Handbook of Advanced Performability Engineering

Krishna B. Misra
Editor

Handbook of Advanced Performability Engineering

Springer

Editor
Krishna B. Misra
RAMS Consultants
Jaipur, Rajasthan, India

ISBN 978-3-030-55731-7 ISBN 978-3-030-55732-4 (eBook)
https://doi.org/10.1007/978-3-030-55732-4

This Springer imprint is published by the registered company Springer Nature Switzerland AG
The registered company address is: Gewerbestrasse 11, 6330 Cham, Switzerland

Dedication

This Handbook is dedicated to the hundreds of thousands of victims around the world who lost their lives and loved ones to the COVID-19 pandemic. Also, this Handbook is as well dedicated to the "Corona Warriors"—the healthcare workers, the doctors, nurses and the supporting staff—who worked selflessly and tirelessly to try to save as many precious human lives as possible.

This Handbook is also dedicated to the younger generation of my family—Nikhil Misra, Meera and Meesha Trivedi, and Cyrus, Anushka and Xenia Chinoy—who can look forward to the benefits that may accrue from the concepts presented in this Handbook.

Foreword

The editor of the present *Advanced Handbook of Performability Engineering*, Prof. Krishna B. Misra, a retired eminent professor of the Indian Institute of Technology, who took to reliability more than half a century ago and is a renowned scholar of reliability. Professor Misra was awarded a plaque by IEEE Reliability Society, in 1995, "in recognition of his meritorious and outstanding contributions to Reliability Engineering and furthering of Reliability Engineering Education and Development in India". Upon his retirement in 2005 from I1T, Kharagpur, where he established the first ever Reliability Engineering Centre in India and the postgraduate course in Reliability Engineering in 1982, he launched the International Journal of Performability Engineering from India in 2005 and served as its Editor-in-Chief until December, 2015. The journal is now being published from USA. In 2014, he started a Book Series on Performability Engineering published jointly by Scrivener and John Wiley & Sons, USA. Ten books under this series have already been published so far.

Two years after successfully establishing the International Journal of Performability Engineering, Prof. Misra took up the responsibility of editing the *Handbook of Performability Engineering*, which was published by Springer in 2008. This version of the handbook received an overwhelming response, with close to 500,000 chapters downloads till 2019 since its publication. At the same time, several new concepts and interpretations have been introduced in performability engineering over the years, hence the timely publication of the *Advanced Handbook of Performability Engineering*, which reflects the changing scenario of the twenty-first century's holistic view of designing, producing and using products, systems or services which satisfy the performance requirements of a customer to the best possible extent.

The word performability reflects an amalgamation of reliability and other reliability-based performance attributes, such as quality, availability, maintainability, and sustainability. Therefore, performability can be considered as the best and most appropriate means to extend the meaning of effectiveness and overall performance of a modem complex and complicated system in which mechanical, electrical and biological elements become increasingly harder to differentiate.

Having reviewed the contents of the present handbook, I find that it clearly covers the entire canvas of performability up to the present: quality, reliability, maintainability, safety and sustainability, including a revised look at assessment and improvement of existing performability parameters like reliability, multi-state performability, analysis of Nonlinear Dynamic Systems, Distributed systems and performability of social robots and models for global warming. I understand that the motivation of this handbook came from the editorial that Prof. Misra wrote in the inaugural issue of International Journal of Performability in 2005.

The handbook addresses how today's systems need to be not only dependable but also sustainable. Modern systems need to be addressed in a practical way instead of simply as a mathematical abstract, often bearing no physical meaning at all. In fact, performability engineering not only aims at producing products, systems and services that are dependable but also involves developing economically viable and safe processes of modern technologies, including clean production that entails minimal environmental pollution. Performability engineering extends the traditionally defined performance requirements to incorporate the modern notion of requiring optimal quantities of material and energy in order to yield safe and reliable products that can be disposed of without causing any adverse effects on the environment at the end of their life cycle.

The chapters included in this handbook have undergone a thorough review and have been carefully devised. These chapters collectively address the issues related to performability engineering. I expect the handbook will create an interest in performability and will bring about the intended interaction between various players of performability engineering.

I am glad to write the Foreword again for the *Advanced Handbook of Performability Engineering* and firmly believe this handbook will be widely used by the practicing engineers as well as serve as a guide to students and teachers, who have an interest in conducting research in the totality of performance requirements of the modern systems of practical use. I would also like to congratulate Prof. Misra once again for taking the bold initiative of editing this timely volume.

June 2020

Way Kuo
President and University Distinguished Professor
City University of Hong Kong
Kowloon, Hong Kong

Preface: The Editor's Journey to Performability Engineering

I would like to take the opportunity to share a few reflections and observations on my professional pursuits and my work in the hope that this may inspire and motivate the readers of this volume, particularly the younger researchers, on whose shoulders rest the responsibility for continuing and expanding further critical work in the area of performability. In the long run, it is they who will be the beneficiaries of the advances that ultimately accrue from research and innovation in the field of performability engineering, leading to sustainable development.

Initial Phase of My Journey

I have long been an admirer of the engineering perfection of German technology and was inspired by the phenomenal improvement in the quality and reliability of Japanese products during the mid-twentieth century. These, among other factors, led me in 1967 to choose reliability engineering for my career with thoughts of trying to emulate the same success in India in order to improve Indian manufactured products, which I felt was vital for improving the economy of the country.

Trained as an engineer, and with the limitations of available resources in a developing country like India, I tried to confine myself to theoretical research in order to develop simple and efficient methods for assessing system reliability as well as the design of systems for which engineers are responsible. I felt that this was an important, but often overlooked, aspect of design, since engineers often work on the "feel" of the systems, and prefer simple and quicker solution to their problem and less on the mathematical rigours involved with the solution of their problems.

Therefore, in search of developing simple methods for evaluating system reliability, I chanced to apply a graph theory approach (which I had learned as an electrical engineering graduate) to system reliability problems. In this manner, I was able to develop simple topological methods based on graph theory to evaluate system reliability for all kinds of system configurations. This included techniques, such as the inspection method, to evaluate system reliability, since I always believed that by studying the inherent characteristics or properties of a given problem, one can develop a simple and effective method of solution. These methods and approaches including network transformation, decomposition and recursive techniques are recorded in [1]—the book I authored in 1992. I then tried to employ this same strategy of developing simple and efficient methods, including some heuristic methods, to the problem of reliability design of engineering systems. Some of these methods are included in the first book on system reliability optimization by Tillman et al. [6]. More information on the work done in the area of reliability design of various types of systems can be found in Chaps. 32 and 33 of my book [5]. The concept of a general purpose dedicated gadget, known as a "reliability analyser", was also proposed during this period in [2]. Later on, in order to provide a resource book for engineers wanting to undergo a training programme in reliability engineering, I also authored a book [4] for such trainees, which was used for many in-house training programmes in India and abroad.

To pursue the field of reliability design and safety of nuclear plants, I went to Germany on a Senior Humboldt Fellowship in 1973–1974 to work at the Laboratorium für Reaktoregelungen und Anlagensicherungen (GRS-Garching, near Munich)—an institute led by Prof. Dr. A. Birkhofer. A nuclear power plant generally consists of subsystems that may employ any of the partial, standby and active redundancies. The design of such a mixed redundancy system, often under some techno-economic constraints, had been previously considered to be quite complex. I was able, however, to provide a simple solution to this problem in a paper published in IEEE PAS Transactions [7]. It was also while working at GRS-Garching that I visualised the usefulness of fuzzy sets theory in the risk assessment of nuclear power plants, since the perceived risk of nuclear plants was always adjudged to be higher than the statistical risk, which had become a deterrent to the development of nuclear power for many countries. I was then able to publish my ideas on this matter in some papers related to the reliability and safety of nuclear power plants. To pursue this concept further, I joined Prof. Dr. H. J. Zimmermann's Institute on Humboldt Fellowship in order to work in the area of the application of fuzzy sets to risk and reliability problems in nuclear plants. Eventually, I was invited to work at Kernforschungszentrum (Reactor Research Centre), Karlsruhe, Germany during 1987–1988, as a Guest Professor. While working at Karlsruhe, my work at this centre [2] resulted in the development of a methodology and a code to carry out Level-I risk assessment studies of nuclear power plants using fuzzy sets theory and I published two papers, including one in the journal on fuzzy set theory and systems.

In 1992, I was appointed Director-Grade-Scientist by the Council of Scientific and Industrial Research (CSIR) at the National Environmental Engineering Research Institute (NEERI), Nagpur, to work on risk problems particularly related to nuclear power plants. It is in this capacity that I was first exposed to the concern of environmental risk problems and to the concept of sustainable development. It is at this point in my career when I began to feel that reliability alone could not be the sole criteria of judging the performance of products, systems and services. It became clear to me that the process of manufacturing, use and disposal of products influence the environment around us and in order to judge the overall performance, one must not be concerned only with the use phase but take the entire lifecycle phases into consideration of a product, system or service. It was during this period that I felt the necessity of editing the book, *Clean Production: Environment and Economic Perspectives*, [3] published by Springer in 1996.

The Concept of Performability

The necessity of considering the environmental impact of products, systems or services [3] and to the assessment of their holistic performance over their entire lifecycle ultimately led me to the development of the concept of performability [5]. In order to further propagate the concept of performability engineering, I launched the *International Journal of Performability Engineering*, under the auspices of Ramsconsultants, from Jaipur, India in 2005. I functioned as the journal's Editor-in-Chief, along with a well-known international team of researchers on the journal's Editorial Board, from July 2005 until December of 2015. Since January of 2016, the journal continues to be published from the USA under a new publisher.

Since the concept of performability and sustainable development are closely inter-linked, I also felt it was vitally important to promote the concept to a wider section of the engineering community, including planners, designers, manufacturers, researchers, technologists and users. As such, I published the first book on performability engineering [5] in 2008. This book, titled *Handbook of Performability Engineering*, published by Springer, London, consisted of 76 chapters with 100 contributors (myself included) and touched on all aspects of performability. The response to this book was overwhelming.

The following figure, adapted from [5], represents the concept of performability as I introduced it in 2005. The model reflects a holistic view of the design, production, use, and disposal of products, systems or services, and includes not only their basic operational requirements—to the best possible extent with minimum cost—but their overall sustainability.

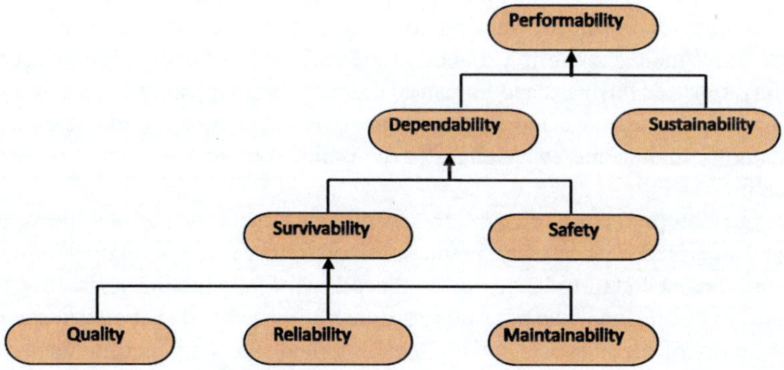

Concept of performability (adapted from Fig. 1.2 of [5])

The sustainability of products, systems and services requires that a minimum of material and energy are consumed during their production, use and disposal, and that they produce minimal waste in order to create the least possible environmental impact. Considering the many-faceted challenges faced by humanity in the twenty-first century in terms of satisfying human needs, conserving resources and protecting the environment, it is my opinion that performability is one of the most critical aspects of high-quality and responsible engineering.

In order to further bolster the importance and propagation of performability engineering, I also initiated a book-series on the topic in 2014 comprising (to date) ten books authored by eminent researchers in the field. The series is a joint publication of Scrivener Publishing LLC, and John Wiley & Sons, USA.

Advancing the Field of Performability

I have been encouraged by the overwhelming worldwide interest in the area of performability (as evidenced by the popularity of the *Handbook* [5], the *Journal* and other publications), and I feel that I must continue to attempt to inspire interest and research in this important engineering concept. Accordingly, I felt that it was time to review the advances that have taken place in performability engineering since the publication of the Handbook [5] over a decade ago.

During this period, several new ideas, concepts, disciplines and areas of application in performability have been introduced. One example is the trend towards automation—termed Industry 4.0 (I4)—which some have deemed a fourth industrial revolution, which encompasses smart manufacturing and data exchange in manufacturing technologies and processes, and include cyber-physical systems, cloud computing, cognitive computing and artificial intelligence, and so on. This trend has impacted several other application areas, such as railways (coining the term Railway 4.0), which includes the digital transformation strategy for railways to create an intelligent business ecosystem for customer demand-driven services. Another important area of application is in time-varying communication systems and software development. Asset management is yet another area which has become important under prevailing dynamic business and industrial scenarios with advanced condition monitoring tools with predictive and prognostic analytics. The presence of a large amount of uncertainty in data and every phase of the evaluation process is still another area of concern.

In performability engineering itself, several new concepts, interpretations, extensions and a revised look at assessment and improvement of existing performability parameters have been introduced during this period. These include reliability, multi-state performability, analysis of nonlinear dynamic systems, distributed systems and performability of social robots and models for global warming, all of which advance the state-of-the-art in performability.

Moving Forward

With so many advancements and expansions, it was considered prudent, therefore, to take stock of what has changed during this intervening period and to revisit the *Handbook of Performability Engineering* once again. To realize this goal, I invited several renowned authors to contribute to this new volume. After a careful selection of 35 peer-reviewed chapters that touch upon most of the above-mentioned areas of performability engineering, the result is the *Advanced Handbook of Performability Engineering* and it is hoped that this Handbook will, one again, find an appreciative audience, and that it will inspire the next generations of researchers and engineers.

Jaipur, India

Krishna B. Misra
kbmisra@gmail.com
http://www.ramsconsutants.org

References

1. Misra, K. B. (1992). *Reliability analysis and prediction: A methodology oriented treatment*. Amsterdam: Elsevier Science Publisher.
2. Misra, K. B. (Ed.) (1993). *New trends in system reliability evaluation.* Amsterdam: Elsevier Science Publisher.
3. Misra, K. B. (Ed.) (1996). *Clean production: Environment and economic perspectives*. Heidelberg, Berlin: Springer.
4. Misra, K. B. (2011). *Principles of reliability engineering*. Sweden: Lülea Technical University Press.
5. Misra, K. B. (Ed.) (2008), *Handbook of performability engineering*. London: Springer. (There have been 495,445 chapter downloads up to 2019 according to Springer Link)
6. Tillman Frank, A., Kuo, W., Hwang, C.-L. (1980). *Optimization of system reliability*. New York: Marcel Dekker Inc.
7. Misra K. B. (1975). Optimum reliability design of a system containing mixed Redundancies. *IEEE Transactions on Power Apparatus and Systems, PAS 94*(3), 983–993.

Acknowledgements

The *Advanced Handbook of Performability Engineering* was conceived in the Fall of 2019, followed shortly thereafter by the eruption of the COVID-19 global pandemic. Governments worldwide enacted lockdowns, which included closures of educational institutions and businesses, as well as home quarantines for much of the global population.

In light of these highly unprecedented events, the editor would like to thank all of the contributors to this Handbook who, in spite of all of the accompanying concerns and difficulties related to the Corona virus threat, generously devoted their time and effort to the development of this Handbook. The editor would also like to thank the peer reviewers who offered their valuable time to offer their comments, which helped to enhance the quality of this work.

The editor would also like to acknowledge the support of his wife, Veena Misra; daughter, Vinita Chinoy and son-in-law, Rayomond Chinoy; son, Vivek Misra and daughter-in-law, Dr. Michelle Misra; as well as my second daughter, Kavita Trivedi and son-in-law, Manoj Trivedi. My beloved family whole-heartedly supported my determination to bring out this Handbook, in spite of the difficult conditions and in addition to the problems related to my debilitating health.

My special thanks are due to my granddaughter, Anushka Chinoy, and daughter, Vinita Chinoy, who assisted me in my editorial work.

Last, but not least, my sincere thanks are due to Dr. Anthony Doyle of Springer, London, Dr. Dieter Merkle of Springer, Nature, and the Production Editor, Vidyaa Shri Krishna Kumar and Mr. Madanagopal of Springer, Nature, who helped me to develop and produce this Handbook.

Krishna B. Misra
Editor

Obituary

Unfortunately, Prof. Ilya Gertsbakh, the co-author of Chap. 10 of this Handbook, died before the publication of this Handbook. Professor Gertsbakh was a well-renowned expert in Reliability, Applied Statistics and other related fields. His skills did not confine to the world of science only, though. Being an extremely versatile person, he was also very approachable. Those who knew him would speak fondly of his warm personal touch and the ability to make everyone feel comfortable around him. Our sincere condolences go out to his close family and friends.

Editor

Contents

1 **Assessment of Sustainability is Essential for Performability Evaluation** 1
Krishna B. Misra

2 **Performability Considerations for Next-Generation Manufacturing Systems** 41
Bhupesh Kumar Lad

3 **Functional Safety and Cybersecurity Analysis and Management in Smart Manufacturing Systems** 61
Kazimierz T. Kosmowski

4 **Extending the Conceptualization of Performability with Cultural Sustainability: The Case of Social Robotics** 89
John P. Ulhøi and Sladjana Nørskov

5 **Design for Performability Under Arctic Complex Operational Conditions** 105
Abbas Barabadi and Masoud Naseri

6 **Dynamic Multi-state System Performability Concepts, Measures, Lz-Transform Evaluation Method** 133
Anatoly Lisnianski and Lina Teper

7 **On Modeling and Performability Evaluation of Time Varying Communication Networks** 161
Sanjay K. Chaturvedi, Sieteng Soh, and Gaurav Khanna

8 **Characteristics and Key Aspects of Complex Systems in Multistage Interconnection Networks** 191
Indra Gunawan

9 **Evaluation and Design of Performable Distributed Systems** 211
Naazira B. Bhat, Dulip Madurasinghe, Ilker Ozcelik, Richard R. Brooks, Ganesh Kumar Venayagamoorthy, and Anthony Skjellum

10 Network Invariants and Their Use in Performability Analysis 229
Ilya Gertsbakh and Yoseph Shpungin

11 The Circular Industrial Economy of the Anthropocene and Its Benefits to Society 249
Walter R. Stahel

12 Sustainment Strategies for System Performance Enhancement 271
Peter Sandborn and William Lucyshyn

13 Four Fundamental Factors for Increasing the Host Country Attractiveness of Foreign Direct Investment: An Empirical Study of India 299
Hwy-Chang Moon and Wenyan Yin

14 Structured Approach to Build-in Design Robustness to Improve Product Reliability 319
Vic Nanda and Eric Maass

15 Time Series Modelling of Non-stationary Vibration Signals for Gearbox Fault Diagnosis 337
Yuejian Chen, Xihui Liang, and Ming J. Zuo

16 Risk-Informed Design Verification and Validation Planning Methods for Optimal Product Reliability Improvement 355
Zhaojun Steven Li and Gongyu Wu

17 Efficient Use of Meta-Models for Reliability-Based Design Optimization of Systems Under Stochastic Excitations and Stochastic Deterioration 383
Gordon J. Savage and Young Kap Son

18 Dynamic Asset Performance Management 403
Aditya Parida and Christer Stenström

19 Asset Management Journey for Realising Value from Assets 429
Gopinath Chattopadhyay

20 Reliability-Based Performance Evaluation of Nonlinear Dynamic Systems Excited in Time Domain 451
Achintya Haldar and Francisco J. Villegas-Mercado

21 Probabilistic Physics-of-Failure Approach in Reliability Engineering 479
Mohammad Modarres

22 Reliability and Availability Analysis in Practice 501
Kishor Trivedi and Andrea Bobbio

23 WIB (Which-Is-Better) Problems in Maintenance Reliability Policies ... 523
Satoshi Mizutani, Xufeng Zhao, and Toshio Nakagawa

24 A Simple and Accurate Approximation to Renewal Function of Gamma Distribution ... 549
R. Jiang

25 Transformative Maintenance Technologies and Business Solutions for the Railway Assets ... 565
Uday Kumar and Diego Galar

26 AI-Supported Image Analysis for the Inspection of Railway Infrastructure ... 597
Joel Forsmoo, Peder Lundkvist, Birre Nyström, and Peter Rosendahl

27 User Personalized Performance Improvements of Compute Devices ... 615
Nikhil Vichare

28 The Neglected Pillar of Science: Risk and Uncertainty Analysis ... 633
Terje Aven

29 Simplified Analysis of Incomplete Data on Risk ... 651
Bernhard Reer

30 Combining Domain-Independent Methods and Domain-Specific Knowledge to Achieve Effective Risk and Uncertainty Reduction ... 679
Michael Todinov

31 Stochastic Effort Optimization Analysis for OSS Projects ... 697
Yoshinobu Tamura, Adarsh Anand, and Shigeru Yamada

32 Should Software Testing Continue After Release of a Software: A New Perspective ... 709
P. K. Kapur, Saurabh Panwar, and Vivek Kumar

33 Data Resilience Under Co-residence Attacks in Cloud Environment ... 739
Gregory Levitin and Liudong Xing

34 Climate Change Causes and Amplification Effects with a Focus on Urban Heat Islands ... 761
Alec Feinberg

35 On the Interplay Between Ecology and Reliability ... 785
Ali Muhammad Ali Rushdi and Ahmad Kamal Hassan

Acronyms

AI	Artificial Intelligence
AS	Alarm System
AVM	Attacker Virtual Machine
BCM	Business Continuity Management
BD	Big Data
BPCS	Basic Process Control System
CCF	Common Cause Failure
CRA	Co-Resident Attack
CS	Cybersecurity
DC	Diagnostic Coverage
DCS	Distributed Control System
DMZ	DeMilitarized Zone
DSA	Different Servers Allocation
DSS	Decision Support System
DTNs	Delay-Tolerant Networks
E/E/PE	Electrical/Electronic/Programmable Electronic
EAL	Evaluation Assurance Level
ERP	Enterprise Resource Planning
EUC	Equipment Under Control
EWA	Early WARNING AGENT
FA	Free Allocation
FANETs	Flying Ad hoc Networks
FCE	First Co-residence Event
FRs	Fundamental Requirements
FS	Functional Safety
HFT	Hardware Fault Tolerance
HIS	Human System Interface
HMI	Human Machine Interface
IACS	Industrial Automation and Control System
ICS	Industrial Control System
IIoT	Industrial Internet of Things
IoT	Internet of Things

ISMS	Information Security Management System
IT	Information Technology
KPI	Key Performance Indicator
LAN	Local Area Network
M2M	Machine to Machine (communication)
MANETs	Mobile Ad hoc Networks
OT	Operational Technology
pdf	Probability Density Function
PFDavg	Probability of Failure on Demand average
PFH	Probability of dangerous Failure per Hour
PL	Performance Level
PLC	Programmable Logic Controller
PLr	Performance Level required
PM	Preventive Maintenance
RAMS&S	Reliability, Availability, Maintainability, Safety and Security
RMS	Resource Management System
SAL	Security Assurance Level
SAR	Security Assurance Requirement
SCADA	Supervisory Control And Data Acquisition
SDP	Sum-of-Disjoint Products
SF	Safety Function
S_{FF}	Safe Failure Fraction
SIEM	Security Information and Event Management
SIL	Safety Integrity Level
SIL CL	Safety Integrity Level CLaimed
SILr	Safety Integrity Level required
SIS	Safety Instrumented System
SL	Security Level
SMS	Smart Manufacturing System
SRCS	Safety Related Control System
SRS	Safety Related System
TAGs	Time Aggregated Graphs
TSE	Timestamped Edge
TS-MCS	Timestamped Minimal Cut Set
TS-MPS	Timestamped Minimal Path Set
TVCNs	Time Varying Communication Networks
*np*TVCN	Non-predictable TVCN
*p*TVCN	Predictable TVCN
UVM	User Virtual Machine
VANETs	Vehicular Ad hoc Networks
VM	Virtual Machine
VPC	Virtual Private Cloud
WAN	Wide Area Network

Chapter 1
Assessment of Sustainability is Essential for Performability Evaluation

Krishna B. Misra

Abstract Performability of a product, a system or a service has been defined by this author (Misra in Inaugural Editorial of International Journal of Performability Engineering 1:1–3, 2005 [1] and Misra in Handbook of performability engineering, Springer, London, 2008 [2]) as an attribute of the holistic performance reckoned over its entire life cycle ensuring not only high dependability (quality, reliability, maintainability and safety) but also sustainability. Sustainability is a characteristic specific to a product, system or service. At the same time, a dependable product, system or service may not be sustainable. It may also be necessary here to point out that without dependability, sustainability is meaningless. Therefore, both dependability as well as sustainability attributes should be considered and must be evaluated in order to evaluate performability of a product. All other attributes of the definition of performability have been defined and can be computed except sustainability. In order to evaluate performability, it is therefore essential to define and compute sustainability. For developing sustainable products, systems and services in the twenty-first century, it is essential that we should be able to define precisely and quantify sustainability since one cannot improve what cannot be measured or assessed. The objective of the present chapter is to understand the implications of sustainability in order to facilitate computation of sustainability and thereby the performability. The purpose of 13 chapters in the Handbook (Misra in Handbook of performability engineering, Springer, London, 2008 [2]) by the author was to provide detailed introduction to each constituent elements of the definition of performability, namely quality, reliability, maintainability, safety and sustainability, and these chapters were received very well by the international academic community as is evident from Table 1.1. This was done with the intent to evoke interest among researchers across the world in the concept of performability leading to a way to compute or assess performability. But this did not happen in the past 12 years after the publication of the Handbook in 2008. The main impediment in this effort is the procedure to evaluate sustainability.

K. B. Misra (✉)
RAMS Consultants, Jaipur, India
e-mail: kbmisra@gmail.com

K. B. Misra (ed.), *Handbook of Advanced Performability Engineering*,
https://doi.org/10.1007/978-3-030-55732-4_1

1.1 Introduction

The intense technological development has led to the destruction of pristine environmental of the Earth and has led to ever-increasing pollution, resulting in extinction of some species [3, 7] on Earth that may even be further accelerated by global warming in the future. Technology has continually affected human society and its surroundings. On one hand, technological progress has helped boost economies of competing nations and bring prosperity to the people; at the same time, very many technological processes have resulted in producing undesirable by-products (solid, gaseous or fluid), known as pollutants which has caused severe degradation of Earth's pristine environment. On the other hand, ever-increasing demand of products due to exponential growth of the world population is causing fast depletion of finite natural resources of the Earth. Although new technologies can help overcome some of these problems, but implementation of several new technologies influences the values of a society and often raises new ethical questions. This is true even for biotechnology or nanotechnology as well.

Realizing the gravity of problem caused by the growth of the world population, over-exploitation of resources on one hand, and their wastage on the other hand, could lead humans to surpass the carrying capacity of the Earth [4], more than 1600 scientists, including 102 Noble laureates collectively signed a Warning to Humanity in 1992, which read as follows:

> No more than one or few decades remain before the chance to avert the threats we confront, will be lost and the prospects for humanity immeasurably diminished... A new ethics is required- a new attitude towards discharging responsibility for caring for ourselves and for Earth ... this ethics must motivate a great movement, convincing reluctant leaders, reluctant governments and reluctant people themselves to affect the needed changes.

Hundreds of papers and reports have appeared on the subject of sustainable development and sustainability but it still remains largely as a concept and no clear outline exists how this can be measured and physically realized.

In twenty-first century, global prosperity [5] would depend increasingly on using Earth's resources, wisely, more efficiently, distributing them more equitably, while reducing the wastages and in fact reducing their overall consumption levels as well. Unless we can accelerate this process, we cannot achieve the goal of sustainable development. Humans would always need products, systems and services all the time to fulfil the basic requirements of life, and all these need to be sustainable in order to achieve the objective of sustainable development.

Economic implications of sustainable development would not only include costs of development but also overall welfare costs. For example, mitigation options in the energy sector may be classified into those that improve energy efficiency and those that reduce the use of carbon-intensive fuels. Energy efficiency improvement reduces reliance on energy supply and it is likely to improve a nation's energy security. Switching to low carbon energy supply sources is the other mitigation category in the energy sector which reduces air pollution with significant GHG benefits.

Another economic consideration that will be important for developing sustainable products, system and services is to utilize the obsolete products at the end of their life through recycling, reuse or remanufacturing. If obsolete materials are not recycled, raw materials have to be processed afresh to make new products. This represents a colossal loss of resources in terms of energy used at every stage of material extraction and transportation, and environmental damage caused by these processes is substantial.

1.2 Concept of Performability

Performability can be called as an attribute of performance assessment of any product, system or service and even of human beings. Performance improvement can be seen as an improvement in input requirements, for example, reduced working capital, material, replacement/reorder time and set-up requirements. Yet another improvement can be seen in the throughput and is often judged by process efficiency, which is measured in terms of time, resource utilization and wastage caused. Lastly, improvement can also be seen in output requirements, often viewed in terms of cost/price, quality, in functionality or longevity [6, 7], durability or even in safety. Humans have been striving to achieve excellence in performance in all their areas of activity. Be it planning, design, execution, manufacturing, or even using a product, system or a service.

The existence of various performance attributes can be traced to the follow-up work in the area of reliability, as no concerted effort was made to standardize the definitions of various terms in use in relation to performance evaluation.

If one takes the Webster definition of performability, it is the ability (this ability expressed in terms of probability just as in case of reliability) to perform under given conditions. Therefore, based on the key terms, "perform" and "ability", performability can be interpreted as performance (which may include reliability, safety, risk, human performance) under given conditions. One must not forget that the given conditions could be normal, abnormal environment conditions and extreme environment conditions. It is in this more general context that the term "performability" is used in this chapter which would not only take into consideration the entire gambit of performance attributes but includes the sustainability aspect of products and systems performance in twenty-first century perspective. In other words, it should represent the holistic performance.

Figure 1.1 presents the concept of performability as introduced by the author in 2005 and reflects a holistic view of designing, producing, using and disposing products, systems or services, which will satisfy not only the basic operational requirements to the best possible extent but are also sustainable. However, it is essential that we quantify or assess performability before it can be used as a criterion of design of a product, system or service. It is more than a decade since the concept was introduced, but so far no progress has been made to compute or assess it.

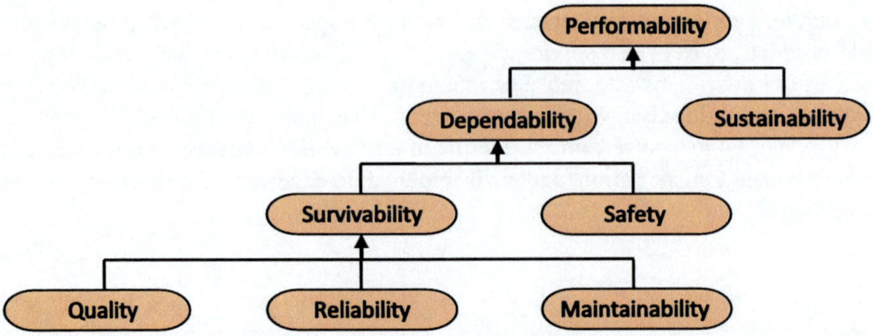

Fig. 1.1 Concept of performability (adapted from Fig. 1.2 of [2])

Peter Ferdinand Drucker, an Austrian-born American legendary management consultant, had famously said, "*What Gets Measured Gets Improved*", some 40 years ago. The sentence is a famous quote by Peter Drucker from his 1954 book titled, "*The Practice of Management*". Since then the quote has been repeated so many times that it has almost become a proverb, adage or slogan. Therefore, unless we can develop a way to assess or measure performability, we will not be able to improve it.

Performability and sustainable development are closely inter-linked. It will be our effort to bring out these linkages between them. As of now, we have been designing products, systems or services based only on the criteria of their dependability as depicted in Fig. 1.1. However, this attribute is very much influenced by the design, raw material, fabrication, techniques and manufacturing processes and their control and finally by the usage, which is the realm of sustainability. Therefore, one needs to lay emphasis on sustainability and needs to design a product, system or service which includes this attribute also.

We will see soon that all the major attributes of performability of Fig. 1.1 along with technology are inter-related with sustainability. This is shown in Fig. 1.4 and a true optimal design of a product, system or service would necessarily be the one which considers performability as the design criteria, which includes the attribute of sustainability within its ambit.

As stated earlier, the author has described more explicitly all the performability attributes in the form of 13 distinct chapters of the Handbook of Performability Engineering [2] in order to understand their implications. For the ease of reference for a reader, the information on these chapters is provided in Table 1.1. The Research gate of Germany publishes every week the statistics of readership of chapters, papers and books authored by researchers published world over, and Table 1.1 displays the number of readers who have read the chapters authored by the author as published in the Handbook of Performability Engineering. Springer Links reports that there have been more than 495,445 chapters download of the Handbook of Performability Engineering since its publication by Springer in 2008.

In fact, Table 1.1 indicates the interest the researchers or readers have shown in the concept of performability engineering. Another inference, one can derive from

Table 1.1 Statistics on chapters by the author published in Handbook of Performability Engineering

From [2] Handbook of Performability Engineering, Springer, London (2008)		
Title of the chapter	Chapter #	Total reads
Performability Engineering: An Essential Concept in the twenty-first Century, pp. 1–12	Chapter 1	50
Engineering Design: A Systems Approach, pp. 13–24	Chapter 2	860
Dependability Considerations in the system Design, pp. 71–80	Chapter 6	140
Quality Engineering and Management, pp. 157–170	Chapter 12	**21199**
Reliability Engineering: A Perspective, pp. 253–288	Chapter 19	**9651**
Tampered Failure Rate Load-Sharing Systems: Status and Perspectives, author with S.V. Amari and Hoang Pham, pp. 289–306	Chapter 20	327
Optimal System Reliability Design, author with Bhupesh Lad and M. S. Kulkarni, pp. 493–514	Chapter 32	483
MIP: A Versatile Tool for Reliability Design of a System, author with S. K. Chaturvedi, pp. 515–526	Chapter 33	56
Risk Analysis and Management: An Introduction, pp. 661–674 (Safety)	Chapter 41	**3179**
Maintenance Engineering and Maintainability: An Introduction, pp. 747–764	Chapter 46	**10434**
Sustainability: Motivation and Pathways for Implementation, pp. 835–848	Chapter 51	**515**
Applications of Performability Engineering Concepts, pp. 971–980	Chapter 60	62
Epilogue—A Peep into the Future, pp. 1239–1250	Chapter 76	18

Source Researchgate, Germany (As accessed on October 15, 2020)

Table 1.1 is that the number of reads of a chapter against each attribute reflects the importance that readers place on the attributes of quality, reliability, maintainability, safety and sustainability. It must be realized that performability engineering not only aims at developing products, systems and services that are dependable but involves developing economically viable and safe processes (clean production and clean technologies) that would entail minimal environmental pollution, require minimum quantities of raw material and energy, and yield safe products of acceptable quality and reliability that can be disposed of at the end of their life without causing any adverse effects on the environment.

Let us first examine the implication of sustainability.

1.3 Implications of Sustainability

The key issues associated with the implementation of sustainability characteristic appear to revolve around:

- The need to conserve essential natural resources, minimize the use of materials, develop renewable sources of energy and avoid over-exploitation of vulnerable resource reserves.
- The need to minimize the use of processes and products that degrade or may degrade the environmental quality.
- The need to reduce the volume of waste produced by economic activities entering the environment.
- The need to conserve and minimize the use of energy.
- The need to reduce or prevent activities that endanger the critical ecological processes on which life on this planet depends.

From an engineer's point of view, to produce sustainable products, system and services would require that we minimize the use of materials and energy. At the same time, we must also ensure that wastages of materials and effluents (solid, liquid or gaseous) produced during entire life-cycle activities (Fig. 2 of [4]) starting from extraction, manufacturing, use and disposal phases are minimal. The throughput at every stage of life cycle is shown in Fig. 1.2.

It is necessary that material and energy utilization and wastage be kept minimal during the actual use of products, systems and services, as well. Energy produced at any stage can be utilized fruitfully elsewhere. In other words, material and energy audit is absolutely necessary in order to produce sustainable products and systems. This will ensure minimal environmental degradation during production. Obviously, less material and energy consumption—either through dematerialization, reuse or recycle or through proper treatment (clean up technology)—would lead to a lesser degree of environmental degradation. This is also necessary that material and energy

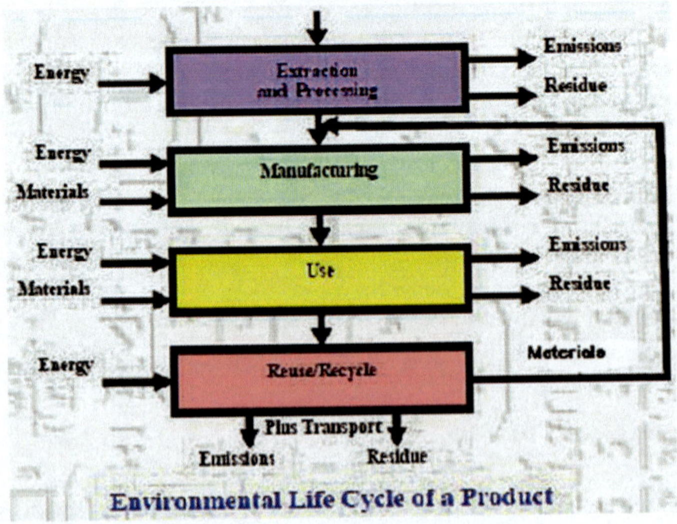

Fig. 1.2 Throughput at every stage of life cycle of a product (adapted from [4])

requirements and wastages are minimized during the actual use of products, systems and services.

1.3.1 Dematerialization

The dematerialization of a product means to use less material to deliver the same level of functionality to a user. A material can be anything from an unprocessed raw material to a finished product. The UNEP defines dematerialization as "the reduction of total material and energy throughput of any product and service, and thus the limitation of its environmental impact. This includes reduction of raw materials at the production stage, of energy and material inputs at the use stage, and of waste at the disposal stage".

Dematerialization considers, besides waste, the natural resources involved in the products' life cycle [3]. It literally means the use of less material. It entails actions at every stage of the production and consumption phase, which include:

- Resource savings in material extraction
- Improved eco-design of products
- Technological innovations in the production process
- Environmentally friendly consumption
- Recycling of waste, etc.

Dematerialization strategy basically translates into:

- The conception, design and manufacture of a smaller or lighter product
- The replacement of material goods by non-material substitutes
- The reduction in the use of material systems or of systems requiring large infrastructures.

According to [8], in the computer industry, silicon wafers are now increasing at the rate of 10–15% per year in size to reduce material losses in cutting. If one considers that, about 400 acres of silicon wafer material is used per year by IBM Corporation alone at a cost of about $100 million per acre with a processed wafer costing approximately $800. Although the volume of cuttings of silicon wafer does not create a waste disposal problem but it do create an environmental problem as their manufacture involves the handling of hazardous chemicals. This is also an interesting example of how the production volume tends to generate demand of large plastic and metal boxes to keep cool the microchips made with the wafers, whereas the world's entire annual chip production can fit inside one 747 jumbo jet. This way miniaturization partially offsets the gains of dematerialization.

1.3.2 Minimization of Energy Requirement

According to the International Council of Chemical Associations (ICCA) report (December 6, 2012) energy-saving products installed in homes in the United States prevented nearly 283 million tons of CO_2 emissions in 2010—equivalent to the greenhouse gas emissions of 50 million passenger vehicles. Studies also show that if this trend continues, more than 7 billion tons of emissions can be avoided by 2050 in the United States alone—equivalent to the CO_2 emissions of more than 1.2 billion passenger vehicles.

Eco-labelling is a promising market-based approach for improving the environmental performance of products through consumer choice. While eco-labelling by itself is not new, eco-labelling to promote energy efficiency or sustainability is a more recent phenomenon.

Five such energy-labelling programmes in the United States are in vogue: Green Seal, Scientific Certification Systems, Energy Guide, Energy Star and Green-e. Of these, the first four certify energy-efficient appliances while the last one certifies renewable electricity. Energy Guide and Energy Star are government-run programmes, and the rest are privately administered.

1.3.3 Minimization of Waste

Waste minimization must consider the full life cycle of a product, starting right from the conception stage to achieve a reduction in total amount of waste produced. Sometimes scraps can be immediately reincorporated at the beginning of the manufacturing line so that they do not become a waste product. Some industries routinely do this; for example, paper mills return any damaged rolls to the beginning of the production line, and in the manufacture of plastic items, off-cuts and scrap are reincorporated into new products. Such innovations help reduce waste material or scraps.

Steps can be taken to ensure that the number of reject batches is kept to a minimum. This is achieved by using better quality control procedures. In fact, waste can be reduced by improving quality and durability of a product so that over a given period of time, it results in less wastage. Waste of energy over the use period forms a part of waste consideration.

Sometimes waste product of one process becomes the raw material for a second process. Waste exchanges represent another way of reducing waste disposal volumes for waste that cannot be eliminated. Recycle and reuse are discussed in the next section.

1.3.4 End-of-Life Treatment

From both environmental as well as economic considerations, the end-of-life treatment of products and systems is now becoming the liability of the manufacturers and distributors eventually. The WEEE directive of European Union is the first step in that direction at least in the electrical and electronic sector. The WEEE directive (2002/96/EC) as passed by European Community is aimed to prevent waste electrical and electronic equipment from ending up in landfills and to promote the level of recycling and reuse in electrical and electronic sector. This directive requires all manufacturers and importers of electric and electronic equipment to meet the cost of collection, treatment and recovery of their waste electrical and electronic equipment at the end of their useful life. Design for end-of-life requires manufacturer to reclaim responsibility for their products at the end-of-life. The alternatives to landfill or incineration include maintenance, recycling for scrap material and remanufacturing. This is shown in Fig. 1.3, adapted from [1].

Remanufacturing, Recycling and Reuse [1]:

While maintenance extends product life through individual upkeep or repair on failures, remanufacturing is a batch process of production involving disassembly, cleaning, refurbishment and replacement of worn-out parts, in defective or obsolete products. However, scrap-material recycling involves separating a product into its constituent materials and reprocessing the material.

Remanufacturing involves recycling at parts level as opposed to scrap-material level. It is actually in effect recycling of materials while preserving value-added components. Remanufacturing also postpones the eventual degradation of the raw materials through contamination and molecular breakdown, which are the characteristics of scrap-material recycling. Since remanufacturing saves 40–60% of the cost of manufacturing a completely new product and requires only 20% energy, several big companies are resorting to remanufacturing. Xerox is an example in this case. IBM also established a facility in Endicott, New York as a reutilization and remanufacturing centre. UNISYS and Hewlett Packard also use this strategy. It must, however,

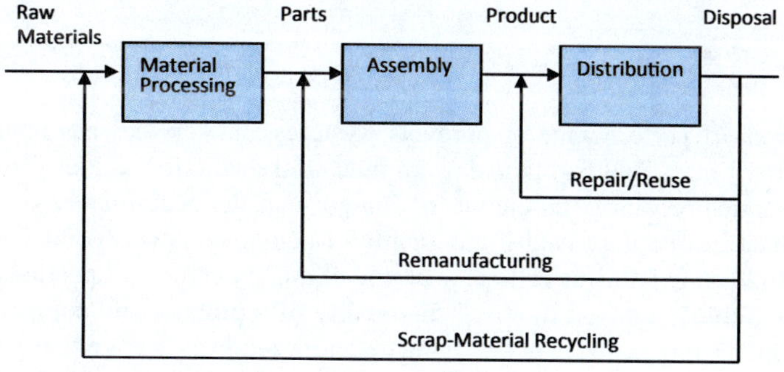

Fig. 1.3 End-of-life options (adapted from [1])

be stated that remanufacturing is not suitable for all types of products, it is appropriate only for those products that are technologically mature and where a large fraction of product can be used after refurbishment.

1.4 Attributes of Dependability and Their Relationship with Sustainability

High dependability of any product, system or service would necessarily demand its better performance over the mission time, characterized by high levels of attributes such as quality, reliability, maintainability and safety. These attributes among themselves are also closely related and govern the overall dependability of a product, system or service. It is also necessary to understand this inter-relationship between these factors and also their relationship with sustainability so that one not only minimizes the chances of occurrence of any untoward incident at the design and fabrication stage but also minimizes the chances of occurrences and the consequences of such an event during system operation and use phase. A brief discussion of these attributes here will not be out of place, although detailed discussion can be found in the chapters listed in Table 1.1.

Sustainability of a product is influenced by a number of factors such as quality, design (size, complexity, reliability, production cost), maintainability (whether the product is to be repaired or replaced) and safety of the product, and the technology employed.

These factors influence one another as is shown in Fig. 1.4. There exist linkages between the attributes of the dependability (namely quality, reliability, maintainability, safety) and technology with sustainability. This is shown in Fig. 1.4. While the linkages between the former attributes are bidirectional (A, B, C, D and E) but that between the attributes and sustainability have been shown as unidirectional (a, b, c, d and e) since we are interested only in how the attributes of dependability affect sustainability.

1.4.1 Quality

Improvement of performance of products, systems and services has always been the concern of man right from the days of industrial revolution. Initially, in the first half of twentieth century, the engineers thought that the performance of a system or equipment can be improved if the quality of components was good. The period between 1920 and 1940 was called the period of quality control inspection since the inspectors were designated to check the quality of a product and compare it with a standard. Discrepancies, if noticed, the deficient products were either rejected or reworked. Therefore, quality has become a worldwide concern of manufacturers.

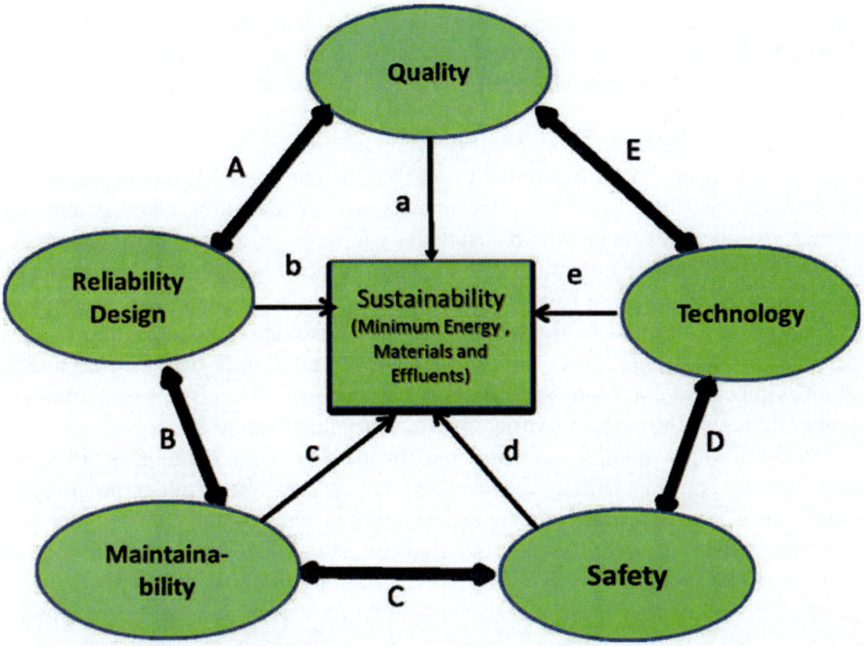

Fig. 1.4 Inter-relationship of attributes of dependability and technology with sustainability

The processes, however, became more and more complex and side-by-side statistical aspects of quality control were also being developed. Shewhart [9] can be said to have laid down the foundation of using control charts to control the variables of a product. Acceptance sampling plans were developed to replace the 100% inspection, and 1930s saw an extensive use of sampling plans in industries. This gradually laid the foundation of statistical quality control and the period 1940–1960 has been called as the period of statistical quality control or SQC [10]. During 1930s, a customer or a user was happy to accept a product as long as it was supported with a warranty. A customer may also have some protection in law, so that he may claim redress for failures occurring within the warranty period. However, there is no guarantee of performance, particularly outside the warranty period. Outside the warranty period, it is only the customer who is left to suffer the consequences of failures. The manufacturer at most may suffer a loss of reputation and possibly future business.

However, the word quality has had different connotations when used by different people and its definition has also undergone several changes and its meaning extended over time but most general definition of quality of a product is a measure of the degree of conformance to applicable design specification and workmanship standards. Thus quality of a product is a concern of a manufacturer and is considered satisfactory if the product is able to satisfy the requirements of consumer. It can be definitely called as an attribute that is generally used to reflect the degree of perfection in manufacturing of a product. It is easy to realize that this degree of perfection is inversely proportional to variability present in the process. Since all manufacturing processes involve materials, men and machines, they all have some element of inherent variability in addition to attributable variability, which can be controlled to an irreducible economic minimum.

Reducing variability in production is synonymous with improving quality of a product. The source of variation due to machine is the natural limits of capability that every process has, which is also known as process/machine capability and any attempt to reduce this range would cost heavily in terms of money. If the process is incapable of acceptable operation within design limits, then we have the option of separating non-conforming from conforming products, using more precise process or change in the design of the product or system in order to achieve an optimum design at minimum total cost. The third source of variation is man himself and is the most important contributor to variability. In fact, man's decisions or actions do directly influence the extent of variability to a very large extent.

The definition of quality does not concern itself with the element of time and would not say if product will retain its quality over a period of time under the specified conditions of use since quality tests either pass a product or fail it. Therefore, the need to have the requirement of a time-based concept of quality was felt. This led to the definition of reliability which emphasizes the ability that a product will perform a specified function over a specified time without failure under the specified conditions of use.

Further detailed discussion of quality and its historical development can be found in Chap. 12 (pages 157–170) of [2].

1.4.1.1 Quality and Sustainability are Related

Since sustainability is characterized by dematerialization, minimization of energy and effluents (wastages), quality can affect sustainability. The dematerialization of a product literally means less material is used to deliver the same level of functionality to the user. Dematerialization [4] therefore affects sustainability, as use of less material would result in a smaller quantity of waste generated during the production as well as the use phase of an industrial product. But if smaller and lighter product is inferior in quality, then more units would be required to be produced, and the net result could be a greater amount of waste generated in both production and consumption phases. In fact, dematerialization can be defined as the change in the amount of waste generated per unit of an industrial product, taking into account overall production and consumption. Westerman [11] indicated that an automobile tire designed for a service life of 100,000 miles could reduce solid waste from tires by 60–75%. Therefore, quality of a product directly affects sustainability. This linkage is shown by "**a**" in Fig. 1.4. Of course, other effective tyre waste reduction strategies may include tire rethreading and recycling.

1.4.2 Reliability

Reliability [6] is a design function and requires skill, indigenous knowhow and experimentation, and of course, it can be designed right at the design stage. Implicit in the probabilistic definition of reliability is the environmental conditions under which the product is supposed to work or used and the period of time (mission time) over which the product is supposed to work satisfactorily that makes the task of ensuring reliability a challenging task. For example, equipment may be exposed to a combination of environments such as temperature, humidity, altitude, shock and vibration, while it is being transported. These environmental conditions are not included in the definition of quality. Sometimes, more than one environmental factor may be acting on parts or the equipment. These combined or concurrent environments may be more detrimental to reliability of a product than the effects of these individual environments separately. Any superposition of effects of individual environmental factors cannot predict the resulting influence of a combination of environmental factors on the reliability or performance of the equipment. In fact, the severity of combination may be much more than individual effects added together. For example, the percentage of failures caused individually by temperature may be 40% of all failures and humidity (19%) but humidity combined with temperature can cause 75% of all failures.

However, quality and reliability are inter-related as shown through link "**A**" in Fig. 1.4. Bad quality or inferior workmanship does definitely shorten the life of a product and thus affects reliability. High reliability could be as a result of high quality but converse is not true. A product may have high quality but may not have high reliability.

About 80% of time, poor performance can be attributed to poor design. Over 90% of field failures result from poor design and most of the product recalls have their origin in faulty design and majority of the law suits are filed on account of improper design and 70–75% of product costs are function of design. Therefore, if any phase in the entire life cycle of a product that has maximum impact on reliability, it is the design phase. Design requires ingenuity and creativeness and a designer ought to know what has been already tried before and did not work. Generally, a product fails prematurely because of inadequate design features, manufacturing and parts defects, human error or external conditions (environmental or operating) that may exceed the designed values. Inadequate attention to the design effort could result in a faulty hardware and retrofits cannot compensate for the faulty design and may be quite costly. There is no substitute to a good design and, it is one of the major responsibilities of the top management to provide a highly competent design team to bring out a reliable product or system.

The detailed discussion of reliability and system reliability design can be found in Chaps. 2, 19, 20, 32 and 33 of [2].

1.4.2.1 Reliability and Sustainability are Related

A good reliability design results in prolonging the lifespan of a product and hence would ensure less adverse effect on the environment over a given period of time. In other words, this would improve sustainability (linkage "**b**" in Fig. 1.4). Cost is implicit with high reliability, since to produce reliable product, we may have to incur increased cost of design and manufacturing. The poor deficient performance attributes not only affect the life-cycle costs but also have effects in terms of environmental consequences. Degraded performance attributes do reflect more on the material and energy demand and wastes and cause more environmental pollution when reckoned over a given period of time. Thus reliability influences sustainability to a large extent.

1.4.3 Maintainability

The post Second World War period saw trade-off between reliability and maintainability, and availability became an important attribute of maintained components and systems. Maintenance is considered as another important factor of product performance after reliability. Broadly speaking, maintenance is a process of keeping an equipment or unit in its operational condition either by preventing its transition to a failed state or by restoring it to an operating state following a failure. Maintenance in reality compliments and compensates for reliability and is an effective tool for enhancing the performance of repairable products or systems. It may be observed that the cost of design and development, manufacturing, and maintenance costs are inter-dependent. For example, a highly reliable product will have lower maintenance costs.

Generally, there are three types of maintenances in use [5, 6], viz., preventative (PM), corrective (CM) and predictive maintenance (PdM). PM is a schedule of planned or scheduled maintenance actions aimed at preventing an equipment failure before it actually occurs and to keep the unit in working condition and/or extend equipment's life. It is performed on a regular basis. Scheduled maintenance can either involve system restoration or a scheduled replacement. Restoration refers to restoring the system to a normal state by disassembling, cleaning or renovating the system at a specified time with the aim of preventing fault occurrences over the wear-out period. Scheduled replacement is carried out when the old and in-service parts or components are scheduled to be replaced in a certain cycle. At the scheduled replacement time, the old parts or components are replaced regardless of the reliability at that time, so is mainly applicable to parts or components with known usefulness lifespans.

PM is designed to enhance the equipment reliability by replacing worn components before they actually fail and this includes activities like equipment checks, partial or complete overhauls at specified periods. Scheduled maintenance is time-based maintenance and is conducted on the basis of previously developed schedules.

Generally, scheduled maintenance involves shutting down the system to check, disassemble or replace components at regularly timed intervals to prevent breakdowns, secondary damage or operating losses. An ideal preventive maintenance programme is one which prevents all equipment failures before they actually occur. If an item has an increasing failure rate, then PM programme is likely to improve system availability. Otherwise, the costs of PM might actually outweigh its benefits. Also, it should be explicitly clear that if an item has a constant failure rate, then PM will have no effect on the item's failure occurrences. Long-term benefits of preventive maintenance include improved system reliability (link "**B**" in Fig. 1.1), decreased cost of replacement, decreased system downtime and better spares inventory management.

Predictive maintenance (PdM) or condition-based maintenance (CBM) is carried out only after collecting and evaluating enough physical data on performance or condition of equipment such as temperature, vibration or particulate matter in oil and so on, by performing periodic or continuous (online) equipment monitoring. This type of maintenance is generally carried out on mechanical systems where historical data is available for validating the performance and maintenance models for the systems and the failure modes are known. Therefore, an important precondition for the application of PdM or CBM is that the system has observable information and that the information is directly related to the fault occurrence. The CBM helps to find a specific fault pattern in the system and should be able to check with appropriate means and parameters, and the potential fault state can also be determined. Of course, there must be a reasonable time interval between the potential fault time and the functional failure time so as to be able to carry out necessary maintenance.

Corrective maintenance (CM) consists of the actions taken to restore a failed equipment or system to operational state. It may include the following steps: fault location, fault isolation, decomposition, replacement, reassembly, adjustment and testing. This maintenance usually involves replacing or repairing the component that caused the failure of the overall system. CM can be performed only at unpredictable intervals because the item's failure time is not known a priori. An item becomes operational after CM or repairs have been performed.

Among the most commonly used parameters that are generally used to reflect performance of maintained products, systems or services are:

- Mean time to repair (MTTR)
- Mean time between failures (MTBF)
- Steady-state availability.

MTTR reflects how good the system's maintainability is, and is a measure of system's ability to perform maintenance to restore assets to a specified condition. It provides a measure of the average time in hours required to restore an asset to its full operational condition after a failure. It can be computed by dividing the total repair time spent by the number of repairs or replacements.

However, MTBF provides a measure of an asset's reliability and can be computed by dividing the total operating time of an asset by the number of failures over a given period of time.

Another parameter used to reflect performance of a maintained system is steady-state availability, which is also called as uptime ratio and is calculated by dividing MTBF by (MTBF + MTTR), namely

$$A = \text{MTBF}/(\text{MTBF} + \text{MTTR}).$$

Further detailed discussion of maintainability can be found in Chap. 46 (pages 746–764) of [2].

1.4.3.1 Maintainability and Sustainability are Related

Small or lighter products or units with low cost of production and quality are generally replaced and not repaired upon failure and this would eventually lead to producing more units for operational requirement which affects sustainability (link "c" in Fig. 1.4) since it will generate more waste although the waste generated per unit may be low.

Sustainability has become an important criterion of design for institutional and commercial buildings in the last few decades. Engineers and architects continuously strive to reduce energy requirements of buildings in order to conserve valuable natural resources and reduce air pollution. However, with the use of unproven building materials and systems to achieve, these goals may create long-term issues for maintenance and these can have a significant impact on maintenance and operational costs.

Similarly, heating, ventilation and air conditioning (HVAC) systems definitely require sustainability improvements in many facilities due to the high costs involved for their installation, operation and maintenance. A properly designed and installed HVAC system can provide years of comfort for occupants, lower energy bills and improved water consumption. But a lack of proper planning can lead to increase in the material costs for preventive maintenance, energy costs and occupant comfort. In recent years, building automation systems (BAS) have become more complex in nature and they have come to be more than just HVAC controls. Today's BAS consists of electrical-power monitoring, lighting controls, condition-based monitoring, access control, and audio/visual system control. They enable technicians to optimize these facility systems to reduce energy use and maintenance costs.

It is needless to stress that regular maintenance also improves safety of an equipment of system (link "C" in Fig. 1.4). Especially, the high-risk systems such as nuclear power plants and very many chemical plants warrant operational safety of the highest order. To achieve this objective such plants need regular inspection and maintenance. Any failure in such a plant may be financially or economically disastrous and environmentally damaging. For example, in 1984, the methyl isocyanate gas leakage in Bhopal Union Carbide plant in India [12] was the worst industrial catastrophe in the history that resulted in immediate death of 2259 persons and some 8000 in the first weeks of disaster and another 8000 died later on from gas-related diseases.

1.4.4 Safety

Safety is another attribute of dependability, just as quality, reliability and maintainability, and allows a system or a product to function under predetermined conditions of use with some acceptable risk. All technological systems have hazards associated with them. A hazard is an implied threat or danger of possible harm. Stimulus is required to trigger a hazard which could be a component failure, operator's failure, maintenance failure or combination of events and conditions which may include an external event as well. The Fukushima Daiichi nuclear disaster that occurred on 11 March 2011 has been one of the gravest disaster in the history in recent times in which three of the plant's six nuclear reactors had melt down. In Fukushima case the stimulus was an external event since the disaster occurred when the plant was hit by a 13–15 m maximum height tsunami triggered by an earthquake of the magnitude 9.0. The plant started releasing substantial amounts of radioactivity on 12 March 2011 becoming the largest nuclear incident since the Chernobyl disaster in April 1986. In August 2013, it was felt that the massive amount of radioactive water is among the most pressing problems affecting the clean-up process, which may even take decades. As of 10 February 2014, some 300,000 people were evacuated from the area.

With the release of Prof. Rasmussen's WASH 1400 report [13] in 1975, safety (in probabilistic terms) became an important design parameter. Safety is planned, disciplined and systematically organized, and the before-the-fact activity is characterized by the identify–analyse–control strategy. Safety is designed into a system or product before it is produced or put into operation. Safety analysis can be of two categories: qualitative and quantitative methods. Both approaches are used to determine dependence between individual components failures with a hazard at system level. Qualitative approaches are used to assess "What possibly can go wrong, such that a system hazard may occur?", while quantitative methods provide estimations about probabilities, rates and/or severity of consequences. Hazard analysis is the cornerstone of safety design. Anticipating and controlling hazards, which may involve risk of loss of life or assets, is the main concern in system safety design.

The two conventional methods of safety analysis are called *failure mode and effects analysis* and *fault tree analysis*. Failure mode and effects analysis (FMEA) is a bottom-up, inductive analytical method which may be performed at either the functional or piece-part level. For functional FMEA, failure modes are identified for each function in a system or equipment item, usually with the help of a functional block diagram. For piece-part FMEA, failure modes are identified for each piece-part component (such as a resistor or a diode, etc.). The effects of the failure mode are described, and assigned a probability based on the failure rate and failure mode ratio of the function or component. When FMEA is combined with criticality analysis, it is known as *failure mode, effects, and criticality analysis* or simply FMECA. On the other hand, *fault tree analysis* (FTA) is a top-down, deductive analytical method. In FTA, initiating primary events such as component failures, human errors and external events are combined through Boolean logic gates to an undesired top event which is usually a system-level event such as an aircraft crash or nuclear reactor core melt. The main objective of system safety analysis here is to ensure that top event is made less probable, and also to verify whether the planned safety goals have been achieved. An earliest study using this technique on a commercial nuclear plant was the WASH-1400 study, also known as the reactor safety study or simply the Rasmussen report.

A fault tree is a logical inverse of a success tree which is basically related to reliability block diagram. Also, an FTA may be *qualitative* or *quantitative*. While the *qualitative FTA* is used to determine minimal cut sets when failure and event probabilities are unknown, the *quantitative FTA* is used to compute top event probability. If a minimal cut set obtained from qualitative FTA contains a single base event, then the top event may be caused by a single failure. Quantitative FTA usually requires computer software, and several software are available to carry out quantitative analysis.

Sometimes, an event tree is also used. An event tree starts from an undesired initiating event (such as loss of critical supply and component failure) and follows possible further system events through a series of final consequences. As each new event is considered, a new node on the tree is added with a split of probabilities of taking either branch. The probabilities of a range of "top events" arising from the initial event can be then seen. In certain situations, both fault trees and event trees can be used.

Further details are available in Chap. 41 (pages 661–674) of [2].

1.4.4.1 Safety and Sustainability are Related

Safety and health are related to sustainability as both of them concern themselves with similar objectives, that is, eliminate incidents, waste and overall losses, improve operational excellence, conduct business in a way that protects human, natural resources and reduce the environmental footprint. Like other attributes, safety also affects sustainability (link "**d**" of Fig. 1.4). This is a very interesting linkage. In a study, Evans [14] found that, in a single-car crash, the unbelted driver of a car weighing about 2000 lb is about 2.6 times as likely to be killed as is the unbelted driver of an approximately 4000-lb car. Also, it was found that the driver of a 2000-lb car crashing into another 2000-lb car is about 2.0 times as likely to be injured seriously or fatally as is the driver of a 4000-lb car crashing into another 4000-lb car. These results indicate that dematerialization alone cannot be a sufficient criterion for product design.

Another example of how sustainability consideration can affect safety: Many cities have switched to LED bulbs in their traffic lights because they use 90% less energy than the old incandescent lamps, last longer and save money. But their great advantage can also be their drawback particularly in winter or in cold countries. They do not waste energy by producing heat and therefore these bulbs don't generate enough heat to melt snow and can become crusted over in a snowstorm—a problem being blamed for many accidents as drivers cannot see traffic lights clearly. Therefore, system design must be carried out depending upon a situation and not just with the consideration of going green.

1.5 Technology Affects Sustainability

Technology also affects sustainability (linkage "**e**" in Fig. 1.4) and we have been able to achieve dematerialization and reduction in energy use as well as wastage in many areas through improved technology. If we look at the evolution of data storage system, there has been tremendous capacity improvement besides achieving sustainability through minimization of materials, energy use and wastage over a short period of time. For example,

- The first hard drive was developed by IBM in 1956 which was of the size of 2 refrigerators and had only a capacity of 5 MB.
- The first 8-inch floppy disk could be developed again by IBM in 1971 which had storage capacity of only 80 KB and was read-only. Its capacity was improved later to 4 MB.
- The first hard drive was developed again by IBM in 1980 and it could store up to 2.52 GB. The drive resembled like an engine and had the size of a refrigerator.
- Sony developed compact disc (CD) in 1980.
- In 1981, 3.5″ floppy disc was developed and there was a reduction in size and improved protection.
- The world's first CD ROM was developed in 1985 which provided compact storage capacity up to 900 MB.
- In 1990, the magneto-optical disc appeared and a special magneto-optical drive was developed to retrieve data from 3.5″ or 5.25″ discs.
- Later on, a compact disc rewritable CD ROM appeared in the market.
- The first DVD ROM appeared in 1996 with storage capacity up to 4.7 GB.
- In 1997, multi-media cards (MMC) using flash memory card were developed by Siemens and SanDisk.
- The first USB flash appeared in 2001, which would store up to 8 MB.
- Secure digital card (SD) with flash memory of 512 MB appeared measuring 32 × 32 × 2.1 mm.
- Blu-Ray (the next-generation optical disc) appears in 2003, which can be used for storing high definition videos.
- Modern USB drives come with a maximum capacity of 1 TB.
- Hitachi develops the world's first 1 TB in 2007 compact in size and a capacity of 1024 GB.
- Samsung develops world's largest capacity solid-state drive (SSD) in 2016. It has a capacity of up to 15 TB and measures only 2.5″ in size.

From the foregoing statements, it is amply clear that while the storage capacity kept increasing, the size and the energy and requirement decreased with every new developed technology. So also the quantum of electronic waste decreased for the same performance which helped improving the sustainability.

1.5.1 Technological Innovations also Affect Sustainability

Improved technology or technological innovations can also help improve sustainability in many ways, for instance:

- By using technological innovation like catalytic converters, we can make vehicular emission, which contributes to 25% of world's total CO_2 which is the single major factor leading to global warming, completely free of gases causing air pollution and carbon loads.
- There are many synthetic biofuels being used for use in vehicles, like ethanol, which is ethyl alcohol and is most often used as a biofuel in vehicles. Dimethyl ether (DME) is being developed as a synthetic second-generation biofuel (BioDME), which is manufactured from lignocellulosic biomass. There are several other sources of fuel for vehicles, like CNG (compressed natural gas), biogas, biodiesel, hydrogen, liquid nitrogen and so on.
- Since we have a limited reserve of gasoline on planet Earth, we need to build cars that will use rechargeable lithium-ion batteries in electric vehicles eliminating air pollution completely. Lithium-ion batteries are commonly used for portable electronics and electric vehicles and are growing in popularity for military and aerospace applications.
- If we were to increase the number of telephones by using old-fashioned standard phones, we would need many kilometres of wire to connect all those phones, and the copper for the wires will have to be mined. The process of mining uses a huge amount of fuel energy and causes considerable amount of pollution of land, water and air. On the other hand, when we use wireless cell phones, we don't need wires, and we can save all that fuel and pollution. Fortunately, this revolution has already taken place.
- In fact, new sustainable and non-polluting technologies have promise of reducing energy requirements of products and systems considerably. It has happened in case of microminiaturization of electronic devices. A laptop today consumes very less power than a system of 1960 which used tubes or a transistorized device of 1970–80 s that uses more power and less reliable. Why this will not happen if move over to the use of nano-devices?
- Technologies like genetic engineering, biotechnology, nanotechnology hold the key to developing sustainable products, systems and services. In fact, all future technological pathways would aim to minimize if not reverse the damage that has already been done to the earth's environment by the last industrial revolution.

Therefore, it is quite understandable that several possibilities exist for using technology to our advantage to prevent pollution and wastage of resources to help improve sustainability.

1.5.2 Technology also has a Lifespan

Shai [15] treats technology as a unique entity; similar to a "product" to which all the measures, interactions, processes and behaviours apply. In his opinion, a technology is born, strives to spread and circulate, reaches maturity and dies. Alternatively, a technology emerges or evolves, adopted, self-organizes to a temporal equilibrium, declines or loses its domination and becomes extinct. This life-cycle phenomenon applies to technologies as it does to any other known entity in nature. Therefore, the reliability theories and concepts can be equally applied to technology as an entity. Shai considers technology as a combination of three ingredients, namely raw materials to be manipulated, tools for production and use and lastly of the adequate manpower skills. He considers each technology as a unique combination of these elements. This set of elements is also suitable for the description of technologies in the post production period. A user needs to have an adequate skill to operate the product, and he also needs special tools to perform maintenance actions and in some cases may need materials to energize the operation. These three ingredients constitute the building blocks or DNA of a technology, sufficient to qualify technology as an entity.

Kemp [16], however, argues that sustainability has to be assessed on the basis of a system, and on the aspects of use, not on a technology basis. Fossil fuel technologies are generally viewed as non-sustainable because they depend on non-renewable resources (gas, oil and coal) whose combustion produces greenhouse gases as well as other emissions. For stationary sources, however, carbon emissions can be captured and stored for reuse at a later time. Fossil fuels can thus be made more sustainable. They can even be reused. Similarly, renewable energy technologies like solar, wind, hydro and biomass are frequently referred to as sustainable energy technologies. Even photovoltaic electricity, which is the cleanest source of electric energy, is not completely free of effects on the environment. As the raw materials for PV systems are shipped to factories, completed products must be transported from factories to consumers. At the end of its lifetime, they must also be safely disposed or given a new useful use. Sustainability can thus not be used as a label for a particular technology.

1.6 Costs Considerations

We have already seen that cost is associated with all attributes of performability, be it quality, reliability, maintainability or safety and even technology used for a product, system or service. To achieve high performance, we may have to incur increased cost of design and manufacturing, using and disposing (linkages **a**, **b**, **c**, **d** and **e** of Fig. 1.4). The poor deficient performance attributes not only affect the life-cycle costs but also have effects in terms of environmental consequences. Degraded performance attributes do reflect more on the material and energy demand and wastes and cause more environmental pollution when reckoned over a given period of time. It may also be mentioned here that a designer shall have to account for various costs associated with end-of-life options such as recycling and remanufacturing, which include the first cost, recycling cost and the cost of failure during disassembly and reassembly. The first cost is the cost of manufacturing and the first assembly. Recycling cost includes the cost of extracting material or cost of separating parts of different materials. Both maintenance and remanufacturing involve disassembly and reassembly and part reuse and failures can occur during these phases. Therefore, the consequences of the above failures are weighted by their probabilities of occurrence. For example, rivets and welds usually get destroyed during the disassembly. The one part of the cost includes the cost of a part getting damaged during assembly or disassembly. The other part of the cost includes the cost of damage done to a part when fastener is extracted. Maintenance costs are the costs associated with disassembly or assembly, whereas the remanufacturing cost is the total cost under all the mentioned heads. Reuse is an option to recycling because it extends the lifespan of a device. Devices still need eventual recycling, but by allowing others to purchase used electronics, recycling can be postponed and value gained from device use.

Another question that often arises is what the "true" cost of consumption and processing of the generated waste is to society. What is the true cost of burning fossil fuel for transportation particularly when the finiteness of resources and consequent long-term damage to the environment are considered? Should not the high-grade resources be made available at higher cost so that the profits may be reinvested towards development of the capital and the knowledge to permit the use of lower-grade resources and the development of technological substitutes later on? What is the actual disposal cost of industrial wastes? To what extent is the cost of waste collection be subsidized by different societies and different segments of a society? Would a higher cost for garbage collection effectively encourage recycling, sorting recyclable materials at the generation source, and dematerialization? Would it encourage more illegal dumping? Can society truly afford to continue functioning in its present "throwaway" mode of products such as watches, radios, flashlights, light bulbs, cameras and calculators?

So far, classical economic theories [4] have treated nature as a bottomless well of resources and infinite sink for waste and this notion will have to be discarded. Environmental costs, that is, the cost of preventing or controlling pollution and ecological disruption, have never been internalized. In fact, it is our incapability to deal with economic nature of environmental pollution that has been largely responsible in destroying Earth's ecological systems. It is time that we need to pass on the hidden environmental costs incurred on resource exploitation onto the consumer or to the user, for preserving our environment for future generations. The internalization of hidden costs of environment preservation will have to be accounted for, sooner or later, in order to be able to produce sustainable products in the long run. It is therefore logical to add these hidden costs to the cost of acquisitioning a product, a system or a service.

1.7 Possible Strategies of Design for Performability

From the definition of performability (Fig. 1.1), it is obvious that a performable product, system or service would be the one which is both *dependable* as well as *sustainable*. As mentioned earlier, a product can be made dependable but may not be sustainable or the other way round. Having just one of these two attributes, a product cannot be called as performable. We must ensure that both attributes are present in a performable entity. Therefore, when we intend to improve performability of an entity; be it a product, system or service, we must maximize sustainability as well as dependability. Dependability comprises reliability, maintainability and safety which are probabilistic and can be maximized with respect to cost of achieving them and literature is abundant with such optimization problems and their solution techniques. We will briefly discuss these here. But we will also enunciate a methodology by which we can optimize sustainability.

1.7.1 Designing for Dependability

Designing a product, system or service for dependability (reliability, maintainability, safety) is not a new problem as we have been doing it over several decades [17–36]. We have maximized product or system reliability with respect to cost of developing the product or system under various conditions.

Reliability, maintainability and safety design are the areas of dependability design, which allow more effective use of resources, at the same time helps decrease the wastage of scarce finances, material and manpower. There are several alternatives available to improve the system dependability. The most known approaches are:

1. Reduction of the complexity of the system.
2. Use of highly reliable components through component improvement programmes.
3. Use of structural redundancy.
4. Putting in practice a planned maintenance, repair schedule and replacement policy.
5. Decreasing the downtime by reducing delays in performing the repair. This can be achieved by optimal allocation of spares, choosing an optimal repair crew size, and so on.

The product improvement programme requires the use of improved packaging, shielding techniques, derating and so on. Although these techniques result in a reduced failure rate of the component, they nevertheless require more time for design and special state-of-the-art production. Therefore, the cost of part improvement programme could become high and may not always be an economical way of system performance improvement. Here the often system optimization design problem would become a reliability allocation problem since components at various stages may have different cost of improving the component.

On the other hand, the use of structural redundancy at subsystem level, keeping system topology intact, is an effective and cheapest means of improving the system reliability to any desired level. Structural redundancy can be of the form of partial redundancy (k-out-of-m type), active redundancy (1-out-of m type), or standby redundancy in which case a switch is used to replace the failed unit by a healthy unit. Here again there could be k units in parallel along with m–k spare units which take the position of one by one upon the failure of a unit to keep in operation k number of units at any time. Redundancy allocation is the most economical method of improving system reliability.

The use of Lawler and Bell's [17] algorithm for reliability design was first introduced by Misra [18]. Subsequently, this algorithm came to be widely used for a variety of reliability design problems. However, it suffered from a major limitation of computational difficulty caused by a substantial increase in the number of binary variables [18].

Misra [19] solved m-constraints design problem by simultaneously solving m-problems of single constraint. This procedure helps generate many feasible solutions to the design problem which can offer a designer an optimal or a near-optimal solution corresponding to flexible resources [19, 20].

Another new development, in the field of reliability design, took place when Misra and Ljubojevic [21] for the first time demonstrated that to obtain a globally optimum design, optimization of system reliability must be done using both component reliability and redundancy level as decision variables in the problem. They formulated the design problem as a mixed-integer programming problem, and solved it by a simple technique. The reliability literature till then offered abundant methods for the optimal design of a system under some constraints. In most of the papers, the problem considered is: given reliabilities of each constituent component and their constraint-type data, optimize the system reliability. This amounts to the assignment of optimal redundancies to each stage of the system, with each component reliability specified. But this was a partial optimization of the system reliability since at the design stage a designer has the option of choosing component reliability improvement as well as a recourse to use of redundancy. A true optimal system design must explore these two alternatives explicitly. The paper [21] demonstrated the feasibility of arriving at an optimal system design using this concept. For simplicity, only a single-cost constraint was used. A typical cost–reliability curve was used to illustrate the approach. However, the approach was more general and could be extended to any number or type of constraints.

Misra [22] also suggested a formulation for a maintained (repairable) system design; reliability and maintainability designs are usually carried out right at the design stage, and failure and repair rates are allocated to each component of the system in order to maximize its availability and/or reliability. For such systems it becomes imperative to seek an optimal allocation for spare parts while maximizing availability/reliability subject to some techno-economic constraints on cost, or other resource and so on.

Literature was full of many sophisticated methods of optimizing system reliability subject to some given techno-economic constraints. Chronologically, the first review was published by Misra [23] in 1975. Subsequent reviews have been published by Tillman et al. [24], Misra [25], and by Kuo and Wan [26] in 2007 and later on again by Misra et al. [27] in 2008.

Misra [28] was also the first to introduce the formulation where mixed type of redundancies (namely active, partial or standby) are found in the optimal reliability design of a system. Prior to the publication of [28], the reliability design formulations invariably considered only active redundancies in redundancy allocation design problems.

Misra [22] also provided formulation for design of maintained system and Sharma and Misra [29] proposed a formulation for an optimization problem involving three sets of decision variables, viz., redundancy, spares and number of repair facilities, simultaneously. The necessity of a proper trade-off, to achieve an optimum reliability design of a system, is stressed in the paper. Both reliability and availability can be considered as design criteria to arrive at an optimum configuration. Nonlinear constraints are permitted. Here again, MIP [25, 31] was shown to be the most effective method to solve the design problems.

The dimensionality difficulty of [17] was overcome by Misra in [30] when he proposed a search procedure similar to [17] in integer domain (and not in zero–one variables domain as is done in Lawler Bell's algorithm) for optimal design of a system which may employ any general type of redundancy, that is, standby, partial or active. In fact, this simple and efficient algorithm, called MIP algorithm (Misra integer programming algorithm [31]) can solve any integer programming problem. It is based on lexicographic search, and MIP requires only functional evaluations and carries out a limited search close to the boundary of resources. It can handle various system design problems with any type of objective and constraints (nonlinear/linear functions) and does not impose any convexity and concavity conditions on functions for optimality condition in which the decision variables are restricted to take integer values only. The method is applicable for both small and large problems and its universality was demonstrated in [31], where MIP search method was applied to integer programming problems which need not be of separable form and may have any arbitrary form of functions.

Misra and Sharma [32] employed a new MIP search algorithm to attempt several system reliability design problems. It provides an advantage of exploring all the feasible design solutions near the constraint boundary and eliminates many of the unwanted feasible points usually searched with L-B algorithm. The MIP algorithm is conceptually simple and efficient for solving many design problems. In Sharma and Misra [33], a more general case of formulation of optimization was considered in which both component reliability and redundancy allocation as decision variables along with mixed type of redundancies at subsystem level are considered. Multi-criteria optimization problems also can be easily solved using MIP. It has been proved to be useful in configuration designing of computer communication system. In fact, MIP [34] provides more general approach of optimization even in case of non-series parallel system configuration as well as to design such configuration with given linear or nonlinear constraints as well as with multiple choices available of component reliability. MIP [34] also provides a useful approach in parametric optimization problems. Among others, problem of where to allocate redundancy, problem of mix of component (that allows for selection of multiple component choice, with different attributes, for each subsystem) and modular or multi-level redundancy allocation are some of the important issues in reliability optimization problems that can be solved by MIP [34].

There have been other approaches [35, 36] for system design, like there has been research on reliability optimization of systems that consist of multi-state system [35]. Unlike two-state systems, multi-state systems assume that a system and its components may take more than two possible states (from perfectly working to completely failed). A multi-state system reliability model provides more flexibility for modelling of system conditions than a two-state reliability model [36].

For probabilistic risk and safety, fuzzy set theory [37] provides the most appropriate framework for its assessment and optimization. Misra and Weber [38] used fuzzy set theory to compute probabilistic risk for carrying out level 1 study of risk in case of nuclear power plant case. This approach can also be used to optimize safety using fault tree methodology [39].

1.7.2 Designing for Sustainability

Designing for sustainability requires selecting a measure by which the sustainability can be reckoned. At present, the sustainability aspect of performance is not quantifiable in probabilistic terms just as the dependability is, but may be in near future that is possible to do. Once that is done, it might become possible to aggregate all attributes in some way to define overall design criterion in probabilistic terms and optimization could then be carried out with respect to cost or any other aspect of life-cycle attribute.

Fortunately, we have life-cycle assessment (LCA) methodology, also called life-cycle analysis methodology that can provide us a measure for assessing environmental impacts associated with all stages of the life cycle of a product, system, process or service (from cradle to grave). For instance, in the case of a manufactured product, environmental impacts can be assessed from raw material extraction stage (mining, drilling, etc.), processing reusable materials, metal melting and so on, through the product's manufacture (ensuring that processes are not polluting or harmful to employees), distribution (materials used in packaging are environmentally friendly and are recyclable) and during the product use, to the recycling or final disposal of the materials composing it (to grave). The end-of-life treatment of a product is also important as some products may produce dangerous chemicals into environment (air, ground and water) if they are disposed of in a landfill. The design of products should be such that the overall energy consumption throughout the product's life. The products should also be designed such that they could be easily disassembled at the end of their life and parts be reused if required and minimum energy is used to achieve this.

An LCA study involves a thorough inventory of the energy and materials that are required across the industry value chain of the product, process or service, and calculates the corresponding emissions to the environment. LCA thus assesses cumulative potential environmental impacts. The aim in improving sustainability is to improve the overall environmental profile of a product. An LCA can be used to forecast the impacts of different production alternatives on the product and thus helps choose the most environmental-friendly process. A life-cycle analysis can serve as a tool to determine the environmental impact of a product or process. Proper LCAs can help a designer compare several different products according to several categories, such as energy use, toxicity, acidification, CO_2 emissions, ozone depletion, resource depletion and many others. By comparing different products, designers can make decisions about which environmental hazard to focus on in order to make the product more environmentally friendly.

Widely recognized procedures for conducting LCAs are included: International Organization for Standardization (ISO), in particular, in ISO 14040 and ISO 14044, which have become the reference standard for several other international standards based on the life-cycle concept. Based on the ISO 14040-Environmental Management—Life Cycle Assessment, Principles and Framework (ISO 14040: 2006) and ISO 14044-Environmental Management, Life Cycle Assessment—Requirements and Guidelines (ISO 14044 2006), recent developments led to a spin-off standard like carbon footprinting (ISO 14067:2018). This document specifies principles, requirements and guidelines for the quantification and reporting of the carbon footprint of a product (CFP), in a manner consistent with International Standards on life-cycle assessment (LCA) (ISO 14040 and ISO 14044).

But at this moment perhaps carbon footprint can help us compute the degree of sustainability since all materials, energy use, processes and even wastages in the form of solids, fluids or gases do create carbon footprints when reckoned over the life cycle of the product; it may help us provide a measure of sustainability. Therefore, maximizing sustainability may be equivalent to minimizing the carbon footprint of a product, system or service through all stages of life cycle, as shown in Fig. 1.2 of a product, system or service.

1.7.3 Carbon Footprint: A Possible Measure of Sustainability

Carbon footprint is the total amount of greenhouse gases (GHG) produced directly or indirectly during the different stages in the life cycle of products and services. It is calculated by summing the emissions resulting from every stage of a product or service's lifetime (material production, processing, transportation, manufacturing, sale, use phase and end-of-life disposal). This is also known as *cradle-to-grave* product carbon footprint (PCF), which sums up the emissions from the extraction of raw materials needed to generate the final product, through manufacturing of precursors and the product itself, down to the use phase and disposal of the product.

There is also a *cradle-to-gate* PCF, which considers only carbon footprint (CF) from extraction of raw materials to production.

All materials leave a carbon footprint during manufacturing and use or disposal. Carbon footprint can help in assessing the efficacy of the strategy of dematerialization process as well as of energy saving and waste reduction, and thus can serve as the measure sustainability effort. Carbon footprint is normally expressed in equivalent tons of carbon dioxide (CO_2), after summing up all the GHGs produced at each stage of the life of a product; the PCF can also be expressed as grams or kilograms of CO_2e per unit of product. For example, the carbon footprint of a 330 ml can of Coke [40] that has been purchased, refrigerated, consumed and then recycled by a consumer in the UK is 170 g CO_2e. The outcome of these calculations is often referred to as product carbon footprints (PCFs), where carbon footprint is the total amount of GHGs produced for a given activity where a product is some goods or a service that is marketed. The development of public and international PCF standards is at a very early stage. The first one to cover a wide range of diverse products, PAS 2050, was published in October 2008 by the British Standards Institute and the Carbon Trust, while the International Organization for Standardization only started to develop a carbon footprint of products standard (ISO/NP 14067-1/2) in late 2008 [40]. Several organizations have calculated carbon footprints of various products [41–50]. The US Environmental Protection Agency has even assessed the carbon footprints of paper, plastic (candy wrappers), glass, cans, computers, carpet and tyres, and so on.

Even energy generation or its use creates a carbon footprint. Studies show that hydroelectric, wind and nuclear power always produced the least CO_2 per kilowatt-hour of any other electricity sources. Even renewable electricity generation sources, like wind power and hydropower, emit no carbon from their operation, but leave a footprint during construction phase and maintenance during their operations.

Material and energy are also consumed during each stage of life cycle. Therefore, carbon footprinting can provide us a common measure of evaluation of sustainability. As mentioned earlier, the measure of carbon footprint can help us in assessing the degree of sustainability we can achieve. In other words, we can determine the degree of dematerialization we can achieve, that is, the amount of material required in developing a product and how much waste will be created (solid or gaseous) and how much polluting a process used in manufacturing of a product is, and finally how much energy is being used during its manufacturing or during its use. All measured in terms of carbon footprints and by summing up the entire carbon footprint over the life cycle, that is,

$$C_{\mathrm{T}} = C_{\mathrm{ex}} + C_{\mathrm{M}} + C_{\mathrm{use}} + C_{\mathrm{dis}} \tag{1.1}$$

where C_{T} is the total carbon footprint created during the entire life cycle, and C_{ex}, C_{M}, C_{use} and C_{dis} are the carbon footprints created during the extraction, manufacturing, use and disposal phases of the life cycle of the product, system or service, respectively.

It may, therefore, be possible to arrive at an optimal design of a product, system or service with respect to sustainability criterion.

1.7.3.1 Carbon Footprint During Extraction (C_{ex})

There are several activities involved in the extraction of the minerals. The first one is the process of extracting the ore from rock using a variety of tools and machinery. The next is processing, during which recovered minerals are processed through huge crushers or mills to separate commercially valuable minerals from their ores. The ore is then transported to smelting facilities either through trucks or belt conveyors. Smelting involves melting the concentrate in a furnace to extract the metal from its ore. The ore is then poured into moulds, producing bars of bullion, which are then ready for sale or use. The last stage in mining operations is closure and reclamation. Once a mining site has been exhausted of reserves, the process of closing the site occurs, dismantling all facilities on the property. The reclamation stage is then implemented, returning the land to its original state. All these activities are energy-intensive and create a substantial amount of carbon footprint. Therefore, C_{ex} shall be the total amount of carbon footprint created during extraction phase and must be assessed reasonably.

It may be worthwhile to mention here that environmental impacts of methods of extraction or mining can be very much different and should be carefully chosen to produce minimum CF. For example, lithium production from hard mineral ore uses large amounts of energy and chemicals and involves significant land clearing. Production from brine ponds, where water is evaporated from high-lithium salty groundwater, is thought by some researchers to be preferable environmentally, but it still uses large amounts of water and toxic chemicals, which can pose risks to water supply.

1.7.3.2 Carbon Footprint During Manufacturing (C_M)

Wang et al. [41] show how carbon emissions during manufacturing phase can be computed. The total carbon emission of a manufacturing phase, C_{MT}, can be calculated as the sum of the carbon emissions generated from various processes and can be expressed as follows:

$$C_{MT} = \sum_{i=1}^{r}(C_{Me}(i) + C_{Mm}(i)) + \sum_{j=1}^{s} C_{Mf}(j) \tag{1.2}$$

where $C_{Me}(i)$ is the directed energy-related carbon emission of the ith unit process in the manufacturing process plan, $C_{Mm}(i)$ is the material-related carbon emission of the ith unit process in the process plan of r processes of manufacturing and $C_{Mf}(j)$ is the indirect energy-related carbon emission of the jth zone in manufacturing plant with s zones. The carbon energy emissions related to direct energy consumption depend upon the state of equipment and process parameters, viz.,

$$E_{total}(i) = E_{tip}(i) + E_{idle}(i) + E_{basic}(i)$$

$$= \int_0^{T_{tip}} P_i d_t + P_{idle} \cdot T_{idle} + P_{basic} \cdot T_{basic}, \quad (1.3)$$

where for the *i*th unit process, $E_{total}(i)$ is the total energy consumed and $E_{tip}(i)$ is the energy associated with the productive part of the operation (e.g., energy during cutting of the material) which is a function of operating parameters, $E_{idle}(i)$ is the "idle" energy consumed when there is no active processing, $E_{basic}(i)$ is the *basic* energy consumed by fundamental activities of the manufacturing equipment, T_{tip} is total processing time, P_{idle} is the total power during *idling*, T_{idle} is the time in a state of being in idle state, P_{basic} is the basic power, and T_{basic} is the time in a basic state. According to Hu et al. [45], if P_i is the total power during the process of cutting material, it is sum of standby power P_u, cutting power P_c and additional load loss power P_a (the certain power loss generated by the load of workpieces). Thus,

$$P_i = P_u + P_c + P_a. \quad (1.4)$$

The energy from each unit process can then be converted to a carbon emission using the following equation:

$$C_e(i) = E_{total}(i) * \alpha_{elec} \quad (1.5)$$

where C_e (i) is the carbon emission of the energy consumed in the *i*th unit process and α_{elec} is the conversion factor for electricity to carbon emissions. α_{elec} depends upon the generation utility service and seasonal changes.

Similarly, material-related carbon footprint of a manufacturing unit in (1.2) can be calculated as follows:

$$C_{Mm}(i) = \sum_{k=1}^{w} m_k(i)\alpha_k(i) \quad (1.6)$$

where $C_{Mm}(i)$ is the total carbon emissions of the material flow for the *i*th unit process in the process plan, $m_k(i)$ is the mass for the *k*th type of material used in the *i*th unit process, and $\alpha_k(i)$ is the conversion factor for the *k*th type of material to carbon emission in the *i*th process.

The carbon footprint caused by auxiliary resource and energy consumptions of ancillary equipment in manufacturing plant in *j*th zone can be calculated by multiplying the total indirect energy consumed in zone *j* with the conversion factor for energy to carbon emissions (i.e., α_{elec}), as expressed in (1.6):

$$C_{Mf}(j) = E_f(j) \cdot \alpha_{elec} \quad (1.7)$$

and (1.7) for a total of *s* zones of manufacturing plant will be given by:

$$\sum_{j=1}^{s} C_{Mf}(j) = \sum_{j=1}^{s} E_f(j) \cdot \alpha_{elec} \quad (1.8)$$

1.7.3.3 Carbon Footprint During Distribution of Finished Product

Finished product is often moved by manufacturers to retailors either by air, land or sea transport. Because of this diversity, a simplified Eq. (1.9) based on PAS 2050 [42] and ISO 14067 [43] can be used to estimate the CF related to distribution. This estimation considers measurements of different types of transport to ship the product to its final destination.

$$C_e = \Sigma m \times d \times \alpha f \quad (1.9)$$

where C_e is the total GHG emissions in kg CO_2eq related to the shipment and m refers to the mass (kg) of the products when they are transported by air or land. When considering sea transportation, m would refer to the volume occupied in m^3. d is the distance travelled during transport, and αf is a specific emission factor (in g CO_2eq/km) that considers relevant load factors to allocate CO_2eq emissions.

1.7.3.4 Carbon Footprint During Use of Product (C_{use})

Carbon footprint (CF) during use phase of a product comprises mainly due to the energy consumed by the product and the wastage created during the use phase of a product.

The estimation of CF in the use phase for an electronic device is based on PAS 2050 [42] and ISO 14067 [43]. A simple Eq. (1.10) proposed in these standards estimates the annual average of energy emission factors for a specific country, considering the average energy consumption of the device in that country. The equation for the estimation:

$$C_s = W \times T \times \alpha_f \quad (1.10)$$

where C_s is the total GHG emissions during their useful life (g CO_2eq); W is the average electric power consumption in kWh, T is the average time in hours the device is working; which based on the design; and α_f is the specific emission factor that depends on how electricity is generated in the country where the device is located (g CO_2eq/kWh).

The energy consumption of some products such as automobiles, appliances and electronic devices is usually much higher in their use phase than in their production or in the extraction and processing of materials used to make them.

A substantial amount of PFC is created by products of general use. For example, it has been estimated [40] that the consumable goods and appliances that an average consumer in the UK buys and uses account for 20% of UK total carbon emissions (not counting the energy to run them), of which food and non-alcoholic drinks, at 9%, comprise the largest category (Carbon Trust 2006).

Consumers have started showing interest in PCF information and would possibly prefer carbon-labelled products and firms over others, while other things being equal and would be willing to pay a premium price for products with lower footprints than the ones not much different from organic price premium. However, they are sceptical and show a preference for a third-party verification for claims made by low CF products. Since PCF is based on LCA, it is likely to have a higher degree of credibility with consumers than any other sort of claims made by manufacturers or retailers in relation to the climate-change attributes of products.

The energy during the use phase of a product can be cut down using the concept of energy sufficiency actions which comprises the actions which reduce energy demand by changing the quantity or quality of services that we can get from products, such as:

- Reducing the duration/frequency of usage, and using products differently
- Unplugging a product instead of leaving it on standby
- Using a tablet instead of a computer to surf on the internet
- Dimensional better sizing of energy-using products to match people's true needs
- Collaborative increased sharing of products
- Sharing a Wi-Fi access.

1.7.3.5 Carbon Footprint During Disposal (C_{dis})

When a product reaches the end-of-life, it can be either disposed of or recycled, or reused. Methods to reckon emissions in this phase are based on PAS 2050 [42] and depend on the final treatment given to the product. If the product is disposed of as waste in a landfill, it will not generate GHG emissions because it is not composed of organic matter. The remaining disposal methods, namely reuse and recycling, would however generate emissions.

Reuse refers to considering a new use to some part of the product instead of discarding it. PAS 2050 [42] provides an expression to simplify the calculation of GHG emissions using (1.11):

$$\text{GHG emissions} = (a + f)/b + c + d + e \tag{1.11}$$

Recycling transforms matter using energy demand to execute it. PAS 2050 [42] offers different options to assess emissions related with this process, depending on the transformation that the product undergoes. One method considers that the recycled material does not maintain the same inherent properties as the virgin material input; for this case, emissions are estimated using (1.12).

The other method considers that recycled material maintains the same inherent properties as the virgin material input; for this case, emissions are estimated using (1.13). Selection of the method depends on the knowledge of the material and its capacity to be recycled.

$$E = (1 - R_1)E_V + R_1 E_R + (1 - R_2)E_D \tag{1.12}$$

$$E = (1 - R_2)E_V + R_2 E_R + (1 - R_2)E_D \tag{1.13}$$

where R_1 is the proportion of recycled material input; R_2 is the proportion of material in the product that is recycled at end-of-life; E_R are emissions and removals arising from recycled material input per unit of material; E_V are emissions and removals arising from virgin material input per unit of material; and E_D are emissions and removals arising from disposal of waste material per unit of material. As mentioned before, in these stages, the location of the product is important to estimate the CF.

When considering the recycling process, data availability of EF per material in each location limits the estimation of GHG emissions. Moreover, comparison of GHG emissions is complex because of the differences encountered in the collection and recycling methods of materials according to the technological level applied [44] for different locations. However, different methods to obtain recycling EF per material are based on ISO 14067 [43] and PAS 2050 standards, which is why emission factors are directly applied in the proposed methodology and EF per material is based on ISO 14067 and PAS 2050 standards, which is why emission factors are directly applied in the proposed methodology.

1.7.4 *Use of Low Carbon Technologies*

We have seen that for better sustainability, a low carbon footprint is essential. It is also widely accepted that for a swift change to meet the Paris Agreement's goals of limiting global warming to below 2 °C, low-carbon technologies will be needed for the world even to strive for 1.5 °C. This will trigger a strong demand for a wide range of base and precious metals, such as aluminium, silver, steel, nickel, lead and zinc along with cobalt, lithium and rare earth elements (REEs), which are a group of 17 chemically similar elements, each having unique properties, making them useful for a wide range of technologies from low-energy lighting and catalytic converters to the magnets used in wind turbines, electric vehicles and computer hard-drives. Neodymium (REE) is relatively abundant in the Earth's crust, but difficult to find in good concentrations to make it economic to mine. The demand for neodymium will become more prevalent if direct-drive technology is used for offshore wind power, since it uses neodymium in its permanent magnets.

Nickel is another ingredient needed for batteries and is expected to form a large demand for future batteries. Nickel is already widely used in stainless steel production. Manganese is also used in batteries, and is an essential ingredient in steel. Copper is widely used as a conductor for power, as well as general wiring, motors and so on. Both copper and manganese are among the most widely extracted metals in the world. Clean technologies also rely on lithium and cobalt. Lithium is crucial ingredient of lithium-ion batteries which are used in smartphones to electric vehicles, but which now pose the biggest demand from consumers. The lithium-ion batteries are used by car makers, including Tesla, BMW, Ford and Nissan. Cobalt, a silver-grey metal produced mainly as a by-product of copper and nickel mining, is another essential component of the cathode in lithium-ion batteries. It is also used in several industrial and military applications. Many other metals are used to a larger or smaller extent in clean-energy production and low-carbon technology. Indium and gallium, for example, are used in the coatings of photovoltaic film and have also been identified by the EU report as critical materials.

A 2015 UNEP report reckoned "cradle-to-grave" GHG emissions of clean-energy sources and are commonly 90–99% lower than for coal power. The report observed that wind, solar PV, concentrated solar-thermal, hydro and geothermal sources of power generated less than 50 g of CO_2 equivalent per kilowatt hour (gCO_2e/kWh).

1.8 Hypothesizing Dependable and Sustainable Designs

Our concern in performability engineering is not limited to designing products, systems and services for performance in conventional sense of dependability but also consider optimizing the processes that create them. It is not difficult to visualize that by employing the strategy of dematerialization, minimum energy and minimum waste while maximizing the yield and developing economically viable and safe processes (clean production and clean technologies),we will have minimal adverse effect on the environment during production, use and disposal (at the end of the life) phases of life cycle. This is basically the goal of performability engineering.

The author, therefore, like to propose a two-stage procedure for designing optimal products, systems and services with respect to performability which not only would be dependable but also be sustainable, or in other words, will have high performability over the specified period of time, usually the life cycle. There are two alternatives to improve sustainability. One way is to use better technology requiring minimum material and energy. The other way to improve sustainability is to use of improved processes resulting in minimum waste. This can probably be done in the following stages, namely,

1. In the first stage, we can develop a base case design with minimum cost using alternative technologies solely from the dependability (reliability, maintainability and safety) consideration. These designs should meet all the technical and functional and operational specifications of the product, system or service being

designed under the specified conditions of use. We arrange these designs with alternate technologies (say, three) D_1, D_2 and D_3 in increasing costs, remembering that the costs associated with these designs (C_1, C_2 and C_3) are the minimum costs for the corresponding technologies D_1, D_2, D_3, such that $C_1 < C_2 < C_3$. Now to start the design iteration, we select the technology D_1 with minimum cost of C_1.

2. In the second step of design, we need to assess the degree of sustainability of the design D_1 arrived at in the first step by calculating the carbon footprints of the product for all the processes used to produce the base case design considering the carbon footprints over of the products life cycle under use and disposal phase.
3. Next, we need to explore the possibility of improving upon the sustainability of designs arrived at in the first step by reviewing the technology used and hardware design which will offer an advantage of minimization of material and energy requirement of the product being designed along with the advantage of minimization of waste generated by this product during its manufacturing and use and disposal stages over the entire life-cycle period. This can be tried using several alternative technologies, substitute materials or processes. Several alternative designs using various technologies can be evaluated.
4. Let CF_1, CF_2 and CF_3 be the carbon footprints using designs D_1, D_2 and D_3, whichever design among the D_1, D_2, D_3 offers the advantage of better sustainability, that is, minimum carbon footprint over the life cycle that is adopted as the base case design replacing the earlier one, selected in Step 1.
5. We can now check upon dependability. If the optimum design arrived at in Step 4 offers a better dependability or at least equal to earlier dependability, the design will be done. In other words, we have a product design which satisfies the criterion of maximized dependability with minimum cost as well as optimized sustainability. In other words, it offers a design with minimum carbon footprint. Otherwise, we move to step 3 to start next iteration.

Lack of data available with the author does not permit actual quantitative evaluation of the optimum design. It is expected that those organizations or institutes with reliable carbon footprint data under their possession will take the lead in this direction. It is just an elementary step towards it.

Sustainability requires that the processes used in manufacture of products, system or services must be clean and non-polluting. Again a reward/or penalty can be introduced in assessing the cleanliness of the processes. The sustainability criterion also requires that the product process should be waste free or should have means of utilizing the waste created by the production processes. Sustainability criteria also require that the energy requirement for the production process as well as during the product maintenance should be minimum and may possibly use clean energy sources. Reuse and recycle possibilities should be rewarded suitably in the design model.

The alternative modern technologies, such as nanotechnology etc. also widen the possibilities of such a realization of a product design. In fact, the concept presented in this chapter may help provide new ideas of research for design of high performability systems, products and services. This chapter only discusses the concept of such a possibility and what may come in the twenty-first century for researcher is to explore further.

1.9 Conclusion

This chapter has explored the possibility of assessing sustainability using the concept of carbon footprints that all materials, energy use and processes create so that future products, systems, and services can be optimally designed with minimum carbon footprint or high performability. It is expected that these designs shall be more environmental friendly as well as economic over their entire life cycle, while offering high performance. The concept of using performability as design criterion is far wider than just reliability or dependability, which has been used by engineers and designers in the past without bothering the environmental impact the design will create in the long run. However, there is much research needed to be done in this area. It is hoped that this chapter would help generate interest and develop new ideas in this area.

Acknowledgements In presenting the state-of-the-art, it is usually necessary to discuss and describe the work done by several authors and researchers, published in the literature. The author would like to record his appreciation and thanks to all those, whose work finds a place of mention in this chapter.

References

1. Misra, K. B. (2005, July). Inaugural Editorial of International Journal of Performability Engineering, *1*(1), 1–3.
2. Misra, K. B. (Ed.). (2008). *Handbook of performability engineering*. London: Springer (76 Chapters, 1318 pages, 100 authors).
3. Misra, K. B. (2013). Sustainable designs of products and systems: A possibility. *International Journal of Performability Engineering, 9*(5), 175–190.
4. Misra, K. B. (Ed.). (1996). *Clean production: Environment and economic perspectives*. Heidelberg, Berlin, Germany: Springer.
5. Misra, K. B. (Ed.). (1993). *New trends in system reliability evaluation*. Elsevier.
6. Misra, K. B. (1992). *Reliability analysis and prediction: A methodology oriented treatment*. Elsevier.
7. Sandra, P. (1994). *Carrying capacity: Earth's bottom line, in "State of the World"*. New York: W. W. Norton & Co.
8. Herman, R., Ardekani, S. A., & Ausubel, J. H. (1990). Technological forecasting and social change, 38, 333–347.
9. Shewhart, W. A. (1931). *Economic control of quality of manufactured product*. New York: Van Nostrand.

10. Feigenbaum, A. V. (1983). *Total quality control* (3rd ed.). New York: McGraw-Hill.
11. Westerman, R. R. (1978). Tires: Decreasing solid wastes and manufacturing throughput. Report EPA-600/5-78-009. Cincinnati, Ohio: U.S. Environmental Protection Agency.
12. Sriramachari, S. (2004, April). The Bhopal gas tragedy: An environmental disaster. *Current Science, 86*(7), 905–920.
13. Rasmussen, N. (1975). US Nuclear Safety Study, WASH-1400 Report, US Nuclear Regulatory Commission.
14. Evans, L. (1985). Car size and safety. Results from analyzing U.S. accident data. In *The Proceedings of the Tenth International Conference on Experimental Safety Vehicles*, Oxford, U.K., July 1–5, 1985. Washington, D.C.: U.S. Government Printing Office, pp. 548–555.
15. Shai, Y. (2015, March). *Reliability of technologies*. Ph.D. Thesis, submitted to of the Technion Israel Institute of Technology, Haifa, Israel.
16. René, K. (2008, September). *Sustainable technologies do not exist*. Paper for DIME Conference on "Innovation, Sustainability and Policy", Bordeaux, September 11–13, 2008.
17. Lawler, E. L., & Bell, M. D. (1966). A method of solving discrete optimization problems. *Operations Research, 14,* 1098–1112.
18. Misra, K. B. (1971, September). A method of solving redundancy optimization problems. *IEEE Transaction On Reliability, R-20*(3), 117–120.
19. Misra, K. B. (1972, February). A simple approach for constrained redundancy optimization problem. *IEEE Transaction on Reliability, R-21*(1), 30–34.
20. Misra, K. B. (1972). Reliability optimization of a series-parallel system-part I: Lagrange multiplier approach, part: II maximum principle approach. *IEEE Transaction on Reliability, R-21*(4), 230–238.
21. Misra, K. B., & Ljubojevich, M. (1973, December). Optimal reliability design of a system: A new look. *IEEE Transaction on Reliability, R-22*(5), 255–258.
22. Misra, K. B. (1974). Reliability design of a maintained system. *Microelectronics and Reliability, 13*(6), 495–500.
23. Misra, K. B. (1975). On optimal reliability design: A review. *IFAC Proceedings, 8,* 27.
24. Tillman, F. A., Hwang, C. L., & Kuo W. (1977). Optimization of system reliability with redundancy—A review. *IEEE Transactions on Reliability, R-26*(3), 148–155.
25. Misra, K. B. (1986). On optimal reliability design: A review. *System Science, 12*(4), 5–30.
26. Kuo, W., & Wan, R. (2007). Recent advances in optimal reliability allocation. *IEEE Transaction on System, Man, and Cybernetics-Part a: System and Humans, 37*(2), 143–156.
27. Lad, B. K., Kulkarni, M. S., & Misra, K. B. (2008). Optimal reliability design. In: K. B. Misra (Ed.), *Handbook of performability engineering*. London: Springer.
28. Misra, K. B. (1975, June). Optimum reliability design of a system containing mixed redundancies. *IEEE Transactions on Power Apparatus and Systems, PAS 94*(3), 983–993.
29. Sharma, U., & Misra, K. B. (1988). Optimal availability design of a maintained system. *Reliability Engineering & System Safety, 20*(2), 147–159.
30. Misra, K. B. (1991). Search procedure to solve integer programming problems arising in reliability design of a system. *International Journal of Systems Science, 22*(11), 2153–2169.
31. Misra, K., & Misra, V. (1993). Search method for solving general integer programming problems. *International Journal of Systems Science, 24*(12), 2321–2334.
32. Misra, K.B., & Sharma, U. (1991). An efficient algorithm to solve integer programming problems arising in a system reliability design. *IEEE Transactions on Reliability, 40*(1), 81–91.
33. Misra, K. B., & Sharma, U. (1991, December). Multicriteria optimization for combined reliability and redundancy allocation in systems employing mixed redundancies. *Microelectronics Reliability, 31*(2–3), 323–335.
34. Chaturvedi, S. K., & Misra, K. B. (2008). MIP: A versatile tool for reliability design of a system. In K. B. Misra (Ed.), Handbook of performability engineering. London: Springer.
35. Ramirez-Marquez, J. E., & Coit, D. W. (2004). A heuristic for solving the redundancy allocation problem for multi-state series-parallel systems. *Reliability Engineering and System Safety, 83,* 341–349.

36. Tian, Z., & Zuo, M. J. (2006). Redundancy allocation for multi-state systems using physical programming and genetic algorithms. *Reliability Engineering and System Safety, 91,* 1049–1056.
37. Misra, K. B. (1993). Use of Fuzzy sets theory, Part I (45 pages); Misra, K. B., & Onisawa, K. Fuzzy Sets Applications, Part II (33 pages) In K. B. Misra (Ed.), *New trends of in system reliability evaluation.* Amsterdam: Elsevier Science, pp. 503–586.
38. Misra, K. B., & Weber, G. G. (1989, December). A new method for Fuzzy fault tree analysis. *Microelectronics Reliability, 29*(2), 195–216.
39. Misra, K. B., & Weber, G. G. (1990, September). Use of Fuzzy set theory for level-I studies in probabilistic risk assessment. *Fuzzy Sets and Systems, 37*(2), 139–160.
40. Bolwig, S., & Gibbon, P. (2009). Counting carbon in the marketplace: Part 1—Overview paper: Report for the OECD. Trade and Agriculture Directorate; Joint Working Party on Trade and Environment, p. 25.
41. Wang, Y., Zhang, H., Zhang, Z., & Wang, J. (2015). Development of an evaluating method for carbon emissions of manufacturing process plans. *Discrete Dynamics in Nature and Society, 2015,* Article ID 784751, p. 8.
42. British Standard Institute. (2011). *Publicly Available Specification (PAS) 2050: Specifications for the assessment of the life cycle greenhouse gas emissions of goods and services; BSI Report PAS 2050.* British Standard Institute: London, UK.
43. ISO. (2013). Greenhouse Gases—Carbon Footprint of Products—Requirements and Guidelines for Quantification and Communication; ISO 14067; ISO: Geneva, Switzerland.
44. Yanjia, W., & Chandler, W. (2010). The Chinese nonferrous metals industry—Energy use and CO_2 emissions. *Energy Policy, 38,* 6475–6484.
45. Hu, S., Liu, F., He, Y., & Hu, T. (2012). An On-line approach for energy efficiency monitoring of machine tools. *Journal of Cleaner Production, 27,* 133–140.
46. Klemeš, J. J. (Ed.). (2015). Assessing and measuring environmental impact and sustainability. Butterworth-Heinemann. Elsevier.
47. https://www.co2list.org/files/carbon.htm "CO_2 released when making & using products". Retrieved October 27, 2009.
48. Center for Sustainable Systems, University of Michigan. (2019). "Carbon Footprint Factsheet." Pub. No. CSS09-05.
49. U.S. EPA. (2017). *Greenhouse gas equivalencies calculator.*
50. Quintana-Pedraza, G. A., Vieira-Agudelo, S. C., & Muñoz-Galeano, N. (2019, August). A cradle-to-grave multi-pronged methodology to obtain the carbon footprint of electro—Intensive power electronic products. *Energies, 12*(17), 3347.

Krishna B. Misra (born 1943 in India) is an electrical engineering graduate of 1963 from Maharaja Sayajirao University of Baroda, India. He also received his M.E. (power systems), Ph.D. (reliability) of the University of Roorkee (now IIT Roorkee) in 1966 and 1970, respectively, which he joined as faculty in 1966 and where he became a full professor in 1976. He delivered lectures and conducted training programmes in Indian industries to popularize the reliability concepts and also helped Department of Science and Technology (DST), New Delhi to set up an NCST working group on reliability implementation programme in 1976, which submitted two reports to Government of India in 1978 and were accepted by DST. However, the change of administration quietly led to shelving the reports after the Task Force committee for implementation was constituted by the earlier DST administration. The self-reliance about which the Indian Government is talking about now in the wake of corona threat would have been put into action, had the government not shelved the follow-up actions in 1978.

He then turned to academia to advance this discipline in institutions and published extensively in international journals. He has served on the editorial boards of several reputed international journals on quality, reliability, and safety, including IEEE Transactions on Reliability. In 1983, he was awarded the first Lal C. Verman award by Institution of Electronics and Telecommunication Engineers, New Delhi for his contributions to Reliability. Dr. Misra received a plaque in 1995 from the IEEE Reliability Society, USA for his contributions to research in reliability and promoting reliability education in the area of reliability in India.

In 1980, he moved to IIT Kharagpur, where he started the first ever postgraduate programme on reliability engineering in India in 1982 and established Reliability Engineering Centre with grants from Ministry of Human Resource & Development in 1983. This centre produced several illustrious postgraduates in reliability engineering, who are working in various prestigious organizations in India and abroad. In 1992, he was appointed Director-Grade-Scientist by Council of Scientific and Industrial Research (CSIR) at the National Environmental Engineering Research Institute (NEERI), Nagpur, where he felt necessity of integrating performance of products be judged to include environmental impact along with reliability. He was also appointed Director of North Eastern Regional Institute of Science and Technology (NERIST) by the Ministry of Human Resource Development, Government of India in 1995 and served this Institute from 1995 to 1998. After his retirement from IIT Kharagpur in 2005, he founded RAMS Consultants, Inc. in Jaipur, India and launched the first International Journal of Performability Engineering which he edited and published under the aegis of Rams consultants in India from 2005 to 2015. He eventually sold its publication rights to a US company, where it is still being published.

Dr. Misra also worked in Federal Republic of Germany at four different reputed institutions with well-known professors.

Prof. Misra has had four decades of teaching, consultancy and research experience in reliability. He has published many papers and five books, including a Handbook of Performability Engineering, which was published in 2008 by Springer, London. He is also the Series Editor of books under Performability Engineering being published jointly by Scrivener—John Wiley & Sons, USA.

Chapter 2
Performability Considerations for Next-Generation Manufacturing Systems

Bhupesh Kumar Lad

Abstract Globally, the manufacturing industry is gearing up for the next level of industrial revolution, and it is called smart manufacturing or Industry 4.0. This chapter aims to discuss various aspects of performability for next-generation manufacturing systems. "Intelligence" is identified as an essential dimension of performability for such systems. Various elements of this new dimension are discussed, and the associated technologies are mapped. New business models that utilize the performability of the next-generation manufacturing systems are presented. Finally, a new philosophy, namely, "Manufacturing by Mass," is built to capitalize the full potential of intelligent factories.

Keywords Industry 4.0 · Performability · Smart manufacturing · Business models · Intelligent manufacturing · Manufacturing by mass

2.1 Introduction

Performability is an aggregate attribute measuring the designer's entire effort in achieving sustainability for a dependable product (Misra [1]). Performability deliberates on sustainability along with other factors like quality, reliability, maintainability, and safety, as shown in Fig. 2.1. Hence, it reflects an all-inclusive aspect of performance of any product or system.

One of the important areas of application of performability concepts is in the field of manufacturing. Performability considerations in manufacturing are involved from two different angles, viz., performability of manufacturing systems and performability of products produced through the manufacturing system. Though the performability is generally studied from the point of view of products, the concept of performability is equally important for manufacturing systems. Moreover, it is difficult to separate these two angles of performability in manufacturing. For example, the performability of products involves factors like sustainability and quality, which are

B. K. Lad (✉)
Indian Institute of Technology Indore, Indore, India
e-mail: bklad@iiti.ac.in

K. B. Misra (ed.), *Handbook of Advanced Performability Engineering*,
https://doi.org/10.1007/978-3-030-55732-4_2

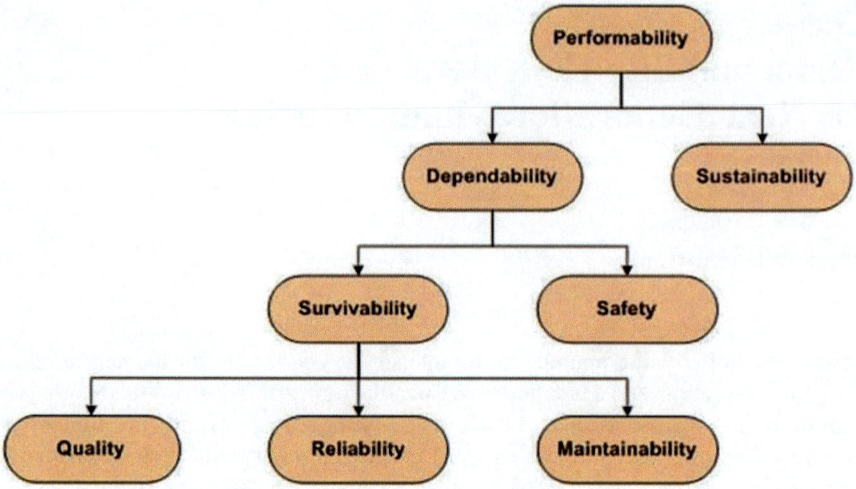

Fig. 2.1 Performability of system (Misra [1])

directly related to the production processes. Similarly, the performability of manufacturing systems is directly linked with the performability of products produced through the system. For example, Lad et al. [2] highlighted the link between product quality and machine tool reliability. The same was then used for reliability and maintenance-based design of machine tools.

The focus of the present chapter is on the performability of manufacturing systems. Many studies focus on reliability, maintainability, maintenance, and life-cycle cost of the machine tools, thereby addressing dependability aspect of the performability of manufacturing systems (Lad et al. [3]). Sustainability considerations have also been studied for manufacturing industries (Baas [4]). The manufacturing industry is going through a paradigm shift powered by information and communication technology (ICT) and artificial intelligence (AI). The fundamental shift in the manufacturing paradigm calls for a new dimension to be added in the traditional performability matrix. The present chapter does not intend to elaborate on any of the traditional dimensions of performability for manufacturing systems. Instead, it aims to identify and elaborate on the implication of the new dimension of performability for the next-generation manufacturing systems.

2.2 Evolution of Manufacturing Systems

The manufacturing sector is going through fundamental changes by the fusion of industrial production and information and communication technology (ICT). This paradigm shift is referred to as the fourth industrial revolution. Figure 2.2 shows the evolution of manufacturing systems.

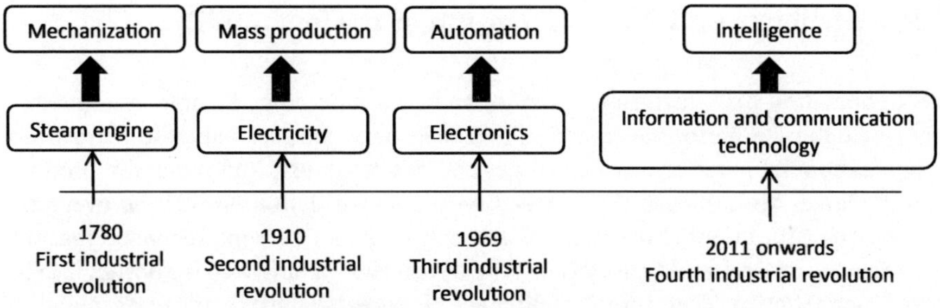

Fig. 2.2 Evolution of manufacturing systems

During first to third industrial revolution the manufacturing sector has shown a significant increase in productivity and efficiency. Today, the fourth industrial revolution or Industry 4.0 has become a hotspot for global manufacturing industries. It is also called smart manufacturing or industrial internet. Industry 4.0 is making it possible to connect machine-to-machine and machine-to-human due to the convergence of the physical and the virtual (cyberspace) world in the form of cyber-physical systems (CPS). It is converting the factories into fully connected and flexible systems called smart factories. The fourth industrial revolution is expected to provide an unprecedented leap in productivity and efficiency of manufacturing systems. An increase in operational efficiency will positively affect the performability of the manufacturing system as it results in a smaller environmental footprint and, in turn, greater environmental sustainability. Moreover, unlike all previous revolutions, which only released human physical power, the fourth industrial revolution is expected to augment, if not fully relieve, the human thinking power that is intelligence and innovatively change the entire manufacturing paradigm (Li et al. [5]). Accordingly, the new industrial paradigm is transforming the ways products are conceptualized, design, produced, sold, and used, which in turn has brought new opportunities for organizations. In such situations, conventional business models for manufacturing systems may not be sufficient to meet the current challenges and utilize the opportunities offered by the new industrial paradigm. Consequently, new and adapted business models are needed (Ibarra et al. [6]). However, any such new business models will require rethinking on performance engineering for new generation manufacturing systems. New dimension is needed to be incorporated in the existing performability matrix, and available technologies need to be mapped with this new dimension. This new dimension of the performability matrix is discussed in the next section.

2.3 Intelligence: A New Dimension of Performability

Next-generation manufacturing systems can be considered as flexible systems that can self-optimize performance across a broader network, self-adapt to learn from new conditions in real or near real-time, and autonomously run the entire production processes (Burke et al. [7]). Based on the above definition and the overview of Industry 4.0 presented in the previous section, it can be comprehended that the performance of the smart factory is heavily dependent on how well we design intelligence in the system. Therefore, "intelligence" is added to the existing performability matrix for the manufacturing systems. Figure 2.3 shows the updated performability matrix. The intelligence can be further achieved by designing the following elements into the next-generation manufacturing systems.

- System visibility
- Flexibility
- Networkability
- Real-time decision-making.

These elements are discussed in the following sub-sections.

2.3.1 System Visibility

Information plays a vital role in achieving the required level of intelligence for any manufacturing system. Information is generated based on the data related to the system and give system visibility and digital presence. For example, today's

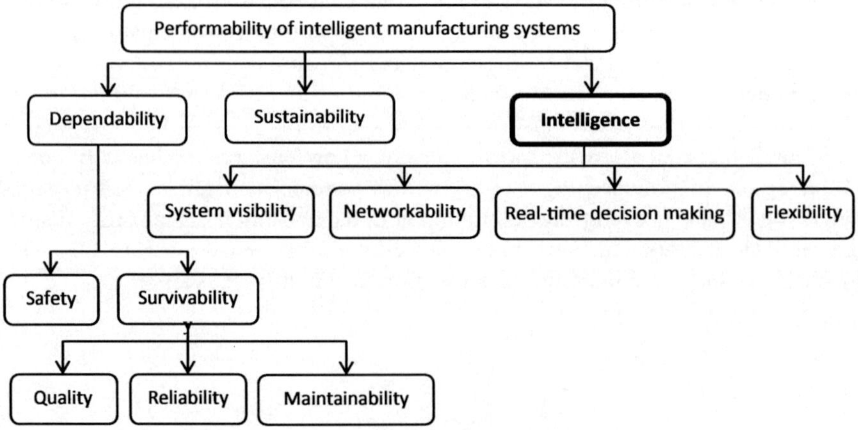

Fig. 2.3 Performability of next generation manufacturing systems

smart products can provide their identity as well as their current status and lifecycle history. This information may be used for optimizing use and maintenance and deciding end-of-life strategy for the product. Traceability of products sometimes may be mandated by the customers, especially for industrial products. Therefore, next-generation manufacturing systems should have the capability of collecting and storing required good quality data. This requires identification and installation of the right sensors for process monitoring like pressure sensor, temperature sensor, accelerometer, etc.; extracting data from machine controllers; employing sensors for tracking of materials, like RFID tags, bar code, etc. Obviously, data collection consumes energy and resources. Hence, identifying the right information and its use will be critical for making a sustainable smart manufacturing system. For example, there is no value in putting sensors into machine tools if operation and maintenance teams do not use these data in product improvement or proactive maintenance planning of machine tools. These data need to be converted into useful information which may be utilized later for decision-making. Descriptive, diagnostic, and predictive analytics, including big data analytics, are essential parts of system visibility. The ability of the smart factory to predict future outcomes, based on historical and real-time data, is crucial in optimizing asset performances in terms of uptime, yield, quality, and prevent safety issues.

Digital twin technology is used for digitizing the asset, including predictive capabilities. This is the starting point for manufacturers to jump from automation (third-generation factories) to intelligence (next-generation smart factories). A digital twin of an industrial asset can be defined as a dynamic virtual replica of the physical asset, which should ideally showcase identical behavior to the physical asset when observed under identical conditions. It is essentially a computer algorithm, modeled upon the entire functioning logic and real-life behavior of the asset, possessing the characteristics of being aware, social, adaptable, and intelligent. Digital twins incorporate big data analytics, artificial intelligence (AI), and machine learning (ML) to provide system visibility and create an interface for communication and further decision-making in smart factory.

Industries exercise various options of processing data, like performing analytics at the edge, at a local server, or a cloud. One needs to think about the most efficient and safest way to achieve real-time capability and help in improving the outcomes of manufacturing systems. For example, edge devices help in getting real-time insights into the manufacturing operations by processing the data analytics close to where the data is born. All these may help in getting useful information like remaining life of the component, machine overall equipment effectiveness (OEE), and so on. This information helps make more-informed business decisions (perspective analytics) in a complex industrial environment. Also, technologies like virtual reality (VR) and augmented reality (AR) are adding new dimension for operational visibility and supporting the organizations on operational activities and operators' training. For example, with the help of VR and AR technologies, an operator can see the performance of the machines while walking along with production facility on the shop floor. If required, the operator can even adjust the machine without physically touching it. Thus, such bi-directional solutions not only enhance system visibility but also create an interface for communication and further decision-making in smart factory.

2.3.2 *Flexibility*

In terms of the design of machine tools, in contrast to the conventional CNCs that are general-purpose machines, such smart machines are expected to be flexible, and reconfigurable, which use mechanical control, hydraulic/pneumatic, and electrical modules to achieve rapid adaptability for a customized range of operational requirements. Flexibility in manufacturing system makes it possible to reconfigure itself and quickly adjust production capabilities and capacities in the event of sudden changes in the market. A flexible manufacturing system can execute many decisions without human intervention. Such machines can even take action when requirements change. The development and supply of machine tool systems that can fulfill the utmost important requirements, such as flexibility, reliability, and productivity for mass production are necessary (Mori et al. [8]). Designing different machining processes, using the same machine tool, can reduce the total energy consumption during the manufacturing process, the need for more substantial floor space in the plant, and the cost per part (Shneor [9]). This not only will improve operational efficiency but also add in sustainability of next-generation manufacturing systems. Additive manufacturing technology, agile and collaborative robotic systems, smart materials, etc., are promising technologies for achieving the required level of flexibility coupled with high-precision and repeatability. Such technologies help improve responsiveness and innovation. Besides, greater process autonomy helps reduce human effort and fatigue, which in turn positively impacts industrial safety.

2.3.3 *Networkability*

At the operational level, the next-generation manufacturing systems can be seen as a process of connected business optimization in real time. Performability of such a process largely depends on the seamless integration or networkability of all devices into a vast manufacturing infrastructure. Interoperability and security are the two of the most essential elements of the networkability. In order to achieve interoperability, it is important to store the data in some standard format that can be utilized by various stakeholders. ZVEI and Platform Industry 4.0 standardization committees and standards like eCl@ss or IEC 61,360 talk about Industry 4.0 semantics (ZVEI [10]).

As mentioned in Sect. 2.3.1, data related to the assets are used for the creation of digital twins. These digital twins act on behalf of the physical machines and make decisions in smart factories. One of the key challenges in the creation of digital twin is the interoperability between the applications managing the manufacturing system and making the assets discoverable in the Industry 4.0 network (Chilwant et al. [11]). Asset administration shell (AAS) technology is evolving as a possible solution to address this challenge. AAS can be viewed as the bridge between a tangible asset and the IoT world or, in other words, the data model which is based on the digital twin. Besides, a seamless connection requires data exchange between products from different manufacturers and across operating systems. Data exchange standards like OPC unified architecture (OPC-UA) are used for a safe, reliable, manufacturer, and platform-independent industrial communication (OPC [12]).

Connected assets make the manufacturing vulnerable to hacking attacks. Intellectual properties and business secrets like design, process flow, process parameters, business models, etc., are susceptible to such attacks. Smart factories need to develop and implement effective, adaptive, and autonomous cyber defense and response mechanisms to secure vital cyber-physical data flows within a manufacturing system. For example, block chain is one of the technologies which brings distributed peer-to-peer network architecture to improve the security and scalability of cloud manufacturing. With the application of smart contracts (SC), block chain technology can be employed to provide a fast and secure platform for machine-to-machine (M2M) communication (Christidis et al. [13]).

2.3.4 Real-Time Decision-Making

The performance of smart factories depends on how quickly and proactively it reacts to the change in demand, machine health, inventory, lead time, etc. Thus, real-time decision-making capability is an essential element of the performability of the next-generation manufacturing system. Such systems are virtually a network of connected assets represented by respective digital twins. It involves both internal and external assets. Thus, there are many connected operations in such systems. Real benefits of Industry 4.0 can be realized only if these networked operations are optimized jointly. However, despite the interdependencies, conventionally, many of these operations are treated independently. For example, production planning, maintenance planning, quality planning, etc. are usually done in isolation (Kumar et al. [14]). This necessitates a managerial-level round table discussion for fine-tuning of multiple interdependent decisions before implementations. This brings in subjectivity, delayed decisions and may lead to sub-optimal solutions. Moreover, in smart manufacturing, advanced data analytics aims to provide shop floor decisions without human intervention. Thus, managerial-level coordination for effective execution of individual decisions will be out of trend (Kumar et al. [14]). Thus, Industry 4.0 calls for joint optimization of connected operations within and outside the factory.

Though imperative, integration brings in computational complexity, which poses significant challenge in terms of responsiveness of the value chain. Responsiveness is the second important requirement from any smart factory. This can be understood with a simple example of a two-machine sequencing problem. Let there be 7 jobs to

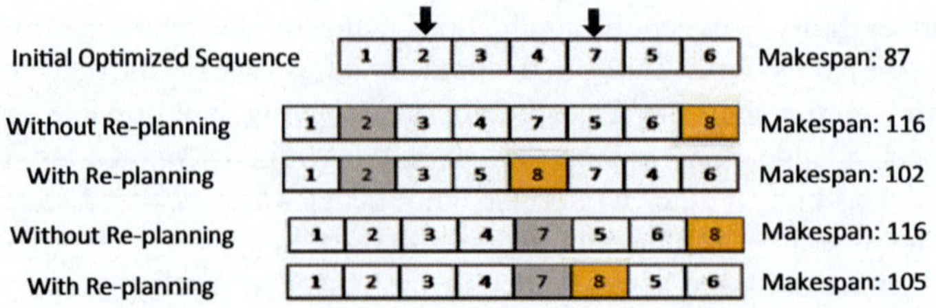

Fig. 2.4 Importance of dynamic planning

be processed on two machines in series. Let the processing time of these jobs on two machines are such that sequence that minimizes makespan, obtained using one of the heuristics like Jonhson's rule, is 1–2–3–4–7–5–6. Let the same sequence is implemented on the shop floor. Let us consider two cases of possible disturbance in the system. The disturbance is caused when a new job (say job 8) enters into the system. Two cases are considered. One when the demand for job 8 comes when the system is processing job 2, and second when the system is processing job 7. The conventional approach is to schedule the new job, i.e., job 8, after all the preplanned jobs are finished, i.e., at the end of the optimal sequence, viz., after the completion of job 6 in this example. An alternate approach is to quickly replan the sequence whenever the disturbance occurs. Figure 2.4 shows a comparison of both the approaches. It can be seen that even for such a small problem, dynamic replanning helps in improving the system performance. Such changes are widespread in any real-life industrial systems. It creates opportunities for performance improvement. However, it requires quick and real-time decision-making. If the decision-making is not autonomous, quick, and real-time, then the conventional approach, viz., "not disturbing current schedule," may win the preference of the managers, which in turn may result in sub-optimal performance. Next-generation manufacturing systems have the potential to capture such missing opportunities at a much larger scale.

Traditionally used solutions for decision-making in manufacturing industries do not address the above requirements adequately. Most of these algorithms fail to provide a real-time solution when integrated with existing enterprise resource planning (ERP) or manufacturing execution system (MES). Moreover, they are not developed considering the intelligence of the system in mind. For example, traditional scheduling algorithms neither utilize M2M communication nor intelligence available with individual machines. Thus, a novel approach is required to deal with two conflicting challenges, viz., integration and responsiveness of decision-making for the next-generation manufacturing systems. Moreover, such solutions should be based on machine-to-machine communication, intelligence available with individual assets, predictive analytics, etc. Distributed or decentralized decision-making from the point of view of computation can solve the problem of responsiveness. Joint planning from the point of view of functional consideration can solve the problem of integration. Thus integrated yet distrusted approaches may be useful. Distributed simulation helps in distributing time-consuming computation that comes with the

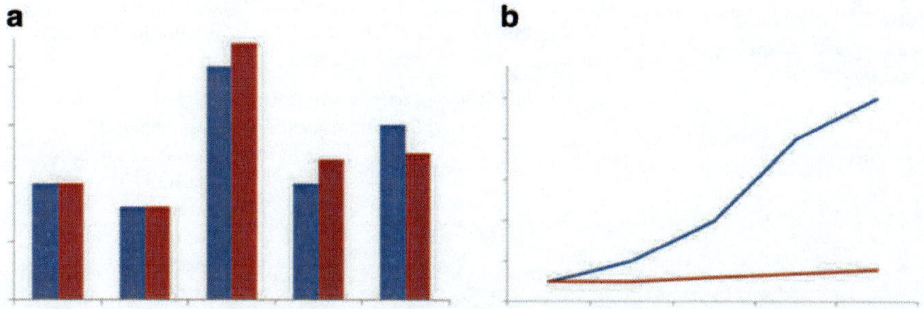

Fig. 2.5 **a** Comparison of objective function value (vertical axis) for decentralized (right bar) and centralized (left bar) approach for increasing problem size (horizontal axis) **b** comparison of computation time (vertical axis) for decentralized (bottom line) and centralized (upper line) approach for increasing problem size (horizontal axis). *Note* Objective function is a negative function (i.e., minimization problem)

stochastic simulation of an intricate model, over multiple computing systems. Each entity of the enterprise is simulated as a virtual entity (digital twin) on a different computer. This allows for a finer level of detailing while modeling the entity, without compromising on the efficiency in terms of computation time.

Figure 2.5a, b show the expected performance of such decentralized approaches. Figure 2.5b shows the required responsiveness of a distributed approach with increasing problem size. Figure 2.5a shows the expected quality of the solution. It shows that for smaller size problem decentralized approach may give similar performance as that given by a centralized approach. For larger size problems, though, the performance of the decentralized approach may not be as optimal as that given by a centralized approach; however, quick response (Fig. 2.5b) may help in capturing more market share and, in turn, will make the decentralized results comparable or even better. Interestingly, for a very large problem size, the decentralized approach may give better result, as the centralized approach may not be able to provide optimal results due to problem complexity. On the other hand, decentralized approach may provide better solution as the benefit of more effective local level optimization may dominate in the overall solution quality. Approaches developed by Kumar et al. [14] and Upasani et al. [15] confirm similar results. More research can be focused in this direction to automate the level of integration and distribution for various problems in connected business optimization.

2.4 Performability-Technology Mapping

Designing the performability into the manufacturing systems requires adopting various technologies. Table 2.1 shows the mapping of technology with the performability elements. A brief overview is already presented in Sects. 2.3.1–2.3.4. The present chapter does not intend to discuss these technologies in detail. It is important

Table 2.1 Performability elements and technology mapping

Performability elements	Performability requirements	Technology aspects
System visibility	Process monitoring Product traceability	Digital twin Big data and industrial analytics Sensors systems Advanced human–machine interface Artificial intelligence and machine learning Computer vision Augmented reality and virtual reality Edge devices Smart sensors
Networkability	Interoperability Network safety	Asset administration shell Cloud computing IoT Cybersecurity MTconnect, OPC UA standard, etc Block chain
Flexibility	Reconfigurability Automation	Reconfigurable machine tools Additive manufacturing Agile and collaborative robotic systems Advanced/smart materials
Real-time decision-making	Integrated decision-making Decentralized decision-making	Distributed decision-making Agent-based decision-making Integrated decision making Simulation modeling Augmented reality and virtual reality

to mention here that it is often not possible or practical to immediately acquire or introduce all these technologies in the existing system. Moreover, these technologies adoption incur significant cost. A long-term technology roadmap may be required to meet the goal of the business. Thus, the performability elements and associated technologies need to be linked with business goals.

2.5 Business Models for Intelligent Manufacturing Systems

Intelligent manufacturing systems powered by enhanced system visibility, flexibility, networkability, and real-time decision-making offer many benefits. Some of these are as follows:

- Improved productivity,
- Improved asset optimization,
- Reduced operating cost,
- Improved quality of products.

Besides, the intelligent factory offers novel business opportunities. Until industries explore and adopt these new business opportunities, it is not possible to extract full benefits of performability of the next-generation manufacturing systems. Novel business models are required to explore these new business opportunities. Some of the novel business models are given by the following:

- Servitization,
- Co-creation,
- Dynamic pricing,
- Mass customization,
- Manufacturing as a service.

2.5.1 *Servitization*

Traditionally, manufacturing industries were focused on selling the products. Lately, companies realized that customers do not always need ownership of the product. Many times, they are only interested in functions provided by the products. Realizing this, manufacturing companies started focusing on satisfying the customers' needs by selling the function of the product rather than the product itself, or by increasing the service component of a product offer. Servitization is thus defined as the processes to shift from selling products to selling integrated products and services that deliver value in use (Baines et al. [16]). It is also called product service system (PSS). Servitization helps in increased differentiation in the market and continued revenue generation. Some of the examples of servitization are given in Table 2.2.

Performability characteristics of next-generation manufacturing systems play a greater role in making PSS or servitization a sustainable model. For example, a gas turbine manufacturer who wants to provide long-term availability contract along with the product, i.e., gas turbine, would need real-time data extraction and descriptive, predictive analytics to optimize its maintenance plans, and earn maximum profit from the contract. This requires the manufacturing companies to invest in system visibility and networking technologies discussed earlier.

By using a service to meet some needs rather than a physical object, more needs can be met with lower material and energy requirements. For example, the "Pay

Table 2.2 Examples of servitization

Company name	Product	PSS offerings
Xerox (Kowalkowaski et al. [17])	Office equipment	Pay-per-use model (1996) Annuity-based business models (2002) Xerox splits into two companies: one hardware-centric and one service-centric (2016)
Rolls-Royce (Ostaeyen [18])	Aircraft engine	Power-by-the-hour service package, whereby maintenance, repair, and overhaul are charged at a fixed price per hour of flight to the customers (i.e., airline companies)
Philips lighting (Ostaeyen [18])	Lighting systems	Selling a promised level of luminance in a building, according to a Pay per Lux concept

per Lux" concept helps the company in providing the exact amount of light for workspaces that employees need when using them for specific tasks, i.e., no more and no less. Thus, it consumes optimal energy for the specific requirements of the customer. This will lead to lower environmental impact over the life cycle.

Servitization requires coordinated efforts by various stakeholders like industry, government, and civil society to create and to facilitate the establishment and smooth functioning of such systems. For example, many electronic products like mobile phones, computers, etc., today have a very short life expectancy and users tend to change them very frequently as soon as technology changes. This creates huge e-waste. PSS can play a significant role in managing this e-waste while creating new business opportunities for companies. Companies may sell such products or ownership of such products for a shorter duration along with their replacement services. Companies may reduce the cost of reliability growth due to lesser life expectancy of the products. Alternatively, companies may reuse some of the parts in other products, thus reduce the cost of manufacturing. Benefits to the customers are that they do not need to pay more for short-duration use of the product and also do not have to worry about the disposal of e-waste that they generate. If the government makes strict norms for e-waste disposal, the customer will be motivated to go for such product service offerings of the manufacturer.

2.5.2 Co-Creation

Co-creation is the process of involving customers, suppliers, and various other stakeholders at different stages of the value-creation process. Though the co-creation concept is not new for industries, its application was not very common in practice. Table 2.3 presents some closely matching models used by some of the companies.

Table 2.3 Co-creation example

Company	Concept used
LEGO (Manufacturers of toys)	Lego created an online platform called "LEGO Ideas," where customers can submit their designs. Some of the highly voted designs are selected for production and worldwide sale (LEGO [19])
Made.com (E-retail furniture company)	Made Talent Lab of Made.com company hosts an annual online contest called "Made Emerging Talent Award," in which budding new designers can submit their work for other designers and customers to vote on. Design that gets the highest votes is produced and sold (MADE.COM [20])
BMW (Automobile manufacturer)	In 2010 BMW co-creation Lab allowed consumers to get closely involved in the design process from start to finish (BMW [21])

Digitization and networking capability present in the smart industry has the potential to take the co-creation business model to the next level. For example, social networking platforms, advanced user-friendly 3D models creation tools, cloud computing, 3D printing technology, virtual and augmented reality, etc. can diminish the boundary between the manufacturing industry and its customers and suppliers. This may take innovation to new heights.

2.5.3 Dynamic Pricing

As the name suggests, dynamic pricing is a strategy in which product price gets updated, depending on the market demand. This is widely used in e-commerce sectors. There are few evidences of the implementation of dynamic prices in manufacturing sectors. For example, Dell Computers offers dynamic pricing based on parameters such as demand variation, inventory levels, or production schedule (Biller et al. [22]). However, the integration of pricing, production, and distribution decisions in manufacturing environments is still in its early stages (Biller et al. [22]). Technologies like internet of things (IoT), data analytics, and real-time decision-making, coupled with direct to customer (DTC) concept are creating opportunities for manufacturing industries to earn more profit by optimizing their product price dynamically. For example, a connected factory can dynamically update the price of the product if it can quickly optimize its value chain and estimate its operations cost by considering real-time demand, inventory, supplier discount, etc. Thus, faster and integrated decision-making is the key to success for such a business model.

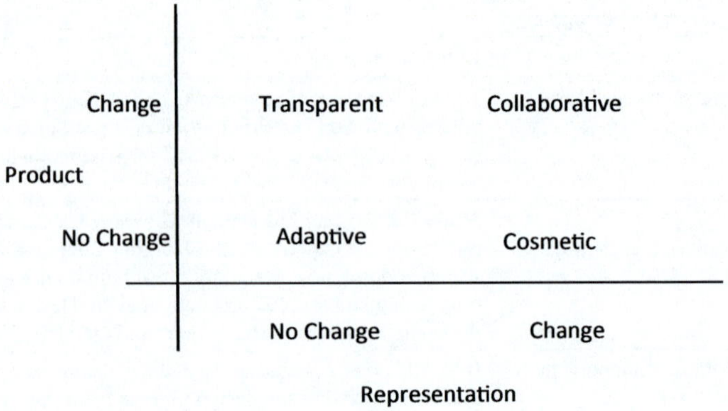

Fig. 2.6 Four approaches for customization

2.5.4 *Mass Customization*

Mass customization offers a higher level of customization along with low unit costs associated with mass production. It allows a manufacturer to customize certain features of a product while still keeping costs closer to that of mass-produced products. James et al. [23] presented four different ways to achieve mass customization, viz., collaborative, adaptive, cosmetic, transparent. These approaches differ on whether the customization is offered on product features or its representation (Fig. 2.6). Flexibility of manufacturing systems is important in realizing this business model.

2.5.5 *Manufacturing as a Service*

Manufacturing as a Service (MaaS) is a concept where manufacturers share their manufacturing equipment via internet to produce goods. It is also called cloud manufacturing (Tao et al. [24]). It is similar to the concept of cloud-based services, for example, Google's Gmail service, where a company uses such services but doesn't buy or maintain its servers. Thus, the server cost is shared across all the customers of the cloud services. Similarly, manufacturing companies that provide manufacturing as service make their facility available for other manufacturers (customers). Thus, customers do not retain full ownership of all the assets they need to manufacture their products. The cost of ownership of such assets, viz., cost of machines, maintenance, software, networking, and more, is distributed across all customers. A properly designed MaaS model can help in reducing manufacturing costs. Manufacturers will be able to offer more customization options to customers by taking advantage of flexibility offered by shared manufacturing facilities. The performance of such a model relies on real-time insight into the status of manufacturing equipment. Performability enabler technologies like sensors, data analytics, internet of things (IoT), and cloud computing has the potential to bring this revolutionary change in the manufacturing business. For example, AI can help the manufacturers to identify the right service provider, right design, right material, etc. to reduce the cost of the product.

2.6 Manufacturing by Mass Philosophy for Intelligent Manufacturing

The next-generation manufacturing systems suggest an integration of shop floor decisions and insights with the rest of the supply chain and broader enterprise through an interconnected information and operations technologies (IT/OT) landscape (Burke et al. [7]). Such systems designed with performability characteristics discussed in the preceding sections can fundamentally change manufacturing business and enhance relationships with suppliers and customers. It can give birth to a new philosophy in manufacturing called "Manufacturing by Mass." Here, the term "Mass" stands for "Customers." The philosophy suggests that customers can play an active role in the connected business processes of the smart factory. It not only can design or customize the product but also can actively participate in the business processes like planning of procurement of raw materials, internal shop floor planning, outsourcing decisions, etc. In contrast to the conventional business philosophies where customer is always a paying entity, in this new philosophy, the customer can even earn revenue while purchasing or producing its product from any manufacturing facility. From the point of view of the production system, it can be looked at as a high variety-high volume system, i.e., mass customization as well as mass production. The same is shown in Fig. 2.7 using a volume–variety curve along with the evolution of manufac-

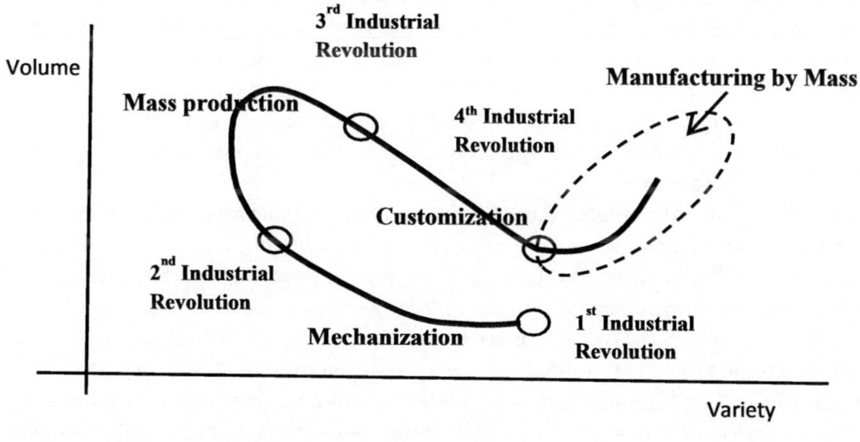

Fig. 2.7 Evolution of manufacturing on volume-variety curve

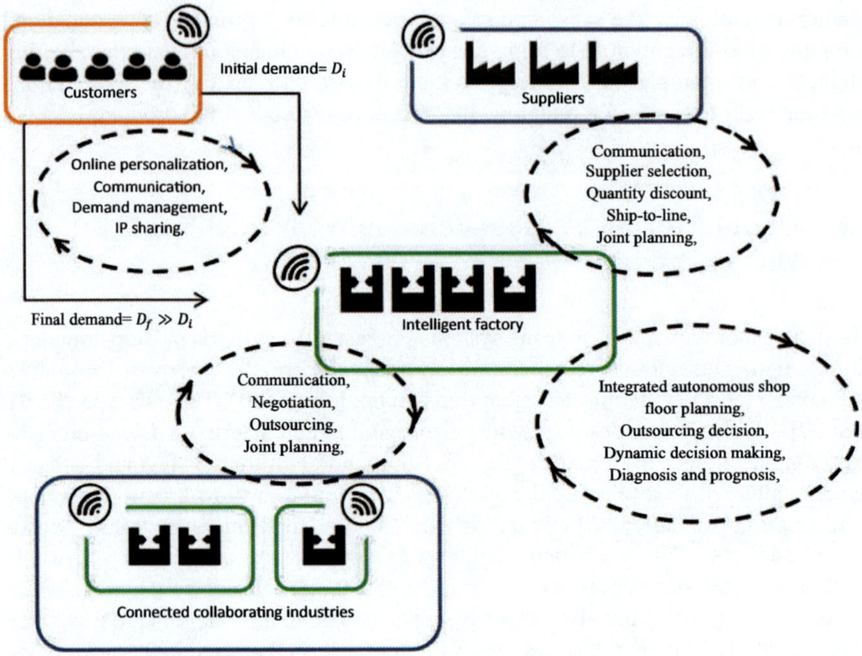

Fig. 2.8 Manufacturing by mass

turing system. Though conflicting, it is achievable scenario for the next-generation manufacturing systems.

To understand the above philosophy, the following example is added, and the same is depicted in Fig. 2.8. In this, a user communicates directly with a digital agent of a smart factory using a humanoid communication platform. Such platforms are akin to the social networking platforms used by humans. Such platforms, apart from humans, also have digital agents of machines and other industrial entities in the network. Let a customer creates a personalized or highly customized design of a product utilizing an interactive design platform powered with artificial intelligence and finalizes a feasible design for production in coordination with the industrial digital agents in the network.

Digital agent of the smart factory identifies a group of machines and other production facilities and raw material suppliers required for the production of the designed product. The digital twins of these machines form a group and perform its internal operations planning and supply planning through the autonomous decision-making and quickly provide a cost estimate to the customer. Being a personalized product, the cost is expected to be very high. Industry agent in such situation explores the global customer pool through the humanoid platform and analyzes their past purchasing behavior and other online activities and identifies potential customers for the designed product. Data analytics can easily cluster these customers into different categories like the most probable customer, the least probable customers, etc. For all possible cases, industry agents will quickly perform integrated planning of inventory, production scheduling, batch sizing (in case machines process other products also), etc. considering its capacity. It may also use cloud manufacturing to identify other connected industries, if the demand exceeds its internal capacity. Consequently, the industry can decide quantity to be produced and the corresponding number of customers to be contacted through the online platform. Industry agent pings to those many customers and receive some confirmed orders. The offered cost for the product is expected to be significantly lower as the same is required in bulk now. As a result, many of the interested customers are expected to accept the proposal. The result is that a highly customized product is produced in mass. In a way, the digital network is connecting customers with other customers and offering benefits of mass production by creating a win–win situation for all.

Also, if the design is novel, the industry may even explore the options of sharing intellectual property (IP) benefits to the original customer who designed the product in consultation with the digital agent of the industry. Thus, customers can even earn revenue while participating in the business network. The above scenario may be an extreme case, and not all the industries may require this extreme. For example, instead of entirely personalized products, it can be just a customized product based on the available options provided by the company. In such cases, IP sharing may not be involved. Similarly, all industries may not go for cloud manufacturing and restrict the production and demand management to its internal capacity. Industries may still follow the rest of the procedure mentioned above to attract new demand and produce that customize product in bulk. Secondly, all the technologies and business models mentioned in earlier sections in this chapter are essential to achieve the Manufacturing by Mass philosophy. In essence, Manufacturing by Mass is a philosophy to realize the benefits of the performability of next-generation manufacturing systems.

2.7 Conclusions

With the change in the industrial scenario, customers' expectations, and the emergence of new technologies, the performability matrix for the manufacturing systems requires update. This chapter identifies "intelligence" as the new dimension of performability for the next-generation manufacturing systems, and elaborates on its various elements. It also emphasizes that the link between performability, technology and business model is key for achieving ultimate benefits of the performability of the manufacturing systems. It is explained that if this link is adequately established, then it may help in evolving a new philosophy in manufacturing.

Figure 2.9 summarizes the overall conclusion of the chapter. The present chapter is expected to ignite new discussions and research in the area of the performability of manufacturing systems and its role in evolving new philosophy in manufacturing.

Fig. 2.9 Broader perspective of performability

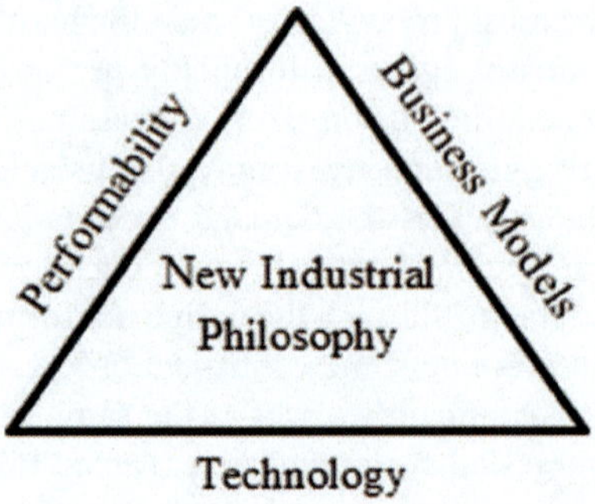

References

1. Misra, K. B. (2008). Performability engineering: An essential concept in 21st century, Chapter 1. In K. B. Misra (Ed.), *Handbook of Performability Engineering* (pp. 1–12). London: Springer.
2. Lad, B. K., & Kulkarni, M. S. (2013). Reliability and maintenance based design of machine tools. *International Journal of Performability Engineering, 9*(3), 321–332.
3. Lad, B.K, D. Shrivastava, M.S. (2016). Kulkarni, *Machine Tool Reliability*, Scrivener Publishing LLC.
4. Baas, L. (2008). Cleaner production and industrial ecology: A dire need for 21st century manufacturing. Chapter 11. In K. B. Misra (Ed.), *Handbook of Performability Engineering* (pp. 139–156). London: Springer.
5. Li, H. X. (2017). and H. *Si, Control for Intelligent Manufacturing: A Multiscale Challenge, Engineering, 3*(5), 608–615.
6. Ibarra, D., Ganzarain J., Igartua, J. I. (2018). Business model innovation through industry 4.0: A review. *Procedia Manufacturing,* 22, 4–10.
7. Burke, R., Mussomeli, A., Laaper, S., Hartigan, M., Sniderman, B. (2020). *The Smart Factory: Responsive, Adaptive, Connected Manufacturing. A Deloitte Series on Industry 4.0*, Deloitte Development LLC, 2017. Downloaded from, https://www2.deloitte.com/content/dam/insights/us/articles/4051_The-smart-factory/DUP_The-smart-factory.pdf, on April 24, 2020.
8. Mori, M., Fujishima, M. (2009). Reconfigurable machine tools for a flexible manufacturing system, Chapter 1. In *Changeable and Reconfigurable Manufacturing Systems* (pp. 101–109). London: Springer.
9. Shneor, Y. (2018). Reconfigurable machine tool: Cnc machine for milling. *Grinding and Polishing, Procedia Manufacturing, 21,* 221–227.
10. ZVEI. (2020). *Which Criteria Do Industry 4.0 Products Need To Fulfill*? German Electrical and Electronic Manufacturers' Association. Germany, 2017, downloaded from, https://www.plattform-i40.de/PI40/Redaktion/EN/Downloads/Publikation/criteria-industrie-40-products.pdf?__blob=publicationFile&v=5, on April 24, 2020.
11. Chilwant, N., & Kulkarni, M. S. (2019). Open asset administration shell for industrial systems. *Manufacturing Letters, 20,* 15–21.
12. OPC. (2020). OPE *Unified Architecture: Interoperability for Industry 4.0 and Internet of Things*. White paper, downloaded from, https://opcfoundation.org/wp-content/uploads/2016/05/OPC-UA-Interoperability-For-Industrie4-and-IoT-EN-v5.pdf, on March 28, 2020.
13. Christidis, K., & Devetsikiotis, M. (2016). Blockchains and smart contracts for the internet of things. *IEEE Access, 4,* 2292–2303.
14. Kumar, S., Manjrekar, V., Singh, V., & Lad, B. K. (2020). Integrated yet distributed operations planning approach: A next generation manufacturing planning system. *Journal of Manufacturing Systems, 54,* 103–122.

15. Upasani, K., Bakshi, M., Pandhare, V., & Lad, B. K. (2017). Distributed maintenance planning in manufacturing industries. *Computers and Industrial Engineering, 108,* 1–14.
16. Baines, T. S., Lightfoot, H. W., Benedettini, O., & Kay, J. M. (2009). The servitization of manufacturing. *Journal of Manufacturing Technology Management, 20*(5), 547–567.
17. Kowalkowaski, C., Gebauer, H., Kamp, B., & Parry, G. (2017). Servitization and deservitization: Overview, concept, and definitions. *Industrial Marketing Management, 60,* 4–10.
18. Ostaeyen, J.V. (2013) *.Analysis of The Business Potential of Product-Service Systems for Investment Goods.* Thesis, KU Leuven—Faculty of Engineering Science, ISBN. 978–94–6018–805–3, pp. 2.
19. LEGO. (2020). *Lego Ideas.* Accessed from https://ideas.lego.com/, on March 28, 2020.
20. MADE.COM. (2020). *Made Talent Lab.* Access from www.made.com, on March 28, 2020.
21. BMW. (2020). *BMW co-creation Lab.* Accessed from https://consumervaluecreation.com/tag/bmw/, on March 28, 2020.
22. Biller, S., Chan, L. M. A., David, S. l., Swann, J. (2005). Dynamic pricing and the direct-to-customer model in the automotive industry. *Electron Commerce Research,* 5, 309–334.
23. James, H. G., Joseph, B. (2020). *The four Faces of Mass Customization,* Harvard Business Review: Operations Management, January-February 1997 issue, Downloaded from, https://hbr.org/1997/01/the-four-faces-of-mass-customization, on March 28, 2020.
24. Tao, F., Zhang, L., Liu, Y., Cheng, Y., Wang, L., & Xu, X. (2015). Manufacturing service management in cloud manufacturing: overview and future research directions. *Journal of Manufacturing Science and Engineering, 137*(4), 040912.

Bhupesh Kumar Lad, Ph.D. is an Associate Professor in the Discipline of Mechanical Engineering at the Indian Institute of Technology Indore, India. He received Ph.D. degree in the area of Reliability Engineering from the Department of Mechanical Engineering at the Indian Institute of Technology Delhi, India, in 2010. Bhupesh worked with GE Global Research Center, India, as a Research Engineer from 2010 to 2011. Bhupesh is one of the founding members and mentor of a startup company, namely Techwarium India Private Limited. He is a leading researcher and one of the members of Industry Academia Consortium on Smart Manufacturing (IndAC-SM). The consortium includes researchers from IIT Indore, IIT Mumbai, University of Cambridge UK, and practitioners from various industries. Bhupesh has published various research papers in peer-reviewed journals and conferences. He is one of the authors of the book- Machine Tool Reliability. He is investigator of various research projects funded by national and international funding agencies. He received the Hamied-Cambridge Visiting Lecture Fellowship of the University of Cambridge, UK, in 2016. His primary research interest includes smart manufacturing, reliability engineering, and prognosis.

Chapter 3
Functional Safety and Cybersecurity Analysis and Management in Smart Manufacturing Systems

Kazimierz T. Kosmowski

Abstract This chapter addresses some of the issues of the integrated functional safety and cybersecurity analysis and management with regard to selected references and the functional safety standards: IEC 61508, IEC 61511, ISO 13849-1 and IEC 62061, and a cybersecurity standard IEC 62443 that concerns the industrial automation and control systems. The objective is to mitigate the vulnerability of industrial systems that include the information technology (IT) and operational technology (OT) to reduce relevant risks. An approach is proposed for verifying the performance level (PL) or the safety integrity level (SIL) of defined safety function, and then to check the level obtained taking into account the security assurance level (SAL) of particular domain, for example, a safety-related control system (SRCS), in which the given safety function is to be implemented. The SAL is determined based on a vector of fundamental requirements (FRs). The method uses defined risk graphs for the individual and/or the societal risk, and relevant risk criteria, for determining the performance level required PL_r or the safety integrity level claimed SIL CL, and probabilistic models to verify PL/SIL achievable for the architecture of the SRCS considered.

Keywords Smart manufacturing systems · Industry 4.0 · Information technology · Operational technology · Safety-related control systems · Functional safety · Cybersecurity

3.1 Introduction

Nowadays, manufacturers face ever-increasing variability demands for innovative products, greater customization, smaller lot sizes and viable in practice supply-chain changes. However, disruptions also occur causing production delays and manufacturing losses. In many industrial sectors various hazards and threats are present or

---

K. T. Kosmowski (✉)
Faculty of Electrical and Control Engineering, Gdansk University of Technology, Gdansk, Poland
e-mail: k.kosmowski@upcpoczta.pl

K. B. Misra (ed.), *Handbook of Advanced Performability Engineering*,
https://doi.org/10.1007/978-3-030-55732-4_3

emerge that contribute significantly to the business and insurance risks [1]. Manufacturers to be successful have to choose and incorporate technologies that help them quickly adapt to dynamic changes in business environment while maintaining high product quality and optimizing the use of energy and resources to limit environmental emissions and pollutions. Such technologies form the core of emerging, information-centric, and the so-called smart manufacturing systems (SMSs) that should be designed and operated to maximize the business potential, in particular the use, flow and re-use of data throughout the enterprise and between cooperating companies [2].

The SMS design and operation principles, and business expectations, are similar to those that stand behind the Industry 4.0 technological concept being in dynamic development [3]. These concepts include new interesting ideas, models, solutions and tools, related to the information technology (IT) and the operational technology (OT), ranging from innovative software supporting business planning and management, using the artificial intelligence (AI) and big data (BD) applications, and the cloud technology (CT), to innovative production and maintenance supporting software tools, and advanced automation solutions, for example, AutomationML concept based on mechatronic metamodels [4]. More and more important functions are to be assigned to the distributed industrial control systems (ICS), operating often in sophisticated computer networks, to be designed using the wire and wireless technologies for communications.

The CT is a relatively new technology of increasing interest that has significant potential to support the effectiveness of the SMSs operating in changing business environment. This technology in principle supports the implementation of advanced internet technologies, currently in dynamic development and use, known as the internet of things (IoT) and the industrial internet of things (IIoT) [5]. Nowadays, the factory automation and process control systems, networks and protocols within the OT are increasingly merged with those of IT. Requirements formulated for the OT and IT are in principle different, but the networks and protocols for communication in the SMS must allow for effective and safe convergence of the IT and OT systems [6], especially when a concept of machine-to-machine (M2M) communication techniques is applied in the industrial interconnected systems.

Therefore, the questions may be raised concerning the security issues of such technical solutions in the context of the reliability and safety requirements. Lately, considerable efforts have been undertaken by the research community to identify existing and emerging problem areas [7, 8], point out more important issues that require further research to support the development and implementation in industrial practice of advanced safety and cybersecurity requirements and technologies [9]. These aspects are considered in some publications from the point of view of technology resilience, in particular, a cyber resilience that should be carefully reviewed in the computer systems and networks to be designed or modernized [10].

The expectations of the industry are high and some institutions have been involved in practically oriented research to propose new solutions for implementation in the industrial hazardous plants [11–13]. Proposing integrated safety and security analysis methodology to support managing of hazardous systems is undoubtedly challenging.

It concerns especially the systems to be designed to achieve possibly high functional safety and cybersecurity goals of relevant domains to be managed in life cycle [14]. It depends on decisions and actions undertaken by responsible management and engineering staff in given industrial company and is influenced significantly by awareness of the safety and security culture to be carefully shaped in time [15].

The complexity of industrial systems and networks, sometimes without clear hierarchy in information flow for controlling various processes, operating in changing internal and external environment, emerging of new hazards and threats, can make some additional challenges to reach, in practice, high level of system reliability and safety [16, 17]. No less important in such systems are the security-related issues, especially those influencing potentially the risk of high consequence losses [18, 19]. An important issue in industrial practice is the business continuity management (BCM) [20] that requires careful consideration of various aspects within an integrated RAMS&S (reliability, availability, maintainability, safety and security) framework. In such analyses the risk evaluation and management in life cycle is of special interest for both the industry and insurance companies [21]. Such issues are of significant interest also in the domain of performability engineering that have been stimulated by Misra for years [22].

In this chapter an approach is proposed for integrated functional safety and cybersecurity analysis and management in the SMSs and hazardous plants in the context of the design and operation of the industrial automation and control systems (IACSs) [14, 23]. The idea of the SMSs assumes the openness of markets and flexible cooperation of companies worldwide. It could not be effective without relevant international standardization. However, some problems have been encountered in industrial practice due to too many existing standards that have been published by various international organizations. Unfortunately, the contents of some related standards were not fully coordinated or require updating. It concerns, in particular, the IT and OT design principles in relation to the IACS functionality and architecture requirements with regard to the safety and security aspects [2, 6].

The main objective of this chapter is to outline a conceptual framework for integrated analyses of the functional safety solutions according to generic functional safety standard IEC 61508 (7 parts) [24], and the IACS cybersecurity, outlined in IEC 62443 (14 parts) [23]. For reducing vulnerability of the IT and OT systems and reduce risks of hazardous events, especially of high consequences, a set of seven fundamental requirements (FRs), defined in the IEC 62443-1 standard, is taken into account to determine the SAL of the domain considered.

The method proposed uses the individual and/or societal risk graphs for determining the performance level required (PL_r) [25], the safety integrity level required (SIL_r) [24, 26] or the safety integrity level claimed (SIL CL) [27] of consecutive safety functions to be defined in the analyses. These levels are then verified to indicate the PL or SIL to be achieved in designed SRCS of architecture proposed, in which particular safety function is to be implemented. For that purpose, relevant probabilistic model of SRCS is developed with regard to potential common cause failures (CCFs), when the redundancy of hardware is proposed. Then, the verified SIL is validated with regard to determined SAL of relevant domain, for example, the domain of SRCS in which particular safety function is implemented, including internal and external communications.

In the analyses and assessments to be carried out, both quantitative and qualitative information available is used, including expert opinions. The analyses and assessments are based on classes defined or categories distinguished. For related evaluations some performance indicators are of interest, in particular the so-called key performance indicators (KPIs) defined, for example, in the standard [28] and numerous publications [e.g. 1].

3.2 Architectures and Conceptual Models of Complex Manufacturing Systems

3.2.1 Manufacturing System General Architecture

Opinions are expressed, based on evidence from industrial systems and networks, that the SMSs are driving unprecedented gains in production agility, quality, and efficiency across manufacturers present on local and global markets, improving both short-term and long-term competitiveness. Specifically, the SMSs use the information and communication technologies along with advanced software applications to achieve the following main goals [2]:

- support intelligent marketing for better production planning,
- develop innovative technologies and products,
- optimize the use of labour, material, and energy to produce customized, high-quality products for the long-term or just-in-time delivery,
- quickly respond to the market demands and supply chains with support of advanced logistics system.

Various categories of computer applications are used in industrial practice for supporting in achieving these goals including [2, 14]: ERP (enterprise resource planning), CRM (customer relationship management), SCM (supply chain management), MES (manufacturing execution system), CMM (computerized maintenance management), PLM (product lifecycle management) and so on.

The ability of potentially disparate systems to gather and exchange the production and business data rests critically on information technology and related standards that enable communication and services for running, supervising and coordinating effectively various processes in normal, transient and abnormal conditions. It becomes evident that a manufacturer's sustainable competitiveness depends on its capabilities with respect to cost, delivery, flexibility and quality, but also the reliability, safety and security of processes and assets.

The SMS's technical and organizational solutions should maximize those capabilities and profits by using advanced technologies that promote rapid flow and widespread use of digital information within and between manufacturing systems [2].

However, it is necessary to consider and assess various risks during the SMS design and its operation to reduce significant risks of potential major losses. It should be supported by the insurer having experience and knowledge gathered from industrial practice [1].

An example of the complex system consisting of the OT, IT and CT networks illustrating generally their functional and architectural issues of convergence is shown in Fig. 3.1. The OT is in the process of adopting the same network technologies as defined in the IT system at an increasing rate, so these two systems begin to merge together. It is expected that the use of the CT in favour of IT and OT will make additional business models and automation structures possible and profitable. Combining these domains is often referred to as the internet of things (IoT) or industrial internet of things (IIoT) [6]. However, such merging can cause some cybersecurity-related problems in relevant domains that require special treatment in the design and in the operation of the IT and OT systems, especially when using the CT network is to be considered [6].

An approach is proposed below for integrated functional safety and cybersecurity evaluation aimed at indicating rational solutions in the context of reducing relevant risks. In the functional safety approach the safety functions [12, 16] are defined to be implemented within the SRCSs, for example, the basic process control system (BPCS) [24], the safety instrumented system (SIS) in process industry [26] or in the machinery sector using, for example, the safety programmable logic controller (PLC) or the relay logic solutions [25, 27] (see the OT part in Fig. 3.1). Adoption of the same networks within the OT and IT systems may be justified regarding costs, but requirements concerning the functional safety and cybersecurity in the domains

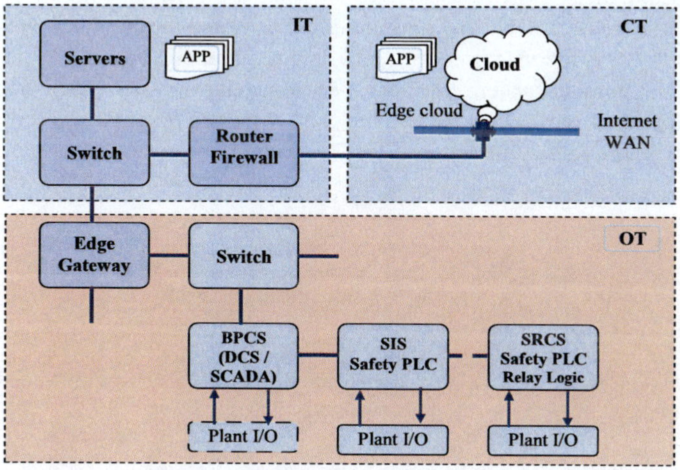

Fig. 3.1 Architectural relations of basic networks consisting of the operational technology (OT), information technology (IT), and cloud technology (CT) (based on [6])

of OT and IT are usually different, which might lead to new challenges in bridging these different technological worlds [6].

3.2.2 Traditional Reference Model of the Manufacturing System

A traditional reference model is based on the ISA99 series of standards derived from the generic model of ANSI/ISA-95.00.01 (Enterprise-Control System Integration), and represents the manufacturing system as the connection of following functional and logical levels (Fig. 3.2):

Level 0—Manufacturing processes: It includes the physical processes and basic process equipment, sensors and actuators, equipment under control (EUC) [24] that are the elements of safety-related system (SRS) for implementing the safety function (SF); these devices are periodically tested and subjected to the preventive maintenance (PM);

Level 1—Basic control: This level includes: local area network (LAN) controller, input/output (I/O) devices, communication conduits, and the PLCs; the devices of this level contribute to the continuous control, discrete/sequence control, or batch control;

Level 2—Area control: This level allows to implement functions for monitoring and controlling the physical process; it consists of LAN and local elements of the control and protection systems, human–machine interface (HMI) on local equipment panels;

Level 3—Site manufacturing and control: For example, the distributed control system (DCS)/supervisory control and data acquisition (SCADA) software that includes: a human–system interface (HSI), an alarm system (AS) and a decision support system (DSS) for the site control human operators; at this level the manufacturing execution system (MES) is placed;

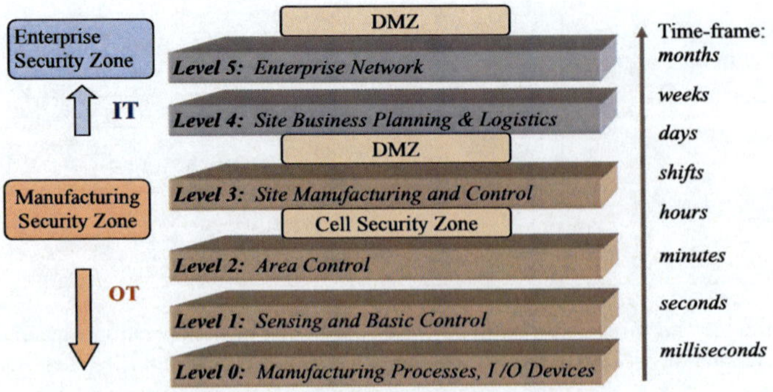

Fig. 3.2 Traditional reference model of the SMS based on ANSI/ISA95 standard

Level 4—Enterprise business planning and logistics: This level is characterized by the business planning and related activities, including logistics, using often the enterprise resource planning (ERP) system to manage and coordinate effectively business and enterprise resources required in manufacturing processes;
Level 5—Enterprise network: At this level additional external functions are to be realized, for example, business and logistics-related support by the CT applications.

Levels 0–3 are to be designed and operated with regard to relevant technical and functional requirements and specifications assigned to the OT network. The levels 4 and 5 are essential parts of the IT network. The purposeful and reliable system-oriented functional convergence of these networks has to include the functional safety and cybersecurity-related aspects. Nowadays, in case of the SMS, an intensive use of the cloud technology is of interest in industrial plants.

In such open manufacturing system, the safety and security aspects require special attention of the designers and operators [14, 29]. From the information security point of view an important role is to be assigned to the cell security zone (CSZ) and the demilitarized zone (DMZ) placed in Fig. 3.2. The safety and security issues, in particular the functional safety and cybersecurity solutions, obviously require careful treatment and management in life cycle.

Many internal and external influences, hazards and threats should be considered in the operation process of the OT and IT systems. Basic features of these system are illustrated in Fig. 3.3. Expected lifetime of the OT system is often to be evaluated in the range of 10–20 years, but only 3–5 years in the case of IT [30]. For characterizing of the OT an AIC (availability, integrity and confidentiality) triad is usually used to prioritize basic safety and security requirements, but a confidentiality, integrity, and availability (CIA) triad is to be assigned to the IT network.

The SMS's reliability, safety and security is influenced by external and internal factors, including human and organizational factors [15]. For high reliability and availability of the OT system an operational strategy should be carefully elaborated

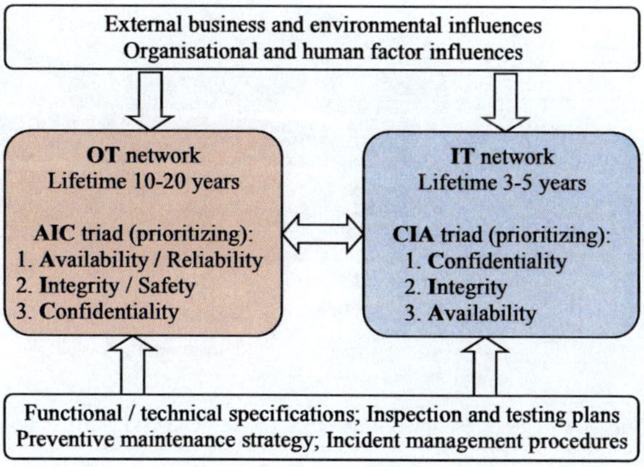

Fig. 3.3 Basic features concerning the OT and IT systems

that includes: inspection, testing, preventive maintenance plans and incident management procedures [21] to reduce the risk of major consequences due to potential hazardous events.

3.2.3 RAMI 4.0 Reference Architecture Model

Another recently published reference architecture model is the RAMI 4.0 (Reference Architectural Model for Industry 4.0), developed to support relevant business-oriented decision-making in practical applications [3, 31]. It seems to be also useful for the reliability, safety and security-related systemic analysis and management in the SMS [14]. This model describes the key elements of manufacturing system based upon the use of structured layers with distinguishing three axes:

- Architecture axis (see Fig. 3.4) of six different layers indicating the information depending view from the assets to business;
- Process axis (value stream) for including the various stages within the life of assets and the value-creation process based on IEC 62890;
- Hierarchy axis (hierarchy levels) for assigning the functional models to individual levels based on IEC 62264 and IEC 61512.

Some remarks concerning the security aspects are as follows:

- Layers—security-related aspects apply to all different levels; the risk evaluation has to be considered for the object/assets as a whole;
- Value stream—the owner of the object must consider security across its entire life-cycle;

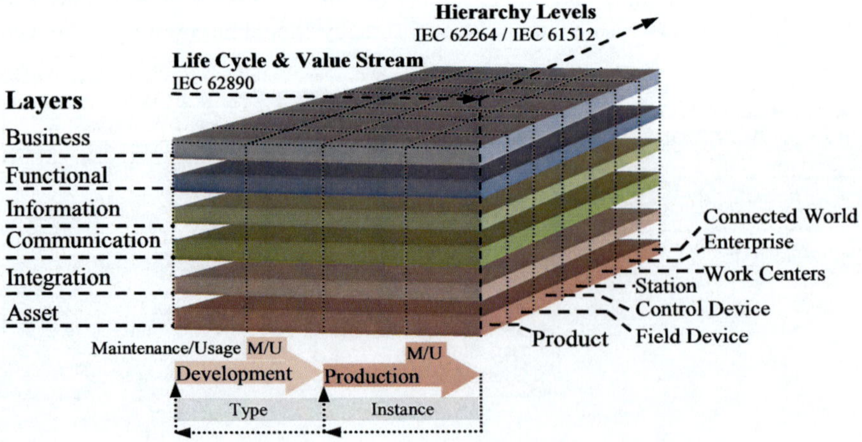

Fig. 3.4 The reference architecture model RAMI 4.0 for Industry 4.0 concept (based on [31])

- Hierarchy levels—all objects/assets are subjected to the security considerations (based on the risk evaluation) and need to possess or provide relevant security characteristics for fulfilling their tasks, thanks to applying appropriate protections.

Opinions are expressed that new opportunities are opened up by the Industry 4.0 idea, but also bring a host of challenges. Security by design, for instance, becomes an indispensable element in designing within Industry 4.0 concept. In some cases, security will be an enabler of new business models [31]. Security-related requirements can act in many cases as a skeleton that carries and holds together all of the structural elements within RAMI 4.0 model and, as a result, the design of the Industry 4.0 components and interrelated systems.

The security-related aspects can also play a role at relevant points of intersection between the various levels. This means that requirements shall be derived for some points of intersection by more specific analyses. The solutions have to be found for these requirements based on new capabilities of the Industry 4.0 components involved in the specific application in question. The manufacturers, integrators and asset owners should all be involved in implementing a holistic safety and security concept that brings the technical and organizational measures together [14, 31].

3.2.4 *Knowledge and Standards Supporting the SMS Operational Analyses Including Functional Safety and Cybersecurity Aspects*

Designing and operating of the SMS require a wide knowledge and considerable efforts. A rational way to deal with relevant issues is at least to consider existing standards. Examples of standards to be of interest in developing operational models of the SMS and the IACS are listed in Table 3.1. In Table 3.2, selected standards and publications useful for supporting the functional safety and cybersecurity analysis based on relevant risk analysis and management methods are collated.

Due to a considerable number of existing standards, the problem lays in purposeful selection of relevant standards, reports and publications, depending on the objectives of analyses. Some of these standards and publications, developed by various organizations to support the design and operation of the SMS or hazardous industrial plant, include mainly the functionality aspects of the IACS, and also some aspects to be included in related reliability, safety and security analyses. The objective is to improve functionality and to limit risks related to production goals with regard to criteria defined.

Nevertheless, a considerable research effort is still necessary to be undertaken directed towards development and successful implementation methods useful for the integration of existing methods and models. As it was mentioned, this chapter is directed towards integration of the functional safety and cybersecurity analyses of the SRCS as a part of the IACS.

Table 3.1 Selected standards useful for developing the operational models of the SMS and its IACS

Topic	Related standards	Remarks
Administration shell	IEC 62794 TR	Reference model for representation of production facilities (digital factory)
	IEC 62832	Industrial process measurement, control and automation—Digital factory framework
Life cycle and value stream	IEC 62890	Life cycle status
Hierarchy levels	IEC 62264/IEC 61512	
	ANSI/ISA 95	Enterprise control system levels
Configuration	IEC 6104 EEDL	Process control and electronic device description language (EDDL)
	IEC 6523 FDT	Information technology, Organization identification schemes
Engineering, data exchange	IEC 61360/ISO 13584 IEC 61987 IEC 62424 IEC 6214 ISO/IEC 20248	Standard data elements Data structures and elements Between P&ID tools and PCE-CAE tools For use in industrial automation systems Automatic identification and data capture
Communication	IEC 61784-2 IEC 61158 IEC 62351	Real-time ethernet (RTE) Industrial communications networks Power system information infrastructure
Condition monitoring	VDMA 24582	Fieldbus neutral reference architecture for condition monitoring in factory automation
OPC UA AutomationML	IEC 62541	Open platform communications unified architecture
	IEC 62714	The automation mark-up language

3.3 Functional Safety Analysis and Management in Life Cycle

3.3.1 Safety Functions for the Risk Reduction

The functional safety is defined as a part of general safety of an industrial hazardous plant installation or manufacturing machinery, which depends on a proper response of the SRCS during abnormal situation or accident to avoid or limit undesirable consequences. The functional safety methodology has been formulated in the generic standard IEC 61508 [24] and is appreciated in the design and operation of the electric/electronic/programmable electronic (E/E/PE) systems. Different names of the SRCS are used in various industrial sectors, for example, the safety instrumented systems (SIS) in case of the process industry sector [26], or the safety-related electrical control system (SRECS) for machinery [27]. Such systems are to be designed to

Table 3.2 Selected standards and publications useful for functional safety and cybersecurity analyses including the risk evaluation and management

Topic	Related standards and publications	Remarks
Risk management	ISO 31000 ISO 31010 ISO/IEC 27001	Risk management—guidelines Risk assessment techniques Information security management systems
	ISO/IEC 27005	Information security risk management
Functional Safety SIL—*safety integrity level* PL—*performance evel*	IEC 61508 ISO 13849-1 (PL) IEC 62061 IEC 61511	Generic standard FS of SRCS Machinery Production lines/systems Process industry
IACS cybersecurity SL—*security level* SAL—*security assurance level*	IEC 62443	Computer systems/networks security
	ISO 22100-4 DTR	Safety of machinery—security aspects
	VDI 2182	IT security for industrial automation
	IEC 63074 CD1	Security aspects of SRCS
	IEC 62351-12 TR	Security recommendation for power systems
Smart manufacturing/ Information security and risk management	NIST IR 8107	Standards for smart manufacturing systems
	NIST SP 800-30	Guide for risk assessments
	NIST SP 800-39	Managing information security risk
	NIST SP 800-53	Security and privacy control
	NIST SP 800-82	ICS security
	NIST SP 800-171	Protecting controlled information

perform specified safety functions to ensure that evaluated risk is reduced to the level specified for the particular industrial installation, and then maintained at a specified tolerable level during the life cycle of the system [16, 32].

Two different requirements are to be specified to ensure appropriate level of functional safety [24]:

- the requirements imposed on the performance of particular safety function being designed for the hazard identified,
- the safety integrity requirements, that is, the probability that the safety function will be performed in a satisfactory way when potential hazardous situation occurs.

The safety integrity is defined as the probability that a safety-related system, such as the E/E/PE system or SIS, will satisfactorily perform defined safety function under all stated conditions within given period of time. For the safety-related system, in which defined safety function is to be implemented, two probabilistic criteria are defined as presented in Table 3.3 for four categories of the SIL [24, 26], namely:

- the probability of failure on demand average (PFD_{avg}) of the SRCS in which particular safety function is to be implemented, operating in a low demand mode, or
- the probability of a dangerous failure per hour (PFH) of the SRCS operating in a high demand or continuous mode.

The SIL requirements for SRCS to be designed for implementing specified safety function stem from the results of the risk analysis and assessment to reduce sufficiently the risk of losses taking into account specified risk criteria, namely for the individual risk and/or the group or societal risk [24]. If the societal risk is of interest, the analyses can be generally oriented on three distinguished categories of losses, namely [16, 24]: Health (H), Environment (E) or Material (M) damage, then the safety integrity level required (SIL_r) for particular safety function is determined as follows:

$$SIL_r = \max\left(SIL_r^H,\ SIL_r^E,\ SIL_r^M\right) \tag{3.1}$$

In case of the machinery only the individual risk is to be considered, and then the performance level required (PL_r) [25] or the safety integrity level claimed (SIL CL) [27] is determined. The SRCS of machinery operates in a high demand or continuous mode, and therefore the PFH probabilistic measure (per hour) is to be evaluated and then assessed against relevant interval criteria.

Figure 3.5 illustrates these interval criteria of PFH in the context of risk graph for determining PL_r according to ISO 13849-1, and a method for determining SIL CL described in IEC 62061. The risk related to identified hazards is to be evaluated taking into account a measure of harm severity (S) that could result from that hazard, and the probability of occurrence of that harm. According to the ISO standard 12100 and ISO 22100 [33], the PFH is influenced by an exposure measure (F) of the person(s) to the hazard considered, the occurrence rate of hazardous event resulting, and the possibility (P) to avoid or limit the harm.

Thus, the PL_r for a safety function considered is determined according to the left side risk graph in Fig. 3.5, taking into account specific parameters to be evaluated

Table 3.3 Safety integrity levels and probabilistic criteria to be assigned to safety-related systems operating in a low demand mode or high/continuous mode

SIL	PFD_{avg}	PFH [h^{-1}]
4	$[10^{-5}, 10^{-4})$	$[10^{-9}, 10^{-8})$
3	$[10^{-4}, 10^{-3})$	$[10^{-8}, 10^{-7})$
2	$[10^{-3}, 10^{-2})$	$[10^{-7}, 10^{-6})$
1	$[10^{-2}, 10^{-1})$	$[10^{-6}, 10^{-5})$

during the risk analysis [14, 25]. The PL_r categories, denoted from a to e, are related to required levels of the risk reduction, being highest in case of category e, which is equivalent to SIL CL 3 according to IEC 62061 [27].

Having the PL_r or SIL CL determined as described above, the relevant level has to be verified, whether it can be achieved by the SRCS of architecture proposed by the system designer, in which particular safety function will be implemented. The verification of the SRCS is based on the PFH probabilistic measure evaluated using appropriate probabilistic model. The result obtained is compared with the interval criteria presented in Fig. 3.5, and verified level PL or SIL is indicated that should be equal or higher than required.

For instance, if the PL (e.g. PL e) or SIL (e.g. SIL 3) obtained are equal or higher than the PL_r ($PL \geq PL_r$) or SIL CL ($SIL \geq SIL\ CL$), respectively, than the architecture proposed can be accepted. Otherwise, it is necessary to propose modified architecture and repeat the verification process as described above. It is worth to mention that the architecture includes the hardware, software and human component. The verification and validation procedure has to be carried out for each safety function considered to be implemented in the SRCS [25, 27].

3.3.2 Issues of the Safety Integrity Level Verification

As it was mentioned above, generally the SIL verification can be carried out for two categories of the operation mode, namely: (1) low operation mode, or (2) high or

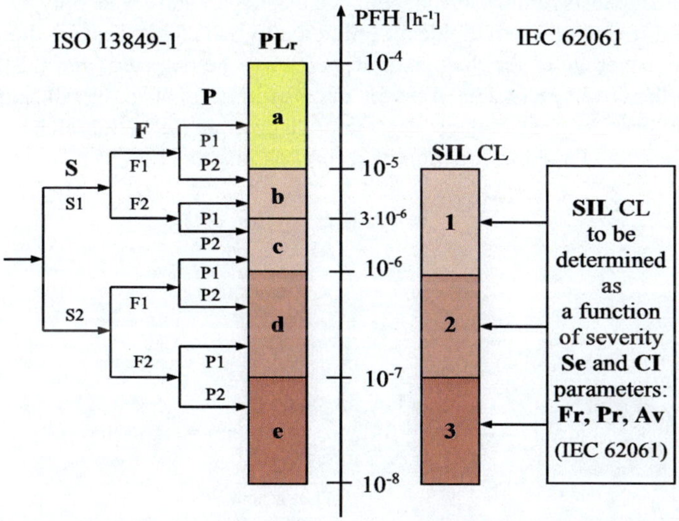

Fig. 3.5 Risk graphs for determining required performance level PL_r or safety integrity level claimed SIL CL (based on standards [25, 27])

continuous mode of operation [24, 34]. The former is characteristic for the process industry [26], and the latter is typical for the machinery [27] or the railway transportation systems, and also for monitoring and the real-time control of any installation using the DCS/SCADA technology.

Typical hardware architecture of an E/E/PE system [16], shown in Fig. 3.6, consists usually of three subsystems: (A) sensors and input devices (transducers, converters etc.), (B) logic device (e.g. safety PLC or safety relay modules) and (C) actuators, that is, the EUC and other output devices.

Such safety-related system constitutes a specific architecture of the hardware, software modules and communication conduits. The logic device comprises typically a safety PLC with its input and output modules. The subsystems shown in Fig. 3.6 can be generally of KooN configuration, for example, 1oo1, 1oo2 or 2oo3. Their hardware fault tolerance (HFT) is understood as ability of the subsystem to perform a required function in the presence of faults or errors [24]. The HFT (0, 1, 2) is an important parameter to be considered in final verification of the subsystem's SIL for the evaluated value of a safe failure fracture (S_{FF}).

Any redundant system, for example, the SRCS, is prone to a common cause failure (CCF) that can contribute significantly to decreasing its dependability due to potential failure mechanisms depending on the site-specific influence factors. The CCF is a failure resulting in one or more events, causing coincident failures of two or more channels in a multiple channel system, leading to the system failure. The multiple failures may occur simultaneously or over a period of time. Various probabilistic models are proposed to deal with CCF in safety-related systems, in particular the E/E/PE systems or SIS [24]. The CCF contribution in the PFD_{avg} or PFH is usually incorporated using the β-factor method [34].

If diagnostic tests run in each channel that can detect and reveal only a fraction of the failures, it is justified to divide all failures into two categories: (1) those that lie outside the coverage of the diagnostic tests (cannot be detected) and (2) those that lie within the coverage (detected by the diagnostic tests). The overall failure event probability per time unit of the subsystem is dangerous (D) failure due to potential failures including CCF is a function of several parameters [24, 34]

$$\mathrm{PF}_{\mathrm{D}}^{\mathrm{CCF}} = f(\lambda_{\mathrm{Du}}\beta, \lambda_{\mathrm{Dd}}\beta_{\mathrm{D}}, \ldots) \tag{3.2}$$

where:

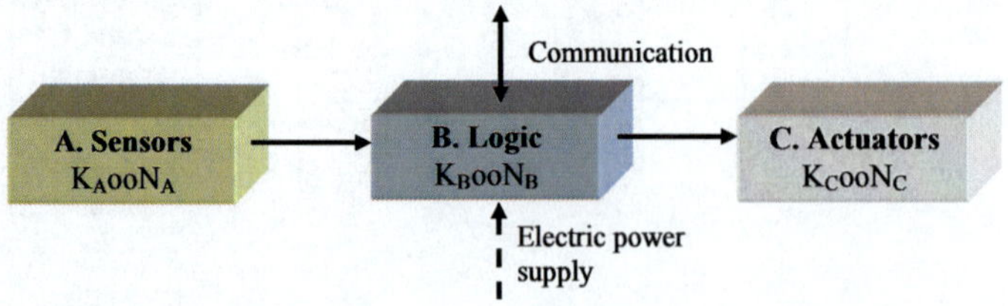

Fig. 3.6 General architecture of the E/E/PE system or SIS for implementing the safety function

Table 3.4 Proposal for evaluation values of β or $β_D$ for subsystems [24]

Score for S or S_D	Values of β or $β_D$ for the logic subsystem (%)	Values of β or $β_D$ for the sensors or actuators (%)
≥120	0.5	1
[70, 120)	1	2
[45, 70)	2	5
<45	5	10

- $λ_{Du}$ is the rate of danger (D) undetected (u) failure in a single channel, influencing the probability of failures that lie outside the coverage of the diagnostic tests; β is the common cause failure factor for undetectable dangerous faults, which is equal to the overall β-factor that would be applicable in the absence of diagnostic testing;
- $λ_{Dd}$ is the rate of a danger (D) detected (d) failure in a single channel, influencing the probability of failures that lie within the coverage of the diagnostic tests, $β_D$ is the common cause failure factor for detectable dangerous faults; as the repetition rate of the diagnostic testing is increased, the value of $β_D$ falls increasingly below β.

In given subsystem probabilistic modelling of the value of β is determined for the score $S = X + Y$ to be evaluated for factors specified in the standard IEC 61508 and the value of $β_D$ is evaluated for the score $S_D = X (Z + 1) + Y$ as it is presented in Table 3.4. These scores are evaluated respectively for the logic subsystem, and for the subsystems of sensors and actuators (final elements). In evaluating scores for *X* and *Y*, the following factors should be taken into consideration [24]:

(1) Separation/segregation,
(2) Diversity/redundancy,
(3) Complexity/design/application/maturity/experience,
(4) Assessment/analysis and feedback of data,
(5) Procedures/human interface,
(6) Competence/training/safety culture,
(7) Environmental control,
(8) Environmental testing.

Each of these factors is divided into several sub-attributes with specified sub-scores to be added to obtain final score, respectively for *X* and *Y*, and finally for *S* and S_D. The value of *Z* in calculating S_D depends on the diagnostic test interval and the diagnostic coverage (DC). For instance, in case of the subsystem of sensors or actuators, if DC ≥ 99% and the diagnostic test interval is between 2 h and 2 days, it is suggested: $Z = 1.5$. If the test interval is greater than 1 week, then $Z = 0$ [24].

Thus, the values of β and $β_D$ parameters used in the probabilistic modelling of subsystems depend significantly on factors specified in IEC 61508 and the expert opinions collected during the functional safety analysis of the E/E/PE system or SIS. In publication [34] two examples are presented of the SIL verification of given SRCS architecture using the probabilistic models of subsystems with regard to the CCF analysis. The architectural constrains with regard to the safe failure fraction (S_{FF}) for subsystems were also considered. It seems to be justified to assume that some categories of factors specified above are also relevant in case of the cybersecurity analysis.

3.4 Cybersecurity of the Safety-Related Control System

The security-related remote attacks are becoming increasingly important threats to the IT and OT systems, especially the IACS operating within industrial networks of hazardous plants [6, 8, 23] and the SMSs characterized in this chapter and publications [2, 3, 14]. The internal or external threats can initiate an IT or OT security-related incidents with the potential to adversely impact the SRCS and machinery operations. Vulnerability understood as a security-related weakness of the IT and/or OT networks that can be exploited by various threats to trigger hazardous events making losses. It is an important issue to be adequately treated in the BCM [20].

A threat may be either passive or active. In case of the passive threat the agents usually gather information by casual communications with employees and contractors. Examples of active threats are as follows [19, 33]: database injection, spoofing and impersonation, phishing, malicious code, Denial of Service (DoS), escalation of privileges, physical destruction, etc. The analyses should be also carried out to identify the SRCS vulnerability that can be exploited by threats, potentially impacting the safety of entire manufacturing system.

The IT security risks shall be mitigated through the combined efforts of component suppliers, the machinery manufacturer, the system integrator, and the machinery end user [23]. Generally, the potential responses to the security risks should take following steps [33]:

(a) eliminate the security risk by design (avoiding vulnerabilities);
(b) mitigate the security risk by risk reduction measures (limiting vulnerabilities);
(c) provide information about the residual security risk and the measures to be adapted by the user.

The standard IEC 62443 [23] proposes an approach to deal systematically with the security aspects of the IACS. Four security levels (SLs) are defined that are understood as a confidence measure that the IACS is free from vulnerabilities and it functions in an intended manner. In the standard IEC 63074 [19] these levels are also proposed to deal with the SRCS security of machinery.

The SL is related to seven foundational requirements (FRs):

FR 1—Identification and authentication control (IAC),
FR 2—Use control (UC),
FR 3—System integrity (SI),
FR 4—Data confidentiality (DC),

FR 5—Restricted data flow (RDF),
FR 6—Timely response to events (TRE), and
FR 7—Resource availability (RA).

Thus, instead to express the SL as a single number, it is proposed to apply a related vector of seven FRs specified above. Such vector is proposed for describing the security requirements for a zone, conduit, component or system. It may contain the integer numbers of SL from 1 to 4 or 0 to be assigned to consecutive FRs. A general format of the security assurance level (SAL) is defined as follows [23]:

$$\text{SL - ? ([FR], domain)} = \left[\text{IAC UC SI DC RDF TRE RA}\right] \quad (3.3)$$

where: SL-? = (required) the SL type-possible formats are: SL-T = Target SAL, SL-A = Achieved SAL, and SL-C = Capabilities SAL vector; [FR,] = (optional) field indicating the FR that the SL value applies; domain = (required) is applicable domain that SL applies—this may be procedure, system or component—when applying the SL to a systems, it may be for instance: Zone A, Machinery B, Engineering Workstation, etc.

For instance, according to the standard [23], it can be written as follows:

(a) SL-T (Control System Zone) = [2 2 0 1 3 1 3],
(b) SL-C (Engineering Workstation) = [3 3 2 3 0 0 1],
(c) SL-C (RA, Safety PLC) = 3; in this example only the RA component is specified, instead of a seven-dimensional SAL vector SL-C.

Thus, three type of vectors describing SL_i for consecutive FR_i of particular domain are distinguished:

- SL-T (Target SAL)—the desired levels of security;
- SL-C (Capability SAL)—the security level that device can provide when properly configured;
- SL-A (Achieved SAL)—the actual level of security of a particular device.

The SL_i numbers provide a qualitative information addressing relevant protection scope of the domain or zone considered, for example, for the IACS or the SRCS as its part, as presented in Table 3.5.

Table 3.5 Security levels and protection description of the IACS domain [19, 23]

Security levels	Description
SL 1	Protection against casual or coincidental violation
SL 2	Protection against intentional violation using simple means with low resources, generic skills and low motivation
SL 3	Protection against intentional violation using sophisticated means with moderate resources, IACS specific skills and moderate motivation
SL 4	Protection against intentional violation using sophisticated means with extended resources, IACS specific skills and high motivation

For instance, in the case of FR 1—identification and authentication control (IAC)—the security levels shall be interpreted in a following way "Identify and authenticate the SRCS users by mechanisms against" [19]:

- causal and coincidental access by unauthorized entities (SL 1),
- intentional unauthorized access by entities using simple means (SL 2),
- intentional unauthorized access by entities using sophisticated means (SL 3),
- intentional unauthorized access by entities using sophisticated means with extended resources (SL 4).

For improving the SRCS security it is suggested to elaborate guidance (the instruction handbook) for the end user that includes the following issues [19, 33, 35]:

(A) Restriction of logical/physical access to the IT systems with potential influence on safety, for example, using internal IT systems with risk reduction measures, such as firewalls, antivirus tools, etc.; providing authentication and access control mechanisms, such as card readers, physical locks, according to specifications of manufacturer or integrator; disabling all unused external ports/interfaces and services, etc.;

(B) Detection and reaction on IT-security incidents with potential influence on safety, for example, checking regularly means for detecting failed IT system components or unavailable service according to the specifications of the machine/component manufacturer; being responsive for vulnerabilities resulting from a new IT security threat and potential attack;

(C) In case of remote maintenance and service, for example, using provided means for setting up and ending a remote access session according to the specifications of the machine/component manufacturer; using encryption means for initiating a remote service according to the specifications of the machine/component manufacturer; watching any remote access session with a restriction of duration for remote access, and so on.

Such topics should be included and carefully treated in a security information and event management (SIEM) to be developed and used proactively in practice according to requirements given in ISO/IEC 27001 [36], and supported by the information security risk management as suggested in ISO/IEC 27005 [37]. Its specific requirements to be formulated should include the target SAL (SL-T) and then verified as achieved SAL (SL-T) taking into account the capability SAL (SL-C) of technology applied. Defined system requirements (SRs) and specific requirement enhancements (REs) for consecutive FRs to be fulfilled at relevant SLs from 1 to 4 are specified in the IEC 62443 standard [23] and a recent publication [14].

3.5 Integrated Functional Safety and Cybersecurity Analysis and Management

The IEC 62443 [23] series of standards consists of 14 parts but some of them are still in development. The main objective of this series is to cover important topics of the IACS security entirely. In the second edition of the generic functional safety standard IEC 61508 [24] it is suggested to use the IEC 62443 standard to deal with the cybersecurity issues at the design stage and operation of the programmable safety-related control systems. Up to now, though, the IEC 61508 and IEC 62443 standards have been rather loosely linked [29]. As it was mentioned, also in case of the SRCS of machinery there is a need to deal more systematically with security issues, as it has been lately emphasized [19, 33].

It is worth to mention that the SRCS security level to be achieved depends strongly on the quality of an information security management system (ISMS) established in industrial practice. The objective of the ISMS is to monitor, continuously control, maintain and, wherever justified, improve the IT and OT security. The IEC 62443 standard is based on general requirements and stipulations of the ISO/IEC 17799 and ISO/IEC 27000 series, especially as regards basic security requirements [36]. Due to complex and dynamic internal and external conditions making technical specifications related to the IT and OT security solutions for implementing in industrial practice is quite challenging.

An important task to be undertaken is the risk evaluation and management, as it is postulated both in ISO/IEC 27001 [36] and ISO/IEC 27005 [37]. It includes the consideration of all functional components of the information system including the hardware (HW) and software (SW), communication conduits and relevant human/organizational issues, especially those related to the IT and OT safety and security. Opinions are expressed that the quantitative risk evaluation is very difficult due to the complexity of the IT and OT system and many influencing factors involved. The credibility of such evaluation depends on a framework adapted and availability of data, and expert opinions concerning specific domain to be evaluated.

Opinions are also expressed that the CIA triad (confidentiality, integrity, availability) is a justified order of requirements in the IT security analysis (see Fig. 3.3), but in case of OT a reversed triad, namely AIC (availability, integrity, confidentiality) is more appropriate. As it was mentioned above the domain SAL defined in IEC 62443 is to be evaluated using the vector of seven FRs, as explained by the formula (3.3). So, there are some doubts how to match these two kinds of requirements in the security-related analyses. It seems to be reasonable that the fundamental requirements of IAC, UC, SI and TRE should be mapped to integrity (I), RA to availability (A), and DC, RDF to confidentiality (C) [14, 29].

Additional issue, worth to be explained in context of the cybersecurity evaluation, is related to the definition of seven evaluation assurance levels (EALs) in the so-called common criteria standard (IEC 15408) [38] that are to be applied in defining the IT security requirements. As explained above only four SLs are defined in IEC 62443. This issue was discussed in the publication [39] in the context of generic functional

Table 3.6 Proposed correlation between SIL and SAL [18]

Safety integrity level (SIL)	Security assurance level (SAL)	Explanation
SIL 1	SAL 1	SAL assignment is based on asset owner's assessment
SIL 2	SAL 2	
SIL 3 and SIL 4	SAL 3	Reserved for total system failure
	SAL 4	Reserved for loss of life

safety standard IEC 61508 [24], in which also four SILs are distinguished (see Table 3.3). So, the problem is encountered how to integrate these concepts in the integrated functional safety and cybersecurity analysis.

In the publication [18] the correlation between SIL and SAL is proposed as it is shown in Table 3.6. Similar correlation can be proposed for the SRCS of machinery; however, remembering that in the machinery sector the highest SIL to be evaluated is SIL 3 (see Fig. 3.5).

In view of the above we propose an approach for integrated functional safety and cybersecurity analysis based on a framework of existing concepts and accepted models suitable to apply the quantitative and qualitative information available, similarly as in the knowledge-based systems [14, 40]. We start from defining the safety functions with regard to hazards and threats identified and then evaluate required risk reduction regarding the risk criteria defined as it was described above in item 3.3.1. It allows to determine: the required safety integrity level SIL_r according to IEC 61508 according to the formula (3.1), or the safety integrity level claimed SIL CL (IEC 62061), or the required performance level PL_r (ISO 13849-1) as it is shown in Fig. 3.5.

As it is known, the levels: the safety integrity level required SIL_r (1, 2, 3 or 4) [24], SIL CL (1, 2 or 3) [27], or the performance level required PL_r (a, b, c, d or e) [25], are related to the required risk reduction with regard to relevant individual or social risk criteria [16]. For instance, the average probability of failure on demand PFD_{avg} (see Table 3.3) is related to the risk reduction measure as its reciprocal.

The PL_r or SIL_r or SIL CL determined for particular safety function has to be then be verified using probabilistic model of the SRCS of architecture proposed at the design stage (see the left site blocks for functional safety evaluation in Fig. 3.7). Such architecture includes generally the hardware configuration and requirements concerning software [24]. Parallelly, the security-related evaluation is to be carried out as it is shown in Fig. 7 (the right side) for cybersecurity evaluation. The integrated functional safety and cybersecurity analysis are repeated when justified to enable a rational management of the SRCS domain in life cycle.

Additional issue to be considered is associated with expressing SAL as a single number to be assigned to the security level achieved SL-A for given domain, as it is outlined in the formula (3.3), according to the standard IEC 62443. It would lead to sometimes disputable requirement that the security levels SL_i would be the same for each FR_i. For instance, confidentiality plays in some cases a minor role for

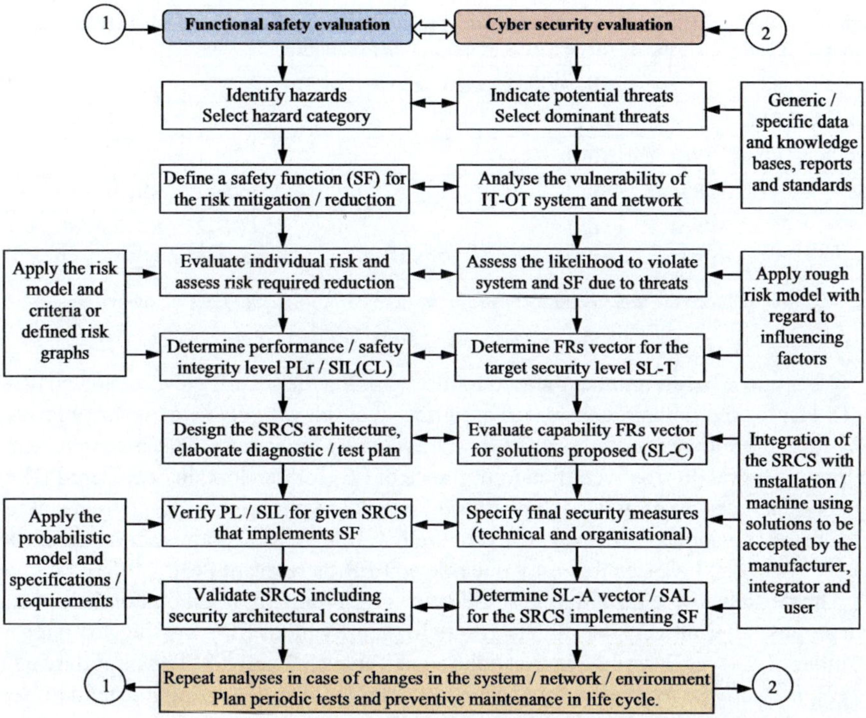

Fig. 3.7 Integrated functional safety and cybersecurity analysis for the SRCS domain

safety-related control system and encryption of all data might lead to complications in testing and the time response longer than required. So generally, different levels of SL_i may be assigned to seven consecutive elements of the FR vector.

This problem was noticed and discussed by Braband in the publication [29]. Only in simple cases of equal levels SL_i for consecutive FR_i (i from 1 to 7) determining SAL of domain of interest (e.g. the SRCS) is straightforward, for instance, SAL 1 = [1 1 1 1 1 1 1]. Generally, the SL_i can be different depending on the security technology applied or FR_i relevance for the domain considered. So, he suggests to use some security profiles, for example, for particular zones or conduits. However, it might also lead to a number of profiles, difficult for evaluation and security-related decision-making.

In our earlier publications [39] it was assumed that resulting SAL for the domain considered can be determined based on dominant FR_i and some common sense rules, in a similar way as in the methodology outlined in the IEC 15408 (common criteria) standard [38]. In this methodology seven evaluation assurance levels (EALs) are distinguished, related to classes of the security assurance requirements (SARs) and defined scope of fulfilling relevant requirements.

Table 3.7 Proposed correlation between security index SI^{Do} or SAL for the domain to be evaluated and final SIL to be attributed to the SRCS of hazardous installation

Security index	SIL verified according to IEC 61508[a]			
SI^{Do} and SAL	1	2	3	4
$SI^{Do1} \in [1.0, 1.5)$/SAL 1	SIL 1	SIL 1	SIL 1	SIL 1
$SI^{Do2} \in [1.5, 2.5)$/SAL 2	SIL 1	SIL 2	SIL 2	SIL 2
$SI^{Do3} \in [2.5, 3.5)$/SAL 3	SIL 1	SIL 2	SIL 3	SIL 3
$SI^{Do4} \in [3.5, 4.0]$/SAL 4	SIL 1	SIL 2	SIL 3	SIL 4

[a]verification includes the architectural constrains with regard to S_{FF} and HFT of subsystems

We propose below another method for determining the security level achieved SL-A (SAL) for the domain considered assuming that the weights w_i of security levels SL_i for consecutive (and relevant) FR_i are evaluated by experts. These weights can differ in general due to diversified importance of FR_i for the domain considered. The method includes cases in which not all fundamental requirements FR_i are relevant to the domain considered. It is suggested in the IEC 62443, as explained in the formula (3.3). There can be cases that only one relevant FR_i is relevant [23].

Thus, instead of determination of SAL for given domain based on dominant FR_i we propose alternatively to evaluate a domain security index SI^{Do} and then to assign a number of SAL as described in first column of Tables 3.7 and 3.8. The importance I_i of FR_i is evaluated by experts for specific domain, for example, using integer number on the scale from 1 to 5 (or 1–10), and 0 if FR_i is not relevant, and then the weight w_i of given FR_i is calculated according to following formula

$$w_i = \frac{I_i}{\sum_{i=1}^{7} I_i} \tag{3.4}$$

The security index SI^{Do} for the domain (Do) and determined security level SL_i (the integer number from 1 to 4, or 0 if FR_i is not relevant) for relevant (Re) fundamental requirements (FR_i) is evaluated as follows

$$SI^{Do} = \sum_{i \in Re} w_i SL_i \tag{3.5}$$

Four intervals of the domain security index SI^{Do} (from SI^{Do1} to SI^{Do4}) are proposed in the first column of Tables 3.7 and 3.8 for assigning the category number of SAL from SAL 1 to SAL 4. Such approach corresponds to attributing SAL for the domain in our earlier publications, based on dominant SL_i for relevant fundamental requirements FR_i.

Proposed correlations between security index to be assigned to the domain SI^{Do} or SAL and final SIL attributing to the SRCS in hazardous installation are presented in

Table 3.7. It was assumed that SIL has been verified according to IEC 61508 including such aspects as the common cause failures (CCFs) in probabilistic modelling, and the architectural constrains regarding the safe failure fraction (S_{FF}) and the hardware fault tolerance (HFT) of subsystems [24, 34].

Table 3.7 can be used to support the function safety and cybersecurity-related decision-making. For instance, if safety integrity level required, obtained from the risk assessment, is SIL_r 3, and it was positively verified according to IEC 61508 for the SRCS as SIL 3, we select the column with number 3. The SAL of the domain should be at least SAL 3 to attribute finally SIL 3 to the SRCS in which relevant safety function is implemented. If the SAL determined in the security analysis of domain considered would be lower (e.g. SAL 2), then the analyst should improve the system security (lowering its vulnerability) to increase SL_i of relevant FR_i to obtain at least SAL 3.

Other correlations are proposed in Table 3.8 for finally attributing the SIL or PL to the SRCS according to, respectively, IEC 62061 [27] or ISO 13849-1 [25]. Similarly, as it was explained above, if required performance level would be SIL CL 2 (or PL_r d), and such level were positively validated as SIL 2 (or PL d) the column 2 (d) of Table 3.8 is selected for the security validation. To obtain PL d the security assurance level should be at least SAL 2. If SAL would be lower (SAL 1) the security of SRCS should be improved to increase SL_i of relevant FR_i to obtain at least SAL 2, and finally validated safety integrity level SIL 2.

A case study was carried out concerning a modern end impregnation line used to treat yarns made of polyamide, polyester, viscose and other raw materials, so they are suitable for applications in tires [14]. A safety function of the pull roll section monitoring and door locking of the installation was analyzed. The performance level required PL_r was determined using a risk graph in Fig. 3.5 for following parameters indicated by a safety engineer for following path: S2, F1, and P2, leading to PL_r d.

The verification of the PL requires probabilistic modelling of the SRCS of known architecture. For HFT = 1, verified performance level obtained is PL e. Taking into

Table 3.8 Proposed correlation between security index SI^{Do} or SAL for the domain evaluated and final SIL (PL) to be attributed to the SRCS of machinery

Security index	SIL (PL) verified according to IEC 62061[a] (ISO 13849-1)			
SI^{Do} and SAL	(a)	1 (b/c)	2 (d)	3 (e)
$SI^{Do1} \in [1.0, 1.5)$/SAL 1	SIL—(PL a)	SIL 1 (PL b/c)	SIL 1 (PL b/c)	SIL 2 (PL d)
$SI^{Do2} \in [1.5, 2.5)$/SAL 2	SIL—(PL a)	SIL 1 (PL b/c)	SIL 2 (PL d)	SIL 2 (PL d)
$SI^{Do3} \in [2.5, 3.5)$/SAL 3	SIL—(PL a)	SIL 1 (PL b/c)	SIL 2 (PL d)	SIL 3 (PL e)
$SI^{Do4} \in [3.5, 4.0]$/SAL 4	SIL—(PL a)	SIL 1 (PL b/c)	SIL 2 (PL d)	SIL 3 (PL e)

[a]verification includes the architectural constrains with regard to S_{FF} and HFT of subsystems

account the domain of SRCS in which the safety function is implemented the vector of SL-A was evaluated as follows: [3 2 3 2 2 3 2]. Assuming that weights of all SL_i are equal ($w_i = 1/7$) and using the Eq. (3.5), the result obtained is $SI^{Do} = 2.43$, that is, SAL 2. Looking at the column 3 (e) of Table 3.7 the final performance level validated with regard to the security requirements is PL d, the same as required performance level PL_r. For the case of hardware fault tolerance HFT = 0 (series configuration of the SRCS), the verified performance level obtained was PL c, lower than required performance level PL_r d. Thus, applying of the redundancy in the SRCS is necessary and the domain security assurance level SAL 2.

3.6 Conclusions

Unprecedented development of the smart manufacturing systems (SMSs) is observed that have the significant potential to make innovative production more profitable and improve business processes. Advanced technologies are under development in area of the internet of things (IoT) and industrial internet of things (IIoT) that offer new manufacturing possibilities, but require also effective monitoring and the control systems having sufficiently high reliability, safety, and security characteristics. These characteristics are especially important when hazardous installations of industrial plants are evaluated to elaborate effective management strategy in life cycle.

Traditionally, the industrial manufacturing system includes the information technology (IT) and the operational technology (OT). Lately, using the cloud technology (CT) is often considered as an external network being important for distributed manufacturing and coordinated management. Advanced automation and control systems are also in development based, for example, on OPC UA and AutomationML concepts that offer new manufacturing solutions and production flexibility. However, it causes also some problems to be solved that include the reliability, safety and security properties, crucial for the business continuity management (BCM) to mitigate the risks of abnormal situations and major accidents contributing to high losses.

Selected design and operational aspects of the OT and IT networks have been overviewed and discussed in this chapter in the context of functionality and architecture of the industrial automation and control systems (IACS). Emphasis was put on the functional safety and cybersecurity of the industrial control systems and networks. These issues are becoming crucial, because the IACS that includes the safety-related control system (SRCS) plays a key role in innovative high-quality manufacturing, especially in so-called smart manufacturing systems (SMSs) of Industry 4.0.

In this chapter a method is proposed for integrated functional safety and cybersecurity analysis, with regard to the concepts outlined in the generic functional safety standard IEC 61508 (7 parts) and the cybersecurity standard IEC 62443 (14 parts). To limit the vulnerability of the IT and OT systems and networks, and the SRCS to be designed and operated to reduce relevant risks, a set of security-related fundamental requirements (FRs) defined in IEC 62443-1 is considered in the analyses and evaluations.

The method proposed uses the individual and/or societal risk graphs for determining the performance level required (PL_r) or the safety integrity level required (SIL_r) or the safety integrity level claimed (SIL CL) of consecutive safety functions defined in the analyses. These levels are then verified to indicate that the required PL or SIL is achievable in the designed SRCS of architecture proposed, in which particular safety function is to be implemented. For that purpose relevant probabilistic models of the SRCSs are developed with regard to potential common cause failures (CCFs), when a hardware redundancy is to be applied. Then, the verified SIL is validated with regard to determined SAL of the domain of interest, for example, the SRCS domain in which particular safety function is implemented, including internal and external communications.

The dependability of the SRCS performing the safety-related functions can be influenced both by technical factors, including requirements concerning hardware (HW) and software (SW), and also the human and organizational factors [1, 15, 17]. These aspects require further research, especially in the context of the design and operation of high complexity manufacturing systems, including the functional safety and cybersecurity aspects with regard to the defence in depths (D-in-D) concept and related strategy to be elaborated and applied in particular industrial plant or smart manufacturing system, characterized by the venture capital, production capacity, existing or emerging hazards and threats that influence various risks in changing environment.

References

1. Kosmowski, K. T., & Gołębiewski, D. (2019). Functional safety and cyber security analysis for life cycle management of industrial control systems in hazardous plants and oil port critical infrastructure including insurance. *Journal of Polish Safety and Reliability Association, 10*(1), 99–126.
2. Lu, Y., Morris, K. C., & Frechette, S. (2016). Current standards landscape for smart manufacturing systems. Systems Integration Division Engineering Laboratory, NISTIR 8107.
3. Li, S.W. et al. (2017). *Architecture alignment and interoperability, an industrial internet consortium and platform industrie 4.0.* IIC:WHT:IN3:V1.0:PB:20171205.
4. Vathoopan, M. Walzel, H., Eisenmenger, W., Zoitl, A., & Brandenbourger, B. (2018). AutomationML mechatronic models as enabler of automation systems engineering: Use-case and evaluation. In *Proceedings of the IEEE 23rd International Conference on Emerging Technologies and Factory Automation (ETFA)*, IEEE 2018.
5. Kivelä, T., Golder, M., & Furmans, K. (2018). Towards an approach for assuring machinery safety in the IIoT-age. *Logistics Journal: Proceedings.*
6. Felser, M., Rentschler, M., & Kleinberg, O. (2019). Coexistence standardisation of operational technology and information technology. *Proceedings of the IEEE.*
7. MERgE. (2016). *Safety & security, recommendations for security and safety co-engineering.* Multi-Concerns Interactions System Engineering ITEA2 Project No. 11011.

8. SESAMO. (2014). *Integrated design and evaluation methodology. Security and safety modelling*. Artemis JU Grant Agr., No. 2295354.
9. EC. (2013). *Cybersecurity strategy of the European Union—An open, safe and secure cyberspace*. https://ec.europa.eu/digital-single-market/en/news/eu-cybersecurity-plan-protect-open-internet-and-online-freedom-and-opportunity-cyber-security, access: April 2020.
10. CISA. *Assessments: Cyber Resilience Review (CRR)*, us-cert.gov/resources/assessments, access: April 2020.
11. ENISA. (2016). *Communication network dependencies for ICS/SCADA Systems*. European Union Agency for Network and Information Security.
12. HSE-1. (2015). *Cyber Security for Industrial Automation and Control Systems (IACS)*, Health and Safety Executive (HSE) Interpretation of Current Standards on Industrial Communication Network and System Security, and Functional Safety.
13. HSE-2. (2016). *Cyber Security for Industrial Automation and Control Systems (IACS)*, Health and Safety Executive (HSE) Report for Chemical Explosives and Microbiological Hazard Division (CEMHD) and Energy Division, Electrical Control and Instrumentation (EC&I) Specialist Inspectors.
14. Kosmowski, K. T., Śliwiński, M., & Piesik, J. (2019). Integrated functional safety and cybersecurity analysis method for smart manufacturing systems. *TASK Quarterly, 23*(2), 1–31.
15. Kosmowski, K. T., & Śliwiński, M. (2016). Organizational culture as prerequisite of proactive safety and security management in critical infrastructure systems including hazardous plants and ports. *Journal of Polish Safety and Reliability Association, 7*(1), 133–145.
16. Kosmowski, K. T. (2013). *Functional safety and reliability analysis methodology for hazardous industrial plants*. Gdansk University of Technology Publishers.
17. Nardello, M., Møller, C., & Gøtze, J. (2017). *Organizational learning supported by reference architecture models: industry 4.0 laboratory study*. Complex Systems Informatics and Modeling Quarterly (CSIMQ) Article 69, Issue 12.
18. Holstein, D. K., & Singer, B. (2010). *Quantitative security measures for cyber & safety security assurance*. ISA: Presented at ISA Safety & Security Symposium.
19. IEC 63074. (2017). *Security aspects related to functional safety of safety-related control systems*. International Electrotechnical Commission.
20. ISO 22301. (2012). *Societal security—Business continuity management—Requirements*. International Organisation for Standardisation.
21. Gołębiewski, D., & Kosmowski, K. T. (2017). Towards process based management system for oil port infrastructure in context of insurance. *Journal of Polish Safety and Reliability Association, 8*(1), 23–37.
22. Misra, K. B. (Ed.). (2008). *Handbook of performability engineering*. London: Springer.
23. IEC 62443. (2018). *Security for industrial automation and control systems*. Parts 1–14 (some parts in preparation). International Electrotechnical Commission.
24. IEC 61508. (2016). *Functional Safety of Electrical/Electronic/Programmable Electronic Safety-Related Systems*, Parts 1–7. International Electrotechnical Commission.
25. ISO 13849-1. (2015). *Safety of machinery—Safety-related parts of control systems—Part 1: General principles for design*. International Organisation for Standardisation.
26. IEC 61511. (2016). *Functional safety: Safety Instrumented Systems for the Process Industry Sector*. Parts 1–3. International Electrotechnical Commission.
27. IEC 62061. (2005). *Safety of machinery—Functional safety of safety-related electrical, electronic and programmable electronic control systems*. International Electrotechnical Commission.
28. ISO 22400. (2014). *Automation systems and integration—Key performance indicators (KPIs) for manufacturing operations management*, Parts 1 and 2. International Organisation for Standardisation.
29. Braband, J. (2016). *What's Security Level go to do with Safety Integrity Level?* 8th European Congress on Embedded Real Time Software and Systems (ERTS 2016), hal-01289437, Toulouse.
30. IS. (2019). *Industrial security*. Siemens, siemens.com/industrial-security. Access: July, 2019.

31. RAMI 4.0. (2016). *Reference architecture model Industrie 4.0*. DIN SPEC 91345.
32. Kosmowski, K. T. (2006). Functional safety concept for hazardous system and new challenges. *Journal of Loss Prevention in the Process Industries, 19*(1), 298–305.
33. ISO 22100-4. (2018). *Safety of machinery—Relationship with ISO 12100, Part 4: Guidance to machinery manufacturers for consideration of related IT-cyber security aspects*, International Organisation for Standardisation.
34. Kosmowski, K. T. (2018). Safety integrity verification issues of the control systems for industrial power plants. In *Advanced solutions in diagnostics and fault tolerant control* (pp. 420–433). Springer International Publishing AG.
35. Malm, T., Ahonen, T., Välisalo, T. (2018). *Risk assessment of machinery system with respect to safety and cyber-security*. Research Report-VTT-R-01428-18.
36. ISO/IEC 27001. (2013). *Information technology—Security techniques—Information security management systems—Requirements*.
37. ISO/IEC 27005. (2018). *Information technology—Security techniques—Information security risk management*.
38. ISO/IEC 15408. (2009). *Information technology, Security techniques—Evaluation criteria for IT security*. Part 1–3.
39. Kosmowski, K. T., Śliwiński, M., Barnert, T. (2006). Functional safety and security assessment of the control and protection systems. *European Safety & Reliability Conference*, ESREL 2006, Estoril. Taylor & Francis Group, London.
40. Kosmowski, K. T., Śliwiński, M. (2015). Knowledge-based functional safety and security management in hazardous industrial plants with emphasis on human factors. In *Advanced systems for automation and diagnostics*, PWNT, Gdańsk.

Kazimierz T. Kosmowski is a Professor at the Gdansk University of Technology, Poland, Department of Electrical and Control Engineering. His scientific interest includes the reliability theory, and the safety and security in technical systems, in particular the functional safety and cybersecurity of industrial control systems. He is involved in teaching Masters courses and training courses for engieers from the industry within a state certification program of persons responsible for functional safety. He contributed to a number of international, state and university projects, and visited a number of universities and research institutes in Poland, Japan, Austria, Germany and Switzerland. He is the author of five books, seven book chapters and over two hundred peer reviewed papers on various aspects of reliability, safety and security in technical systems, including the human reliability aspects. He is a board member of the Polish Safety and Reliability Association.

Chapter 4
Extending the Conceptualization of Performability with Cultural Sustainability: The Case of Social Robotics

John P. Ulhøi and Sladjana Nørskov

Abstract A more comprehensive conceptualization of performability, beyond pure economic, technological, and environmental performance, is needed. Adopting and using a technological innovation in its socio-cultural context is likely to have performative impacts well beyond techno-economic and environmental conditions. Examples, as discussed in this chapter, include changes of human and social behavior conditions following from the adoption of social robotics. Reviewing recent developments in social robotics and the adoption of this technology in professional activities, this chapter argues that contemporary conceptualization of performability is incapable of capturing all important conditions and therefore needs to be extended to include *cultural sustainability*. Borrowing from theory on technology and innovation development, impact, responsibility, and living labs allows us to lay some preliminary stepping stones toward an extended conceptualization of performability and how such technology can be tested in the right context. Before closing, the chapter briefly sketches out avenues for future research.

Keywords Performability · Psycho-social performability · Socio-cultural performability · Innovation · Social robots · Responsible research and innovation · Technology assessment · Living labs

This work is co-sponsored by a Carlsberg Foundation Semper Ardens Grant [CF16-0004]. Any opinions, findings, conclusions and/or recommendations expressed in this work are those of the author(s) and do not necessarily reflect the views of either the sponsor or the employer(s) of the author(s). The usual disclaimers apply.

J. P. Ulhøi (✉)
Department of Management, Aarhus University, Aarhus, Denmark
e-mail: jpu@mgmt.au.dk

S. Nørskov
Department of Business Development and Technology, Aarhus University, Herning, Denmark

K. B. Misra (ed.), *Handbook of Advanced Performability Engineering*,
https://doi.org/10.1007/978-3-030-55732-4_4

4.1 Introduction

The etymological meaning of performability refers to the ability to perform, that is, the execution, accomplishment of an action, task, and/or function. Performability is thus a key concept when trying to understand how a specific technology or product performs in terms of its techno-economic, ecological, and psycho-social performability. If one or more of these performative dimensions are omitted, a true representation of a product's or a technology's overall performability is not provided. Visible and quantifiable factors have been much in focus when trying to come to grips with a given technology's or technological artefact's expected performability. Not surprisingly, the relative straightforwardness associated with metricating and thus measuring the techno-economic and/or environmental performability of a new technology has attracted key interest when developing tools to document the performability. Efforts devoted to documenting the expected performability of a given technological innovation, however, seldom consider possible unintended performability effects. Moreover, such effects often surface (some time) after the technology and the artefact has been developed and marketed.

The contemporary definition of performability includes a number of diverse yet related areas like quality, reliability, maintainability, safety, risk, environmental impacts, and sustainability related to the performance of a product, system, or service [1, p. xi]. Possible unintended and/or undesirable performability-related effects of technological innovations should not be left out of consideration. An example of the effects could be a situation (e.g., a disabled citizen being serviced by a robot), where the roboticized services (whenever needed) are provided at the expense of privacy (if the nature and frequency of service are recorded). The various performative properties of a new technology and/or artefact are, however, on the one hand affected by human values, preferences, and human reactions, and on the other by energy and raw materials, technology, methods, and production processes (during design and development as well as during the use of the products, systems, or services).

As emphasized in a recent World Economic Forum White Paper, the present technological and economic progress can no longer be assumed to be automatically in alignment with social progress [2]. Roboticists are no longer only involved in "pure" technological engineering. They also increasingly seem to be involved in cultural engineering. Robotics appears to be approaching a level of development, where responsible decision-makers cannot afford to be concerned with only documenting the issues related to immediately quantifiable properties of new technology. Developers of social robots are increasingly confronted by possible psycho-social and socio-cultural effects of new technology that cannot be swept under the carpet. Lack of information on any important performative aspects and/or effects of new technology and associated moral responsibilities limits the decision-makers' ability to make informed decisions about technology, which is particularly questionable as new robotic technologies tend to increasingly blur the lines between humans and technology. As an example of the latter, this chapter examines and discusses the recent development of social robotics and its increasing human-like properties and capacities. The latter capacities suggest that this technology is capable of affecting humans beyond their work activities and/or functions in qualitatively different ways that transcend what the technology was developed for in the first place. Social robotics is thus an obvious case of technological development that is not only altering the technology itself, but is in fact capable of engineering the psycho-social context and conditions in which the technology is being used, that is, cultural engineering.

Social robots refer to a technology that contains a robot and a social interface, which is designed in a way that causes users to attribute social qualities to the robot and to perceive the robot as a socially interactive agent [3]. Social robots are capable of expressing and/or perceiving emotions, communicate via dialogue, establish/maintain social relationships, display personality, use natural cues such as gaze and gestures, and so on [4, p. 145]. Such features lead humans to perceive social robots differently than other technologies [5]. In fact, people tend to interact with social robots as with social others [6]. Research in psychology shows that users of social robots are capable of developing a significant attachment to social robots. This is because robots are able to create an impression of mutual relating, thus triggering people's desire to nurture robots [7, 8]. Darling [5] argues that our relationships with robots are distinctly stronger than our relationships with other technologies or objects. This is, she suggests, due to three factors: (i) physicality of social robots, that is, social robots are a part of our physical rather than virtual space, (ii) perceived autonomous movement of social robots, and (iii) social behavior of this type of robots. This development has led to an increasing research interest in the human side of social robotics, including how humans respond to, perceive and interact with robots; how they develop social and emotional relationships with robots; how human–robot interactions and relationships affect human cognition and emotions; and how they shape cultural and social arrangements. Social robotics applications have been explored for a variety of purposes, for example, in therapy [9], education [10], creativity [11], reduction of perceptual biases [12, 13], and so on. While research shows promising effects of robots in these areas, social robotics is also related to a number of challenges from a user perspective, for example, issues related to responsibility [14], liability [15], privacy [16], and socio-emotional wellbeing of people [17], which are further discussed in this chapter.

Culture engineering refers to the situation where the roboticist intentionally decides how humans can interact with the robot through the subsequent design and functionality choices leading to the social robot in question, thus extending the programmer role from (merely) a coder to a creator [18]. In addition to being creators, the developers are also "*imagineers*" [19, p. 1], not only of technologies and their applications but also of cultural and social arrangements [20]. Given these acknowledged roles, roboticists have been criticized for facilitating "the transcendence of ethnocentrism, paternalism and sexism, and their associated power relations" [21, p. 28]. Unintentional reproduction of cultural stereotypes related to gender in robotics development in Japan has, for instance, been documented and discussed (e.g., [20, 22]). The evolution and adoption of robotics is thus not only limited to promoting techno-economic performance but also to (un)intentional effects on psycho-social and socio-cultural performance. Infringing users' physical, social, and psychological privacy adds to the pool of potentially undesirable performative effects associated with technological development, which becomes even more problematic if psycho-social implications are difficult to assess and/or prevent [23]. In pace with the increasing sophistication and complexity associated with social robotics technologies, there is thus a corresponding need to assess human rights and socio-emotional wellbeing directly against potential harm associated with such technologies [17, 24].

We use caregiving as an application area for social robotics to illustrate why social robotics presents itself as a particularly relevant technology category in which to include psycho-social and socio-cultural dimensions of performability. Based on this example we will advance arguments in favor of expanding the definition of performability to address uncovered context-related psychological, social, and cultural conditions. We will show how social robotics holds potential for not only adding or replacing designed functionality but also for unintentionally altering the quality and outcomes of social interactions in organizations and why such effects also need to be encapsulated by conceptualization of performability. The remainder of the chapter is organized as follows. In the following section, we discuss technology development in the light of responsibility and control issues. This is important as social robotics seems to have reached a level of technological progress in terms of anthropomorphic as well as social properties, and thus has the capacity to transgress the human and social activity domain. We briefly sketch out some robotic cases of professional applications, where the use of social robots is growing. We then articulate and discuss social robotics in the context of responsible research and innovation, and performability. Before closing, we conclude and identify avenues for future research.

4.2 Technology Engineering, Responsibility, and Control

Although engineers implicitly and/or explicitly apply visions and value judgements about intended use and associated autonomy, transparency, and/or fairness when developing technology [25], the wider uses of a given technology will not be fully known until long after having been introduced to the market. Johnson [14] recently made an interesting analysis of a possible responsibility gap that may arise in the case of autonomous technologies that perform tasks without direct human control or intervention. She challenges the existing views on responsibility for behaviors of autonomous technologies by arguing that they are based on incorrect assumptions. Let us take a look at why existing views of technological responsibility have been criticized. One of those views argues that since it is not possible to predict the behavior or exert complete control of autonomous agents, it is not possible to hold humans responsible (e.g., [18]). Another view posits that the control requirement is not a necessary criterion for holding humans responsible, as there are many situations in which humans are held responsible for outcomes that are out of their control, for example, strict liability [26]. Yet another view contends that in spite of their inability to control the behavior of autonomous technologies, engineers should be held responsible due to professional responsibility [27].

The underlying assumption in these views, according to Johnson [14], is the fact that humans cannot be held responsible "because of the nature of technology" [p. 714]. This assumption is inaccurate, she explains, because it relies on a "narrow and deficient view of technological development" [ibid., p. 711] in which the progression and chronology of developmental steps is determined solely by the nature of the particular technology. According to these lines of reasoning, engineers have no choice but to follow the given, inevitable developmental logic. As pointed out by Johnson [14], however, it is not only the nature of technology, but also social norms and expectations that influence the technology development and thus the responsibility arrangements. The development of technologies is subject to a negotiation process of many different stakeholders, and it is therefore the outcome of human choices. Consequently, responsibility is embedded in social relationships, which means that humans are responsible for the behavior of (autonomous) technologies. Although she does not take a stance on whether humans should be held responsible, she maintains that "[t]here are good reasons for staying with human responsibility, namely to keep the pressure on developers to ensure the safety and reliability of such devices" [14, p. 714].

While society is aimed at ensuring that new products and technologies do not cause unacceptable safety and/or health risk hazards, much less attention seems to be directed toward possible psycho-social hazards following from the adoption and application of social robots. Assessment schemes and/or techniques have been developed to address safety, reliability, quality, and/or environmental properties of technology. Further, more recently local and/or national regulations have also been supplemented by supranational guidelines (c.f., e.g., [28]). Public authority is thus undertaking the responsibility for possible risks and safety concerns [29]. While such general guidelines certainly can serve as a good starting point, we still lack a way to systematically assess how the introduction of a new technology, that is, a social robot, may affect the ethics, quality, and outcomes of social interactions *before* the robots are developed and used at a larger (social) scale. This is an important endeavor, because neither the development nor the subsequent application of new technology is neutral. Designers' personal values are not isolated from the design phases. Rather, user-related decisions are being taken on their behalf during the design and development processes. Design choices are privileged choices that are not available to public scrutiny. Those choices also contain designers' and developers' social, cultural, and user assumptions and economic considerations which are inscribed in the software and hardware [22, 30]. The user phase itself also involves values that affect freedom, autonomy, transparency, and fairness [25]. Technology has been characterized as a "moral mediator" that affects how humans interact with the world [31]. Designers, engineers, and firms are making critical design choices which have implications well beyond the techno-economic performance.

We therefore argue that the existing conceptualization of performability is in need of an extension. First, existing institutionalized performability-related regulations do not cover all important dimensions. Second, market players incentivized by micro-economically exploitable opportunities cannot be expected to pay attention to possible psycho-social (un)desirability of a given technological innovation. In a free market economy, by definition, technological development is governed by what is (i) technologically possible (from the point of view of science) and (ii) commercially viable (from the point of view of profit). Neither of these incentives are related to societal desirability (or the opposite). If it is technologically possible (and in alignment with existing regulations), and if there is a perceived market, then technology is likely to be developed. Market or consumer preferences, in turn, are determined by individual taste and perceived value, which is based on available information and existing experiences. Third, and more recently, environmental consideration has emerged as an important criterion not previously included in the "weighing and metrication process", thus paving the way for the inclusion of softer dimensions of performability.

Fully comprehending and controlling all the important performability-related impact of a new technology, however, is far from being straightforward. When discussing the possibility of exerting control of the technological development, the situation has been referred to as a choice between dilemmas, with reference to the Collingridge Dilemma [32]. Controlling a technology is easier and cheaper during its development, while it becomes costly and slow once the technology has been commercialized and integrated into the economic and social fabrics of society [32]. The dilemma describes the inherent tradeoff between being able to anticipate the wider impact of a specific technology and the possibility of correcting the technology's development trajectory [ibid.].

4.3 The Emergence of Social Robots in Eldercare

To add a practical dimension to our discussion, we will use the example of social robotics. In spite of industrial robots having been around for decades, the professional application of social robots in business sectors is much more recent. Below, we will briefly discuss how social robots have been used in the eldercare sector, and lay the groundwork for explaining why additional performability dimensions ought to be considered in the case of this technology.

Considering the growing challenges that follow from a drastic increase in global aging demographics and a declining caregiver-to-senior ratio, using assistive robots is becoming increasingly attractive in relation to care provision in the eldercare sector [33]. A recent review revealed that robots are used for five different purposes: affective therapy, cognitive training, social facilitator, companionship, and physiological therapy [34]. In *affective therapy*, the review showed that group interventions involving a social robot seem to be better at generating positive emotions, while one-on-one interventions seem to be more effective at mending negative emotions. The majority of the studies on *cognitive training* (e.g., improving working memory) that Abdi et al. [34] identified also showed positive effects on cognitive functions of elderly subjects. However, the authors note that the lack of objective outcome measures and control groups in some of the identified studies limits the value of their results. Furthermore, *social robots as facilitators* of social interactions were found to be able to improve sociability of elderly with the care staff and/or fellow residents in all the identified studies. The review further identified only three studies related to *robots as companions* which showed that robots were able to reduce loneliness in elderly subjects. One of the studies showed, for instance, that some subjects also became emotionally engaged with the robotic companion [35]. The interviews additionally revealed that residents enjoyed sharing, interacting with, and talking about the robot. Finally, the review found that *physiological effects* of interacting with social robots include the ability of robots to decrease systolic and diastolic blood pressure (where only systolic decreases were sustained over time) and improved physiological reactions to stress. Nonetheless, the design of these studies made them vulnerable to several confounders (e.g., increased interaction with other residents during the course of the studies), thus limiting the clinical interpretability and reliability of the results. While this review showed that social robots may hold value for eldercare, some of the identified studies suffer from methodological limitations such as a lack of objective outcome measures, lack of control groups, a possible cultural bias (one-third of the studies were conducted in Japan), and a small number of participants [34, 36].

Despite the methodological insufficiencies and limited generalizability, these studies help point out why psycho-social dimensions are relevant to consider and include in the conceptualization of performability. One reason is because the studies confirm that people get attached to social robots. This is in line with existing research in psychology [17]. Based on her research on social robots as companions, Turkle [7] warns that social robots could fundamentally change the nature of social relations, because they "do not teach us what we need to know about empathy, ambivalence, and life lived in shades of gray" [pp. 9–10], which may lead to a deterioration of the socio-emotional wellbeing of people. Along the same veins of reasoning, Darling [5] argues that social robots may impact the way human individuals treat each other, which makes it relevant to consider and discuss how social robotics may affect the quality and outcomes of social interactions, and how this could expand the conceptualization of performability. When developing new technologies, it is therefore not sufficient to ask "what technologies can do *for us*", but rather "what they may do *with us*" [37, p. 119].

Research on telepresence robot solutions for assisting elderly people in their daily activities and for supporting professional caregivers in their work found that the users expressed positive perceptions of the technology and a willingness to use it for both social and professional purposes [38]. Considering that this target group is particularly vulnerable and highly dependent on the assistance of others, it appears relevant to test the technology for its broader performability dimensions—before a wider diffusion and adoption have taken place—to understand how the technology affects both work and social relations. While it probably may seem more agreeable to many observers to accept social robots being used instrumentally in physical caregiving for routine care tasks, more concern may be likely to surface if human care providers are replaced with social robots in psycho-socially related care provision situations that require emotional and personal involvement (e.g., [39]). Securing meaningful human contact is relevant for the psychological wellbeing of people and for the development of their social and emotional skills [17].

Regardless of how effective robots are at caregiving, Vallor [37] maintains that a moral dilemma will be permanently attached to their use. Caregiving is a moral skill, and to acquire this kind of skills, one needs to practice it while at the same time receiving meaningful feedback. As an example of unintentional consequences of the professional use of social robots, in this case social robots would reduce the opportunity for professionals to practice and exercise caregiving skills, thus leading to what Vallor [37] terms moral deskilling. She argues that "moral skills appear just as vulnerable to disruption or devaluation by technology-driven shifts in human practices as professional or artisanal skills [...] because moral skills are typically acquired in specific practices which, under the right conditions and with sufficient opportunity for repetition, foster the cultivation of practical wisdom and moral habituation that jointly constitute genuine virtue" [p. 109]. Replacing human caring relations with robots rather than assisting them may not only deprive the care receivers of a human touch, eye contact, and conversation, but it will also take away the possibility to cultivate these profoundly important skills. Differently put, there are plausible arguments for considering *cultural sustainability* in the conceptualization of performability as a subdimension of the already included sustainability dimension. According to Misra [1], sustainability "focuses on providing the best outcomes for both human and natural environments now, and indefinitely into the future" [p. 843]. Based on the above discussion, we propose that cultural sustainability is particularly concerned with the human behavioral and value-related aspect of the current sustainability dimension. We therefore suggest that achieving cultural sustainability means ensuring that any new technology, which may potentially transform or disrupt human practices, values, and norms, does so in a way that leads to moral, social, and emotional upskilling or reskilling rather than deskilling of humans.

4.4 Some Stepping Stones Toward an Extension

The application of social robotics in eldercare discussed in the previous section illustrates that this technology holds a potential for changing the way humans interact with one another (e.g., people now confide in robots rather than in their peers or caregivers). A few existing studies, however, focus on only a few performability dimensions of the technology in question and have limited generalizability. So, while they suggest that social robots can have positive effects (e.g., lower depression) on the narrowly selected user segments, there is a need for shedding more light on how social robots may affect these and other parts of the population and the involved processes of interactions in the longer run.

Recently, literature on social robots and possible unintended consequences has started to address the issue of surveillance (privacy intrusion) and social bonding associated with this technology, thus challenging social and informational privacy

[23, p. 413]. A few observers [40] have argued that health services are inherently social entities which deploy technologies such as robots. In consequence, they suggest to go beyond perceiving robots as technical devices and instead focus on the sociological dimension of robots which might help to understand how the use of robotics in delivering health care affects the working and social relationships at hospitals, as robots can simultaneously reconfigure services and influence their scope and ability to add value.

Mazmanian et al. [41], for example, studied the impact of the use of this relatively simple technology among consultants and documented that this technology had contradicting effects on their work life. Studies have documented that robotic telepresence for intensive care at a hospital department intensified coordination outcome both positively and negatively, which in turn had contrary implications for subsequent coordination [42], while other researchers have examined the use of robot technology in pharmacies revealing how robots influenced the work practices and boundary relations of disparate occupational groups [43]. More recently, concerns have been voiced from within the business community itself regarding whether businesses should produce and/or market products that are capable of irreversibly changing social norms [44]. There is an emerging acknowledgment of the wider responsibility and desirability of robotic innovations and the fact that new, smart technologies, such as social robotics, can have unplanned effects on, for instance, the work environment and its associated interactions and relationships.

Theoretically, some conceptual "stepping stones" are needed which can bridge technology innovation, impact, and responsibility. Traditionally, however, literature on research and innovation management has not paid much attention to ethical concerns associated with technological innovations [45] until the emergence of responsible research and innovation (RRI). This broader conceptualization of innovation activities has been defined as "a transparent, interactive process by which societal actors and innovators become mutually responsive to each other with a view to the (ethical) acceptability, sustainability and societal desirability of the innovation process and its marketable products (in order to allow a proper embedding of scientific and technological advances in our society)" [29, p. 19]. Drawing on literature on technology assessment, Stilgoe et al. [46] refer to four important dimensions of responsible innovation: anticipation, reflexivity, inclusion, and responsiveness [p. 1570] which allow technologists to build on past experiences "rather than reinventing responsibilities for each particular emerging technology" [p. 1577].

Although an interest for RRI can be traced back many years [47], it is only recently that this area in the research and innovation fields has begun to surface as a new subfield [48–50]. Responsible innovation [51] or responsible research and innovation is defined as a "transparent, interactive process by which societal actors and innovators become mutually responsive to each other with a view to the (ethical) acceptability, sustainability and societal desirability of the innovation process and its marketable products" [29, p. 19]. Stretching this a bit further, RRI engulfs a technology dimension (possible impact following from the design decisions), a product dimension (possible impact following from the production), and a usage dimension (possible impact following from the usage). The relevance of bringing responsible innovation into the focus here is that it allows for reflecting on what kind of future development society and organizations want technology to provide, including what specific challenges and needs must be met by technology and its underlying values and the extent to which such enabled futures are democratic [52]. Increasing culturally engineered effects of new ICTs, for example, gives reason for growing ethical concern. Ethical concern may lead to the wrong action, or an action may lead to consequences that may harm society in the long run [45]. Apart from ethical concerns, dilemmas may also be expected to arise, if, for example, a new robot technology, which holds promise for increased society protection and/or safety, is usable only at the expense of sacrificing existing levels of individual privacy.

In order to ensure that ethical and legal issues are properly addressed, Liu and Zawieska [24] recently recommended that responsibility considerations should be included at each stage of design, development, and deployment of robot technology. The authors further identified a knowledge gap between (i) circumstantial responsibility and (ii) conceptual responsibility. Circumstantial responsibility, they argue, relates to the actual context in which robotics is being used. This type of responsibility is thus intimately related to control, predictability, and foresight. The problem is, however, that the developer of the robot, in particular when the robot is an autonomous artificial intelligence-based agent, has no possibility of predicting its future behavior.

An obvious place to consider addressing responsibility would be in the early research and development (R&D) stages of new technology. While modeling and forecasting can be useful for specific analyses, they are nothing more than useful means to produce indicative rather than reliable predictive results. Moreover, the inherent complexity and dynamics related to the techno-economic, psycho-social, and socio-cultural factors make traditional cross-sectional or controlled lab setups less relevant. Rather than revisiting the future-oriented methods, we therefore believe that an obvious place to start addressing responsibility would be the prototyping stage and related prototyping techniques. First, this is where the applicative dimension is in focus. Second, this stage typically represents the part of the product development process, in which the largest sunk costs of R&D are surfacing [53]. Third, at the prototyping stage, the issue of responsibility regarding the wider use and/or societal desirability has also received surprisingly little attention. In their review of the literature, they found that the engineering design literature did not seem to pay much interest to this human and social aspect. Menold and Simpson [53] further found that most of the literature suffered from a lack of interest into physical prototypes that allowed for incorporating user feedback and concern. To meet this gap, they point toward user-centered design as a method which offers an opportunity to integrate the desirability, feasibility, and viability dimensions during the design process. We need, however, to reconsider and revise existing approaches to consumer satisfaction to ensure that future approaches can capture effects beyond those which are immediately intended for application.

An interesting evaluative approach to consider here is the social experimentation approach or the living lab approach. Dell'Era and Landoni [54] define a living lab as "a design research methodology aimed at co-creating innovation through the involvement of aware users in a real-life setting" [p. 139]. Their literature review led them to conclude that the living lab methodology is applicable to both examine many different user needs and to provide context as a critical element of the design process, which allows users to interact with the new artefact in their daily lives. Social experiments represent an approach where society is perceived as a kind of "living" laboratory that allows for experimenting with new technologies [55]. The social dimension of experiments, he argues [p. 64], both refers to the location (i.e., happens *in* society), the consequences (i.e., *on* society), and the experimenter (i.e., done *by* society). It thus follows that social experiments are different from scientific experiments and foresight studies, as they are tested under real conditions (which normally cannot be reproduced in scientific laboratory settings). Social experiments, he argues, can be used as a model for moral experiments with new technology discovering new moral issues caused by the technology, finding out how to specify existing normative standards to assess these moral issues involved, and figuring out new normative standards needed to deal with associated new moral issues [56].

Van de Poel [55] further emphasizes that using society as a laboratory implies responsible experimentation, which in turn raises both epistemological and ethical concerns. Where the former relates to the preparation of the experiments (to secure reliability and relevance), the latter concerns the possible negative impact on society. Their study of the urban smart energy campus identified three potential obstacles associated with this approach. One concern relates to the "messy" co-creation agency involved which allows society to "speak back" and/or to disrupt along with the technological transformation (rather than being "freezed" during the investigations). A second worry refers to the implicit tensions associated with the open-ended experimentation and pressures to show success. A third concern relates to the potentially conflicting needs of the local socio-cultural specificity and the need for scalability and generalizability. In a related recent study, Engels et al. [57] refer to the societal test of technology as "test beds" and "living labs" and describe these approaches as experimental, co-creative approaches to innovation policy. Based on two case-studies, they are found to be capable of offering useful test beds that can be used for true societal tests of the possible desirability (or undesirability) of emerging technological transformation. Canzler et al. [58] emphasize that such "living labs" are not only restricted to offering spaces of experimentation and co-creation. Such labs also enable the emergence of institutional public–private formations between separated actors and/or policy areas.

4.5 Conclusion and Implications

In pace with the increasing digitalization, the demarcation line between technology and users has become increasingly blurred. Adopting and using a technological innovation in a psycho-social and socio-cultural context, we asserted, is likely to cause performative impacts beyond techno-economic and environmental effects directly associated with the artefact, some of which may negatively affect the long-term performability of the technology while others may act as a brake toward the adoption of the technology. Examples of such cultural effects include changes of the human and social behavior conditions beyond the activity replaced or supplemented by technology. In developing support for this assertion, we analyzed a practical example of professional use of social robotics in eldercare. Our analysis showed that the application and wider performability of social robotics hold a potential for affecting the quality and outcomes of social interactions in ways that may neither be intended nor appreciated. Moreover, our examination determined that the wider performability aspects of social robotics cannot be adequately accounted for by the contemporary conceptualization of performability. To rectify this situation, we have extended the sustainability dimension of performability to include cultural sustainability.

In our search for some preliminary conceptual stepping stones, we have (re)visited theory that addresses social robotics, technology, innovation, responsibility, and living labs. While theory offers useful conceptualizations, which allow for including wider human and social responsibility into a performability context, there still remains a need for accounting for such endeavor. This brings us to address avenues for future research in the field of performability. As documented by Lutz et al. [23] empirical studies of social robotics applications are few and based on small samples thus limiting the comparability and generalizability of results. Social robotics research and development needs to broaden its scope to include tests of robots in real-life settings, including human–robot interactions as well as teams of humans and robots in different contexts and performing different tasks. As pointed out by Magrani [59], there is a need for new ontological and epistemological lenses to conceptualize and understand the increasingly human-like robots, not as technical devices but as moral agents embedded into existing socio-technical systems and capable of interacting with human beings in both private and public domains. Therefore, more studies are needed that go beyond only focusing on the intended performability of the robot technology used. To help expose some of the not yet covered and often unintended aspects of the performability of social robotics, we suggested that the application of social experiment research should be considered.

Our examination of social robotics from a performability perspective also reveals some practical implications associated with the adoption of an extended conceptualization of performability. Implementation of major technological changes in organizations are likely to meet unexpected friction (e.g., related to work relations) and/or social problems (e.g., related to privacy intrusion), which in turn may lead to less successful implementation, that is, only realizing a part of its full potential. Lack of attention to the human factor, however, is not only critical for preventing undesirable human or social consequences from new robotics technology, it also holds a key to unlock some of the problems related to insufficient implementation and/or realization of the wider potential of the robotics technology in question. Studies of advanced manufacturing technology adoption, for example, documented decades ago that management variables most related to the human factor can distinguish firms which are successful in adopting the technologies from the less successful ones [60].

References

1. Misra, K. B. (Ed.). (2008). *Handbook of performablity engineering*. London: Springer-Verlag.
2. Philbeck, T., Davis, N., & Larsen, A. M. (2018). *Values, ethics and innovation. Rethinking technological development in the fourth industrial revolution*. White Paper of the World Economic Forum.
3. Hegel, F., Muhl, C., Wrede, B., Hielscher-Fatabend, M., & Sagerer, G. (2009, February). *Understanding social robots*. In *Proceedings of 2009 Second International Conferences on Advances in Computer-Human Interactions* (pp. 169–174), retrieved from: https://aiweb.techfak.uni-bielefeld.de/files/2009%20hegel%20ACHI.pdf on December 23, 2019.
4. Fong, T., Nourbakhsh, I., & Dautenhahn, K. (2003). A survey of socially interactive robots. *Robotics and Autonomous Systems, 42*(3–4), 143–166.
5. Darling, K. Extending Legal protection to social robots: The effects of anthropomorphism, empathy, and violent behavior towards robotic objects (April 23, 2012). In C. Robot Law, & Kerr, F. (Eds.), *We Robot Conference 2012*, University of Miami, 2012, retrieved from SSRN: https://ssrn.com/abstract=2044797 or https://doi.org/10.2139/ssrn.2044797.
6. Breazeal, C., Takanishi, A., & Kobayashi, T. (2008). Social robots that interact with people. In B. Siciliano & O. Khatib (Eds.), *Springer handbook of robotics* (pp. 1349–1369). Berlin, Heidelberg: Springer.
7. Turkle, S. (2006). A nascent robotics culture: New complicities for companionship. *AAAI Technical Report Series*, retrieved from https://web.mit.edu/~sturkle/www/nascentroboticsculture.pdf.
8. Turkle, S. (2007). Authenticity in the age of digital companions. *Interactions Studies, 8*(3), 501–517.
9. da Silva, J. G. G., Kavanagh, D. J., Belpaeme, T., Taylor, L., Beeson, K., & Andrade, J. (2018). Experiences of a motivational interview delivered by a robot: Qualitative study. *Journal of Medical Internet Research, 20* (5), e116.
10. Belpaeme, T., Kennedy, J., Ramachandran, A., Scassellati, B., & Tanaka, F. (2018). Social robots for education: A review. *Science Robotics, 3*(21), 1–9.
11. Kahn, P. H., Kanda, T., Ishiguro, H., Gill, B. T., Shen, S., Ruckert, J. H., & Gary, H. E. (2016). Human creativity can be facilitated through interacting with a social robot. *11th ACM/IEEE International Conference on Human-Robot Interaction (HRI)*, 173–180.
12. Nørskov, S., Damholdt, M. F., Ulhøi, J. P., Jensen, M. B., Mathiasen, M. K., Ess, C. M., & Seibt, J. (2019). Fairness perceptions in job interviews: Using a teleoperated robot as a fair proxy. In *35th European Group for Organizational Studies (EGOS) Colloquium*. Edinburgh, United Kingdom.
13. Seibt, J., & Vestergaard, C. (2018). Fair proxy communication: Using social robots to modify the mechanisms of implicit social cognition. *Research Ideas and Outcomes, 4*, e31827.
14. Johnson, D. G. (2015). Technology with no human responsibility? *Journal of Business Ethics, 127*(4), 707–715.
15. Hubbard, F. P. (2016). "Sophisticated robots": Balancing liability, regulation, and innovation. *Florida Law Review, 66*(5), 1803–1872.
16. Lutz, C., & Tamò, A. (2015). *RoboCode-ethicists: privacy-friendly robots, an ethical responsibility of engineers?* In *Proceedings of the 2015 ACM Web Science Conference*, Oxford (pp. 1–12).
17. Turkle, S. (2011). *Alone together: Why we expect more from technology and less from each other*. New York: Basic Books.
18. Matthias, A. (2004). The responsibility gap: Ascribing responsibility for the actions of learning automata. *Ethics and Information Technology, 6*, 175–183.
19. Rossini, M. (2003). *Science/fiction: Imagineering Posthuman bodies*. Paper presented at Gender and Power in the New Europe, the 5th European Feminist Research Conference, retrieved from https://www.researchgate.net/publication/250814946_Science_Fiction_Imagineering_Posthuman_Bodies.

20. Robertson, J. (2017). *Robo Sapiens Japanicus: Robots, gender, family, and the Japanese Nation.* Berkeley, CA: University of California Press.
21. Robertson, J. (2010). Gendering Humanoid Robots: Robo-Sexism in Japan. *Body & Society, 16*(2), 1–36.
22. Šabanović, S. (2014). Inventing Japan's 'robotics culture': The repeated assembly of science, technology, and culture in social robotics. *Social Studies of Science, 44*(3), 342–367.
23. Lutz, C., Schöttler, M., & Hoffmann, C. P. (2019). The privacy implications of social robots: Scoping review and expert interviews. *Mobile Media & Communication, 7*(3), 42–434.
24. Liu, H.-Y., & Zawieska, K. (2017). From responsible robotics towards a human rights regime oriented to the challenges of robotics and artificial intelligence. *Ethics and Information Technology*. Published online November 2017.
25. Martin, K., Shilton, K., & Smith, J. (2019). Business and ethical implications of technology introduction to the symposia. *Journal of Business Ethics, 160*(2), 307–317.
26. Santoro, M., Marino, D., & Tamburrini, G. (2008). Learning robots interacting with humans: From epistemic risk to responsibility. *AI & Society, 22*(3), 301–314.
27. Nagenborg, M., Capurro, R., Weber, J., & Pingel, C. (2008). Ethical regulations on robotics in Europe. *AI & Society, 22*(3), 349–366.
28. High-Level Expert Group on AI (HLEG AI). *Ethics guidelines for trustworthy AI* [Report/Study]. Brussels: European Commission, 2019, retrieved from https://ec.europa.eu/futurium/en/ai-alliance-consultation/guidelines#Top.
29. von Schomberg, R. (2013). A vision of responsible innovation. In R. Owen, M. Heintz, & J. Bessant, (Eds.), *Responsible innovation: Managing the responsible emergence of science and innovation in society*. London: Wiley, retrieved from https://www.pacitaproject.eu/wp-content/uploads/2014/04/von-Schomberg-RRI-owenbookChapter.pdf.
30. Faraj, S., Pachidi, S., & Sayegh, K. (2018). Working and organizing in the age of the learning algorithm. *Information and Organization, 28*, 62–70.
31. Verbeek, P. (2011). *Moralizing technology: Understanding and designing the morality of things*. Chicago; London: The University of Chicago Press.
32. Collingridge, D. (1980). *The social control of technology*. London: Francis Pinter.
33. Moro, C., Lin, S., Nejat, G., & Mihailidis, A. (2019). Social robots and seniors: A comparative study of the influence of dynamic social features on human-robot interaction. *International Journal of Social Robotics, 11*(5), 5–24.
34. Abdi, J., Al-Hindawi, A., Ng, T., & Vizcaychipi, M. P. (2018). Scoping review on the use of socially assistive robot technology in elderly care. *British Medical Journal Open, 8*(2), 1–20.
35. Robinson, H., Broadbent, E., & MacDonald, B. (2015). Group sessions with Paro in a nursing home: Structure, observations and interviews. *Australasian Journal on Aging, 35*(2), 106–112.
36. Broekens, J., Heerink, M., & Rosendal, H. (2009). Assistive social robots in elderly care: A review. *Gerontechnology, 8*(2), 94–103.
37. Vallor, S. (2015). Moral deskilling and upskilling in a new machine age: Reflections on the ambiguous future of character. *Philosophy & Technology, 28*(1), 107–124.
38. Koceski, S., & Koceska, N. (2016). Evaluation of an assistive telepresence robot for elderly healthcare. *Journal of Medical Systems, 40*(5), 1–7.
39. Coeckelbergh, M. (2010). Health care, capabilities, and AI assistive technologies. *Ethical Theory and Moral Practice, 13*(2), 181–190.
40. Oborn, E., Barrett, M., & Darzi, A. (2011). Robots and service innovation in the health care. *Journal of Health Services Research & Policy, 16*(1), 46–50.
41. Mazmanian, M., Orlikowski, W. J., & Yates, J. (2013). The autonomy paradox: The implications of mobile email devices for knowledge professionals. *Organization Science, 24*(5), 1337–1357.
42. Beane, M., & Orlikowski, W. J. (2015). What difference does a robot make? The material enactment of distributed coordination. *Organization Science, 26*(6), 1553–1573.
43. Barrett, M., Oborn, E., Orlikowski, W. J., & Yates, J. (2012). Reconfiguring boundary relations: Robotic innovations in pharmacy work. *Organization Science, 23*(5), 1448–1466.
44. Kiron, D., & Unruh, G. (2009). Even if AI can cure loneness—Should it? *MIT Sloan Management Review, 60* (2), 1–4, retrieved from https://mitsmr.com/2ovm04.

45. Nathan, G. (2015). Innovation process and ethics in technology: An approach to ethical (responsible) innovation governance. *Journal of Chain and Network Science, 15*(2), 119–134.
46. Stilgoe, J., Owen, R., & Macnaghten, P. (2013). Developing a framework for responsible innovation. *Research Policy, 42*(9), 1568–1580.
47. Genus, A., & Stirling, A. (2018). Collingridge and the dilemma of control: Towards responsible and accountable innovation. *Research Policy, 47,* 61–69.
48. Grunwald, A. (2011). Responsible innovation: Bringing together technology assessment, applied ethics, and STS research. *Enterprise and Work Innovation Studies, 7,* 9–31.
49. Guston, D. H. (2006). Responsible knowledge-based innovation. *Society,* 19–21.
50. Owen, R., Macnaghten, P., & Stilgoe, J. (2012). Responsible research and innovation: From science in society to science for society, with society. *Science and Public Policy, 39*(6), 751–760.
51. Owen, R., Maynard, D. B. T., & Depledge, M. (2009). Beyond regulation: Risk pricing and responsible innovation. *Environmental Science & Technology, 43*(18), 6902–6906.
52. Owen, R., Stilgoe, J., Macnaghten, P., Gorman, M., Fisher, E., & Guston, D. (2013). A framework for responsible innovation. In R. Owen, J. Bessant, & M. Heintz (Eds.), *Responsible innovation: Managing the responsible emergence of science and innovation in society*. Chichester: Wiley (pp. 27–50).
53. Menold, J., & Simpson, T. W. (2016). The prototype for X (PFX) framework: assessing the impact of PFX on desirability, feasibility, and viability of end designs. In *Proceedings of the ASME 2016 International Design Engineering Technical Conferences and Computers and Information Engineering Conference*, August 21–24, Charlottte, North Carolina, USA.
54. Dell'Era, C., & Landoni, P. (2014). Living lab: A methodology between user-centred design and participatory design. *Creativity and Innovation Management, 23*(2), 137–154.
55. van de Poel, I. (2017). Society as a laboratory to experiment with new technology. In D. M. Bowman, E. Stokes, & A. Rip (Eds.), *2017, Embedding new technologies into society* (pp. 61–87). Singapore: Pan Stanford Publishing Pte. Ltd.
56. van de Poel, I. (2017b). Moral experimentation with new technology. In I. van de Poel, L. Asveld, & D. C. Mehos (Eds.), (2017), *New perspectives on technology in society* (pp. 59–79). London: Routledge.
57. Engels, F., Wentland, A., & Pfotenhauer, S. M. (2019). Testing future societies? Developing a framework for test beds and living labs as instruments of innovation and governance. *Research Policy, 48,* 1–11.
58. Canzler, W., Engels, F., Rogge, J.-C., & Simon, D. (2017). From "living lab" to strategic action field: Bringing together energy, mobility, and information technology in Germany. *Energy Research & Social Science, 27,* 25–35.
59. Magrani, E. (2019). New perspectives on ethics and the laws of artificial intelligence. *Internet Policy Review, 8*(3), 1–17.
60. Co, H. C., Patuwo, B. E., & Hu, M. Y. (2010). The human factor in advanced manufacturing technology adaption: An empirical analysis. *International Journal of Operations & Production Management, 18*(1), 106–187.

John P. Ulhøi Ph.D., serves as Professor of OMT (since 1998), Department of Management, Aarhus University. He earned his Ph.D. from Aarhus School of Business (1991), specialized in technology and innovation management. His research appears in journals such as Journal of Business Venturing, Entrepreneurship, Theory and Practice, Journal of Organizational Behavior, Managerial & Decision Economics, Business Strategy and the Environment, Technological Forecasting and Social Change, Management Decision, Scandinavian Journal of Management. He serves as Vice-President of The European Doctoral Association in Management and Business Administration and as Board Member of The Nordic Academy of Management. He is Associate Editor of Scandinavian Journal of Management (Elsevier) and Journal of Global Entrepreneurship (Springer).

Sladjana Nørskov, Ph.D., serves as Associate Professor, Department of Business Development and Technology, School of Business and Social Sciences, Aarhus University. She specializes in innovation management, organizational behaviour and social robotics. Her work has been published in international peer-reviewed journals such as *Creativity and Innovation Management, Information Technology and People, International Journal of Innovation Management, Journal of Consumer Marketing,* and *European Journal of Innovation Management.*

Together with her co-authors she was recently awarded the best annual paper award by *Creativity and Innovation Management* for their research on how to automatize identification of new product ideas in online communities by using machine learning.

Chapter 5
Design for Performability Under Arctic Complex Operational Conditions

Abbas Barabadi and Masoud Naseri

Abstract Complex operational conditions such as those in the Arctic regions can affect the performability and its integrated elements in various ways. Historical performability data such as failure and repair data play important roles in performability assessment. Such data should reflect the real conditions that equipment and human experience during operations. However, in practice, in some applications, there are not many efforts for collecting, reporting, and analyzing the performability data together with all associated influencing factors, which are the parameters of the complex operational conditions affecting the performability of a system. A case in point is the Arctic offshore, where compared to normal-climate regions, the performability data and associated influencing parameters (e.g. environmental conditions) are scarce. Hence, operations in such a complex environment are associated with a great deal of uncertainties. Such uncertainties can lead to unforeseen failures or in some cases to expensive over-designed concepts. One of the main reasons for lack of performability data is that most of available databases are not prepared originally for performability analysis of systems in complex operational conditions. For example, OREDA database, which is a database for failure and repair data of different components of oil and gas facilities in the Norwegian Continental Shelf, focuses only on reliability and maintainability that are two pillars of performability concept and thus the required data for other performability elements, including quality, safety, and sustainability, are not addressed accordingly. This chapter discusses the effects of complex operational conditions of the Arctic on the performability of offshore facilities. It also discusses the challenges of available methods for performability data collection, and thereafter, it introduces a methodology based on expert judgments for performability assessment of systems operating in the Arctic.

Keywords Arctic complex operational conditions · Arctic offshore · Expert judgments · Performability · Cold-climate regions

---

A. Barabadi (✉) · M. Naseri
UiT the Arctic University of Norway, 9037 Tromsø, Norway
e-mail: abbas.b.abadi@uit.no

K. B. Misra (ed.), *Handbook of Advanced Performability Engineering*,
https://doi.org/10.1007/978-3-030-55732-4_5

5.1 Introduction

Industrial activities including oil and gas, mining, wind farms, fishing, and so on are increased significantly in the Arctic region. The Arctic region is characterized by a range of harsh and severe climatic conditions and sensitive environment, with less-developed infrastructure and in far distances from main hubs. The severe and complex operational condition in the Arctic can significantly affect the performability of equipment. For example, low temperatures and icing can change the properties of materials such as changing the plastic behavior of polymers and make them brittle, a phenomenon that increases the failure rate. In such conditions, the crew need protective equipment against the cold which are often heavy clothes that can affect the human performance and consequently decrease the maintainability performance of equipment. Moreover, logistics is a challenging task in the Arctic which can be associated with long downtime due to uncertainties in harsh weather conditions. In such conditions, one of the main questions which should be answered is: what should the equipment be designed for? The concept which is used to design for should be comprehensive such that it reflects the robustness of the system against all sources of stress in order to reduce the failure rate and increase the equipment maintainability to reduce the consequences of the failures, while minimizing the repair and maintenance resources. Moreover, considering the sensitivity of the Arctic area, it should be a green design concept. Although these concepts might sometimes seem conflicting, the performability acts as an umbrella to cover all these essential needs for design and operation in complex Arctic regions. Hence, design for performability (Fig. 5.1) can play a significant role as a decision-support tool for decision-makers (e.g. managers, engineers, stakeholders) who deal with various different challenges in order to meet the varying demands of internal and external customers, regulation, production control, and the optimization of processes.

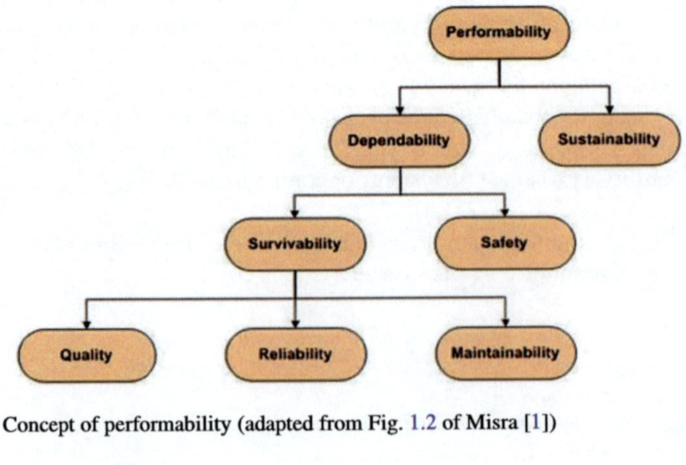

Fig. 5.1 Concept of performability (adapted from Fig. 1.2 of Misra [1])

However, compared to normal and temperate-climate regions, there exist little literature published on experience related to the operation and design to be used in the Arctic region. For instance, while oil and gas industry has gained extensive knowledge and experience by operating in the North Sea, the amount of knowledge and experience for operations in the Barents Sea with severe Arctic condition is very limited [2], Freitag and McFadden [3], Larsen and Markeset [4]. In this chapter, first, we discuss the main elements of the Arctic conditions and how Arctic operational conditions can affect different elements of performability; thereafter the need for establishing a comprehensive performability data collection is discussed. Expert judgments play an important role for performability design in the Arctic operational condition where there is limited experience and information. Hence in the next part, the expert judgment application for the performability design in the Arctic is discussed.

5.2 Arctic Operational Conditions and Their Effects on Performability

Studies have shown that the challenges faced by designer and operators in the Arctic can be grouped into three main groups: (i) harsh climatic conditions and sensitive environments; (ii) less-developed infrastructure; (iii) long distance to the market and main industrial hubs. These challenges lead to a great deal of uncertainties if the technologies solutions for industrial activities available in normal-climate regions are used in Arctic without being reassessed and modified accordingly [2].

In general, Arctic regions are sparsely populated areas with less-developed infrastructure [3, 4]. Moreover, the industrial activities taking place in the Arctic offshore are usually located in remote locations and far away from suppliers, manufacturers, and well-stablished ports and hubs. Compared to normal-climate regions, there are few weather stations in the Arctic and thus weather forecasts for Arctic offshore locations are limited [5–7].

Arctic regions are associated with low air and sea surface temperatures that vary considerably over the region and throughout the year. For instance, the air temperature in the Barents Sea varies greatly from summer to winter and is associated with high temporal variability during winter mainly due to the flow of water masses with different temperatures (see Fig. 5.2), latitudinal changes in solar radiation rates, and the presence of sea ice in the northern and north-eastern areas and usually open-waters in the west and southwest regions [8–10]. For example, while the annual minimum air temperature in the southern parts of the Barents Sea varies from -9–6°C, the minimum air temperature in the vicinity of Shtokman and Prirazlomnoye fields in the eastern and south-eastern parts is approximately -28°C and -48°C, respectively [9, 10].

Snow deposition and icing are the other characteristics of the Arctic regions. The metamorphose process of deposited snow over time that changes the snow into an

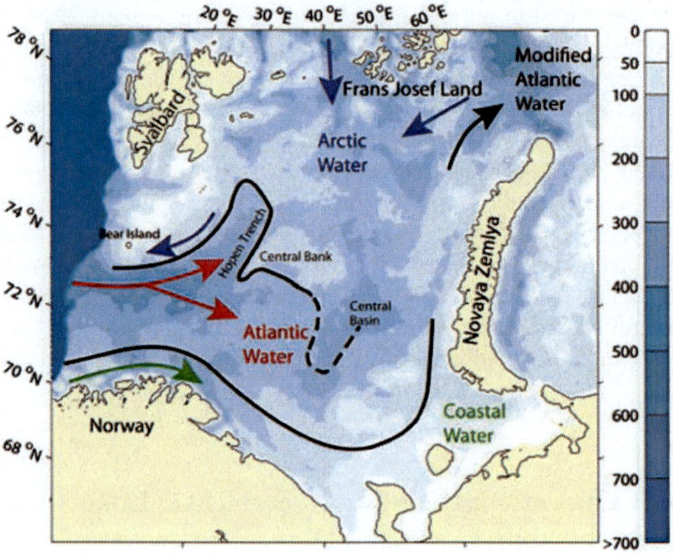

Fig. 5.2 Schematic of main water masses in the Barents sea—black lines represent frontal areas [11]

assemblage of roughly spherical ice grains, which increases the density of deposited snow to that of ice, is along with a rise in deposited snow strength and rigidity, a phenomenon which is known as hardening process [3, 12]. Icing has two main types depending on its water source, namely atmospheric icing and sea spray icing. Atmospheric icing includes "all processes, where drifting or falling water droplets, rain, drizzle or wet snow in the atmosphere freeze or stick to any object exposed to the weather" [13]. The type of atmospheric icing, and thus its mechanical properties depends on wind speed, air temperature, and freezing process itself [12, 13].

Sea spray icing (see Fig. 5.3) is usually considered the most dangerous icing type which forms as the water influxes, forming ship–wave interactions or wind-blown

Fig. 5.3 Vessel–wave interaction and resulting sea spray icing on the deck [17]

water droplets generated from whitecaps on the ocean surface, freeze on the surface of the vessels, and equipment onboard [14]. Sea spray ice accretion rate and its possible location on a vessel or platform depend on a number of factors, including meteorological and oceanographic conditions (such as air temperature, wind speed, wave height, wave period, sea surface temperature, and atmospheric pressure), shape and location of the equipment onboard, characteristics and type of the surface, design characteristics of the platform or vessel, and so on [8, 15, 16]. Understanding the icing process and gaining knowledge on the parameters including its rate, together with knowledge on mechanical characteristics of spray ice provide us with necessary information for designing and implementing anti-icing and de-icing strategies [16].

Polar low pressures, which form when a system of cold polar air moves over relatively ice-free warmer waters [18], are common meteorological phenomena in some Arctic seas including in the Barents Sea from September to early summer. Polar low pressures have a relatively short lifespan of 6–48 h from initiation to decay, with a diameter of 200–1000 km. They are associated with sudden weather changes in terms of storm force winds, high waves up to 15 m, decreased air temperature to −30°C, considerable snow and ice showers with reduced visibility down to less than 50 m [18, 19]. Sea ice and presence of iceberg are other features of the Arctic offshore. The distribution and extension of sea ice, and the drifting patterns of sea ice floes are mainly governed by the flow of different water masses, air temperature, wind, and current speed. Thus, the sea ice extent varies greatly throughout the year. For instance, while the maximum ice extent in the southern Barents Sea occurs in March and its minimum extent happens in September and October, the northern and north-western parts are usually covered by ice during summer [20]. An important feature of the Arctic seas is marginal ice zone where there are various ice floes of different sizes and masses in transition areas between open sea and continuous ice cover. The area between the ice floes is occupied by either brush ice or open water [20, 21]. The main issue is the retreat of the sea ice edge time due to the global warming impact, as shown in Fig. 5.4(A) and 5.4(B), where the frequency of sea ice extent over the Barents Sea is depicted with at least 40% ice concentration over the period October 1980 to May 1981 and October 2011–May 2012, respectively. Loss of sea

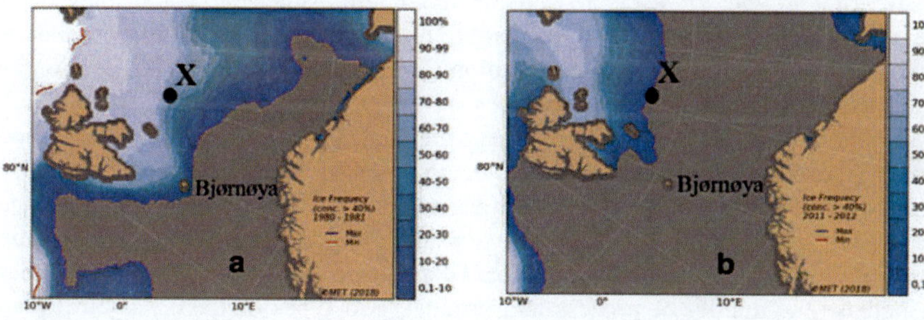

Fig. 5.4 Frequency of 40% sea ice concentration in the western and central barents sea. **a** October 1980 to May 1981; and **b** October 2011 to May 2012— adopted from [8]

ice leads to more complex operational conditions. It contributes to increase in sea surface temperature and significant wave heights, particularly in areas close to the ice edge [8], leading to changes in design philosophies and approaches; a shift from design for sea ice to design for spray icing in that location over time (see Fig. 5.4).

Reduced visibility due to fogs and polar nights are other characteristics of the Arctic. Fog is defined as suspended water droplets or ice particles near the Earth's surface that can lead to reduction of horizontal visibility below 1 km [22]. Arctic fogs are divided into four main types: advection fog, radiation fog, steam fog or Arctic smoke, and ice fog [23], of which advection fog that occurs when relatively warm and moist air flows over a cold surface is the most common type especially in the seas. Fog is observed most often in the Arctic basin and over the Arctic seas with an average annual number of days with fog of 80–100. The highest number of days with fog is about 140 days per year that can occur in the central parts of the Arctic [23]. Although fogs do not last for very long, their duration can reach sometimes as long as 72 h [23]. The duration and frequency of Arctic fogs often correlate with ice concentration, wind direction, and wind speed [24]. The highest frequency of fogs occurs over sea with a 70–90% ice cover. Fog frequency varies throughout the year as well due to low absolute humidity of water masses. This leads to the highest frequency of 65–80% in the Barents Sea in the summer to about 5–10% in winter [23], Proshutinsky et al. [24].

The above-mentioned characteristics of the Arctic regions add additional challenges for the design, construction, and operation of the facilities in the Arctic. In order to develop solutions and to tackle such challenges, the first step is to understand the mechanisms that such complex environmental conditions impact the operations, activities, and performability of the facilities. Therefore, it is of utmost important to understand how the Arctic harsh and complex operational conditions can affect the performability of equipment.

5.2.1 *Quality*

Quality of a product is a measure of its degree of conformance to applicable design specifications and workmanship standards [1]. To have a high-quality product, a high-quality input material and information should be used in high-quality process which is run by high-quality employees on the controlled environment. In such conditions, all variables which can change the quality of product can be controlled properly. Variation in input parameters of a process including inherent variability and attributable variability leads to defect in the final product which can reduce the quality rate significantly. Operational conditions under which equipment and operators are working may have significant contributions to such variations. In order to increase the quality of process and product, two types of quality control plan should be implemented, including design quality and manufacturing quality. The aim of design quality is to confirm the equipment, material, and operator's integrity over the life cycle of the facility. Hence, it requires an early assessment for material selection and performance analysis of equipment and human. For this, a set of tests such as accelerated life testing needs to be established to check the quality of design. In the accelerated life testing the equipment are tested on stress conditions (e.g. temperatures, voltage, vibration rate, pressure, etc.) higher than the design operational conditions to uncover potential modes of failure in short amount of time. Thereafter, the results will be extrapolated to predict their behavior under the design conditions. Hence, in order to establish such a type of tests, a comprehensive understanding about the involved physical environments and the stresses that are applied in real-world conditions should be available for designers [25]. As mentioned, due to the lack of weather forecasting infrastructures and technical experience as well as rapid effect of climate change on the Arctic, such information is not usually available. It can lead to significant uncertainties in design quality. Moreover, establishing such type of testes for all equipment will be very expensive and maybe impossible.

In manufacturing quality, by using some statistical tools such as statistical control process and by establishing some program such as lean management, total quality management, and Six Sigma, we try to control the variation during the operation phase and thus to increase the quality of products. Considering the fact that the most important contributor to variability is human himself, hence the competence development and crew training are the main core elements of such programs. Arctic operational conditions may increase the stress on human significantly that leads to deteriorations in human's physical and cognitive performance. For example, studies show that in outdoor work in the winter, cold stress frequently reduces working ability by 70% for short periods [26, 27]. Long period of exposure to the cold results in decreased cognitive performance, injury, hypothermia, loss of sensitivity, and reduced manual dexterity and grip [28, 29]. These conditions can directly influence the variability of human's decisions and human reliability to a very large extent. In this regard, the impact of Arctic conditions on equipment units and on human is discussed in more details in Sects. 5.2.2 and 5.2.3, when reliability and maintainability under Arctic conditions are addressed.

5.2.2 *Reliability*

Arctic environmental conditions negatively affect the reliability of equipment units. For instance, low temperatures change the properties (e.g. ductile/brittle behavior) of metals, polymers, and plastics, as well as rheological properties (e.g. viscosity) and chemical composition of the fluids such as lubricants and crude oils [3]. For instance, increase in oil viscosity at low temperatures requires more pumping power, leading to more energy consumption and even more failures in pumping facilities and hoses (see Fig. 5.5) [30]. The rate of corrosion-induced failures increases in low temperatures. Moreover, one may also consider the reduced performance of the operation or maintenance crew [32], which potentially results in reduced operational or equipment reliability by increasing human error [12]. Functionality of electrical and electronic devices, such as cables, wires, switches, pushbuttons, lighting elements,

Fig. 5.5 Burst hose as a result of excessive pressure from pumping highly viscous oil [31]

and gauges, can be impaired at low temperatures because of material deterioration. Moreover, as temperature decreases, resistance and capacitance of conductors can change and thus lead to potential changes in electrical properties of electrical and electronic components [33].

The impact of low temperatures can negatively affect the psychological and physical performance of the crew and thus lead to an increase in human error. The combination of low temperatures, high waves, and wind speeds may lead to severe icing storms. Wind action on iced structures such as antennas is different from the un-iced ones, which is due to the larger drag coefficient for ice-covered structures. Asymmetric icing and snow accumulation, which can happen due to the changes in prevailing icing direction, can unbalance the forces exerted on equipment and thus threatens its stability or increase the fatigue failure possibility [12, 13]. Loads imposed by ice on equipment (e.g. small diameter tubes, chains, ropes, pipes, connections) and shelter ceilings may cause damage and malfunction [14]. A major concern in reliability of equipment onboard when it concerns the presence of sea ice is the structure vibration due to the crushing of sea ice and its resulting intensive shaking of the deck, which may cause hazardous working conditions for the crew and increase human error. Platform vibration may also decrease equipment reliability by, for instance, inducing fatigue failure. Falling of objects due to significant shaking of platform can deteriorate operational safety and may harm the crew [35, 36].

5.2.3 Maintainability

The harsh and complex operational conditions of the Arctic often impact the maintainability of equipment units in various ways, including imposing delays in logistics operations, deteriorating the performance of the maintenance and repair crew, and finally through affecting the failed equipment and technical aspects of the maintenance or repair [12, 26].

Less-developed infrastructure and remote distances from suppliers and market can affect overall support strategies and logistics, such as transportation of equipment, modules, repair crew, and spare parts. This can result in extended plant downtimes due to the unavailability of materials, tools, spare parts, and personnel [7]. Interruptions in offloading spare parts due to high winds and wind-speed-induced limitations associated with crane operations can add to equipment downtime and reduce equipment maintainability. Such disruptions may also occur because of high waves and forces exerted by sea ice on supply vessel and thus pose delays on maintenance tasks and intervention operations [12]. In polar nights and fogy days, the impaired visibility can delay the logistics operations, for instance, by interrupting spare delivery plans [32]. Uncertainties in weather conditions and difficulties in predicting polar low pressures [12] contribute to prolonged delays in logistics operations and thus contribute to extended equipment downtime.

Poor visibility can lead to extended active repair times as well; for instance, by making it difficult to read technical data and manuals that further increase the propensity to miss something perform incorrect repair and maintenance [26, 32]. Accretion of ice on failed equipment reduces accessibility to the equipment and thus interrupts operations and maintenance tasks and increase equipment downtime, for instance, by increasing the time required for disassembling, fault isolation, replacement and removal time of failed components, and reassembling. Sensors on test equipment (e.g. temperature sensors, accelerometers, etc.) can be affected by different types of ice, leading to measurement errors in inspections and repairs process [26, 32]. Maintenance supervisors estimate that a 30% saving in overall maintenance time could be achieved if access to equipment were ideal or unrestricted [26]. Similar issues can arise due to the accumulation of wind-blown snow in low-velocity areas.

The combined impact of low temperatures and high wind speeds that result in low wind chill index [37, 38] negatively impact the physical and psychological performance [29] of the repair crew and thus equipment maintainability by increasing equipment active repair time. Such a process might be associated with crew's loss of strength, mobility, and balance due to low temperatures, together with confusion and impaired consciousness [29]. In such conditions, the maintenance and repair crew should wear warm clothes and gloves. However, although thermal protective clothing may mitigate the neurophysiologic responses, it can negatively affect manual performance due to a decrease in mobility and inability to perceive external elements or cues [26]. Moreover, studies show that when a person who is fully dressed in Arctic clothing is exposed to extremely cold air temperatures, a significant reduction in performance is still observed when compared with a person working in normal temperatures [29].

5.2.4 Safety

Safety is defined as "freedom from unacceptable risk" (i.e. risk which is not tolerable) [40]. In other words, safety can refer to a situation that could have negative consequences, such as harm to humans, environment, economic loss, which implies that safety can be seen as the capacity of a unit to avoid an endangering of persons, environment, or the facilities, for specified time and conditions. In order to analyze the risks of operations, one should account for hazards, the likelihood of the occurrence of such hazards, and the consequences should such hazards occur [41]. Accounting for the uncertainties with the type of hazard that might occur, the probability of its occurrence and the extent and severity of the outcomes is of crucial importance in any risk and safety management [42].

The risks of operations and activities in the Arctic regions with harsh and complex environmental conditions is higher than those in normal-climate areas for three main reasons, including (i) increased probability of failure, (ii) increased severity of negative consequences, and (iii) increased number of failure scenarios. Such higher risks can lead to reductions in safety levels of operations and activities [12]. Some failure scenarios are unique to the Arctic and cold-climate regions, and thus are not experienced in other regions. For example, the forces of drifting sea ice and icebergs on platforms and vessels increases the probability of failure of mooring lines and thus loss of station keeping systems. Structural safety of the vessels and platforms that are threatened by iceberg collision and sea-ice build-up around the platform structures is another example of unique failure scenario in the Arctic. Failure of components and delays in operations and activities due to the negative impact of atmospheric icing and spray icing are other examples of hazards and failure scenarios that are unique to the Arctic and cold-climate regions. Moreover, the sensitive Arctic environment, its remoteness, and less-developed infrastructure can contribute to increase in the severity of failure consequences, especially if we consider crises and large failure scenarios, that lead to platform evacuation, search and rescue, and oil spill clean-up. Poor satellite coverage in northern latitudes and thus less reliable telecommunication means pose limitations on communication and data transfer that can negatively affect transferring real-time technical advice and remote support for decision-making onboard during emergency situations.

Lack of data and relatively less industrial experience (compared to normal-climate regions) adds to the risks associated with Arctic industrial activities, especially through three important types of failure scenarios, namely (i) unknown unknown (i.e. situations where the actual future hazardous event is not a part of the set of events discussed in risk assessment because no one knew about it), (ii) unknown known (i.e. situations where the actual future hazardous event is not part of the set of events that are discussed in risk assessment, because someone knew but not those performing the risk assessment), and (iii) event with negligible probability (i.e. situations where the subjective probability of a particular failure scenario is considered negligible by those performing the risk assessment) [42, 43].

Moreover, considering the relationship between safety and risk, one may conclude that the performance of the established safety procedures, which are basically active/passive risk reducing barriers is severely impaired by the harsh and complex conditions of the Arctic. For instance, reduced visibility (e.g. because of fogs) and darkness, through promoting human being's ocular inability to distinguish objects given limited brightness and contrast, increase the chances of judgment errors and thus rise the probability of accidents [22]. Nascimento and Majumdar [44] reported that the helicopter fatal accident rate during night is 15 times higher than that during daytime. Such a difference is attributed to the visual perception and decision-making in degraded visual environment. Spray icing on platforms and vessels can severely threaten the stability of vessels and platforms leading to capsizing and loss of lives [8]. Reduced visibility and degraded visual environment because of snow showers and fogs can threaten the safe evacuation and its followed search and rescue operation. Huge spray ice accumulation on the windward side of the platforms and vessels can cause an imbalance in the structure, leading to problems in heaving and thus platform's motion characteristics. Falling ice and compacted snow during thawing, and slippery surfaces because of ice or hardened snow can cause injuries. Loss of accessibility to doors, stairways, pathways, helicopter pad, and escape routes, in addition to safety equipment, lifeboats, and fire-fighting equipment due to ice and snow accumulation threatens the safety of the crew [34], Crowley [45]. The safety functions of electrical and electronic equipment are of vital concern, especially if they are installed in areas with potential leakage of explosive and hazardous gases. The ability of materials to withstand potential gas explosions can be impaired at low temperatures [33]. Potential build-up of static charges on plastic surfaces in cold environment due to low humidity increases the possibility of explosions in case of hydrocarbon gas leakage. It may also cause problems for devices such as analogue meters with plastic faces by giving incorrect or erratic readings and thus affect the operation of sensitive controllers, shutdown systems, and alarms [33].

5.2.5 *Sustainability*

Given limited resources on planet, any design we should try to meet the needs of the present without compromising the ability of future generations to meet their own needs [1]. Performability by considering the sustainability as one of its concepts tries to reach this goal. To reach a sustainable design, we should minimize the footprint of technology and human activities by minimizing the material and energy usage throughout their entire life cycle. The material which is going to be used should be green materials and be highly recyclable at the end of their life. In order to reach this goal, the efficiency of the energy and material usage should be increased by merging creativity in design, economics, manufacturing, and policy. Design for sustainability preserves ecosystem integrity and promotes human health and happiness.

Arctic operational conditions can significantly increase the material and energy usage. For example, icing as a common phenomenon in the Arctic region can provide a lot of challenges for operators and the equipment. Ice can reduce the quality of communication tools and sensors. For instance, in icing conditions, wind speed errors can be as high as 30% [46]. It can increase the vertical load and it can change the dynamic characteristics of structures. Hence, to avoid these adverse effects of icing, mechanical or electrical anti-icing and de-icing measures need to be taken into consideration [12, 26, 47]. Such measures negatively affect the sustainability by increasing energy and material consumption. Moreover, current practices for de-icing are very expensive. For example, a study shows 5% of the cost of a 600 kW wind turbine should be allocated for the anti-icing and de-icing systems [46], or, for a windmill farm with medium icing severity, with an average of 30 icing days per year, the anti-icing and de-icing system payback time can be 5 years [48]. To avoid this and increase the sustainability of the technologies, some new creative solutions should be developed. Anti-icing coating is one of the promising solutions which can increase the satiability of design significantly. It decreases the ice accretion rate by ensuring high degree of water repellency, delays any ice nucleation, and slows ice adhesion. However, the durability of such coating is a requested research subject to ensure lifelong functionality. The other way is to design equipment in such a way that the ice accumulation is reduced; for example, as the diameter of the subjects is incised the ice accretion rate will be reduced. Or they can be situated in protected locations, so that sea spray and weather cannot reach them. This may be accomplished by using fully enclosed spaces, semi-enclosures, and recesses with removable "curtains" in front or similar [49].

5.3 Performability Data—Current Situation and Future Needs

Performability tries to integrate different aspects of equipment performance into a holistic and comprehensive concept. Hence, making the right decision with respect to the performability design, performability optimization, and performability monitoring requires accurate predictions of failure time, repair time, defect rate, human performance, energy and material usage, as well as all accident and incident consequences, and so on. This can be achieved by an effective performability analysis, which in lower levels needs an effective reliability, maintainability, quality, safety, and sustainability analysis on both component and system levels.

To have an effective input from performability analysis, it is necessary to establish a process so that the right person at the right time has access to the right data, which are collected and reported in the right format. The right person for an effective performability analysis is someone who has comprehensive understanding of (i) the methodology, data and information needed for model building, (ii) the properties of different models, and (iii) the tools and techniques to determine whether a particular model is appropriate to model a given dataset. Recording the data in an unsuitable format, such as a qualitative format, makes them difficult to analyze. The data need to be stored in systems that make them easy to retrieve, analyze, and draw conclusions on a continuous basis. Moreover, timely data can provide the required information for a reliable and cost-effective design.

However, the most challenging part of such a process is collecting the right data, which can reflect the real world. Performability analysis greatly relies on historical data at the component level and well-established knowledge regarding the interconnections between different components which build up the whole system. In order to have a valid performability analysis, collected data must be able to provide a clear understanding of technical characteristics of the equipment, all sources of stress, operating and environmental conditions, component potential failures and failure routes, common causes and special cause variation effect on quality, failure consequences, as well as maintenance history and bill of usage material. These data and other relevant data constitute the performability data that actually vary based on the type of the system. For example, for an isolated item in a control office environment, the ambient conditions (e.g. ambient temperature) can be considered as identical and there is no need to collect them. However, for an outdoor pump, the ambient conditions can change over time, and hence they need to be collected in the performability database. In order to have valid and high-quality performability data under complex operational conditions such as those in the Arctic regions, it may be necessary to collect operational data for some years. Analyzing such data by an appropriate model can provide us with a clear and comprehensive understanding of the performability of equipment in designated operating conditions.

At the current stage due to the limited industrial activities in the Arctic region as well as less-developed infrastructures such as limited numbers of weather stations, the performability data are not available to a large extent. This can increase the uncertainties associated with performability analysis and consequently performability design as well as the costs of investment, operation, and maintenance. There are vast changes in the operational conditions in the Arctic region throughout the year and from one year to another it can cause a significant fluctuation on stress levels on human and equipment. Such unforeseen fluctuations in stress can change the performability characteristics and it may cause a catastrophic consequence. On the other hand, the climate change that has drastic effects on Arctic climatic conditions introduces another challenge for designer as it can make the limited available historical data collected over previous years less reliable for future applications.

The current practice is to use the available database such as OREDA [50] for performability analysis of equipment to be operated under Arctic conditions. These databases are restricted to the area south of the polar circle where the operational conditions are very different from those found in the Arctic region. The use of such data for performability analysis of Arctic equipment without considering the Arctic operational conditions leads to unreliable results.

Performability data (e.g. time to failure, time to repair, etc.) are often collected from multiple and distributed facilities and operational units working under different conditions. For instance, a specific type of pump may be installed in different places of a specific plant that experiences different flow rates, different working pressures and temperatures, and chemical composition of the fluids might be different (i.e. pumps are operating under different conditions exposed to different stresses). However, it is a common practical approach to collect their failure and repair data in a single database where their different operational conditions are not explained correctly. These differences in operating conditions are also known as covariates, stressors, or risk factors that can introduce heterogeneity into the data. However, in many reliability studies, datasets are assumed to be homogeneous, with the failure data being independent and identically distributed [51].

In a broad sense, covariates can be categorized into two different groups, namely observed covariates and unobserved covariates. Observed covariates are those factors which may have an influence on the performability characteristics of an item, and their associated values are collected and recorded in a database. Examples of observed covariates are the surrounding operating condition (e.g. weather data, temperature, humidity, dust, etc.), condition-indicating parameters (e.g. vibration and pressure), human aspects (e.g. the skill of the operators and maintenance crew), and organizational parameters such as organization culture and norms, training program, in addition to the implemented design modifications, and the history of the repair activities carried out on the system (e.g. type of repair, number of the repairs, etc.). Based on the effect of covariates on performability characteristics of an item, they can be divided into two groups: (i) categorical covariates and (ii) continuous covariates. The categorical covariates are qualitative variables and often have binary or multiple categories (e.g. effect of icing can be coded as no-icing, light-icing, moderate-icing, and heavy-icing). Continuous covariates have a defined scale and can be quantified, which can change linear or nonlinear [2]. Moreover, covariates can be time-dependent and time-independent. In the time-dependent covariates their effects on performability change over time.

Unobserved covariates are typically unknown, or their associated values are not collected properly, or they are missing in the databases. For example, if a pump has a soft foot problem, then it will put the bearing in an over-stressed situation. Hence it should be considered as a covariate in performability analysis. In the case that there is no information regarding soft foot in the performability database of the bearing, an unobserved covariate should be defined to capture the effect of soft foot on the performability of the bearing. As unobserved covariates are typically unknown, they cannot be explicitly included in the performability analysis. Observed and unobserved covariates lead to observed and unobserved heterogeneity.

In most of performability analyses, not only the effects of observed and unobserved covariates are neglected [51], but also the effects of observed covariates are not adequately addressed and quantified (Barabadi et al.). Therefore, it is not possible to extrapolate the result of analyses to a wide range of operational conditions. This issue highlights the limited application of a lot of available databases such as OREDA (OREDA Participants [50] in the performability design for new operational condition such as those in Arctic regions [52]. In the OREDA database, only operating time has been recorded and the other influence factors have not been collected. However, if the associated covariates with each time between failure and time to repair data are recorded then reliability and maintainability can be modelled as a function of time and observed and unobserved covariates. Thereafter, the results can be extrapolated to the new operational conditions. Such results will provide a basis to make decisions with respect to the design and operation of the production facility and technology under new environmental conditions.

Most of available data collection systems are not designed for performability analysis purposes and they are not collected in detail when it concerns the failure mechanism, failure model, failure consequences, and, more specifically, the operational conditions. A review of available standard for data collection such as ISO 14224 [53] shows that it does not cover the collection of important factors, such as the operational conditions, as being a minimum requirement for an effective performability analysis. This issue is a considerable drawback, especially for applications in a complex operational environment such as the Arctic where modelling of the effect of observed and unobserved covariates are of utmost importance. After collecting the performability data and their associated covariates, an appropriate tool should be used to analyze the data. For example, covariate-based model such as frailty model, proportional hazard model, proportional covariate model, accelerated failure time model and stress-strength model are some of the models which can be used to quantify the effect of covariates on the performability and its constituting elements [2]. Proportional hazard models and proportional covariate models are built based on the assumption that the hazard/repair rate of an item is the product of a baseline hazard/repair rate and a positive functional term that describes how the hazard/repair rate changes as a function of covariates. The baseline hazard/repair rate is assumed to be identical and equal to the total hazard rate when the observed and unobserved covariates have no influence on the failure pattern. Recently, some studies in the reliability field have used the frailty model to model the effect of unobserved covariates on some concepts of performability including the maintainability and reliability [51].

Hence, based on above discussion, establishing a correct, comprehensive, and suitable data collection system and selecting a suitable model for data analysis are important requirements for the performability design and performability optimization under Arctic conditions. Such a database should include all relevant potential covariates that can affect the performability characteristics of the item. Based on the ISO 31000 risk management process [41], the performability assessment process can be developed. Figure 5.6 shows the proposed approach for performability data collection and assessment. Performability assessment is the overall process of performability analysis and performability evaluation. Performability analysis is the process to comprehend and to determine the level of performability, based on selected data and information including the observed and unobserved covariate effects. Performability evaluation is the process of comparing the results of performability analysis against criteria or objectives and thus identifying areas for improvement. After performability evaluation, if needed a performability enhancement should be established to modify performability which can be focused on different concepts of performability such as reliability or safety. The focused area needs to be identified based on the result of performability analysis as well as the design criteria. As Fig. 5.6 shows at the first step the scope, context, and criteria should be identified, thereafter using the stablished context the performability data including their associated covariates need

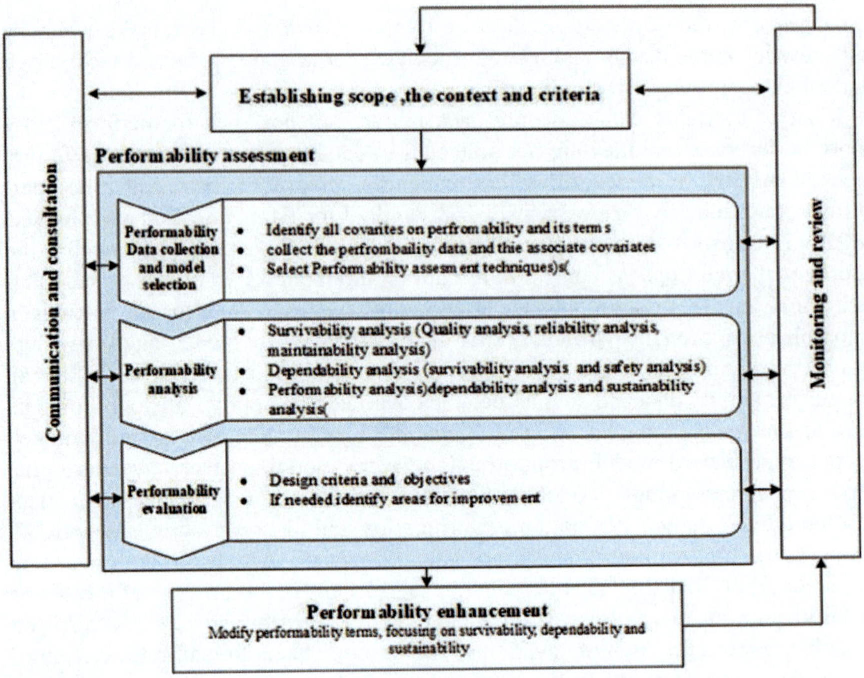

Fig. 5.6 Performability assessment process in analogy with ISO 31000 risk management process

to be collected. Moreover, in this step based on the nature of collected data, an appropriate statistical tool should be selected to estimate the performability characteristics of the item.

5.4 Expert Judgments for Performability Assessment in the Arctic Region

The methods that are available for quantifying the impact of operational conditions on the performance of the equipment units and crew, such as those developed based on proportional hazard models [54, 55] and accelerated life models [56, 57] rely on wide range of performability data in addition to data and information about underlaying operational conditions throughout the equipment life. Nowadays, the fourth industrial revolution, the internet of things, and big data provide us with opportunities to make production systems and services more efficient, more flexible, and more resilient. This has come with the advancements in knowledge, methods, and techniques, the increase in information sharing, data availability, and computational capabilities in addition to new opportunities of development for the analysis and assessment of risk. This leads us to a new era where knowledge, information, and data available for analyzing and characterizing hazards, modelling, and computing risks continue to grow [58]. The vast extent of digitization in recent years also provides us with a huge amount of data that can be used for real-time decision-making involved in maintenance planning and optimization, resource allocation, manufacturing cost reduction, and finally for sustainable design manufacturing and operation [59].

However, real-time data collection methodologies are usually available in operational phase. During the design phase, the main issue is the lack of extensive industrial experience in the Arctic regions compared to normal-climate regions. For example, although oil and gas industry has considerable experience in offshore oil spill clean-up, the only experience relating to that in the Arctic is mainly limited to the coastline cleaning after the grounding of Exxon Valdez oil tanker in 1988 in Alaska. As another example, one can consider the Norwegian Continental Shelf, wherein oil and gas industry has extensive knowledge and experience in the North Sea, its experience in the Barents Sea is limited to the south-western parts [52]. In this regard, oil and gas companies adapt a step-by-step approach where the industrial activities are currently limited to south-western parts. However, when it concerns Arctic tourism and cruise ships sailing in the Arctic Ocean, one should design evacuation facilities in such a way that they withstand the harsh and severe operational conditions of the Arctic offshore, as discussed in Sects. 5.2.2–5.2.4. To this, aim, the lack of industrial experience, and thus lack of data remains the crucial issue [32, 52]. This issue is addressed in the conceptual model presented in Fig. 5.7 in more details. To cope with such shortcoming, expert judgment process can be applied as an alternative method. Expert opinions have been widely used in various fields such as supply chain and traffic network risks [60], chemical process plants [61], human reliability analysis

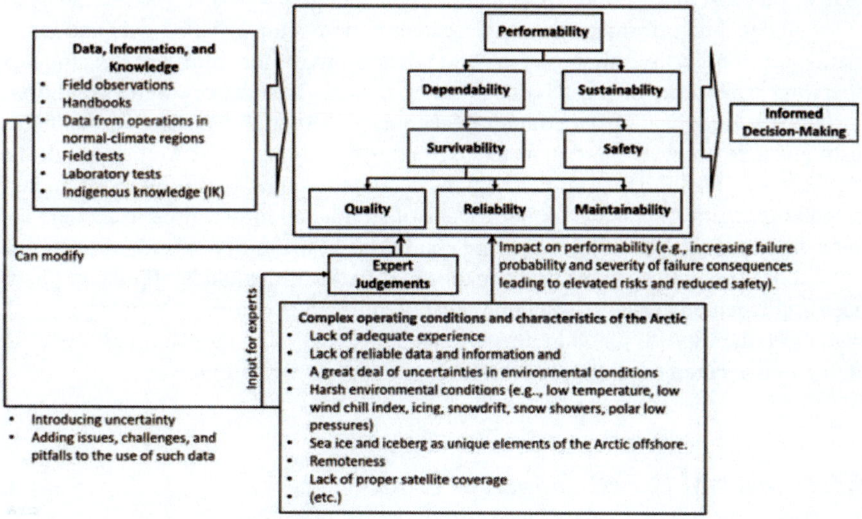

Fig. 5.7 A conceptual model for integrating expert judgments in performability analysis of operations and facilities in the arctic

[39, 62], oil and gas industry [52], to name a few. Expert opinions are often used for different reasons, including, but not limited to [63–68]

- regulation and management of industrial activities from risk and safety perspectives
- when other sources of data such as measurements and observations are not available
- when existing data are sparse, not reliable, questionable, or only indirectly applicable
- when estimates on new, rare, complex, or poorly understood phenomena are required
- predicting future events when good data are not available
- interpreting and integrating existing data that could range from qualitative data to quantitative data
- determining the state-of-the-art and what is currently known, what is not known, and what the gaps are and what should be learnt.

Given above discussion, the application of expert judgments for industrial activities in the Arctic regions are thus thoroughly justified due to, for instance, different operational conditions compared to normal-climate regions, lack of adequate data and information, new, complex, and poorly understood phenomena affecting the performability of the facilities, and integrating existing data (from normal-climate regions) with the new data.

Figure 5.7 illustrates a conceptual model, where the expert judgments are used in the context of performability analysis of equipment units in the Arctic. The output of performability analysis is often some inputs for making informed decisions for design, manufacturing, operation, and maintenance of facilities. Such an analysis relies on data and information that are scarce in the Arctic. Existing data and information are not usually suitable as they do not account for the impact of the complex operating conditions of the Arctic. In other words, elements of Arctic conditions introduce a great deal of uncertainty in such data. Although laboratory tests, fields tests, and field observations connected to existing operations in the Arctic provide the industry with useful data, but their extent is limited. Under these situations, expert judgments can be employed to modify existing data and information to include the impact of the Arctic complex operating conditions or to modify existing models and approaches for analyzing the performability of the facilities. Such approaches have been implemented in analyzing equipment reliability [52], equipment downtime [69], system availability [57], and to develop risk index [26], and human reliability [29], taking into account the impact of Arctic harsh operating conditions.

5.4.1 Formal Expert Judgment Process

Expert judgments are usually referred to the state of knowledge of experts when they reflect upon or answer to a question at the time of response. Expert judgments, also known as educated guess, expert opinions, and expert forecasts are expressions of opinions that experts make in responding to a technical problem based on their knowledge and experience [63], Meyer [66]. A technical problem, which is usually formed as a question, is the problem that the experts are asked about their opinions. Expert judgments also include the experts' mental processes of assumptions, definitions, biases, source of information, and algorithms, through which they arrive to the stated opinions and formulated answers [70]. Such a process is associated with a great deal of uncertainty and different sorts of bias. In this regard, it is of utmost importance to follow a formal process for acquiring and aggregating expert judgments. A formal expert judgment process has usually three main phases including expert selection, expert judgment elicitation, and expert judgment aggregation [63, 65–67, 70–72].

5.4.2 Expert Selection

An expert can be defined as "a person who has background in the subject area and is recognized by his or her peers or those conducting the study as qualified to answer questions" [66]. Such a definitions considers, among others, three main criteria, for experts, namely:

- Having background pertinent to the subject area
- Being recognized by his/her peers
- Being qualified and having a desired level of expertise (both substantive and informative expertise).

These criteria that are, to some extent, subjective introduce some uncertainty in deciding who the expert is. O'Hagan and Buck [73] considers an expert, as the person "who has great knowledge of the subject matter", or even as "the person whose judgements are to be elicited, whatever their actual degree of expertise". The expertise, however, should not be limited to the knowledge about the technical problem, but also on the response mode and how the person organizes and uses his/her knowledge in addition to person's mental processes for making assumptions, definitions, and algorithms for expressing his/her opinions [70], O'Hagan and Buck [73].

It is widely accepted that a panel of expert should have a pool of diverse background in such a way that a balanced set of viewpoints is achieved, and excessive influence of a single individual or single viewpoint is avoided. Moreover, while choosing experts with similar disciplinary backgrounds might cause problems if the experts are asked about their opinions that might go beyond their immediate expertise [65, 73], and selecting a panel of experts with very diverse background might make reaching a consensus difficult, if a consensus is desired. Aggregating expert opinions using weighted averaging methods that are based on assigning weighting factors to expert also becomes challenging when dealing with a group of experts with different, if not contradicting, views and expertise. In relation to the application of expert judgments for performability analysis of technical systems and facilities operating in the Arctic, one should make sure of including some key areas of expertise such as:

- The concept of performability and its elements (i.e. quality, reliability, maintainability, safety/risk, and sustainability (referring to substantive expertise))
- The technical aspects of the operation/activity/facility (referring to substantive expertise)
- The operational conditions—meteorological and atmospheric conditions of the location of interest (referring to substantive expertise)
- The mechanisms through which operational conditions can impact performability (referring to substantive expertise)
- Uncertainty (its concept, representation, and characterization), judgment, and decision-making (referring to normative expertise).

5.4.3 Expert Judgment Elicitation

Elicitation refers to the process of "obtaining experts' subjective opinions through specifically designed methods of communication" [66]. The elicitation process and the information and the assumptions that experts consider for expressing their opinions affect the elicited opinions [63, 70]. The quantities of interest together with the underlaying assumptions of the problem should be carefully and clearly defined and communicated with the experts [65, 66]. Expert judgments are affected by the response mode and the process through which expert opinions are elicited. The method for aggregating expert opinions can also determine the elicitation approach. O'Hagan and Buck [73] reviews various models and frameworks for eliciting expert judgments, a common element of which is the response mode. There are various response modes for eliciting expert opinions such as single probability values, set of probability values, probability distribution, quantiles of a distribution, parameters of a distribution, etc. that to a great extent depend on the objective of the study [63, 66, 67, 70].

Experts usually prefer to present their opinions along with the associated uncertainties in the form of, for instance, mean and variance, distribution, and percentiles. However, the eliciting probability distribution is an inherently imprecise process, mainly due to two factors: (i) it is difficult for the experts to give numerical values for the probabilities, quantiles, standard deviation, or parameters of a distribution, and (ii) experts can only provide a finite number of probability judgments that makes it difficult to determine an empirical probability distributions [63, 66, 70, 73]. In other words, the use of probability distributions does not eliminate the uncertainties associated and thus does not guarantee a perfect representation of an expert's uncertainty. Cooke [67] proposes a method, known as Cooke's performance-based method, for elicitation and aggregation of expert opinions, where he tackles the issue of experts' uncertainties as well as the normative and substantive expertise of the experts by eliciting expert distributions for some calibration questions [67, 68]. A weighting factor will then be computed for each expert that will be in weighted arithmetic averaging approach for combining expert distributions of the technical problem.

5.4.4 Expert Judgment Aggregation

Expert judgment aggregation may refer to the procedure by which expert opinions are combined in order to form an overall opinion to be further used as an answer to the predefined technical problem [66]. There are various ways to aggregate expert judgments, which can be grouped into behavioral and mathematical approaches.

In behavioral approaches, experts interact with each other and decision-maker, for instance, in face-to-face group meetings, where they assess technical problem or even simply discuss relevant issues and ideas with only informal judgmental assessment. In these settings the aim is to reach an agreement or consensus within the group of experts. Solutions given by the group may require that the experts compromise in some ways in order to reach an agreement. While the main advantage of this approach is that the decision-maker does not need to combine and aggregate expert opinions, its main drawbacks is group polarizing by some experts [71], Clemen and Winkler [72].

Mathematical aggregation approaches, which consist of processes or analytical models that operate on each expert's probability distributions to combine them for obtaining a single probability distribution to be used by the decision-maker [72], can be mainly grouped into axiomatic approaches and Bayesian approaches.

In Bayesian paradigm of combining expert judgments, if n experts provide information $g_1, g_2, \ldots, g_n$ to decision-maker regarding some quantity of interest, or a technical problem solution, θ, then the decision-maker's probability distribution of θ, denoted by p^* is obtained by applying Bayes' theorem to update a prior distribution $p(\theta)$ [64], Clemen and Winkler [72]:

$$p^* = p(\theta | g_1, g_2, \ldots, g_n) \propto p(\theta) L(g_1, g_2, \ldots, g_n | \theta) \tag{5.1}$$

where L represents the likelihood function that is associated with the experts' information. The challenging part of combining expert judgments using Bayesian aggregation method is the assessment of the likelihood function $L(g_1, g_2, \ldots, g_n|\theta)$. More information on Bayesian aggregation of expert's distribution are given in approaches [64, 67, 71, 72, 74].

Axiom-based aggregation methods, which are the earlier methods of combining expert judgments, are based on postulating certain properties that the combined distribution should follow and then deriving the functional form of the combined distribution [72]. Linear opinion pool or a weighted arithmetic averaging approach is a common and yet less-mathematically complex aggregation method. Let $p(\theta)_i, i = 1, \ldots, N$ be the probability distribution of the quantity of interest given by expert i, with N being the total number of experts. The combined expert opinions, denoted by $p(\theta)_{\mathrm{DM}}$, is obtained by taking the weighted arithmetic average of expert opinions [63], Clemen and Winkler [72]:

$$p(\theta)_{\mathrm{DM}} = \sum_{i=1}^{N} w_i p(\theta)_i \tag{5.2}$$

where w_i is the non-negative normalized weighting factor assigned for expert i, such that $\sum_{i=1}^{N} w_i = 1$. A detailed discussion on weighted average techniques is given by [63, 67, 68, 70–72, 75, 76].

One of the main challenges of using weighted arithmetic average aggregation method is the curial issue of assigning or computing the weights w_i. French [74] lists several obstacles to the approaches used for computing or assigning weights to experts, namely expert calibration, expert honesty, correlation among experts, and relative expertise of the experts. Genest and McConway [76] reviews different works on assigning expert weights and summarize them into, namely,

- Weights as veridical probabilities—the decision-maker's distribution is generated by one of the assessors' distributions, and the weight represents the probability that the assessors' distribution is the true distribution.
- Outranking probabilities—the weights are interpreted as the probability that the next prediction made using an expert distribution outperforms predictions made from other experts' distributions.
- Weights derived from scores—the weights are computed by applying strictly proper scoring rules in order to ensure that an individual's probability assessment correspond to his or her judgments.
- Minimum variance weights—the weights are assigned by minimizing the variance of the composite forecaster (i.e. the combined distribution).
- Weights as a measure of correlation—the weights are computed by considering the dependence among the assessors' sources of information.
- Self-assigned weights—the decision-maker asks the experts to select their own weights.

To tackle the issue of assigning weights, Cooke [67] introduces a performance-based aggregation approach, also known as Cook's classical approach, which is used to compute expert weights based on their performance on a set of so-called calibration or seed variables. In this method, weights are computed based on calibration and information scores that each expert receives according to his/her 5th, 50th, and 95th quantiles on seed variables, whose realizations (i.e. true values) are available post hoc. The information score is defined as a measure of the degree to which the expert's distribution is concentrated around the realizations of uncertain variables and the calibration is a measure of how well the uncertain quantities of the realizations are independent and identically distributed with hypothetical density $p = (0.05, 0.45, 0.45, 0.05)$. A detailed description of this approach is given in [67, 77–79] and its application for system reliability and production performance analysis is given in [52, 56].

5.5 Conclusions

Complex Arctic operational conditions can affect the equipment and human performance in various ways. They can increase the failure rate, power losses, life cycle costs, repair time, and safety hazards. Taking into consideration these types of effects, the designed systems must be dependable and safe as well as economically viable and it should minimize environmental pollution quantity of used raw materials and energy. Designing for performability contains appropriate approaches that can enable us to meet these important goals. In design for performability the main objective is to optimize reliability, maintainability, quality, safety, and risk analysis, as well as sustainability of selected technical solution simultaneously. This provide necessary information for selection, developing, optimization, and monitoring the most appropriate technology. Design for performability in complex Arctic operational conditions requires a range of statistical and simulation tools to be used and is dependent on a large amount of data and information. However, currently, there are not sufficient amount of performability data, including reliability, maintainability data, and information for industrial activities in the Arctic, which are essential inputs for an accurate performability analysis and assessment. This chapter has reviewed the effects of Arctic operational conditions on performability elements and has then showed that most of available databases are not suitable for performability analysis of equipment units in the Arctic regions. Thereafter, considering the urgent needs for establishing a correct, comprehensive, and suitable data collection system, it discussed the main elements of such a data collection system including the identification and then collecting all performability covariates. Expert judgments are effective ways to reduce the uncertainties associated with the design for performability while there is no high-quality historical data, with the Arctic being a case in point. Hence in the last part of the chapter, a conceptual model for integrating expert judgments in performability analysis of operations and facilities in the Arctic is developed.

References

1. Misra, K. B. (2008). *Handbook of performability engineering*. London: Springer.
2. Barabadi, A. (2011). *Production performance analysis: Reliability, maintainability and operational conditions*. Stavanger: University of Stavanger, Stavanger.
3. Freitag, D. R., & McFadden, T. T. (1997). *Introduction to cold regions engineering*. New York: ASCE Press.
4. Larsen, A. C., & Markeset, T. (2007). Mapping of operations, maintenance and support design factors in arctic environments. In T. Aven & J. E. Vinnem (Eds.), *Risk, reliability and societal safety*. London: Taylor & Francis.
5. Gudmestad, O. T., & Karunakaran, D. *Challenges faced by the marine contractors working in western and southern barents sea*. OTC Arctic Technology Conference; 3–5 December, Houston, Texas, USA2012.
6. Zaki, R. (2015). *Drilling waste minimization in the barents sea*. Tromsø: UiT—The Arctic University of Norway.

7. Kayrbekova, D., Barabadi, A., & Markeset, T. (2011). Maintenance cost evaluation of a system to be used in arctic conditions: A case study. *Journal of Quality in Maintenance Engineering, 17*(4), 320–336.
8. Naseri, M., & Samuelsen, E. M. (2019). Unprecedented vessel-icing climatology based on spray-icing modelling and reanalysis data: A risk-based decision-making input for arctic offshore industries. *Atmosphere, 10*(4), 197.
9. ISO. ISO 19906. (2010). *Petroleum and natural gas industries—arctic offshore structures.* Geneva: ISO.
10. Nikiforov, S., Dunaev, N., & Politova, N. (2005). Modern environmental conditions of the pechora sea (climate, currents, waves, ice regime, tides, river runoff, and geological). In H. A. Bauch, Y. A. Pavlidis, Y. I. Polyakova, G. G. Matishov, & N. Koç (Eds.), *Pechora sea environments: past, present and future.* Bremerhaven, Germany: Berichte zur Polar-und Meeresforschung.
11. Årthun, M., Bellerby, R. G. J., Omar, A. M., & Schrum, C. (2012). Spatiotemporal variability of air-sea co2 fluxes in the barents sea, as determined from empirical relationships and modeled hydrography. *Journal of Marine Systems, 98–99*(September), 40–50.
12. Naseri, M., & Barabady, J. (2016). On RAM performance of production facilities operating under the barents sea harsh environmental conditions. *International Journal of System Assurance Engineering and Management, 7*(3), 273–298.
13. ISO. ISO 12494. (2001). Atmospheric Icing of Structures. Geneva: ISO.
14. Ryerson, C. C. (2008). *Assessment of superstructure ice protection as applied to offshore oil operations: Safety problems, hazards, needs, and potential transfer technologies.* New Hampshire, USA: US Army Engineer Research and Development Center.
15. Samuelsen, E. M., Edvardsen, K., & Graversen, R. G. (2017). Modelled and observed sea-spray icing in arctic-norwegian waters. *Cold Regions Science and Technology, 134,* 54–81.
16. Rashid, T., Khawaja, H. A., & Edvardsen, K. (2016). Review of marine icing and anti-/de-icing systems. *Journal of Marine Engineering & Technology, 15*(2), 79–87.
17. Toomey, R. M., Lloyd, M., House, D. J., & Dickins, D. (2010). *The ice navigation manual.* Edinburgh, UK: Witherby Seamanship International Ltd.
18. Hamilton, L. (2004) The polar low phenomenon. Group for Earth Observation (GEO) Quarterly. 10–2.
19. Rasmussen, E. A., & Turner, J. (2003). *Polar lows—mesoscale weather systems in the polar regions.* Cambridge: Cambridge University Press.
20. Løset, S., Shkhinek, K., Gudmestad, O. T., Strass, P., Michalenko, E., Frederking, R., et al. (1999). Comparison of the physical environment of some arctic seas. *Cold Regions Science and Technology, 29*(3), 201–214.
21. Løset, S., Shkhinek, K., Strass, P., Gudmestad, O. T., Michalenko, E. B., & Kärnä, T. (1997). Ice conditions in the barents and kara seas. 16th *International Conference on Offshore Mechanics and Arctic Engineering.* April 13–18, Japan: Yokohama.
22. Holton, J. R., Curry, J. A., & Pyle, J. A. (2002). *Encyclopedia of atmospheric sciences.* London: Academic Press.
23. Przybylak, R. (2016). *The climate of the arctic* (2nd ed.). Cham: Springer.
24. Proshutinsky, A. Y., Proshutinsky, T., & Weingartner, T. (1999). INSROP working paper no. 126—environmental conditions affecting commercial shipping on the northern sea routes. Lysaker: International Northern Sea Route Programme.
25. Bagdonavicius, V., & Nikulin, M. (2001). *Accelerated life models: modeling and statistical analysis.* Boca Raton: Chapman & Hall/CRC.
26. Barabadi, A., Garmabaki, A. H. S., & Zaki, R. (2016). Designing for performability: an icing risk index for arctic offshore. *Cold Regions Science and Technology, 124,* 77–86.
27. Anttonen, H., & Virokannas, H. (1994). Assessment of cold stress in outdoor work. *Arctic Medical Research, 53*(1), 40–48.
28. Holmér, I. (1994). Cold stress: part i—guidelines for the practitioner. *International Journal of Industrial Ergonomics, 14*(1–2), 139–149.

29. Mäkinen, T. M., Palinkas, L. A., Reeves, D. L., Pääkkönen, T., Rintamäki, H., Leppäluoto, J., et al. (2006). Effect of repeated exposures to cold on cognitive performance in humans. *Physiology & Behavior, 87*(1), 166–176.
30. Gao, Y., & Li, K. (2012). New models for calculating the viscosity of mixed oil. *Fuel, 95,* 431–437.
31. NOAA. (2010). *Characteristics of response strategies: a guide for spill response planning in marine environment.* Seattle: National Oceanic and Atmospheric Administration (NOAA).
32. Markeset, T., Saeland, A., Gudmestad, O., & Barabady, J. (2015). Petroleum production facilities in arctic operational environments. In A. Bourmistrov, F. Mellemvik, & A. Bambulyak (Eds.), *International arctic petroleum cooperation: barents sea scenarios* (pp. 184–203). London: Routledge.
33. Keane, B., Schwarz, G., & Thernherr, P. (2013). Electrical equipment in cold weather applications. *Petroleum and Chemical Industry Technical Conference* (PCIC); September 23–25, Chicago: IEEE.
34. Ryerson, C. C. (2011). Ice protection of offshore platforms. *Cold Regions Science and Technology, 65*(1), 97–110.
35. Wright, B., & Timco, G., (Eds.). (1994). *A review of ice forces and failure modes on the molikpaq.* Proceeding of the 12th International Symposium on Ice, Trondheim.
36. Zhang, D., & Yue, Q. (2011). Major challenges of offshore platforms design for shallow water oil and gas field in moderate ice conditions. *Ocean Engineering, 38*(10), 1220–1224.
37. Osczevski, R., & Bluestein, M. (2005). The new wind chill equivalent temperature chart. *Bulletin of the American Meteorological Society, 86*(10), 1453–1458.
38. Bluestein, M., & Quayle, R. (2003). Wind chill. In J. R. Holton, J. A. Curry, & J. A. Pyle (Eds.), *Encyclopedia of atmospheric sciences* (pp. 2597–2602). Oxford: Academic Press.
39. Noroozi, A., Abbassi, R., MacKinnon, S., Khan, F., & Khakzad, N. (2014). Effects of cold environments on human reliability assessment in offshore oil and gas facilities. *Human Factors, 56*(5), 825–839.
40. ISO. ISO Guide 73. (2009). *Risk management—vocabulary—guidelines for use in standards.* Geneva: ISO.
41. ISO. ISO 31000. (2009). *Risk management—principles and guidelines.* Geneva: ISO.
42. Aven, T. (2008). *Risk analysis: Assessing uncertainties beyond expected values and probabilities.* West Sussex: Wiley.
43. Aven, T. (2012). *Foundations of risk analysis.* West Sussex: Wiley.
44. Nascimento, F. A. C., Majumdar, A., Ochieng, W. Y., & Jarvis, S. R. (2012). A multistage multinational triangulation approach to hazard identification in night-time offshore helicopter operations. *Reliability Engineering and System Safety., 108,* 142–153.
45. Crowley, J.D. (Eds.). (1988). *Cold water effects upon marine operations.* Proceedings OCEANS' 88—A Partnership of Marine Interests, Oct 31–Nov 2: IEEE.
46. Laakso, T., & Peltola, E. (Eds). (2005). *Review on blade heating technology and future prospects.* BOREAS VII Conference Proceedings; Mar 7–8.
47. Farzaneh, M., Volat, C., & Leblond, A. (2008). *Anti-Icing and de-icing techniques for overhead lines.* In: M. Farzaneh (Ed.). *Atmospheric Icing of power networks* (pp. 229–68). Dordrecht: Springer Netherlands.
48. Tammelin, B., Säntti, K., Dobech, H., Durstewich, M., Ganander H., & Kury, G., et al. (2005). *Wind turbines in icing environment: improvement of tools for siting, certification and operation-NEW ICETOOLS.* Finnish Meteorological Institute.
49. DNV. (2013). *DNV-OS-A201: Winterization for cold climate operations.* Høvik: Det Norske Veritas (DNV).
50. OREDA Participants. (2009). *Offshore reliability data handbook* (5th ed.). Trondhim: OREDA Participants.
51. Zaki, R., Barabadi, A., Qarahasanlou, A. N., & Garmabaki, A. H. S. (2019). A mixture frailty model for maintainability analysis of mechanical components: A case study. *International Journal of System Assurance Engineering and Management, 10*(6), 1646–1653.

52. Naseri, M., & Barabady, J. (2015). An expert-based model for reliability analysis of arctic oil and gas processing facilities. *Offshore Mechanics and Arctic Engineering., 138*(5), 051602.
53. ISO. *ISO 14224.* (2006). *Petroleum, petrochemical and natural gas industries—collection and exchange of reliability and maintenance data for equipment*. Geneva: ISO.
54. Gao, X., Barabady, J., & Markeset, T. (2010). An approach for prediction of petroleum production facility performance considering arctic influence factors. *Reliability Engineering and System Safety, 95*(8), 837–846.
55. Barabadi, A., & Markeset, T. (2011). Reliability and maintainability performance under arctic conditions. *International Journal of System Assurance Engineering and Management., 2*(3), 205–217.
56. Naseri, M., & Barabady, J. (2016). An expert-based approach to production performance analysis of oil and gas facilities considering time-independent arctic operating conditions. *International Journal of System Assurance Engineering and Management, 7*(1), 99–113.
57. Naseri, M., Baraldi, P., Compare, M., & Zio, E. (2016). Availability assessment of oil and gas processing plants operating under dynamic arctic weather conditions. *Reliability Engineering and System Safety, 152*(August), 66–82.
58. Zio, E. (2018). The future of risk assessment. *Reliability Engineering and System Safety, 177,* 177–190.
59. Ghobakhloo, M. (2020). Industry 4. digitization, and opportunities for sustainability. *Journal of Cleaner Production, 252,* 119869.
60. Nogal, M., Morales Nápoles, O., & O'Connor, A. (2019). Structured expert judgement to understand the intrinsic vulnerability of traffic networks. *Transportation Research Part A: Policy and Practice., 127,* 136–152.
61. Wang, D., Zhang, P., & Chen, L. (2013). Fuzzy fault tree analysis for fire and explosion of crude oil tanks. *Journal of Loss Prevention in the Process Industries, 26*(6), 1390–1398.
62. Kuselman, I., Pennecchi, F., Epstein, M., Fajgelj, A., & Ellison, S. L. R. (2014). Monte carlo simulation of expert judgments on human errors in chemical analysis—a case study of ICP–MS. *Talanta, 130,* 462–469.
63. Ayyub, B. M. (2001). *elicitation of expert opinions for uncertainty and risks.* Boca Raton: CRC Press.
64. Kelly, D., & Smith, C. (2011). *Bayesian inference for probabilistic risk assessment: a practitioner's guidebook.* London: Springer Science and Business Media.
65. Otway, H., & von Winterfeldt, D. (1992). Expert judgment in risk analysis and management: process, context, and pitfalls. *Risk Analysis, 12*(1), 83–93.
66. Meyer, M. A., & Booker, J. M. (2001). *Eliciting and analyzing expert judgement—A practical guide.* Philadelphia: Society for Industrial and Applied Mathematics.
67. Cooke, R. M. (1991). *Experts in uncertainty: Opinion and subjective probability in science.* New York: Oxford University Press.
68. Cooke, R. M., & Goossens, L. L. H. J. (2008). TU delft expert judgment database. *Reliability Engineering and System Safety, 93*(5), 657–674.
69. Naseri, M. On maintainability of winterised plants operating in arctic regions. *Proceedings of the ASME 2017 36th International Conference on Ocean,* Offshore and Arctic Engineering. June 25–30, Trondheim, Norway.
70. Dias, L. C., Morton, A., & Quigley, J. (2018). *Elicitation—the science and art of structuring judgement.* Gham: Springer.
71. Simola, K., Mengolini, A. M., & Bolado, Lavin R. (2005). *Formal expert judgement—an overview.* Petten: Directorate-GeneralJointResearchCentre (DGJRC) Institute for Energy.
72. Clemen, R. T., & Winkler, R. L. (2007). Aggregating probability distributions. In W. Edwards, R. F. Miles Jr., & D. Von Winterfeldt (Eds.), *Advances in decision analysis: From foundations to applications* (pp. 154–176). Cambridge: Cambridge University Press.
73. O'Hagan, A., Buck, C. E., Daneshkhah, A., Eiser, J. R., Garthwaite, P. H., Jenkinson, D. J., et al. (2006). *Uncertain judgements: Eliciting expert probabilities.* West Sussex: Wiley.
74. French, S. (1985). Group consensus probability distributions: A critical survey. In J. M. Bernardo, M. H. D. Groot, D. V. Lindley, & A. F. M. Smith (Eds.), *Bayesian statistics* (pp. 183–201). North Holland: Elsevier.

75. Pulkkinen, U. (1993). Methods for combination of expert judgements. *Reliability Engineering and System Safety, 40*(2), 111–118.
76. Genest, C., & McConway, K. J. (1990). Allocating the weights in the linear opinion pool. *Journal of Forecasting, 9*(1), 53–73.
77. Cooke, R. M., & Goossens, L. H. J. (2000). Procedures guide for structural expert judgement in accident consequence modelling. *Radiation Protection Dosimetry, 90*(3), 303–309.
78. Bedford, T., & Cooke, R. (2001). *Probabilistic risk analysis: Foundations and methods.* Cambridge: Cambridge University Press.
79. Cooke, R., Mendel, M., & Thijs, W. (1988). Calibration and information in expert resolution: A classical approach. *Automatica., 24*(1), 87–93.

Abbas Barabadi, Ph.D. Dr. Abbas Barabadi works as Professor in Safety and Technology at the Department of Technology and Safety at the University of Tromsø—The Arctic University of Norway (UiT). He is also serving as a Visiting Associate Professor at Luleå University of Technology (LTU), Sweden and as Adjunct Associate Professor at the University Centre in Svalbard (UNIS), Norway. He holds a B.Sc. in Mining Engineering from the University of Tehran, Iran, in the area of Mining Engineering. He received his M.Sc. in Mining Engineering from the Azad University of Science and Technology of Tehran, Iran. After working for a number of years as a Technical Office Manager and Senior Expert at Jajarm Bauxite Mine, Iran, he obtained a Ph.D. degree in Offshore Technology from the University of Stavanger (UiS). The Ph.D. project was a collaboration between UiS and UiT-The Arctic University of Norway. During the Ph.D. study, his office was located at the Department of Engineering and Safety, UiT that financed the project. His research interests include reliability and risk analysis and management, operation and maintenance engineering, cold climate engineering, technology and innovation in the Arctic region, Arctic operations, production performance analysis and management, risk-based inspection, and risk-based maintenance. He has published around 80 papers in international journals and conference proceedings.

Masoud Naseri, Ph.D. Dr. Masoud Naseri has received his Ph.D. in Natural Science in 2016 majored in risk and reliability form UiT-The Arctic University of Norway. During his Ph.D. study, he focused on risk and safety of technological developments, and operation and maintenance of technical and engineering facilities in the Arctic by considering the adverse impact of environmental conditions. He has gained experience and knowledge on risk assessment and risk analysis, expert judgements, offshore operations, and technologies, as well as reliability, availability, maintainability and safety modelling, simulation and analysis. He is currently enrolled as an Associate Professor in the B.Sc. study programme International Emergency Preparedness in Department of Technology and Safety at the UiT The Arctic University of Norway. He received his M.Sc. and B.Sc. in Petroleum Engineering in 2011 and 2008, prior to a one-year research assistantship at Middle East Technical University, Turkey. He has authored/co-authored 25 peer-reviewed articles published in international journals and conference proceedings.

Chapter 6
Dynamic Multi-state System Performability Concepts, Measures, Lz-Transform Evaluation Method

Anatoly Lisnianski and Lina Teper

Abstract In the chapter a performability concept for dynamic multi-state system as an extension of multi-state system reliability is considered. Steady-state and instantaneous (dynamic) indices for performability estimation in real-world multi-state systems are presented. The main obstacle in assessment of these indices is a "curse of dimensionality"—a huge number of system's states even for relatively simple multi-state system. In order to overcome on these difficulties, in this chapter modern mathematical method is considered—L_z-transform—for evaluation of dynamic performability indices (measures) for multi-state systems. Numerical example is presented in order to illustrate the approach.

Keywords Dynamic performability · Multi-state system · Markov process · L_z-transform · Performability measures

6.1 Introduction

All technical systems are designed to perform their intended tasks in a given environment under given conditions. Many systems can perform their task with various distinctive levels of efficiency, which is usually referred to as system performance rates. A system that can have a finite number of performance rates is called a multi-state system (MSS) [15]. Usually MSS is composed of elements that in their turn can be multi-state. Actually, a binary system is the simplest case of an MSS having two distinguished states (perfect functioning and complete failure).

There are many different situations in which a system should be considered as MSS. It may be power system, where performance is interpreted as power generating capacity; computer system, where data processing speed is treated as system performance; and so on. Many detailed examples of technical MSSs can be found in Lisnianski et al. [11].

A. Lisnianski (✉) · L. Teper
Reliability and Risk Management Centre, Shamoon College of Engineering, RAFAEL, Haifa, Israel
e-mail: lisnianski@bezeqint.net

K. B. Misra (ed.), *Handbook of Advanced Performability Engineering*,
https://doi.org/10.1007/978-3-030-55732-4_6

The term performability was first used in 1980 by Meyer [13] in the context of evaluation of highly reliable aircraft control computer systems. By using this term, he reflected a set of such systems' characteristics such as availability and maintainability. In this case performability was treated as an "umbrella" concept for all these properties. The definition of term performability was further extended by Misra [14] to include attributes of dependability and sustainability where dependability includes attributes of quality, reliability, maintainability, and safety. In other words, the performance was considered in totality over the entire life cycle of a product, system, or service. However, here we shall use the term performability in the sense of customer satisfaction from the system operation. It means that the following issue should be analyzed: How the system satisfies demands of its customer? This concept was suggested by Young and Kapur in [20]. Therefore, our MSS's performability measure should be considered as measures of customer demands satisfaction or even part of dependability attributes, namely reliability and availability (maintainability).

It should, however, be noted that there is a substantial difference between MSS performance and MSS performability measures. *A system performance is usually a physical parameter. Performability is a system property that characterizes a customer satisfaction from the system's functioning.* For example, for a power system such physical parameter as a generating capacity is usually treated as a system output performance. Such parameters as expected energy not supplied to consumers, loss of load probabilities, *and so on* that characterize a customer satisfaction from the system operation are measures of performability. Evaluation of such performability measures in dynamic modes when MSS has thousands and even millions of possible states is not a trivial job. It is so, because in dynamic modes each element should be presented by using discrete-state continuous-time (DSCT) stochastic process. If, for example, in the system there are eight components and each component has five states, then a state–space diagram for the entire system will have $5^8 = 390,625$ states. Because of the huge number of states and transitions, such model can be built only by using a special code (program) that should be developed for each case. Then a system of 3,900,625 differential equations should be solved in order to assess the performability measures. It requires enormous efforts. In this chapter we consider a modern method that can be used in such cases in order to overcome these difficulties. This method is called L_z-transform method. The method is an extension of widely known universal generating function approach that was primarily suggested by Ushakov [18]. L_z-transform was primarily introduced in [8] and its brief description and possible applications will be presented in the following sections of the chapter. Here only Markov stochastic processes will be considered.

6.2 Generic MSS Model and Its Evolution in State Space Associated Performability Measures

The generic MSS model [11, 15] should include models of the performance stochastic processes

$$G_j(t),\ j = 1, 2, \ldots, n \tag{6.1}$$

for each system element *j*, and the system structure function that produces the stochastic process corresponding to the output performance of the entire MSS is given by

$$G(t) = f(G_1(t), \ldots, G_n(t)) \tag{6.2}$$

The MSS behavior is characterized by its evolution in the space of states.

Since the system functioning is characterized by its output performance $G(t)$, the state acceptability at any time instant t depends on this value. In some cases, this dependency can be expressed by the acceptability function $F(G(t))$ that takes non-negative values if and only if the MSS functioning is acceptable. This takes place when the efficiency of the system functioning is completely determined by its internal state. For example, only the states where a network preserves its connectivity are acceptable. Other states are unacceptable. Usually unacceptable states are interpreted as system failure states, which when reached, imply that the system should be repaired or discarded.

Much more frequently, the system state acceptability depends on the relation between the MSS performance and the desired level of this performance (demand) that is determined by the customer. In general, the demand $W(t)$ is also a random process. Below we shall consider such a case when the demand can take discrete values from the set $\boldsymbol{w} = \{w_1, \ldots, w_M\}$. Often the desired relation between the system performance and the demand can be expressed by the acceptability function $F(G(t), W(t))$. The acceptable system states correspond to $F(G(t), W(t)) \geq 0$ *and the unacceptable states correspond to* $F(G(t), W(t)) < 0$. *The last inequality defines the* MSS failure criterion.

In many practical cases, the MSS performance should exceed the demand level determined by the customer. In such cases the acceptability function takes the form $F(G(t), W(t)) = G(t) - W(t)$.

The system behavior during the operation period can be characterized by the possibility of entering the subset of unacceptable states more than once. The case when MSS can enter this subset only once usually corresponds to unrepairable deteriorating systems. For repairable systems the transitions between subsets of acceptable and unacceptable states may occur an arbitrary number of times.

Note that in some cases it may be impossible to divide MSS's state–space to acceptable and unacceptable states. Only some functional associated with two stochastic processes $G(t)$ and $W(t)$ may be of interest in order to define MSS failure. For example, MSS failure may be defined as an event, when functional $J = \int_0^T \propto [G(t), W(t)]dt \geq J_0$ will be greater than or equal to some specified value J_0 and $\alpha(*)$ is defined as some arbitrary function. For example, for power system, where $G(t)$ and $W(t)$ are treated as generating capacity and load (demand, which is required by consumers), such functional J is interpreted as an energy not supplied to consumers, when $\alpha(*)$ is defined such as the following:

$$\propto (t) = W(t) - G(t), \text{if } W(t) - G(t) \geq 0$$

and

$$\propto (t) \equiv 0, if\, W(t) - G(t) < 0.$$

According to customer requirements, this functional J may also characterize the number of system entrances the set of some specified states (where, for example, $G(t) < W(t)$) during some time period [0,T]), the accumulated time of system staying in this set of states, the accumulated performance deficiency during time period [0,T], and so on.

In general, a value of functional J, which is defined by the customer requirements for two stochastic processes $G(t)$ and $W(t)$, is considered as MSS performability measure. In the next subsection we introduce different measures of MSS performability.

6.2.1 MSS Performability Measures

To numerically characterize MSS behavior from a performability point of view one has to determine the MSS performability measures (indices). In general case, these indices are different modifications of functional J and they are based on considering the system evolution in the time domain. In this case the relation between the system's output performance and the demand represented by the two corresponding stochastic processes must be studied. Figure 6.1 shows an example of the behavior of the MSS performance and a demand as the realizations of the stochastic processes [11].

When one considers an MSS evolution in the space of states during the system operation period T, the following measures are usually of interest from a customer point of view:

Time to failure, T_f is the time from the beginning of the system life up to the instant when the system enters the subset of unacceptable states for the first time.

Time between failures, T_b is the time between two consecutive transitions from the subset of acceptable states to the subset of unacceptable states.

Number of failures, N_T is the number of times the system enters the subset of unacceptable states during the time interval $[0,T]$.

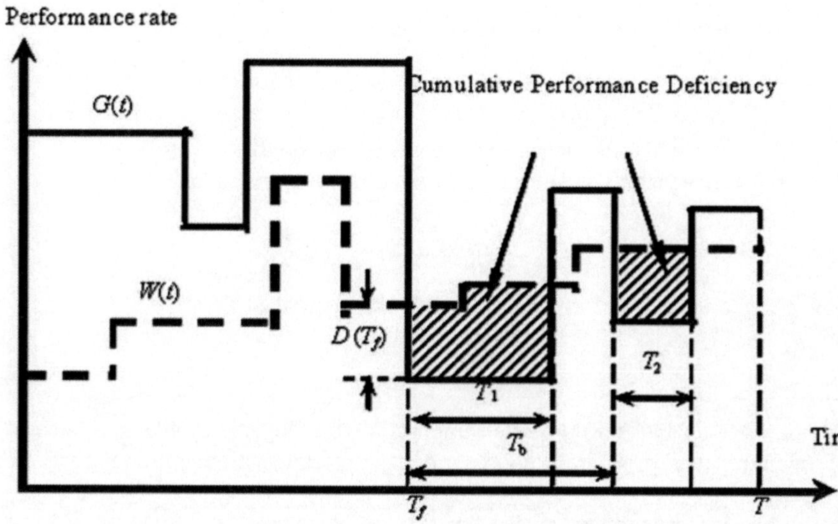

Fig. 6.1 MSS behavior as relation between two stochastic processes—mss output performance $G(t)$ and demand $W(t)$

These measures are the measures of an MSS reliability.

In Fig. 6.1 one can see an example of realizations of two stochastic processes $G(t)$ and $W(t)$. Assume that according to customer requirement the system performance value should exceed the value of demand: $F(G(t), W(t)) = G(t)\text{-}W(t) > 0$. In this case, the first time that the process $G(t)$ down crosses the level of demand, $W(t)$ determines the time to MSS failure. This time is designated as T_f. The random variable T_f is characterized by the following indices:

Probability of a failure-free operation or reliability function $R(t)$ is the probability that T_f is greater than or equal to the value t ($t > 0$), where in the initial state (at instant $t = 0$) MSS is in one of the acceptable states:

$$R(t) = \Pr\{T_f \geq t | F(G(0), W(0)\} \geq 0. \tag{6.3}$$

Mean time to failure (MTTF) is the mean time up to the instant when the system enters the subset of unacceptable states for the first time:

$$E\{T_f\}. \tag{6.4}$$

From now on $E\{*\}$ will be used as an expectation symbol.

The same two indices can be defined for the random variable T_b:

The probability that the time between failures is greater than or equal to t is:

$$\Pr\{T_b \geq t\}. \tag{6.5}$$

The mean time between failures (MTBF):

$$E\{T_b\} \tag{6.6}$$

The reliability indices associated with the random variable N_T are:
The probability that N_T is not greater than some specified number n:

$$Pr\{N_T\} \leq n \tag{6.7}$$

The expected number of system failures in the interval $[0,T]$:

$$E\{N_T\}. \tag{6.8}$$

Measures (6.7) and (6.8) are often important when logistic problems related to MSS operation are considered (e.g. in order to determine the required number of spare parts).

MSS instantaneous (point) availability $A(t, w)$ is the probability that the MSS at instant $t > 0$ is in one of the acceptable states

$$A(t, w) = \Pr\{F(G(t), W(t)) \geq 0\}. \tag{6.9}$$

MSS availability in the time interval $[0,T]$ is defined as:

$$A_T = \frac{1}{T}\int_0^T A(t, w)\mathrm{d}t. \tag{6.10}$$

The random variable A_T represents the portion of time when the MSS output performance rate is in an acceptable area. This index characterizes the portion of time when the MSS output performance rate is not less than the demand. For example, in Fig. 6.1

$$A_T = (T - T_1 - T_2)/T. \tag{6.11}$$

The expected value of A_T is often used and is named demand availability [1]:

$$A_D = E\{A_T\}. \tag{6.12}$$

For large t ($t \to \infty$), the system initial state has no influence on its availability. Therefore, the steady-state (stationary or long-term) MSS availability $A_\infty(w)$ for the constant demand level $W(t) = w$ can be determined on the base of the system steady-state performance distribution:

$$A_\infty(w) = \sum_{k=1}^{K} p_k 1(F(g_k, w) \geq 0) \tag{6.13}$$

where

$$1(F(g_i, w) \geq 0) = \begin{cases} 1, & \text{if } F(g_i, w) \geq 0, \\ 0, & \text{if } F(g_i, w) < 0, \end{cases} \tag{6.14}$$

and $p_k = \lim_{t \to \infty} p_k(t)$ is the steady-state probability of the MSS state k with the corresponding output performance rate g_k.

In the case where $F(G(t), W(t)) = G(t) - W(t)$ we have $F(g_k, w) = g_k - w$ and, therefore,

$$A_\infty(w) = \sum_{k=1}^{K} p_k 1(g_k \geq w) = \sum_{g_k \geq w} p_k. \tag{6.15}$$

In power systems this index is called as loss of load probability [2].

In order to obtain the indices that characterize the average MSS output performance, one can use the performance expectation. The mean value of MSS instantaneous output performance at time t is determined as:

$$G_{\text{mean}}(t) = E\{G(t)\}. \tag{6.16}$$

The average MSS expected output performance for a fixed time interval $[0,T]$ is defined as:

$$G_T = \frac{1}{T} \int_0^T G_{\text{mean}}(t) \mathrm{d}t. \tag{6.17}$$

Observe that the mean MSS performance does not depend on demand.

In some cases, a conditional expected performance is used. This index represents the mean performance of MSS on condition that it is in acceptable states.

It is often important to know the measure of system performance deviation from a demand when the demand is not met. In the special case where $F(G(t), W(t)) = G(t) - W(t)$, the instantaneous performance deviation can be represented as:

$$D(t, w) = \max\{W(t) - G(t), 0\} \tag{6.18}$$

and is called the instantaneous performance deficiency at instant t. For example, in power systems $D(t,w)$ is interpreted as a generating capacity deficiency.

The average MSS expected performance deficiency for a fixed time interval $[0,T]$ is defined as follows:

$$D_T = \frac{1}{T}\int_0^T D(w,t)\mathrm{d}t \tag{6.19}$$

The cumulative performance deficiency for a fixed interval [0,*T*] is defined as follows:

$$D_{\sum^T} = \int_0^T D(t,w)\mathrm{d}t. \tag{6.20}$$

For example, in power systems it is an energy not supplied to consumers during time interval [0,*T*]. (In Fig. 6.1 the cumulative performance deficiency is the sum of the dashed areas).

In some cases, the instantaneous performance deficiency makes no sense as the system uses storage facilities to accumulate a product. The deficiency appears not when the system performance does not meet the demand, but rather when the accumulated performance in interval [0,*T*] is less than the accumulated demand at this interval. This takes place in oil and gas transmission systems with intermediate reservoirs. The accumulated performance deficiency in this case takes the following form:

$$D_{\sum^T} = \int_o^T (W(t) - G(t))\mathrm{d}t = \int_0^T W(t)\mathrm{d}t - \int_0^T G(t)\mathrm{d}t. \tag{6.21}$$

Computation of most of the above-mentioned performability indices is quite a difficult problem, especially in dynamic modes. In this chapter we shall consider for this purpose a modern mathematical method—L_z-transform.

6.3 L_z—Transform and Inverse L_z—Transform: The Method Description

6.3.1 $\mathbf{L_z}$-*Transform*

We consider a discrete-state continuous-time (DSCT) Markov process [17] $X(t) \in \{x_1, \ldots, x_K\}$, $t \geq 0$, which has *K* possible states *i* (*i* = 1, ...,*K*), where performance level associated with any state *i* is x_i. This Markov process is completely defined by a set of possible states $\mathbf{x} = \{x_1, \ldots, x_K\}$, transitions intensities matrix $\mathbf{A} = \|a_{ij}(t)\|$, $i, j = 1, \ldots K$ and by initial states probability distribution that can be presented by corresponding set

$$\mathbf{p}_0 = [p_{10} = \Pr\{X(0) = x_1\}, \ldots, p_{K0} = \Pr\{X(0) = x_K\}]. \tag{6.22}$$

From now on, we shall use for such Markov process the following notation by using triplet:

$$X(t) = \{\mathbf{x}, \mathbf{A}, \mathbf{p}_0\}. \tag{6.23}$$

Definition *L_z-transform of a discrete-state continuous-time Markov process $X(t) = \{x, A, \mathbf{p_0}\}$ is a function of* u$(z, t, \mathbf{p_0})$ *defined as*

$$L_z\{X(t)\} = \mathrm{u}(z, t, \mathbf{p}_0) = \sum_{i=1}^{K} p_i(t) z^{x_i} \tag{6.24}$$

where $p_i(t)$ is a probability that the process is in state i at time instant t for any given initial state probability distribution $\mathbf{p_0}$, and z is a complex variable.

6.3.1.1 Existence and Uniqueness of L_z-Transform

Each discrete-state continuous-time Markov process $X(t) = \{\mathbf{x}, \mathbf{A}, \mathbf{p_0}\}$ (where transition intensities $a_{ij}(t)$ are continuous functions of time) under certain initial conditions has only one (unique) L_z-transform u$(z, t, \mathbf{p_0})$ and each L_z-transform u$(z, t, \mathbf{p_0})$ has only one corresponding DSCT Markov process $X(t)$ developing from these initial conditions $\mathbf{p_0}$.

We'll formulate this as an ***existence and uniqueness property*** of L_z-transform. It was proven by Lisnianski [8].

Remark 1 In reliability interpretation, L_z-transform may be applied to an aging system and to a system at burn-in period as well as to a system with constant failure and repair rates.

6.3.1.2 Main Properties of L_z-Transform

Property 1 Multiplying DSCT Markov process on constant value a is equal to multiplying corresponding performance level x_i at each state i on this value:

$$L_z\{aX(t)\} = \sum_{i=1}^{K} p_i(t) z^{a x_i} \tag{6.25}$$

Property 2 L_z-transform from a single-valued function $f(G(t), W(t))$ of two independent DSCT Markov processes $G(t)$ and $W(t)$ can be found by applying Ushakov's universal generating operator Ω_f to L_z-transform from $G(t)$ and $W(t)$ processes over all time points $t \geq 0$

$$L_z\{f(G(t), W(t))\} = \Omega_f(L_z\{G(t)\}, L_z\{W(t)\}). \qquad (6.26)$$

The property provides L_z-transform application to multi-state system reliability analysis. Computation procedures for operator Ω_f have been established for many different structure functions f [11]; Levitin [7]. An example of application of these procedures will be presented in the following section.

6.3.1.3 Inverse L_z-Transform

Inverse L_z-transform was introduced by Lisnianski and Ding [10], and here we present its brief description.

Definition. Let a function

$$\mathrm{u}(z, t, \mathbf{p_0}) = \sum_{i=1}^{K} p_i(t) z^{x_i}, \qquad (6.27)$$

be L_z-transform of some ***unknown*** discrete-state continuous-time Markov process $X(t) = \{\mathbf{x}, \mathbf{A}, \mathbf{p_0}\}$. Here $p_i(t)$, $i = 1, \ldots, K$ is a probability that Markov process $X(t)$ is in state i at time instant $t \geq 0$, x_i is the performance in this state, $\mathbf{p_0}$ vector of the process states probabilities at initial time instant $t = 0$, and z is a complex variable.

Based on a given L_z-transform (6.27) of some ***unknown*** DSCT Markov process $X(t)$, inverse L_z-transform

$$L_Z^{-1}\left\{\sum_{i=1}^{K} p_i(t) z^{x_i}\right\} \qquad (6.28)$$

reveals the underlying Markov process $X(t)$. Therefore, the following definition can be written:

$$L_Z^{-1}\{L_Z\{X(t)\}\} = X(t) = \{\mathbf{x}, \mathbf{A}, \mathbf{p_0}\}. \qquad (6.29)$$

if all transition intensities in matrix $\mathbf{A}$ are continuous function of time.

In other words, based on a given L_z-transform of some DSCT Markov process $X(t)$, inverse L_z-transform reveals (or uncovers) the underlying Markov process $X(t)$.

As it was stated above, "to reveal (uncover) underlying Markov process" means to determine for this process: a set of possible states $\mathbf{x}$; a transition intensities matrix $\mathbf{A}$; a vector of initial conditions $\mathbf{p_0}$.

6.3.1.4 Computational Procedure for Determining Inverse L_z-Transform

From computational point of view, the problem of finding inverse L_z-transform can be summarized as the following:

It is given L_z-transform of some unknown Markov process $X(t)$

$$L_z\{X(t)\} = \mathrm{u}(z, t, \mathbf{p}_0) = \sum_{i=1}^{K} p_i(t) z^{x_i}. \tag{6.30}$$

Based on this expression for L_z-transform one should reveal (uncover) the underlying Markov process $X(t)$, or in other words, to find the set of states $\mathbf{x}$, the set of initial conditions $\mathbf{p_0}$, and the matrix A of transitional probabilities of the process.

In reliability interpretation we consider the case when each MSS's component at time $t = 0$ may be in any arbitrary state and any MSS's state with performance lower than any specified demand level is treated as MSS's failure. We suppose that MSS consists of n independent components, where each component i is described by corresponding Markov stochastic process $X_i(t)$. Usually MSS's structure function f, which defines MSS output stochastic process $X(t)$ is known and given by the following expression

$$X(t) = f(X_1(t), X_2(t), \ldots, X_n(t)), \tag{6.31}$$

where $X_i(t) \in \{x_{i1}, x_{i2}, \ldots, x_{ir_i}\}$ is a discrete state continuous time Markov process that describes performance behavior of component i.

Notice that L_z-transform (6.30) is obtained under specified initial conditions for all system's components. We designate these conditions by using the following notation

$$X_1(0) = x_{1i}, i \in \{1, \ldots r_1\}, \ldots, X_n(0) = x_{nk}, k \in \{1, \ldots, r_n\}, \tag{6.32}$$

where $r_m, m \in \{1, \ldots, n\}$ is a number of performance levels for every component m.

Thus, the problem is to uncover (reveal) the underlying process $X(t)$ based on a given information regarding the system structure function (6.31), L_z-transform of process $X(t)$ (6.30) and given initial conditions (6.32) for each MSS's component.

*Determining set of states **X** and set of initial conditions.*

From expression (6.30), one knows a number of states K of resulting Markov process and performance in each state i corresponding to value x_i. Thus, one determines a set of states $\mathbf{x} = \{x_1, x_2, \ldots, x_K\}$ for underlying process $X(t)$.

The initial conditions for MSS are obtained by the given initial conditions of its components (6.32). Thus, the initial state of the entire system will be defined by the

MSS's structure function (6.31), where the corresponding performances of components are in their initial states. If the initial conditions of all system's components are given by (6.32), then according to MSS's structure function (6.31), the initial state of the entire MSS is the following

$$X(0) = f(x_{1i}, x_{2l}, \ldots, x_{nk}) = x_j, \; j \in \{1, 2, \ldots, K\}. \tag{6.33}$$

Thus, the following initial conditions are determined for entire MSS

$$\mathbf{p_{0j}} = [\Pr\{X(0) = x_1\} = 0, \Pr\{X(0) = x_2\} = 0, \ldots,$$
$$\Pr\{X(0) = x_j\} = 1, \ldots, \Pr\{X(0) = x_k\} = 0], \; j \in \{1, 2, \ldots, K\}. \tag{6.34}$$

It means that at instant $t = 0$ the system is in state j, with performance x_j, $j \in [1, \ldots, K]$.

In order to emphasize the fact that L_z-transform (3.36) is obtained for the given initial states of all MSS's components (6.32) (which then provide initial condition $\mathbf{p_{0j}}$ (6.34) for the entire system), we will use the following designation for the given L_z-transform of MSS's resulting (output) performance process $X(t)$:

$$L_Z\{X(t)\} = \sum_{i=1}^{K} p_i^{(j)}(t) z^{x_i}, \; j \in \{1, 2, \ldots, K\}. \tag{6.35}$$

Determining matrix ***A***

The resulting stochastic process $X(t)$ is a Markov process that has K states and from expression (3.41), we know all its state probabilities $p_i^{(j)}(t)$, $i = 1, \ldots, K$ under the condition that the process begins from state j at time instant $t = 0$.

Generally, probabilities for each of K states can be found by solving the following system of differential equations [17] under given initial conditions $\mathbf{p_{0j}}$ (in matrix notation)

$$\frac{\mathrm{d}\mathbf{p}(t)}{\mathrm{d}t} = \mathbf{p}(t)\mathbf{A}, \tag{6.36}$$

where

$\mathbf{p}(t) = [p_1(t), p_2(t), \ldots, p_K(t)]$ is row-vector of state probabilities;

$\frac{\mathrm{d}\mathbf{p}(t)}{\mathrm{d}t} = \left[\frac{\mathrm{d}p_1(t)}{\mathrm{d}t}, \frac{\mathrm{d}p_2(t)}{\mathrm{d}t}, \ldots, \frac{\mathrm{d}p_K(t)}{\mathrm{d}t}\right]$ is row-vector of state probabilities' derivatives;

A—transition intensities matrix (that in our case is still unknown)

$$\mathbf{A} = \begin{bmatrix} -(a_{12} + \ldots + a_{1K}) & a_{12} & \ldots & a_{1K} \\ a_{21} & -(a_{21} + \ldots + a_{2K}) & \ldots & a_{2K} \\ \ldots & \ldots & \ldots & \ldots \\ a_{K1} & a_{K2} & \ldots & -(a_{K1} + \ldots + a_{K,K-1}) \end{bmatrix}. \tag{6.37}$$

We designate the solution of system (6.36) under initial condition $\mathbf{p_{0\,j}}$ (the system is in state $j, j = 1,2, \ldots, K$ at instant $t = 0$) as $\mathbf{p_{Aj}}(t) = \left[p_{A1}^{(j)}(t), p_{A2}^{(j)}(t), \ldots, p_{AK}^{(j)}(t)\right]$.

Ideally, matrix A should be determined in order to provide the following equations

$$p_{A1}^{(j)}(t) = p_1^{(j)}(t), p_{A2}^{(j)}(t) = p_2^{(j)}(t), \ldots, p_{AK}^{(j)}(t) = p_K^{(j)}(t), j = 1, \ldots, K. \tag{6.38}$$

It means that state probabilities of the resulting stochastic process that can be found as a solution of system (6.36) for initial conditions $\mathbf{p_{0j}}$, and state probabilities from known L_z-transform of this process (expression (3.36)) should be the same. In practice, we should find matrix A in order to provide a minimal difference between all probabilities $p_{Ai}^{(j)}(t)$ and $p_i^{(j)}(t)$, $i = 1,2,\ldots K$ for the given $j = 1, \ldots, K$ at any time instant t.

A numerical solution to this problem can be obtained by using genetic algorithm (GA) that was implemented by Lisnianski and Ding [10]. In accordance with GA in each genetic cycle, we randomly generate all $(K - 1) \cdot (K - 1)$ coefficients a_{ij} in matrix A (6.37). We should generate only $(K - 1)$ omponents in each row, because the diagonal component a_{ii} in each row i in matrix $\mathbf{A}$ is defined as minus the sum of all other components in this row and should not be generated.

Finally, we should find set of coefficients that minimize the accumulated relative error—the measure of difference between solutions of system (3.42) and probabilities, which are presented in L_z-transform (expression 3.36). This error will be used in GA as a measure of accuracy.

For the given initial conditions $\mathbf{p_{0j}}$, the accumulated relative error, which should be minimized, is the following:

$$\text{ERR}_j = \sum_{i=1}^{N_p} w_1 \frac{\left|p_{1A}^{(j)}(t_i) - p_1^{(j)}(t_i)\right|}{p_1^{(j)}(t_i)} + \ldots + \sum_{i=1}^{N_p} w_K \frac{\left|p_{KA}^{(j)}(t_i) - p_K^{(j)}(t_i)\right|}{p_K^{(j)}(t_i)} \tag{6.39}$$

where N_p—number of time points t_i, for which the probability values are compared; $w_1, \ldots, w_K$ are defined as weights of relative error for state 1 and state K, respectively.

The weights of relative errors in some states can be set as small values if their absolute errors are quite small. In order to determine the number of time points N_p, one should analyze functions $p_i^{(j)}(t)$ from the given L_z-transform. Usually around 1000 time points should be sufficient for the calculation of accumulated relative error

(6.39). Approximately, a quarter of these points may be taken within a steady-state mode and three quarters within transient interval of functions $p_i^{(j)}(t)$.

6.3.2 MSS Model for the Method Application

We consider a multi-state system consisting of n multi-state components. Any component j in MSS can have k_j different states corresponding to different performance, represented by the set $\mathbf{g_j} = \{g_{j1}, \ldots, g_{jk_j}\}$, where g_{ji} is the performance rate of component j in the state I, $i \in \{1, 2, \ldots, k_j\}$. The generic MSS model consists of the performance stochastic processes $G_j(t) \in \mathbf{g_j}$, $j = 1, \ldots, n$ for each system component j, and the system structure function that produces the stochastic process corresponding to the output performance of the entire MSS: $G(t) = f(G_1(t), \ldots, G_n(t))$. At first, a model of stochastic process should be built for every multi-state component in order to define output performance stochastic process for the entire MSS.

6.3.2.1 Model of Repairable Multi-state Element

Markov performance stochastic process for each component j can be represented by the triplet $G_j(t) = \{\mathbf{g_j}, \mathbf{A_j}, \mathbf{p_{j0}}\}$, where $\mathbf{g_j}, \mathbf{A_j}, \mathbf{p_{j0}}$ are defined by the following:

- $\mathbf{g_j} = \{g_{j1}, \ldots, g_{jk_j}\}$ - set of possible states;
- $\mathbf{A_j} = \left(a_{\mathrm{lm}}^{(j)}(t)\right)$, $l, m = 1, \ldots, k; j = 1, \ldots n$, transition intensities matrix (for aging elements $a_{\mathrm{lm}}^{(j)}(t)$ are increasing functions of time);
- $\mathbf{p_{j0}} = \left[p_{10}^{(j)} = \Pr\{G_j(0) = g_{10}\}, \ldots, p_{k_j0}^{(j)} = \Pr\{G_j(0) = g_{k_j0}\}\right]$ initial states probability distribution.

The following system of differential equations can be written for the state probabilities [17]

$$\begin{cases} \frac{\mathrm{d}p_{\mathrm{j}k_j}(t)}{dt} = \sum\limits_{e=1}^{k_j-1} a_{\mathrm{e}k_j}^{(j)}(t)p_{\mathrm{je}}(t) - p_{\mathrm{jk}}(t)\sum\limits_{e=1}^{k_j-1} a_{k_je}^{(j)}(t) \\ \frac{\mathrm{d}p_{\mathrm{ji}}(t)}{\mathrm{d}t} = \sum\limits_{e=i+1}^{k_j} a_{\mathrm{ei}}^{(j)}(t)p_{\mathrm{je}}(t) + \sum\limits_{e=1}^{i-1} a_{\mathrm{ei}}^{(j)}(t)p_{\mathrm{je}}(t) - p_{\mathrm{ji}}(t) \\ \left(\sum\limits_{e=1}^{i-1} a_{\mathrm{ie}}^{(j)}(t) + \sum\limits_{e=i+1}^{k_j} a_{\mathrm{ie}\ i,e}^{(j)}(t)\right) \\ \qquad \text{for } 1 < i < k_j \\ \frac{\mathrm{d}p_{j1}(t)}{\mathrm{d}t} = \sum\limits_{e=2}^{k_j} a_{e1}^{(j)}p_{\mathrm{je}}(t) - p_{j1}(t)\sum\limits_{e=2}^{k_j} a_{1e}^{(j)} \end{cases} \tag{6.40}$$

By solving this system (6.40) under initial conditions

$$\mathbf{p}_0^{(j)} = \left[p_{10}^{(j)} = \Pr\left\{G_j = g_{10}\right\}, \ldots, p_{k_j0}^{(j)} = \Pr\left\{G_j(0) = g_{k_j0}\right\}\right] \tag{6.41}$$

one can find (for each element j in the MSS) state probabilities as functions of time:

$$p_{\mathrm{ji}}(t) = \Pr\left\{G_j(t) = g_{ji}\right\}, i \in \left\{1, \ldots, k_j\right\}, \; j \in \{1, \ldots, n\}. \tag{6.42}$$

Based on solution (6.40) we can obtain L_z-transform $L_z\left\{G_j(t)\right\}$ of a discrete-state continuous-time (DSCT) Markov process $G_j(t)$

$$L_z\left\{G_j(t)\right\} = \sum_{i=1}^{k_j} p_{\mathrm{ji}}(t) z^{g_{ji}}, \tag{6.43}$$

where $p_{\mathrm{ji}}(t)$ is a probability that the process $G_j(t)$ is in a state with performance g_{ji} at time instant $t \geq 0$ for a given initial states probability distribution $\mathbf{p}_0^{(j)}$, and z in general case is a complex variable. So, for each of the MSS's element the system of differential Eq. (6.40) should be solved under a given initial condition (6.41) and a corresponding L_z-transform (6.43) should be found.

6.3.2.2 Entire Multi-state System Model

A logical arrangement of the elements in the system is defined by the system structure function $f(G_1(t), \ldots, G_n(t))$. The output performance distribution for the entire MSS *at each time instant t* should be defined based on previously determined states probabilities (6.42) for all elements and a logical arrangement of the elements in the system is defined by the system structure function $f(G_1(t), \ldots, G_n(t))$. At this stage L_z-transform and Ushakov's universal generating operator provide the corresponding computations. L_z-transform of the output stochastic process for the entire MSS can be defined based on previously determined L_z-transform for each component j and system structure function f, which produces the output stochastic process of the entire MSS based on stochastic processes of all MSS's elements:

$$G(t) = f(G_1(t), \ldots, G_n(t)) \tag{6.44}$$

In order to find L_z-transform of the MSS's output performance, Markov process $G(t)$, which is the single-valued function (6.44) of n independent DSCT Markov processes $G_j(t), j = 1, \ldots, n$, one can apply Ushakov's universal generating operator (UGO) [18] to all individual L_z-transforms $L_z\left\{G_j(t)\right\}$ over all time points $t \geq 0$.

$$L_z\{G(t)\} = \Omega_f(L_z\{G_1(t)\}, \ldots, L_z\{G_n(t)\}) = \sum_{i=1}^{K} p_i(t) z^{g_i}. \tag{6.45}$$

The technique of Ushakov's operator applying is well established for many different structure functions f [11, 7].

If all components in some MSS are described by Markov process, the entire MSS is described by Markov process too. So, the resulting process $G(t)$ is a Markov process. But after like term collection (summarizing all terms with same powers of z in expression $L_Z\{G(t)\}$), one will have a new expression for L_z-transform, which is corresponding with new stochastic process $G_I(t)$. This new stochastic process $G_I(t)$ can be considered as the process, which was obtained from the primary Markov process $G(t)$ by lumping (uniting) all states with the same performance. In general case lumpability conditions [6] are not fulfilled for this process and the resulting process $G_1(t)$ (after like term collection) is not Markov. This fact does not change the computation of availability and performability indices according to expressions (6.9)–(6.21), because all states probabilities for output stochastic process are known. But it will be very important and should be taken into account when reliability function and mean time up to failure will be calculated.

Therefore, after like terms collection in expression (6.45) one will have L_z-transform for new process $G_1(t)$ with restricted number of states $K_1 < K$, which in general case is not Markov process

$$L_z\{G_1(t)\} = \sum_{i=1}^{K_1} p_i(t)z^{g_i} \tag{6.46}$$

The possibility of like terms collection is one of the main advantages of UGF and L_z-transform method, because in many practical cases it helps to restrict drastically a number of states in resulting stochastic process.

In order to use this important advantage and remain in Markov framework we will deal with new Markov process $G_{IM}(t)$, which will be equivalent to the process $G_1(t)$ in sense of equality of all probabilities of staying in states with same performances over all time points $t \geq 0$. In other words, L_z-transform for the process $G_{IM}(t)$ is equal to L_z-transform of the process $G_1(t)$

$$L_z\{G_{1M}(t)\} = L_z\{G_1(t)\} = \sum_{i=1}^{K_1} p_i(t)z^{g_i} \tag{6.47}$$

This process $G_{IM}(t)$ will be called as *approximating Markov process* for primary non-Markov output stochastic process $G_1(t)$.

6.3.3 Calculation of Dynamic Availability and Performability Measures

If L_z-transform (6.46) of output stochastic process $G_1(t) \in \{g_1, \ldots, g_{K1}\}$ is known, then important system's performability measures can be found.

The system availability at instant $t \geq 0$ is given by:

$$A(t) = \sum_{g_i \geq 0} p_i(t). \tag{6.48}$$

The system instantaneous mean expected performance at instant $t \geq 0$ is

$$E(t) = \sum_{i=1}^{K_1} p_i(t) g_i. \tag{6.49}$$

The system average expected performance for a fixed time interval [0,T] is

$$E_T = \frac{1}{T} \int_0^T E(t)\mathrm{d}t. \tag{6.50}$$

The system instantaneous performance deficiency is

$$D(t) = \sum_{i=1}^{K_1} p_i(t)\min(g_i, 0) \tag{6.51}$$

The system accumulated performance deficiency for a fixed time interval [0,T] is

$$D_\int = \int_0^T D(t)\mathrm{d}t = \sum_{k=1}^{K_1} \min(g_i, 0) \int_0^T p_i(t)\mathrm{d}t. \tag{6.52}$$

In order to find $A(t)$, $E(t)$, E_T, $D(t)$, $D_\int$, one doesn't need to know approximating Markov process. The exact values of these measures are obtained from expression (6.46) because all states probabilities $p_i(t)$, $i \in [1, K_I]$ for an output of non-Markov stochastic process $G_I(t)$ are known.

But in order to find MSS reliability function $R(t)$, one has to uncover approximating Markov process G_{IM}(t) for output stochastic process $G_1(t)$.

Inverse L_z-transform can uncover an underlying Markov process, when L_z-transform of this process is known. Based on the revealed (uncovered) output process, the MSS reliability function and mean time to failure (MTTF) can be found.

Applying inverse L_z-transform (L_Z^{-1}-transform) to L_z-transform in expression (6.47), one can reveal the underlying approximating Markov process $G_{IM}(t)$:

$$L_Z^{-1}\{\sum_{i=1}^{K_1} p_i(t)z^{g_i} = G_{1M}(t) = \{\mathbf{g}, \mathbf{A}, \mathbf{p_0}\}. = \{\mathbf{g}, \mathbf{A}, \mathbf{p_0}\}. \tag{6.53}$$

The approximating Markov process has K_1 states $\mathbf{g} = \{g_1, \ldots, g_{K1}\}$ that are arranged in the ascending order $g_1 \le g_2 \le \ldots \le g_{K1}$. Reliability function *R*(*t*) is treated as probability that the process $G_{IM}(t)$, which begins at $t = 0$ from state *j* will downgrade the below specified demand level $w_r$at time instant *t*.

In order to find *R*(*t*) all states with performance lower than $w_r$should be united in one absorbing state and all transitions from this absorbing state to any other states should be constrained. If $g_k < w_r$ and $g_{k+1} \ge w_{\text{req}}$, then all states 1, 2, …, *k* should be united in one absorbing state and all transitions from this absorbing state to any other states should be constrained. It means that in matrix **A** of the revealed process all elements in rows with numbers equal or lower than *k* should be zeroed. We designate the matrix as $\mathbf{A_0}$

$$\mathbf{A_0} = \begin{bmatrix} 0 & 0 & \ldots & 0 \\ 0 & 0 & \ldots & 0 \\ \ldots & \ldots & \ldots & \ldots \\ 0 & 0 & \ldots & 0 \\ a_{k+1,1} & a_{k+1,2} & \ldots & a_{k+1,K} \\ \ldots & \ldots & \ldots & \ldots \\ a_{K,1} & a_{K,2} & \ldots & a_{K,K} \end{bmatrix}. \tag{6.54}$$

Reliability function *R*(*t*) may be found after solving the following system of differential equations in matrix notation

$$\frac{\mathbf{dp}(t)}{dt} = \mathbf{p}(t)\mathbf{A}_0, \tag{6.55}$$

where

$\mathbf{p}(t) = [p_1(t), p_2(t), \ldots, p_K(t)]$ is row-vector of state probabilities,

$\frac{\mathbf{dp}(t)}{dt} = \left[\frac{dp_1(t)}{dt}, \frac{dp_2(t)}{dt}, \ldots, \frac{dp_K(t)}{dt}\right]$ is row-vector of state probabilities' derivatives.

Then the reliability function can be obtained

$$R(t) = \sum_{i=k}^{K} p_i(t). \tag{6.56}$$

Based on the reliability function we can obtain the mean time to failure (MTTF) as time up to the first entrance into united absorbing state

$$\text{MTTF} = \int_0^\infty R(t)\text{d}t. \tag{6.57}$$

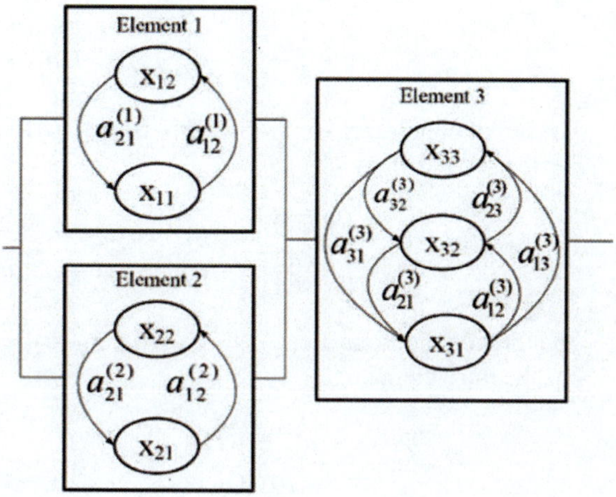

Fig. 6.2 MSS's Structure and State-transition Diagram of its Elements

6.3.4 *Illustrative Example*

Consider an MSS, which consists of three multi-state elements. The MSS's structure, state-transition diagrams of the elements, and the corresponding performance levels are presented in Fig. 6.2.

One can see that $G_1(t) = \{g_{11}, g_{12}\} = \{0,1.5\}$, $G_2(t) = \{g_{21}, g_{22}\} = \{0, 2\}$, $G_3(t) = \{g_{31}, g_{32}, g_{33}\} = \{0, 1.8, 4\}$. Failure rates and repair rates are the following:

$$a_{21}^{(1)} = 7\,\text{year}^{-1}, a_{12}^{(1)} = 100\,\text{year}^{-1}, a_{21}^{(2)} = 10\,\text{year}^{-1}, a_{12}^{(2)} = 80\,\text{year}^{-1},$$
$$a_{32}^{(3)} = 10\,\text{year}^{-1}, a_{31}^{(3)} = 3\,\text{year}^{-1}, a_{21}^{(3)} = 7\,\text{year}^{-1}, a_{13}^{(3)} = 0\,\text{year}^{-1},$$
$$a_{12}^{(3)} = 120\,\text{year}^{-1}, a_{23}^{(3)} = 110\,\text{year}^{-1}.$$

The MSS structure function is given by the expression

$$G(t) = f(G_1(t), G_2(t), G_3(t)) = \min\{G_1(t) + G_2(t), G_3(t)\}. \tag{6.58}$$

The system has to satisfy a constant demand $w_{\text{req}} = 1$. The system failure is treated as an output performance downgrading under this demand. The problem is to calculate the system availability and reliability for time period T as well as mean time to failure.

Solution Applying the procedure described above, we proceed as follows. According to the Markov method we build the systems of differential equations for each element (using the state-transitions diagrams presented in Fig. 6.2):

For the first element

$$\begin{cases} \frac{dp_{11}(t)}{dt} = -a_{12}^{(1)} p_{11}(t) + a_{21}^{(1)} p_{12}(t), \\ \frac{dp_{12}(t)}{dt} = -a_{21}^{(1)} p_{12}(t) + a_{12}^{(1)} p_{11}(t). \end{cases} \tag{6.59}$$

The initial conditions are: $\mathbf{p_{10}} = \{p_{11}(0), p_{12}(0)\} = \{0, 1\}$.
For the second element:

$$\begin{cases} \frac{dp_{21}(t)}{dt} = -a_{12}^{(2)} p_{21}(t) + a_{21}^{(2)} p_{12}(t), \\ \frac{dp_{22}(t)}{dt} = -a_{21}^{(2)} p_{22}(t) + a_{12}^{(2)} p_{21}(t). \end{cases} \tag{6.60}$$

The initial conditions are: $\mathbf{p_{20}} = \{p_{21}(0), p_{22}(0)\} = \{0, 1\}$.
For the third element:

$$\begin{cases} \frac{dp_{31}(t)}{dt} = -a_{12}^{(3)} p_{31}(t) + a_{21}^{(3)} p_{32}(t), \\ \frac{dp_{32}(t)}{dt} = a_{32}^{(3)} p_{33}(t) - (a_{21}^{(3)} + a_{23}^{(3)}) p_{32}(t) + a_{12}^{(3)} p_{31}(t), \\ \frac{dp_{33}(t)}{dt} = -a_{32}^{(3)} p_{33}(t) + a_{23}^{(3)} p_{32}(t). \end{cases} \tag{6.61}$$

The initial conditions are: $\mathbf{p_{30}} = \{p_{31}(0), p_{32}(0), p_{33}(0)\} = \{0, 0, 1\}$.

After numerical solution of these three systems of differential equations under corresponding initial conditions by using MATLAB, one obtains L_z-transforms for three processes:

Process $G_1(t) : \mathbf{g}_1 = \{g_{11}, g_{12}\} = \{0, 1.5\}$, $\mathbf{p}_1(t) = \{p_{11}(t), p_{12}(t)\}$, $\mathbf{p_{10}} = \{p_{11}(0), p_{12}(0)\} = \{0, 1\}$.

The associated L_z-transform: $L_z\{G_1(t)\} = \sum_{i=1}^{2} p_{1i}(t) z^{g_{1i}}$.

Process $G_2(t) : \mathbf{g}_2 = \{g_{21}, g_{22}\} = \{0, 2.0\}$, $\mathbf{p}_2(t) = \{p_{21}(t), p_{22}(t)\}$,

$$\mathbf{p_{20}} = \{p_{21}(0), p_{22}(0)\} = \{0, 1\}.$$

The associated L_z-transform: $L_z\{G_2(t)\} = \sum_{i=1}^{2} p_{2i}(t) z^{g_{2i}}$.

Process $G_3(t) : \mathbf{g}_3 = \{g_{31}, g_{32}, g_{33}\} = \{0, 1.8, 4.0\}$, $\mathbf{p}_3(t) = \{p_{31}(t), p_{32}(t), p_{33}(t)\}$,

$$\mathbf{p_{30}} = \{p_{31}(0), p_{32}(0), p_{33}(0)\} = \{0, 0, 1\}.$$

The associated L_z-transform: $L_z\{G_3(t)\} = \sum_{i=1}^{3} p_{3i}(t) z^{g_{3i}}$.

Now by using Ushakov's operator Ω_f over all L_z-transforms of individual elements we can obtain L_z-transform $L_z\{G(t)\}$ associated with output performance stochastic process $G(t)$ of the entire MSS:

$$L_z\{G(t)\} = \Omega_f(L_z\{G_1(t)\}, L_z\{G_2(t)\}, L_z\{G_3(t)\}), \tag{6.62}$$

where the system structure function is as follows:

$$G(t) = f(G_1(t), G_2(t), G_3(t)) = \min\{G_1(t) + G_2(t), G_3(t)\}. \tag{6.63}$$

Based on the known rules for series-parallel MSS [11], after like terms collection we finally obtain

$$L_z\{G(t)\} = \sum_{i=1}^{5} p_i(t) z^{g_i}, \tag{6.64}$$

where

$$\begin{aligned}
&g_1 = 0, \quad p_1(t) = p_{11}(t)p_{21}(t) + p_{31}(t)p_{12(t)} + p_{31}(t)p_{11}(t)p_{22}(t);\\
&g_2 = 1.5, \ p_2(t) = p_{12}(t)p_{21}(t)(p_{32}(t) + p_{33}(t));\\
&g_3 = 1.8, \ p_3(t) = p_{32}(t)p_{22}(t);\\
&g_4 = 2.0, \ p_4(t) = p_{33}(t)p_{11}(t)p_{22}(t);\\
&g_5 = 3.5, \ p_5(t) = p_{33}(t)p_{12}(t)p_{22}(t).
\end{aligned}$$

These two sets $\mathbf{g} = \{g_1, \ldots, g_5\} = \{0, 1.5, 1.8, 2.0, 3.5\}$ and $\mathbf{p}(t) = \{p_1(t), \ldots, p_5(t)\}$ define performance rates and states probabilities of output *non-Markov* performance stochastic process $G_1(t)$ for the entire MSS.

The failure is treated as the system performance degradation lower than $w_{\text{req}} = 1$. So, summarizing all probabilities with z-powers greater than or equal to 1, we obtain the MSS instantaneous availability $AV(t)$

$$AV(t) = \sum_{i=2}^{5} p_i(t) = 1 - p_1(t). \tag{6.65}$$

The calculated MSS instantaneous availability $AV(t)$ is presented in Fig. 6.3.

Now by using inverse L_z-transform the underlying approximating Markov process $G_{IM}(t)$ can be revealed.

As one can see from the obtained L_z-transform, the underlying approximating Markov output process has five states

$$\mathbf{g} = \{g_4, g_2, g_3, g_4, g_5\} = \{0, 1.5, 1.8, 2.0, 3.5\}.$$

The corresponding states probabilities are as follows

$$\mathbf{p}(t) = \{p_1(t), p_2(t), p_3(t), p_4(t), p_5(t)\}$$

The initial state is the best state with performance g_5.

In general case, states probabilities for five-state approximating Markov process can be obtained from the following system of ordinary differential equations:

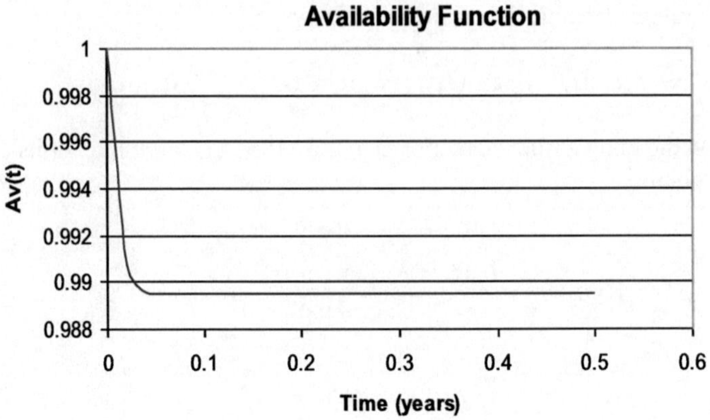

Fig. 6.3 MSS Instantaneous Availability

$$\begin{cases} \frac{dp_{1A}(t)}{dt} = -\left(\sum_{i=2}^{5} a_{1i}\right) p_{1A}(t) + \sum_{i=2}^{5} a_{i1} p_{iA}(t), \\ \frac{dp_{2A}(t)}{dt} = -\left(\sum_{i=1,i\neq 2}^{5} a_{2i}\right) p_{2A}(t) + \sum_{i=1,i\neq 2}^{5} a_{i2} p_{iA}(t), \\ \frac{dp_{3A}(t)}{dt} = -\left(\sum_{i=1,i\neq 3}^{5} a_{3i}\right) p_{3A}(t) + \sum_{i=1,i\neq 3}^{5} a_{i3} p_{iA}(t), \\ \frac{dp_{4A}(t)}{dt} = -\left(\sum_{i=1,i\neq 4}^{5} a_{4i}\right) p_{4A}(t) + \sum_{i=1,i\neq 4}^{5} a_{i4} p_{iA}(t), \\ \frac{dp_{5A}(t)}{dt} = -\left(\sum_{i=1}^{4} a_{5i}\right) p_{5A}(t) + \sum_{i=1}^{4} a_{i5} p_{iA}(t). \end{cases} \tag{6.66}$$

Under the given initial conditions: $p_{1A}(0) = p_{2A}(0) = p_{3A}(0) = p_{4A}(0) = 0$, $p_{5A}(0) = 1$. (Symbol A means "approximating").

The solution of this system $p_{1A}(t), p_{2A}(t), p_{3A}(t), p_{4A}(t), p_{5A}(t)$ should be numerically closed to probabilities $p_1(t), p_2(t), p_3(t), p_4(t), p_5(t)$ that were found above by using L_Z-transform for non-Markov process $G_I(t)$. In accordance with the GA procedure, transition intensities a_{ij} should be found in order to minimize the following error for five-state Markov process:

$$\text{Err} = \sum_{i=1}^{N_p} \frac{|p_{1A}(t_i) - p_1(t_i)|}{p_1(t_i)} + \ldots + \sum_{i=1}^{N_p} \frac{|p_{5A}(t_i) - p_5(t_i)|}{p_5(t_i)}. \tag{6.67}$$

In each GA cycle in this example the solution $p_{1A}(t), p_{2A}(t), p_{3A}(t), p_{4A}(t), p_{5A}(t)$ should be obtained for the period of 0.15 year. During 0.15 year, the transient mode for the solution $p_1(t), p_2(t), p_3(t), p_4(t), p_5(t)$

will be completely over (finished) and the process will be in steady-state. For the comparison, we shall take 1000 points—one point for 0.00015 year.

The population size in the GA is 100. The offspring will mutate with probability, which avoids premature convergence to a local optimum and facilities jumps in the solution space. The mutation probability is 0.005. The convergence criterion in the proposed GA is set as satisfying both a minimal number of genetic cycles (500 cycles) and a number of genetic cycles without improving the solution performance (50 cycles). The GA converges to optimal solutions by performing about 700 genetic cycles.

So, the underlying Markov process $G_{IM}(t)$ was completely revealed:

$$G(t) = \{\mathbf{g}, \mathbf{A}, \mathbf{p}_0\},$$

where

$$\mathbf{g} = \{g_1, \ldots, g_5\} = \{0, 1.5, 1.8, 2.0, 3.5\},$$

$$\mathbf{A} = \begin{bmatrix} -295 & 95 & 120 & 80 & 0 \\ 10.63 & -157.51 & 96.88 & 0 & 50.0 \\ 8.0 & 50.0 & -299.09 & 191.09 & 50.0 \\ 10.07 & 0 & 10.75 & -408.32 & 387.5 \\ 0 & 11.31 & 12.75 & 7 & -31.06 \end{bmatrix}, \tag{6.68}$$

$$\mathbf{p}_0 = \{0, 0, 0, 0, 1\}.$$

In order to find the reliability function $R(t)$, all transitions from worst state 1 should be constrained. It means that all a_{1i}, $i = 2, \ldots, K$ should be zeroed. Therefore, we will have the matrix

$$\mathbf{A}_0 = \begin{bmatrix} 0 & 0 & 0 & 0 & 0 \\ 10.63 & -157.51 & 96.88 & 0 & 50.0 \\ 8.0 & 50.0 & -299.09 & 191.09 & 50.0 \\ 10.07 & 0 & 10.75 & -408.32 & 387.5 \\ 0 & 11.31 & 12.75 & 7 & -31.06 \end{bmatrix}. \tag{6.69}$$

Therefore, reliability function may be obtained as follows:

$$R(t) = \sum_{i=2}^{5} p_i(t), \tag{6.70}$$

where functions $p_i(t)$ are obtained by solving the system of ordinary differential equations

$$\frac{d\mathbf{p}(t)}{dt} = \mathbf{p}(t)\mathbf{A}_0 \tag{6.71}$$

under initial conditions $\mathbf{p_0} = \{0, 0, 0, 0, 1\}$.

For the case that was considered in this methodical example, calculation may be done by using straightforward Markov method, because the number of states is not great.

In Fig. 6.4 one can see graphs of reliability functions, calculated by using inverse L_z-transform and by a conventional straightforward Markov method, which is presented in [11].

As one can see the reliability curves representing those two solutions are positioned so close together that the difference between them cannot be distinguished visually. (The difference is in the fourth digit after the decimal point).

Now mean time to failure can be obtained

$$\text{MTTF}_L = \int_0^\infty R_L(t)dt = 0.569\,\text{years},$$

where

$R_L(t)$ is reliability function that was computed by using inverse L_z-transform,

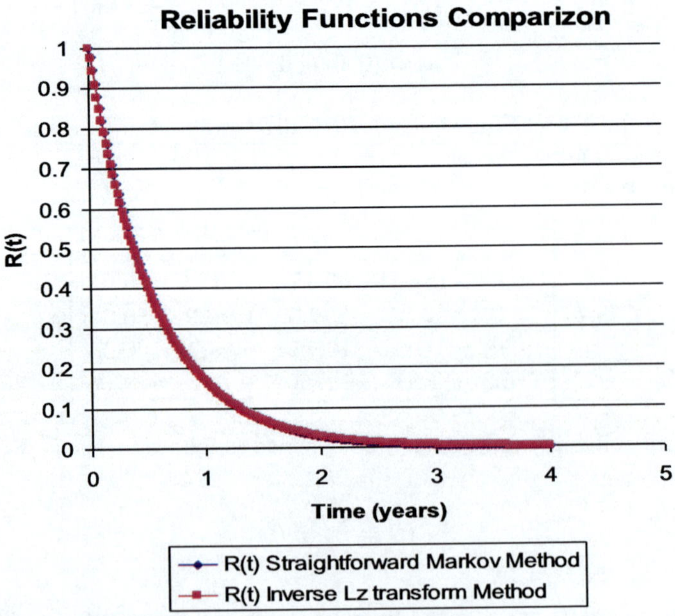

Fig. 6.4 Graphs of Reliability Functions calculated by using Straightforward Markov Method and by using Inverse L_z-transform

$MTTF_L$ is mean time to failure that was computed by using inverse L_z-transform

Notice that mean time to failure $MTTF_M$ calculated by using straightforward Markov method is almost the same

$$\mathrm{MTTF}_M = \int_0^\infty R_M(t)dt = 0.568\,\text{years}.$$

The error for $MTTF_L$ calculation is less than 0.2% which is a very good accuracy for reliability problems where high uncertainty in failure data is usually expected.

It should be noticed that resulting output stochastic process was found by using L_z-transform after like term collection. Approximating Markov process was revealed by inverse L_z-transform and has only five-state Markov process, which was built by using straightforward Markov method in (Lisninaski et al. 2010) in order to perform reliability analysis and for this example has 12 states. Therefore, even for this simple methodological example the computational complexity decreases when the suggested method is used.

6.4 Application Experience

Based on L_z-transform method wide range of MSS dynamic performability assessment and optimization problems can be solved. In [16], L_z-transform method was applied in order to find an optimal age replacement policy for MSS. By using L_z-transform the state probabilities were computed for the entire MSS and the expected cost and profit functions were derived. Finally, cost minimization or profit maximization policy was determined. Such practical problem as availability assessment for aging refrigeration system for the big supermarket was solved in [4]. The system is enough complex—it has 2048 states and its analysis by using classical straightforward Markov methods or simulation is very difficult. The problem was solved by using L_z-transform method and it has been proven that in order to provide the required reliability level, replacement of aging mechanical part in compressors should be performed after 9.5 years.

More complicated system was considered by Frenkel et al. [3], where instantaneous mean cooling performance was assessed for the water-cooling system of magnetic resonance imaging medical equipment. The system has 3840 states and the problem solving without using L_z-transform method is impossible in practice. In [5], availability was evaluated for this water-cooling system.

Short-term reliability evaluation for power station by using L_z-transform method was presented in [9]. Power station consisting of some power generating units is naturally a MSS, because every unit is multi-state element. Evaluation of such important reliability indices as power system availability, expected capacity deficiency, the expected energy not supplied to consumers is considered. Based on these short-term reliability indices in this paper corresponding operative decisions for units' dispatch were suggested.

Based on L_z-transform in [12], the method for Birnbaum importance assessment of aging MSS was suggested. Dynamic Birnbaum importance evaluation for aging MSS is especially important, because the relative importance of MSS's components is changing over the time and strongly depends on system demand. Reliability function evaluation based on inverse L_z-transform was considered in [19] for power system. In this paper the risk function for real-world power station has been computed. Based on this units' dispatch was suggested.

6.5 Conclusions

In this chapter performability measures for dynamic multi-state systems were considered. It was shown that L_z-transform and inverse L_z-transform can be successfully applied to dynamic performability analysis of multi-state system. It was demonstrated that the L_z-transform method is well formalized and a suitable tool for practical application in performability engineering for real-world MSSs analysis. By using this method, it is possible to overcome the main obstacle in a problem of performability assessment of dynamic MSS—a huge number of system's states (dimension curse).

L_z-transform is not a universal generating function; it is a new special mathematical object. Based on L_z-transform it is possible to utilize Ushakov's universal generating operator in order to perform performability analysis for MSS in transient modes where initial conditions have a great impact on performability, for aging MSS, MSS under stochastic demand and so on.

The method provides drastic decrease of computational burden compared with straightforward Markov method and Monte-Carlo simulation. Its application is especially effective for MSSs with complex structure function and many redundant elements, which have many equal performance levels. In the chapter a brief overview of successful applications of L_z-transform method to performability analysis of real-world industrial systems is presented.

References

1. Aven, T., & Jensen, U. (1999). *Stochastic models in reliability*. NY: Springer-Verlag.
2. Billinton, R., & Allan, R. (1996). *Reliability evaluation of power systems*. New York: Plenum Press.
3. Frenkel, I., et al. (2013). Assessing water cooling system performance: L_Z-transform method. *Proceedings of 2013 International Conference on Availability, Reliability and Security*. IEEE. doi:10.1109 ARES 2013.97, pp. 737–742.
4. Frenkel, I., et al. (2012). Availability assessment for aging refrigeration system by using *Lz*-transform. *Journal of Reliability and Statistical Studies, 5*(2), 33–43.
5. Frenkel, I., et al. (2014). On the L_Z-transform application for availability assessment of an aging multi-state water cooling system for medical equipment. In I. Frenkel, A. Lisnianski, A.

Karagrigoriou, & A. Kleiner (Eds.), *Applied reliability and risk analysis: Probabilistic models and statistical inference* (pp. 60–77). New York: Wiley.
6. Kemeny, J., & Snell, J. (1960). *Finite Markov Chains*. New York: Van Nostrand.
7. Levitin, G. (2005). *The universal generating function in reliability analysis and optimization*. London: Springer.
8. Lisnianski, A. (2012). L_Z—transform for a discrete-state continuous-time markov process and its applications to multi-state system reliability. In: A. Lisnianski, & I. Frenkel (Eds.). *Recent advances in system reliability. Signatures, multi-state systems and statistical inference* (pp. 79–96). London: Springer.
9. Lisnianski, A., & Ben Haim, H. (2013). Short-term reliability evaluation for power stations by using L_Z-transform. *Journal of Modern Power System and Clean Energy, 1*(2), 110–117.
10. Lisnianski, A., & Ding, Y. (2014). Inverse L_Z-transform for a discrete-state continuous-time marrkov process and its application to multi-state system reliability analysis. In: I. Frenkel, A. Lisnianski, A. Karagrigoriou, & Kleiner A. (Eds.). *Applied reliability and risk analysis: probabilistic models and statistical inference* (pp. 43–58). New York: Wiley.
11. Lisnianski, A., Frenkel, I., & Ding, Y. (2010). *Multi-state system reliability analysis and optimization for engineers and industrial managers*. London: Springer.
12. Lisnianski, A., Frenkel, I., & Khvatskin, L. (2015). On birnbaum importance assessment for aging multi-state system under minimal repair by using L_Z-transform method. *Reliability Engineering and System Safety, 142*, 258–266.
13. Meyer, J. (1980). On evaluating the performability of degradable computing systems. *IEEE Transactions on Computers, 29*(8), 720–731.
14. Misra, K. (2008). *Performability engineering: An essential concept in the 21st century*. In: K. Misra (Ed.), *Handbook of performability engineering* (PP. 1–12). Springer.
15. Natvig, B. (2011). *Multistate systems reliability*. New York: Wiley, Theory with Applications.
16. Sheu, S.-H., & Zhang, Z. (2013). An optimal age replacement policy for multi-state systems. *IEEE Transactions on Reliability, 62*(3), 722–735.
17. Trivedi, K. (2002). *Probability and statistics with reliability, queuing and computer science applications*. New York: Wiley.
18. Ushakov, I. (1986). A universal generating function. *Soviet Journal of Computer and Systems Sciences, 24*, 37–49.
19. Lisnianski, A., & Ding, Y. Using inverse L_Z-transform for obtaining compact stochastic model of complex power station for short-term risk evaluation. *Reliability Engineering and System Safety*. Accepted for publication.
20. Young, K., Kapur, K. (1995 Oct). *Customer driven reliability: Models, testing and improvement*, FORD Robustness Reliability Symposium, Dearborn, Michigan.

Anatoly LISNIANSKI, Ph.D. received his M.Sc. degree in Electrical Engineering from the University of Information Technologies, Precision Mechanics and Optics, Sankt-Petersburg, Russia; and Ph.D. degree in Reliability in 1984 from Federal Scientific and Production Centre "Aurora" in Sankt-Petersburg. Up to 1989, he was a senior researcher at this Centre. Starting in 1991 he was working in the Reliability Department of the Israel Electric Corporation as an engineer and then as a senior engineering expert. He is also a scientific supervisor of Reliability and Risk Management Centre in Shamoon College of Engineering. He is specialized in reliability assessment and optimization for complex technical systems. He is the author and co-author of more than 200 journal papers, two books, some book chapters, and inventions in the field of reliability and applied statistics. He is a senior member of IEEE and a member of Israel Statistical Association.

Lina TEPER, Ph.D. received M.Sc. degree from Technion – Israel Institute of Technology in 2001 and Ph.D. in 2007. Working at RAFAEL from 2005 at Reliability Analysis Center – and from 2013 as Head of Department. She focused her work on the Reliability Mathematics, System Safety, Metrology and System Engineering. Along with the position of Chief Research Engineer she continues academic activity and published papers on Metrology and Reliability Engineering.

Chapter 7
On Modeling and Performability Evaluation of Time Varying Communication Networks

Sanjay K. Chaturvedi, Sieteng Soh, and Gaurav Khanna

7.1 Introduction

In the past few years, networks with dynamic connectivity have gained research interests from engineering community and computer scientists. The widespread interest in the domain led to the development of fixed infrastructure-less wireless networks, like mobile ad hoc networks (MANETs), vehicular ad hoc networks (VANETs), flying ad hoc networks (FANETs) and delay-tolerant networks (DTNs) [1, 2]. We henceforth refer to all such networks as time varying communication networks (TVCNs) as their topologies change as a function of time. A topology change in a TVCN can be attributed to a variety of *intrinsic* (predictable and inherent) interruptions, such as node mobility, and/or *extrinsic* (unpredictable) interruptions, such as shadowing that occurs in wireless channel, and hardware failures [3]. Note that topology changes in these networks are not considered as an anomaly, rather regarded as an intrinsic feature [4]. Further, in general, TVCNs exhibit extremely long delays and show intermittent connectivity as nature of the system. More specifically, TVCNs seldom have *end-to-end* multi-hop paths between any node pair and utilize *device-to-device* communication via *store-carry-and-forward* mechanism for data transmission [5]. Thus, TVCNs may actually be disconnected at every time instant; however, data transmission can be made possible via routes available over time and space.

Primarily, TVCNs were designed to provide internet services in remote areas of developing countries [6], wildlife monitoring [7] and battlefield reconnaissance

S. K. Chaturvedi (✉) · G. Khanna
Subir Chowdhury School of Quality and Reliability, Indian Institute of Technology Kharagpur, Kharagpur, India
e-mail: skcrec@hijli.iitkgp.ac.in

S. Soh · G. Khanna
School of Electrical Engineering, Computing and Mathematical Sciences, Curtin University, Perth, Australia

K. B. Misra (ed.), *Handbook of Advanced Performability Engineering*,
https://doi.org/10.1007/978-3-030-55732-4_7

[1]. However, the realization of potential and opportunities of these networks led to more advanced networks like low earth orbiting (LEO) satellite networks, deep space networks (DSN), interplanetary networks (IPN) and the networks of unmanned aerial vehicles (UAVs) [8, 9]. Among these applications, movement of animals acting as *data ferry* during wildlife monitoring presents an instance of non-predictable TVCN (*np*TVCN). In contrast, LEO satellite networks and fleet of buses with predestined trajectories and schedules have somewhat predictable network dynamics over time [10]. Such TVCNs are referred to as *predictable* TVCNs (*p*TVCNs). Thus, specifically, a TVCN is an umbrella term depicting both *np*TVCN and *p*TVCN.

It is important to note here that although a plethora of works exists in the literature which deals with topology control [11], routing [12] and trace collection [1, 13] in TVCNs, however, assessment of their reliability is still rarely explored and needs considerable attention. One main reason for the deficient *state-of-the-art* works for the reliability evaluation of TVCNs is due to dynamically changing topology. Interested readers may refer [14] to learn about major differences between both static networks and TVCNs. Further, due to the dynamic topology changes, the definitions and notions of conventional minimal path/cut sets and spanning trees/arborescences developed for evaluating static networks' reliability become mostly inapplicable in their usual form in *p*TVCNs. Thus, there is a need of substantial modification and extension of these notions to *p*TVCNs as well.

Recently, the concept of minimal path set and cut set of static networks was extended to *p*TVCNs; see references [15, 16, 17]. Similarly, this chapter extends the concept of spanning arborescences to TVCNs, for data collection, that is, convergecasting. More specifically, the chapter presents two types of timestamped spanning arborescences, viz., timestamped *valid* spanning arborescences and timestamped *invalid* spanning arborescences, of TVCNs. A timestamped spanning arborescence is a spanning arborescence wherein each of its constituting edge accompanies a contact, representing its active time. Thus, a timestamped *valid* spanning arborescence, *aka* time-ordered spanning arborescence, is a timestamped spanning arborescence in which each edge is *time-ordered.* This means that in a *time-ordered* spanning arborescence traversal over edges is possible as we only move forward in time. On the contrary, a timestamped spanning arborescence is a timestamped *invalid* spanning arborescence if its edges are not time-ordered; thus, traversal over its edges is impossible. Note that here onwards we refer to terms minimal path/cut sets, spanning trees/arborescences, and timestamped (*valid*) spanning arborescences as paths/cuts, trees/arborescences, and timestamped (*valid*) arborescences, respectively. Further, note that we have used terms timestamped *valid* arborescence and time-ordered arborescence interchangeably throughout the chapter.

The chapter is organized as follows: Sect. 7.2 gives an overview of the existing modeling techniques to represent different features of TVCNs, that is, mobility, connectivity, data communication and topology. Section 7.3 reviews the available techniques for performability evaluation of *end-to-end* connected TVCNs. Section 7.4 discusses techniques developed for the performability evaluation of *device-to-device* connected *p*TVCNs. Section 7.5 presents a novel technique to enumerate *all* time-ordered arborescences, converging to each sink node. Section 7.6

presents two applications of the enumerated arborescences: (i) to enumerate all time-ordered paths between a specified source–destination, that is, (s, d) pair of nodes using all timestamped arborescences converging to sink node v_d, and (ii) to evaluate reliability metrics $R_c(v_d)$ and $R_c(K)$ using all time-ordered arborescences. The section also evaluates another reliability metric, viz., $R(s, d)$ using all time-ordered paths. Section 7.7 presents analysis of simulation results obtained from experiment over ten arbitrary TAGs. Finally, we draw the chapter conclusions and some future scope in Sect. 7.8.

7.2 Modeling TVCNs

This section presents some existing models for representing various features of TVCNs.

7.2.1 Mobility Models

A mobility model of TVCN aims to capture the time varying speed and direction of mobile nodes. Although literature [18, 19] shows many mobility models, yet there is no panacea. The main reason is the extremely complex and often non-deterministic nature of real-world mobility pattern of human beings, animals and/or vehicles on which wireless sensors are mounted, and the vast diversity of areas of application. A facile classification of mobility models would include: (i) trace-based mobility models and (ii) synthetic mobility models [14]. A trace-based mobility model is developed by monitoring and extracting features from the real movement patterns of users carrying mobile nodes; thus, represents reality. It is worth mentioning here that in the past, mostly data traces have been collected by deploying mobile devices in a small region, usually a university campus or a conference room. Further, note that the task of data trace collection requires a long period of time, say six months to one year, to collect a good amount of traces which are devoid of any biased data. On the contrary, a synthetic mobility model depicts randomly generated movements and creates synthetic traces. Note that a synthetic mobility model requires complex mathematical modeling, but it can be easily applied to an arbitrary number of nodes and over a large scale. Avid readers may refer [1, 13, 20, 21] and the references therein for detailed surveys on mobility models, software tools for synthetic mobility modeling and real-world trace repositories.

7.2.2 Connectivity Models

The traditional routing protocols designed for MANETs assumed existence of an *end-to-end* connectivity between (s, d) pair of nodes. However, in practice, such protocols fail to deliver data if an *end-to-end* connectivity/path is not found. The reason is because if a network gets partitioned, then the traditional ad hoc routing protocols fail to interconnect such partitions [12]. Moreover, with the increasing hop count between s and d, the end-to-end connected path tends to become unstable due to frequent disconnection of path caused by the movements of intermediate nodes [22]. Thus, the lack of continuous end-to-end connectivity between devices gave an impetus to the development of *device-to-device* connectivity for opportunistically routing data from one device to another via *store-carry-and-forward* mechanism. More specifically, the DTN group under Internet Research Task Force (IRTF) addressed the issues of intermittent connectivity and partitioned networks via their proposed store-carry-and-forward paradigm. In addition, the group also resolved the needs of the overlay architecture by using an addressing scheme that exploits the late binding of addresses [23]. It is worthy to note here that in store-carry-and-forward paradigm of data routing a next hop may not be immediately available to the current mobile device for forwarding data packets. Thus, it necessitates the current device to store data packets, maybe for a considerable duration, until it gets an opportunity to forward the packets to some other device. This renders transmissions between intermediate devices to be independent of each other. Therefore, the notion of store-carry-and-forward, accomplished by *device-to-device* communication, mitigates the effect of hop count to a large extent [22]. This chapter covers some works from the literature, in Sects. 7.3 and 7.4, which consider end-to-end connectivity and device-to-device communication, respectively, for TVCN performability evaluation.

7.2.3 Communication Models

The two primarily used communication models in TVCNs are *convergecasting* and *broadcasting*. Network convergecasting model is for collecting data from some (all) nodes toward a *sink/destination* node [24], *for example*, the reception of data gathered by LEO satellites from different sites across the world, including remote places which are inaccessible for ground-based data acquisition center, at ground station [25]. On the other hand, broadcast deals with data or information dissemination from a *root/source* node to all other nodes. Some instances which require broadcasting in pTVCNs are maneuvering, tracking, software update and maintenance of satellites from the Earth-based ground station [26]. It is worth mentioning here that both convergecasting and broadcasting models are equally important. However, most of the works in the literature focus on either of these, mainly because they both represent complementary problems. These models entail generation of efficient arborescence(s) which consume minimum energy/cost, ensure collision-free communication and/or provide maximum reliability, and so on. The recent works considering convergecast [27] and broadcast [28, 29] in *p*TVCNs only aim to find a single time-ordered arborescence satisfying the constraints on energy/cost and/or time. In contrast, Sect. 7.5 of this chapter presents in detail a novel approach to enumerate all time-ordered arborescences for convergecasting in *p*TVCNs.

7.2.4 Topology Models

TVCN topology modeling is crucial to understand the underlying dynamics between nodes and the overall performance of the network. Among other models presented in [14, 30], *random geometric graph* (RGG), *time-aggregated graph* (TAG), *space–time graph* and *line graph* have been excessively utilized by the researchers, and are of particular interest for this chapter. In the upcoming paragraph, we discuss salient features of each of them.

(1) Gilbert's model [31], commonly known as *random geometric graph* (RGG), is used to study the creation or snapping of edges in wireless networks, thereby their topology changes. The model considers that $\#V$ devices are placed uniformly at random in an area of $[0, 1]^2$; where V is a set of devices and $\#V$ represents the number of devices in set V. Among $\#V$ devices, any two devices in the network can communicate with each other if their Euclidean distance is less than or equal to the transmission range r of the devices. This model is widely utilized as a simplified topological model for wireless sensor networks and MANETs.
(2) *Time aggregated graph* (TAG) is used to model the changes in a spatio-temporal network, for example, road networks, over time by collecting the node/edge attributes into a set of time series [32]. Figure 7.1(i) shows a *p*TVCN with a period of four time slots. More specifically, G_1 to G_4 in Fig. 7.1(i) (a–d), respectively, represent four sequential snapshots of the network taken at four different slots of time viz., t_1, t_2, t_3 and t_4, where $t_1 < t_2 < t_3 < t_4$. Note that t_0 represents origin of the network evolution, and *slot length* is given by the difference between end time and start time of a slot. Thus, these snapshots

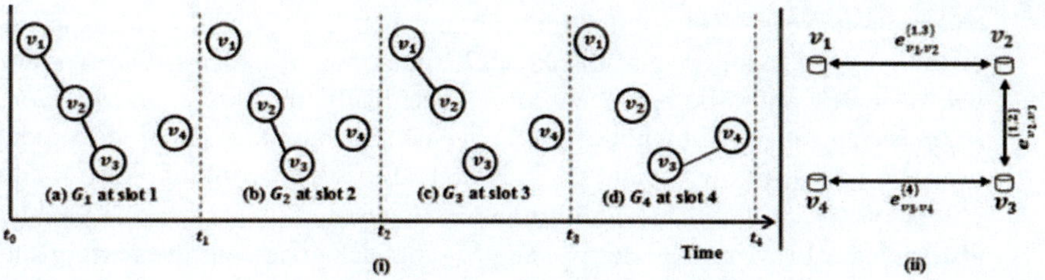

Fig. 7.1 (i) A sequence of four snapshots at four different time slots representing a *p*TVCN; (ii) TAG representation of the network evolution shown in (i)

are a sequence of static graphs representing interactions between four different nodes at four different instants of time. Figure 7.1(ii) shows the TAG representation of the network evolution depicted in Fig. 7.1(i). More specifically, in Fig. 7.1(ii) each edge accompanies a time-series representing edge activity schedule. For example, observe from Fig. 7.1(i) that edge between node pair (v_2,v_3) is active in slots 1 and 2. Figure 7.1(ii) shows the same bidirectional edges as $e^{\{1,2\}}_{v_2,v_3}$. Here, superscript denotes an ordered set $\{1, 2\}$ of timestamps associated with bidirectional edge e_{v_2,v_3}; thereby, represents edge's activity schedule. Thus, TAG model can compactly depict a *p*TVCN topology, while preserving all temporal information. Note that conventional *static* graph representation cannot effectively model time varying topology conditions, as it will obscure important temporal details of the network. For example, the static graph of the TAG shown in Fig. 7.1(ii) can be visualized by neglecting the contacts over each edge. Thus, from this static graph, one can wrongly adjudge that there exists a path between nodes (v_4,v_2) via node v_3, that is. $(e_{v_4,v_3}.e_{v_3,v_2})$. However, when we consider temporal information corresponding to each edge, such a path can never exist. This happens because here each edge accompanying a *single* timestamp, called as *timestamped edge* (TSE), will not lead to any time-ordered path aka *timestamped minimal path set* (TS-MPS) [17]. For example, timestamped path $(e^4_{v_4,v_3}.e^1_{v_3,v_2})$ is *invalid* with respect to time or can never exist as TSE $e^4_{v_4,v_3}$ is active in slot four; however, by that time its successive TSE $e^1_{v_3,v_2}$, having timestamp 1, ceases to exist, thereby making data transmission impossible between the specified pair of nodes. Thus, it is important to note that as we cannot traverse backwards in time, the time-order of interaction between nodes is critical in *p*TVCNs to facilitate a TS-MPS. Further, the notion of TSEs, along with TAG, effectively models *store-carry-and-forward* mechanism of data transmission between mobile devices. To illustrate this, consider a TS-MPS, viz., $(e^1_{v_1,v_2}.e^1_{v_2,v_3}.e^4_{v_3,v_4})$ between nodes (v_1,v_4). Observe that data can be transmitted from node v_1 to v_3 via node v_2 in slot 1. The received data is then stored at node v_3 in slots 2 and 3 and is later forwarded to node v_4 in slot 4. Similarly, we can model time-ordered arborescences converging (diverging) to (from) sink (source) in *p*TVCNs. Due to the effectiveness of TAG model, we utilize it later in Sect. 7.5.

(3) A *space–time graph* [11, 15] combines a sequence of static graphs as shown in Fig. 7.1 (i), to represent them as a directed graph defined in both spatial and temporal space. Thus, the model captures both space and time dimensions of the *p*TVCN topology and displays *all* time-ordered paths, between every pair of nodes, and arborescences. More specifically, to capture *transmission*, *reception* and *storage* at a node $v_i \in V$ in each time slot $t_x \in [1, \tau]$, the model uses two nodes, viz., $v_i^{t_x,\mathcal{T}}$ and $v_i^{t_x,\mathcal{R}}$. Note that τ is the period of *p*TVCN and $v_i^{t_x,\mathcal{T}}$ $(v_i^{t_x,\mathcal{R}})$ represents data *transmitting* (*receiving*) node v_i in slot t_x. The model also includes two virtual nodes v_i^0 and $v_i^{\tau+1}$ for each node v_i as the starting and ending points, respectively, of the time span. Further, two types of edges that

is spatial and temporal edges, are added in each layer of space–time graph; see dotted box for each time slot in Figs. 7.2 and 7.3.

A horizontal line, that is, temporal edge, either represents storage of packets at a node or a virtual edge connecting two consecutive time slots. On the other hand, a slanted line, that is, spatial edge, within a slot represents exchange of packets between two nodes at the given time instant. More specifically, in Fig. 7.2, a bidirectional horizontal edge, $\overleftrightarrow{v_i^{t_x,\mathcal{T}} v_i^{t_x,\mathcal{R}}}$ within slot t_x is a temporal edge, which represents storage of packets at node v_i in slot t_x. This bidirectional edge model is helpful in representing multi-hop communication amongst nodes because if the model uses unidirectional horizontal temporal edge $\overrightarrow{v_i^{t_x,\mathcal{T}} v_i^{t_x,\mathcal{R}}}$, then only one-hop transmission would be possible within any slot (see Fig. 7.3). The reason is that in the latter case any node cannot behave as a transmitter and receiver simultaneously. A horizontal edge between slots t_x and t_{x+1}, that is, a temporal

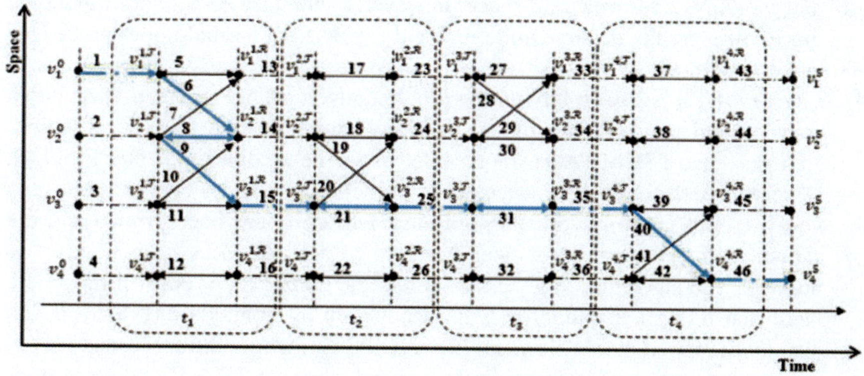

Fig. 7.2 Space–time graph, of *p*TVCN in Fig. 7.1(i), with multi-hop communication capability

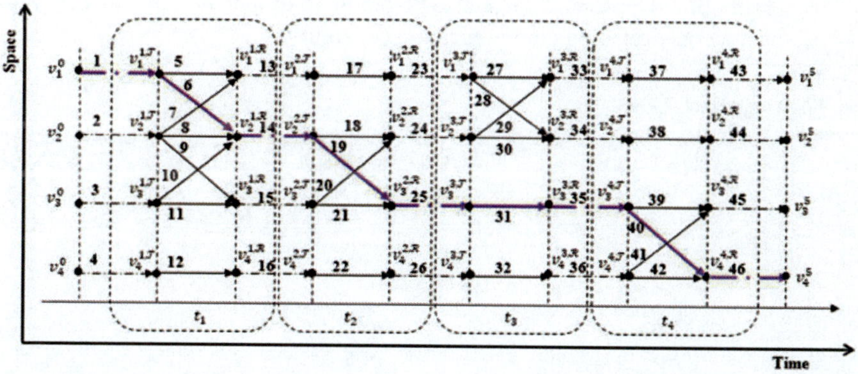

Fig. 7.3 Space–time graph, of *p*TVCN in Fig. 7.1(i), with one-hop communication capability

edge $\overrightarrow{v_i^{t_x,\mathcal{R}} v_i^{t_{x+1},\mathcal{T}}}$ is a virtual edge connecting two consecutive time slots. A non-horizontal edge inside slot t_x is a spatial edge $\overrightarrow{v_i^{t_x,\mathcal{T}} v_k^{t_x,\mathcal{R}}}$, which represents forwarding of packet(s) from node v_i to its neighbor v_k in time slot t_x. For better understanding of this representation, let us again consider Fig. 7.1(ii) and observe the TS-MPS between nodes (v_1,v_4), viz., $(e^1_{v_1,v_2}.e^1_{v_2,v_3}.e^4_{v_3,v_4})$. Figure 7.2 has an equivalent path as (1–6–8–9–15–21–25–31–35–40–46), wherein edges 6, 9 and 40 correspond to TSEs $e^1_{v_1,v_2}$, $e^1_{v_2,v_3}$ and $e^4_{v_3,v_4}$, respectively. All other edges in the path are either virtual edges or edges representing data storage at node itself. In contrast, if only one-hop communication is possible within a slot, then Fig. 7.3 is used to model TS-MPS. For instance, see highlighted path (1–6–14–19–25–31–35–40–46) depicting $(e^1_{v_1,v_2}.e^2_{v_2,v_3}.e^4_{v_3,v_4})$ in Fig. 7.1(ii). The reader can verify by visual inspection that Fig. 7.1(ii) and Fig. 7.2 can represent both TS-MPS $(e^1_{v_1,v_2}.e^1_{v_2,v_3}.e^4_{v_3,v_4})$ and $(e^1_{v_1,v_2}.e^2_{v_2,v_3}.e^4_{v_3,v_4})$, while Fig. 7.3 cannot represent TS-MPS $(e^1_{v_1,v_2}.e^1_{v_2,v_3}.e^4_{v_3,v_4})$. Although space–time graphs can represent *p*TVCNs over time and space, however, as the size of network increases, space–time graphs become quite unwieldy with many virtual nodes and edges in the structure.

(4) *Line graph* [3] is a useful modeling tool to convert a TAG representation into a conventional static graph without loss of any temporal reachability information. Note that each TSE of TAG is represented as a node in a line graph (see Fig. 7.4). Observe that the line graph successfully models connectivity over time between nodes v_1 and v_4. It is worthy to note that leaf $e^3_{v_1,v_2}$ has been pruned and not shown in the diagram as it does not lead to any TS-MPS from node v_1 to node v_4. The algorithm to generate such a line graph can be seen from [3, 17]. Note that this graph model is helpful not only in enumerating TS-MPS, but also *timestamped minimal cut set* (TS-MCS) [17], whose failure leads to (s, d) pair disconnection. For example, in Fig. 7.4, the two TS-MPS between node pair (v_1,v_4) can be observed as $(e^1_{v_1,v_2}.e^1_{v_2,v_3}.e^4_{v_3,v_4})$ and $(e^1_{v_1,v_2}.e^2_{v_2,v_3}.e^4_{v_3,v_4})$, while the three TS-MCS are $e^1_{v_1,v_2}$, $e^4_{v_3,v_4}$ and $(e^1_{v_2,v_3}.e^2_{v_2,v_3})$. Further, without loss of generality, line graph model can also be modified to include time latency, that is, to show only one hop communication capability.

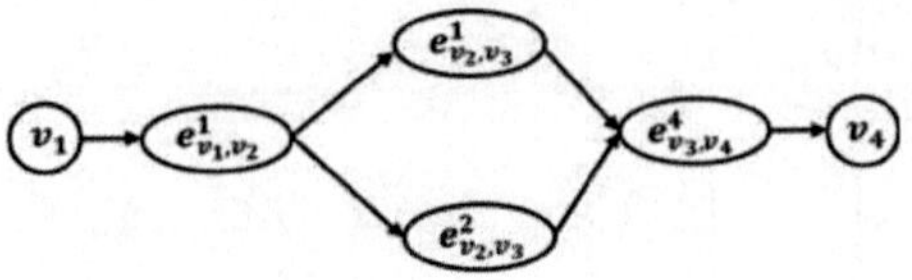

Fig. 7.4 Line graph of TAG shown in Fig. 7.1(ii)

7.3 Performability Evaluation: End-to-End Connected TVCNs

A MANET is a TVCN in which each node maintains a *routing* table. Using information in the routing table, a source node sends data to a destination node via some other nodes acting as routers, thereby ensuring *end-to-end* connectivity/communication [33]. References [34, 35] review various routing schemes developed for MANETs. However, in harsh environmental conditions it is difficult to maintain accurate routing tables owing to frequent disconnection of *end-to-end* path(s). More specifically, the reliability of MANETs, with end-to-end connectivity, decreases in case of low node density and/or high mobility. The reason is because low density and/or high moving speed of the nodes make it difficult to construct appropriate routing tables [33]. Thus, *reliability* assessment of TVCNs is an important aspect for their performance evaluation. The reliability of a TVCN is the success probability of a packet sent by a source node to be received by a destination node under variable topology conditions [36, 37]. Authors in [36, 38, 39] proposed a Monte-Carlo simulation-based approach to assess the reliability of MANETs. In [40], the authors proposed a critical node detection-based approach for the reliability evaluation of large-scale MANETs. In [41], the authors proposed to use *logistic regression* to evaluate the reliability of MANETs. A stochastic edge failure model was used in [42] for the reliability evaluation of wireless multi-hop networks. In [43], the authors utilized *universal generating function* to assess MANET reliability. However, the aforementioned works do not model store-carry-and-forward mechanism of data transmission. Thus, in general, the above works are not *viable* for most real-world applications of *p*TVCNs, which by design have sparse and intermittent connectivity, and use device-to-device communication methodology. Note that *p*TVCNs seldom have an end-to-end connectivity, and can provide an acceptable network reliability and throughput over time via store-carry-and-forward mechanism. In this context, next section discusses recently developed techniques for analyzing the performance of device-to-device connected TVCNs.

7.4 Performability Evaluation: Device-to-Device Connected TVCNs

As discussed in [12], device-to-device communication methodology increases network robustness in the presence of disruptions. Besides, the mechanism also decreases the impact of number of hops on data transmission [22]. Moreover, by using device-to-device communication it is possible to find opportunistic routes over time for information exchange. In [27], the authors utilized device-to-device communication to address topology control problem in network convergecasting applications. More specifically, the work in [27] uses space–time graph to model topology changes, and then presents three heuristic algorithms for constructing a sparse *p*TVCN by enumerating time-ordered minimum cost arborescence that also satisfies time delay constraint. Among the three polynomial time algorithms presented in [27], two are adaptations of well-known Kruskal's and Prim's algorithm [44] used to find a minimum cost tree in static networks. Similarly, topology control problem in [11] maintains cost-efficient and connected *p*TVCN topology for supporting data exchange between all pairs of nodes. In contrast, in [45], the authors considered arborescence packing problem in dynamic setting and presented an algorithm to find the desired number of edge-disjoint time-ordered arborescences in an acyclic temporal network. However, literature lack works that enumerate *all* converging and/or diverging time-ordered arborescences which can be utilized to evaluate the reliability of convergecasting and/or broadcasting in *p*TVCNs as is evaluated in static networks using arborescences. Authors in [15–17] present techniques for enumerating *all* TS-MPS and TS-MCS in *p*TVCNs. More specifically, they present three techniques (Cartesian product-based [15], connection matrix-based [16] and line graph-based [17]) to enumerate all TS-MPS and two techniques (TS-MPS inversion-based [15] and line graph-based [17]) to enumerate all TS-MCS between a specified (s, d) node pair. Later the enumerated TS-MPS/TS-MCS were used to evaluate two-terminal reliability of a *p*TVCN via sum-of-disjoint products (SDP) technique [46, 47]. Authors in [48] presented algorithms for finding shortest TS-MPS in a temporal network. Similarly, [49] presented algorithms for finding shortest, fastest and foremost TS-MPS in *p*TVCNs. Authors in [50] investigated utility of opportunistic store-carry-and-forward mechanism to conserve energy in multi-hop cellular networks, which integrate both cellular networks and device-to-device communication. In [3], the authors presented a new survivability framework for *p*TVCNs and extended the concept of maxFlow and minCut of static networks to *p*TVCNs. They also showed that Menger's theorem only conditionally holds in *p*TVCNs. The upcoming section of this chapter presents a novel algorithm for enumerating all time-ordered arborescences for network convergecasting.

7.5 Time-Ordered Arborescences Enumeration Method

Let TS-CA$_d$ (TS-VCA$_d$) be the set of all timestamped arborescences (timestamped valid arborescences) converging to sink node $v_d \in V$, where V represents set of nodes in a *p*TVCN. We use TS-CA$_d^i$ (TS-VCA$_d^i$) to denote the ith timestamped arborescence (timestamped valid arborescence) in TS-CA$_d$ (TS-VCA$_d$). Next, let TS-CA$_{All}$ (TS-VCA$_{All}$) be the set of all timestamped arborescences (timestamped valid arborescences) in a *p*TVCN. We use #TS-CA$_d$, #TS-VCA$_d$, #TS-CA$_{All}$ and #TS-VCA$_{All}$ to represent the number of arborescences in TS-CA$_d$, TS-VCA$_d$, TS-CA$_{All}$ and TS-VCA$_{All}$, respectively. It is worthy to note here that #TS-VCA$_d$ is no larger than #TS-CA$_d$, for each sink $v_d \in V$. This is because each set TS-VCA$_d$ is obtained from respective set TS-CA$_d$. Besides, the elements in both TS-CA$_d$ and TS-VCA$_d$ are mutually exclusive of the elements in TS-CA$_j$ and TS-VCA$_j$, respectively, where $v_d \neq v_j$ and $\{v_d, v_j\} \in V$. Thus, we have #TS-VCA$_{All} = \sum_{d=1}^{\#V}$ (#TS-VCA$_d$),

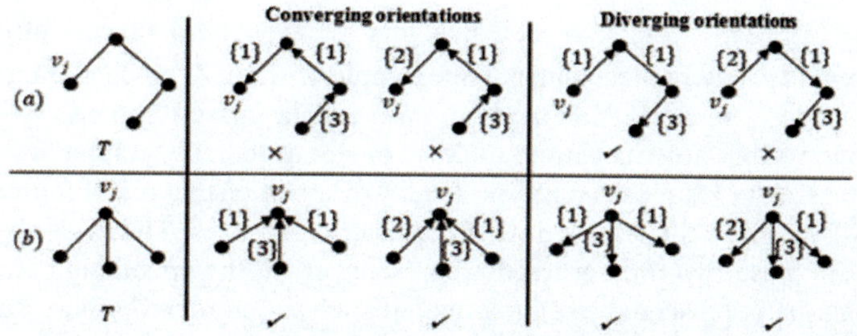

Fig. 7.5 An illustration of tree and corresponding timestamped arborescences

$\#\text{TS-CA}_{All} = \sum_{d=1}^{\#V} (\#\text{TS-CA}_d)$, and $\#\text{TS-VCA}_{All} \leq \#\text{TS-CA}_{All}$. For a given set of sink nodes $K \subseteq V$, a *p*TVCN is said to be *#K-connected* if it has at least one time-ordered arborescence, that is, $\#\text{TS-VCA}_d \geq 1$, for each sink node $v_d \in K$. Further, this chapter calls a *#K-connected p*TVCN as a *global-connected p*TVCN when $\#K = \#V$. In other words, a *p*TVCN can be considered as global-connected for data collection if it has at least one time-ordered arborescence for each node of the network acting as sink.

Let us apprehend with examples—why the conventional definition of arborescence needs modification for *p*TVCNs. Let T be a tree in a graph and v_j a vertex of T. It is well-known from graph theory [51] that T has exactly one orientation that is an arborescence converging to (diverging from) vertex v_j. However, in *p*TVCNs it may be or may not be true for arborescences due to the presence of TSEs. To understand this notion clearly, let us first consider the tree shown in Fig. 7.5a. If the timestamps on the three consecutive edges from vertex v_j are: ({1,2}-{1}-{3}), then as shown there are two timestamped arborescences converging to vertex v_j, that is, $\#\text{TS-CA}_j = 2$. However, none of them is time-ordered as the timestamps on edges directed toward sink vertex v_j are not in non-decreasing order, so, $\#\text{TS-VCA}_j = 0$.

Next, consider Fig. 7.5b, in which all edges are directly connected to vertex v_j. In this case, irrespective of the timestamps, both converging arborescences are valid; thus, $\#\text{TS-CA}_j = 2$ and $\#\text{TS-VCA}_j{}' = 2$. The inclusion of TSEs has similar effect on diverging orientation(s) as illustrated in Fig. 7.5. Therefore, unlike trees (arborescences) for undirected (directed) static networks, we can infer that some timestamped arborescences may not be time-ordered. Thus, although there may be a large number of timestamps over each edge of a TAG, they are not always fruitful for data collection and dissemination. Hence, careful selection and utilization of potential contacts can result in *p*TVCN topology optimization. Now, reconsider the TAG of Fig. 7.1(ii) and observe the arborescence $(e_{v_4,v_3}.e_{v_1,v_2}.e_{v_2,v_3})$ converging to sink node v_3. Here, each edge also has timestamps associated with it which indicate their activity schedule, for example, e_{v_4,v_3} is active in slot 4, e_{v_1,v_2} is active in slots 1 and 3, while e_{v_2,v_3} becomes active in slots 1 and 2. Thus, upon considering all combinations of timestamps associated with each comprising edge of $(e_{v_4,v_3}.e_{v_1,v_2}.e_{v_2,v_3})$, we will obtain four timestamped arborescences converging to sink node v_3, viz., $\text{TS-CA}_3^1 = (e^4_{v_4,v_3}.e^1_{v_1,v_2}.e^1_{v_2,v_3})$, $\text{TS-CA}_3^2 = (e^4_{v_4,v_3}.e^1_{v_1,v_2}.e^2_{v_2,v_3})$, $\text{TS-CA}_3^3 = (e^4_{v_4,v_3}.e^3_{v_1,v_2}.e^1_{v_2,v_3})$ and

TS-CA$_3^4$ = $(e^4_{v_4,v_3}.e^3_{v_1,v_2}.e^2_{v_2,v_3})$. Recall that this representation successfully depicts *store-carry-and-forward* mechanism. For example, in TS-CA$_3^2$ node v_1 transfers data packets to node v_2 in slot 1, then node v_2 carries data forward to node v_3 in slot 2. At last, node v_4 forwards its data to node v_3 in slot 4, thereby, completing the data collection at sink v_3 from each node of the network. Notice that out of four only two, that is, TS-CA$_3^1$ and TS-CA$_3^2$ are time-ordered arborescences. Thus, our objective is to enumerate these two time-ordered arborescences. In the upcoming paragraph, a novel technique is presented to enumerate all such time-ordered arborescences for each sink node $v_d \in V$ in a pTVCN.

The presented approach to enumerate all time-ordered arborescences adapts and extends Tutte's Matrix Tree Theorem [52], well-known for enumerating all arborescences of a static directed graph. The approach consists of three steps:

Step 1: Generate directed multigraph from TAG.
Step 2: Generate TS-CA$_{All}$ by using Tutte's Matrix Tree Theorem on directed multigraph obtained in Step 1.
Step 3: Generate TS-VCA$_{All}$ from TS-CA$_{All}$ by discarding all timestamped arborescences which are not time-ordered, that is, are invalid.

We illustrate the steps using the TAG of Fig. 7.1(ii). Step 1 expands each edge of TAG into multiple TSEs. For example, edge $e^{\{1,2\}}_{v_2,v_3}$ becomes $e^1_{v_2,v_3}$, $e^2_{v_2,v_3}$, $e^1_{v_3,v_2}$ and $e^2_{v_3,v_2}$. Similar transformation of other edges of TAG finally results in directed multigraph shown in Fig. 7.6.

Step 2 applies Tutte's Matrix-Tree Theorem [52] on the directed multigraph of Fig. 7.6 to enumerate all timestamped arborescences converging to sink node v_d, for all sink node $v_d \in V$. More specifically, Step 2 consists of two enumeration steps, which are:

(1) *Generate Laplace matrix L^- of the directed multigraph by filling it with indeterminates*: This is accomplished by filling the (i, i)- th entry of Laplace matrix L^- with variables representing the outgoing edges from i^{th} node. The remaining entries of L^- are completed by placing a negative sign and filling in the variables representing each edge from node v_i to v_j. Thus, we obtain

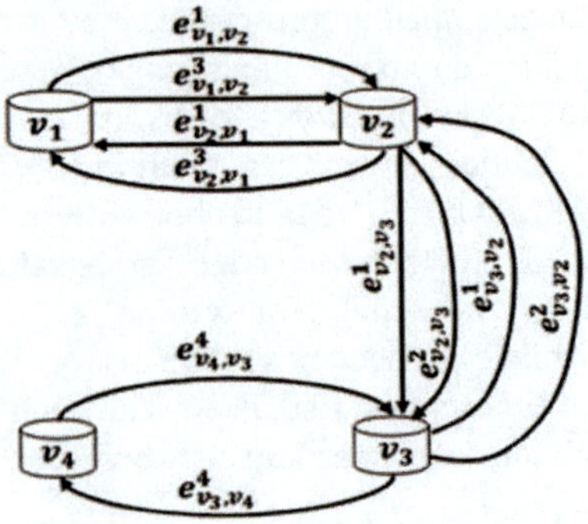

Fig. 7.6 Directed multigraph representation of TAG of Fig. 7.1(ii)

$$L^- = \begin{bmatrix} (e^1_{v_1,v_2}+e^3_{v_1,v_2}) & -(e^1_{v_1,v_2}+e^3_{v_1,v_2}) & 0 & 0 \\ -(e^1_{v_2,v_1}+e^3_{v_2,v_1}) & (e^1_{v_2,v_1}+e^3_{v_2,v_1}+e^1_{v_2,v_3}+e^2_{v_2,v_3}) & -(e^1_{v_2,v_3}+e^2_{v_2,v_3}) & 0 \\ 0 & -(e^1_{v_3,v_2}+e^2_{v_3,v_2}) & (e^1_{v_3,v_2}+e^2_{v_3,v_2}+e^4_{v_3,v_4}) & -e^4_{v_3,v_4} \\ 0 & 0 & -e^4_{v_4,v_3} & e^4_{v_4,v_3} \end{bmatrix}$$

(2) *Generate reduced Laplace matrix $\hat{L}^-_d$ for all sink node $v_d \in V$, and calculate their determinant*: The reduced Laplace matrix $\hat{L}^-_d$, for sink v_d, is obtained by deleting d^{th} row and column from L^-. The determinant of $\hat{L}^-_d$, that is, $\left|\hat{L}^-_d\right|$ results in a polynomial. Each non-vanishing monomial in this polynomial has coefficient one and corresponds to a timestamped arborescence converging to sink node v_d. For example, for $d = 1$, that is, v_1 as sink, the reduced Laplace matrix $\hat{L}^-_1$ is:

$$\hat{L}^-_1 = \begin{bmatrix} (e^1_{v_2,v_1}+e^3_{v_2,v_1}+e^1_{v_2,v_3}+e^2_{v_2,v_3}) & -(e^1_{v_2,v_3}+e^2_{v_2,v_3}) & 0 \\ -(e^1_{v_3,v_2}+e^2_{v_3,v_2}) & (e^1_{v_3,v_2}+e^2_{v_3,v_2}+e^4_{v_3,v_4}) & -e^4_{v_3,v_4} \\ 0 & -e^4_{v_4,v_3} & e^4_{v_4,v_3} \end{bmatrix}$$

Next, we calculate $\left|\hat{L}^-_1\right|$. Note that here vertical bars mean "determinant of".

$$\begin{aligned}
\left|\hat{L}^-_1\right| &= (e^1_{v_2,v_1}+e^3_{v_2,v_1}+e^1_{v_2,v_3}+e^2_{v_2,v_3}) \cdot \begin{vmatrix} (e^1_{v_3,v_2}+e^2_{v_3,v_2}+e^4_{v_3,v_4}) & -e^4_{v_3,v_4} \\ -e^4_{v_4,v_3} & e^4_{v_4,v_3} \end{vmatrix} \\
&\quad + (e^1_{v_2,v_3}+e^2_{v_2,v_3}) \cdot \begin{vmatrix} -(e^1_{v_3,v_2}+e^2_{v_3,v_2}) & -e^4_{v_3,v_4} \\ 0 & e^4_{v_4,v_3} \end{vmatrix} + 0 \cdot \\
&\quad \begin{vmatrix} -(e^1_{v_3,v_2}+e^2_{v_3,v_2}) & (e^1_{v_3,v_2}+e^2_{v_3,v_2}+e^4_{v_3,v_4}) \\ 0 & -e^4_{v_4,v_3} \end{vmatrix} \\
&= (e^1_{v_2,v_1}+e^3_{v_2,v_1}+e^1_{v_2,v_3}+e^2_{v_2,v_3})[e^1_{v_3,v_2}.e^4_{v_4,v_3} \\
&\quad + e^2_{v_3,v_2}\cdot e^4_{v_4,v_3} + e^4_{v_3,v_4}\cdot e^4_{v_4,v_3} - e^4_{v_4,v_3}\cdot e^4_{v_3,v_4}] \\
&\quad + (e^1_{v_2,v_3}+e^2_{v_2,v_3})\cdot[-e^1_{v_3,v_2}\cdot e^4_{v_4,v_3} - e^2_{v_3,v_2}\cdot e^4_{v_4,v_3}] \\
&= e^1_{v_3,v_2}\cdot e^4_{v_4,v_3}\cdot e^1_{v_2,v_1} \\
&\quad + e^2_{v_3,v_2}\cdot e^4_{v_4,v_3}\cdot e^1_{v_2,v_1} + e^1_{v_3,v_2}\cdot e^4_{v_4,v_3}\cdot e^3_{v_2,v_1} \\
&\quad + e^2_{v_3,v_2}\cdot e^4_{v_4,v_3}\cdot e^3_{v_2,v_1} + e^1_{v_3,v_2}\cdot e^4_{v_4,v_3}\cdot e^1_{v_2,v_3} \\
&\quad + e^2_{v_3,v_2}\cdot e^4_{v_4,v_3}\cdot e^1_{v_2,v_3} + e^1_{v_3,v_2}\cdot e^4_{v_4,v_3}\cdot e^2_{v_2,v_3} \\
&\quad + e^2_{v_3,v_2}\cdot e^4_{v_4,v_3}\cdot e^2_{v_2,v_3} - e^1_{v_3,v_2}\cdot e^4_{v_4,v_3}\cdot e^1_{v_2,v_3} \\
&\quad - e^1_{v_3,v_2}\cdot e^4_{v_4,v_3}\cdot e^2_{v_2,v_3} - e^2_{v_3,v_2}\cdot e^4_{v_4,v_3}\cdot e^1_{v_2,v_3} \\
&\quad - e^2_{v_3,v_2}\cdot e^4_{v_4,v_3}\cdot e^2_{v_2,v_3}.
\end{aligned}$$

Simplifying the above expression results in a polynomial with four monomials (see Eq. (7.1)).

$$\left|\hat{L}_1^-\right| = [e^4_{v_4,v_3}.e^1_{v_3,v_2}\cdot e^1_{v_2,v_1} + e^4_{v_4,v_3}.e^2_{v_3,v_2}\cdot e^1_{v_2,v_1} + e^4_{v_4,v_3}.e^1_{v_3,v_2}\cdot e^3_{v_2,v_1} + e^4_{v_4,v_3}\cdot e^2_{v_3,v_2}\cdot e^3_{v_2,v_1}]. \quad (7.1)$$

It can be easily verified that each monomial represents a timestamped arborescence converging to sink node v_1 and thus #TS-CA$_1 = 4$. Similarly, repeating Step 2 with all other nodes of the network, acting as sink, we obtain TS-CA$_{All}$. Table 7.1 collates TS-CA$_d$ for each sink $v_d \in V$. Therefore, #TS-CA$_{All}$ $= 16$ upon considering all four sink nodes.

In Step 3, each of the timestamped arborescence generated in Step 2 is assessed for their validity with respect to time. The step is implemented by Algorithm 1 which discards all timestamped arborescences that are not time-ordered. The details of Algorithm 1 are as follows.

In Line 1, Algorithm 1 creates an array TS-VCA$_{All}$ to store set TS-VCA$_d$ for all sink nodes $v_d \in V$. Lines 2-20, generate set TS-VCA$_d$ for each sink node $v_d \in V$. Line 3 creates an array TS-VCA$_d$ to store all time-ordered arborescences TS-VCA$_d^i$

Table 7.1 Timestamped converging arborescences of *p*TVCN shown in Fig. 7.1(ii) for different sink nodes

$\hat{L}_1^- = \begin{bmatrix} (e^1_{v_2,v_1}+e^3_{v_2,v_1}+e^1_{v_2,v_3}+e^2_{v_2,v_3}) & -(e^1_{v_2,v_3}+e^2_{v_2,v_3}) & 0 \\ -(e^1_{v_3,v_2}+e^2_{v_3,v_2}) & (e^1_{v_3,v_2}+e^2_{v_3,v_2}+e^4_{v_3,v_4}) & -e^4_{v_3,v_4} \\ 0 & -e^4_{v_4,v_3} & e^4_{v_4,v_3} \end{bmatrix}$
TS-CA$_1$ $= \{(e^4_{v_4,v_3}.e^1_{v_3,v_2}.e^1_{v_2,v_1}), (e^4_{v_4,v_3}.e^2_{v_3,v_2}.e^1_{v_2,v_1}), (e^4_{v_4,v_3}.e^1_{v_3,v_2}.e^3_{v_2,v_1}), (e^4_{v_4,v_3}.e^2_{v_3,v_2}.e^3_{v_2,v_1})\}$
$\hat{L}_2^- = \begin{bmatrix} (e^1_{v_1,v_2}+e^3_{v_1,v_2}) & 0 & 0 \\ 0 & (e^1_{v_3,v_2}+e^2_{v_3,v_2}+e^4_{v_3,v_4}) & -e^4_{v_3,v_4} \\ 0 & -e^4_{v_4,v_3} & e^4_{v_4,v_3} \end{bmatrix}$
TS-CA$_2$ $= \{(e^4_{v_4,v_3}.e^1_{v_3,v_2}.e^1_{v_1,v_2}), (e^4_{v_4,v_3}.e^1_{v_3,v_2}.e^3_{v_1,v_2}), (e^4_{v_4,v_3}.e^2_{v_3,v_2}.e^1_{v_1,v_2}), (e^4_{v_4,v_3}.e^2_{v_3,v_2}.e^3_{v_1,v_2})\}$
$\hat{L}_3^- = \begin{bmatrix} (e^1_{v_1,v_2}+e^3_{v_1,v_2}) & -(e^1_{v_1,v_2}+e^3_{v_1,v_2}) & 0 \\ -(e^1_{v_2,v_1}+e^3_{v_2,v_1}) & (e^1_{v_2,v_1}+e^3_{v_2,v_1}+e^1_{v_2,v_3}+e^2_{v_2,v_3}) & 0 \\ 0 & 0 & e^4_{v_4,v_3} \end{bmatrix}$
TS-CA$_3$ $=$ $\{\boxed{(e^4_{v_4,v_3}.e^1_{v_1,v_2}.e^1_{v_2,v_3})}, \boxed{(e^4_{v_4,v_3}.e^1_{v_1,v_2}.e^2_{v_2,v_3})}, (e^4_{v_4,v_3}.e^3_{v_1,v_2}.e^1_{v_2,v_3}), (e^4_{v_4,v_3}.e^3_{v_1,v_2}.e^2_{v_2,v_3})\}$
$\hat{L}_4^- = \begin{bmatrix} (e^1_{v_1,v_2}+e^3_{v_1,v_2}) & -(e^1_{v_1,v_2}+e^3_{v_1,v_2}) & 0 \\ -(e^1_{v_2,v_1}+e^3_{v_2,v_1}) & (e^1_{v_2,v_1}+e^3_{v_2,v_1}+e^1_{v_2,v_3}+e^2_{v_2,v_3}) & -(e^1_{v_2,v_3}+e^2_{v_2,v_3}) \\ 0 & -(e^1_{v_3,v_2}+e^2_{v_3,v_2}) & (e^1_{v_3,v_2}+e^2_{v_3,v_2}+e^4_{v_3,v_4}) \end{bmatrix}$
TS-CA$_4$ $=$ $\{\boxed{(e^1_{v_1,v_2}.e^1_{v_2,v_3}.e^4_{v_3,v_4})}, \boxed{(e^1_{v_1,v_2}.e^2_{v_2,v_3}.e^4_{v_3,v_4})}, (e^3_{v_1,v_2}.e^1_{v_2,v_3}.e^4_{v_3,v_4}), (e^3_{v_1,v_2}.e^2_{v_2,v_3}.e^4_{v_3,v_4})\}$

$\in$ TS-VCA$_d$. Lines 4-18 traverse over each timestamped arborescence TS-CA$_d^i$ $\in$ TS-CA$_d$ to check its validity with respect to time. Line 5 initializes a variable **notValid** with 'FALSE' to initially consider a TS-CA$_d^i$ to be valid. The validity of this TS-CA$_d^i$ is verified later by using **notValid**. The algorithm also defines an array **nodeTime**[1, 2,…, $\#V$] to fill it with the timestamp at which the starting node of each TSE in TS-CA$_d^i$ transmit data; see Lines 6-8. Note from Line 5 that initially each entry of **nodeTime** contains timestamp one more than the period of the pTVCN, that is, $(\tau + 1)$. For example, from Eq. (7.1), consider $(e^4_{v_4,v_3}.e^1_{v_3,v_2}.e^1_{v_2,v_1})$ converging to sink node v_1. For this timestamped arborescence, **nodeTime** will contain 4 at location 4, 1 at location 3 and 1 at location 2 corresponding to the starting node of each TSE. Note that here location corresponding to sink node v_1 will contain five as the period τ of pTVCN is four. This is done as sink node will not transmit any data to other nodes. Thus, here **nodeTime** = (5, 1, 1, 4). The content of **nodeTime** is utilized for discarding invalid TS-CA$_d^i$. More specifically, in Lines 9-14, Algorithm 1 iterates over each TSE of TS-CA$_d^i$ and checks whether all edges are time-ordered or not; see Line 10. If any violation is found, the algorithm sets **notValid** to TRUE and stops checking the remaining edges; see Lines 11–12. Thus, TS-CA$_d^i$ is not added to the TS-VCA$_d$. Otherwise, TS-CA$_d^i$ is time-ordered and is added to TS-VCA$_d$; see Lines 15-17. For previous example having **nodeTime** = (5, 1, 1, 4), for TSE $e^4_{v_4,v_3}$, $t = 4$ and **nodeTime** [3] = 1, and thus **nodeTime** [3] = 1 < t = 4. So, variable **notValid** becomes TRUE. It means that data transmitted by node v_4 to node v_3 at timestamp four cannot be further collected at node v_2 because node v_3 and v_2 were connected at timestamp one, prior to the activation of link between nodes v_4 and v_3. Thus, $(e^4_{v_4,v_3}.e^1_{v_3,v_2}.e^1_{v_2,v_1})$ is adjudged as invalid and is not added to TS-VCA$_1$. Notice that for sink node v_1 each timestamped arborescence is invalid, that is, #TS-VCA$_1$ = 0. Thus, to illustrate an example of a time-ordered arborescence, consider $(e^1_{v_1,v_2}.e^1_{v_2,v_3}.e^4_{v_3,v_4})$ converging to sink node v_4. Here, **nodeTime** = (1, 1, 4, 5). Observe that each TSE $e^1_{v_1,v_2}$, $e^1_{v_2,v_3}$ and $e^4_{v_3,v_4}$ is time-ordered, that is, does not satisfy the condition in Line 10; thus, variable **notValid** remains FALSE. Therefore, the timestamped arborescence is valid and added to TS-VCA$_4$. Note that Table 7.1 shows all time-ordered arborescence in each TS-VCA$_d$, $\forall\, v_d \in V$, encapsulated inside box. So, we have #TS-VCA$_2$ = 0, #TS-VCA$_3$ = 2 and #TS-VCA$_4$ = 2. Thus, in this example, #TS-VCA$_{All}$ = 4. Note that as there is no time-ordered arborescence converging to sink node v_1 and v_2 in TAG of Fig. 7.1(ii), this implies that nodes v_1 and v_2 cannot collect data from all other nodes. Thus, the network is not *global-connected* for data collection.

Algorithm 1

Input: TS-CA$_{All}$; **Output:** TS-VCA$_{All}$

```
TS-VCA_All ← { }
for (d ← 1 to #V)
  TS-VCA_d ← { }
  for (i ← 1 to #TS-CA_d)
    Set notValid ← FALSE and initialize each entry in nodeTime[1, ..., #V] to (τ + 1)
    for each e^t_{v_a,v_b} ∈ TS-CA^i_d
      nodeTime[a] ← t
    end for
    for each e^t_{v_a,v_b} ∈ TS-CA^i_d
      if (nodeTime[b] < t)
        notValid ← TRUE
        break
      end if
    end for
    if (notValid = FALSE)
      Insert TS-CA^i_d into TS-VCA_d
    end if
  end for
  Insert TS-VCA_d into TS-VCA_All
end for
```

7.6 Applications of Enumerated Timestamped Spanning Arborescences

This section shows how the enumerated timestamped arborescences converging to sink node v_d can be used to (i) enumerate all (*s*, *d*) TS-MPS, and (ii) evaluate reliability metrics $R_c(v_d)$ and $R_c(K)$. Then, it shows how to evaluate $R(s, d)$ using all (*s*, *d*) TS-MPS enumerated in (i).

7.6.1 *All (s, d) TS-MPS Enumeration from* TS-CA$_d$

This section presents an algorithm to enumerate a set of all TS-MPS, denoted by TS-MPS$_{s,d}$, between a specified node pair (*s*, *d*) of a *p*TVCN from set TS-CA$_d$ for node $v_s \neq v_d$ and $\{v_s, v_d\} \in V$. Note that by definition, there is a timestamped path from a source node v_s to sink node v_d in a timestamped arborescence TS-CA^i_d ∈ TS-CA$_d$. Further, each TS-CA^i_d contains at most one TS-MPS from source v_s to destination node v_d. Let TS-MPS$^i_{s,d}$ be the TS-MPS generated from TS-CA^i_d. It is important to note here that the set TS-VCA$_d$ should not be used for generating set TS-MPS$_{s,d}$. The reason is because all TS-CA$_d^i$ ∈ TS-CA$_d$ may contain at least one TS-MPS between node pair (s, d), while there may be zero TS-VCA$_d^i$ ∈ TS-VCA$_d$; thereby, making it impossible to generate any TS-MPS from TS-VCA$_d$. Next, we present Proposition 7.1, which is used by our proposed all (s, d) TS-MPS enumeration algorithm.

Proposition 7.1 A TS-MPS$_{s,d}^i$ ∈ TS-MPS$_{s,d}$ can be generated from TS-CA$_d^i$ ∈ TS-CA$_d$ with time complexity of $O(\#V)$.

Proof: We need two main steps to find a TS-MPS$_{s,d}^i$ from TS-CA$_d^i$: (i) search for source node v_s among $\#V$ nodes in TS-CA$_d^i$, and (ii) check if all TSE(s) of the path from source v_s to destination node v_d are time-ordered. Using a linear search, one can find node v_s for step (i) in $O(\#V)$. For step (ii), in the worst case, the path contains at most $(\#V - 1)$ number of TSEs. Thus, time-ordering of TSEs can be validated by traversing each TSE from v_s to v_d in TS-CA$_d^i$, which requires $O(\#V)$. Hence, the time complexity to generate TS-MPS$_{s,d}^i$ from TS-CA$_d^i$ is $O(\#V + \#V) = O(\#V)$. **Q.E.D.**

Note that a TS-CA$_d^i$ may not contain TS-MPS$_{s,d}^i$ when the path from v_s to v_d is not time-ordered. For example, consider generating TS-MPS$_{1,3}$ from the four timestamped arborescences in TS-CA$_3$, shown in Table 7.1, that is, TS-CA$_3^1$ = $(e^4_{v_4,v_3}.e^1_{v_1,v_2}.e^1_{v_2,v_3})$, TS-CA$_3^2$ = $(e^4_{v_4,v_3}.e^1_{v_1,v_2}.e^2_{v_2,v_3})$, TS-CA$_3^3$ = $(e^4_{v_4,v_3}.e^3_{v_1,v_2}.e^1_{v_2,v_3})$ and TS-CA$_3^4$ = $(e^4_{v_4,v_3}.e^3_{v_1,v_2}.e^2_{v_2,v_3})$. The last two arborescences render invalid paths $(e^3_{v_1,v_2}.e^1_{v_2,v_3})$ and $(e^3_{v_1,v_2}.e^2_{v_2,v_3})$, respectively. Thus, one cannot generate TS-MPS$_{1,3}^3$ and TS-MPS$_{1,3}^4$ from TS-CA$_3^3$ and TS-CA$_3^4$, respectively. On the other hand, we have TS-MPS$_{1,3}^2$ = $(e^1_{v_1,v_2}.e^2_{v_2,v_3})$ because TSE $e^1_{v_1,v_2}$ appears earlier than TSE $e^2_{v_2,v_3}$. More specifically, as there are two TS-MPS between node pair (1, 3), viz., TS-MPS$_{1,3}^1$ = $(e^1_{v_1,v_2}.e^1_{v_2,v_3})$ and TS-MPS$_{1,3}^2$ = $(e^1_{v_1,v_2}.e^2_{v_2,v_3})$; thus, #TS-MPS$_{1,3}$ = 2.

Proposition 7.1 can be used to find all TS-MPS$_{s,d}^i$ ∈ TS-MPS$_{s,d}$ using all TS-CA$_d^i$ ∈ TS-CA$_d$. More specifically, we present the following Algorithm 2 to enumerate, for a given (s, d) node pair, all TS-MPS$_{s,d}^i$ ∈ TS-MPS$_{s,d}$ using all TS-CA$_d^i$ ∈ TS-CA$_d$.

Algorithm 2

Input: TS-CA$_d$, source node v_s; **Output:** TS-MPS$_{s,d}$

1. TS-MPS$_{s,d}$ ← { }
2. **for** each TS-CA$_d^i$ ∈ TS-CA$_d$
3.   **if** TS-CA$_d^i$ contains TS-MPS$_{s,d}^i$ from v_s to v_d
4.   **if** TS-MPS$_{s,d}^i$ ∉ TS-MPS$_{s,d}$
5.   Insert TS-MPS$_{s,d}^i$ into TS-MPS$_{s,d}$
6.   **end if**
7.   **end if**
8. **end for**

In Line 1, Algorithm 2 creates an empty set TS-MPS$_{s,d}$ to store all (s, d) TS-MPS. Lines 2-8 iterate over each TS-CA$_d^i$ ∈ TS-CA$_d$ in order to find all (s, d) TS-MPS. Line 3 uses Proposition 7.1 to find a (s, d) TS-MPS from each TS-CA$_d^i$.

It is worth mentioning here that there can be many duplicate TS-MPS, which may appear from different $\text{TS-CA}_d^i \in \text{TS-CA}_d$. For example, consider generating TS-MPS between node pair (2, 4) from arborescences $\text{TS-CA}_4^1 = (e^1_{v_1,v_2}.e^1_{v_2,v_3}.e^4_{v_3,v_4})$ and $\text{TS-CA}_4^3 = (e^3_{v_1,v_2}.e^1_{v_2,v_3}.e^4_{v_3,v_4})$ converging to sink node v_4; see Table 7.1. Then, Line 3 would generate $\text{TS-MPS}^1_{2,4} = (e^1_{v_2,v_3}.e^4_{v_3,v_4})$ and $\text{TS-MPS}^3_{2,4} = (e^1_{v_2,v_3}.e^4_{v_3,v_4})$, respectively, from the two arborescences. Notice that the generated TS-MPS are identical to each other, that is, are duplicate. Line 4 checks for duplicating TS-MPS, and Line 5 puts each unique TS-MPS into $\text{TS-MPS}_{s,d}$. We use the following illustration to show steps of the Algorithm 2.

Illustration Let us find all (3, 1) TS-MPS for the TAG shown in Fig. 7.1(ii). Here sink node is v_1, while source node is v_3. From Table 7.1, $\#\text{TS-CA}_1 = 4$, viz., $\text{TS-CA}_1^1 = (e^4_{v_4,v_3}.e^1_{v_3,v_2}.e^1_{v_2,v_1})$, $\text{TS-CA}_1^2 = (e^4_{v_4,v_3}.e^2_{v_3,v_2}.e^1_{v_2,v_1})$, $\text{TS-CA}_1^3 = (e^4_{v_4,v_3}.e^1_{v_3,v_2}.e^3_{v_2,v_1})$ and $\text{TS-CA}_1^4 = (e^4_{v_4,v_3}.e^2_{v_3,v_2}.e^3_{v_2,v_1})$. The algorithm begins with $\text{TS-CA}_1^1 = (e^4_{v_4,v_3}.e^1_{v_3,v_2}.e^1_{v_2,v_1})$; see Line 2. Line 3 finds a (3, 1) TS-MPS, viz., $\text{TS-MPS}^1_{3,1} = (e^1_{v_3,v_2}.e^1_{v_2,v_1})$ in $O(\#V)$; see Proposition 7.1. As initially $\text{TS-MPS}_{3,1} = \{\}$, so Lines 4-6 add $\text{TS-MPS}^1_{3,1}$ to $\text{TS-MPS}_{3,1}$. Thus, now $\text{TS-MPS}_{3,1} = \{(e^1_{v_3,v_2}.e^1_{v_2,v_1})\}$. Next, the algorithm's Line 2 uses $\text{TS-CA}_1^2 = (e^4_{v_4,v_3}.e^2_{v_3,v_2}.e^1_{v_2,v_1})$. Using TS-CA_1^2, Line 3 finds path $(e^2_{v_3,v_2}.e^1_{v_2,v_1})$, which is not time-ordered and hence discarded. So, the algorithm goes back to Line 2 to begin with $\text{TS-CA}_1^3 = (e^4_{v_4,v_3}.e^1_{v_3,v_2}.e^3_{v_2,v_1})$. From TS-CA_1^3, Line 3 finds $\text{TS-MPS}^3_{3,1} = (e^1_{v_3,v_2}.e^3_{v_2,v_1})$. The algorithm at this stage checks whether $\text{TS-MPS}^3_{3,1}$ is unique or the same TS-MPS already exists in $\text{TS-MPS}_{3,1}$. This means that if each TSE of $\text{TS-MPS}^3_{3,1}$ is same as of any TS-MPS already in set $\text{TS-MPS}_{3,1}$, then $\text{TS-MPS}^3_{3,1}$ will be considered as duplicate and will be discarded. For example, the only TS-MPS, up to this stage, in $\text{TS-MPS}_{3,1}$, that is, $\text{TS-MPS}^1_{3,1} = (e^1_{v_3,v_2}.e^1_{v_2,v_1})$, is not same as $\text{TS-MPS}^3_{3,1} = (e^1_{v_3,v_2}.e^3_{v_2,v_1})$. Thus, $\text{TS-MPS}_{3,1}$ is updated to store $\{(e^1_{v_3,v_2}.e^1_{v_2,v_1}), (e^1_{v_3,v_2}.e^3_{v_2,v_1})\}$. At last, algorithm uses $\text{TS-CA}_1^4 = (e^4_{v_4,v_3}.e^2_{v_3,v_2}.e^3_{v_2,v_1})$. Line 3 finds $\text{TS-MPS}^4_{3,1} = (e^2_{v_3,v_2}.e^3_{v_2,v_1})$ from TS-CA_1^4. Now, it can be observed that $\text{TS-MPS}^4_{3,1}$ is unique and therefore Algorithm 2 updates $\text{TS-MPS}_{3,1}$ to store $\{(e^1_{v_3,v_2}.e^1_{v_2,v_1}), (e^1_{v_3,v_2}.e^3_{v_2,v_1}), (e^2_{v_3,v_2}.e^3_{v_2,v_1})\}$. It is important to note here that unlike static networks, $\#\text{TS-MPS}_{3,1} = 3$ is not equal to $\#\text{TS-MPS}_{1,3} = 2$, shown before. This happens because of the presence of timestamps with edges. Now, let us see Proposition 7.2.

Proposition 7.2 Time complexity to generate set $\text{TS-MPS}_{s,d}$ from set TS-CA_d is $O\left(\frac{\#\text{TS-MPS}_{s,d}(\#\text{TS-MPS}_{s,d}-1)}{2}(\#V)\right)$.

Proof For each $\text{TS-CA}_d^i \in \text{TS-CA}_d$, Line 3 of Algorithm 2 takes $O(\#V)$; see Proposition 7.1. However, the TS-MPS generated in Line 3 needs to be further assessed to verify whether it is unique or duplicate. More specifically, the first TS-MPS is always unique and needs no comparison; thus, is added directly to $\text{TS-MPS}_{s,d}$. The TS-MPS generated next is compared with first TS-MPS stored in $\text{TS-MPS}_{s,d}$. Similarly, the third TS-MPS is compared with the first two, and so on. Note that each comparison needs element by element inspection, which in worst case requires $O(\#V)$.

Thus, in total it takes [0 + 1 + … + (#TS-MPS$_{s,d}$− 1)] × (#V) comparisons to obtain set TS-MPS$_{s,d}$. In other words, time complexity to obtain TS-MPS$_{s,d}$ is $O\left(\frac{\#\text{TS-MPS}_{s,d}(\#\text{TS-MPS}_{s,d}-1)}{2}(\#V)\right)$. **Q.E.D.**

7.6.2 *Reliability Evaluation*

This section presents three reliability metrics, viz, $R_c(v_d)$, $R_c(K)$ and $R(s, d)$, of *p*TVCNs. Their respective definitions and evaluation method are explained in the upcoming paragraphs.

(i) **v_d-Convergecast Reliability**

Reliability metric $R_c(v_d)$ depicts reliability of convergecast to sink node v_d of *p*TVCN. More specifically, $R_c(v_d)$ depicts a probability that data packets disseminated by all other nodes except sink node v_d are successfully collected at sink. To illustrate reliability $R_c(v_d)$, consider Fig. 7.1(ii) with sink node v_3; thus, $R_c(v_3)$ is the probability of at least one of the two time-ordered arborescences, that is, $\left(e^4_{v_4,v_3}.e^1_{v_1,v_2}.e^1_{v_2,v_3}\right)$ and $\left(e^4_{v_4,v_3}.e^1_{v_1,v_2}.e^2_{v_2,v_3}\right)$, being operational. We propose a two-step approach for computing the reliability $R_c(v_d)$. Step 1 requires enumeration of all time-ordered arborescences converging to sink node v_d. For Step 2, we propose using any SDP technique, for example, [46, 47] to convert the obtained set of time-ordered arborescences into its compact reliability expression. For example, disjointing the two time-ordered arborescences converging to sink node v_3 results in the $R_c(v_3)$ expression as:

$$R_c(v_3) = p_{e^4_{v_4,v_3}} p_{e^1_{v_1,v_2}} p_{e^1_{v_2,v_3}} + \left(1 - p_{e^1_{v_2,v_3}}\right) p_{e^4_{v_4,v_3}} p_{e^1_{v_1,v_2}} p_{e^2_{v_2,v_3}} \tag{7.2}$$

Here, $p_{e^t_{v_a,v_b}}$ denotes the probability of success of a TSE $e^t_{v_a,v_b}$. Note that $R_c(v_3)$ expression in Eq. (7.2) can have dissimilar value of $p_{e^t_{v_a,v_b}}$ for different pair of nodes and also for different timestamp t between a specified pair of nodes. For example, $p_{e^1_{v_1,v_2}}$ may not be equal to $p_{e^1_{v_2,v_3}}$, and $p_{e^1_{v_2,v_3}}$ and $p_{e^2_{v_2,v_3}}$ may also be different as they appear at different instants of time between a given node pair (v_2,v_3). Assuming each TSE has an equal communication probability of 0.9, then the above expression results in $R_c(v_3) = 0.8019$. Similarly, we can evaluate the value of $R_c(v_d)$ for other sink nodes. Note that $R_c(v_1) = R_c(v_2) = 0$ as there is no time-ordered arborescence converging to sink node v_1 and v_2.

(ii) ***K*-Convergecast Reliability**

K-convergecast reliability, that is, $R_c(K)$ represents a probability that data packets disseminated by all nodes of *p*TVCN are successfully collected at *all* sink nodes in set *K*. Note that the number of nodes in set *K* ranges from 1 to #*V*, and thus $R_c(K)$

$= R_c(v_d)$ when $\#K = 1$. To illustrate $R_c(K)$ consider Fig. 7.1(ii) with $K = \{v_3, v_4\}$. As shown in Table 7.1, there are two time-ordered arborescences converging to both nodes v_3 and v_4. For this example, $R_c(K)$ computes the probability that at least one of the two time-ordered arborescences converging to v_3 *and* one of the two time-ordered arborescences converging to v_4 are operational, for example, $(e^4_{v_4,v_3}.e^1_{v_1,v_2}.e^1_{v_2,v_3})$ and $(e^1_{v_1,v_2}.e^1_{v_2,v_3}.e^4_{v_3,v_4})$. Equivalently, for this example, $R_c(K)$ gives the probability that one of four $(= 2 \times 2)$ possible combinations of time-ordered arborescences converging to v_3 and v_4 is operational. In the following, we describe in detail the two-step approach for evaluating $R_c(K)$.

(1) Find all possible *irredundant* and *minimal* combinations of time-ordered arborescences using one arborescence from each TS-VCA$_K$ for all $K \in [1, 2, \ldots, \#V]$. For TS-VCA$_1$ = {TS-VCA$^1_1$, TS-VCA$^2_1$, TS-VCA$^3_1$, …,TS-VCA$^k_1$ }, TS-VCA$_2$ = {TS-VCA$^1_2$, TS-VCA$^2_2$, TS-VCA$^3_2$, …,TS-VCA$^l_2$ }, …, and TS-VCA$_{\#V}$ = {TS-VCA$^1_{\#V}$, TS-VCA$^2_{\#V}$, TS-VCA$^3_{\#V}$, …, TS-VCA$^m_{\#V}$ }, we obtain at maximum $z = k \times l \times \ldots \times m$ irredundant and minimal combinations represented by C$_Q$, where $1 \leq Q \leq z$. A combination is *irredundant* if it is unique and *minimal* if there exists no superset of this combination, for example, C_1 = {TS-VCA$^1_1$ ∪ TS-VCA$^1_2$ ∪ … ∪ TS-VCA$^1_{\#V}$ }, C_2 = {TS-VCA$^1_1$ ∪ TS-VCA$^1_2$ ∪ … ∪ TS-VCA$^2_{\#V}$} and so on. Thus, for $K = \{v_3, v_4\}$, we have at maximum $2 \times 2 = 4$ irredundant and minimal combinations. From Table 7.1, observe that each time-ordered arborescence has three TSEs, so each of the four combinations can have a maximum length of six TSEs, if each of them is unique. However, only two of the four combinations are irredundant and minimal, that is, $\{e^4_{v_4,v_3}.e^1_{v_1,v_2}.e^1_{v_2,v_3}.e^4_{v_3,v_4}\}$ and $\{e^4_{v_4,v_3}.e^1_{v_1,v_2}.e^2_{v_2,v_3}.e^4_{v_3,v_4}\}$, each of which has length four. The remaining two combinations of length five are redundant and superset of the two irredundant and minimal combinations.

(2) Generate the desired reliability expression and/or value from the combinations obtained in (1) using any SDP approach [46, 47]. For example, $R_c(K)$ expression, for $K = \{v_3, v_4\}$, obtained by disjointing the two combinations via SDP is shown in Eq. (7.3)

$$R_c(K) = p_{e^4_{v_4,v_3}} p_{e^1_{v_1,v_2}} p_{e^1_{v_2,v_3}} p_{e^4_{v_3,v_4}} + \left(1 - p_{e^1_{v_2,v_3}}\right) p_{e^4_{v_4,v_3}} p_{e^1_{v_1,v_2}} p_{e^2_{v_2,v_3}} p_{e^4_{v_3,v_4}}. \tag{7.3}$$

If each $p_{e^t_{v_a,v_b}} = 0.9$, then Eq. (7.3) results in $R_c(K) = 0.72171$. Note that, in general, generating irredundant and minimal combinations in (1) is impractical as it becomes computationally intractable for *p*TVCNs having large number of time-ordered arborescences converging to each sink node $v_d \in K \subseteq V$. Further, inversion of the irredundant and minimal combinations of arborescences to obtain their dual and to use them to compute well-known Esary and Proschan's [53] lower bound of reliability is also intractable. Finally, as the combinations may have high degree of dependence among each other, due to the presence of multiple common TSEs, the Bonferroni inequalities [54] also cannot be utilized to find bounds, as they are expected to result in loose inequalities.

(iii) (s, d) Node Pair Reliability

The two-terminal reliability of pTVCN related to node pair (s, d), that is, $R(s, d)$, is the probability that data packets sent from source node v_s will be successfully received by sink node v_d. Thus, it depicts success probability of at least one of the TS-MPS between (s, d) node pair. To evaluate reliability $R(s, d)$ of pTVCNs, we follow a two-step approach similar to the one used to evaluate $R_c(v_d)$. The steps are: (1) Enumerate all TS-MPS for a specified (s, d) node pair, for $v_s \neq v_d$ and $v_s, v_d \in V$, and (2) Apply any SDP approach, for example, [46, 47] on all TS-MPS enumerated in (1) to make them disjointed and to obtain a compact reliability expression. The R(3,1) expression for node pair (3, 1) of the TAG in Fig. 7.1(ii) is given in Eq. (7.4), and we have $R(3,1) = 0.972$ if each $p_{e^t_{v_a,v_b}}$ is 0.9.

$$R(3,1) = p_{e^1_{v_3,v_2}} p_{e^1_{v_2,v_1}} + \left(1 - p_{e^1_{v_2,v_1}}\right) p_{e^1_{v_3,v_2}} p_{e^3_{v_2,v_1}} + \left(1 - p_{e^1_{v_3,v_2}}\right) p_{e^2_{v_3,v_2}} p_{e^3_{v_2,v_1}}. \tag{7.4}$$

7.7 Simulation Results

The algorithms have been implemented using Java code supported on JDK 8 and above, and the tests were run on IntelliJ IDEA Community edition version 2018.3.6 running on PC with the following configuration: (1) Processor: Intel (R) Core (TM) i7-7700 CPU @ 3.60 GHz, (2) RAM: 16.00 GB, and (3) System Type: 64-bit Operating System, x64-based processor.

To analyze the performance of the two proposed algorithms, ten arbitrary TAGs, as shown in Fig. 7.7, have been utilized. Note that period of each TAG is assumed to be four discrete units of time, the number of nodes vary between four and seven, and $p_{e^t_{v_a,v_b}} = 0.90$ for each TSE $e^t_{v_a,v_b}$. Table 7.2 presents #TS-CA$_d$ and #TS-VCA$_d$, for all sink $v_d \in V$; see row 1 and 2, respectively, for each TAG of Fig. 7.7. For each given TAG, we have the same #TS-CA$_d$ for each $v_d \in V$ because we assumed bidirectional contacts. Besides, Table 7.2 also analyzes the obtained reliability $R_c(v_d)$; see row 3 for each TAG of Fig. 7.7. The cell corresponding to #TS-CA$_d$ (#TS-VCA$_d$) and total for each TAG depicts #TS-CA$_{All}$ (#TS-VCA$_{All}$). Note that for each example TAG, #TS-CA$_{All}$ > #TS-VCA$_{All}$, which supports our discussion in Sect. 7.5. As discussed in Sect. 7.6.2, evaluation of $R_c(K)$ is often intractable, hence the metric is not computed here. It is worth mentioning that, in general a large #TS-VCA$_d$ for a sink node v_i as compared to v_j does not guarantee $R_c(v_i) > R_c(v_j)$, where $v_i \neq v_j$. The reason is because there may exist more multiple common TSEs in time-ordered arborescences converging to sink node v_i as compared to at node v_j. For example, in TAG 4, nodes v_1 and v_4 have #TS-VCA$_d$ equal to 121 and 60, respectively; however, $R_c(v_1) < R_c(v_4)$. Table 7.2 also shows that each TAG is globally connected as there is at least one time-ordered arborescence converging to each sink node $v_d \in V$. In

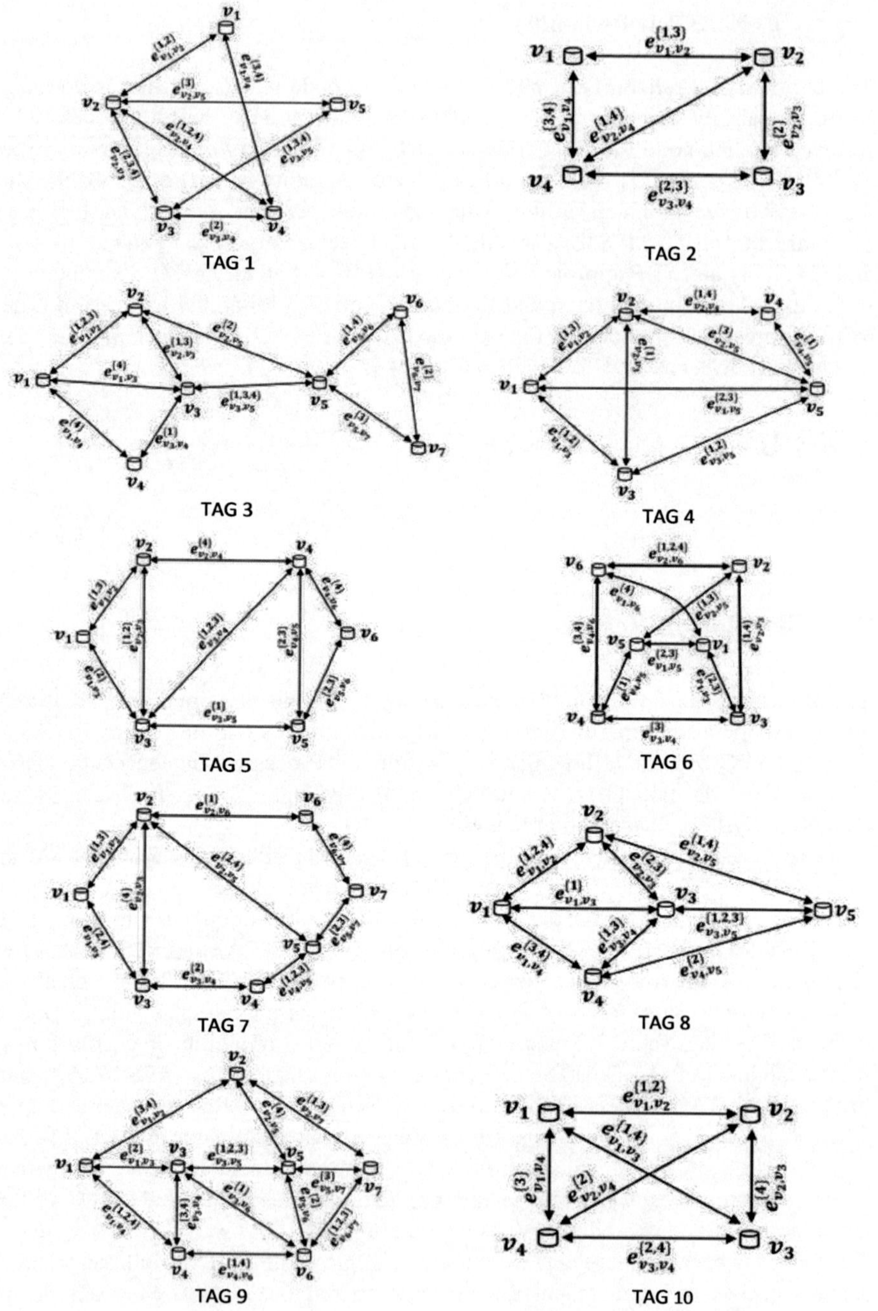

Fig. 7.7 Ten arbitrary TAG examples

Table 7.2 Timestamped arborescences and reliability from different TAGs of Fig. 7.7

TAG	Metric	Sink node							Total
		v_1	v_2	v_3	v_4	v_5	v_6	v_7	
1	**#TS-CA$_d$**	364	364	364	364	364	NA	NA	1820
	#TS-VCA$_d$	31	120	103	73	70	NA	NA	397
	$R_c(v_d)$	0.9779	0.9985	0.9977	0.9886	0.9959	NA	NA	NA
2	**#TS-CA$_d$**	44	44	44	44	NA	NA	NA	176
	#TS-VCA$_d$	23	19	9	34	NA	NA	NA	85
	$R_c(v_d)$	0.9968	0.9959	0.9775	0.9986	NA	NA	NA	NA
3	**#TS-CA$_d$**	675	675	675	675	675	675	675	4725
	#TS-VCA$_d$	76	6	108	38	116	59	23	426
	$R_c(v_d)$	0.9216	0.6488	0.9436	0.8379	0.9525	0.9191	0.8487	NA
4	**#TS-CA$_d$**	262	262	262	262	262	NA	NA	1310
	#TS-VCA$_d$	121	138	41	60	126	NA	NA	486
	$R_c(v_d)$	0.9899	0.9987	0.9876	0.9955	0.9899	NA	NA	NA
5	**#TS-CA$_d$**	539	539	539	539	539	539	NA	3234
	#TS-VCA$_d$	5	130	20	221	40	165	NA	581
	$R_c(v_d)$	0.7150	0.9680	0.9365	0.9949	0.9717	0.9947	NA	NA
6	**#TS-CA$_d$**	1230	1230	1230	1230	1230	1230	NA	7380
	#TS-VCA$_d$	288	220	163	180	27	379	NA	1257
	$R_c(v_d)$	0.9970	0.9971	0.9959	0.9875	0.9560	0.9984	NA	NA
7	**#TS-CA$_d$**	648	648	648	648	648	648	648	4536
	#TS-VCA$_d$	25	75	57	9	64	15	24	269

(continued)

Table 7.2 (continued)

TAG	Metric	Sink node v_1	v_2	v_3	v_4	v_5	v_6	v_7	**Total**
	$\boldsymbol{R_c(v_d)}$	0.8593	0.8790	0.8695	0.7652	0.8771	0.7780	0.8545	NA
8	**#TS-CA$_d$**	661	661	661	661	661	NA	NA	3305
	#TS-VCA$_d$	212	197	119	197	131	NA	NA	856
	$\boldsymbol{R_c(v_d)}$	0.9999	0.9998	0.9986	0.9999	0.9995	NA	NA	NA
9	**#TS-CA$_d$**	12750	12750	12750	12750	12750	12750	12750	89250
	#TS-VCA$_d$	1531	1728	523	1199	1261	607	326	7175
	$\boldsymbol{R_c(v_d)}$	0.9996	0.9997	0.9953	0.9984	0.9976	0.9959	0.9871	NA
10	**#TS-CA$_d$**	51	51	51	51	NA	NA	NA	204
	#TS-VCA$_d$	22	19	35	27	NA	NA	NA	103
	$\boldsymbol{R_c(v_d)}$	0.9968	0.9887	0.9995	0.9985	NA	NA	NA	NA

NA: Not applicable

Table 7.3 TS-MPS and two-terminal reliability results from different TAGs of Fig. 7.7

TAG	**(*s*, *d*)**	**#TS-MPS$_{s,d}$**	***R*(*s*, *d*)**	**(*d*, *s*)**	**#TS-MPS$_{d,s}$**	***R*(*d*, *s*)**
1	(1, 5)	20	0.9962	(5, 1)	13	0.9779
2	(1, 3)	4	0.9784	(3, 1)	6	0.9969
3	(1, 7)	8	0.9528	(7, 1)	4	0.9331
4	(1, 5)	12	0.9999	(5, 1)	12	0.9999
5	(1, 6)	29	0.9950	(6, 1)	2	0.7151
6	(1, 6)	9	0.9986	(6, 1)	9	0.9991
7	(1, 7)	6	0.9735	(7, 1)	6	0.9566
8	(1, 5)	18	0.9999	(5, 1)	27	0.9999
9	(1, 7)	14	0.9979	(7, 1)	32	0.9999
10	(1, 4)	8	0.9997	(4, 1)	5	0.9980

addition, for each TAG of Fig. 7.7, Table 7.3 analyzes the obtained reliability *R*(*s*, *d*) between specified pair of nodes (*s*, *d*). The results verify the fact that it is not necessary that #TS-MPS$_{s,d}$ is equal to #TS-MPS$_{d,s}$. Note that even if #TS-MPS$_{s,d}$ is equal to #TS-MPS$_{d,s}$, it is not necessary that $R(s, d) = R(d, s)$. The reason is because there may exist more multiple common TSEs in TS-MPS between (*s*, *d*) as compared to between (*d*, *s*). For example, in TAG 6, #TS-MPS$_{1,6}$ = #TS-MPS$_{6,1}$ = 9, but $R(6,1) > R(1, 6)$.

7.8 Conclusions and Future Scope

This chapter has reviewed some recently developed models to study and analyze various facets of TVCNs. It then extended the usual notion of arborescences to timestamped arborescences for network convergecasting, that is, timestamped valid and invalid arborescences, and showed their differences. Further, the chapter has presented a method to enumerate all timestamped valid arborescences of a *p*TVCN. It has also defined the connectivity in *p*TVCNs, and presented an approach to enumerate all TS-MPS between a specified (*s*, *d*) node pair in a network using all timestamped arborescences converging to sink v_d. Finally, this chapter proposed three reliability metrics for *p*TVCNs, viz., $R_c(v_d)$ and $R_c(K)$ that have been evaluated using time-ordered arborescences, and $R(s, d)$ evaluated using all (*s*, *d*) TS-MPS. This work can be extended to efficiently find tight bounds on the reliability metric $R_c(K)$ as its exact evaluation is still intractable. Further, the presented notions and algorithms can be utilized for designing feasible contact plans, optimal topology, and executing upgrades of *p*TVCNs.

Notation

τ	Periodicity of a *p*TVCN
$e^t_{v_i,v_j}$	A TSE from node v_i to v_j at timestamp *t*
$p_{e^t_{v_i,v_j}}$	Probability of success of the TSE $e^t_{v_i,v_j}$
(s, d)	Source–destination node pair
V	Set of TVCN nodes
$\#V$	Number of nodes in set V
TS-CA$_d$	A set of all timestamped arborescences converging to sink $v_d \in V$
TS-VCA$_d$	A set of all timestamped valid arborescences converging to sink $v_d \in V$
TS-CA^i_d	*i*th timestamped arborescence in set TS-CA$_d$
TS-VCA^i_d	*i*th timestamped valid arborescence in set TS-VCA$_d$
TS-CA$_{All}$	A set of TS-CA$_d \forall v_d \in V$
TS-VCA$_{All}$	A set of TS-VCA$_d \forall v_d \in V$
#TS-CA$_d$	Number of timestamped arborescence(s) in set TS-CA$_d$
#TS-VCA$_d$	Number of timestamped valid arborescence(s) in set TS-VCA$_d$
#TS-CA$_{All}$	Total number of timestamped arborescences in a TAG
#TS-VCA$_{All}$	Total number of timestamped valid arborescences in a TAG
L^-	Laplace out-degree matrix of a multigraph
$\widehat{L}^-_d$	Reduced L^- matrix obtained by deleting *d*th row and column
$\lvert\widehat{L}^-_d\rvert$	Determinant of $\widehat{L}^-_d$
$R_c(v_d)$	Reliability of convergecast to sink node $v_d \in V$
$R_c(K)$	Reliability of convergecast to *all* sink nodes in set $K \subseteq V$
$R(s, d)$	Reliability of successful transmission between (s, d) node pair
TS-MPS$_{s,d}$	A set of all TS-MPS between (s, d) node pair
TS-MPS$^i_{s,d}$	A TS-MPS in TS-MPS$_{s,d}$ generated from timestamped arborescence TS-CA^i_d
n	Number of servers in the cloud computing system
k	Number of UVMs created by cloud RMS
m	Number of AVMs created by cloud RMS
v, *u*	Probability of AVM's success in stealing, corrupting data of UVM residing in the same server
q(*n*, *k*, *h*)	Probability that exactly *h* out of the *n* servers host UVMs
p(*n*, *k*, *m*, *x*)	Probability that any UVM and AVM co-reside in *x* out of the *n* servers
g(*n*, *m*, *h*, *x*)	Conditional probability that in *x* servers AVMs co-reside with UVMs given UVMs reside in *h* servers
w(*n*, *k*, *m*)	Probability that at least one of UVMs co-resides with AVMs when *k* UVMs and *m* AVMs are randomly distributed among *n* servers
z(*n*, *k*, *m*)	Probability that all UVMs co-reside with AVMs when *k* UVMs and *m* AVMs are randomly distributed among *n* servers
s(*n*, *k*, *m*, *u*)	Probability that AVM succeeds in corrupting user's data in at least one out of *n* servers given that *k* UVMs and *m* AVMs are created and

	AVM succeeds in corrupting UVM residing in the same server with probability u
$f(n, k, m, v)$	Probability that AVMs steals the entire user's data given that k UVMs and m AVMs are created in cloud system having n available servers and AVM succeeds in stealing the data of UVM residing in the same server with probability v
A	Attack (request to create AVMs) rate
$\pi(t, A, m)$	Probability that m attacks happen in time interval $[0, t)$
$z(n, k, A, t)$	Probability that during time t since the FCE the attacker succeeds to get its AVMs co-residing with all the UVMs
T_D	Random time needed by EWA for detecting the attack
T_A	Random time needed by AVM to gain access to data of co-resident UVMs
$f_D(t), f_A(t)$	*pdf* of attack detection T_D, T_A

References

1. Khanna, G., & Chaturvedi, S. K. (2018). A comprehensive survey on multi-hop wireless networks: milestones, changing trends and concomitant challenges. *Wirel Pers Commun, 101*, 677–722.
2. Soh. S., Rai, S., & Brooks, R. R. (2008). Performability issues in wireless communication networks. *Handbook of Performability Engineering* (pp. 1047–1067), Springer.
3. Liang, Q., & Modiano, E. (2017). Survivability in time-varying networks. *IEEE Transactions on Mobile Computing, 16*, 2668–2681.
4. Casteigts, A., et al. (2012). Time-varying graphs and dynamic networks. *International Journal of Parallel, Emergent and Distributed Systems, 27*, 387–408.
5. Casteigts, A., et al. (2011). Time-varying graphs and dynamic networks. In H. Frey, X. Li, & S. Ruehrup (Eds.), *Ad-hoc, mobile, and wireless networks* (pp. 346–359). Berlin Heidelberg: Springer.
6. Pentland, A., Fletcher, R., & Hasson, A. (2004). DakNet: Rethinking connectivity in developing nations. *Computer, 37*, 78–83.
7. Juang, P., et al. (2002). Energy-efficient computing for wildlife tracking: design tradeoffs and early experiences with ZebraNet. *ACM SIGARCH Comput Archit News, 30*, p. 12.
8. Zhang, W., et al. (2019). Efficient topology control for time-varying spacecraft networks with unreliable links. *Int J Distrib Sens Netw, 15*(9).
9. Bekmezci, I., Sahingoz, O. K., & Temel, Ş. (2013). Flying ad-hoc networks (FANETs): A survey. *Ad Hoc Networks, 11*, 1254–1270.
10. Xuan, B. B., Ferreira, A., & Jarry, A. (2003). Evolving graphs and least cost journeys in dynamic networks. In *WiOpt'03: Modeling and Optimization in Mobile, Ad Hoc and Wireless Networks* (p. 10). Sophia Antipolis, France.
11. Huang, M., et al. (2013). Topology control for time-evolving and predictable delay-tolerant networks. *IEEE Transactions on Computers, 62*, 2308–2321.
12. Raffelsberger, C., & Hellwagner, H. (2014). Combined mobile ad-hoc and delay/disruption-tolerant routing. In S. Guo, J. Lloret, P. Manzoni, et al. (Eds.), *Ad-hoc, mobile, and wireless networks* (pp. 1–14). Cham: Springer International Publishing.
13. Baudic, G., Perennou, T., & Lochin, E. (2016). Following the right path: Using traces for the study of DTNs. *Computer Communications, 88*, 25–33.

14. Khanna, G., Chaturvedi, S. K., & Soh, S. (2020). Time varying communication networks: Modelling, reliability evaluation and optimization. In M. Ram & H. Pham (Eds.), *Advances in reliability analysis and its applications* (pp. 1–30). Cham: Springer International Publishing.
15. Chaturvedi, S. K., Khanna, G., & Soh, S. (2018). Reliability evaluation of time evolving Delay Tolerant Networks based on Sum-of-Disjoint products. *Reliab Eng Syst Saf, 171,* 136–151.
16. Khanna, G., Chaturvedi, S. K., & Soh, S. (2019). On computing the reliability of opportunistic multihop networks with Mobile relays. *Qual Reliab Eng Int, 35,* 870–888.
17. Khanna, G., Chaturvedi, S. K., & Soh, S. (2020). Two-terminal reliability analysis for time-evolving and predictable delay-tolerant networks. *Recent Adv Electr Electron Eng Former Recent Pat Electr Electron Eng, 13,* 236–250.
18. Santi, P. (2012). *Mobility models for next generation wireless networks: Ad hoc, vehicular and mesh networks.* Wiley.
19. Batabyal, S., & Bhaumik, P. (2015). Mobility models, traces and impact of mobility on opportunistic routing algorithms: A survey. *IEEE Commun Surv Tutor, 17,* 1679–1707.
20. Aschenbruck, N., Munjal, A., & Camp, T. (2011). Trace-based mobility modeling for multi-hop wireless networks. *Computer Communications, 34,* 704–714.
21. Munjal, A., Camp, T., & Aschenbruck, N. (2012). Changing trends in modeling mobility. *J Electr Comput Eng, 2012,* 1–16.
22. Matis, M., Doboš, L., & Papaj, J. (2016). An enhanced hybrid social based routing algorithm for MANET-DTN. *Mob Inf Syst*, pp. 1–12.
23. Zhang, Z. (2006). Routing in intermittently connected mobile ad hoc networks and delay tolerant networks: Overview and challenges. *IEEE Commun Surv Tutor, 8,* 24–37.
24. Du, D., & Hu, X. (2008). *Steiner tree problems in computer communication networks,* World Scientific.
25. Fraire, J. A., Madoery, P., & Finochietto, J. M. (2017). Contact plan design for predictable disruption-tolerant space sensor networks. In H. F. Rashvand & A. Abedi (Eds.), *Wireless sensor systems for extreme environments* (pp. 123–150). Chichester, UK: Wiley.
26. Maini, A. K., & Agrawal, V. (2014). *Satellite technology: Principles and applications* (3rd ed.). Chichester, West Sussex: Wiley.
27. Chen, H., Shi, K., & Wu, C. (2016). Spanning tree based topology control for data collecting in predictable delay-tolerant networks. *Ad Hoc Networks, 46,* 48–60.
28. Bhadra, S., & Ferreira, A. (2003). Complexity of connected components in evolving graphs and the computation of multicast trees in dynamic networks. In *Ad-Hoc, Mobile, and Wireless Networks* (pp. 259–270). Springer.
29. Bhadra, S., & Ferreira, A. (2002). *Computing multicast trees in dynamic networks using evolving graphs.* RR-4531, INRIA.
30. Holme, P. (2015). Modern temporal network theory: A colloquium. *European Physical Journal B: Condensed Matter and Complex Systems, 88,* 1–30.
31. Díaz, J., Mitsche, D., & Santi, P. (2011). Theoretical aspects of graph models for MANETs. *Theoretical Aspects of Distributed Computing in Sensor Networks* (pp. 161–190). Springer.
32. George, B., & Shekhar, S. (2008). Time-aggregated graphs for modeling spatio-temporal networks. *Journal on Data Semantics XI*. Springer, pp. 191–212.
33. Kawamoto, Y., Nishiyama, H., & Kato, N. (2013). Toward terminal-to-terminal communication networks: A hybrid MANET and DTN approach. In *18th IEEE International Workshop on Computer Aided Modeling and Design of Communication Links and Networks (CAMAD)* (pp. 228–232).
34. Abolhasan, M., Wysocki, T., & Dutkiewicz, E. (2004). A review of routing protocols for mobile ad hoc networks. *Ad Hoc Networks, 2,* 1–22.
35. Alotaibi, E., & Mukherjee, B. (2012). A survey on routing algorithms for wireless Ad-Hoc and mesh networks. *Computer Networks, 56,* 940–965.
36. Khanna, G., Chaturvedi, S. K., & Soh, S. (2019). Reliability evaluation of mobile ad hoc networks by considering link expiration time and border time. *Int J Syst Assur Eng Manag, 10,* 399–415.
37. Misra, K. B. (Ed.). (2008). *Handbook of performability engineering*. London: Springer.

38. Cook, J. L., & Ramirez-Marquez, J. E. (2008). Mobility and reliability modeling for a mobile ad hoc network. *IIE Transactions, 41,* 23–31.
39. Padmavathy, N., & Chaturvedi, S. K. (2013). Evaluation of mobile ad hoc network reliability using propagation-based link reliability model. *Reliab Eng Syst Saf, 115,* 1–9.
40. Ahmad, M., & Mishra, D. K. (2012). A reliability calculations model for large-scale MANETs. Int J Comput Appl, *59.*
41. Singh, M. M., Baruah, M., & Mandal, J. K. (2014). Reliability computation of mobile ad-hoc network using logistic regression. In *Eleventh International Conference on Wireless and Optical Communications Networks (WOCN)* (pp. 1–5). IEEE.
42. Egeland, G., & Engelstad, P. (2009). The availability and reliability of wireless multi-hop networks with stochastic link failures. *IEEE Journal on Selected Areas in Communications, 27,* 1132–1146.
43. Meena, K. S., & Vasanthi, T. (2016). Reliability analysis of mobile ad hoc networks using universal generating function: Reliability analysis of manet using ugf. *Qual Reliab Eng Int, 32,* 111–122.
44. Clark, J., & Holton, D.A. (1991). *A first look at graph theory.* World Scientific.
45. Kamiyama, N., & Kawase, Y. (2015). On packing arborescences in temporal networks. *Inf Process Lett, 115,* 321–325.
46. Rai, S., Veeraraghavan, M., & Trivedi, K. S. (1995). A survey of efficient reliability computation using disjoint products approach. *Networks, 25,* 147–163.
47. Misra, K. B. (1993). *New trends in system reliability evaluation.* Elsevier Science Ltd.
48. Mertzios, G. B., Michail, O., & Spirakis, P. G. (2019). Temporal network optimization subject to connectivity constraints. *Algorithmica, 81,* 1416–1449.
49. Xuan, B. B., Ferreira, A., & Jarry, A. (2003). Computing shortest, fastest, and foremost journeys in dynamic networks. *Int J Found Comput Sci, 14,* 267–285.
50. Coll-Perales, B., Gozalvez, J., & Friderikos, V. (2016). Energy-efficient opportunistic forwarding in multi-hop cellular networks using device-to-device communications. *Trans Emerg Telecommun Technol, 27,* 249–265.
51. Tutte, W. T. (1984). *Graph theory.* Menlo Park, Calif: Addison-Wesley Pub. Co., Advanced Book Program.
52. Aigner, M. (2007). *A course in enumeration.* Springer Science & Business Media.
53. Hsieh, Y.-C. (2003). New reliability bounds for coherent systems. *Journal of the Operational Research Society, 54,* 995–1001.
54. Schwager, S. J. (1984). Bonferroni sometimes loses. *American Statistician, 38,* 192–197.

Sanjay K. Chaturvedi is currently working as a Professor and Head at Subir Chowdhury School of Quality and Reliability, Indian Institute of Technology, Kharagpur (WB), India. Please visit http://www.iitkgp.ac.in/department/RE/faculty/re-skcrec. For publications and citation details, visit *Google Scholar*. He has published papers in most of the leading international journals of his research interests, and executed several consultancy projects of private and Govt. organizations. He is also on the review panel of IEEE Transactions on Reliability, IJPE, IJQRM, IJ System Science, IJ Failure Analysis. He is on the editorial board of International Journal of Mathematical, Engineering and Management Sciences (IJMEMS). He is also a senior member to IEEE and has also served as Co-Editor-in-Chief to IJPE.

Sieteng Soh received the B.S. degree in Electrical Engineering from the University of Wisconsin-Madison in 1987, and M.S. & Ph.D. in Electrical Engineering from the Louisiana State University-Baton Rouge in 1989 and 1993, respectively. From 1993 to 2000, he was with the Tarumanagara University, Indonesia. He is a Senior Lecturer with the School of Electrical Engineering, Computing and Mathematical Sciences at Curtin University, Perth, Western Australia. His research interests include P2P systems, computer network, network reliability, and parallel and distributed processing. He is a member of the IEEE.

Gaurav Khanna received his B.Tech. degree in Electronics and Communication Engineering from GBTU, Lucknow, India, in 2011 and M. Tech. from School of Information and Communication Technology, Gautam Buddha University, Greater Noida, India, in 2015. He is associated with Subir Chowdhury School of Quality and Reliability, Indian Institute of Technology Kharagpur, India as a Ph.D. scholar since 2015. He is concurrently enrolled with the School of Electrical Engineering, Computing and Mathematical Sciences at Curtin University, Australia, under a Dual Doctoral Degree programme. His research interests include ad hoc networks, delay-tolerant networks and reliability engineering.

Chapter 8
Characteristics and Key Aspects of Complex Systems in Multistage Interconnection Networks

Indra Gunawan

Abstract Multistage Interconnection Networks (MINs) have been used extensively to provide reliable and fast communication with effective cost. In this paper, four types of systems, characteristics and key aspects of complex systems, are discussed in the context of MINs. Shuffle-Exchange Networks (SEN), a common network topology in MINs, is analysed as a complex system. Different perspectives on how MINs possess all characteristics of complex systems are discussed and therefore it is managed as complex systems accordingly.

Keywords Complex systems · Multistage interconnection networks (MINs) · Shuffle-Exchange networks (SEN)

8.1 Introduction

Although computer processing power has increased tremendously in the last few decades, the demand for processing power far exceeds the processing power that is currently available [1, 2, 28]. Thus there is a need for improved techniques that will deliver higher computer processing power to satisfy the needs of processor-intensive applications such as engineering and science simulations. This project looks at the notion of interconnection network as a means to fulfilling the demand for higher computer processing power.

Interconnection network technology is used to link together multiple processor-memory modules or computers in order to share resources, exchange data or to achieve parallel processing capability. Interconnection networks are applied in many fields such as telephone switches, supercomputers with multiprocessor and wide area networks [12].

I. Gunawan (✉)
The University of Adelaide, Adelaide, SA 5005, Australia
e-mail: indra.gunawan@adelaide.edu.au

K. B. Misra (ed.), *Handbook of Advanced Performability Engineering*,
https://doi.org/10.1007/978-3-030-55732-4_8

Interconnecting large number of processors and memory modules that allow communication among processors and processor-memory modules or in a communication network is a complicated task as issues such as connectivity, latency, bandwidth, cost, scalability and reliability need to be addressed. Numerous approaches had been proposed, ranging from single bus to fully connected architecture [24, 30]. Although the single-bus architecture can be easily implemented, its scalability highly depends on the bandwidth and requires arbitration of the bus usage, which somehow disallows parallel communication among processors or with memory modules. While a fully connected network might meet the need of parallel communication, it is difficult to rescale as it requires a large number of connection lines that make it impractical to be implemented in the real world.

Other alternatives of interconnection networks include crossbar network, hypercube network, tree network and multistage interconnection networks. Each of these networks has its own advantages and disadvantages. Thus, it is difficult to justify which is the best interconnection network. A rule of thumb in choosing the best interconnection network is to define the objectives and requirements that are needed in the system and then select the best method of interconnection that would satisfy most of the requirements.

Multistage Interconnection Networks (MINs) as its name implies is made of stages of crossbar switches, which are linked together in certain patterns to provide the needed interconnection between input devices and output devices. MINs have been used extensively in circuit switching networks and later in packet switching networks with the introduction of buffered switches. Examples of a multiprocessor system that implements MIN are ultracomputer and IBM RP3. Besides its application in a multiprocessor system, MINs are also used in communication networks such as ATM switches and Gigabit Ethernet, which are the forms of optical networks [12].

The number of stages, types of switches and interconnections among the network switches/stages determine the MIN configuration. There exist many types of MINs with different topologies such as multistage cube network (generalised cube topology), shuffle exchange network, gamma network, delta network, Tandem-Banyan networks and multilayer MINs [20, 25, 28].

MINs reliability evaluation has been cited in past researches [3–6, 8, 11, 26, 32]. Numerous methods are applied to compute the reliability of network systems [18, 34, 35]. In general, network reliability can be analysed in three main areas; terminal, broadcast and all-terminal network reliability. Terminal reliability is defined as probability of the existence of at least one fault-free path between a designated pair of input and output terminals (two terminals). Terminal reliability is commonly used as a robustness indicator of MIN. Broadcast reliability signifies a MIN ability to broadcast data from a given input terminal to all the output terminals of the network. When a connection cannot be made from a given input to at least one of the output terminals, the network is assumed to have failed. All-terminal reliability (or network reliability) represents the probability of the existence of a connection between each input to all outputs (all-terminal).

The research is initiated with the literature review of the past research that has been done in the area of interconnection network systems. Then, the proposed methodology to improve the performance of network systems is demonstrated. New configurations of network systems are proposed and the reliability performance of the systems is analysed. Finally, the algorithms to compute the terminal, broadcast and network reliability and the simulation of the interconnections in the network systems are discussed.

Depending on the availability of paths to establish new connections, MINs have been traditionally divided into three classes [12]:

1. *Blocking*. A connection between a free input/output pair is not always possible because of conflicts with the existing connections. Typically, there is a unique path between every input/output pair, thus minimising the number of switches and stages. However, it is also possible to provide multiple paths to reduce conflicts and increase fault tolerance. These blocking networks are also known as multipath.
2. *Nonblocking*. Any input port can be connected to any free output port without affecting the existing connections. Non-blocking networks have the same functionality as a crossbar. They require multiple paths between every input and output, which in turn leads to extra stages.
3. *Rearrangeable*. Any input port can be connected to any free output port. However, the existing connections may require rearrangement of paths. These networks also require multiple paths between every input and output but the number of paths and the cost are smaller than in the case of non-blocking networks.

Non-blocking networks are expensive. Although they are cheaper than a crossbar of the same size, their cost is prohibitive for large sizes. The best-known example of non-blocking multistage network is initially proposed for telephone networks. Rearrangeable networks require less stages or simpler switches than a non-blocking network. Rearrangeable networks require a central controller to rearrange connections and were proposed for array processors. However, connections cannot be easily rearranged on multiprocessors because processors access the network asynchronously. Therefore, rearrangeable networks behave like blocking networks when accesses are asynchronous.

Depending on the kind of channels and switches, MINs can be split into two classes [24]:

1. *Unidirectional MINs*. Channels and switches are unidirectional.
2. *Bidirectional MINs*. Channels and switches are bidirectional. This implies that information can be transmitted simultaneously in opposite directions between neighbouring switches.

Additionally, each channel may be either multiplexed or replaced by two or more channels. In the latter case, the network is referred to as dilated MIN. Obviously, the number of ports of each switch must increase accordingly.

8.2 Unidirectional Multistage Interconnection Networks

The basic building blocks of unidirectional MINs are unidirectional switches. An a × b switch is a crossbar network with a inputs and b outputs. If each input port is allowed to connect to exactly one output port, at most min $\{a, b\}$, connections can be supported simultaneously. If each input port is allowed to connect many output ports, a more complicated design is needed to support the one-to-many or multicast communications. In the broadcast mode or one-to-all communication, each input port is allowed to connect to all output ports. Figure 8.1 shows four possible states of a 2 × 2 switch. The last two states are used to support one-to-many and one-to-all communications.

In MINs with N inputs = M outputs, it is common to use switches with the same number of input and output ports, i.e. $a = b$. If $N > M$, switches with $a > b$ will be used. Such switches are also called concentration switches. In the case of $N < M$, distribution switches with $a < b$ will be used.

It can be shown that with N input and output ports, a unidirectional MIN with $k \times k$ switches require at least $\log_k N$ stages to allow a connection path between any input port and any output port. By having additional stages, more connection paths may be used to deliver a message between an input port and an output port at the expense of extra hardware cost. Every path through the MIN crosses all the stages. Therefore, all the paths have the same length.

In general, the topological equivalence of MINs can be viewed as follows: Consider that each input link to the first stage is numbered using a string of n digits $s_{n\text{-}1}s_{n\text{-}2}\ldots s_1s_0$, where $0 \leq s_i \leq k-1$, for $0 \leq i \leq n-1$. The least significant digit s_0 gives the address of the input port at the corresponding switch and the address of the switch is given by $s_{n\text{-}1}s_{n\text{-}2}\ldots s_1$. At each stage, a given switch is able to connect any input port with any output port. This can be viewed as changing the value of the least significant digit of the address. In order to connect any input to any output of the network, it should be possible to change the value of all the digits. As each switch is only able to change the value of the least significant digit of the address, connection patterns between stages are defined in such a way that the position of digits is permuted, and after n stages, all the digits have occupied the least significant position.

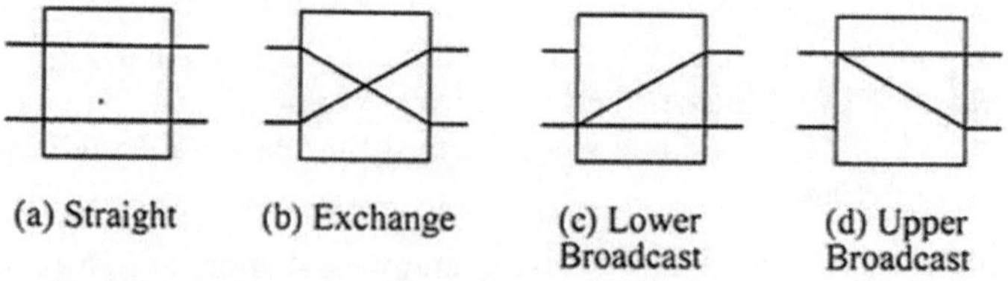

Fig. 8.1 Four possible states of a 2 × 2 switch

8.3 Bidirectional Multistage Interconnection Networks

A bidirectional switch supports three types of connections: forward, backward and turnaround. Figure 8.2 illustrates a bidirectional switch in which each port is associated with a pair of unidirectional channels in opposite directions. This implies that information can be transmitted simultaneously in opposite directions between neighbouring switches. As turnaround connections between ports at the same side of a switch are possible, paths have different lengths. An eight-node butterfly bidirectional MIN (BMIN) is illustrated in Fig. 8.3. For ease of explanation, it is assumed that processor nodes are on the left-hand side of the network.

Paths are established in BMINs by crossing stages in forward direction, then establishing a turnaround connection and finally crossing stages in backward direction. This is usually referred to as turnaround routing. Figure 8.4 shows two alternative paths from node S to node D in an eight-node butterfly BMIN. When crossing stages in forward direction, several paths are possible. Each switch can select any of its

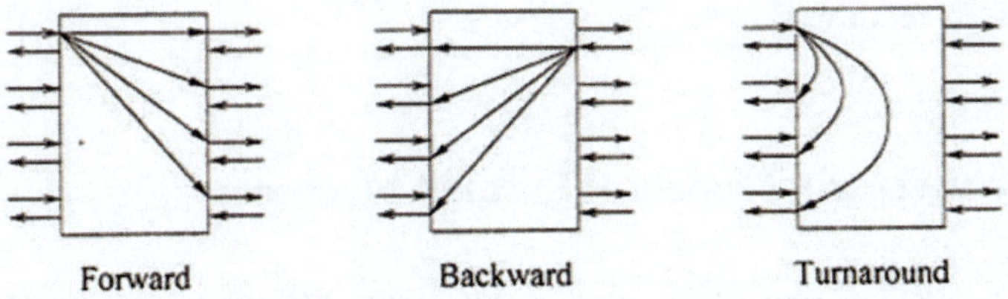

Fig. 8.2 Connections in a bidirectional switch

Fig. 8.3 An eight-node butterfly bidirectional MIN

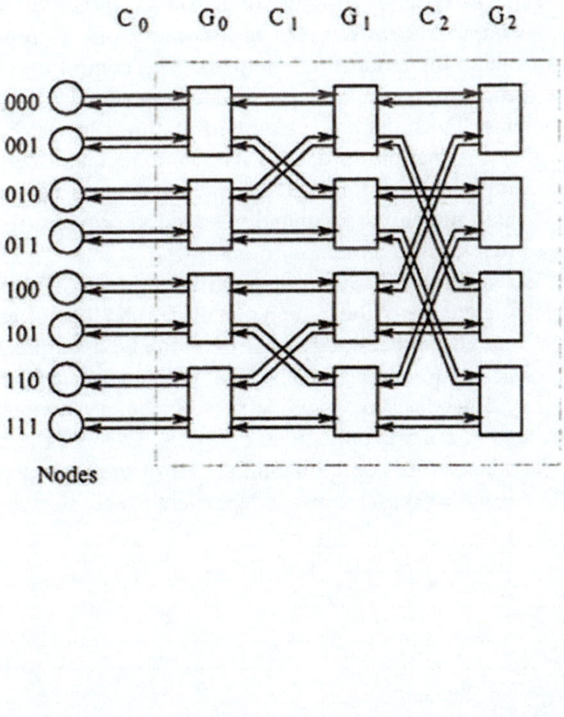

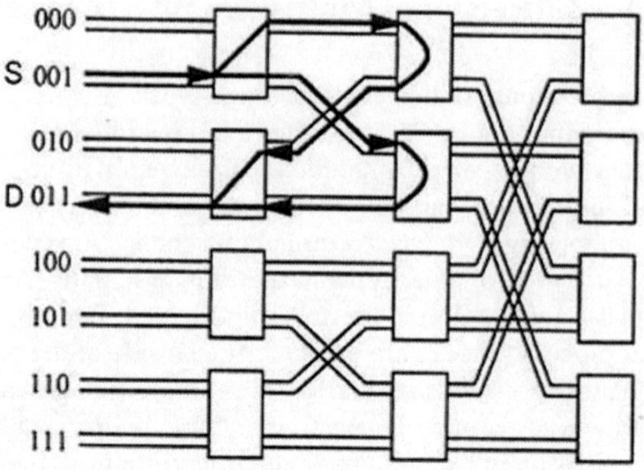

Fig. 8.4 Alternative paths in an eight-node butterfly bidirectional MIN

output ports. However, once the turnaround connection is crossed, a single path is available up to the destination node. In the worst case, establishing a path in an n-stage BMIN requires crossing $2n-1$ stages.

8.4 Architectural Models of Parallel Machines

There are a variety of ways to organise the processors, memories and interconnection network in a large-scale parallel processing system. In this section, models of a few of the basic structures are briefly introduced:

1. *SIMD Systems.* A model of an *SIMD* (*single instruction stream—multiple data stream*) *system* consists of a control unit, N processors, N memory modules and an interconnection network. The control unit broadcasts instructions to the processors, and all active processors execute the same instruction at the same time. Thus, there is a single instruction stream. Each active processor executes the instruction on data in its own associated memory module. Thus, there are multiple data streams. The interconnection network, sometimes referred to as an alignment or permutation network, provides for communications among the processors and memory modules.
2. *Multiple-SIMD Systems.* A variation on the SIMD model that may permit more efficient use of the system processors and memories is the *multiple-SIMD system*, a parallel processing system that can be dynamically reconfigured to operate as one or more independent SIMD subsystems of various sizes. A multiple-SIMD system consists of N processors, N memory modules, an interconnection network and C control units, where $C < N$. Each of the multiple control units can be connected to some disjoint subset of the processors, which communicate over sub-networks, creating independent SIMD subsystems of various sizes.

3. *MIMD Systems.* In contrast to the SIMD system, where all processors follow a single instruction stream, the processors in a parallel system may each follow its own instruction stream, forming an *MIMD (multiple instruction stream—multiple data stream) system.* One organisation for an MIMD system consists of *N* processors, *N* memory modules and an interconnection network, where each of the processors executes its own program on its own data. Thus there are multiple instruction streams and multiple data streams. The interconnection network provides communications among the processors and memory modules. While in an SIMD system all active processors use the interconnection network at the same time (i.e. synchronously), in an MIMD system, because each processor is executing its own program, inputs to the network arrive independently (i.e. asynchronously).
4. *Partitionable SIMD/MIMD Systems.* A fourth model of system organisation combines the features of the previous three. A *partitionable SIMD/MIMD system* is a parallel processing system that can be dynamically reconfigured to operate as one or more independent SIMD and/or MIMD subsystems of various sizes. The *N* processors, *N* memory modules, interconnection network and C control units of a partitionable SIMD/MIMD system can be partitioned to form independent subsystems as with multiple-SIMD systems. Furthermore, each processor can follow its own instructions (MIMD operation) in addition to being capable of accepting an instruction stream from a control unit (SIMD operation). Thus, each subsystem can operate in the SIMD mode or the MIMD mode. The processors can switch between the two modes of parallelism from one instruction to the next when performing a task, depending on which is more desirable at the time.
5. *System Configurations.* With any of these four models, there are two basic system configurations. One is the *PE-to-PE configuration,* in which each processing element or *PE* (formed by pairing a processor with a local memory) is attached to both an input port and an output port of an interconnection network (i.e. PE *j* is connected to input port j and output port *j*). This is also referred to as distributed memory system or private memory system. In contrast, in the *processor-to-memory configuration,* processors are attached to one side of an interconnection network and memories are attached to the other side. Processors communicate through shared memories. This is also referred to as a global memory system. Hybrids of the two approaches are also possible, such as using a local cache in a processor-to-memory system. Which configuration or hybrid of them is 'best' for a particular system design is a function of many factors, such as the types of computational tasks for which the system is intended (e.g. are most data and/or programs shared by all processors or local to each processor), the operating system philosophy (e.g. will multitasking be done within each processor to hide any latency time for network transfer delays when fetching data) and the characteristics of the processors and memories to be used (e.g. clock speed, availability of cache). Beware of the term *shared memory* as applied to these parallel systems. Some researchers use this term to refer to the way in which a system is physically constructed (i.e. processor-to-memory configuration) and others use it to refer to the logical addressing method.

8.5 Terminology

Many interconnection networks for large-scale multiprocessor computer systems have been proposed. Of these, Multistage Interconnection Networks (MINs) offer a good balance between cost and performance. In this section, the common terminologies in MINs such as switches, links, ports, crossbar, fault-tolerant, topology, routing tag, path and connection are discussed.

Figure 8.5 shows MIN hardware in the most general terms; dots indicate items that may repeat. MINs are composed of a collection of *switches* and *links* between switches. A signal may enter or leave a network through a *port*. A network with A input ports and B output ports is an $A \times B$ network.

A switch may be viewed as a very simple network. Switches are multiport devices; the number of ports and the port-to-port connections supported within a switch vary among switch designs. A *crossbar* switch can simultaneously connect, in any pattern, a number of input/output port pairs equal to the minimum of the number of inputs and the number of outputs. A *selector* switch connects only one of its inputs to one of its outputs at a time.

The term *network component* may denote any element of the structure of a network. An interconnection network may consist of a single *stage* or bank of switches and may require that data pass through the network more than once to reach its destination. A MIN is constructed from two or more stages of switches, and typically is designed, so that data can be sent to the desired destination by one pass through the network.

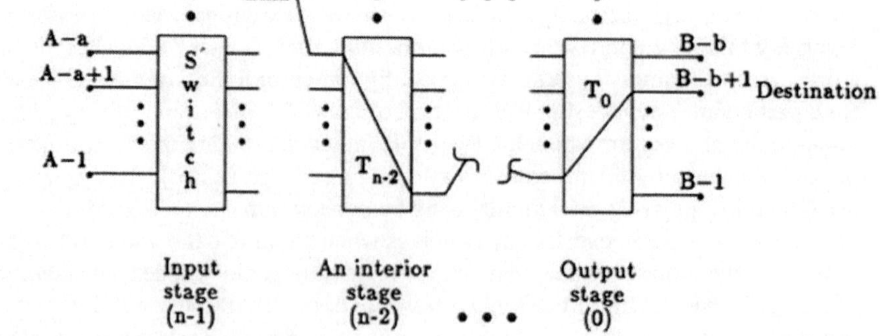

Fig. 8.5 A generic MIN diagram detailing one path

MINs can be considered from either a topological (graphical) or algebraic viewpoint. The *topology* of a network is the pattern of connections in its structure, where the pattern can be represented by a graph. Topology is determined by switch design and the pattern of links. Different MINs are often compared graphically because comparison by topology is independent of hardware. When one network is said to be an instance of another, it is the network graphs being compared. Nodes in the graph of a MIN can be numbered, and then a MIN can be described in terms of the algebraic relations among the nodes. The algebraic model is useful in discussing control and communication routing strategy.

There are three basic forms of connection through a network. A *one-to-one connection* passes information from one network port, the *source*, to another network port, the *destination.* The exact route taken by the information is its *path.* Multiple one-to-one connections may be active simultaneously. A *permutation connection* is a set of one-to-one connections such that no two one-to-one connections have the same source or destination. Such connections are meaningful only in the context of networks with equal number of sources and destinations. Information flow from one source simultaneously to two or more destinations is supported by a *broadcast connection*, and the route taken is a *broadcast path.*

Routing tags are way of describing a path through a network and providing for distributed network control. For MINs, tags often take the form of a multidigit integer, each successive digit encoding the setting for the switch in the next stage along a desired path. Control is distributed if devices using the network generate their own routing tags and network switches can set themselves based on tag information. Figure 8.5 shows a switch in stages $n-1$, $n-2$ and 0 being set, respectively, by tag digits T_{n-1}, T_{n-2} and T_0. Routing tags are particularly important for fault-tolerant MINs since they should be able to specify a functioning path if one exists; tag limitations translate into fault-tolerance limitations.

There are three methods for sources to generate routing tags that specify a fault-free path. With *non-adaptive routing,* a source learns of a fault only when the path it is attempting to establish reaches the faulty network component. Notice of the fault is sent to the source, which tries the next alternative path. This approach requires little hardware but may have poor performance. There are two forms of *adaptive routing.* With *notification on demand,* a source maintains a table of faults it has encountered in attempting to establish paths and uses this information to guide future routing. With *broadcast notification* of a fault, all sources are notified of faulty components as they are diagnosed.

A fault-free path needs not be specified by a source if routing tags can be modified in response to faults encountered as a path is followed or established. This *dynamic routing* can be accomplished in MINs constructed of switches capable of performing the necessary routing tag revisions.

The following section explores the network topologies in Multistage Interconnection Networks (MINs). Various inherent properties include path establishment, distributed routing tag control and partitionability. In general, the multistage networks have analogous, but not identical, properties. The standard networks and the hardware modifications made to provide redundancy, from less to more extensive, are introduced. Many possible techniques exist, including adding an extra stage of switches, varying switch size, adding extra links and adding extra ports. Finally, some irregular MINs that have different connection patterns between stages are also discussed.

8.6 Fault-Tolerant

A *fault-tolerant* MIN is one that provides service, in at least some cases, even when it contains a faulty component or components. A fault can be either permanent or transient, unless stated otherwise, it is assumed that faults are permanent. Fault tolerance is defined only with respect to a chosen *fault-tolerance model*, which has two parts. The *fault model* characterises all faults assumed to occur, stating the failure modes (if any) for each network component. The *fault-tolerance criterion* is the condition that must be met for the network to be said to have tolerated a given fault or faults.

Fault models may or may not correspond closely to predicted or actual experience with MIN hardware. In particular, a fault model may be chosen with characteristics that simplify reliability analysis, even if those characteristics depart widely from reality (such as assuming certain network components never fail). While fault-tolerance criteria typically closely reflect the normal (fault-free) operational capability of a network, this need not be so. The variability of fault-tolerance models hinders comparison of the engineering characteristics of fault-tolerant MINs.

A network is *single-fault tolerant* if it can function as specified by its fault-tolerance criterion despite any single fault conforming to its fault model. More generally, if any set of *i* faults can be tolerated, then a network is *i-fault tolerant*. A network that can tolerate some instances of *i* faults is robust although not *i*-fault tolerant.

Many fault-tolerant systems require fault diagnosis (detection and location) to achieve their fault tolerance. Techniques such as test patterns, dynamic parity checking and write/read-back/verify are used in various MINs. Techniques for fault-tolerant design can be categorised by whether they involve modifying the topology (graph) of the system. Three well-known methods that do not modify topology are error-correcting codes, bit-slice implementation with spare bit slices, and duplicating an entire network (this changes the topology of the larger system using the network). These approaches to fault tolerance can be applied to MINs. A number of techniques have also been developed tailored closely to the nature of MINs and their use.

As there are many unknown parameters in MINs, it is important to analyse MIN from complex systems approach. In the next sections, type of systems is described, characteristics of complex systems are discussed, key aspects of complexity are presented and finally a case study on SEN is analysed.

8.7 Type of Systems

Snowden and Boone [29] developed a hierarchy containing the following four type of systems:

(a) Simple
Known; in simple projects, operations are predictable and repeatable.
(b) Complicated
Known Unknown; in the case of complicated projects, problems are of coordination or specialised expertise: management is essentially linear.
(c) Complex
Unknown Unknown; in complex projects, it is not possible to know and understand all the features within any given situation, and there are ambiguity and uncertainty.
(d) Chaotic
Unclear boundary; in chaotic situations, there are high turbulence and no clear cause and effect relationships. Decisions need to be made quickly.

8.8 Characteristics of Complex Systems

There are many different typologies of system complexity. Boardman and Sauser [7] identified five characteristics of complex systems as presented below:

(i) Emergence; new properties develop through evolution
(ii) Autonomy; the ability of a system to make independent choices
(iii) Connectivity; the ability of a system to stay connected to other constituent systems
(iv) Diversity; evidence of heterogeneity between systems
(v) Belonging; systems have the right and ability to choose to belong to the system.

8.9 Aspects of Complexity

Emergence:

Emergence occurs as system characteristics and behaviours emerge from simple rules of interaction. Individual components interact and some kind of property emerges, something you could not have predicted from what you know of the component parts. Emergent behaviour then feeds back to influence the behaviours of the individuals that produced it [33, 27].

The emergent properties or patterns and properties of a complex system that emerge can be difficult to predict or understand by separately analysing various 'causes' and 'effects', or by looking just at the behaviour of the system's component parts. Emergent properties can be seen as the result of human action and not of human design.

Examples of emergent properties are structure, processes, functions, memory, measurement, creativity, novelty and meaning. While the nature of the entities, interactions and environment of a system are key contributors to emergence, there is no simple relationship between them. Emergence has been used to describe features such as social structure, human personalities, the internet, consciousness and even life itself. As one lucid account has it [23].

Nonlinearity:

Complexity science generally accepts that human systems do not work in a simple linear fashion as feedback processes between interconnected elements and dimensions lead to relationships that see change that is dynamic, non-linear and unpredictable [16]. Non-linearity is a direct result of the interdependence between elements of complex systems.

Linear problems can be broken down in a reductionist fashion, with each element analysed separately and the separate aspects can be recombined to give the right answer to the original problem. In a linear system, the whole is exactly equivalent to the sum of the parts.

Phase space:

The phase space of a system is the set of all the possible states—or phases—that the system can occupy. These states or phases can be mapped. The phase space is identified by noting all the dimensions that are relevant to understanding the system, then determining the possible values that these dimensions can take. This range of possible spaces can then be represented in either graphical or tabular form. Such a representation can be useful as a way to describe complex systems 'because it does not seek to establish known relationships between selected variables, but instead attempts to shed light on the overall shape of the system by looking at the patterns apparent when looking across all of the key dimensions [27]. Identifying patterns of interaction across different elements and dimensions of such systems in valuable. Phase space can be used to show how a system changes over time and the constraints that exist to change in the system.

Strange attractors, edge of chaos and far from equilibrium

The concept of phase space and attractors are central to understanding complexity, as complexity relates to specific kinds of system trajectories through phase space over time. The behaviour of complex systems can at first glance appear to be highly disordered or random. Moreover, these systems move through continually new states, with change as a constant in a kind of unending turbulence. However, there is an underlying pattern of order that is recognisable when the phase space of the system is mapped, known as a strange attractor [27].

Strange attractors show how complex systems move around in phase space, in shapes which resembles two butterfly wings. A complex system—such as the three-body planetary system, or the weather—would move around one loop of the attractor, spiralling out from the centre. When it got close to the edge of the 'wing' it would move over to the other 'wing' and spiral around again. Complex systems can have a chaotic dynamic and develop through a series of sudden jumps [13]. Such a jump, usually referred to as a bifurcation, is an abrupt change in the long-term behaviour of a system, to some critical value. As one gets close to the bifurcation points—which may be seen as those points where the system moves from one wing of the attractor to the other, the values of fluctuations increase dramatically [13], [27].

Adaptive agents:

All living things are adaptive agents. Individual people are adaptive agents, so are the teams they work in, and so are organisations. Some complex systems are said to be adaptive or evolving when individual adaptive agents respond to forces in their environments via feedback. Regardless of size and nature, adaptive agents share certain characteristics, in that they react to the environment in different ways [17]. Some adaptive agents may also be goal directed; still more may attempt to exert control over their environment in order to achieve these goals. Agents may have goals that can take on diverse forms, including desired local states; desired end goals; rewards to be maximised and internal needs (or motivations) that need to be kept within desired bounds. They can sense the environment and respond through physical or other behaviours or actions. They may also have internal information processing and decision-making capabilities, enabling them to compare the environmental inputs and their own behavioural outputs with their goals. They may anticipate future states and possibilities, based on internalised models of change (which may be incomplete and/or incorrect); this anticipatory ability often significantly alters the aggregate behaviour of the system of which an agent is part. They may also be capable of abstract self-reflection and internally generated sources of unpredictable conduct [15, 27].

Self-organisation:

Self-organisation is a form of emergent property and supports the notion that complex systems cannot be understood in terms of the sum of its parts, since they may not be understood from the properties of individual agents and how they may behave when interacting in large numbers. Racism provides an example as a result of segregated neighbourhoods in that racial attitudes develop. The economy is a self-organising system.

Mitleton-Kelly [22] adds that emergent properties, qualities, patterns or structures, arise from the interaction of individual elements; they are greater than the sum of the parts and may be difficult to predict by studying the individual elements. Emergence is the process that creates new order together with self-organisation. Mitleton-Kelly [22] reminds us that Checkland defines emergent properties as those exhibited by a human activity system as a whole entity, which derives from its component activities and their structure, but cannot be reduced to them [9]. The emphasis is on the interacting whole and the non-reduction of those properties to individual parts.

Co-evolution:

When adaptable autonomous agents or organisms interact intimately in an environment, such as in predator-prey and parasite-host relationships, they influence each other's evolution. This effect is called co-evolution, and it is the key to understanding how all large-scale complex adaptive systems behave over the long term. Each adaptive agent in a complex system has other agents of the same and different kinds as part of its environment. As the agent adapts to its surroundings, various elements of its surroundings are adapting to it and each other. One important result of the interconnectedness of adaptive bodies is the concept of co-evolution. This means that the evolution of one domain or entity is partially dependent on the evolution of other related domains or entities [19].

Fitness:

Complexity theory defines fitness as the ability to cope with complexity. To survive challenges and make the most of opportunity, a fit organism can process information about and deal with many variables. The theory posits that all life forms exist on a spectrum ranging from instability (chaos) to ultra stability (ordered hierarchy). Fitness is found in the middle ranges of this spectrum between rigid order and chaos—not in a crystal, where every atom resides in an ordered hierarchy; nor in gases whose molecules move at random. Move too far towards either pole, and you lose fitness [10].

Fitness landscapes:

Work in biology on fitness landscapes is an interesting illustration of competitive co-evolution [19]. A fitness landscape is based on the idea that the fitness of an organism is not dependent only on its intrinsic characteristics, but also on its interaction with its environment. The term 'landscape' comes from visualising a geographical landscape of fitness 'peaks', where each peak represents an adaptive solution to a problem of optimising certain kinds of benefits to the species. The 'fitness landscape' is most appropriately used where there is a clear single measure of the 'fitness' of an entity, so may not always be useful in social sciences [27].

8.10 Case Study: Shuffle-Exchange Networks

The shuffle-exchange multistage interconnection network (SEN) is one network in a large class of topologically equivalent MINs that include the omega, indirect binary n-cube, baseline and generalised cube. Figure 8.6 is an example of an 8×8 SEN. Each switching element (SE), the basic building block of a SEN, can be viewed as a 2×2 SEN. The SE can either transmit the inputs straight or has crossed connections.

A SEN has $N = 2^n$ inputs, termed sources (S), and 2^n outputs termed destinations (D). There is a unique path between each source–destination pair. The SEN has n stages and each stage has N/2 switching elements. The network complexity, defined

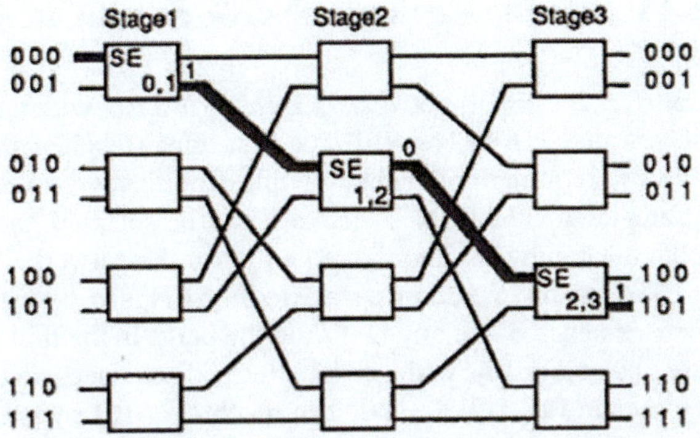

Fig. 8.6 8 × 8 Shuffle-exchange multistage interconnection network

as the total number of switching elements in the MIN, is $(N/2)$ $(\log_2 N)$, which for the 8 × 8 SEN is 12 SE's. The position of switching element *i* in stage *j* is represented by $SE_{i,j}$.

The SEN is a self-routing network. That is, a message from any source to a given destination is routed through the network according to the binary representation of the destination's address. For example, from S = 000 sends a message to $D = 101$, the routing can be described as follows: $S = 000$ presents the address of D = 101 plus the message for *D* to the *SE* in stage 1 to which $S = 000$ ($SE_{0,1}$) is connected. The first bit of the destination address (1) is used by $SE_{0,1}$ for routing. So output link 1 of $SE_{0,1}$ is used. At $SE_{1,2}$ the second bit of *D* (0) is used and output link 0 of $SE_{1,2}$ is chosen. Finally, at $SE_{2,3}$ the third bit of *D* (1) is used and output link 1 of $SE_{2,3}$ is selected. Figure 8.7 shows this *S–D* connection.

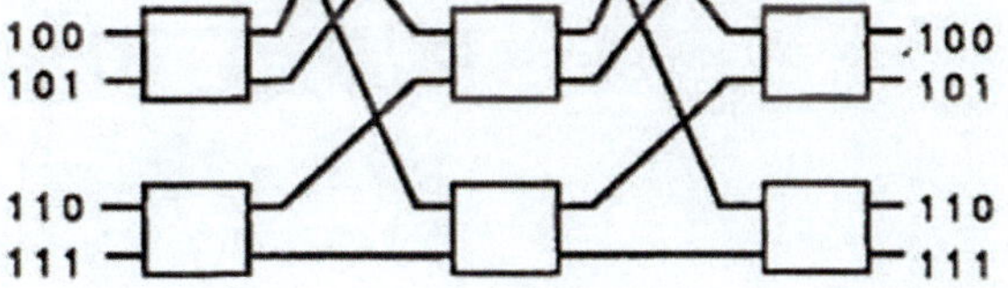

Fig. 8.7 Routing for communication between S = 000 and D = 101

8.11 Shuffle-Exchange Network with an Additional Stage

An $N \times N$ SEN + network is an $N \times N$ SEN with an additional stage. Figure 8.8 shows an 8×8 SEN + [6]. The first stage (labelled stage 0) is the additional stage and will require implementation of a different control strategy. While several control strategies for the SEN + network can be selected, the strategy chosen may affect both the bandwidth and the reliability of the network.

The reason for adding a stage to the SEN is to allow two paths for communication between each S and any D. While the paths in the first and last stages of the SEN + are not disjoint, the paths in the intermediate stages do traverse disjoint links. As can be seen in Fig. 8.9, $S = 000$ can reach $D = 101$ by two paths. In this case, the path redundancy is achieved in the SEN + at the expense of one extra stage added to the SEN.

The control strategy allows a switching element in stage 0 to use the T (straight) setting until a failure in a SE along the path from a given S to a given D is detected. At that time, the SE in stage 0 is placed in the X (exchange) setting for all future accesses between that S–D pair. In this way, it is shown that two paths between each S–D pair given that the failures occur only in the intermediate stages of the SEN +.

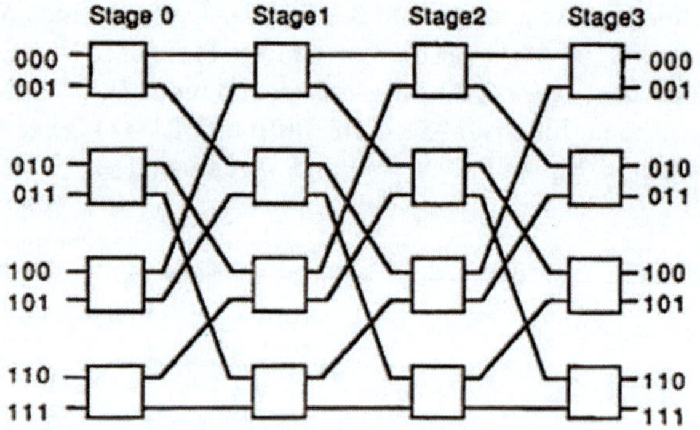

Fig. 8.8 8×8 SEN with an extra stage

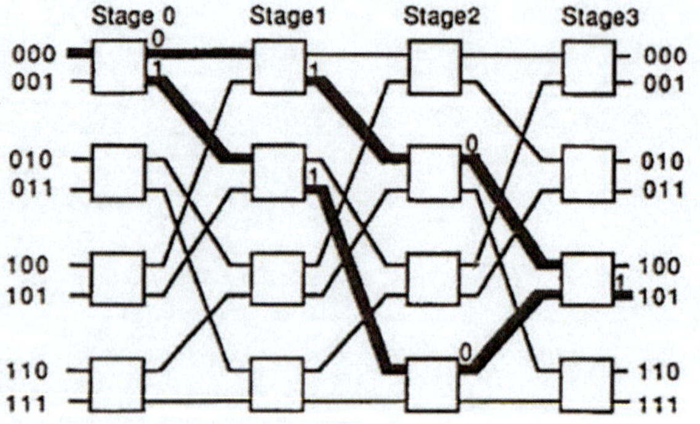

Fig. 8.9 Two paths for routing communication between S = 000 and D = 101 in the 8×8 SEN+

It is recognised that in actual implementations, the network should be reconfigured to reduce congestion.

Figure 8.8 shows that the switch complexity (the total number of switching elements in the network) for the 8×8 SEN + is 16. In general, the switch complexity for the $N \times N$ SEN + is $N/2$ ($\log_2 N + 1$). Thus, the additional cost of the SEN + is $N/2$ switches or a fractional increase in $1/\log_2$ N is small for a large N. In the next chapter, how much the increase in the redundancy improves the reliability of the SEN will be evaluated.

Shuffle-exchange networks have been widely considered as practical interconnection systems due to their size of its switching elements and uncomplicated configuration. Shuffle-exchange network (SEN) is a network among a large class of topologically equivalent MINs that includes omega, indirect binary n-cube, baseline and generalised cube [14].

A shuffle-exchange network is a unique path MIN [6, 8, 21, 31]. Therefore, there is only a single path between a particular input and a particular output. In this type of network, all switching elements are critical and assumed as series connection. The switching element (SE) can either transmit the inputs straight through itself or has cross-connections. The number of switches per stage and the number of links and the connection between stages are consistent. There is an eight-input/eight-output shuffle-exchange network with three stages, 12 switches (SEs), and 32 links.

SEN system has N inputs and N outputs. It has $n = \log_2 N$ stages and each stage has N/2 switching elements. In general, the network complexity for $N \times N$ SEN is $N/2$ ($\log_2 N$).

The following characteristics of complex systems and key aspects of complexity related to MINs/SEN:

Emergence: System failure cannot be predicted and switching elements are independent to each other. Therefore, there is no specific pattern on how connections are delivered between input and output stages in network systems.

Autonomy: Each type of MINs, in this case, SEN has its own network topology, which is the connections of all switching elements in the system. Every switching element also works independently and does not rely on other components.

Connectivity: Switching elements are set up in a network system through a number of stages. In 8×8 SEN above, it consists of 2×2 switching elements in three stages.

Diversity: There is flexibility in the arrangement of switching elements and number of stages. As shown in SEN, 2×2 switching elements are employed in all stages. In other MINs, different types of switching elements can be integrated.

Belonging: SEN is a part of MINs as it has the characteristics of interconnection network systems that interconnect a set of processors and a set of memory modules.

Nonlinearity: The type of switching elements, the number of switching elements and the number of stages vary. There is no linearity in terms of these topologies in network systems.

Phase space: MINs/SEN consists of stages of switching elements and this is a phase space to connect switching elements in input and output stages. There is a liberty to choose the type of switching elements and how connections will be arranged.

Strange attractors: It is clear to note that there are paths to connect various interconnections as described in terminal paths (between one to one switching element), broadcast paths (one to all switching elements) and network paths (all to all switching elements).

Adaptive agents: Different size of switching elements and number of stages are implemented in network systems to optimise terminal, broadcast and network reliability.

Self-organisation: The characteristics of MINs/SEN should be seen as how the whole network systems work and cannot be judged by the way individual switching element operates.

Co-evolution: Different types of switching elements can be integrated as they are designed to provide the fast and effective communication.

Fitness: All switching elements are set up to deliver optimum connection between input and output stages. Links and number of stages are interconnected in network systems to achieve this goal.

Fitness landscapes: The topology of network systems is measured on how reliable network systems are in terms of terminal, broadcast and network reliability. Other parameters include network complexity, system throughput, failure rate and cost.

8.12 Conclusion

Multistage Interconnection Networks (MINs) connect input devices to output devices through a number of switch stages, where each switch is a crossbar network. The number of stages and the connection patterns between stages determine the routing capability of the networks. MINs were initially proposed for telephone networks and later for array processors. In these cases, a central controller establishes the path from input to output. In cases where the number of inputs equals the number of outputs, each input synchronously transmits a message to one output, and each output receives a message from exactly one input. Such unicast communication patterns can be represented as a permutation of the input addresses. For this application, MINs have been popular as alignment networks for storing and accessing arrays in parallel from memory banks. Array storage is typically skewed to permit conflict-free access, and the network is used to unscramble the arrays during access. These networks can also be configured with the number of inputs greater than the number of outputs and vice versa. On the other hand, in asynchronous multiprocessors, centralised control and permutation routing are infeasible. In this case, a routing algorithm is required to establish the path across the stages of a MIN.

In this paper, characteristics of complex systems are mapped in SEN, which is a common network topology in MINs. This analysis describes key aspects of complex systems that are belonged to MINs. It is expected that the observation provides a clear description on the way SEN operates in complex environments. The key aspects of complex systems are important to be incorporated in complex system analysis to achieve final results effectively.

As described above for SEN case study, it is important to recognise the characteristics and key aspects of complex systems to fully understand the system behaviour. All these parameters have been discussed in detail in the context of MINs.

References

1. Abd-El-Barr, M., & Abed, O. (1995). Fault-tolerance and terminal reliability for a class of data manipulator networks. *Computer*, 225–229.
2. Adams, G. B., III, Agrawal, D. P., & Siegel, H. J. (1987). A survey and comparison of fault-tolerant multistage interconnection networks. *IEEE Transactions on Computers, 20*(6), 14–27.
3. Ball, M. O. (1986). Computational complexity of network reliability analysis. *IEEE Transactions on Reliability, R-35*(3).
4. Berge, C. (1973). *Graphs and Hypergraphs*, North-Holland.
5. Bertsekas, D., & Gallager, R. (1987). *Data networks.* NJ: Prentice-Hall.
6. Blake, J.T., & Trivedi, K. S. (1988). Reliabilities of two fault-tolerant interconnection networks. *Proceeding of the Eighteenth International Symposium on Fault Tolerant Computing*, 300–305.
7. Boardman, J., & Sauser, B. (2008). Systems thinking RC press, Boca Raton.
8. Booting, C., Rai, S., & Agrawal, D. P. (1994). Reliability computation of multistage interconnection networks. *IEEE Transactions on Reliability, 38*(1), 138–145.
9. Checkland, P. B. (1981). *Systems Thinking.* Systems Practice: John Wiley.
10. Clemens, W. C. (2001). Complexity theory as a tool for understanding and coping with ethnic conflict and development issues in post-soviet Eurasia. *International Journal of Peace Studies,, 6*(2), Autumn/Winter.
11. Colbourn, C. J. (1987). *The combinatorics of network reliability.* NY: Oxford University Press Inc.
12. Duato, J., Yalmanchili, S., & Ni, L. M. (1997). *Interconnection networks an engineering approach.* Los Alamitos, CA: IEEE Computer Society.
13. Feigenbaum, M. (1978). Quantitative universality for a class of nonlinear transformations. *Journal of Statistical Physics, 19*(1), 25–52.
14. Gunawan, I. (2008). Reliability analysis of shuffle-exchange network systems. *Reliability Engineering and System Safety, 93*(2), 271–276.
15. Harvey, D. (2001). Chaos and complexity: Their bearing on social policy research. *Social Issues, 1*(2), http://www.whb.co.uk/socialissues/index.htm.
16. Ireland, V. (2014). Complex project management 1 notes, The University of Adelaide.
17. ISCID. (2005). Encyclopaedia of science and philosophy, http://www.iscid.org/about.php.
18. Jolfaei, N.G., Jin, B., Gunawan, I., Vanderlinden, L., & Jolfaei, N.G. (2019). Reliability modelling with redundancy—A case study of power generation engines in a wastewater treatment plant. *Quality and Reliability Engineering International.*
19. Kauffman, S. (1995). *At home in the universe: The search for the laws of self-organisation and complexity.* London: Penguin Books.
20. Lee, K. Y., & Hegazy, W. (1986). The extra stage gamma network. *Computer*, 175–182.
21. Menezes, B. L., & Bakhru, U. (1995). New bounds on the reliability of augmented shuffle-exchange networks. *IEEE Transactions on Computers, 44*(1), 123–129.
22. Mitleton-Kelly, E. (2003). Ten complex systems and evolutionary perspectives on organisations: Complex systems and evolutionary perspectives on organisations: The application of complexity theory to organisations, Elsevier.
23. Newell, D. (2003). *Concepts in the study of complexity and their possible relation to chiropractic healthcare in Clinical Chiropractic, 6,* 15–33.
24. Ni, L. M. (1996). Issues in designing truly scalable interconnection networks. *Proceedings of the 1996 ICPP Workshop on Challenges for Parallel Processing*, 74–83.

25. Parker, D. S. & Raghavendra, C. S. (1984). The gamma network. *IEEE Transactions on Computers, C-33*(4), 367–373.
26. Provan, J. S. (1986). Bounds on the reliability of networks. *IEEE Transactions on Reliability, 35,* 260–268.
27. Ramalingam, B., Jones, H., Reba, T., & Young, J. (2008). Exploring the science of complexity ideas and implications for development and humanitarian efforts, London: ODI, Working Paper 285.
28. Siegel, H. J. (1985). *Interconnection networks for large scale parallel processing: Theory and case studies.* Lexington, MA: Lexington Books.
29. Snowden, D. J. & Boone, M. E. (2007). The leaders framework for decision making, Harvard Business Review, 69–76.
30. Thurber, K. J. (1979). Parallel processor architectures—Part 1: General purpose systems. *Computer Design, 18,* 89–97.
31. Trahan, J. L., Wang, D. X., & Rai, S. (1995). Dependent and multimode failures in reliability evaluation of extra-stage shuffle-exchange MINs. *IEEE Trans Reliability, 44*(1), 73–86.
32. Trivedi, K. S. (1982). *Probability and statistics with reliability.* Prentice-Hall, Englewood Cliffs, NJ: Queuing and Computer Science Applications.
33. Urry, J. (2003). *Global complexity.* Cambridge: Blackwell Publishing Ltd.
34. Zarghami, S. A., & Gunawan, I. (2019). A fuzzy-based vulnerability assessment model for infrastructure networks incorporating reliability and centrality. *Engineering, Construction and Architectural Management.*
35. Zarghami, S. A., Gunawan, I., & Schultmann, F. (2019). Exact reliability evaluation of infrastructure networks using graph theory. *Quality and Reliability Engineering International.*

Indra Gunawan is the Discipline Head of Management and Associate Professor in Complex Project Management in the Adelaide Business School, Faculty of the Professions, the University of Adelaide, Australia. He received his Ph.D. in Industrial Engineering and M.Sc. in Construction Management from Northeastern University, USA. Prior to joining the University of Adelaide, he was the post-graduate program coordinator for Maintenance and Reliability Engineering at Monash University. Previously, he has also taught in the Department of Mechanical and Manufacturing Engineering at Auckland University of Technology, New Zealand and worked as the Head of Systems Engineering and Management program at Malaysia University of Science and Technology (in collaboration with the MIT, USA). His current research interests include system reliability modelling, maintenance optimisation, project management, applications of operations research and operations management. He is actively involved in the Asset Management Council, a technical society of Engineers Australia.

Chapter 9
Evaluation and Design of Performable Distributed Systems

Naazira B. Bhat, Dulip Madurasinghe, Ilker Ozcelik, Richard R. Brooks, Ganesh Kumar Venayagamoorthy, and Anthony Skjellum

Abstract Performability measures system performance including quality, reliability, maintainability and availability over time, regardless of faults. This is challenging for distributed systems, since the internet was designed as a best-effort network that does not guarantee that data delivery meets a certain level of quality of service. In this chapter, we explain the design, test and performability evaluation of distributed systems by utilizing adversarial components. In our approach, the system design uses adversarial logic to make the system robust. In system test, we can leverage existing, powerful attacks to verify our design by using existing denial of service (DoS) attacks to stress the system.

Keywords Performability · Robust control · Game theory · Blockchain · Denial of service

9.1 Introduction

Critical infrastructure is geographically distributed and vulnerable. A performable system withstands and mitigates unfavourable conditions. Distributed system components share data and information. In practice, sensor malfunctions, data communication hijacking or external disturbances can occur. The system has to tolerate disturbances in operation. To design and test these systems, we integrate adversarial logic into our approach. Game theory is the mathematics used to model conflict between

N. B. Bhat · D. Madurasinghe · R. R. Brooks (✉) · G. K. Venayagamoorthy
Holcombe Department of Electrical and Computer Engineering,
Clemson University, Clemson, South Carolina, USA
e-mail: rrb@g.clemson.edu; rrb@acm.org

I. Ozcelik · A. Skjellum
SimCenter, University of Tennessee at Chattanooga, Chattanooga, Tennessee, USA

I. Ozcelik
Department of Computer Engineering, Recep Tayyip Erdogan University, Rize, Turkey

K. B. Misra (ed.), *Handbook of Advanced Performability Engineering*,
https://doi.org/10.1007/978-3-030-55732-4_9

rational decision-makers. In this chapter, we present a game theory-based approached for DCS design. Game analysis techniques improve performability.

We model the system as a Two-person Zero-sum (TPZS) game. We introduce different disturbances and countermeasures. We start by finding the best metric(s) to measure network performance. Zero sum means that the disturbances and countermeasures have no cooperation, the success of a disturbance is equal to the lack of performance of the countermeasure, and vice versa. Two person means we consider only affects on one system and cooperation between disturbance and countermeasure is impossible. If more than one metric is used, we either use a weighted sum of the individual metrics or we could always use the value of the worst metric (this is known as the H-infinity metric).

A game between these parties is established. For each component, we determine how it could malfunction to most severely sabotage the system. We then implement these behaviours. We then look at the other components to find how they could best counter these malfunctions to create a set of countermeasures.

We then create a set of simulations or experiments where we measure the results of each malfunction against each countermeasure. This provides a payoff matrix that we can use to analyze the different possible deviations from our initial design. The final resulting system should tolerate disturbances, which improves the performability of the system.

9.2 Game Theory-Based Robust Control Design

Game theory is the mathematics of competition scenarios. It models conflicts and cooperation between intelligent rational decision-makers. Game theory used general mathematical techniques to analyze the interactions in between parties that influence each other (Myerson [1]). It has been widely used in social science (Shubik [2]), economics (Leven [3], Chen [4]) and engineering (Trestian [5], Changwon [6]) applications. In our approach, we consider system deviations by positing that the system will fail in the worst possible way. We then embed in the system countermeasures that allow it to successfully adapt to these adverse conditions.

The rest of this section describes the mathematical tools we need to analyze the system. In Sect. 1.1.1, we describe the Z-test that tells us from our experimental analysis whether or not the effects of two specific malfunctions or countermeasures differ significantly. We then look in Sect. 1.1.2 as to whether one malfunction, or countermeasure, is always superior to another. If one approach is always superior (dominant), then we can discard the other one and simplify our design space. If we are lucky, the system recursively simplifies itself and we end up with a saddle point. For the saddle point, there is one failure mode that is more significant than all the others and it is tied to single countermeasure that best counters it. If this is not the case, Sect. 1.1.3, then the worst possible failure is a randomized combination of individual faults. Luckily, our approach finds this worst condition and also provides the best-randomized set of countermeasures to minimize the disturbance.

9.2.1 Z-Test

The Z-test is a statistical analysis based on the difference between the means of the test sample set and a mean of the population (Li [7]). This is a hypothesis test used for larger sample set (Bo [8]). Based on the central limit theorem, when the number of samples becomes larger, the sample average follows a Gaussian distribution with mean equal to the mean of the distribution and standard deviation, where σ is the standard deviation of the distribution and *n* is the number of the samples. The Z-test can be conducted on distributions, which Gaussian and the standard deviation are known (Hedge [9]). In our study, we use Z-test to compare two strategies in the game to find the dominant strategy under define significance level, which is useful in finding the saddle point of large payoff matrix. The selected level of significance (α) is 0.05. Considering two-tail tests, under the given significance level, we identify whether selected strategy (*R*) is dominant, is dominated or cannot conclude the dominance status compared with another strategy (*S*) based on the Z-score calculated using (9.1) and referring to Tables 9.1 and 9.2.

Table 9.1 Example game between players A & B

		Player B			
		J	K	L	M
Player A	P	$PO_{(PA)}$	$PO_{(PB)}$	$PO_{(PC)}$	$PO_{(PD)}$
	Q	$PO_{(QA)}$	$PO_{(QB)}$	$PO_{(QC)}$	$PO_{(QD)}$
	R	$PO_{(RA)}$	$PO_{(RB)}$	$PO_{(RC)}$	$PO_{(RD)}$
	S	$PO_{(SA)}$	$PO_{(SB)}$	$PO_{(SC)}$	$PO_{(SD)}$

Table 9.2 Dominance strategy selection conditions

Z-score	Conclusion
($Z_{(R,S)}$ >= 1.96) for all Player B strategies	*S* is dominated by *R*
($Z_{(R,S)}$ >= 1.96) for one or more Player B strategies & (1.96 > ($Z_{(R,S)}$ >= −1.96) for rest of the Player B strategies	*S* is dominated by *R*
(1.96 > $Z_{(R,S)}$ >= −1.96) for all Player B strategies	Uncertain
$Z_{(R,S)}$ values are on all three regions for all Player B strategies	Uncertain
(−1.96 > $Z_{(R,S)}$) for one or more Player B strategies & ($Z_{(R,S)}$ >= 1.96) for rest of the Player B strategies	Uncertain
(−1.96 > $Z_{(R,S)}$) for all Player B strategies	*R* is dominated by *S*
(−1.96 > $Z_{(R,S)}$) for one or more Player B strategies & (1.96 > $Z_{(R,S)}$ >= −1.96) for rest of the Player B strategies	*R* is dominated by *S*

$$Z_{(R,S)} = \frac{\bar{X}_R - \bar{X}_S}{\sqrt{\frac{\sigma^2{}_R}{n_R} - \sqrt{\frac{\sigma^2{}_S}{n_S}}}} \tag{9.1}$$

9.2.2 Dominant Strategies

Consider the game in Table 9.1. Consider two strategies *R* and *S* of Player A. Let us assume according to Table 9.2, the *R* strategy dominates the *S* strategy. We can remove the *S* strategy (a row in the example game) from the payoff matrix, since it is never a better choice than *R*. The same approach can be used for the Player B (column-wise).

9.2.3 Mixed Equilibria

John Forbes Nash Jr. won the Nobel Prize, by proving that every finite game must have at least one Nash equilibrium (Nash [10]). But there are instances where a game is not having a pure strategy Nash equilibrium. The matching penny game is an example. Hence, John Forbes Nash Jr. divides Nash equilibrium into two types, Pure Strategy Nash Equilibrium and Mixed Strategy Nash Equilibrium. To solve a mixed strategy and find Nash Equilibrium, we use a mixed strategy algorithm. Mixed strategy algorithm is based on probability distribution of pure strategies.

Let A choose option A and B with probability of p and $(1-p)$. Then, the following statements are derived and illustrated in Fig. 9.1's graph:

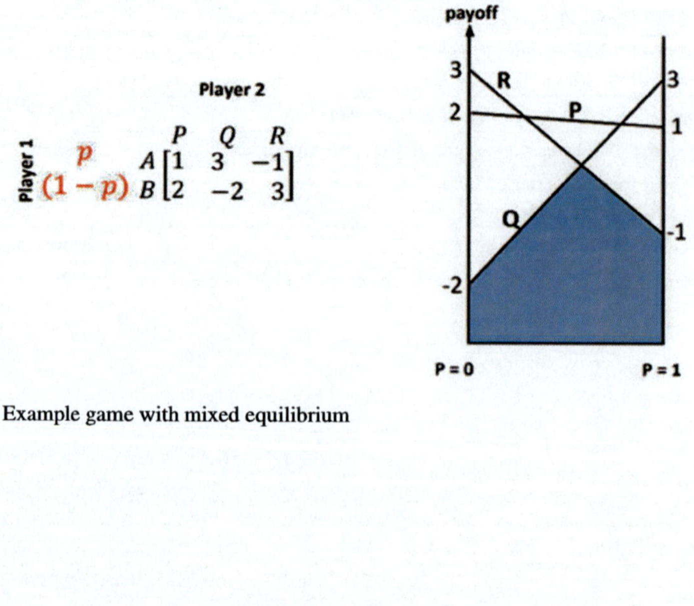

Fig. 9.1 Example game with mixed equilibrium

1. When player 2 chooses option P the payoff of player 1 $= p + 2(1 - p)$.
2. When player 2 chooses option Q the payoff of player 1 $= 5p - 2$.
3. When player 2 chooses option R the payoff of player 1 $= 3 - 4p$.

A mixed equilibrium game is shown in Fig. 9.1. Player 1's optimal strategy can be found from the graph, which is to play option A with 0.5556 probability and option B with 0.4444 probability. The payoff of the game is 0.7778.

The procedure of the design approach process is shown in Fig. 9.2. This is the generic model that can be used for any distributed control system application.

9.3 Evaluating System Robustness

Once a distributed system has been designed to maintain performability, it becomes important to verify that the system is in fact successful. To do this, we consider common approaches for disabling or degrading distributed systems. Luckily, criminals have been attacking systems on the internet for decades. Over time, they have created a set of common attacks that are used to destroy internet applications. The most robust attacks can be classified as denials of service. Our performability verification leverages decades of work by these malicious actors to be certain that the system continues to provide service even under the harshest circumstances.

The most successful tools are generally referred to as Distributed Denial of Service (DDoS) attacks (Ozcelik [11]). To verify our systems, we therefore integrate successful DDoS methods into our verification suite. This section looks at system verification for distributed ledger technologies.

Blockchain distributed ledgers have become one of the most frequently considered solutions for ensuring security of the storage of data and its transfer through decentralized, peer-to-peer networks in recent years. Blockchains are a data structure based on shared, distributed and fault-tolerant database that every participant in the network can have access to, but none of those can tamper with it. Being a cryptographic-based distributed ledger, trusted transactions are enabled among untrusted participants in the network using the blockchain technology. Blockchains assume that malicious nodes are present in the participating network but rely on the computational capabilities of the honest nodes to ensure that the exchanged information is resilient to manipulation. The absence of a centralized entity speeds up the entire process. Owing to the cryptographic structure of the blockchain, it is challenging to alter it. Based on these features, blockchains have drawn attention from a wide range of stakeholders including academics, healthcare and other government agencies.

Blockchain has seen a surge in interest among researchers, software developers and other industry practitioners because of security features like immutability it offers (Kan [12], Miller [13], Fiaidhi [14], Samaniego [15]). However, an important aspect for which the blockchain-based networks need to be tested is their vulnerability to D/DoS attacks. If successful, the entire data stored on the application can be rendered

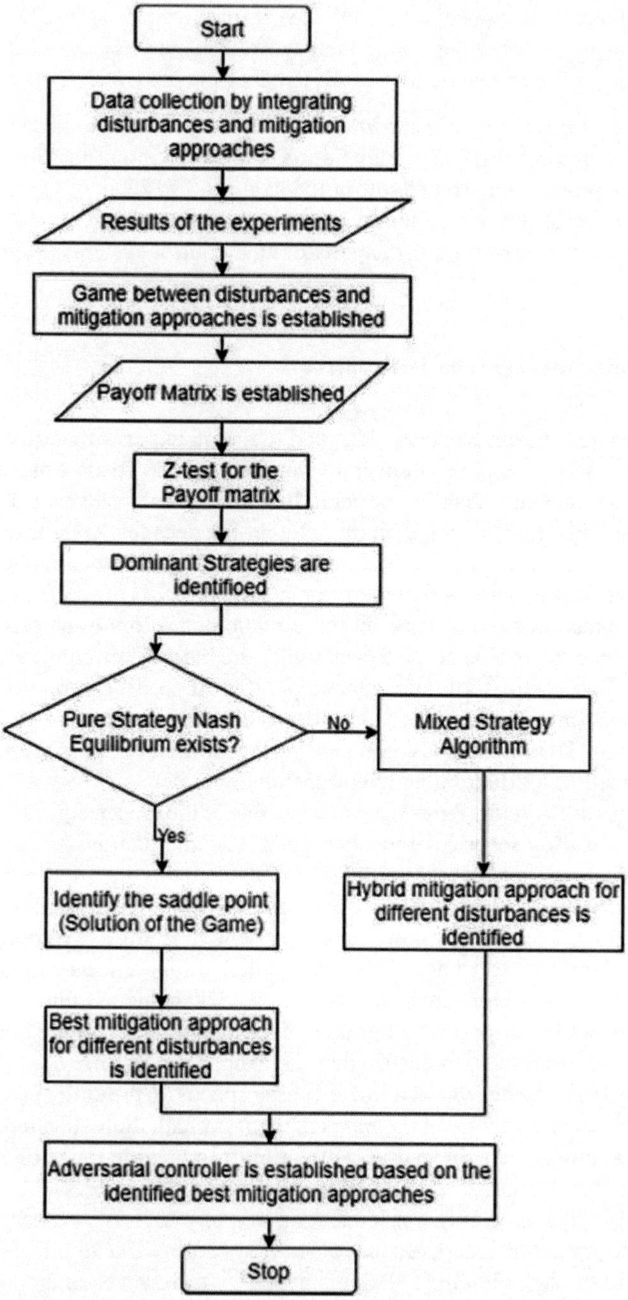

Fig. 9.2 Flow of the design process

useless and unavailable to its owners. This is even more critically important for time-critical applications such as healthcare. Therefore, network attacks on blockchain-based structures are to be taken seriously regardless of the intrinsic security properties of this data structure.

9.3.1 *Case study—A Blockchain-Based (Distributed) System*

The communication protocol used in our proposed distributed system is TCP. Therefore, we exploit the vulnerabilities of the TCP protocol for stress-testing our network. One of the most effective attacks against TCP is SYN floods. Hence, we focus on evaluating the performance of our proposed blockchain network in the presence of the SYN flood attack and its variants. It is described in Sect. 9.4.2 how the three-way TCP handshake is exploited to make the SYN floods effective. For our experiments, the attack is launched using IP-spoofing, randomized IP spoofing and the local area network denial-of-service (LAND) attack technique. The tool used to generate the attack traffic is Hping3, and the network analyzer used is Wireshark.

The tests have been conducted offline in the Network Security Lab at Clemson University's ECE department. The reason for offline testing is to ensure that the attack traffic does not disrupt the infrastructure of the university campus and that the effects of the attacks are contained within the lab.

Our distributed network uses Castro and Liskov's practical Byzantine-fault-tolerant (pBFT) algorithm to reach consensus. pBFT can tolerate a maximum value of faulty nodes (f) equal to less than $N/3$, where N is the total number of participating miners in each round. For our experiments, we used $N = 5$; and $f = 1$. The system architecture is shown in Fig. 9.3.

A brief description of the architecture components is as follows:

1. Clients:
   Clients submit data or transactions to the participating miners for registration over the blockchain. These are the users of our blockchain-based technology.

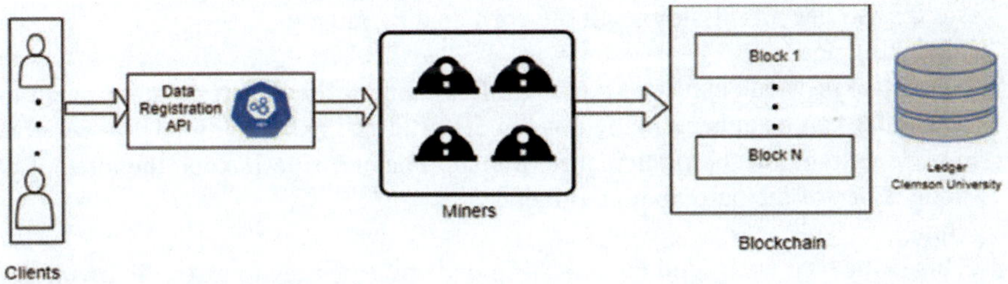

Fig. 9.3 Our distributed system architecture

2. Transactions:
Transactions are the data or files stored on the blockchain. These are the backbone of the provenance. Transactions can reference previous transactions if they are not the first transaction themselves, or they can be genesis events, (i.e. the first data collected from a particular use case of our blockchain network). We store the SHA-3 hash of the transaction on the server instead of the transaction itself. This helps to drastically reduce the size of the blockchain.
3. Blocks:
Blocks are one of the prime components of the system. A sequence of verified blocks forms the blockchain. The current block consists of hash of the previous block. This property makes the blockchain immutable. Blocks are added to the blockchain by miners or entities authorized to participate in the mining round.
4. Servers:

The servers are locally maintained and will hold the raw data comprising the ledgers in which the blockchains are held.

9.3.1.1 Experimental Setup

For the experiments, we have configured seven virtual machines (VMs) on seven different host machines—one on each host. A brief description of the components of the experiment is as follows:

1. Miners:
There can be N participating miners. For our experiments, we use $N = 5$, four of which are honest, and one of which is
malicious. Miners receive transactions submitted by the clients. Based on the algorithm described above, the selected miner mines the next block and appends it to the current length of the blockchain.
2. Client:
Clients submit their files to the miners with the intention of making their data secure and immutable. For our experiment, we have initially configured a single client node, which is submitting xml files with a wait time of 3 s between each successive file submission to all the participating miners.
3. Attacking node:
The attacking node can send attack traffic to any of the N participating miners. This node can be either a malicious miner participating in the mining process or a node external to the participating miners. For our experiments, the attacking node is one of the participating miners.
4. Hping3:
This is the tool we use on the attacking node to generate attack traffic owing to its versatility and simple usage.
5. Wireshark:
This is the tool we use as a network analyser on the victim's machine.

Blockchain update requests could be sent by either a new miner who attempts to join the mining network and needs to retrieve the length of blockchain mined so far before being able to participate in subsequent rounds or by a new client who joins using IP address of any miner but needs to know to the peer list of miners such that they are able to send transactions to all the participating miners simultaneously. For our experiments, we assume that the blockchain request is sent by a new miner who wants to join the mining network.

9.3.1.2 Results

To evaluate the performance of the blockchain-based network under the influence of each attack, we determine the amount of time the network takes to send a response to the blockchain update request by the requesting miner. This is the most strenuous step in the functioning of this distributed network. If the network is able to withstand torture testing in this step, we assume that it would perform well in rest of the steps. The miner who sends the response to the blockchain request is randomly selected and the selection is equiprobable. For reference, we also checked how much time it takes for the new miner to get the response when the network is not under the influence of any attack. It should be noted that SYN cookies are enabled on each of the machines involved in our experiments.

For all the cases, each of our mining rounds was successfully executed and the miner was randomly selected to produce the next block. The results for the response times are summarized in Table 9.3 and Fig. 9.4.

Table 9.3 Response times for different attack cases

Case	No. of attack packets sent		Response time of the victim [seconds]	
			When miner under attack gets selected to send the response to the blockchain request	When miner that is not under attack gets selected to send the response to the blockchain request
No attack	–		–	5
SYN flood with IP spoofing	Spoofed IP reachable	612195	20	5
	Spoofed IP unreachable	1643336	48	5
SYN flood with randomized IP spoofing		926612	30	5
LAND attack		1567141	47	5

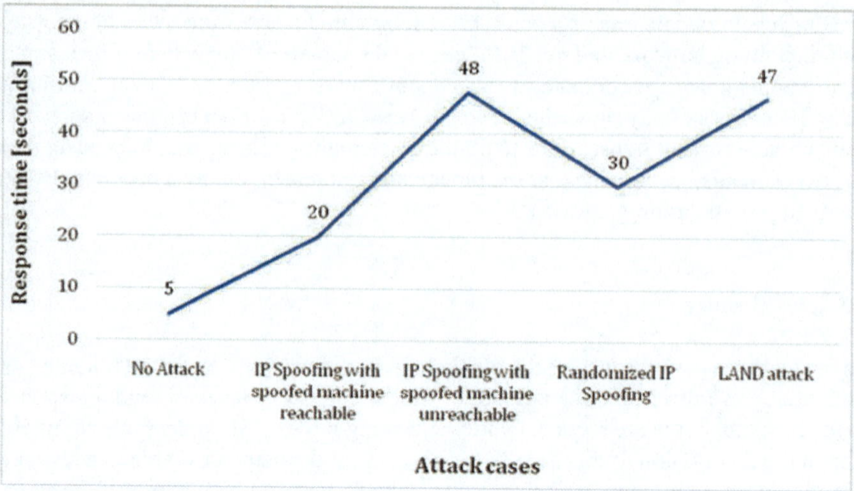

Fig. 9.4 Graphical representation of response times in case of no attack and different sub-cases of the SYN flood attack

9.3.1.3 Analysis

From results in Sect. 2.1.2, we observe that the network performance changes under the influence of an attack by a considerable extent. When the miner who gets selected to send the response to the blockchain request is under attack, the response time increases. The reason for the increase in response time is the incomplete TCP connection requests in the victim's TCP queue as a result of a large number of SYN packets sent by the attacker. On receiving the SYN packets, the victim responds with the SYN-ACK packet and keeps waiting for the final ACK packet from the node that initiated the SYN request. Until the TCP queue is reaped from the incomplete connection entries, the victim's resources get throttled. As a result, miner cannot process the requests coming from legitimate users while it is attacked. This is explained in detail in Sect. 9.3. The attack with maximum impact is the SYN flood using IP spoofing with spoofed machine not reachable. The maximum number of attack packets is sent in this case to the victim. The reason lies in the TCP parameter 'syn-ack-retries'. The victim is configured to keep retransmitting the SYN-ACK packet in response to the initial SYN request for a certain number of times. Since the attacker employed a spoofed IP address, and kept that machine powered off, the victim has to spend an additional time in trying to reach the machine first.

In all the attack cases with the victim being the selected miner to send the response, the performance of the network is degraded. This raises an important question about the reliability of distributed systems. Even though blockchain claims to ensure security and immutability, delay in the response of the network can cause many harmful implications. The stakes are higher for critical applications such as banking, healthcare and other important infrastructure.

However, an important observation is made when an unattacked miner gets selected to send the response to the blockchain request. The response time in this case is the same as when no attack is launched. This means potentially the attacker failed to launch a successful DoS attack on our network in this case. This implies that the probability of users experiencing denial-of-service while using our network is equal to the probability that an attacked miner gets selected to send a blockchain update response.

As mentioned earlier, the miner selection is random and equiprobable. Thus, the selection probability follows uniform distribution. So we can mathematically express this as:

$$\begin{aligned} & P(\text{Successful DoS attack on system}) \\ & = P(\text{Selected min er to send blockchain updates be the victim}) \end{aligned}$$

$$P(\text{Selected miner to send blockchain updates to be the victim}) = 1/N$$

Therefore, for our experiments, where $N = 5$:

$$P(\text{Successful DoS attack on system}) = 1/5$$

As N increases, the probability of a successful DoS attack on our distributed system decreases for a constant number of miners under attack simultaneously.

9.3.1.4 Inference

This raises an important question about the reliability of distributed systems. Even though blockchains claim to ensure security and immutability, delay in the response of the network can cause many harmful implications. The stakes are higher for critical applications such as e-banking, healthcare and other important infrastructure. Thus, distributed systems such as blockchains do not necessarily offer a considerable amount of availability and reliability unless designed to perform robustly.

9.4 Denial of Service

A Denial of Service (DoS) attack intentionally disables a system or a service to its legitimate users. Criminals generally perform these attacks by targeting the limited resources of a system. If an attacker uses more than one node to perform these attacks, it is called Distributed Denial of Service (DDoS) attack.

Targeting scarce system and network resources are common approaches used to perform a denial of service attack. These attacks are called resource-starvation attacks. The attacker may target system resources such as memory, disk space, CPU

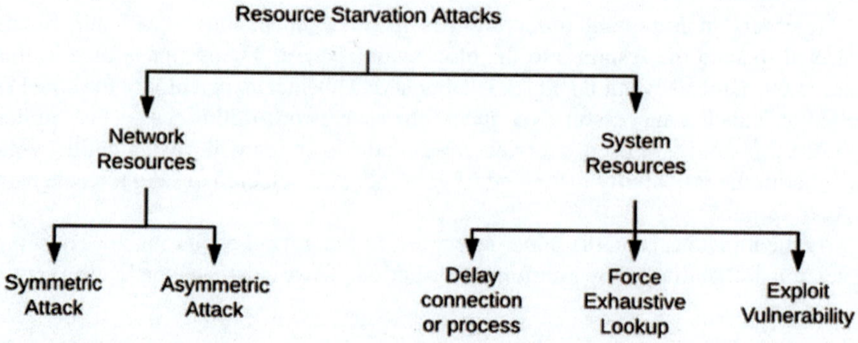

Fig. 9.5 Resource starvation (D)DoS attacks based on their main target

time or network bandwidth. Resource starvation attacks are easy to perform but hard to mitigate. Attackers make attack detection difficult by spoofing attack packets and utilizing multiple compromised nodes to generate the attack traffic. Additionally, using public network services, such as DNS and NTP, as a proxy to reflect and amplify attack traffic is a known approach utilized by attackers.

In this section, we focus on commonly used DDoS attacks targeting system and network resources to disable and/or disrupt distributed systems on the internet. We classify and present these attacks based on the resources they mainly target (See Fig. 9.5).

9.4.1 Network Resources

DDoS attacks targeting network resources are easy to perform and comprise a powerful way of disabling an online system. Attackers generally use network protocols, such as UDP, ICMP, HTTP and DNS, that do not authenticate the sender's identity. They flood the victim network with dummy traffic using zombie agents. The amount of attack traffic generated is proportional to the number of zombies used by the attacker. If zombies send attack traffic directly to the victim, it is called a symmetric DDoS attack. Attackers need a large number of zombie agents to perform an effective symmetric DDoS attack. In an asymmetric DDoS attack case, the attacker reflects and amplifies the attack traffic from an unprotected and misconfigured public network server; such as DNS, NTP and Memcached. Asymmetric DDoS attacks conceal the attackers' identity and amplify the attack strength based on the protocol exploited in the reflecting server (Ozcelik [11]).

9.4.2 System Resources

Hackers also attack limited resources of online systems and services to disable them. The purpose of these attacks is to consume all the CPU time, memory and HDD space of these systems and force them to deny legitimate users. Attackers use three general approaches to accomplish this goal: stalling communication or process to create long queues, forcing system for exhaustive lookups and exploiting vulnerabilities.

The TCP protocol defines how to establish a reliable connection and data exchange between two nodes. Attackers abuse certain protocol steps to stall the communication process during both the connection-establishment and data-exchange phases. In SYN Flood, the attacker sends many SYN packets to start new TCP connections at the victim server, but never follows up to complete the process. Eventually, the attacker uses all the memory space designated for TCP session records and forces the victim to deny service for legitimate users. Similarly, in low and slow communication attacks, the attackers prolong sessions as long as possible. In this case, the attacker establishes the TCP connection, but it slows down the data transfer rate to the minimum level required to keep the connection alive. The victim eventually reaches the maximum possible connection limits and the system denies the rest of the incoming connection requests. Slowloris, RUDY, and Slow read attacks are some of the examples of low and slow attacks.

A system needs to keep track of active TCP connections and perform a lookup to find the destination process when it receives a new packet. This lookup creates a bottleneck during peak hours. Attackers exploit this inherent weakness of TCP protocol in SYN-ACK Flood, ACK & PUSH ACK Flood and Fragmented ACK attacks. In SYN-ACK and ACK & PUSH ACK attacks, the victim server is bombarded with dummy SYN-ACK and ACK packets and overwhelmed with non-existing session lookups. In Fragmented ACK attack, the attacker sends ACK packets larger than network MTU. Therefore, the victim server deals with the defragmentation process in addition to TCP session lookups. Attackers use spoofed TCP FIN and TCP RST packets for the same purpose.

Attackers also exploit vulnerabilities of systems and protocols to perform denial of service. Fragmentation & Reassembly, killapache, and Local Area Network Denial (LAND) attacks are some of the examples in this category. In these attacks, attackers exploit a vulnerability to consume all available resources of the system. Ping of Death is an outdated attack that leveraged a vulnerability in the network stack. The attacker sent packet fragments that were larger than the system could handle after reassembly. Teardrop is also one of the more famous attacks, which targets the TCP/IP reassembly mechanism. In this attack, the attacker specially crafts packet fragments whose offset values overlap. This overwhelms the target during reassembly and causes it to fail. Similarly, a perl script released by a security researcher, whose screen name is Kingcope, sends specially crafted HTTP GET requests to exhaust the CPU and system memory of vulnerable Apache servers (Gulik [16]). In a LAND attack, the attacker specially crafts an SYN packet with the same source and destination

IP address. When the victim responds to the request, it creates an infinite loop that eventually causes the victim to crash.

9.4.3 DDoS Mitigation

For effective DDoS mitigation, attack detection and reaction systems need to work together. While efficient DDoS attack detection can be done at the attack target, reaction systems should be placed closer to the attack source. This requires a distributed system design in DDoS mitigation.

Today, most of the efficient and practical DDoS mitigation systems utilize contemporary networking and cloud technologies. Instead of reacting to DDoS attacks on premise by packet filtering using Firewalls and IDSs, attack reaction systems are moved to the cloud. Many companies, such as Cloudflare, Akamai and Arbor Networks, optimized their cloud infrastructure for DDoS reaction. These infrastructures, also called scrubbing centres, are used to separate attack traffic from the legitimate traffic. These companies claim that they can reroute victim traffic to their scrubbing centre and react to DDoS attacks in almost real time. The cost of this service depends on the size of victim service or network. In 2011, Verisign charged an average of $500,000 annually to large corporations for their DDoS mitigation service (Osborne [17]). In 2017, single attack mitigation was expected to cost around $2.5 Million (Osborne [17]). There are also DDoS mitigation solutions available for small and medium-sized businesses. By using on demand elastic cloud systems, many systems were developed, such as Deflect (Deflect [18]) and DDM (Mansfield [19]), to increase availability and reduce response time of a system. These systems aim to dissipate the overwhelming impact of DDoS attack by increasing the attack surface.

Using game theory, researchers have also proposed moving target-based DDoS mitigation approaches (Brooks [20], Dingankar [21], Venkatesan [22], Wright [23]). Moving Target Defense (MTD) approaches continuously change system configuration to reduce or else to move the attack surface. These configuration changes are described as a two-player game between an attacker and a defender and the most viable change is chosen by defender to make a successful attack difficult (Ozcelik [11]). Although, these approaches are mostly theoretical, they pave the way to build and defend performable distributed systems on the internet.

9.5 Summary

This chapter's goal is to provide practical guidelines for creating performable systems. We provide the following insights:

- Concepts from game theory should be integrated into system design.
- System testing should include intentional disruptions to verify performability.
- The internet can provide many tools that can be leveraged for testing a given system's ability to adapt to disruption.

Using these concepts, it is possible to create systems that adapt well under most condition.

References

1. Myerson, R. B. (1997). *Game theory: Analysis of conflict.* Harvard University Press.
2. Shubik, M. (1982). *Game theory in the social sciences* (Vol. 1). The MIT Press.
3. Leven, S. (1996). Models of market behavior: Bringing realistic games to market. *In IEEE/IAFE 1996 Conference on Computational Intelligence for Financial Engineering (CIFEr)*, (pp. 41–48).
4. Chen, M., Tian, S., & Chen, P. (2015). Evolutionary game analysis between service of public library and the investment of government. In *2015 7th International Conference on Intelligent Human-Machine Systems and Cybernetics* (Vol. 1, pp. 191–194).
5. Trestian, R., Ormond, O.,& Muntean, G. (2012). Game theory-based network selection: Solutions and challenges. *IEEE Communications Surveys Tutorials, 14*(4), 1212–1231, Fourth 2012.
6. Changwon Kim, R. L. (2014). Game theory based autonomous vehicles operation. *International Journal of Vehicle Design, 65*(4), 360–383.
7. Li, C. (2013). Z-test of computer-and-internet-aided multimodal English listening and speaking. In *2013 IEEE Third International Conference on Information Science and Technology (ICIST)* (pp. 98–101).
8. Bo, P., Hai, L., & Xing, J. (2009). A new sampling test method for maximum maintenance time of normal distribution items. In *2009 IEEE 10th International Conference on Computer-Aided Industrial Design Conceptual Design* (pp. 2210–2212).
9. Hegde, V., Pallavi, M. S. (2015). Descriptive analytical approach to analyze the student performance by comparative study using z score factor through r language. In *2015 IEEE International Conference on Computational Intelligence and Computing Research (ICCIC)* (pp. 1–4).
10. Nash, J. F. Equillibrium points in N-person Games. *Proceedings of the National Academy of Sciences of the United States of America, 36*(1), 48–49.
11. Ozcelik, I., & Brooks, R. (2020). *Distributed denial of service attacks: Real-world detection and mitigation.*
12. Kan, L., Wei, Y., Hafiz Muhammad, A., Siyuan, W., Linchao, G., Kai, H. (2018). A multiple blockchains architecture on inter-blockchain communication. In *2018 IEEE International Conference on Software Quality, Reliability and Security Companion (QRS-C)* (pp. 139–145).
13. Miller, D. (2018). Blockchain and the internet of Things in the industrial sector. *IT Professional, 20*(3), 15–18.
14. Fiaidhi, J., Mohammed, S., & Mohammed, S. (2018). EDI with blockchain as an enabler for extreme automation. *IT Professional, 20*(4), 66–72.
15. Samaniego, M., & Deters, R. (2016). Blockchain as a service for IoT. In *2016 IEEE International Conference on Internet of Things (iThings) and IEEE Green Computing and Communications (GreenCom) and IEEE Cyber, Physical and Social Computing (CPSCom) and IEEE Smart Data (SmartData)* (pp. 433–436)
16. van Gulik, D. -W. (2011). Retrieved from https://mail-archives.apache.org/mod_mbox/httpdannounce/201108.mbox/<20110824161640.122D387DD@minotaur.apache.org>.

17. Osborne, C. (2017). *The average DDoS attack cost for businesses rises to over $2.5 million*, ZDNet. Retrieved from https://www.zdnet.com/article/the-average-ddos-attack-cost-for-businesses-rises-to-over-2-5m/.
18. eQualit.ie. *Deflect*, Deflect. Retrieved from https://deflect.ca/#about.
19. Mansfield-Devine, S. (2011). DDoS: Threats and mitigation. *Network Security, 2011*(12), 5–12.
20. Brooks, R. R. (2004). *Disruptive security technologies with mobile code and peer-to-peer networks*. CRC Press.
21. Dingankar, C., & Brooks, R. R. (2007). Denial of service games. In *Proceedings of the Third Annual Cyber Security and Information Infrastructure Research Workshop* (pp. 7–17).
22. Venkatesan, S., Albanese, M., Amin, K., Jajodia, S., Wright, M. (2016). A moving target defense approach to mitigate DDoS attacks against proxy-based architectures. In *2016 IEEE Conference On Communications And Network Security (CNS)*, (pp. 198–206). IEEE.
23. Wright, M., Venkatesan, S, Albanese, M., & Wellman, M. P. (2016). Moving target defense against DDoS attacks: An empirical game-theoretic analysis. In *Proceedings of the 2016 ACM Workshop on Moving Target Defense* (pp. 93–104).

Naazira B. Bhat is a second-year Master's student of Holcombe Department of Electrical & Computer Engineering, Clemson University, USA and is working as a Research Assistant for Dr. Richard Brooks. She is currently working on an NSF-funded project 'Provenance Assurance using Cryptocurrency Primitives' and her work focuses on evaluating the reliability and availability of the proposed blockchain technology when exposed to cyberattacks like the D/DoS. Naazira is currently doing her second Master's degree. She received her first Master's degree from SMVD University, Jammu, India and Bachelor's degree from the University of Kashmir, India with major in Electronics and Communication Engineering. Her research interests include cybersecurity, network traffic analysis, system security and blockchain.

Dulip Madurasinghe is a second-year Ph.D. student of Department of Electrical & Computer Engineering, Clemson University. Research Assistant for Dr. Brooks working on control resiliency analysis based on game theory under the objective of distributed cyber-physical system robustness. His research interests are cybersecurity for smart grid and autonomous vehicle. He was an embedded engineer for Atlas Labs (pty) Ltd. from 2017 to 2018. He received his Bachelor from University of Moratuwa, Sri Lanka major in Electrical Engineering in 2017.

Ilker Ozcelik received the MS degree in electrical engineering from Syracuse University in 2010, and the Ph.D. degree in electrical engineering from the Holcombe Department of Electrical and Computer Engineering, Clemson University, Clemson, South Carolina in 2015. His research interests include network traffic analysis, network security, software-defined networking, blockchain, security and privacy in intelligent systems.

Richard R. Brooks has in the past been PI on research programs funded by the Air Force Office of Scientific Research, National Science Foundation, Department of Energy, National Institute of Standards, Army Research Office, Office of Naval Research and BMW Corporation. These research projects include coordination of combat missions among autonomous combat vehicles (ARO), situation and threat assessment for combat command and control (ONR), detection of protocol tunnelling through encrypted channels (AFOSR), security of intelligent building technologies (NIST), experimental analysis of Denial of Service vulnerabilities (NSF), mobile code security (ONR) and security analysis of cellular networks used for vehicle remote diagnostics (BMW). He received his BA in mathematical sciences from Johns Hopkins University and Ph.D. in computer science from Louisiana State University.

Ganesh Kumar Venayagamoorthy is the Duke Energy Distinguished Professor of Power Engineering and Professor of Electrical and Computer Engineering and Automotive Engineering at Clemson University. Prior to that, he was a Professor of Electrical and Computer Engineering at the Missouri University of Science and Technology (Missouri S&T), Rolla, USA from 2002 to 2011 and Senior Lecturer in the Department of Electronic Engineering, Durban University of Technology, Durban, South Africa from 1996 to 2002. Dr. Venayagamoorthy is the Founder (2004) and Director of the Real-Time Power and Intelligent Systems Laboratory (https://rtpis.org). He holds an Honorary Professor position in the School of Engineering at the University of Kwazulu-Natal, Durban, South Africa. Dr. Venayagamoorthy received his Ph.D. and M.Sc. (Eng) degrees in Electrical Engineering from the University of Natal, Durban, South Africa, in February 2002 and April 1999, respectively. He received his BEng (Honors) degree with a First Class from Abubakar Tafawa Balewa University, Bauchi, Nigeria in March 1994. He holds an MBA degree in Entrepreneurship and Innovation from Clemson University, SC (2016). Dr. Venayagamoorthy's interests are in the research, development and innovation of smart grid technologies and operations, including computational intelligence, intelligent sensing and monitoring, intelligent systems, integration of renewable energy sources, power system optimization, stability and control, and signal processing. He is an inventor of technologies for scalable computational intelligence for complex systems and dynamic stochastic optimal power flow. He has published over 500 refereed technical articles. His publications are cited ~17,000 times with an h-index of 63 and i10-index of 255. Dr .Venayagamoorthy has been involved in over 75 sponsored projects in excess of US $12 million. Dr. Venayagamoorthy has given over 500 invited keynotes, plenaries, presentations, tutorials and lectures in over 40 countries to date. He has several international educational and research collaborations. Dr. Venayagamoorthy is involved in the leadership and organization of many conferences including the General Chair of the Annual Power System Conference (Clemson, SC, USA) since 2013, and Pioneer and Chair/co-Chair of the IEEE Symposium of Computational Intelligence Applications in Smart Grid (CIASG) since 2011. He is currently the Chair of the IEEE PES Working Group on Intelligent Control Systems, and the Founder and Chair of IEEE Computational Intelligence Society (CIS) Task Force on Smart Grid. Dr Venayagamoorthy has served as Editor/Guest Editor of several IEEE Transactions and Elsevier Journals. Dr Venayagamoorthy is a Senior Member of the IEEE and a Fellow of the IET, UK and the SAIEE.

Anthony Skjellum studied at Caltech (BS, MS, Ph.D.). His Ph.D. work emphasized portable, parallel software for large-scale dynamic simulation, with a specific emphasis on message-passing systems, parallel non-linear and linear solvers, and massive parallelism. From 1990 to 1993, he was a computer scientist at the Lawrence Livermore National Laboratory focusing on performance-portable message passing and portable parallel math libraries. From 1993 to 2003, he was on the faculty in Computer Science at Mississippi State University, where his group coinvented the MPICH implementation of the Message Passing Interface (MPI) together with colleagues at Argonne National Laboratory. From 2003–2013, he was professor and chair at the University of Alabama at Birmingham, Department of Computer and Information Sciences. In 2014, he joined Auburn University as Lead Cyber Scientist and led R&D in cyber and High-Performance Computing for over three years. In Summer 2017, he joined the University of Tennessee at Chattanooga as Professor of Computer Science, Chair of Excellence, and Director, SimCenter, where he continues work in HPC (emphasizing MPI, scalable libraries, and heterogeneous computing) and Cybersecurity (with strong emphases on IoT and blockchain technologies). He is a senior member of ACM, IEEE, ASEE, and AIChE, and an Associate Member of the American Academy of Forensic Science (AAFS), Digital & Multimedia Sciences Division.

Chapter 10
Network Invariants and Their Use in Performability Analysis

Ilya Gertsbakh and Yoseph Shpungin

Abstract Network-type systems with binary components have important structural parameters known in literature as Signature, Internal Distribution, D-spectra and BIM-spectra. The knowledge of these parameters allows obtaining the probabilistic description of network behaviour in the process of their component failures, and probabilistic description of such network parameters as resilience, component importance, system failure probability as a function of component failure probability q, and the approximation to reliability if q tends to 0. When the network has many components, the exact calculation of Signatures or D-spectra becomes a very complicated issue. We suggest using efficient Monte Carlo procedures. All relevant calculations are illustrated by examples of networks, including flow in random networks and network structural comparison in the process of network gradual destruction process.

Keywords Network structure invariants · Network resilience · Shock model · Reliability approximation

10.1 Introduction

Reliability and availability [1] are two important probabilistic attributes of performability of any network or system with binary states. This chapter lays down discussion of important structural parameters known as Signature, Internal Distribution, D-spectra and BIM-spectra. Before we do that some definitions are provided to make the discussion meaningful.

Prof. Gertsbakh has sadly passed away prior to the publication of this manuscript.

I. Gertsbakh
Ben Gurion University, Beer Sheva, Israel

Y. Shpungin (✉)
Shamoon College of Engineering, Beer Sheva, Israel
e-mail: yfromb@gmail.com

K. B. Misra (ed.), *Handbook of Advanced Performability Engineering*,
https://doi.org/10.1007/978-3-030-55732-4_10

10.1.1 Networks, Node and Edge Failures. Success Criteria

We meet networks every day and everywhere in our life. For the formal study of network properties, we must operate with abstract models of networks. In further, our principal network model will be a triple $N = (V, E, T)$, where V is the vertex or node set, E is the edge or link set and T is a set of special nodes called terminals, $T \in V$. In simple words, a network is a collection of circles (nodes) and links (edges), i.e. line segments connecting the nodes. Terminals are special nodes that do not fail and they are represented as bold circles, like in Fig. 10.1.

Our exposition will be centred around network behaviour when its elements (nodes and/or links) fail. We will deal with so-called binary elements that can be in two states up and down denoted by 1 and 0, respectively. When speaking about links, link i failure means that this link is erased, i.e. it does not exist.

The state of link $i, i = 1, \ldots, n$ is denoted by binary variable x_i. If $x_i = 1$, link i is up; if $x_i = 0$, link i is down. x_i is often called link indicator variable. In some models, the elements subjected to failure are network nodes (vertices). If the indicator variable of node j is $y_j = 0$, i.e. node j is down, it means that all links incident to node j are erased, but the node itself remains intact. By an agreement, the terminals do not fail.

By network state, we mean the set of all its elements (nodes and edges) that are in up state. We will distinguish network UP (operating) and DOWN (non-operating) states according to a certain criterion.

Below we give several examples of different UP and DOWN criteria. All examples relate to the network are shown in Fig. 10.1. This network has two terminals: 1 and 6.

Terminal Connectivity Criterion. Nodes Unreliable, Edges Reliable
We say that the network is UP if each pair of terminals is connected by a path of non-erased elements. Let node 2 is up, and nodes 3, 4, 5 are in the down state. Then the network is UP, because the node 2 connects terminals 1 and 6. Let now nodes 2 and 4 are down, and nodes 3 and 5—up. Then the network is DOWN.

Terminal Connectivity Criterion. Edges Unreliable, Nodes Reliable
Suppose that edges (1,2), (2,5) and (5,6) are up, and all other edges are down. The network is UP. Let now the edges (1,2) and (1,4) be down, and all other edges are up. The network is DOWN.

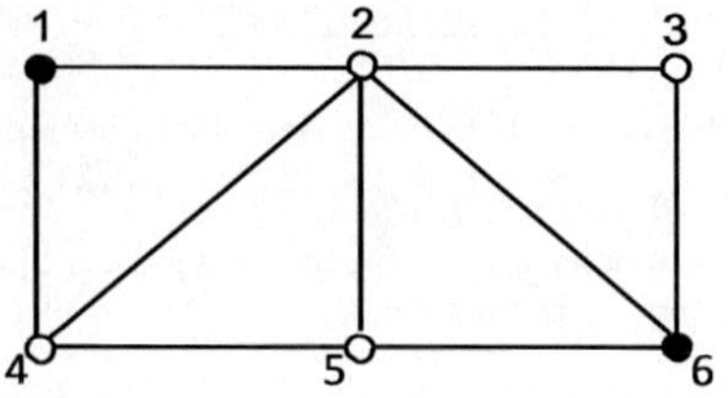

Fig. 10.1 Network with six nodes and nine edges. Nodes 1 and 6 are Terminals

Max Component and Max Cluster Criteria. Nodes Unreliable

For this example, we need the definition of a *component*. A subset $V_1 \subset V$ is called an *isolated component* of N if all nodes of V_1 are connected to each other and there are no edges of type $e = (a, b)$, where $a \in V_1$ and $b \in V - V_1$. An isolated node is considered as an isolated component. The *size* of a component is the number of nodes in it.

We say that the network is in UP state if the maximal component has at least x nodes, where x is some given number. Suppose that $x = 3$. Let the nodes 3 and 5 are up, and the rest of the nodes—down. Then the maximal component consisting of nodes 3, 5, 6 and edges (3,6), and (5,6) is of size 3, and therefore N is UP. Let now nodes 3 and 4 are up, and nodes 2, 5 are down. Obviously N is DOWN.

By the definition, an isolated component of N is called a *cluster* if it contains at least one terminal node. We say that the network is *UP* if it contains a cluster of size at least x, where x is some given number. Let $x = 4$. If the nodes 2 and 5 are up and 3, 4 are down, then we have a cluster of size 4, and N is UP. If only nodes 3 and 5 are up then we have maximal cluster of size 3, and N is DOWN. (Recall that the terminals are always up.)

Further we will use the notions cut and min-cut. Appropriate definition follows.

Definition 1 A subset of network unreliable elements $(c_1, c_2, \ldots, c_k)$ is called a cut if the following condition is satisfied:

If all these elements are in state down, then the network is also in state DOWN.

A cut is called minimal (min-cut) if after removing any element, the new subset is no more a cut.

Consider for example, the network in Fig. 10.1 for the case when the nodes are unreliable. The subset of nodes (2, 3, 5) is cut, but not a min-cut. Indeed, if node 3 is removed, the remaining subset (2, 5) is still a cut. It is obvious that (2,5) is min-cut.

10.2 Destruction Spectrum and Network Reliability

10.2.1 D-Spectrum and CD-Spectrum

Definition 2 Let $\pi = e_{i_1}, e_{i_2}, \ldots, e_{i_n}$ be a permutation of all unreliable elements (edges or nodes). Start with a network with all elements being up and 'erase' the elements in the order they appear in π, from left to right. Stop at the first element e_{i_r} when the network becomes DOWN. The ordinal number r of this element is called the anchor of permutation π and denoted $r(\pi)$.

Remark 1 Note that the anchor value for given π depends only on the network structure and its DOWN definition. It is completely separated from the stochastic mechanism that governs the node or edge failures in a real network destruction process.

Example 1 Consider the network shown in Fig. 10.1. In this network with two terminals 1 and 6, the edges are reliable and nodes 2, 3, 4, 5 are unreliable. Consider an arbitrary permutation π of node numbers, e.g. $\pi = (3, 5, 2, 4)$. We start the destruction process with all nodes in the *up* state. Erase one node after another in the order prescribed by π, from the left to right. The network becomes DOWN after erasing the third node, i.e. node 2. So we have $r(\pi) = 3$.

Definition 3 Let x_i be the number of permutations such that their anchor equals i. The set

$$D = \left\{ d_1 = \frac{x_1}{n!}, \ldots, d_n = \frac{x_n}{n!} \right\} \tag{10.1}$$

is called the D-spectrum of the network.

Remark 2 'D' in Definition 1 refers to the 'destruction' process of erasing network elements from left to right in the permutation π. *D*-spectrum is distribution of the anchor value, and obviously $\sum_{i=1}^{n} d_i = 1$. Numerically, the *D*-spectrum coincides with the so-called Signature introduced first in (Samaniego 1985, see [2]). It was proved there that if system elements fail independently and their lifetimes X_i have identical continuous distribution function $F(t)$, then the system lifetime distribution $F_S(T) = \sum_{i=1}^{n} d_i \cdot F_{i:n}(t)$ where $F_{i:n}(t)$ is the cumulative distribution function of the i th order statistics in random sample $X_1, X_2, \ldots, X_n$.

Example 1 (continued) Table 10.1 shows all 24 permutations of the nodes. The nodes destruction of which caused the failure of the network are marked by asterisk. Directly from this table we get $x_1 = 0, x_2 = 8, x_3 = 10, x_4 = 6$, and *D*-spectrum of the network equals $(d_1 = 0, d_2 = 1/3, d_3 = 5/12, d_4 = 1/4)$.

Definition 4 Let $y_b = \sum_{i=1}^{b} d_i, b = 1, 2, \ldots, n$. Then the set $(y_1, y_2, \ldots, y_n)$ is called the Cumulative *D*-spectrum (CD-spectrum).

Remark 3 Like an anchor, both spectra (*D* and CD) depend only on the network structure and the definition of network DOWN state. That is, they are invariant with respect to the up/down probabilities of the elements.

Table 10.1 All permutations of nodes

Column 1	Column 2	Column 3	Column 4
2,3,4*,5	3,2,4*,5	4,2*,3,5	5,2*,3,4
2,3,5*,4	3,2,5*,4	4,2*,5,3	5,2*,4,3
2,4*,3,5	3,4,2*,5	4,3,2*,5	5,3,2*,4
2,4*,5,3	3,4,5,2*	4,3,5,2*	5,3,4,2*
2,5*,3,4	3,5,2*,4	4,5,2*,3	5,4,2*,3
2,5*,4,3	3,5,4,2*	4,5,3,2*	5,4,3,2*

The following theorem establishes an important combinatorial property of the CD-spectrum.

Theorem 1 *Let $C(i)$ be the number of cut sets of size i in the network. Then*

$$C(i) = y_i \cdot \frac{n!}{i!(n-i)!} \tag{10.2}$$

The proof of Theorem 1 can be found in the textbook [3].

Example 1 (continued)

We get the following CD-spectrum of our network: ($y_1 = 0, y_2 = 1/3, y_3 = 3/4, y_4 = 1$).

Using formula (10.2), we get: $C(1) = 0, C(2) = 2, C(3) = 3, C(4) = 1$.

10.2.2 Network Reliability and CD-Spectrum Monte Carlo

The following theorem gives an expression of network reliability using CD - spectrum.

Theorem 2 *If all $p_i = p$, then network static reliability $R(N)$ can be expressed in the following form:*

$$R(N) = 1 - \sum_{i=1}^{n} y_i \cdot \frac{n!q^i p^{n-i}}{i!(n-i)!} \tag{10.3}$$

It is clear that even for relatively small networks, the exact calculation of network CD-spectrum is extremely difficult. Below we present Monte Carlo algorithm for estimating the CD spectrum.

Algorithm 1: Evaluation of CD-spectrum

1. **Initialize** all a_i to be zero, $i = 1, \ldots, n$.
2. **Simulate** permutation π of all elements.
3. **Find out** the anchor $r(\pi)$.
4. **Put** $a_r = a_r + 1$.
5. **Put** $r = r + 1$. **If** $r \leq n$ **GOTO 4.**
6. **Repeat** 2–5 M times.
7. **Estimate** y_i via $\hat{y}_i = \frac{a_i}{M}$.

Figure 10.2 shows a network with 32 unreliable nodes and 60 reliable edges. Nodes 4, 13, 27, 30 are terminals. Table 10.2 shows CD-spectrum for grid network with unreliable nodes and for terminal connectivity criterion. Table 10.3 shows CD-spectrum for the same network, but with unreliable edges and maximal cluster criterion ($x = 25$). Both spectra were obtained using algorithm 1 with $M = 10{,}000$.

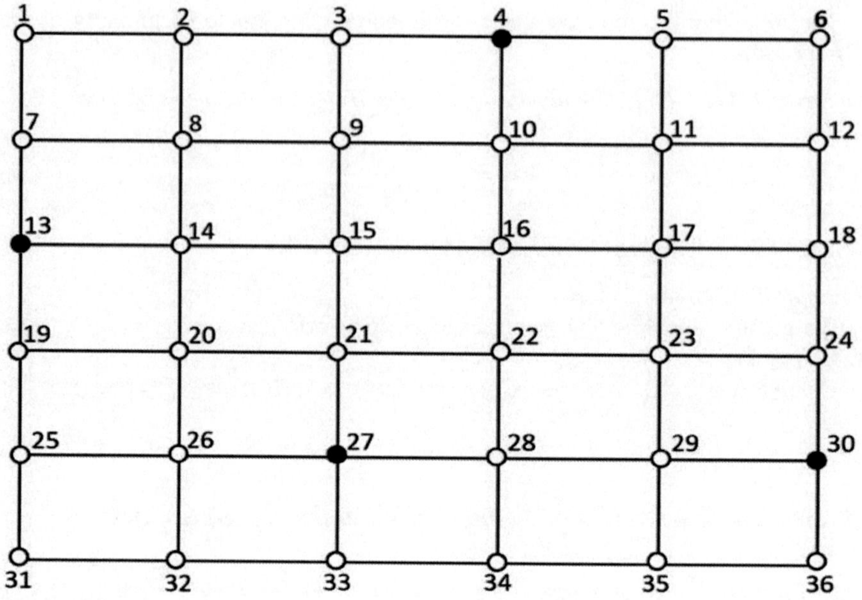

Fig. 10.2 Grid network with 32 unreliable nodes and 60 reliable edges. Nodes 4, 13, 27, 30 are terminals

Table 10.2 Grid CD-spectrum. Nodes unreliable. Terminals $T = (4, 13, 27, 30)$ terminal connectivity criterion

i	y_i	i	y_i	i	y_i	i	y_i
1	0	9	.2069	17	.9664	25	1
2	0	10	.3188	18	.9824	26	1
3	.0011	11	.4490	19	.9924	27	1
4	.0046	12	.5838	20	.9967	28	1
5	.0137	13	.7070	21	.9986	29	1
6	.0352	14	.8078	22	.9996	30	1
7	.0677	15	.8811	23	1	31	1
8	.1210	16	.9322	24	1	32	1

Remark 4 Using CD-Monte Carlo for calculating network reliability has several advantages over other methods, including the following two.

1. Since CD-spectrum is an invariant, **once estimated** it, we can calculate network reliability for any values of p.
2. Using this method, we avoid the so-called rare event phenomenon [3].

Table 10.3 Grid CD-spectrum. Edges unreliable. Terminals T = (4, 13, 27, 30). Maximal cluster criterion, $x = 25$

i	y_i	i	y_i	i	y_i	i	y_i	i	y_i
1	0	9	.0001	17	.0089	25	.2642	33	.9660
2	0	10	.0002	18	.0147	26	.3500	34	.9881
3	0	11	.0002	19	.0227	27	.4510	35	.9971
4	0	12	.0002	20	.0370	28	.5641	36	.9996
5	0	13	.0004	21	.0593	29	.6757	37	1
6	0	14	.0013	22	.0899	30	.7791	38	1
7	0	15	.0026	23	.1321	31	.8657	39	1
8	0	16	.0044	24	.1891	32	.9291	40-60	1

10.2.3 *Two Alternative Methods for Evaluating Network Reliability*

Method 1: Crude Monte Carlo

A common method for evaluating network reliability is the Crude Monte Carlo (CMC) method. We present below the corresponding algorithm.

Algorithm 2: CMC

1. **Set** $Y = 0$
2. **For** each element i, **Simulate** its state with probability p_i
3. **Check** the network state in accordance with given criterion
4. **If** the network state is UP **Then** $Y := Y + 1$
5. **Repeat** steps 2, 3, 4 M times.
6. **Estimate** $\hat{R}$ as $\hat{R} = \frac{Y}{M}$

In many cases, using CMC gives good results, but unlike the CD-Monte Carlo, it has some drawbacks, including the following two.

1. For each p value, it is necessary to restart the simulation process.
2. The main disadvantage of CMC is the presence of a rare event phenomenon. That is, if $p \to 1$, then the relative error $r.e.(\text{CMC}) \to \infty$. Therefore CMC is not suitable for evaluating very reliable networks, which is actually an important practical case.

Method 2: Burtin–Pittel Approximation

Burtin–Pittel approximation provides rather accurate network reliability estimates for the case of a highly reliable network and independent and equal element unreliability $q_i = q$.

Assume that $q \to 0$, that is the network is highly reliable. Let the number of min-cuts of a minimal size r is equal to s.

Table 10.4 Grid network reliability by CMC, Destruction Monte Carlo (DMC) and Burtin–Pittel approximation (B–P)

p/Algorithm	CMC	DMC	B–P
0.7	0.6786	0.678619	0.865
0.8	0.9126	0.909937	0.96
0.9	0.9922	0.991761	0.995
0.95	0.9995	0.999185	0.999375
0.99	1	0.9999944	0.999995

Then by Burtin-Pittel approximation

$$Q(N) = 1 - R(N) \approx s \cdot q^r. \tag{10.4}$$

Note that this approximation was first suggested in a more general form by Burtin and Pittel (see [4]).

We explain this approximation using the example of the network in Fig. 10.1. As we have seen in the above example, the unreliability of this network is $Q(N) = 1 - R(N) = 2q^2p^2 + 3q^3p + q^4$. The main term here (when $q \to 0$) is $2q^2p^2$, where 2 is the number of min-cuts of minimal size. Clearly that $2q^2p^2 = 2q^2(1 + o(1))$ as $q \to 0$.

Consider now the grid network in Fig. 10.2. Obviously, the minimal size of the min-cuts is 3. All min-cuts of size 3 are as follows: (3,5,10), (7,14,19), (24,29,36), (24,29,34), (24,29,35). So in this case, we get $1 - R(N) \approx 5 \cdot q^3$.

Table 10.4 presents grid network reliability for different values of p, calculated using CMC, Destruction Monte Carlo (DMC), both with $M = 10{,}000$, and also Burtin-Pittel approximation (B-P). Comparing CMC and DMC we see a good correspondence up to $p = 0.95$. However, starting from $p = 0.99$ reliability values obtained using CMC with $M = 10{,}000$ will be 1.

Note that here we see the rare event phenomenon.

For example, we want to estimate the reliability of the order of 0.99999 with relative error at least 10%. (Note that rel.arr. (CMC) $= \frac{\sqrt{R}}{\sqrt{M \cdot (1-R)}}$.) Then we get $M = 10{,}000{,}000$.

A more detailed comparison of these methods can be found in [5].

As for B-P method, we see that it gives good approximation starting from $p = 0.9$.

10.3 Network Resilience

One of the important concepts in the analysis of the network behaviour under random attack on its elements is *network resilience*.

Definition 5 *Probabilistic resilience [6]* Assume that network element failures appear in *random order*, i.e. all $n!$ orderings are equally probable.

Let N be a network with n elements. The probabilistic resilience $\mathrm{res}_{\mathrm{pr}}(N; \beta)$ is the largest number of element failures such that N is still UP with probability $1 - \beta$. Formally,

$$\mathrm{res}_{\mathrm{pr}}(N, \beta) = \max\left\{ I : \sum_{i=1}^{I} P(N, i) \leq \beta \right\}.\#$$

The concepts of resilience and CD-spectrum are closely related. From the CD-spectrum, we can get the network resilience for any β.

Consider, for example, CD-spectrum shown in Table 10.2, and let $\beta = 0.01, 0.05, 0.1, 0.3, 0.5$. Then we get:

$$\begin{aligned}
&\mathrm{res}_{\mathrm{pr2}}(N, 0.01) = 2, \mathrm{res}_{\mathrm{pr}}(N, 0.05) = 6,\\
&\mathrm{res}_{\mathrm{pr}}(N, 0.1) = 7, \mathrm{res}_{\mathrm{pr}}(N, 0.3) = 9,\\
&\mathrm{res}_{\mathrm{pr}}(N, 0.5) = 11.
\end{aligned}$$

Consider now CD-spectrum shown in Table 10.3. We get:

$$\begin{aligned}
&\mathrm{res}_{\mathrm{pr2}}(N, 0.01) = 17, \mathrm{res}_{\mathrm{pr}}(N, 0.05) = 20,\\
&\mathrm{res}_{\mathrm{pr}}(N, 0.1) = 22, \mathrm{res}_{\mathrm{pr}}(N, 0.3) = 25,\\
&\mathrm{res}_{\mathrm{pr}}(N, 0.5) = 27.
\end{aligned}$$

Note that resilience is also an invariant of the network, since it depends solely on the network topology and criterion UP/DOWN.

10.4 Birnbaum Importance Measure (BIM) and BIM-Spectrum

In this section, we introduce the Birnbaum Importance Measure (BIM) [3, 7] of network element j, $j = 1, 2, \ldots, k$. Let network reliability $R(p_1, p_2, \ldots, p_k)$ be a function of element reliability p_i. Then BIM of element j is defined as

$$\begin{aligned}
\mathrm{BIM}_j = \frac{\partial R(p_1, p_2, \ldots, p_n)}{\partial p_j} &= R\big(p_1, p_2, \ldots, 1_j, \ldots, p_n\big)\\
&- R\big(p_1, p_2, \ldots, 0_j, \ldots, p_n\big) \qquad (10.5)
\end{aligned}$$

BIM has a transparent probabilistic meaning: it is the gain in network reliability received from replacing a down element j by an absolutely reliable one. BIM_j gives the approximation to the network reliability δR resulted from element j reliability increment by δp_j.

Table 10.5 BIM-spectrum for network in Fig. 10.1

i	z_{i2}	z_{i3}	z_{i4}	z_{i5}
1	0	0	0	0
2	1/3	0	1/6	1/6
3	3/4	1/2	1/2	1/2
4	1	1	1	1

The use of BIM in practice is limited since usually the reliability function $R(p_1, p_2, \ldots, p_k)$ is not available in explicit form. However the BIM-spectrum that we define below allows to estimate the element BIM's without knowing the analytic form of the reliability function [3].

Definition 6 Denote by Z_{ij} the number of permutations satisfying the following two conditions:

(1) If the first i elements in the permutation are down, then the network is DOWN.
(2) Element j is among the first i elements of the permutation.

The collection of $z_{ij} = Z_{ij}/k!$ values, $i = 1, \ldots, k$; $j = 1, \ldots, k$, is called BIM-spectrum of the network. The set of z_{ij} values for fixed j and $i = 1, \ldots, k$ is called BIM$_j$-spectrum, or the importance spectrum of element j.

Example 2 Let us return to the network in Fig. 10.1, and using Table 10.1 calculate one of the Z_{ij} values, say Z_{42}. The permutations that satisfy the condition of the above definition are the following: (2,4,3,5), (2,4,5,3), (4,2,3,5), (4,3,5,3). Table 10.5 presents the BIM-spectrum for our network.

The columns in this table are the BIM$_j$ spectra.

The following theorem [3] demonstrates how BIM$_j$ can be calculated without using the reliability function.

Theorem 3 *BIM$_j$, $j = 1, \ldots, k$, equals,*

$$\mathrm{BIM}_j = \sum_{i=1}^{n} \frac{n!(z_{i,j} \cdot q^{i-1} p^{n-i} - (y_i - z_{i,j}) q^i p^{n-i-1}}{i!(n-i)!} \tag{10.6}$$

Note that $y_k - z_{kj} = 0$, which means that in the second term of the numerator of (10.6) one can assume that index i changes from 1 to $k - 1$.

Remark 5 BIM-spectrum depends only on the network structure and the definition of network DOWN state. That is, this is invariant with respect to the up/down probabilities of the elements.

The exact calculation of BIM-spectra is a formidable task, but we can estimate the spectra using Monte Carlo approach. An appropriate algorithm [3, 5] simultaneously estimates the CD-spectrum and the BIM-spectra for all network elements.

Sometimes in the problems of analysis and design of networks, we do not need to know the values of the importance of the elements. We want to know how the elements are ranked by importance. The following theorem [3, 5] allows us to compare elements without calculating their BIM's.

Theorem 4 *Suppose the BIM's for the network are given. Let us fix two indices α and β, $\alpha \neq \beta$, and the corresponding $Z_{i,\alpha}$ and $Z_{i,\beta}$ values. Then if for all i, $i = 1, \ldots, k$, $Z_{i,\alpha} \geq Z_{i,\beta}$, then $\mathrm{BIM}_\alpha \geq \mathrm{BIM}_\beta$ for all p values.*

For the network in Fig. 10.1, comparing the columns in Table 10.5 we get:

$$\mathrm{BIM}_2 > \mathrm{BIM}_4 = \mathrm{BIM}_5 > \mathrm{BIM}_3$$

Additional information on BIM's can be found in [8].

10.5 Border States

10.5.1 Border States and Reliability Gradient

In this section, we introduce the so-called network *border states* that are closely related to the reliability gradient.

Definition 7 Reliability gradient vector ∇R is defined as,

$$\nabla R = \left[\frac{\partial R}{\partial p_1}, \ldots, \frac{\partial R}{\partial p_k}\right] \tag{10.7}$$

In words: component i of ∇R is BIM_i.

For the following definition, it is more convenient for us to determine the state of the network as a vector of element indicator variables, i.e. state $\mathbf{x} = (x_1, \ldots, x_k)$, where $x_i = 1$, if element i is up, and $x_i = 0$ otherwise.

Definition 8 Network state $\mathbf{x} = (x_1, \ldots, x_k) \in$ DOWN is called the neighbour of the state $\mathbf{y} = (y_1, \ldots, y_k)$ if $\mathbf{x}$ differs from $\mathbf{y}$ in exactly one position. If $\mathbf{y} \in UP$ then we call the $\mathbf{x}$ *border state*. The set of all border states is called the *border set* and denoted as DN*.

Remark 6 It is clear from the definition 8 that the border state and also the border set are network invariants.

Example 3 Consider the network in Fig. 10.1. Its state is determined by the vector of nodes indicators (x_2, x_3, x_4, x_5). (Recall that nodes 1 and 6 are terminals.) For example $(x_2 = 0, x_3 = 0, x_4 = 1, x_5 = 0) \in$ DOWN is the neighbour of $(x_2 = 1, x_3 = 0, x_4 = 1, x_5 = 0) \in$ UP (and also the neighbour of

$(x_2 = 0, x_3 = 0, x_4 = 1, x_5 = 1) \in \text{UP}$). So x is a border state. The border set for our network is

$$\begin{aligned} \text{DN}^* = \{ & v_1 = (0,0,0,0), v_2 = (0,1,0,0), v_3 = (0,0,1,0), \\ & v_4 = (0,0,0,1), v_5 = (0,1,1,0), v_6 = (0,1,0,1).\} \end{aligned}$$

To clarify the connection between border states and gradient, we introduce an *artificial evolution process* [3, 9] on network elements.

Assume that at $t = 0$ each element is down. Element i is born after random time $\tau_i \sim \exp(\mu_i)$. After the 'birth', the element remains up 'forever'. Note that for fixed time t_0, $P(\tau_i > t_0) = q_i = e^{-\mu_i t_0}$.

The following theorem [3, 9] opens the way to calculating the reliability gradient.

Theorem 5 *Let* $P(v; t)$ *be the probability that the network is in state* v *at time* t*. Denote by* $\Gamma(v)$ *the sum of* μ_i *over all set of indices* i *such that* $v + (0, \dots 1_i, 0, \dots 0) \in \text{UP}$. *Formally*

$$\Gamma(v) = \sum_{v \in DN^*, v+(0,\dots,1_i,0,\dots,0) \in UP} \mu_i \tag{10.8}$$

Then the following equation holds:

$$\nabla R \bullet \{q_1\mu_1, \dots, q_k\mu_k\} = \sum_{v \in DN^*} P(v; t)\Gamma(v), \tag{10.9}$$

where by $\bullet$ denoted scalar profuct.

We see from the last equation, that knowing the probabilities of border states, we can calculate the reliability gradient. In most cases, the explicit expression of these probabilities is not available. However, formula 10.9 makes possible using the well-known Lomonosov's algorithm [3, 9] for estimating $P(v; t)$ and ∇R.

Here we restrict ourselves to a brief description of the idea of the Monte Carlo algorithm of estimation the gradient.

First of all, we introduce an evolutionary process on network elements, as described above. We recall that for fixed time t_0 element i is in the state up with probability p_i. Now consider a sequence in an evolution process. This sequence starts from a zero state w_0. Let $\pi = (i_1, i_2, \dots, i_k)$ be some permutation of the network elements, so that i_1 has a minimum birth time, i_2 was born the second, and so on. We associate with this permutation a sequence of network states: a state w_1 in which only i_1 is up, a state w_2 with two elements in up, and so on, up to the first state UP. This sequence of states we call the trajectory $w = (w_0, w_1, \dots, w_s)$. Consider, for example, the network with unreliable nodes is shown in Fig. 10.1. Suppose node 4 is born first, node 5 is born next and node 3 is born third. Then we get the following trajectory:

$$w = \{w_0 = (x_2 = 0, x_3 = 0, x_4 = 0, x_5 = 0),$$

$$w_1 = (x_2 = 0, x_3 = 0, x_4 = 1, x_5 = 0),$$

$$w_2 = (x_2 = 0, x_3 = 0, x_4 = 1, x_5 = 1),$$

$$w_3 = (x_2 = 0, x_3 = 1, x_4 = 1, x_5 = 1) \in UP\}$$

Note that w_2 is one of the border states.

The following is a simplified algorithm for estimating the gradient.

Algorithm 3: Evaluation of Gradient

1. **Put** $\frac{\partial R}{\partial p_1} = 0, i = 1, \ldots, n$.
2. **Generate** trajectory $w = (w_0, w_1, \ldots, w_s)$.
3. **Find** the first j so that w_j is a border state.
4. **Calculate convolution** $\text{Conv} = P(\tau(w_j) \le t_0) - P(\tau(w_{j+1}) \le t_0)$,

   where $\tau(w_j)$ and $\tau(w_{j+1})$ are the birth time of w_j and w_{j+1}, respectively.
   For each $x_i \in \Gamma(w_j)$ **calculate** $\frac{\partial R}{\partial p_1} = \frac{\partial R}{\partial p_1} + \text{Conv}$.
5. **Put** $j = j + 1$. **If** $j < s$ **Goto 4.**
6. **Repeat** 2–5 M times.
7. **For each** $i = 1, \ldots, n$ **put** $\frac{\partial R}{\partial p_i} = \frac{\partial R}{\partial p_i} / M \cdot q_i$.

Detailed explanation of the algorithm as well as an analytical expression for calculating the convolution of exponentials are given in [3].

10.5.2 Border States and Availability

Let us now consider the following dynamic model. Each network element, independently of others, alternates between two states: up and down. When element i is up, it has failure rate λ_i. if it is down, it has repair rate μ_i. In equilibrium element i is up with probability $p_i = \mu_i/(\mu_i + \lambda_i)$. Let T_U and T_D be the average UP and DOWN periods of the network in equilibrium. Our goal is to find these periods.

It is known [4] that the network availability $Av(N)$ can be expressed as follows:

$$Av(N) = R(p_1, p_2, \ldots, p_k) = \frac{T_U}{T_U + T_D} \tag{10.10}$$

The value $\rho = \frac{1}{T_U + T_D}$ is called network DOWN $\to$ UP *transition rate*. The following theorem shows the relationship between the transition rate and the border states.

Theorem 6 It can be shown in [3, 9],

$$\rho = \sum_{v \in DN^*} P(v)\Gamma(v) \tag{10.11}$$

Example 4 Consider the network in Fig. 10.1. Assume that node i has failure rate λ_i and repair rate μ_i. Rewrite the network border set obtained above.

$$DN^* = \{v_1 = (0, 0, 0, 0), v_2 = (0, 1, 0, 0), v_3 = (0, 0, 1, 0),$$
$$v_4 = (0, 0, 0, 1), v_5 = (0, 1, 1, 0), v_6 = (0, 1, 0, 1)\}.$$

Now by (10.11) we get:

$$\rho = P(v_1)\mu_2 + P(v_2)\mu_2 + P(v_3)(\mu_2 + \mu_5)$$
$$+ P(v_4)(\mu_2 + \mu_4) + P(v_5)(\mu_2 + \mu_5) + P(v_6)(\mu_2 + \mu_4).$$

For convenience, let $\mu_i = \mu, \lambda_i = \lambda$, for all $i = 2, 3, 4, 5$. Suppose that $\mu = 4$, $\lambda = 1$. Then $p = 0.8$, $q = 0.2$. Using a little arithmetics, we get $\rho = \mu\big(q^4 + 5pq^3 + 4p^2q^2\big) = 0.544$. Now, easy to get $R(N) = p + p^2 - p^3 = 0.928$. Finally using (10.10) and (10.11), we obtain: $T_U = 1.706$, $T_D = 0.132$.

In the case of a large network, the Lomonosov's algorithm adapted for this purpose is used.

Let Ω be the set of all trajectories. We can rewrite (10.11) in the following form:

$$\rho = \sum_{w \in \Omega}(w)P(v|w)\Gamma(v),$$

where $Pr(w)$ is the probability of the trajectory w (see [3], Chap. 9), and v is the border state determined by the trajectory w. Now, simulating the trajectories and using the corresponding variant of the Lomonosov's algorithm we obtain the availability estimate.

Remark 7 An extremely efficient Lomonosov's algorithm is based on ingenious graph-theoretic construction known as an evolution process on so-called Lomonosov's 'turnip' [3], Chap. 9. This algorithm has a number of useful properties. Let us mention some of them.

1. The algorithm is a highly effective tool for calculating the reliability of monotone systems, for any criteria UP, and for arbitrary (not necessarily equal) element probabilities up.
2. The algorithm avoids the occurrence of a rare event phenomenon. Indeed, a distinctive feature of a rare event is that the relative error in estimating the probability of this event tends to infinity. In the Lomonosov's algorithm, the random

choice of the trajectories does not depend on the probabilities of the elements and this explains the absence of this phenomenon.
3. It can be used to evaluate the mean stationary UP and DOWN periods.
4. It can be used to evaluate reliability gradient ∇R.

Detailed description of the algorithm and its applications can be found in the book [3], Chap. 9.

10.6 Examples

In this section, we present several examples of using the network invariants described above.

10.6.1 Network Reliability Improvement

Consider the network with unreliable nodes in Fig. 10.2. Assume that all nodes are in state up with probability $p = 0.7$. The network reliability (see Table 10.4) equals then $R = 0.6786$. Our goal is to increase network reliability to $R^* = 0.8$. Suppose it is possible to replace several nodes with more reliable ones, say with up probability $p^* = 0.9$, and we are interested in doing minimal number of such replacements. A good heuristic approach to solve this problem is the following.

First, rank all the nodes in descending order of their BIM's. Next, successively replace the nodes with more reliable ones until we get the required reliability.

The calculations performed show that all the nodes can be divided into several groups according to their importance. In particular, the first group consists of one node—29, the second group consists of four nodes: 10, 14, 24 and 28. We write it for clarity as follows:

$$\mathrm{BIM}_{29} > (\mathrm{BIM}_{10} = \mathrm{BIM}_{14} = \mathrm{BIM}_{24} = \mathrm{BIM}_{28})$$

This conclusion is based on the analysis of the network BIM-spectra.

Further, replacing the nodes 29, 10, 14, 24 with more reliable ones, we achieve the desired reliability $R^* = 0.8169$.

Partially, the BIM-spectrum data for the nodes 1, 10, 14, 24, 28, 29 are presented in Table 10.6.

Spectrum values in the range of 20–32 are not shown. These values are almost the same, since the probability of network failure starting from step 20 is very close to 1, and at step 23 is already equal to 1. From the table, we see that the BIM spectrum values of node 29 are greater than those of the other nodes. Spectrum values for nodes 10, 14, 24, 28 are close and intertwined. Node 1 does not belong to the first two groups.

Table 10.6 Grid Network BIM-spectrum. Nodes unreliable. Terminals T = (4,13,27,30)

i	1	3	5	7	10	13	15	17	19
Z_1	0	0	.0010	.0101	.0810	.2144	.3990	.5089	.5923
Z_{10}	0	.0002	.0046	.0230	.1185	.3102	.4280	.5187	.5953
Z_{14}	0	.0002	.0040	.0201	.1217	.3155	.4378	.5266	.5968
Z_{24}	0	.0008	.0069	.0264	.1217	.2965	.4150	.5131	.5899
Z_{28}	0	0	.0020	.0156	.1136	.3103	.4235	.5167	.5928
Z_{29}	0	.0008	.0073	.0274	.1291	.3225	.4394	.5270	.5945

Remark 8 It should be noted that the problem described above can also be solved taking into account the cost of replacing elements. More information on network analysis and optimal network design can be found in [3, 6].

10.6.2 Resilience of Flow Network

In this section, we consider the flow network. These networks are important in many applications. By the definition, flow network is a directed network, where each edge (a, b) has a flow capacity $c(a, b)$. The flow delivered from a to b cannot exceed $c(a, b)$. Denote by s and t the source and sink nodes of the network. Denote by Maxflow the maximal flow from s to t when all edges are up. We say that the network is in DOWN state if its maximal flow is below some fixed level Φ. (Note that there exists an extensive literature with several fast algorithms for finding the maximum flow in networks.)

Let us consider now the network shown in Fig. 10.3. It has 16 reliable nodes and 30 unreliable and directed edges. The nodes 1 and 16 are the source and sink,

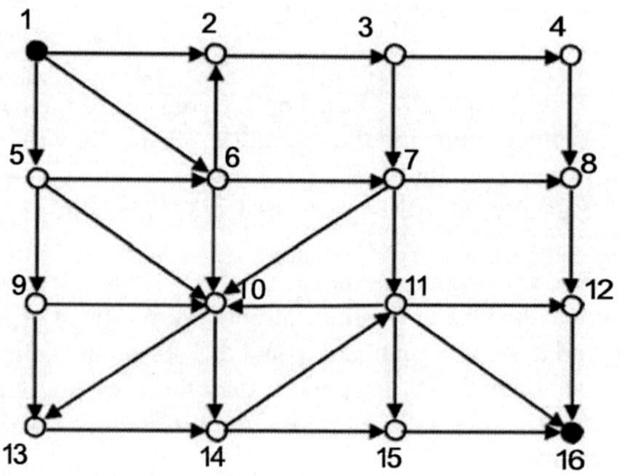

Fig. 10.3 Flow network with 16 reliable nodes and 30 unreliable edges

respectively. The corresponding capacities are given in Table 10.7. The calculated value of Maxflow equals 26.

Table 10.8 presents 14 values of the network spectra for $\Phi = 10$ and $\Phi = 15$. Based on these spectra, we can obtain the network resilience for different values of Φ. $\Phi = 10$ and $\Phi = 15$. Table 10.9 shows resilience for $\Phi = 10$ and $\Phi = 15$, for some values of α.

Remark 9 More detailed information on resilience of flow networks can be found in [10]. Note also that in [5] an example of comparing the resilience of networks with the same number of nodes and edges but with different topological structures is given.

Table 10.7 Edge capacities

(i, j)	c(i, j)	(i, j)	c(i, j)	(i, j)	c(i, j)
(1, 2)	10	(6, 2)	9	(10, 14)	7
(1, 5)	9	(6, 7)	7	(11, 10)	7
(1, 6)	10	(6, 10)	8	(11, 12)	8
(2, 3)	7	(7, 8)	6	(11, 15)	7
(3, 4)	6	(7, 10)	8	(11, 16)	11
(3, 7)	8	(7, 11)	9	(12, 16)	9
(4, 8)	7	(8, 12)	8	(13, 14)	8
(5, 6)	8	(9, 10)	8	(14, 11)	7
(5, 9)	9	(9, 13)	6	(14, 15)	7
(5, 10)	8	(10, 13)	8	(15, 16)	10

Table 10.8 Flow network CD-spectrum for $\Phi = 10$ and $\Phi = 15$

i	y_i (Φ =10)	y_i (Φ =15)	i	y_i ($\Phi = 10$)	y_i ($\Phi = 15$)
1	0	0	8	.5463	.8656
2	.0073	.0550	9	.6739	.9293
3	.0357	.1659	10	.7786	.9678
4	.0923	.3130	11	.8594	.9869
5	.1749	.4865	12	.9147	.9958
6	.2843	.6484	13	.9544	.9989
7	.4130	.7708	14	.9753	.9997

Table 10.9 Comparing resilience of flow network for $\Phi = 10$ and $\Phi = 15$

α	0.05	0.1	0.2	0.3	0.4	0.5	0.6
$\Phi = 10$	3	4	5	6	6	7	8
$\Phi = 15$	1	2	3	3	4	5	5

10.7 Concluding Remarks

Analysing the text above and the example section, we see that network sustainability analysis involves two types of information. Type A information is of non-stochastic nature and is based on network graph description, node, edge, terminal definitions and UP/DOWN definition of network states.

Four structural invariants have been defined in this paper (Signatures or Internal Distributions), CD-spectrum, BIM -spectrum and Border States) representing type A information.

All further analysis of network performance is done by combining structural invariants with information on the stochastic behaviour of network components subject to failure (edges or nodes), in static or dynamic situations. This information we call of type B. A typical example of combining A and B types of information is given in Sect. 10.6.1 on network reliability improvement.

A special 'artificial' variant of B-type information was an assumption that network components subject to failure fail in random and equiprobable manner imitating an external 'shock' situation. This shock model allows defining network resilience parameter and compares networks resilience for various versions of their structure.

In conclusion, let us note that this chapter is based on 'binary' approach to network structure. The book [11] goes further and introduces networks with several DOWN states. This leads to multi-dimensional invariants. Moreover, also the binary nature of failing edges or nodes can also be generalised, see [11] where in addition to up and down states of failed components, an intermediate third 'mid' state has been added.

References

1. Misra, Krishna B. (Ed.). (2008). *Handbook of performability engineering*. London: Springer.
2. Samaniego, F. (2007). *System signatures and their use in engineering reliability*. New York: Springer.
3. Gertsbakh, I., & Shpungin, Y. (2009). *Models of network reliability: analysis*. Combinatorics and Monte Carlo: CRC Press.
4. Gertsbakh, I. (2000). *Reliability theory with applications to preventive maintenance*. Springer-Verlag.
5. Gertsbakh, I., & Shpungin, Y. (2019). Network reliability: a lecture course. Springer Briefs in Electrical and Computer Engineering, Springer.
6. Gertsbakh, I., & Shpungin, Y. (2011). Network reliability and resilience. Springer Briefs in Electrical and Computer Engineering, Springer.
7. Birnbaum, Z. W. (1969). On the importance of different components in multicomponent system. In P. R. Krishnaiah (Ed.), *Multivariate analysis* (pp. 581–592). New York: Academic Press.
8. Gertsbakh, I., & Shpungin, Y. (2012). Combinatorial approach to computing importance indices of coherent systems. *Probability in Engineering and Information Sciences, 26,* 117–128.
9. Elperin, T., Gertsbakh, I., & Lomonosov, M. (1991). Estimation of network reliability using graph evolution models. *IEEE Transactions on Reliability, 40*(5), 572–581.
10. Gertsbakh, I., Rubinstein, R., Shpungin, Y., & Vaisman, R. (2014). Permutation methods for performance analysis of flow networks. *Probability in the Engineering and Information Sciences, 28*(1), 21–38.

11. Gertsbakh, I., Shpungin, Y., & Vaisman, R. (2014). Ternary networks. Springer Briefs in Electrical and Computer Engineering, Springer.

Ilya Gertsbakh was Professor Emeritus in the Department of Mathematics at Ben-Gurion University of the Negev, Israel. He is the author or co-author of eight books, the recent of which is 'Network Reliability: A Lecture Course' with Y. Shpungin as co-author and is published by Springer in 2019. Ilya Gertsbakh has published some 100 papers on topics in operations research, reliability, applied probability and applied statistics. He received M.Sc. degrees in Mechanical Engineering and Mathematics (1961) from Latvian State University in Riga, Latvia, and Ph.D. degree (1964) in Applied Probability and Statistics from Latvian Academy of Sciences, also in Riga.

Yoseph Shpungin is Professor Emeritus in the Software Engineering Department, at Shamoon College of Engineering, Beer Sheva, Israel. He is the co-author of four books and numerous publications in international scientific journals. He received his M.Sc. degree in Mathematics (1969) from Latvian State University in Riga, Latvia, and his Ph.D. degree (1997) in Mathematics from Ben Gurion University, Beer Sheva, Israel.

Chapter 11
The Circular Industrial Economy of the Anthropocene and Its Benefits to Society

Walter R. Stahel

Abstract Circular economy has always been about maintaining the value of stocks, be it natural, human, cultural, financial or manufactured capital, with a long-term perspective. A circular economy has evolved through three distinct phases, which today co-exist in parallel: a bioeconomy of natural materials ruled by Nature's circularity, an anthropogenic phase (Anthropocene: to define a new geological epoch, a signal must be found that occurs globally and will be incorporated into deposits in the future geological record. The 35 scientists on the Working Group on the Anthropocene (WGA) decided at the beginning of 2020 that the Anthropocene started with the nuclear bomb in Hiroshima on 6 August 1945. The radioactive elements from nuclear bomb tests, which were blown into the stratosphere before settling down to Earth, provided this 'golden spike' signal. https://quaternary.stratigraphy.org/working-groups/anthropocene/) characterised by synthetic (man-made) materials and objects and a phase of 'invisible' resources and immaterial constraints. This chapter will focus on how the anthropogenic phase and the 'invisible' resources and immaterial constraints can integrate into a mature circular industrial economy.

Keywords Circular economy · Innovation · Engineering · Circular sciences · Embodied resources · Accountancy · Full producer liability · Waste prevention · Service-life extension of objects · Recovery of atoms and molecules · Intelligent decentralisation · Regional economy

11.1 Introduction

The concept of a circular economy is becoming increasingly popular, to the extent of being in danger to lose its identity. To prevent this, it is crucial to state the common denominators of its different facets. Whereas the initial focus was on a sustainable use of natural resource, a report in 1976 put the emphasis on the issue of substituting manpower for energy, which happens in extending the service-life of objects in

W. R. Stahel (✉)
Faculty of Engineering and Physical Sciences, University of Surrey, Guildford, England
e-mail: wrstahel2014@gmail.com

K. B. Misra (ed.), *Handbook of Advanced Performability Engineering*,
https://doi.org/10.1007/978-3-030-55732-4_11

an ‘economy in loops’ [1]. This also means a considerable reduction of the CO_2 emissions of economic activities, a fact which only caught the attention of politicians 40 years later. In 2010, the Ellen MacArthur Foundation revitalised the concept of a circular economy and proposed a two-fold approach of bio-cycle and a tech-cycle, where the latter copied the findings of the 1976 study.

It is important to state that a circular economy is not a synonym for sustainable development, and it is not the only sustainable or intelligent solution there is. A circular economy of scarcity and necessity is as old as humankind. The circular industrial economy, however, mainly applies to societies of abundance in industrialised economies with markets near saturation, where people are no longer motivated by scarcity and new motivations are needed. A multitude of approaches such as circular societies in industrialising countries, cultural issues and self-help groups (repair cafés) can increase sustainability as much as the circular economy.

11.1.1 Managing the Wealth of Resource Stocks

The common denominator of all facets of the circular economy is the objective to maintain the value of stocks of.

- natural capital: including rocks and minerals, fauna and flora, water and biodiversity. Regenerative management by intention is appropriate for volatile natural resources ruled by nature’s circularity, as in regenerative agriculture and regenerative medicine. But natural biodiversity maintains its value more by human non-action than design,
- human capital, people and their creativity and manual skills,
- cultural capital, the UNESCO world heritage register originally listed only physical assets, but has now been extended to include immaterial assets (scientific knowledge, traditions, music),
- manufactured capital, such as infrastructure, buildings, equipment and goods. Mass-produced man-made objects and synthetic materials of the Anthropocene, which include agrochemicals, pharmaceuticals, plastics and manufactured objects made of these materials, cannot by digested by nature’s circularity and thus imply a man-made liability over their full product-life. They are the main focus of the circular industrial economy,
- financial capital.

The first two of these stocks are of natural origin, but influenced by human activities, the last three are created by people. The new term ‘urban mining’, which refers to stocks of manufactured capital as substitute for virgin resources is misleading as it ignores the fact that any manufactured stock embodies such resources as the water consumed and the CO_2 emitted from the mine to the point of sale. These embodied resources will be preserved through the service-life extension activities of the circular economy, but lost in urban mining. Note that virgin resources do not contain embodied resources.

Growth in the linear industrial economy is defined as an increase in the monetary throughput (flow), measured in the Gross National Product (GDP), which is the basis of today's political decisions. Growth needs to be redefined in a circular economy as an increase in societal wealth, represented as the quality and quantity of all stocks. In order to study the evolution of societal wealth, society needs statistics measuring at least the five stocks listed above [2]. In the absence of stock statistics, we measure 'production' flows but without knowing if this production increases our wealth stocks or makes us poorer. We know how much nations spend on the health systems, but we ignore if as a result people are healthier. If nobody is sick, the health system is bankrupt. Other cultures such as old China had health systems where everybody, patients and doctors, were better off when people were in good health [3].

Nature has been based on circularity since the beginning. 'Living' materials at the end of their useful life become food for other organisms. Other materials like rocks go through very slow erosion processes, which turn stones into gravel and sand, or are in a permanent circular process like water molecules, which evaporate and return as precipitation and can be stocked in glaciers for long periods of time. Circularity is at the basis of natural processes; Nature's resource stock is permanently preserved but constantly changing its form. This has no importance as Nature has no objective and suffers no monetary or time pressures.

Early humankind survived by exploiting both 'living' and inert natural resources. Early crafts were based on a 'bioeconomy' with all materials coming from nature: rocks were transformed into stones and tools; trees into beams and planks; animal skins into clothing. Objects that were disposed of became food for others as part of nature's circularity. Basic infrastructure, such as aqueducts, bridges, roads and fortifications have been built and maintained using natural resources since Roman times as part of an emerging circular society focused on maintaining stocks of assets through appropriate operation and maintenance.

A circular economy emerged with the industrialisation of the bioeconomy[1] starting with quarries, mills and sawmills and the monetarisation of what had been a barter economy Even during the early industrial revolution, which in Europe started around 1800, circularity by nature remained the dominant principle with people and nature living in a circular society where nature provided most resources. Any shortage of food, shelter and personal objects forced people to reuse objects and materials in a circular economy of scarcity, necessity and often poverty. When the climate effects of the 1815 explosion of the volcano Tambora in Indonesia reached the Northern Hemisphere in 1816, it caused a year without summer, where food was scarce and many people in Europe and North America experienced severe famine. This situation still persists in some regions of the world in times of droughts or locust invasions.

The industrial revolution helped society to overcome the general shortage of food, shelter, personal objects and infrastructure by optimising the supply chains of a linear production still based on natural resources. This industrial economy was

[1]The bioeconomy comprises those parts of the economy that use renewable biological resources from land and sea—such as crops, forests, fish, animals and micro-organisms—to produce food, materials and energy. https://ec.europa.eu/research/bioeconomy/index.cfm.

specialised and monetarised. Manufactured objects were sold to owner-users, who at the end of their service-life sold them to 'rag and bone' men for reprocessing. Organic waste was dispersed in nature.

With the growth of cities, the issue of waste management as the final stage of the linear industrial economy became a societal problem, which had to be managed by public authorities. Cities started building water supply systems and sewers, with the latter mostly ending in rivers and lakes, and designated landfills for organic and manufactured waste.

This linear approach, together with the increasing production efficiency of the industrial economy and a growing population, increasingly jeopardised nature's carrying capacity through effects of concentration and saturation. Pollution in rivers from leather and textile factories; ammonia from cemeteries in groundwater streams; manure from animals and sewage in rivers and atmospheric pollution from burning wood and coal were the results. This early linear industrial economy, based on the availability of natural resources and the absorption capacity of nature's circularity, still exists in some regions of the world, but its sustainability is jeopardised by continued population growth.

11.1.2 Managing Human Labour and Water

An industrial society needs to pay attention to two natural renewable resources—human labour and water—which cannot be replaced by manufactured objects and are special for the following reasons:

- Water: quantitatively because there is no resource that can replace it, and qualitatively because clean water is a necessity for the health and survival of people and animals.

  Traditional agriculture has been a major consumer of water, but industrial agriculture for instance to grow cotton and avocados, industrialisation, watershed management (dams), urbanisation and climate change may in the future lead to an increasing battle for access to water. A number of new methods such as drip irrigation, 'solid rain' and salt-tolerant plants may reduce the water demand to produce food. A fully circular use of water is possible for sewage to drinking water, at a high cost and technology input. But the issue remains that there is no other resource that can replace water.
- Labour: because people are a renewable resource, and the only resource with a qualitative edge, which can be greatly improved through education and training, but which rapidly degrades if unused.

  Policymakers should thus give preference to the use of labour over all other resources. And as any renewable resource, labour should not be taxed. This implies that labour (jobs) should be regarded as the basis for qualitative economic growth, instead of considering jobs as the result of economic growth based on higher consumption of material resources. Robots may replace labour for repetitive jobs, but without people there is no economy or society!

The concept of a circular economy of wealth management by maintaining existing capitals or assets greatly contributes to the sustainability of these two resources and allows to reduce climate change. Especially, the strategy of extending the service life of manufactured objects—which the author calls [4] the era of '**R**' for **R**euse, **R**efill, **R**epair, **R**eprogramme and **R**emanufacture—enables to:

- save water by preserving the water used in the original manufacturing process, which remains embodied in objects. This argument is valid for the service-life extension of objects made of manufactured materials (concrete, steel) as well as certain agricultural produce (cotton),
- create jobs of all skill levels locally and regionally because service-life extension activities correspond to a substitution of manpower for energy, in comparison to manufacturing new goods, and are best done where the objects and their owners are located,
- reduce CO_2 emissions because **R** activities only use a fraction of the energy that would have been spent by manufacturing new (replacement) goods, and because **R** activities need only a fraction of the transport, storage and marketing input necessary to commercialise mass-produced objects.

These benefits apply to all manufactured objects of the linear economy. The emergence of new synthetic materials and globalised mass production in the Anthropocene has greatly increased the magnitude of these benefits and given rise to the concept of a modern circular industrial economy as a strategy to manage values and liabilities of the stocks of manufactured objects made of synthetic materials.

11.2 The Circular Industrial Economy of the Anthropocene

The Anthropocene began in 1945. With the first nuclear bomb—brighter than a thousand suns—scientific man took over the command from Nature. Man-made energies and synthetic materials appeared based on scientific progress in physics, metallurgy (metal alloys, stainless steel) and chemistry (plastics, (agro) chemicals, pharmaceuticals like oestrogen), which are unknown to, and indigestible by, nature.

Society—politicians—overlooked the control issue arising from this change of command. Economic man using these synthetic materials now has to take responsibility for end-of-service-life objects in order to 'close the loop' for synthetic materials and other man-made substances and objects. Synthetic materials, unknown to nature, need industrial recovery processes within a circular industrial economy.

At the centre of the circular industrial economy remains the idea of maintaining the value of stocks as assets, be they of natural origin, human nature, cultural, manufactured or financial capital. But with the synthetic materials and resource-intensive production technologies of the Anthropocene, it becomes crucial to also maintain the invisible stocks of embodied resources and such immaterial issues as producer liability.

With hindsight, the rise of the Anthropocene split circularity into two domains:

- a biologic one of natural resources and natural wastes—the bio-cycle—which is not at the centre of analysis of this chapter, because the era of 'R' does not apply—you cannot eat an apple twice. However, with the development of life sciences and gen technologies—the bio-Anthropocene—this domain is undergoing fundamental structural changes,[2]
- a technology one of stocks of manufactured infrastructure, objects and materials—the tech-cycle—which end as wasted resources outside nature's circularity and outside the economy. As they have no positive economic value or ultimate liable owner; witness plastic in the oceans. The tech-cycle demands a change in societal thinking which is at the centre of this chapter. An analysis of the material resources has to be complemented by a consideration of immaterial topics, such as producer liability and invisible resources such as digital data and water consumed and CO_2 emitted in production and now embodied in objects.

At the end of the twentieth century, the technology domain of the Anthropocene expanded by including two new domains:

- the concept of an industrial bioeconomy is gaining increased attention. There are a growing number of technical processes, which are 'mining' nature—processed food and 'artificial meat'; asbestos and carbon fibres, gene-technology applications—the wastes of which may be outside nature's absorption capacity. The bioeconomy definition of the EU, which encompasses all organic matter, raises ethical issues: does this include such 'manufactured nature' as hormones (oestrogen), molecular machines, biosimilars, enzyme engineering, CRISPR and RNA editing?
- data mining in the digital economy—BIG DATA—has become a major new industrial activity. But despite its immaterial character, it is built on numerous objects such as smartphones and a huge physical infrastructure of server farms and communication systems, which are often owned by the data traders. As with synthetic materials, society—politicians—have underestimated the control issue involved in the exploitation of this new resource. Is data a manufactured product? Is the author or the user its owner?

Worldwide, many traditional forms of circularity co-exist today with the circular industrial economy, which itself continues to develop. Its most sustainable form today are the business models of the performance economy, which include systems solutions and sufficiency—creating wealth without resource consumption [5].

[2]The start of the Bio-Anthropocene was probably the discovery of DNA. American biologist James Watson and English physicist Francis Crick discovered the double helix of DNA in the 1950s. But DNA was first identified in the late 1860s by Swiss chemist Friedrich Miescher. Source: Wikipedia, accessed 11 August 2019.

11.3 The Seven Major Challenges of a Circular Industrial Economy

This book identifies seven major challenges to build a circular industrial economy of manufactured infrastructure, objects and materials:

1 CARING: motivating owner-users of objects to enjoy the use of, and take care of, their belongings—in the case of individuals—and assets—in the case of organisations—for as long as possible. All caring activities are labour-intensive and should not be taxed.
This is the era of 'R', of reuse, refill, repair, reprogramme and remanufacture of objects, including the right to do this. Remanufacturing objects to a quality 'better than new' (better than the original manufactured object) is the most labour and skill intensive as well as the most ecologic and profitable strategy of the circular industrial economy.

2 SOLVING the problem of the LEGACY WASTES of the synthetic (man-made) materials of the Anthropocene. Metal alloys, plastics, agrochemicals cannot be undone by nature.
This is the era of 'D', of developing processes and technologies to de-bond, de-link or de-polymerise compound materials in order to recover atoms and molecules for reuse, and to deconstruct infrastructure and buildings to recover materials. Note: The second law of thermodynamics, which limits these material recovery processes, is not applicable to service-life extension in the era of 'R'.

3 LEGISLATING sustainable framework conditions [6] to promote the service-life extension of objects in use and the recovery of used materials. These services are labour-intensive, a renewable resource that should not be taxed. And it cannot be up to municipalities to pay for problems created by profit-making industrial producers.
If waste is defined as objects and materials with no positive economic value and without an ultimate liable owner, legislation can add value (deposit laws) or nominate the producers as ultimate liable owner who has to solve the problem they created.

4 SPREADING the technical and economic KNOWLEDGE of the circular economy to all classrooms, boardrooms and parliaments.
The technical and economic knowledge and its benefits for society are largely unknown today, only known to insiders in SMEs, research institutes and fleet managers.

5 DARING: innovation to develop novel repair technologies, materials and processes in new disciplines, such as circular energy, circular chemistry and circular metallurgy, as well as new types of reusable components and systems solutions, to prevent future legacy wastes.

6 SHARING: a performance economy selling performance (molecules [7]) and goods as a service [3] instead of selling goods demand a stewardship attitude by nation–states, producers, owners and users, and includes big data—no

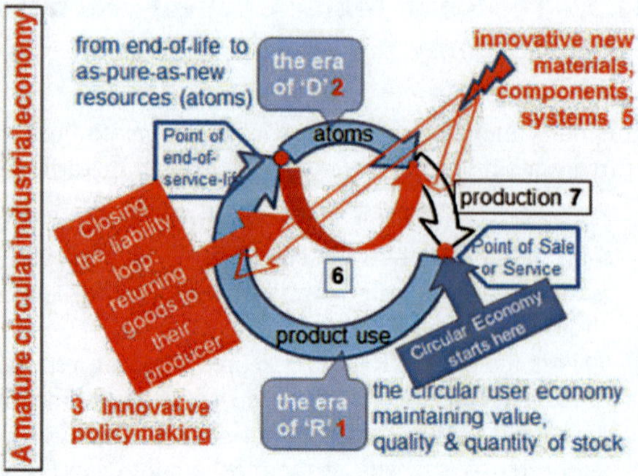

Fig. 11.1 View of a mature circular industrial economy

sharing without caring. Research into the fields of behaviour sciences, personality psychology and 'computational social sciences' will help to tackle issues of sharing and caring.

7 MEASURING THE INVISIBLE: accounting for the consumption of water and energy resources from mine to point of sale, in order to quantify the resources embodied in objects. Novel approaches like Life Cycle Inventory Analysis (LCIA) or Resource Quantity Surveying will speed up the development of tools for this topic (Fig. 11.1).

11.3.1 CARING, the Era of 'R'

The challenge of CARING: motivating owner-users of objects to enjoy the use of, and take care of, their assets (organisations) and belongings (individuals), for as long as possible—the era of 'R', of reuse, repair and remanufacture, including the right to repair. **The era of 'R'**: techno-commercial strategies to keep objects—infrastructure, buildings, goods and components—at the highest level of value and utility through:

- **R**euse,
- **R**efill,
- **R**epair,
- **R**eprogramme,
- **R**emarket,
- **R**emanufacture,
- **R**erefine,
- **R**eprogramme.

Remanufacture, which enables to achieve objects with a quality 'better than new' (better than the original manufactured object), is the most labour and skill intensive as well as the most ecologic and profitable solution of the circular industrial economy.

This challenge is especially important in economies of abundance. In a consumer society, individuals need motivation to look after what they have. In less developed or less industrialised regions, especially after a disaster has stricken, a circular society of caring and sharing or a circular economy of necessity or scarcity may dominate.

The era of 'R' of the circular industrial economy aims to maintain manufactured objects and their components at the highest utility and use-value at all times. Maximising the use-value of manufactured stocks over time and space follows some rules:

- the circular industrial economy is about economics, innovation and competitiveness but is counter-intuitive to manufacturing economics—small and local is beautiful and profitable, instead of bigger and global is more profitable,
- the smaller the loops are, the more profitable and resource efficient they are—the inertia principle [3]: *do not repair what is not broken, do not remanufacture something that can be repaired, do not recycle an object that can be remanufactured*,
- the lower the speed of the loops is, the more resource efficient they are, because of the law of reverse compound interests[3] and the second law of thermodynamics,
- loops have no beginning and no end—newcomers can enter a loop at any point,
- the circular industrial economy substitutes manpower for energy and resources by managing manufactured stocks, whereas the linear industrial economy substitutes energy (machines) for manpower and manages production flows.

The era of 'R' is:

- modern, part of a general twenty-first century trend of intelligent decentralisation,[4] which embraces production and use: 3D printing (to produce cheap spare parts just in time), local production and use (micro-breweries, -bakeries, -hydroelectricity, solar photovoltaic power), decentralised robot manufacturing, urban farming, soda fountains in pubs and at home; they are all local, decentralised, as are the era of 'R' services.
- economically profitable because 'R' activities for mass-produced goods are on average 40% cheaper than equivalent newly manufactured objects with which they compete. This ratio is even higher for customised objects and when the external cost differences with production are taken into consideration: 'R' activities are not subjected to compliance costs (such as proof of the absence of child labour, conflict minerals), carbon taxes or import duties and have a lower risk of environmental impairment liability.
  As 'R' methods differ from those used in the linear industrial economy, profitability can be substantially increased by innovation on a systems level; witness the reusable rockets developed by the start-up companies Blue Horizon and SpaceX's Falcon 9.

[3]Reverse Compound Principle: see Sect. 4.2.

[4]A term first used by Prof. Heinrich Wohlmeyer in Austria.

- ecologically desirable because 'R' activities preserve most embodied resources (energy, material and water), consume only few resources and cause little waste. As they are local, they do not need transport over large distances with intermediate storage, nor shopping centres and flashy packaging.
  Because 'R' activities do not depend on global publicity, they are invisible and silent, in contrast to linear industrial economy. Innovative ways of reaching the objects' owners will gain importance in the shift from a circular (craftsmen) economy to a circular industrial economy.
- socially viable because 'R' activities are labour-intensive services best done locally where the clients are; they demand skilled labour to judge the minimal interventions necessary (the inertia principle); they partly rely on 'silver workers'[5] who know the technology of times gone by; and they nurture a caring attitude towards goods by owners and users, which is absent in the fashion-driven linear industrial economy.
  Each economy is based on trust. For new goods, it is trust into manufacturing quality. For pre-owned objects, innovative ways to create the objects' owners trust are needed.
- labour and skill intensive because each step of 'R' activities involves caring. Each step from the non-destructive collection and value-preserving dismantling of used goods to analysing the repair or remanufacture options of each dismantled component demands a qualitative judgement. The ultimate engineering challenge in remanufacturing is developing innovative cheap repair and remanufacture methods for components destined for scrap; this is also where the highest profits are in remanufacturing.
  In buildings, about a quarter of the labour input but 80% of the material resources consumed to build a structure are stored in its load-bearing structure, the reminder in fixings and equipment. Refurbishing buildings (exchanging fixings and equipment) saves the majority of resources embodied in the structure but may need as much labour as the initial construction [8].

11.3.2 SOLVING the Problem of the Legacy Wastes, the Era of 'D'

The challenge of the era of 'D' is to develop processes and technologies to recover atoms and molecules of high purity from the legacy wastes of synthetic materials (metal alloys, plastics, agrochemicals) of the Anthropocene. However, de-bonding or de-linking compound materials in order to recover atoms and molecules for reuse does not allow recovering the embodied water, energy and CO_2 emissions for most materials. And the second law of thermodynamics (entropy) hampers these material recovery processes.

[5]Elderly workers with the skills and knowledge of technologies and objects of the past.

The era of 'D' of the circular industrial economy has the objective of maintaining the quality (purity) and value of these stocks of atoms and molecules. Today's high-volume low-value recycling technologies, which are often a solution of last resort in an 'out of sight out of mind' approach to dispose of end-of-service-life objects, enable to minimise waste costs but not to recover the material resources. Yet the priority should be to maintain the economic and resource value of the atoms and molecules [9].

The era of 'D' comprises a number of technologies and actions in an attempt to recover atoms and molecules as pure as virgin:

- **De**-polymerise,
- **De**-alloy,
- **De**-laminate,
- **De**-vulcanise,
- **De**-coat objects and
- **De**-construct high-rise buildings and major infrastructure.

Three conditions have to be fulfilled for any activities of the era of 'D' to be effective:

- a non-destructive collection of used goods,
- a sorting of the used goods into components of clean material fractions and
- a continued ownership and liability for objects and embodied materials.

Where this is not the case, for instance, because materials are mixed or dispersed (automobile shredders, rubber wear from tyres, micro-plastics in sun creams) or deliberately disposed into the environment after use (metal drink and oxygen containers), most manufactured objects made of synthetic materials will become a long-time environmental hazard, such as plastic in the oceans.

Natural circularity can only de-bond natural materials, such as foodstuff, wood, wool and iron under favourable conditions and over time. It took 100 years for nature to digest most of the materials, which made up the 'Titanic', but the iron hulk has mostly survived to this day.

11.3.3 LEGISLATING Sustainable Framework Conditions, Innovative Policymaking

Key requirements are to legislate:

- labour-friendly framework conditions, as all activities involving caring—including the circular economy—are labour intensive but consume few resources and
- a full producer liability for used objects and materials with zero value to society and nature, and without an ultimate liable owner: he who created the object or

material has to pay to solve the problem. It cannot be up to municipalities to pay for problems created by profit-making industrial producers.

The sustainability efficiency of the circular industrial economy can be greatly enhanced by closing the invisible liability loops both for objects and materials, in addition to closing the physical loops of objects in the era of 'R' and of molecules in the era of 'D'.

In the linear industrial economy, liability for the use of objects lies with the user-owner of goods: guns do not kill, the person pulling the trigger does. But this manufacturer strategy of the linear industrial economy, to limit producer liability after the point of sale to a short warranty period, has started to fade in the second half of the twentieth century. Nestlé was accused of 'killing babies' by selling milk powder without detailed instructions how to use it; the tobacco industry has been accused of killing smokers through its products; and the asbestos industry has been accused of causing the death of workers handling asbestos cement goods, even decades after the production has ended. This list now includes 'diesel gate' and could spread to immaterial goods, as some people consider "social media as the new cigarette".

This development is not revolutionary. It builds on the 1976 US Resource Conservation and Recovery Act, the 1980 US Superfund legislation and the Polluter Pays Principle (PPP)[6] in Europe at the turn of the twentieth century, when producers were made liable for environmental harm caused.

Full Producer Liability (FPL) goes far beyond the European Extended Producer Responsibility (EPR) legislation,[7] which allows producers to outsource their responsibility to third parties against payment of a small fee. But as these recyclers have no access to producer knowledge and commercial expertise to remarket components at the highest value level—or are not allowed by contractual obligations—they focus on the cheapest recycling or disposal methods.

The present policy framework of the circular industrial economy, which only closes the highly visible material loops of the era of 'R' but neglects the invisible immaterial liability loops, thus misses a major driver to reach sustainability: closing the liability loop through a full producer liability, which will show up in the financial balance sheets of corporations—which is not the case for extended producer responsibility.

Defining waste as 'objects without positive value or ultimate liable owner' opens:

- an industrial solution—use materials with inherent value, such as gold or copper, to give used objects a positive value,
- a political option—deposit laws which give used objects an economic value and
- a policy solution—define the original producer as the ultimate liable owner.

[6]The Polluter Pays Principle is an environmental law to make the party, which is responsible for the pollution, also responsible for paying for the damage done to the natural environment.

[7]EPR legislation, such as the European Union's WEEE directive for waste of electrical and electronic equipment.

Closing the liability loop then means that goods with no value at the end of their service life can be returned to their producer as the ultimate liable owner.[8]

A full producer liability will give producers strong incentives to prevent future liabilities by designing goods both for maximum value at the end of their service life and a minimum liability. A full producer liability furthermore puts manufacturers selling goods on a level with economic actors selling goods as a service, which already today retain the ownership and liability of their objects and materials over the full service-life (see section Caring, Performance Economy).

11.3.4 Spreading the Knowledge

Promoting the knowledge pool of the circular economy—technical, commercial and economic—to class-and boardrooms, to parliaments, academia and technical training institutions, and to new 'R' professions is a major opportunity to speed up the transition to a circular industrial economy.

But only a few of these key messages of the circular industrial economy have to do with engineering. Replacing production with service-life extension activities is part of a new trend of intelligent decentralisation, like 3D print, AI-led robotised manufacturing, micro-breweries and bakeries as well as urban farming, and opens up technology and engineering opportunities.

11.3.5 DARING—Radical Innovation

In order to achieve the vision of a circular industrial economy of zero waste and zero carbon without future legacy 'waste', radical innovation into systems, components and materials to upgrade stocks, and novel materials from new disciplines will be necessary. Cooperation between multitudes of scientific disciplines will speed up this development; these disciplines include:

- Biology, biophysics and biochemistry.
- Chemistry (e.g. circular chemistry, constructed molecules).
- Metallurgy (e.g. circular metallurgy, marking/identifying alloys).
- Circular energy (e.g. hydrogen from green sources).
- Material sciences (e.g. de-bonding alloys, carbon fibre laminates, de-constructing infrastructure and high-rise buildings).
- Space sciences (e.g. the neglected commons).
- Law schools and accountancy (e.g. the definition of waste, 'used product' liability and full producer liability).

[8]The author has derived the concept of the ULO from that of the Ultimate Beneficial Owner (UBO), which was introduced in the USA in the 1970s to reduce tax evasion through chain ownerships of companies in tax havens.

- Macro-economics (input/output models for alternative economies).
- Micro-economics (e.g. ROI for remanufacturing versus manufacturing).
- Systems management (IT, IoT, pharmacogenomics, gene therapy).
- History (e.g. the atomic bomb as the beginning of the Anthropocene and the reasons why its consequences have been overlooked for decades).
- Literature as a means of informing and motivating users (e.g. circular economy examples described in the literature, such as death of a salesman by Miller [10]; Zen and the art of motorcycle maintenance by Pirsig [11].
- Behavioural sciences (e.g. motivating owner-users to enjoy the use of, and care for, their belongings/assets; prevention of vandalism and abuse of shared assets (Tragedy of the Commons),
- Political sciences (facilitating new policymaking on, for instance taxation of labour, resources, emissions).

Based on visions of a sustainable future, pull innovation (building markets using public procurement) can be used to move industry into a desired direction. A Norwegian shipping fleet has ordered zero-emission coastal express vessels using hydrogen and fuel cells for propulsion[9] [12].

Engineering innovation in the era of 'R'

Since the 1990s, techno-economic research with environmental objectives has flourished in areas like Life-Cycle Analysis (LCA), which has a time horizon defined as 'Cradle-to-Grave' [13]. But accepting that a reduction of resource consumption in industrialised countries by 90%—a factor of 10—was desirable and feasible, and analysing its implications for economy and society, was impossible for most experts.[10]

Political interests to reduce end-of-pipe waste volumes guided academic research to look into sectors of the circular economy concerned with finding uses for high volume wastes, for instance from the building industry, recovering building materials as aggregate in concrete.

In a mature circular industrial economy, production becomes a segment of the loop of the circular industrial economy by producing innovation to upgrade and renew the stocks of objects. In construction and the electro-mechanical world, technology upgrades often involve singular components, which can be replaced by new-tech components fulfilling the same function. In vintage cars, mechanical distributors, which need regular adjustments, can be replaced by maintenance-free electronic ones. Transforming a mechanical typewriter into a personal computer does not make

[9]2019 Boreal and Wärtsilä Ship Design have agreed to develop a hydrogen-powered ferry for the Hjelmeland–Skipavik–Nesvik stretch. The ferry will be the first in the world where the vessel will use hydrogen as a fuel. The Norwegian Public Roads Administration has announced a development contract for a hydrogen-powered ferry, which will be put into operation in 2021. The ferry service will be operated by two ferries, one being fully electric and the other hybrid hydrogen electric with 50% of the hydrogen output.

[10]In 2017, the Factor Ten concept was reinvented by the World Business Council of Sustainable Development.

sense, but upgrading mechanical bicycles into e-bikes, by using wheel-integrated electric micro-motors and adding a battery, transforming an original E-type Jaguar into an electric one, or a 60-year-old seaplane into an electric one are feasible and have been done by singular economic actors.

These are hidden business opportunities, which need engineers knowledgeable in new technologies and the existing stock of objects. In today's markets, very few of these go-betweens exist, and most of these business opportunities remain unexploited.

But technology push is forcing economic actors of the linear industrial economy into the performance economy, driven by a trend to replace complex mechanical systems, such as combustion engines with gearboxes, with long-life low-maintenance components, such as electric motors. As long-life low-maintenance components lead to long-life objects, producers start to seize the opportunity to sell goods as a service in order to retain market control and revenue flows.

In the old IT world, hardware and software could be upgraded separately: hardware items were routinely replaced by new more powerful and/or energy-saving components; and software periodically upgraded, often on-line, to make computer systems more resilient. New external hardware like printers and hard disks were mostly compatible with existing equipment like personal computers. The latter could be used 'as is' for a long time, as stand-alone systems working off-line without limits. Ownership and control remained with the owner of the hardware, who bought a software licence. This is still the case for isolated systems like dash cams and portable GPS.

Engineering innovation in the era of 'D'

This is the sector of the circular industrial economy with the biggest potential for improvements through technical R&D. Once the reuse and service-life extension options of the era of 'R' have been exhausted, recovering the stocks of atoms and molecules at their highest utility and value (purity) level for reuse is the best option. This demands sophisticated sorting technologies and processes to dismantle used objects and sort them into clean material fractions, for instance into different alloys of the same metal, and finally, technologies to recover molecules and atoms as pure as virgin resources.

Sorting manufactured materials is a problem that does not exist in mining and opens up new fields of R&D. The same is true for de-bonding manufactured molecules. Innovative economic actors should be in the driver seat of the era of 'D', and governments can support these activities by financing R&D at universities. Science and technology opportunities to recover atoms and molecules are almost unlimited in an open internationally competitive environment, and many solutions will be patentable.

In contrast to the decentral processes in the era of 'R', technologies and actions of the era of 'D' will often be global and include:

- de-bonding molecules, such as to de-polymerise polymers, de-alloy metal alloys, de-laminate carbon and glass fibre laminates, de-vulcanise used tyres to recover rubber and steel, de-coat objects,
- salvaging non-renewable resources, such as phosphorous, from general waste streams,
- de-constructing high-rise buildings and major infrastructure is a different kettle of fish. In Tokyo, the first high-rise building has been de-constructed using a method that enabled recovering not only the equipment and materials used in construction but also the energy originally consumed to hoist things up. Spain has started to dismantle its Yecla de Yeltes hydroelectric dam, the largest de-construction project of its kind ever in Europe; Germany and Switzerland have started deconstructing nuclear power stations.

These examples are just the tip of the iceberg. Some manufacturers like Plasto, a Norwegian company producing equipment for fish farming, have started to take back end-of-service-life objects made of High-density Polyethylene (HDPE) in order to reprocess the material to produce new equipment in a profitable process.

Research into reusing atoms and molecules also opens up new territories in basic sciences. Questions like 'can CO_2 turn from waste into a resource to produce new chemicals, and will this new carbon chemistry ever be able to compete with petro-chemistry?' may find an answer through scientific research. Using Carbon Capture for Utilisation (CCU) to produce hydrogen is a research topic under study in Norway.

In cases where no technologies are found for used materials in the era of 'D', pressure will mount on producers of the linear industrial economy to look for alternative materials or change their business model in order to exploit the economic opportunities of the era of 'R' or the performance economy.

11.3.6 SHARING—The Performance Economy

In the performance economy, economic actors selling results instead of objects retain the ownership of objects and embodied resources and internalise all liability over the full product-life. These economic actors may be manufacturers of technical systems, buildings or equipment, or fleet managers operating them; in both cases, they sell the use of these objects as a service over the longest period of time possible and maximise their profits by exploiting both efficiency and sufficiency solutions (Fig. 11.2).

The performance economy of selling goods and molecules as a service, function guarantees or results and performance is the most sustainable and potentially the most profitable business model of the circular industrial economy because it internalises the costs of product liability, of risk and waste, and thus incorporates a strong financial incentive to prevent losses and waste. The performance economy is potentially highly profitable because it maximises the profit potential by exploiting sufficiency, efficiency and systems solutions but depends on a caring stewardship attitude of users.

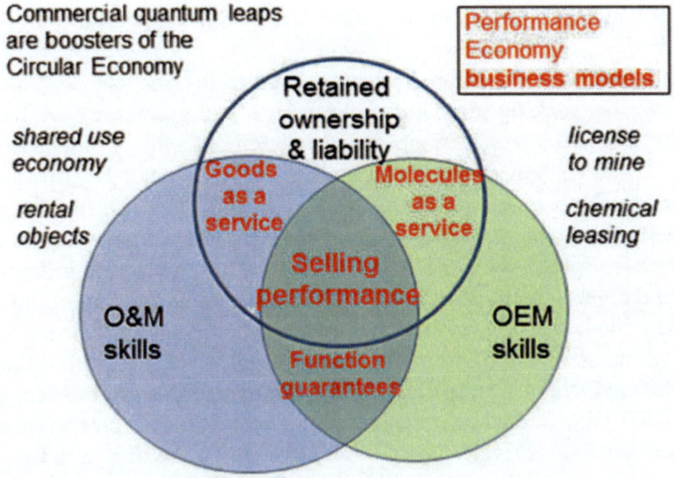

Fig. 11.2 Selling performance instead of selling goods: combining retained ownership with original equipment manufacturer (OEM) and operation and maintenance (O&M) skills

No Sharing Without Caring!

Maintaining ownership of objects and embodied resources creates corporate and national resource security at low cost:

If Producers Retain the Ownership of Their Goods, the Goods of Today Are tomorrow's Resources at yesteryear's Commodity Prices, locally available.

The performance economy redefines the role of the supply side but also implies a radical change of the demand side, from ownership to usership of objects. Is this really new? Aristotle had already stated 2000 years ago that

Real Wealth Lies in the Use, not the Ownership of Goods.

In a rental economy, users do not need capital to buy goods,[11] but they do not profit from capital gains either. Owning goods makes economic sense for individuals for goods that increase in value over time; owning real estate often makes sense, owning a smartphone or washing machine does not. By renting objects, users gain flexibility in use, know the cost of using a product in advance and only pay when using it. Users with a weakness for fashion and constant change can fully live their fancies without causing excess waste, for instance by renting every weekend a different fashionable sports car, costume or handbag.

Fleet managers prefer objects of high quality and low-maintenance cost, focusing on function not short-lived fashion. They also have the knowledge to optimise operation and minimise maintenance costs of the objects in their possession, for example

[11] Quasi rental service activities are sometimes called sharing economy, platform economy (UBER, Airbnb) or Product-Service Systems (PSS), for reasons of brand distinctions rather than factual differences.

through standardised components and technical systems as well as cascading uses of objects.[12]

Witness textile leasing companies renting uniforms and catering for hospital textiles; they start making profits after the fabrics and garments have been in use for more than 3 years. As their operation is geographically limited by transport costs, and the knowledge of their clients' specific needs is vital, they operate through franchising, not globalisation. Real estate owners are often life insurance companies or family trusts interested in long-term value preservation and low operation and maintenance costs, which can best be achieved through a high initial quality of materials and objects, and a familiarity with local customs and conditions.

Detailed knowledge of operation and maintenance is also necessary for facility managers in charge of building or operating complex infrastructures. The French company Eiffage in 2001 signed a 78-year contract to design, finance, build and operate till 2079 the viaduct near Millau, with a maintenance contract running until 2121. The project is a Private Finance Initiative (PFI); the bridge did not cost the French taxpayer a penny, but each vehicle crossing the bridge has to pay a toll (as the bridge deck is more than 200 m above the valley, pedestrians are not allowed to cross in order to prevent suicides). All risks are carried by, and profits go to, Eiffage,[13] who will only know 78 years after the signature of the contract how much loss or profit it has made.

Innovation in the performance economy comes from a shift of focus, from optimising production to optimising the utilisation of objects, and from including the factor time in this optimisation. A focus on the use or utilisation of objects opens up new opportunities, such as long-life goods, multifunctional goods,[14] service strategies and systems solutions. These opportunities are of no interests for manufacturing industries selling objects, whose objective is optimising production up to the point of sale, not product use or duration. Examples are.

- goods sold as a service for *exclusive use* are rental apartments, tools and vehicles for rent, but also public toilets, ISO shipping containers, leased equipment and reusable packaging.
- goods or systems sold as a service for *shared use* are all forms of public transport (busses, trains, aircraft), as well as public swimming pools, concert halls and laundromats.
- molecules as a service are chemical leasing contracts (also called rent a molecule). They enable a precise accounting of the losses of chemicals into the environment between all the actors involved (lessor, lessee, distributors), which may be legally required to establish Toxic Release Inventories. UNIDO promotes

[12]Xerox very early imposed its commonality principle, specifying the same components across its equipment range; Airbus introduced from the beginning a standardised flight deck for all its aircraft, saving airlines O&M costs in crew training, stand-by crews; airlines routinely transform passenger jets into cargo jets to extend their service life.

[13]Pure risks can be insured, of course, but not the entrepreneurial risk.

[14]With the digitalisation of technology, multifunctional goods such as all-in-one printer–copier–scanner–fax machines became standard.

chemical leasing as a strategy for Africa in order to minimise the reuse of toxic packaging waste to store water or food.[15] Dow Chemical, through its subsidiary Safe-Chem, has been a pioneer of leasing solvents.

An international agreement to stop the production of mercury after 2020 could radically change its marketing and lead to closed loop and rental strategies in the future commercialisation of mercury.[16]

Strategies of molecules as a service are not limited to chemicals with a catalytic function or high toxicity. Metal leasing is proposed by two scientists (Hagan [7]) as a strategy for mining companies and governments of mining regions. It would give governments and mining companies a smaller short-term income than selling the minerals but guarantee constant long-term revenues. Developing innovative smart materials for rent is an opportunity open to material companies. "*The UK's Cookson Group developed a composite powder that can be pressed into any form and, when magnetised, becomes an extremely powerful permanent magnet. The two characteristics, easy shaping and magnetisation on demand, make it an ideal material for use as a rotor in small electric motors. After use, this smart material can easily be demagnetised by grinding it back into a powder and then remixed for its next use. To benefit from the successive life cycles of its smart materials, the manufacturer has to retain ownership by, for example, leasing the material to component manufacturers with a return guarantee. In this case, the strategy of selling performance in the Functional Service Economy must be imposed on all levels from material to final product. Otherwise, there is no guarantee that the smart material will be returned to its manufacturer at the end of the product's life.*"[17]

A performance economy selling performance (molecules [7] and goods as a service [3] instead of selling goods, demands a stewardship attitude by nation–states, producers, owners and users and includes big data. The fields of personality psychology [14] and 'computational social sciences' may help to seize these opportunities!

11.3.7 Measuring the Invisible and Immaterial Assets

One common denominator of all circular economy efforts is to prevent waste by maintaining the value and utility of durable objects and perishable produce, and the value and purity of molecules.

The circular economy is like a lake; of course there are currents in lakes but overall the stock of water shows little changes in quality and quantity. This also means that

[15] UNIDO is promoting chemical leasing in Africa to reduce the uncontrolled release of used chemicals into the environment.

[16] For catalytic goods, which are contaminated but not consumed during use, an integrated 'rent-a-molecule' service enables economic actors to create revenue without resource consumption [3, p. 87].

[17] Cookson's 'rent-a-molecule' of smart materials, [3, p. 109].

great care should be taken as any pollution will take a long time to be diluted or washed out naturally. The heritages we leave to our children are stocks and include the CO_2 emissions of past and present generations, which are accumulated as carbon stock in the atmosphere.

The linear industrial economy is like a river; rivers experience great and rapid changes in flow volume (throughput) and any change in quality (pollution) will be rapidly washed downstream and less affect our children's heritage.

The invisible assets of engineering—the heritage to our children—are water and materials consumed, and CO_2 emitted, during mining, manufacturing and distribution activities, embodied in the final objects and preserved as long as the objects exist.

In manufacturing, three quarters of energy is used in the production of basic materials such as cement and steel, only one quarter in producing goods such as buildings or cars. For labour input, the relation is reverse; three quarters are being used in producing the goods [8].

The nature of the service life extension activities of goods—the era of 'R'—is similar to that of production. It uses a lot of manpower but few energy resources but it differs from production in that it preserves all the manufacturing energy embodied in the objects. But today, we do not keep account of water and materials consumed, and CO_2 emitted, during the manufacturing process. In the future, novel approaches such as the 'Madaster' concept [15] may keep track of the inputs of material resources.

But in order to motivate decisions in favour of the circular economy, we need to know the amount of invisible resources embodied in objects. And that means that economic actors will have to measure the water and energy resources consumed from mine to point of sale, and the CO_2 emitted, in order to quantify the invisible resources embodied in each object. New techniques like Life Cycle Inventory Analysis (LCIA) and professions like Resource Quantity Surveyors have to be put in place to do this.

Among the immaterial topics to be considered are financial assets, digital data, liability and behavioural issues. Financial capital may have reached the peak of non-materiality in Bitcoins; if a banker or the client loses the code and password, monies may no longer be accessible, lost in cyberspace.

The 'IT software-data-hardware issue' is another immaterial topic. Tractors, which are mechanically fit but their IT does not work, cannot be repaired if manufacturers do not make available their software source codes, ignoring owner-users' rights to repair. Smartphones, which were upgraded to function less efficiently when the next generation came to market, are now considered as planned obsolescence. The French government has fined Apple 25 million Euros for such an incident, opening the door for class action suits by the owners of the remotely manipulated smartphones.

Liability is an immaterial topic that needs to become visible. Full Producer Liability (FPL) will shift the present extended producer responsibility from the soft Corporate Sustainability Reporting to the Financial Statements, to Balance Sheets of Assets and Liabilities, which are the focus point of financial investors. As FPL has no known amount or deadline, investors will shy away if economic actors do not change their business model from selling goods to the more sustainable alternatives.

11.4 Conclusions

What are the drivers of the circular economy today?

In general:

- restrictions on export/import of mixed waste (the enlarged Basel agreement),
- speeding up the market penetration of intelligent decentralisation, such as 3D print, IA-robot manufacturing, micro-pharmaceutical, -bakeries, -breweries, local reuse–repair–remarket organisations, social reuse and repair networks,
- social bottom-up (Greta) behavioural changes, education and information.

In engineering:

- shift to selling goods as a service (performance economy) as a result of long-life low-maintenance technologies (e.g. electric motors instead of combustion engines),
- changes in energy resources, from oil to electricity (battery or H2-fuel cell?),
- R&D into circular energy (hydrogen fuel cells), circular chemistry and circular metallurgy.

In the European Union:

- politics and research linked to limiting and ultimately stopping carbon emissions,
- changes in legislation, a shift from taxing labour to digital transactions, carbon emissions and resource consumption; a rise in extended producer liability towards FPL,
- bans on the destruction of unsold food and surplus stocks and on returns of internet sales,
- ecodesign directives imposing the right to repair and the obligation for producers to supply mandatory information, tools and spare parts for a period of 10 years,
- the lack of economic feasibility of waste management (recycling).

In Asia:

- in the PRC, the choice of remanufacturing as national industrial strategy,
- in Singapore, the scarcity of resources and of landfills leading to innovation in the reuse of drinking water (NEWater) and preowned objects.

In the USA:

- R&D in university and national laboratories,
- private bottom-up initiatives like Kyle Wien's www.iFixit.com.

In South America:

- business associations, chambers of commerce.

References

1. Stahel, W. R., Reday-Mulvey, G. (1981). The potential for substituting manpower for energy; 1976 report to the Commission of the European Communities, Brussels. Published 1981 as: Stahel, W.R., Reday-Mulvey, G. Jobs for Tomorrow, the potential for substituting manpower for energy, Vantage Press, New York, N.Y. 1981.
2. World Bank Group. (2020). *The Changing Wealth of Nations Report 2018*. https://openknowledge.worldbank.org/bitstream/handle/10986/29001/9781464810466.pdf. Accessed 27 Apr 2020. 09:11 hours.
3. Stahel, W. R. (2010). *The performance economy* (2nd ed.). Houndmills: Palgrave Macmillan.
4. Stahel, W. R. (2019). *The circular economy, a user's guide*. Abingdon: Routledge.
5. Stahel, W. R. (1997). The service economy: Wealth without resource consumption? *Philosophical Transactions a, Royal Society London, 355*(June), 1309–1319.
6. Stahel, W. R. (2013). Policy for material efficiency—sustainable taxation as a departure from the throwaway society. *Philosophical Transactions A of the Royal Society, 371* no. 1986 20110567. https://doi.org/10.1098/rsta.2011.0567.
7. Hagan, A. J., Tost, M., Inderwildi, O. R., Hitch, M., Moser, P. (2019). *The license to mine: Making resource wealth work for those who need it most*. https://www.sciencedirect.com/science/article/abs/pii/S0301420717305445. Accessed 27 Apr 2020 09:12 hours.
8. The case study on the French construction industry in [1].
9. Material economics. (2018). Ett värdebeständigt svenskt materialsystem (Retaining value in the Swedish Materials System). Research study, unpublished.
10. Miller, H. (1976). *Death of a salesman, 1949 play*. London: Published Penguin Plays.
11. Pirsig, R. (1974). *Zen and the art of motorcycle maintenance*. London: William Morris and Company.
12. Ferry shipping news (2918) Feb 08 and May 17. https://www.ferryshippingnews.com/.
13. **ISO** 14044. (2006). Environmental management—Life cycle assessment—Requirements and guidelines. International Standardisation Organisation, Geneva.
14. Dumont, F. (2015). Personality: Historical and conceptual perspectives. In J. D. Wright (editor-in-chief), *International Encyclopedia of the Social & Behavioral Sciences* (2nd ed., Vol. 17, pp. 906–913). Oxford: Elsevier.
15. Oberhuber, S. (2016). *Material matters*. Berlin: ECON im Ullstein Verlag.

Walter R. Stahel M.A. in architecture from ETH Zurich (1971). Visiting Professor, Faculty of Engineering and Physical Sciences, University of Surrey (2005); Full Member of the Club of Rome; Member of the Coordination Group of the European Circular Economy Stakeholder Platform of the European Parliament; Member of the Scientific Advisory Board of Institute CRETUS at Universidade de Santiago de Compostela; Member of the Advisory Board of CRESTING, Circular Economy, University of Hull, http://cresting.hull.ac.uk/. Since 1982: Founder-director of the Product-Life Institute Geneva, the oldest consultancy in Europe devoted to developing sustainable strategies and policies. 1986–2014: Director of risk management research of the Geneva Association. Present activity: keynote speaker and author promoting the opportunities of a circular industrial economy, and of a performance economy selling goods and molecules as a service. Honours: 2020 Thornton Medal of the Institute of Materials, Minerals and Mining (IOM3). Degrees of Doctor honoris causa, University of Surrey (2013) and l'Université de Montréal (2016). Mitchell Prize in Houston, TX (1982, 'The Product-Life Factor'); first prize in the German Future's Society's competition (1978, 'unemployment, occupation and profession').

Chapter 12
Sustainment Strategies for System Performance Enhancement

Peter Sandborn and William Lucyshyn

Abstract "Sustainment" (as commonly defined by industry and government) is comprised of maintenance, support, and upgrade practices that maintain or improve the performance of a system and maximize the availability of goods and services while minimizing their cost and footprint or, more simply, the capacity of a system to endure. System sustainment is a multitrillion-dollar enterprise, in government (infrastructure and defense) and industry (transportation, industrial controls, data centers, and others). Systems associated with human safety, the delivery of critical services, important humanitarian, and military missions and global economic stability are often compromised by the failure to develop, resource, and implement effective long-term sustainment strategies. System sustainment is, unfortunately, an area that has traditionally been dominated by transactional processes with little strategic planning, policy, or methodological support. This chapter discusses the definition of sustainment and the relationship of sustainment to system resilience, the economics of sustainment (i.e., making business cases to strategically sustain systems), policies that impact the ability to sustain systems, and the emergence of outcome-based contracting for system sustainment.

Keywords Sustainment · Cost · Business case · Policy · Complex systems · Maintenance · System health management

12.1 Introduction

Sustainability and its variants have captured the interest of engineering (and other disciplines) for several decades. Even though sustainability and sustainment are sometimes used interchangeably, these words have unique connotations that depend on discipline in which they are used. The focus of this chapter is on the sustainment of complex engineered systems, but let us first look at the most prevalent usages of sustainment [1].

P. Sandborn (✉) · W. Lucyshyn
University of Maryland, College Park, MD, USA
e-mail: Sandborn@umd.edu

K. B. Misra (ed.), *Handbook of Advanced Performability Engineering*,
https://doi.org/10.1007/978-3-030-55732-4_12

Environmental sustainability is "the ability of an ecosystem to maintain ecological processes and functions, biological diversity, and productivity over time" [2]. The objective of environmental sustainability is to increase energy and material efficiencies, preserve ecosystem integrity, and promote human health and happiness through design, economics, manufacturing, and policy.

Economic (business or corporate) sustainability refers to an increase in productivity (possibly accompanied by a reduction of consumed resources) without any reduction in quality or profitability. Business sustainability is often described as the triple bottom line [3]: financial (profit), social (people), and environmental (planet). "Sustainable operations management" integrates profit and efficiency with the stakeholders and resulting environmental impacts [4].

Social sustainability is the ability of a social system to indefinitely function at a defined level of social wellbeing [5]. Social sustainability has also been defined as "a process for creating sustainable, successful places that promote wellbeing, by understanding what people need from the places they live and work" [6]. Social sustainability is a combination of the physical design of places that people occupy with the design of the social world, i.e., the infrastructure that supports social and cultural life.

Technology or system sustainment refers to the activities undertaken to: (a) maintain the operation of an existing system (ensure that it can successfully complete its intended purpose), (b) continue to manufacture and field versions of the system that satisfy the original requirements, and (c) manufacture and field revised versions of the system that satisfy evolving requirements [7]. The term "sustainment engineering" when applied to technology sustainment activities is the process of assessing and improving a system's ability to be sustained by determining, selecting, and implementing feasible and economically viable alternatives [8].

Many specialized uses of sustainability exist,[1] which overlap into one or more of the categories above, including urban sustainability, sustainable living, sustainable food, sustainable capitalism, sustainable buildings, software sustainment, sustainable supply chains, and many others. Technology and system sustainment is the topic of this chapter (starting in Sect. 12.3).

12.2 A General Sustainment Definition

With so many diverse interests using sustainability/sustainment terminology, sustainment can imply very different things to different people. Both sustainment and sustainability are nouns. However, sustainment is the act of sustaining something, i.e., determination and execution of the actions taken to improve or ensure a system's

[1]There are other usages that are not particularly relevant to engineered systems, for example, sustainment and sustainability are used as a general programmatic/practice metric; "sustainability" is a term used to refer to what happens after initial implementation efforts (or funding ends) where sustainability measures the extent, nature, or impact of adaptations to the interventions or programs once implemented, e.g., in health care [9].

longevity or survivability; while sustainability is the ability to sustain something or a system's ability to be sustained. Today, *sustain* is defined as keeping a product or system going or to extend its duration [10]. The most common modern synonym for *sustain* is *maintain*. *Sustain* and *maintain* may be used interchangeably, however, maintaining most often refers to actions taken to correct problems, while sustaining is a more general strategic term referring to the management of the evolution of a system. Basiago [11] points out that sustainability is closely tied to *futurity*; meaning renewed or continuing existence in a future time. To sustain embraces a philosophy in which principles of futurity guide current decision-making.

The first use of the word sustainability in the context of man's future was in 1972 [12, 13], and the term was first used in a United Nations report in 1978 [14]. For the history of the origin and development of socioecological sustainability, see, Refs. [15, 16]. The best-known socioecological definition of sustainability (attributed to the "Brundtland Report" [17]) is commonly paraphrased as "development that meets the needs of present generations without compromising the ability of future generations to meet their own needs." While the primary context for this definition is environmental (and social) sustainability, it has applicability to other types of sustainability. In the case of technology sustainment if the word "generations" is interpreted as the operators, maintainers, and users of a system, then the definition could be used to describe technology sustainment. Unfortunately, the concept of sustainability has been coopted by various groups to serve as a means-to-an-end in the service of special interests and marketing.

At the other end of the spectrum, the US Department of Defense (DoD) defines sustainment as "the provision of logistics and personnel services necessary to maintain and prolong operations through mission accomplishment and redeployment of the force" [18]. Sustainment provides the necessary support to operational military entities to enable them to perform their missions. The second, and perhaps more germane defense definition, is in the systems acquisition context. Once a system is developed and deployed the system operations and support phase consists of two major efforts "sustainment and disposal." How do these definitions relate to the design and production of systems? For many types of critical systems (systems that are used to ensure the success of safety, mission, and infrastructure critical activities), sustainment must be part of the initial system design (making it an afterthought is a prescription for disaster—see Sect. 12.3).

In 1992, Kidd [15] concluded that "The roots of the term 'sustainability' are so deeply embedded in fundamentally different concepts, each of which has valid claims to validity, that a search for a single definition seems futile." Although Kidd was only focused on socioecological sustainability, his statement carries a kernel of truth across the entire scope of disciplines considered in this chapter. Nonetheless, in an attempt to create a general definition of sustainment that is universally applicable across all disciplines, we developed the following. The best short definition of sustainment is the capacity of a system to endure. A potentially better, but longer, definition of sustainment was proposed by Sandborn [19]: "development, production, operation, management, and end-of-life of systems that maximizes the availability of goods and services while minimizing their footprint". The general applicability of this definition is embedded in the following terms:

- "footprint" represents any kind of impact that is relevant to the system's customers and/or stakeholders, e.g., cost (economics), resource consumption, energy, environmental, and human health;
- "availability" measures the fraction of time that a product or service is at the right place, supported by the appropriate resources, and in the right operational state when the customer requires it;
- "customer" is a group of people, i.e., individual, company, geographic region, or general population segment.

This definition is consistent with environmental, social, business, and technology/system sustainment concerns.

12.3 The Sustainment of Critical Systems

Having discussed the general sustainment/sustainability landscape, we now focus on technology/system sustainment, which is the topic of the remainder of this chapter. In this section, we define the type of "systems" we are concerned with and then describe what the sustainment of these systems entails.

Critical systems perform safety-, mission-, and infrastructure-critical activities that create the transportation, communications, defense, financial, utilities, and public health backbone of society.[2] The cost of the sustainment of these systems can be staggering. For example, the global maintenance, repair, and overhaul (MRO) market for airlines is expected to exceed $100B per year by 2026 [20]. Amtrak has estimated its capital maintenance backlog (which includes physical infrastructure and electromechanical systems) in the US Northeast Corridor, alone, at around $21 billion [21]. The annual cost to operate and maintain the Department of Defense vast sustainment enterprise was over $170B in 2011 [22]. The sustainment of critical systems encompasses all the elements in Table 12.1.

While it is easy to map the disciplines listed in Table 12.1 onto managing hardware components and subsystems, sustainment is more than hardware. Critical systems are composed of combinations of: hardware, software, operational logistics, business models, contract structures, and applicable legislation, policy and governance. If any of these system elements fails, the system potentially fails. The term "system resilience," which is the intrinsic ability of a system to resist disruptions, i.e., it is its ability to provide its required capability in the face of adversity, in part encompasses sustainment. In the case of sustainment, we are concerned with adversity from

[2] Another term for these systems is "mission critical". These systems often become "legacy" systems because their field life is so long that during the majority of their life they are based on, or are composed of, out-of-date (old) processes, methodologies, technologies, parts, and/or application software.

Table 12.1 Elements of critical system sustainment

Affordability	Availability	Policy/governance	Mission engineering
Cost–benefit analysis	Readiness	System health management	Modernization/technology insertion
Warranty	Reliability	Upgradability	Logistics[a]
Maintainability	Obsolescence	Open systems	Outcome-based contracts
Viability	Prognostics	Qualification/certification	Sparing
Risk	Testability	Counterfeit management	
Diagnosability	Workforce	Configuration control	

[a]In [18], sustainment is distinguished from logistics, which is "supply, maintenance operations, deployment and distribution, health service support (HSS), logistic services, engineering, and operational contract support"

"aging" issues, both technological and nontechnological. The subsections that follow highlight some less obvious complex system sustainment issues.

12.3.1 *Software Sustainment*

All of the discussion so far can be readily applied to hardware, but sustainment also applies to software (and obviously, systems composed of both hardware and software). In the case of hardware, when a component fails, maintenance personnel can remove the failed component and replace it with a working component. The resolution to a software failure is less straightforward. First, the term "software failure" is more nebulous, and may mean that latent defects ("bugs") in the software have been encountered during operation, that the software has become incompatible with the system it is in due to other software or hardware changes to that system, or a host of other negative system impacts caused by the software, [23].

12.3.2 *Operational Logistics—Supply Chain Sustainment*

The supply chains for complex systems are becoming increasingly volatile and difficult to manage. Consider the F-35 Joint Strike Fighter aircraft, which partners with more than 1200 domestic suppliers and nine "partner countries" to produce "thousands of components from highly sophisticated radar sensors to the aircraft's mid fuselage" [24]. The F-35 manufacturing will continue until at least the mid-2020s and the aircraft must be maintained (i.e., spared) for the next 30+ years; how do you manage the F-35′s complex, multinational supply chain for those 30+ years so that you can keep the aircraft flying?

In short, supply chain sustainment involves managing supply chain risk over potentially long periods of time. This involves the management of sourcing, existing inventories, and disruptions to the supply chain. Unlike cell phones, for example, critical systems generally do not control the supply chain for their components, i.e., the supply chain does not exist for (and is not driven by) the critical system application. Many practices from high-volume industries (e.g., just-in-time and lean inventories), which were created to improve the efficiency of supply chains have increased the supply chain's "brittleness" and, consequently, an enterprise's exposure to supply disruptions [25]. Developing additional sources of supply can help reduce risks, but having them does not necessarily reduce supply chain vulnerabilities. Better options to reduce vulnerabilities may be available by working with the existing suppliers, e.g., using dual sites to assure supply at one site should a disaster strike the other, or making sure that suppliers have plans to address a wide variety of contingencies. Mission, safety, and infrastructure critical systems can complicate support because they require more sophisticated testing to ensure that all system interfaces are properly functioning. Budget constraints coupled with the increasing costs of new systems and personnel are increasing pressure to reduce the physical size of and budgets for support infrastructure.

12.3.3 Operational Logistics—Workforce Sustainment

The sustainment of critical systems is also impacted by the loss of critical human skills that either cannot be replaced or take impractically long times to reconstitute. Critical skills loss [26] becomes a problem for sustaining systems that depend on an aging workforce that has highly specialized, low-demand skills. Critical skills loss occurs when skilled workers retire and there is an insufficient number of younger workers to take their place. This does not occur because of inactivity, poor planning, or a lack of foresight by an organization. Rather, it is simply an inevitable outcome of the dependence on low-demand specialized skills. System sustainment challenges resulting from the loss of critical human skills have been reported in industries that include healthcare, nuclear power, and aerospace. An example is the shortage of mainframe application programmers that are experienced in legacy applications—in this case, the required skills are no longer taught as part of any structured educational program and younger workers are not interested in learning them. For critical systems, the problems can be devastating: "Even a 1-year delay in funding for CVN-76 [aircraft carrier] will result in the loss of critical skills which will take up to 5 years to reconstitute through new hires and training. A longer delay could cause a permanent loss in the skills necessary to maintain our carrier force" [27].

12.3.4 Contract Structure

The long-term contract structures under which critical systems are delivered and supported play an increasingly critical role in defining the strategies that govern how sustainment is performed. In addition to a legacy transactional approach, there are a group of strategies for system support is called outcome-based logistics or contracting (also referred to as "performance contracting," "availability contracting," "contract for availability (CfA)," "performance-based service acquisition (PBSA)," "performance-based logistics (PBL)," and "performance-based contracting"). In outcome-based contracting, a contractor delivers performance outcomes that are defined by performance metric(s) for a system instead of delivering a particular good or service. The mindset behind outcome-based contracts is well summarized by Levitt [28] as, "The customer doesn't want a drilling machine; he wants a hole-in-the-wall." Outcome-based contracts pay the contractor for effectiveness (which can be articulated as availability, readiness, or other performance-related measures) at a fixed rate, penalizing shortcomings in the effectiveness delivered, and/or awarding gains beyond the target goals.

Outcome-based contracting exists because customers that require high availability systems are interested in buying system (in some cases subsystem or component) availability, rather than buying the system. For this type of contract, the customer pays for the outcome delivered, instead of buying the system and paying for system sustainment. Outcome-based contracts include cost penalties for failing to meet specified availability and performance requirements during defined time periods.

Outcome-based contracts make the sustainment community responsible for designing systems (including designing the sustainment of systems) and to coordinate the system design and the design of the contract terms. "For systems managed under outcome-based contracts, contract failure may mean significant money is spent by the customer (potentially the public) for either no outcome or inadequate outcome, or result in the contractor being driven out of business, which can lead to disaster for both parties" [29].

12.3.5 Governance and Policy

When designing and producing complex systems, there are host of technical challenges that must be resolved to meet their sustainment requirements. Although, resolving these engineering issues is necessary, it is generally insufficient to meet these requirements. Most critical systems operate at the intersection of the public and private sectors, where their sustainment is subject to a host public policy as well as business considerations. Moreover, during this era of disruptive technical developments, government policies and business models lag and may in fact impede the use of innovative sustainment practices and processes.

The US DoD, for example, has a host of legislative and regulatory policies that must be considered when performing sustainment tasks. These include legislation that specifies the definition of organic depot maintenance,[3] types and amounts of work that must be performed there, along with guidance on how these depots can form public–private partnerships—these all constrain the sustainment solutions space. Federal acquisition regulations provide detailed guidance on acquisition planning and awarding contracts. There are also a myriad of Department, Military Service, and Agency instructions and regulations that provide guidance on every level of maintenance, supply, and transportation operations. Contracts, for example, maybe restricted in terms of type and contract length, potentially limiting the benefits of outcome-based contracting.

Business models are also fundamentally connected to, and informed by, technological innovation. They serve as the intermediary link between traditional firm performance and operation, enabling firms and organizations to leverage the benefits that technology can offer. As the role of technology increasingly affects the production, supply chains, and system sustainment, these innovations will necessitate changes to the existing business models, particularly as these businesses transition into the digital era. To adapt, businesses will have to strategically pivot their existing models and add a focus on their digital supply chain.

As a result, in order to develop comprehensive sustainment solutions, engineering innovations must be coupled with a consideration of public policy and business challenges. Only then can the full potential of the emerging technologies for the sustainment of complex critical systems be achieved.

12.4 The Economics of Sustainment

Traditionally, for many systems, sustainment is an afterthought. Unfortunately, these systems are often too expensive to replace except under emergency or catastrophic circumstances, and in many cases, the financial resources expended to sustain them over their long lifetimes effectively preclude their replacement. The cost of supporting old systems is not only economic but also safety, resource consumption, and quality of life. For example, imagine a 911 system in a major city that used the latest communications technology (instead of 15-year-old technology)—lives would be saved [30]; or FAA air traffic control systems incorporating the latest technology (rather than 25-year-old technology)—aircraft could fly with reduced separation and more optimal paths, significantly improving efficiency [31, 32]. These systems are too expensive to replace or even update, and therefore they become costly legacy systems that eventually impact people's lives (convenience and most importantly safety).

---

[3] The DoD's military departments own and operate industrial facilities to maintain, repair, and overhaul equipment that are referred to as organic depots.

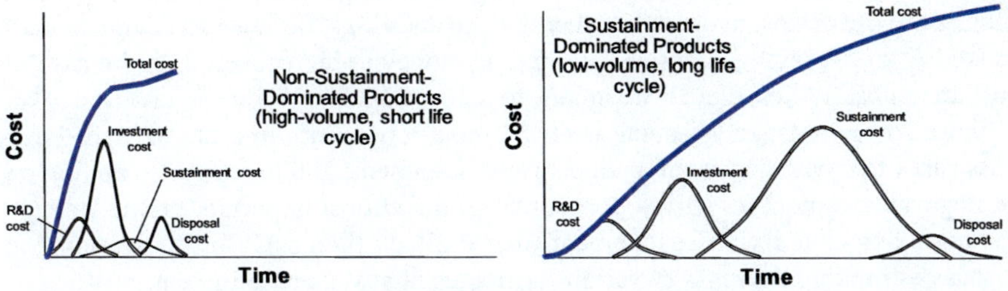

Fig. 12.1 Life-cycle spending profile for high and low volume products, [19]

A sustainment-dominated system is defined as a system for which the life-cycle footprint significantly exceeds the footprint associated with making it [7]. Sustainment-dominated systems are generally manufactured and supported for very long times, are very expensive to replace, and have very large qualification/certification overheads. Figure 12.1 illustrates the difference between sustainment-dominated products and non-sustainment-dominated products. Non-sustainment-dominated products are generally high-volume products sold to the general public that have relatively little investment in sustainment activities (probably only a limited warranty) and the total life cycle of the product (production and support) is short (e.g., a particular model of cell phone). Alternatively, sustainment-dominated products, are low-volume expensive systems, have large sustainment costs and long manufacturing and/or field lives (e.g., an airplane).

Commercial companies that develop critical systems consider operating and support costs integral to their product development decisions. Controlling these costs directly impacts revenues, profits, and market growth. Consequently, they establish product availability, operating, and sustainment costs as key system requirements. As a result, the product developers focus on designing a product that meets the availability requirements, is easy to maintain, and reliable. When we look at government system development, although they may have the same vision, their execution is often flawed. The US DoD's systems often last decades, and their sustainment dominates life-cycle costs (LCC), typically 60–80% of LCC for a system that lasts 30 years [33]. However, when faced with immediate near-term pressures, such as those related to R&D, production, and system acquisition costs, they must make real-time trade-offs against the future impacts those decisions may have on sustainment cost and performance 10, 20, even 30 or more years in the future. While simply understanding these trade-offs can be difficult, justifying and defending a 30–40-year return on investment against the immediate resource demands of today is even more challenging. As a result, addressing sustainment issues is often delayed, until the systems are operational, which is too late.

The value of process, equipment, and yield changes for manufacturing systems are often quantified as cost *savings*. However, the value of sustainment activities is usually characterized as cost *avoidance*. "Cost avoidance is a reduction in costs that have to be paid in the future to sustain a system" [19]. The sustainment community prefers the use of cost avoidance rather than cost savings, because an action characterized as a cost savings implies that there is money to be recovered. In the case of sustainment activities, there is no money to recover. Making business cases based on a future cost avoidance argument is challenging. Therefore, in order to make business cases to create and retain budgets for sustainment; and to support spending on strategic sustainment initiatives, it becomes of the utmost importance to understand the costs associated with sustainment (and the lack thereof).[4] In this section, we discuss estimating the costs of various attributes of system sustainment.

12.4.1 Maintenance Management

Maintenance refers to the measures taken to keep a product in operable condition or to repair it to an operable condition [35]. No one knows how much economies spend on maintenance, partly because most maintenance is performed in-house, not purchased on the market. The best numbers are collected by Canada, where firms spent 3.3% of GDP on repairs in 2016, more than twice as much as the country spends on research and development [36]. "Maintenance lacks the glamour of innovation. It is mostly noticed in its absence." [36].

Fundamentally, maintenance is about money and time. The decision to spend money doing maintenance is based on the value obtained, i.e., money does not have to be spent on maintenance; the system could be simply discarded each time it fails and replaced with a new system. Optimizing the maintenance activities is justified by a combination of economic and availability arguments.

12.4.1.1 Corrective Maintenance

Corrective maintenance (also called "break-fix" or "run-to-failure") primarily depends on the system's reliability. The cost of maintenance in this case is simply the number of system failures that have to be resolved multiplied by the cost of resolving them. Assume that we have a system whose failure rate is constant. The reliability of the system is given by Eq. (12.1) as,

$$R(t) = \mathrm{e}^{-\lambda t} \tag{12.1}$$

where t is time and λ is the failure rate. The mean time between failure (MTBF) for this system is $1/\lambda$. Suppose, for simplicity, the failures of this system are resolved

[4]Sometimes this is referred to as "life-cycle sustainment planning" [34]. The purpose of life-cycle sustainment planning is to maximize readiness by delivering the best possible support outcomes at the lowest Operating and Support (O&S) cost. Programs that emphasize sustainment early in the system life cycle, deliver designs with the highest likelihood of achieving operational performance requirements, and reduced demand for sustainment.

instantaneously at a maintenance cost of $1000/failure. If we wish to support the system for 20 years and the units on λ are failures/year, how much will it cost? Assuming that the discount rate on money is zero, this is a trivial calculation:

$$\text{Total Cost} = 1000(20\lambda) \tag{12.2}$$

The term in parentheses is the total number of failures in 20 years. If $\lambda = 2$ failures per year, the Total Cost is $40,000. If we include a cost of money, i.e., a discretely compounded discount rate (r), the solution becomes a sum, because each maintenance event has a different cost in year 0 dollar,

$$\text{Total Cost} = \sum_{i=1}^{20\lambda} \frac{1000}{(1+r)^{i/2}} \tag{12.3}$$

where $i/2$ is the event date in years.[5] If we assume $r = 8\%$/year, the Total Cost is now $20,021.47 in year 0 dollar.

In reality, the actual event dates in the example presented above are not known (they do not happen at exactly MTBF intervals), rather the time-to-failures are represented by a failure distribution. The failure distribution can be sampled to capture a sequence of failure events whose costs can be summed using Eq. (12.3). See Ref. [19] for an example.

In the simple example described, 20λ in Eq. (12.2) is the number of "spares" needed to support the system for 20 years (if $\lambda = 2$ failures/year then 40 spares are necessary). Sparing analysis, i.e., determining the number of spares required to support a system for a specified period of time to a specified confidence level is central to maintenance planning and budgeting. In general, the number of spares needed can be determined from Ref. [37],

$$\Pr(X \le k) = \sum_{x=0}^{k} \frac{(n\lambda t)^x e^{-n\lambda t}}{x!} \tag{12.4}$$

where

k = number of spares.
n = number of unduplicated (in series, not redundant) units in service.
λ = mean failure rate of the unit or the average number of maintenance events expected to occur in time t.
t = time interval.
$\Pr(X \le k)$ = probability that k is enough spares or the probability that a spare will be available when needed (this is known as the "protection level" or "probability of sufficiency").

[5] The $i/2$ assumes that $\lambda = 2$ and the failures are uniformly distributed throughout the year.

Solving Eq. (12.4) for k gives the number of spares needed. The time interval (t) in Eq. (12.4) can be interpreted several ways. If the spares are permanent than t is the total time that the system needs to be supported. Conversely, if the spares are only required to support the system while the original failed item is being repaired, then t is the time-to-repair the original item.

Renewal functions are another way of estimating spares. A renewal function gives the expected number of failures in an interval. For a constant failure rate, the number of renewals in a period of length t is given by,

$$M(t) = \lambda t \tag{12.5}$$

For other types of time-to-failure distributions (e.g., Weibull), the renewal function may not have a simple closed-form like Eq. (12.5) but can be estimated using,

$$M(t) = \frac{t}{\mu} + \frac{\sigma^2}{2\mu^2} - \frac{1}{2} \tag{12.6}$$

where μ is the mean and σ^2 is the variance of the distribution (this estimation is valid for large t, other approximations exist). For a three-parameter Weibull distribution, the μ and σ^2 are given by,

$$\mu = \gamma + (\eta - \gamma)\Gamma\left(1 + \frac{1}{\beta}\right), \sigma^2 = (\eta - \gamma)^2\left[\Gamma\left(1 + \frac{2}{\beta}\right) - \Gamma^2\left(1 + \frac{1}{\beta}\right)\right] \tag{12.7}$$

where β is the shape parameter, η is the scale parameter, and γ is the location parameter.

$M(t)$ and k are not the same thing. k is the number of spares necessary to satisfy a specified confidence that you have enough spares to last t (i.e., $\Pr(X \leq k)$ in Eq. (12.4)). $M(t)$ is the expected number of spares needed to last for t. Renewal functions are commonly used to estimate warranty reserve funds for a warranty period of t and to estimate maintenance budgets, but if one wants to know how many spares are necessary to satisfy a particular confidence level then a treatment like that in Eq. (12.4) is necessary.

To illustrate the analysis of maintenance costs, consider a bus that is intended to operate for 200,000 miles per year. Reliability analysis indicates that the failure of a critical component follows an exponential distribution with a failure rate of $\lambda = 1.4 \times 10^{-5}$ failures/mile. Assume that it takes 5 days (2740 miles of lost bus usage) and costs \$5,000 each time the component must be replaced when it fails.[6] Assume that the replacement component is "as good as new" and that the failure mechanism

[6]Note, everything in this illustration is in miles rather than time. Mileage can be converted to time if desired, but it is not necessary to do so. We are also assuming that all maintenance is via component replacement, i.e., there is no component repair.

only accumulates damage while the bus is operating (not while it is being repaired). What is the expected maintenance cost for one bus, for 1 year?

The component failures follow an exponential distribution, so we can use Eq. (12.5) to estimate the number of renewals in 1 year (200,000 miles) period. Using Eq. (12.5), we get $M(t = 200{,}000) = 2.8$ renewals/year (repairs in this case). This would be the correct number of repairs if the relevant failure mechanism accumulated damage continuously over calendar time, but because it only accumulates damage when it is operating, this is too large. The time (miles) to perform the corrective maintenance is not zero (the calculation above implicitly assumes it is zero, i.e., it assumes the bus is fixed instantaneously on failure, which it is not). One way to fix this is by adjusting the failure rate,

$$\lambda_{\text{modified}} = \frac{1}{{}^{1}\!/_{\lambda_{\text{original}}} + 2740} = 1.348 \times 10^{-5} \text{ failures/year} \tag{12.8}$$

Equation (12.8) effectively extends the MTBF ($1/\lambda_{\text{original}}$), by the maintenance duration. Using the new value of λ, $M(t = 200{,}000) = 2.697$ renewals.[7]

Now, the annual maintenance cost for a bus is given by,

$$\text{Cost}_{\text{annual}} = c_f M(t) \tag{12.9}$$

where c_f is the cost per maintenance event. For the bus problem, from Eq. (12.9) with $c_f = \$5000$, the annual maintenance cost per bus is \$13,668. The operational availability of the bus is given by,

$$\text{Availability} = \frac{\text{Uptime}}{\text{Uptime} + \text{Downtime}} = \frac{200{,}000 - (2.697)(2740)}{200{,}000} = 0.9631 \tag{12.10}$$

The availability is the fraction of time that the bus is operational.

How many spares do we need to have a 90% confidence that we have enough spares for one bus for 1 year? 2.697 is the expected number of spares (per bus per year). To solve this problem, we need to use Eq. (12.4) with $n = 1$ (one bus). When $k = 3$, the confidence level is $\Pr(X \le k) = 0.69$; to obtain a confidence level greater than 0.9, $k = 5$ spares have to be used, $\Pr(X \le k) = 0.93$ in this case.

12.4.1.2 Preventative Maintenance

Next, we consider preventative maintenance. Preventative maintenance potentially avoids more expensive corrective maintenance. Corrective maintenance is generally

[7] 2.697 is the expected number of spares (per bus per year). If we want to know the corresponding confidence level, or conversely the number of spares needed to meet a given confidence level, we have to solve this problem using discrete-event simulation.

more costly because it occurs at unplanned times making the logistics of repair more difficult and it may cause collateral damage to other system components. To assess the cost of a system with a combination of corrective and preventative maintenance, we define a maintenance cycle length, which is the length of time between maintenance events (corrective or preventative). In terms of this maintenance cycle length, the total maintenance cost per unit time is given by Ref. [38],

$$\begin{aligned}\text{Cost}(t_p) &= \frac{\text{Total expected replacement cost}}{\text{Expected maintenance cycle length}} = \frac{R(t_p)c_p + [1 - R(t_p)]c_f}{R(t_p)t_p + \int_0^{t_p} tf(t)dt} \\ &= \frac{R(t_p)c_p + [1 - R(t_p)]c_f}{\int_0^{t_p} R(t)dt}\end{aligned} \tag{12.11}$$

where

t_p = preventative maintenance time.
c_p = preventative maintenance cost.
c_f = corrective (on failure) maintenance cost.
$R(t)$ = reliability at time t.
1-$R(t)$ = unreliability at time t.
$f(t)$ = PDF of the failure distribution.

The maintenance interval (t_p), is determined by minimizing value of Cost(t_p), i.e., determining the value of t_p that satisfies $\text{dCost}(t_p)/\text{d}t_p = 0$. For the bus problem described in Sect. 12.4.1.1, Cost(t_p) is minimized when $t_p = \infty$, why? An exponential distribution is memoryless, i.e., the failure rate is constant and independent of the age of the system or whether preventative maintenance has been done. In order for preventative maintenance to make sense there must be an increasing failure rate over time, i.e., the system has to age.

To demonstrate preventative maintenance, let's change the example from Sect. 12.4.1.1. Assume that the failure of the component of interest follows a Weibull distribution with $\beta = 2$, $\eta = 74{,}000$ miles and $\gamma = 0$. Assuming just corrective maintenance, and using Eqs. (12.6) and (12.7) with the addition of 2740 miles to μ, the $M(t = 200{,}000) = 2.553$. Let's assume that a scheduled preventative replacement task that takes 1 day (550 miles of lost usage) and costs $2050. In this case, $\text{dCost}(t_p)/\text{d}t_p = 0$ when $t_p = 65{,}500$ miles (solved numerically ignoring the time to perform maintenance). At $t_p = 65{,}500$ miles, Eq. (12.11) gives Cost(t_p) = $0.07056/mile. Using discrete-event simulation, the average number of corrective maintenance events per year per bus is 1.976 and the average number of preventative maintenance events per year per bus is 1.498. The availability in this case, determined via the discrete-event simulation, is 0.9688.[8] The annual cost per bus is given by,

[8]In this case, we assume that the preventative maintenance clock is reset to zero if the bus fails and has a corrective maintenance event prior to t_p. This also assumes the component of interest starts each year good-as-new.

$$\text{Cost}_{\text{annual}} = c_f(1.976) + c_p(1.498) = \$12{,}948 \tag{12.12a}$$

$$\text{Cost}_{\text{annual}} = \text{Cost}(t_p)(200{,}000) = \$14{,}111 \tag{12.12b}$$

Equations (12.12a) and (12.12b) do not result in the same cost. They do not match because the simulation (which is more accurate) accommodates incomplete maintenance cycles (for which the incomplete portion is free).[9]

12.4.1.3 Predictive Maintenance

Preventative maintenance occurs on some predetermined schedule, e.g., every 65,500 miles in the example in Sect. 12.4.1.2. Predictive maintenance occurs when the system needs maintenance based on reliability predictions, the actual condition of the system (condition-based maintenance) or the condition of the system coupled with the expected future environmental stress conditions (prognostics and health management—PHM). In the case of PHM, predictive maintenance cost modeling is based on the prediction of a remaining useful life (RUL). The RUL provides a time period prior to failure in which maintenance can be scheduled to minimize the interruption to system operation.[10]

The economics of predictive maintenance includes predicting the return-on-investment (ROI) associated with investing in predictive maintenance (it may be costly to add and support in systems); and optimizing when to act (and what action to take) when a predicted RUL (including its associated uncertainties) is obtained.

A cost avoidance ROI for PHM can be calculated using Ref. [39],

$$\text{ROI} = \frac{\text{Cost Avoided} - \text{Investment}}{\text{Investment}} = \frac{C_u - C_{\text{PHM}}}{I_{\text{PHM}}} \tag{12.13}$$

where

C_u = life-cycle cost of the system managed using unscheduled (corrective) maintenance.

C_{PHM} = the life-cycle cost of the system when managed using a PHM (predictive) maintenance approach.

I_{PHM} = the investment in PHM when the system is managed using a PHM (predictive) maintenance approach.

[9] If the length (in miles) of the problem is increased, the two models will converge to the same cost.

[10] For example, if an airline had a 24-h RUL prediction (assume there is no uncertainty in this prediction), they could reroute an aircraft to insure that it was at an airport that has the appropriate maintenance resources between midnight and 6 am tomorrow morning to obtain the required maintenance without interrupting any flight schedules.

To illustrate an ROI analysis, consider the bus example from the previous two sections. As part of the business case for the inclusion of PHM into a particular subsystem in the bus, its ROI has to be assessed. Assume the following:

- The system will fail three times per year
- Without PHM, all three failures will result in unscheduled maintenance actions
- With PHM, two out of the three failures per year can be converted from unscheduled corrective to scheduled maintenance actions (the third will still result in an unscheduled maintenance action)
- The cost of an unscheduled maintenance action is \$5000 and takes 5 days of downtime
- The cost of a preventative maintenance action is \$1000 (all repairs, no spares) and takes half a day of downtime
- The recurring cost (per system instance) of putting PHM into the system is \$20,000
- In addition, you have to pay \$2000 per year (per system instance) to maintain the infrastructure necessary to support the PHM in the system
- The bus has to be supported for 25 years.

We wish to calculate the ROI of the investment in PHM relative to performing all unscheduled maintenance. First, consider a case where the discount rate is 0. The analysis is simple in this case,

$$C_u = (25)(3)(\$5000) = \$375{,}000.$$
$$C_{\text{PHM}} = (25)[(1)(\$5000) + (2)(\$1000)] = \$175{,}000.$$
$$I_{\text{PHM}} = \$20{,}000 + (25)(\$2000) = \$70{,}000.$$
$$\text{ROI} = \frac{375{,}000 - 175{,}000}{70{,}000} = 2.86$$

If the discount rate is nonzero, the calculation becomes more involved; for a 5%/year discount rate the solution becomes,[11]

$$C_u = \sum_{i=1}^{25} \frac{(3)(\$5000)}{(1+0.05)^i} = (3)(\$5000)\frac{(1+0.05)^{25} - 1}{(0.05)(1+0.05)^{25}} = \$211{,}409$$

$$C_{\text{PHM}} = \sum_{i=1}^{25} \frac{(1)(\$5000) + (2)(\$1000)}{(1+0.05)^i} = \$98{,}658$$

$$I_{\text{PHM}} = \$20{,}000 + \sum_{i=1}^{25} \frac{\$2000}{(1+0.05)^i} = \$48{,}188$$

[11]There are several implicit assumptions in this analysis including that all charges for maintenance occur at the end of the year (end-of-year convention), that the \$20,000 investment in PHM occurs at the beginning of year 1, and discrete annual compounding. In this case, the values of C_u and C_{PHM} are both year 0 present values.

$$\text{ROI} = \frac{211{,}409 - 98{,}658}{48{,}188} = 2.34$$

In reality, the ROI calculation associated with adding health management to a system is more complex than the simple analysis provided above. For example, predictive maintenance (e.g., PHM), will result in a combination of repairs and replacements with spares. Since the health management system will tell the maintainer to take action prior to the actual failure, some remaining life in the original component will be disposed of, which could eventually translate into the need for more spares. The availability of the system may also be a relevant issue; a simple availability calculation for this case is:

$$A_{\text{no PHM}} = \frac{(24)(7)(365) - (3)(5)(24)}{(24)(7)(365)} = 0.9941,$$

$$A_{\text{PHM}} = \frac{(24)(7)(365) - [(1)(5)(24) + (2)(0.5)(24)]}{(24)(7)(365)} = 0.9977$$

A positive or negative ROI does not make or break a business case, but, being able to assess an ROI is part of making a business case to management or to a customer.

When predictive maintenance is analyzed, the operative question is often when to perform maintenance in response to a predicted RUL. The longer the predicted RUL, the more flexibility the sustainer has to manage the system, but RULs are uncertain and the longer one waits after an RUL indication, the higher the risk of the system failing before the appropriate maintenance resources are available. One method of optimizing the action to take (and when to take it) based on an uncertain RUL is using a maintenance option.

A maintenance option is a real option is defined by Ref. [40] as,

- Buying the option = paying to add PHM to the system (including the infrastructure to support it)
- Exercising the option = performing predictive maintenance prior to system failure after an RUL indication
- Exercise price = predictive maintenance cost
- Letting the option expire = do nothing and run the system to failure then perform corrective maintenance.

The value from exercising the option is the cost avoidance (corrective vs. predictive maintenance) tempered with the potential loss of unused life in system components that were removed prior to failure or the predictive maintenance revenue loss. The predictive maintenance revenue loss is relevant to systems where uptime is correlated to revenue received (e.g., energy generation systems) and is the difference between the cumulative revenue that could be earned by waiting until the end of the RUL to do maintenance versus performing the predictive maintenance at some point that is earlier than the end of the RUL. In summary, the loss that appears in the value calculation is the portion of the system's RUL that is thrown away when predictive maintenance is done prior to the end of the RUL. See Refs. [40, 41] for the analysis of systems with maintenance options.

12.4.2 The Aging Supply Chain

Technology evolution is often driven by high-volume consumer product demands (e.g., cell phones, tablet computers, etc.), not by the type of critical systems defined in Sect. 12.3 (e.g., airplanes, control systems, networks, and power plants). As a result, unless the application is the demand driver it likely lags state-of-the-art technology by 10 or more years. Unfortunately, many of the most affected systems are safety, mission, and/or infrastructure critical so changes cannot be made to hardware or software without very expensive qualification and certification.

For sustainment-dominated systems, an aging supply chain that is not controlled by the application is reality. If we could forecast, plan for, and optimize how we manage aging technology (i.e., "gracefully" age critical systems), billions of dollars could be saved and the public's safety and convenience significantly enhanced.

The aging supply chain often manifests itself as an inability to procure the needed resources to sustain a system because the supply chain has "moved on". Most often those resources are spare parts, however, they can also be human resources (see Sect. 12.3.3), consumable materials needed to support a manufacturing process, equipment needed to manufacturing or test systems, intellectual property rights, and governance.

12.4.2.1 Diminishing Manufacturing Sources and Material Shortages (DMSMS)

DMSMS is defined as the "loss of impending loss of original manufacturers of items or suppliers of items or raw materials" [42], i.e. obsolescence. While there are several types of obsolescence, the most prevalent and relevant form for aging supply chains is procurement obsolescence, i.e., due to the length of the system's manufacturing and support life and possible unforeseen life extensions to the support of the system, the necessary components and other resources become unavailable (or at least unavailable from their original manufacturer) before the system's demand for them is exhausted. For many critical systems, simply replacing obsolete components with newer components is not a viable solution because of high reengineering costs and the potentially prohibitive cost of system requalification and recertification. For example, if an electronic component in the 25-year-old control system of a nuclear power plant fails, an instance of the original component may have to be used to replace it because replacement with a component that has the same form, fit, function and interface that is not an instance of the original component could jeopardize the "grandfathered" certification of the plant.

Electronic components are the most impacted and most managed aging supply chain components. A host of obsolescence mitigation approaches are used ranging from substitute/alternate parts to aftermarket suppliers and emulation foundries. A common mitigation approach is called lifetime buy. Lifetime buys,[12] although simple in concept, can be challenging to optimize and execute. A lifetime buy means making a one-time purchase of all the components that you think you will need forever. The opportunity to make a lifetime buy is usually offered by manufacturers of electronic components prior to part discontinuance (in the form of a published "last order date"). Lifetime and bridge buys play a role in nearly every component obsolescence management portfolio no matter what other reactive, proactive, or strategic management plans are being followed. At its most basic level, a lifetime buy means simply adding up all the projected future demand for the component, adding some "buffer" to that quantity, and buying and storing those components until needed. Unfortunately, everything is uncertain (most notably the demand forecasts) and the cost penalties for buying too few components can be astronomically larger than the penalty for buy too many components.

In this chapter, we not only present a simple lifetime buy quantity optimization treatment but also warn the reader that real lifetime buy optimization is done via stochastic discrete-event simulation for a number of reasons that will be articulated later in this section. The lifetime buy optimization problem is a version of the Newsvendor Problem (a classic optimization problem from operations research). The newsvendor problem seeks the optimal inventory level for an asset, given an uncertain demand and unequal costs for overstock and understock. In Newsvendor problems, the critical ratio is

$$F(Q_{\text{opt}}) = \frac{C_U}{C_O + C_U} \tag{12.14}$$

The factors relevant to solving this problem are:

- $F(Q)$ the cumulative distribution function (CDF) of demand evaluated for a particular lifetime buy quantity of Q.
- C_O the overstock cost—the effective cost of ordering one more unit than what you would have ordered if you knew the exact demand (i.e., the effective cost of one left-over unit that cannot be used or sold).
- C_U the understock cost—the effective cost of ordering one fewer unit than what you would have ordered if you knew the exact demand (i.e., the penalty associated with having one less unit than you need or the loss of one sale you can not make).
- Q the quantity ordered.
- D demand.

The objective is to find the value of Q that satisfies Eq. (12.14), i.e., Q_{opt}.

---

[12] Also called life-of-need, life-of-type, or all-time buys. Alternatively, bridge buys mean purchasing enough parts to last until a planned design refresh point in the future where the part will be designed out.

Consider the bus example, we defined in earlier sections of this chapter. Assume that there will be no future opportunity to procure additional spare parts for the component, we previously considered (with an exponential distribution with $\lambda = 1.4 \times 10^{-5}$ failures/mile). A lifetime buy is offered for this component. How many spare components should be bought per bus now to support 10 years worth of bus operation? Assume that the components cost \$1400 to procure now, but if you run out of components and have to procure them from a third party in the future, they will cost \$20,000 per component. Using Eq. (12.14) with $C_U = \$20{,}000 - \$1400 = \$18{,}600$, and $C_O = \$1400$, $F(Q_{\text{opt}}) = 0.93$. $F(Q)$ is the CDF of the demand, which means that life-cycle cost is minimized by purchasing the number of components gives you 93% confidence that you have enough spares. In the last paragraph of Sect. 12.4.1.1, this problem was worked using Eq. (12.4) and the number of spares that satisfied a 93% confidence was found to be 5 spares/year, therefore $Q_{\text{opt}} = 5$, which indicates that you will need $(5)(10) = 50$ components/bus purchased at the lifetime buy to last 10 years. Note, the actual demand is 2.697 spares/year, Q_{opt} is larger because of the asymmetry in the penalties, for example, if the future cost was \$4500, then the $F(Q_{\text{opt}}) = 0.69$, which corresponds to 3 spares/year.

The treatment of lifetime buy quantity optimization using a Newsvendor approach is elegant, but does not incorporate several key attributes of the problem, most notably Newsvendor solutions do not accommodate time. Time enters into the problem as discounting of the cash flows and in holding costs. The initial purchase of parts happens at time zero and does not need to be discounted, however, the penalties C_O and C_U occur years later when the buy runs out or the support of the system ends. C_O and C_U can be discounted and if one assumes that they would occur at approximately the same future time, then the value of $F(Q_{\text{opt}})$ given by Eq. (12.14) is unaffected. The bigger problem is holding cost.[13] Holding happens continuously until parts are used up—this is a problem that we cannot overcome with the newsvendor solution, and holding costs are not negligible.[14] See Ref. [43] for a more extensive treatment of lifetime buy problems.

Lifetime buys are a common reactive mitigation approach to obsolescence management. Because of the long manufacturing and field lives associated with sustainment-dominated systems, they are usually refreshed or redesigned one or more times during their lives to update functionality and manage obsolescence. Unlike high-volume commercial products in which redesign is driven by improvements in manufacturing, equipment or technology, for sustainment-dominated systems, design refresh is often driven by technology obsolescence that would otherwise render the product unproducible and/or unsustainable. The challenge is to determine the optimum design refresh plan (dates and content) that balances reactive obsolescence mitigation (including lifetime buys) with the large expense of redesign and

[13] There are Newsvendor solutions that include holding costs, however, the holding costs are \$/part (no time involved), so these types of holding costs are not applicable to the lifetime buy problem.

[14] For parts that have to be stored for many years in environmentally controlled inventory facilities, it is not unusual for the holding cost of the parts to be many times larger than the original cost to procure the parts.

requalification. The refresh planning problem can be articulated as finding the Y_R that minimizes, (12.15)

$$C_{Total} = \underbrace{\sum_{i=1}^{N} C_{before_i}}_{\text{Buying components as needed from 0 to their obsolescence date}} + \underbrace{\sum_{i=1}^{N} C_{LTB_i}}_{\text{Lifetime buy of components at their obsolescence date}} + \underbrace{\sum_{i=1}^{N} C_{H_i}}_{\text{Lifetime buy holding cost}} + \underbrace{C_{DR}}_{\text{Design refresh cost (all obsolete components addressed)}} + \underbrace{\sum_{i=1}^{N} C_{after_i}}_{\text{Buying components as needed from } Y_R \text{ to } Y_{EOS}} \tag{12.15}$$

where C are discounted costs, there are N total unique components, with a single design refresh at Y_R. The simplest solution to Eq. (12.15) only includes the second and fourth terms (the component buy to get to the refresh and the refresh costs) for a single ($N = 1$) component is known as a Porter model [44] for which closed-form solutions to this exist [19].

More detailed solutions to Eq. (12.15) exist including discrete-event simulation models that can find multiple refresh optimums and include other reactive mitigation options besides just last-time buys, e.g., Ref. [45]. These solutions can also incorporate various constraints governing when refreshes can and cannot occur [46].

12.4.2.2 Counterfeit Components

The obsolescence of components creates an opportunity for counterfeit components [47]. Counterfeit components are components that are misrepresented to the customer and may have inferior specifications and quality. Counterfeit components can take many forms, they may be used (salvaged) components misrepresented as new, remarked components, manufacturing rejects, components manufactured during factory shutdowns, and others. Whatever the form of the counterfeit, these components are problematic in critical systems. The risk of obtaining counterfeit components increases substantially when components become obsolete and have to be procured from sources that are not the original manufacturer.

12.4.2.3 Sourcing Small Quantities

For lean manufacturing approaches used for high-volume products (e.g., hundreds of thousands to millions of products a year), supply-chain disruptions are usually relatively short in duration (e.g., hours or days). For critical systems that are low volume (e.g., hundreds to a few thousand products a year) manufactured and supported for long periods of time, supply-chain disruptions may have durations of months or even years. Unlike high-volume products, critical systems often do not focus on minimizing the procurement prices of the components, rather, they care more about supply-chain stability because they are often subject to system availability requirements that penalize them if the product does not operate due to a lack of spare components.

High-volume applications commonly use a host of approaches to minimize their sourcing risk including second sourcing, and other strategies. This sourcing strategy decreases the impact of disruptions as component orders can be rerouted to the other suppliers when disruptions occur. For high-volume demand, multisourcing strategies are good for supplier negotiations (manufacturers can put pressure on the price), but for low-volume demand there is often little or no supplier negotiation. For low-volume demand,[15] the additional qualification and support costs associated with a backup source can negate its benefits. Single sourcing is defined as an exclusive relationship between an original equipment manufacturer (OEM) and a single supplier with respect to a specific part. However, while single sourcing minimizes qualification costs and allows for greater supplier–manufacturer coordination, the manufacturer is more susceptible to supplier-specific disruptions.

Buffering involves stocking enough parts in inventory to satisfy the forecasted component demand (for both manufacturing and maintenance requirements) for a fixed future time period so as to offset the impact of disruptions. While buffering can decrease the penalty costs associated with disruption events, there can be negative impacts, e.g., it can delay the discovery of counterfeit components in the inventory. Similarly, long-term storage of components can lead to part deterioration (such as the reduction of important solderability characteristics for electronic parts). For this reason, OEMs that utilize long-term buffering as a disruption mitigation strategy need to employ unique (and potentially expensive) long-term storage techniques that include regular assessment of the status/condition of the buffered components.

The supply chain for critical systems can also be subject to allocation problems. Allocation issues can occur for components that are not obsolete, but have extremely long delivery times (e.g., months to years). This is often due to circumstances that are out of the control of the system sustainers (natural disasters, political unrest, pandemics, etc.) that limit the quantity of components available on the market. When demand significantly exceeds supply, usually the largest customers (e.g., highest-volume customers) are supplied before low-volume customers meaning that critical systems may go to the "back of the line" for their components.[16]

---

[15] As additive manufacturing technologies and processes mature, they will create an alternative path for the production of some low-volume components.

[16] Note, some critical systems, i.e., approved national defense and energy programs may be covered by the Defense Production Act (DPA) and thereby can be given allocation priority. With respect to technology, the DPA was invoked by President Donald Trump for critical technology in the space industry [48] and more recently associated with ventilator manufacturing to combat the COVID-19 pandemic.

12.5 The Role of Policy and Acquisition in Sustainment

The sustainment of complex systems across the span of their life cycle involves a range of planning, implementation, and execution activities. These systems must meet user needs, as evidenced by their availability, effectiveness, and affordability. To achieve the best results, requires that sustainment be considered during all phases of the system's life cycle, particularly during the initial phases of its acquisition. Sustainment professionals need to be involved early in the system's development to influence the system design and support concepts for sustainability, since decisions made early in a program's development will have a profound impact on the system's life-cycle cost.

During these early phases, when examining performance requirements trade-offs (e.g., speed, range, payload), they should be balanced with the system sustainment requirements (e.g., availability, reliability, operating and support costs). These decisions should be based on a business case analysis to identify and compare the various alternatives, then analyze the mission and business impacts, risks, and sensitivities.

Technological trends are also placing increasing emphasis on digital data to support sustainment applications, such as prognostic health monitoring, condition-based maintenance, additive manufacturing, and failure prediction. Consequently, early in the life of programs, acquisition decisions must be made regarding the data collection and data rights.

12.5.1 A Broadened Sustainment Perspective

The concept of sustainability implies that a stakeholder's present needs are met while not placing the future well-being of the stakeholders at risk.

Under the best of circumstances, sustainment provides a framework for assuring the financial, security, and mission-success of an enterprise (where the enterprise could be a population, company, region, or nation). However, today, sustainment is usually only recognized as an organizational goal after it has already impacted the bottom line and/or the mission success of the organization, which is too late. Given that increasingly complex systems are embedded in everything, the sustainment culture needs to change to make it a part of the system's design and planning. Suggestions include [1]:

(1) Design systems for sustainability from the beginning of the system's development.
(2) Developing sustainment requirements and metrics is as critical to a program's success as identifying requirements for cost, schedule, and performance; but, often does not receive the requisite attention.
(3) Socialize the concept of sustainment. Generally, universities are good at preparing students to design new things, but the majority of students receive

minimal exposure to the challenges of keeping systems going or the role that government policies play in regulating sustainment.

- We need to educate students (engineers, public policy, and business) to contribute to the sustainment workforce.
- We need to educate everyone—even the students that will not enter the sustainment workforce need to understand sustainment because all of them will become customers or stakeholders at some level (taxpayers, policy influencers, decision-makers, etc.). The public has to be willing to resource the sustainment of critical systems.

(4) Leverage sustainment to create more resilient systems—resilience is more than just reliable hardware and fault-tolerant software. Resilience is the intrinsic ability of a system to resist disruptions, i.e., it is the ability to provide required capability in the face of adversity, including adversity from nontechnological aging and governance issues. Resilient design seeks to manage the uncertainties that constrain current design practices. From an engineered systems point of view, system resilience requires all of the following:

- reliable hardware and fault-tolerant software;
- resilient logistics (which includes managing changes that may occur in the supply chain and the workforce);
- resilient legislation or governance (rules, laws, policy);
- a resilient contract structure;
- and a resilient business model.

(5) Sustainment is not only an engineering problem. Engineering, public policy, and business must all come together in order to appropriately balance risk aversion with innovation and system evolution.

The world is full of complex systems (communications, transportation, energy delivery, financial management, defense, etc.). Because these systems are expensive to replace, they often become "legacy" systems. At some point, the amount of money and resources being spent on sustaining the legacy system hinders the ability to invest in new systems, creating a vicious cycle in which old systems do not get replaced until they become completely unsustainable or result in a catastrophic outcome.

References

1. Sandborn, P., & Lucyshyn, W. (2019). Defining sustainment for engineered systems—A technology and systems view. *ASME Journal of Manufacturing Science and Engineering, 141*(2).
2. Dunster, J., & Dunster, K. (1996). *Dictionary of natural resource management*. Vancouver: UBC Press.
3. Elkington, J. (1997). *Cannibals with forks: The triple bottom line of 21st century business*. Oxford: Capstone Publishing.

4. Kleindorfer, P. R., Singhal, K., & Van Wassenhove, L. N. (2005). Sustainable operations management. *Production and Operations Management, 14*(4), 482–492.
5. Thwink.org. Social sustainability [Internet]. thwink.org [cited 2018 Jul 26]. Available from: https://www.thwink.org/sustain/glossary/SocialSustainability.htm.
6. Wookcraft, S., Bacon, N., Caistor-Arendar, L., & Hackett, T. (2018). Design for social sustainability [Internet]. London: Social Life Ltd; 2012 [cited 2018 Jul 21]. Available from: https://www.social-life.co/media/files/DESIGN_FOR_SOCIAL_SUSTAINABILITY_3.pdf.
7. Sandborn, P., & Myers, J. (2008). Designing engineering systems for sustainability. In K. B. Misra (Ed.), *Handbook of performability engineering* (pp. 81–103). London: Springer.
8. Crum, D. (2002). Legacy system sustainment engineering. In *Proceedings of the DoD Diminishing Manufacturing Sources and Material Shortages Conference*, New Orleans.
9. Wiltsey-Stirman, S., Kimberly, J., Cook, N., Calloway, A., Castro, F., & Charns, M. (2012). The sustainability of new programs and innovations: A review of the empirical literature and recommendations for future research. *Implementation Science., 7*(17), 17–35.
10. Sutton, P. (2004). What is sustainability? *Eingana, 27*(1), 4:9
11. Basiago, A. D. (1995). Methods of defining 'sustainability'. *Sustainable Development, 3*(3), 109–119.
12. Goldsmith, E., & Allen, R. (1972). A blueprint for survival. *The Ecologist,2*(1).
13. Meadows, D. H., Meadows, D.L., Randers, J., Behrens, III W.W. (1972). *The limits to growth.* New York: Potomac Associates—Universe Books.
14. United Nations Environment Programme. (1978). *Review of the areas: Environment and development, and environment management*. New York: United Nations Environment Programme.
15. Kidd, C. V. (1992). The evolution of sustainability. *Journal of Agricultural and Environmental Ethics, 5*(1), 1–26.
16. Pisani, D. J. A. (2006). Sustainable development—Historical roots of the concept. *Environmental Sciences, 3*(2), 83–96.
17. The World Commission on Environment and Development. (1987). *Our common future.* Oxford: Oxford University Press.
18. US Government Joint Chiefs of Staff. (2017). Joint Publication (JP) 3–0, Joint Operations.
19. Sandborn, P. (2017). *Cost analysis of electronic systems* (2nd ed.). Singapore: World Scientific.
20. IATA Maintenance Cost Task Force. (2017). *Airline maintenance cost executive commentary.* Montreal: International Air Transport Association.
21. Peterman, D.R., & Frittelli, J. (2015). Issues in the reauthorization of Amtrak. *Congressional Research Service.*
22. Applying best business practices from corporate performance management to DoD. Defense Business Board; 2013. DBB Report FY13–03.
23. Depot Maintenance Core Capabilities Determination Process, US Department of Defense. DOD Instruction 4151.20, May 2018, NIST GAO-19–173.
24. Lockheed Martin (2020) Powering job creation for America and its allies. [Internet]. Lockheed Martin Corporation [cited 2020 Apr 23]. Available from: https://www.f35.com/about/economic-impact.
25. Griffin, W. (2008). The future of integrated supply chain management utilizing performance based logistics. *Defense Acquisition Review Journal, 15*(1), 3–17.
26. Sandborn, P., & Prabhakar, V. J. (2015). Forecasting and impact of the loss of the critical human skills necessary for supporting legacy systems. *IEEE Trans on Engineering Management, 62*(3), 361–371.
27. Defense Authorization and CVN-76, The Next Nuclear Aircraft Carrier Import to our Nation. [Internet]. *Congressional Record, 140*(65) [cited 2020 Apr 23]. Available from: https://www.gpo.gov/fdsys/pkg/CREC-1994-05-23/html/CREC-1994-05-23-pt1-PgH68.htm.
28. Levitt, T. (1972). Production-line approach to service. *Harvard Business Review, 50*(5), 41–52.
29. Sandborn, P., Kashani-Pour, A., Goudarzi, N., & Lei, X. (2016). Outcome-based contracts—Toward concurrently designing products and contracts. In *Proceedings of the International Conference on Through-Life Engineering*, Cranfield.
30. Wheeler, T. (2015). The 911 system isn't ready for the iPhone era. *The New York Times.*

31. Breselor, S. (2020). Why 40-year old tech is still running America's air traffic control. [Internet]. Gear; 2015 [cited 2020 Apr 23]. Available from: https://www.wired.com/2015/02/air-traffic-control/.
32. GAO 17-450, Air Traffic Control Modernization, August 2017.
33. Dallost, P. A., & Simick, T. A. (2012). Designing for supportability. Defense AT&L. 34–38.
34. Defense Acquisition University. Defense Acquisition Guidebook. Chapter 4 Life Cycle Sustainment. [cited 2020 Apr 27]. Available from: https://www.dau.edu/guidebooks/Shared%20Documents/Chapter%204%20Life%20Cycle%20Sustainment.pdf.
35. Dhillon, B. S. (1999). *Engineering maintainability*. Houston: Gulf Publishing Company.
36. Economics and the art of maintenance: Repair is as important as innovation. *The Economist*. 2018 Oct 20.
37. Myrick, A. (1989). Sparing analysis—A multi-use planning tool. In *Proceedings of the Reliability and Maintainability Symposium*, Atlanta, pp. 296–300.
38. Elsyed, E. A. (1996). *Reliability engineering*. Reading: Addison Wesley Longman.
39. Feldman, K., Jazouli, T., Sandborn, P. (2009). A methodology for determining the return on investment associated with prognostics and health management. *IEEE Transactions on Reliability, 58*(2), 305 316.
40. Haddad, G., Sandborn, P. A., & Pecht, M. G. (2014). Using maintenance options to maximize the benefits of prognostics for wind farms. *Wind Energy, 17*, 775–791.
41. Lei, X., & Sandborn, P. A. (2016). PHM-based wind turbine maintenance optimization using real options. *International Journal Prognostics and Health Management, 7*, 1–14.
42. Sandborn, P. (2008). Trapped on technology's trailing edge. *IEEE Spectrum, 45*(4), 42–45, 54, 56–58.
43. Feng, P., Singh, P., & Sandborn, P. (2007). Optimizing lifetime buys to minimize lifecycle cost. In *Proceedings of the Aging Aircraft Conference*. Palm Springs.
44. Porter, G. Z. (1998). An economic method for evaluating electronic component obsolescence solutions. Boeing Company.
45. Singh, P., & Sandborn, P. (2006). Obsolescence driven design refresh planning for sustainment-dominated systems. *The Engineering Economist, 51*(2), 115–139.
46. Nelson III, R., & Sandborn, P. (2012). Strategic management of component obsolescence using constraint-driven design refresh planning. *International Journal of Product Life Cycle Management, 6*(2), 99–120.
47. Pecht, M., & Tiku, S. (2006). Bogus: Electronic manufacturing and consumers confront a rising tide of counterfeit electronics. *IEEE Spectrum, 43*(5), 37–46.
48. Presidential Determination to Adequately Provide Critical Technology a Timely Manner Pursuant to Section 4533(a)(5) of the Defense Production Act of 1950". Federal Register, 2017, 82(114).

Peter Sandborn is a Professor in the CALCE Electronic Products and Systems Center and the Director of the Maryland Technology Enterprise Institute at the University of Maryland. Dr. Sandborn's group develops life-cycle cost models and business case support for long-field life systems. This work includes: obsolescence forecasting algorithms, strategic design refresh planning, lifetime buy quantity optimization, return on investment models for maintenance planning, and outcome-based contract design and optimization. Dr. Sandborn is an Associate Editor of the IEEE Transactions on Electronics Packaging Manufacturing and a member of the Board of Directors of the PHM Society. He was the winner of the 2004 SOLE Proceedings, the 2006 Eugene L. Grant, and the 2017 ASME Kos Ishii-Toshiba awards. He has a BS degree in engineering physics from the University of Colorado, Boulder, in 1982, and the MS degree in electrical science and PhD degree in electrical engineering, both from the University of Michigan, Ann Arbor, in 1983 and 1987, respectively. He is a Fellow of the IEEE, the ASME, and the PHM Society.

William Lucyshyn is a Research Professor and the Director of Research at the Center for Public Policy and Private Enterprise, in the School of Public Policy, at the University of Maryland. In this position, he directs research on critical policy issues related to the increasingly complex problems associated with improving public sector management and operations, and how government works with private enterprise. Current projects include public and private sector partnering; transforming Department of Defense logistics and supply chain management; and identifying government sourcing and acquisition best practices. Previously, Mr. Lucyshyn served as a program manager and the principal technical advisor to the Director of the Defense Advanced Research Projects Agency (DARPA) on the identification, selection, research, development, and prototype production of advanced technology projects. Mr. Lucyshyn received his Bachelor's Degree in Engineering Science from the City University of New York in 1971. In 1985, he earned his Master's Degree in Nuclear Engineering from the Air Force Institute of Technology. He was certified Level III, as an Acquisition Professional in Program Management in 1994.

Chapter 13
Four Fundamental Factors for Increasing the Host Country Attractiveness of Foreign Direct Investment: An Empirical Study of India

Hwy-Chang Moon and Wenyan Yin

Abstract Protectionist policies and recent coronavirus outbreak have made it more difficult for host countries to attract Foreign Direct Investment (FDI) and require governments to enhance their country's attractiveness for adapting to this changing environment. In this respect, this study introduces four fundamental factors that improve the inflow of FDI by comparing them with conventional elements that are commonly considered as being positive for such inflows. Unlike traditional factors that particularly stress what resources the host countries must possess in order to attract FDI, the fundamental factors suggested by this study emphasize more the *how* aspects, the effective way to utilize and mobilize a country's available resources. Furthermore, in order to understand better the importance of these factors, it uses India as an illustrative example. The Modi government introduced its "Make in India" policy to enhance its manufacturing sector by attracting FDI, yet such inflows to the manufacturing industries have remained very low. Thus, India requires more systemic measures for improving its business environment. By comparing its FDI attractiveness based on the four factors against nine other Asian economies, this study identifies strengths and weaknesses of India. It then suggests a series of strategic guidelines for enhancing India's FDI attractiveness.

Keywords Foreign direct investment (FDI) · Host country · Attractiveness · India · Make in India

H.-C. Moon (✉) · W. Yin
Seoul Business School, aSSIST University, Seoul, South Korea
e-mail: cmoon@snu.ac.kr

Graduate School of International Studies, Seoul National University, Seoul, South Korea

K. B. Misra (ed.), *Handbook of Advanced Performability Engineering*,
https://doi.org/10.1007/978-3-030-55732-4_13

13.1 Changes in Global Investment Environment and Challenges for Host Countries

The Trump administration's "America First" policies have strengthened the intensity of protectionism and reshoring in the USA. Such measures have had a major impact on the global investment environment, as the USA has long been one of the largest sources for outward Foreign Direct Investment (FDI). The main objective behind this policy is to impose high import duties on Multinational Corporations (MNCs) as a way to induce a transition from the long espoused "export strategy to the USA" to an "FDI strategy in the USA." At the same time, the Trump administration introduced a series of policies to improve the US investment environment by relaxing regulations and improving the efficiency of the government's operations. In fact, not only more US companies operating overseas but also foreign MNCs have invested in the USA in response to this policy approach (Economist [1], Moon and Yin [2]; Wall Street Journal [3]).

The recent United Nations Conference on Trade and Development (UNCTAD) report [4] has acknowledged that US tax reform contributed to the reduction of its outward FDI flows while global FDI inflows have witnessed a decline in both 2018 and 2019. However, MNCs do not just respond to government policies, rather they would make the decision to invest based on whether the host country has sufficient investment attractiveness or not. It may then seem like that the Trump administration pressured many of these MNCs, but a careful analysis of their true motivations will show that their decisions are based more on the fact that the USA enjoys strong investment attractiveness and offers many business opportunities (A.T. Kearney [5]; Moon and Yin [2]).

Global FDI inflows remain flat, with a 1% decline from US$1.41 trillion in 2018 to US$1.39 trillion in 2019 (UNCTAD [4]). FDI flows to developed countries decreased by a further 6% to an estimated US$643 billion, a historically low level, and FDI flows to developing countries remained unchanged compared to the previous year. The weaker macroeconomic performance and uncertain investment policies for firms such as the ongoing USA–China trade tensions were the main reasons behind the global downturn of FDI flows. Furthermore, with the current outbreak and spread of the coronavirus, global FDI flows may shrink by 5–15% and may hit the lowest levels since the Global Financial Crisis of 2008 (UNCTAD [6]). Therefore, it can be expected that competition in the future among countries will become more intense toward attracting FDI.

In order to sustain FDI inflows, the most important policy objective for governments is to enhance investment attractiveness. The locational determinants of FDI flows have long been investigated by previous studies and the conventional view has often emphasized the importance of production costs, labor skills, technical and managerial knowhow, infrastructure adequacy, and institutional quality (Du et al. [7]; Singh [8]). Although these factors do influence the FDI attractiveness of host countries, they may not be always applicable or doable for all countries which are at different development stages. Moreover, some factors often require a long period of time in order for a country to enhance their competitiveness to a level that would satisfy global investors. In this respect, this study[1] seeks to introduce four fundamental determinants that affect FDI attractiveness. Unlike preceding studies that mostly emphasize "what" factors of locational advantages, the four factors stress "how" aspects and well explain why countries endowed with similar resources show better performance in attracting FDI. Therefore, the four factors are particularly useful in providing strategic directions for developing countries to effectively mobilize and integrate the available resources to improve FDI attractiveness.

The rest of this study is organized as follows. It begins by presenting the four fundamental determinants that affect FDI attractiveness of host countries. We then take the case of India. We examine first the status of its FDI inflows and assess the effectiveness of the "Make in India" policy in particular adopted in 2014 to revitalize its manufacturing competitiveness. To better understand the relative strengths and weaknesses of India compared with its Asian counterparts, we then conduct a comparative analysis of nine other Asian economies by comparing the four key determinants. Based on the above investigation and analysis, this study provides a series of policy implications and strategic directions for Indian policymakers by applying the global value chain (GVC) approach, which encompasses not only trade but also various international means including investment and non-equity mode (NEM).

13.2 Four Determinants of FDI Attractiveness[2]

Dunning [9] classified the motivations for MNCs into four categories: resource-seeking, market-seeking, efficiency-seeking, and strategic-asset-seeking FDI. Hence, in order to attract FDI from firms with these four factors, the host country should have advantages of abundant natural resources, large market, cheap labor, or superior technology embedded in a specific field. However, these factors are featured as either inherited advantages or are difficult to emulate for all countries, particularly developing economies.

Moreover, with respect to manufacturing industries, previous studies have found that cheap labor is often an important factor in influencing FDI inflows. However, this stands in contrast to the fact that MNCs' automation rate for the production process is increasing while the proportion of labor costs in total production costs is decreasing. For example, the Taiwanese company Foxconn, which makes half of the world's iPhones, plans to fully automate 30% of its production by 2020, and it has already reduced more than 400,000 jobs by using tens of thousands of robots from

[1]This study was extended and further developed from Moon and Yin's [44] study titled "Chap. 1: Strategic Direction for Promoting FDI in India," which is a part of report entitled, *Policy Recommendation for the Development of Invest India*, prepared by KOTRA, Korea.

[2]The four factors in this part are correlated with the four elements in Moon's [36] Korea's economic development strategy which include agility, benchmarking, convergence, and dedication.

2012 to 2016 (South China Morning Post [10]). Therefore, low labor costs are no longer a critical factor in attracting FDI to the manufacturing sector.

Accordingly, other aspects are required to assess the overall investment attractiveness in a more comprehensive and systematic manner. In the section below, we present four more fundamental and doable factors for host countries to attract FDI, by comparing with the four general factors that are commonly believed to enhance a country's FDI attractiveness.

13.2.1 Cheap Labor Versus Productive Labor

Theoretically, low-cost labor is considered as an important determinant for MNCs to invest abroad in developing countries (Dunning [9]). Thus, they should possess a comparative advantage with labor, particularly with low wages, in order to attract FDI from developed countries. Yet, while developing countries have a comparative advantage of low-cost labor compared with advanced countries, there is no significant difference in wages among developing countries. In this case, given the cheap labor among developing countries, labor productivity becomes a more important determinant for attracting FDI. Empirically, Campos and Kinoshita's [11] study found that there were no significant effects with labor cost on FDI inflows. They argued that labor cost should be adjusted for labor productivity, and low wage rates alone are not a good indicator of labor cost advantages. Other studies (Redding and Venables [12], Ma [13]) found that although MNCs prefer low-cost labor countries, they will not simply move to less developed regions of a certain country, but rather they will tend to seek the regions with a qualified labor force. Our study defined productivity as an indicator for addressing both aspects of speed and precision; yet preceding studies were mostly focused on the speed aspect only.

This logic explains well why Apple and Samsung Electronics selected China and Vietnam, respectively, as the locations for the production and assembly of their smartphones. Although the wages of China and Vietnam are lower than those of some emerging economies such as the "Four Asian Tigers," they are higher than those of other developing countries such as Cambodia, Indonesia, and the Philippines (Moon and Yin [2], Yin [14]). Samsung's smartphone factory is located near Hanoi, the capital of Vietnam, while Apple's smartphone factory is located in Guangdong province. Both are the most expensive regions in the two countries (Yin [14, 15]). Therefore, low wages alone do not adequately explain why Vietnam and China were selected as the manufacturing base for these two large smartphone producers. Instead, the productivity of Chinese and Vietnamese workers is much higher than that of neighboring countries. In Vietnam, the labor cost of unskilled workers is only one-sixth of that for their counterparts in South Korea (Korea, hereafter), but there is no significant difference in labor productivity between Vietnam and Korea (Moon and Parc [16]). On the other hand, Chinese production plants are much bigger and more efficient than their counterparts in the USA, and thus have a high degree of agility to respond quickly to requirements in a changing international environment.

13.2.2 Better Environment Versus Adapting to the Global Standard

Government policies toward attracting FDI appear to emphasize what they have achieved over a certain period. For example, the Modi government in India has set its goal of becoming one of the top 50 places in the world for World Bank's ease of doing business index and has been working to create a good business environment. In order to enhance its attractiveness for foreign investment, India is seeking to improve the business environment by reducing corruption and improving its general infrastructure and has already achieved significant improvement in these fields. In 2020, India was ranked 63 among the list of 190 countries in the ease of doing business index, and this is a significant improvement from its 2014 ranking of 142. This would suggest that the macroenvironment has improved since the Modi government took office in 2014. Furthermore, out of the 12 macroeconomic indicators selected by the *Wall Street Journal* in 2016, India demonstrated a stronger performance across eight indicators when compared with the previous government (Wall Street Journal [17]).

In general, multinational managers often compare foreign countries where they can better exploit local resources and complement their asset portfolios as well as enhance their overall competitiveness (Moon [18]). While it is necessary to regularly improve the business environment and build upon past performances, it is more important to adapt to the international standard in terms of institutional regulations and industrial and living infrastructure. Such a factor can influence the MNCs' overall operational costs directly or indirectly. Notably, in an era where the value chains of firms have become more global and finely sliced up, host countries will be less likely to take on their entire value chain. Instead they will only host part of their value activities which reveal how they must adapt to a changing business environment and follow global best practices in order to ensure smooth and effective linkages among the value activities of MNCs dispersed among different regions of the world. Therefore, it is important for the host country to regularly compare strengths and weaknesses against their rivals, and secure higher investment competitiveness. Moreover, in addition to benchmarking the global best practices in terms of FDI attractiveness, the host country can further improve its attractiveness and outperform its rivals by adding plus alpha to better serve the investment needs of MNCs.

13.2.3 Entire Country Versus Industry Cluster

The mainstream literature of International Business has mainly adopted the entire country as the geographic unit of analysis for locational selection of FDI flows (Qian et al. [19]). However, when an MNC invests in a particular country, it tends to be in a specific area rather than the entire country, so regional competitiveness has become a more influential factor for MNCs when selecting the destination for their overseas investment. Some studies (Alcácer and Chung [20, 21]; Mudambi et al. [22]) have found that the analysis at the county and city level appeared to provide more solid evidence in respect of locational choices for FDI, particularly among high-technology firms. Moreover, from the perspective of enhancing the firms' competitiveness, they have long relied on localized resources and competences for new ideas and technologies which are often generated from interaction and communication among professionals within local communities (Moon [23]; Porter [24]). Therefore, firms within the cluster possess the advantage of accessing and exploiting local resources and are more likely to pursue innovation and competence enhancement strategies than those outside the cluster (Li and Bathelt [25]). In addition, given the context of GVCs, firms prefer regional clusters that have linkages with other clusters around the world. This is due to the fact that firms can benefit from mobilizing and exploiting knowledge and resources located in different regions on a global scope (Alcácer and Chung [20]; Moon and Jung [26]; Yin [8]). Furthermore, from the "doability" aspect of a nation's government, it is more effective to develop competitiveness in a specific region because the larger the country, the more difficult it is to achieve balanced regional development. Therefore, it is more efficient for the government to develop specific clusters that can attract foreign investors by providing good facilities and infrastructure. The cluster dimension of this study emphasizes three aspects, industrial infrastructure, living infrastructure, and international linkages, whereas preceding studies have mostly focused on industrial infrastructure.

The importance of clusters can be seen by examining the geographical distribution of Korea's investment in Vietnam where it is the leading investor. Korean companies invested heavily in clusters in northern Vietnam, near Hanoi, and in clusters in southern Vietnam, near Ho Chi Minh City. Among the northern regions, Bac Ninh Province has attracted the most FDI from Korea, which is the result of the investment by Samsung Electronics and its suppliers in Yen Phong and Que Vo industrial clusters. The second largest area for receiving Korean FDI is Hanoi, followed by Dong Nai, Thai Nguyen, Ho Chi Minh City, Haiphong, and Vung Tau (ASEAN Secretariat and UNCTAD [27]). Six out of the seven top regions for Korean FDI are categorized as Focal Economic Zones, which consist of a number of coastal provinces and major cities in Vietnam. By June 2016, Vietnam has established a total of 324 industrial clusters and 16 special economic zones, which accounted for about 50% of the cumulative FDI to Vietnam (HKTDC [28]). In particular, more than 75% of these were clustered in the Focal Economic Zone.

13.2.4 Education Versus Desire for a Better Life

Labor force can be generally divided into two categories: unskilled and skilled workers. Relatively high-skilled labor is needed to attract investment and such labor force is created by a high level of education. However, developing countries usually have more comparative advantage in low-skilled labor when attracting investment. For low-skilled workers in developing countries, a high level of education will help improve their productivity, but it means that there is a chance that they will be less likely to engage in long-term repetitive tasks such as assembly line production. Furthermore, they might also be sensitive to issues surrounding human rights and social welfare. In fact, many companies are increasing salaries due to repeated union strikes in their factories.

In this respect, when MNCs invest in developing countries, they will usually prefer hardworking, highly motivated workers who can meet the production standards that are required even if their level of education is low. For example, workers in Apple's Chinese assembly plant can work 6 days a week, 12 h a day. In addition, China has the flexibility to mobilize a large number of workers within a short period of time (Moon [29]). Thus, as soon as the parts and components arrive at the Foxconn assembly plant at midnight, 8,000 workers can be quickly assembled from the company's dorms and will begin work after a 30-min break (New York Times [30]). In other words, Chinese workers have a high sense of motivation and can always be put to work in a rapid way.

Vietnam has a high degree of flexibility in terms of working conditions and long-working hours. In terms of the number of working days per year, Vietnam has 302 days, while Korea has 249 days; and it also has longer working days (Vietnam: 2,416 h, Korea: 1,992 h) (Moon and Parc [16]). Of course, in developed countries, such conditions could be criticized for exploiting human rights or poor-working conditions, but in developing countries, such diligence and high motivation can be regarded as a great competitive factor for catching up with developed countries. This advantage influences the decision of MNCs for overseas investment among developing countries with similar labor costs.

Table 13.1 summarizes the comparative analysis examined above between the general understanding of location determinants of FDI and the four fundamental factors. The conventional factors that are commonly regarded as critical in attracting FDI are necessary but not sufficient for improving the attractiveness of the host country. Moreover, preceding studies have emphasized part of the four fundamental factors, but not all of them in a single framework in a comprehensive and systematic way. This study redefines or extends the concept of each of the four factors for their influences in attracting FDI toward the host country as shown in Table 13.1.

Table 13.1 Key factors affecting the attractiveness of FDI

General understanding	Fundamental factors
Cheap labor	Productive labor (Agility: speed and precision)
Better environment	Better than competitors (Benchmarking: learning and plus alpha)
Entire country	Industry cluster (Convergence: related industries, living environment, and international linkages)
Education	Desire for a better life (Dedication: diligence and motivation)

13.3 An Empirical Study of India's FDI Attractiveness

13.3.1 The Performance of India's FDI Inflows and Promotion Policy

India is the eighth largest recipient of FDI. In 2019, it attracted US$49 billion of FDI inflows, which is a 16% increase from the previous year (UNCTAD [6]). As Fig. 13.1 shows, FDI inflows to India declined amid the global economic downturn of 2008, but they have been steadily increasing since 2012. Today they have even surpassed the pre-Global Financial Crisis level, demonstrating an increasing trend over the last decade. Moreover, India was ranked 16 in A.T. Kearney's FDI Confidence Index Top 25 for 2019 that judges which countries are likely to attract the most investment over the next three years (A.T. Kearney [5]). This should be attributed to its rapid economic growth, the government's relaxation of FDI regulations, and a proactive FDI incentive policy. The top five investors in India are Singapore, Mauritius, Netherlands, USA, and Japan (in order) which altogether accounted for 77% of India's FDI inflows for fiscal year 2018–2019 (see Table 13.5), which reveals much about how India is highly dependent on the investment of just a few countries. In terms of sectoral distribution, as of the fiscal year 2018/2019, the service[3] industry received the highest FDI inflows, accounting for 20.6% of the country's total amount. This is followed by computer software and hardware (14.4%) and trade (10.0%) (see Table 13.6). FDI inflows to India are still concentrated on the service sectors, and FDI inflows to other manufacturing sectors, such as automobiles, chemicals, and pharmaceuticals, are still relatively low.

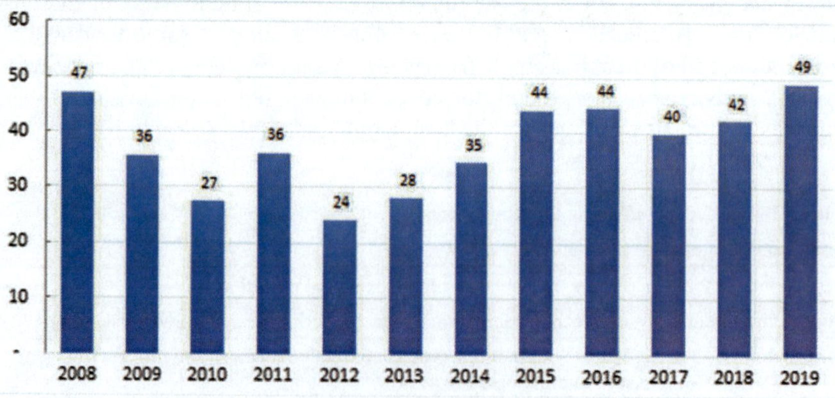

Fig. 13.1 The trend of India's FDI inflows, 2009–2019 (US$ billion). *Source* UNCTAD FDI Statistics, https://unctadstat.unctad.org/EN/; UNCTAD (2020)

[3]Services sector includes Financial, Banking, Insurance, NonFinancial/Business, Outsourcing, R&D, Courier, and Tech.

Despite India's high potential for FDI attractiveness, the actual investment environment is not as attractive relative to its competitors. According to the World Bank's ease of doing business index 2020, the investment environment is considered to be still relatively poor, ranking 63 out of 190 countries, and in some other categories (starting a business, 136; registering property, 154; enforcing contracts, 163) it shows substantial weaknesses. Therefore, in order to attract more FDI, the government should adopt new measures such as deregulation and simplification of procedures. Furthermore, as mentioned above, India's FDI inflows are dependent upon only a few investors, so it is necessary to further increase the range of the total FDI inflows. In the cumulative period of 2000–2019, the share of investment among these top five countries is 69%, and this trend is intensifying.

While FDI inflows in India have been centered on the competitive industries such as services and Information Technology (IT), for a more sustainable future it will be necessary to expand its range to other sectors. Investment in the service and IT sectors accounted for 18 and 10% of the cumulative total for the period 2000–2019, respectively, and the portion of investment for the service sectors has been surging in recent years. The concentration of FDI inflows among a few industries is still high as they are predominantly led by large MNCs rich in capital. At the same time, small businesses are dissuaded due to the country's poor infrastructure. This contrasts with Vietnam, where FDI inflows among both large and small firms have surged recently. For example, an increasing number of Korean SMEs as well as large conglomerates have both invested in Vietnam and are also located in or near the same cluster (ASEAN Secretariat and UNCTAD [27]).

In his inaugural speech in 2016, Modi emphasized the need to attract more FDI through "minimum government, maximum governance," which would be achieved by implementing a series of reforms. He also stressed the importance of revitalizing the economy through improving the business environment. The role of FDI for enhancing economic growth was evident in his desire to increase the range of FDI inflows. Such a policy intends to supplement the lack of capital and technology in India by attracting more investment.

The Modi government seeks to foster India as a global manufacturing center through the "Make in India" campaign launched in 2014. The goal is to increase the share of manufacturing for its total GDP from the current 15 to 25% by 2022. By the fiscal year 2018–2019, while India's service sector has maintained more than 50%, the share of its manufacturing sector has remained at 15%. This is lower than its Asian competitors, such as China (30%), Korea (30%), and Indonesia (24%) (KOTRA [31]).

A key means toward achieving the "Make in India" goal is to attract FDI. To this end, the Indian government introduced a series of policies including (1) creating a favorable environment for businesses such as simplification of complex regulations; (2) building new social infrastructures such as industrial clusters and smart cities; and (3) nurturing 25 key industries including IT, aviation, and renewable energy. In this respect, it will inevitably compete with China and Vietnam in Asia, which already enjoys a high level of competitiveness as bases for global manufacturing.

Despite this approach, recent statistics have indicated that the "Make in India" policy has not achieved the desired results in improving the level of manufacturing. As shown in Fig. 13.2 and Table 13.2, since the government has promoted the "Make in India" policy in 2014, FDI flows have gone more to the service sector than to the manufacturing industry. As of the fiscal year 2018–2019, FDI inflows to four major services (services, telecommunications, computer software, and hardware and trade) accounted for more than 50% of the total FDI inflows to India. By contrast, the portion of the three major manufacturing industries (automotive, chemical, and pharmaceutical) was a bit more than 10%. In addition, as shown in Table 13.2, the ratio of FDI inflows to major service sectors has increased significantly from 40 to 51.2% over the past three years, but the ratio of FDI inflows to manufacturing has in fact decreased from 16.1 to 11%.

Although some manufacturing sectors (e.g., mobile phone production) appeared to be doing well, the key stated outcomes were unlikely to happen by the target year of 2022 (The Hindu Business Line [32]). Recently, The Department-Related Parliamentary Standing Committee on Commerce of India also acknowledged that the FDI inflows in manufacturing is declining, and the low inflow of FDI in the manufacturing sectors fails to achieve the original purpose of Make-in-India scheme (Business Standard [33]). It recommended the government to take further efforts to increase the share of manufacturing sectors in the total FDI inflows.

In order to offer a new approach for India to enhance its FDI attractiveness, we will first examine its relative position in terms of the four fundamental determinants of locational attractiveness of FDI inflows. For this, we have selected nine other Asian economies for comparison from which we can investigate the relative strengths and weaknesses of FDI attractiveness.

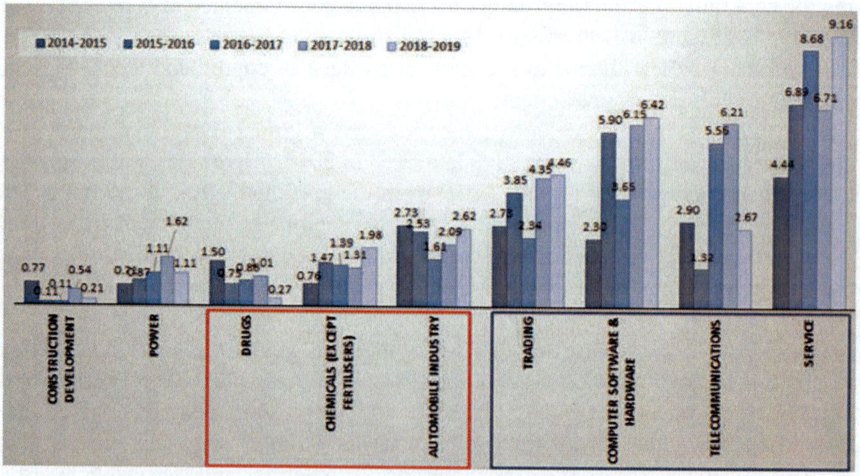

Fig. 13.2 Sectors attracting highest FDI inflows (US$ million). *Source* DIPP FDI Statistics, Development of Industry Policy & Promotion of India, https://dipp.gov.in/publications/fdi-statistics

Table 13.2 Percentage of FDI in core services and manufacturing sectors

	2014/4–2015/3	2015/4–2016/3	2016/4–2017/3	2017/4–2018/3	2018/4–2019/3
4 non-manufacturing sectors	40.0%	44.9%	46.5%	52.2%	51.2%
3 core manufacturing sectors	16.1%	11.9%	8.9%	9.8%	11.0%

Source DIPP FDI Statistics, Development of Industry Policy & Promotion of India, https://dipp.gov.in/publications/fdi-statistics

Table 13.3 Criteria for measurement

Factors	Criteria	Source	Data type
Labor productivity	1.1 Workforce productivity	IMD	Survey
	1.2 Ease of doing business	World Bank	Survey
Best practice adaptability	2.1 Adaptability of government policy	IMD	Survey
	2.2 Firm-level technology absorption	WEF	Survey
Cluster competitiveness	3.1 State of cluster development	WEF	Survey
	3.2 Value chain breadth	WEF	Survey
Goal orientation	4.1 Working hours	IMD	Hard
	4.2 Worker motivation	IPS	Survey

13.3.2 An Empirical Study: Comparative Analysis Between India and Asian Countries

This section highlights the need to quantify the major factors that influence FDI attractiveness as described above by comparing India's competitiveness with nine other economies in Asia. This will be helpful toward understanding India's current position in terms of FDI attractiveness in a more rigorous and systematic manner. The criteria for measuring the four factors were selected from the National Competitiveness Report (e.g., IMD, WEF, and IPS), and statistics published by international organizations (e.g., World Bank) (see Table 13.3). In addition to India, the countries for evaluation include four first-tier Asian newly industrialized economies (NIEs) which are Korea, Taiwan, Hong Kong, and Singapore, and four second-tier NIEs which are Indonesia, Malaysia, the Philippines, and Thailand. China is also included, bringing the total to 10 countries for this comparative analysis.

Among the eight criteria, the reasons for selecting two indicators related to best practice adaptability are as follows. Criterion 2.1 is an indicator that measures the adaptability of the government's policies to changes in the external environment. A higher level of adaptability implies that the government has a strong intention to compete with its rivals. On the other hand, Criterion 2.2 measures the level of firms' acceptance of the latest technology that helps understand its standing relative to its rivals.

Since each individual data contain different units, we had to first standardize them.[4] The indices of the four factors were obtained by calculating the average of the two criteria that belong to them. We then determined the overall FDI attractiveness by calculating the average of the four factors. Based upon this approach, the higher the composite competitiveness index is, the higher the FDI attractiveness. By applying this methodology of measurement and quantification, the results for the investment attractiveness of the 10 economies are summarized in Table 13.4.

[4]The years of data for the eight criteria were 2016 or 2017.

Table 13.4 Results: competitiveness ranking

Country	Overall ranking	Productive labor	Best practice adaptability	Cluster competitiveness	Goal orientation
Singapore	1	2	1	3	2
Hong Kong	2	1	3	1	4
Taiwan	3	3	5	2	1
Malaysia	4	4	2	4	10
China	5	7	6	7	3
Korea	6	5	7	5	8
Thailand	7	6	4	9	5
Indonesia	8	10	8	6	9
India	9	9	10	8	7
Philippines	10	8	9	10	6

The 10 economies in this study are pursuing different strategies toward attracting FDI in the manufacturing sector. Singapore and Hong Kong, which are ranked first and second, respectively, play a role as global or regional hubs. They seek to attract regional or global headquarters of MNCs by engaging in the manufacturing sector. On the other hand, in Taiwan there are a large number of internationally competitive SMEs, and most of them supply high value-added parts and components to global companies. Therefore, Taiwan seeks to attract investment through its connection with the GVC of MNCs. China and the four second-tier NIEs mainly attract FDI for low value-added activities such as assembly.

Korea has a number of internationally competitive global companies. Most of these usually transfer their low value-added production activities to developing countries in order to utilize cheap labor, while concentrating on high value-added activities in Korea. Therefore, the appropriate strategy for Korea would be to attract FDI in high value-added activities, such as R&D centers, rather than low value-added activities. This shows that policies for attracting FDI should be related to the characteristics of GVC in host economies.

On the other hand, all the 10 economies have established domestic clusters and international linkages with neighboring countries, by utilizing the comparative advantage of relevant countries. For example, Singapore, Johor in Malaysia, and Riau in Indonesia have cooperated to develop a transnational growth triangle known as SIJORI and have successfully promoted regional economic cooperation for attracting FDI. In effectively transferring Singapore's existing labor-intensive industries to neighboring countries, it has not only contributed to the advancement of its industrial structure, but Malaysia and Indonesia were also able to achieve economic growth by attracting a large amount of capital and technology know-how. Meanwhile Hong Kong, Korea, and Taiwan were seeking to promote economic development through regional linkages that have been supported by FDI with their bigger neighbor—China.

India is ranked ninth overall and is thus less competitive when compared with other Asian countries. As Fig. 13.3 shows, India is weaker than China across all four

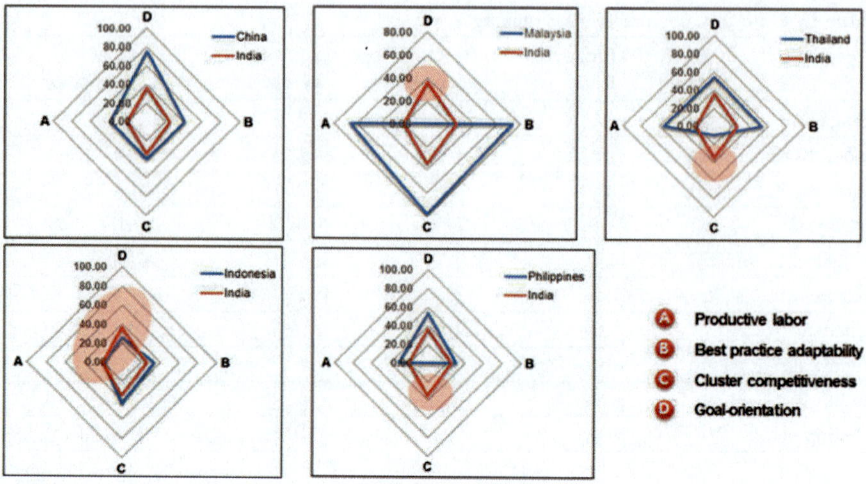

Fig. 13.3 Comparison of the structure of FDI attractiveness

factors. However, compared with Malaysia, Thailand, Indonesia, and the Philippines, it shows competitive advantage in part for these four factors. Specifically, India has a competitive edge for the factor of "goal-orientation" compared to Malaysia and Indonesia, and has a competitive advantage in terms of "cluster competitiveness" compared to Thailand and the Philippines. Therefore, India has a relative superiority in cluster competitiveness and goal orientation compared to the second NIEs, but it is inferior in the other two factors—labor productivity and best practice adaptability.

Here we can see that India's rigid labor market hinders the improvement of its labor productivity. On top of this, the lack of skilled labor due to a high turnover rate, frequent demand for increases in wages, and limited motivations are problems that limit the improvement of labor productivity in India. Much of this is due to the fact there has been no significant change in the labor market system since the economic reforms in 1991. Political difficulties continue to block the amendment of the country's labor laws that could enhance flexibility in the market [40].

Under the GVC context, comprehensive competitiveness consisting of all parties involved in these activities becomes more important than a single firm's competitiveness. India has cheaper labor costs compared with China and some developing countries, but its logistics costs are particularly high. In general, India is responsible for MNCs' assembly in the GVC, or producing low value-added intermediate goods and exporting them to other countries. Therefore, logistics costs arising from linkages with other countries are important because India accounts for only a part of MNCs' entire value chain activities. But the drawback in this case is that India's logistics costs are four to five times higher than international standards (Economist [34]). In addition, the traffic control and management system in India are lagging, which causes a high rate of traffic accidents and consequently increases the cost of doing business in India (Millennium Post [35]).

To address environmental and energy issues as well as the infrastructure, the Indian government announced in 2014 that it planned to create one hundred smart cities across the country by 2022. A solid infrastructure is crucial in attracting FDI, but the more fundamental solution is to reduce unnecessary regulations. According to Moon [36], leading MNCs are more sensitive to excessive regulations than government incentives. This is not only because regulation increases the cost of doing business but it also has a significant negative impact on its current competitive advantage in the host country. Therefore, if local governments sufficiently ensure the basic business activities by reducing regulations, MNCs will be able to make the most of their ownership advantages through investment. Furthermore, they will more likely work with local governments and participate in building infrastructure and improving other economic factors.

13.3.3 Implications for India's FDI Policies

The following presents strategic directions for attracting FDI across the four aspects. The first one is agility. The Modi government has been deregulating various industries over the past three years but there are still many other regulations that hinder investment by MNCs. In particular, labor-related regulations have a negative impact on labor market flexibility and productivity. This highlights the need to improve not only labor productivity but also create a more competitive labor force.

The second one is competitor comparison. Since 1990, India has been steadily pursuing a series of reforms and opening to attract FDI. Notably, the Modi government has implemented more active policies to speed up this process, yet India's FDI attractiveness still lags behind those of its Asian competitors, particularly China. In order to address this issue, systematic developmental strategies are needed by benchmarking specific national and industrial policies in accordance with India's current development stage. In addition, India has a competitive edge in industries such as software, automobiles, aerospace, but it still requires a good foundation for attracting FDI by securing its competitiveness in many other fields.

The third one is upgrade of related industries and living environment. The development of a living environment (software: education, medical, cultural, and entertainment facilities) as well as the industrial infrastructure (hardware: development of physical infrastructure and related industries) are important in developing international-linkage clusters. In order to attract high value-added activities among MNCs, it is important to draw in personnel with world-class skills by being able to provide high-level living and cultural facilities. India has to further consider the international linkage of its industrial clusters with other clusters around the world, thereby facilitating interlinkages of value chain activities spreading around the world.

The fourth one is clear and viable goal setting. As many of India's policies still have high political tendencies, it is important to establish consistent policies with a focus on economic development. As India is a federal state, the power has been decentralized across local governments which usually have various regulations and divergent policies for attracting FDI. Therefore, it is important to have close coordination and cooperation toward achieving intended economic goals, implementing relevant policies, and establishing efficient institutions. The federal government should provide common economic goals and establish institutions that are able to adjust the conflicts and enhance regional cooperation. At the same time, the policies should be formulated and implemented in a way to lower regional transaction costs and increase the overall efficiency through the establishment of regional-linkage clusters.

13.4 Conclusion

With the US protectionist measures, attracting FDI from MNCs is becoming more difficult for other countries. And with the outbreak of the coronavirus and its impact on the global economy, the international investment environment has been very difficult. This has heightened the competition among countries around the world to attract FDI. In order to respond effectively to this challenging environment, this paper introduced four fundamental factors that influence the creation of an attractive environment for FDI. They are productive labor with both speed and precision, best practice adaptability, cluster development, and goal-orientation with diligence and strong motivation. In contrast with general factors such as cheap labor or educated labor force that are commonly believed to influence the FDI inflows, these four factors of this study assume that without superior inherited or created advantages in the resources themselves, countries that are able to mobilize their available resources in an efficient manner will be able to enhance their position compared to their rivals.

For a clearer understanding on the importance of these factors, we take India as an illustrative example. Despite its great potential in attracting FDI, India's current status of investment attractiveness is relatively weak when compared with China in particular. In order to enhance its overall attractiveness toward foreign investors, this study conducted an empirical analysis by comparing India's competitive position against the nine Asian economies. Despite India's relative advantage in some factors such as cluster development and goal orientation, its overall competitiveness in attracting FDI is still not high.

For India to attract FDI effectively in the manufacturing sector, it should improve the competitiveness of the four fundamental determinants suggested in this study. In addition to promoting the "Make in India" policy, India should be linked to the GVC activities of MNCs to improve the productivity and competitiveness of its firms. Furthermore, India should maximize values created in India by broadening the tool of globalization, from trade to FDI and then to more comprehensive value creation mode via GVC. In the end, the scope of competition and cooperation of clusters in India should be extended to globally linked ones.

Table 13.5 Top 10 investors for India's FDI inflows, April 2018–March 2019 (million US$, %)

Country	FDI inflows	Country	FDI inflows
Singapore	16,228 (36.6)	UK	1,351 (3.0)
Mauritius	8,084 (18.2)	UAE	898 (2.0)
Netherlands	3,870 (8.7)	Germany	886 (2.0)
US	3,139 (7.1)	France	France (0.9)
Japan	2,965 (6.7)	Cyprus	296 (0.7)

Source FDI Statistics, Development of Industry Policy & Promotion of India, https://dipp.gov.in/publications/fdi-statistics

Table 13.6 India's FDI inflows by industry, April 2018–March 2019 (million US$, %)

Industry	FDI inflows	Industry	FDI inflows
Service	9,158 (20.6)	Construction	2,258 (5.1)
Computer software & hardware	6,415 (14.5)	Chemicals (other than fertilizers)	1,981 (4.5)
Trade	4,462 (10.1)	Power	1,106 (2.5)
Telecommunications	2,668 (6.0)	Drugs and pharmaceuticals	266 (0.6)
Automobile industry	2,623 (5.9)	Construction development	213 (0.5)

Source FDI Statistics, Development of Industry Policy & Promotion of India, https://dipp.gov.in/publications/fdi-statistics

Appendix

See Tables 13.5 and 13.6.

References

1. *Economist, Ford Motors Courts Donald Trump by Scrapping a Planned Plant in Mexico*, January 5, 2017.
2. Moon, H. C., & Yin, W. (2017). *Korea's Investment Promotion Strategy in Response to the New US Government [in Korean]*, KOTRA report.
3. *Wall Street Journal, Trump Says Apple CEO Has Promised to Build Three Manufacturing Plants in U.S.*, July 25, 2017.
4. UNCTAD. (2020a). *Investment Trends Monitor*, January 20, 2020.
5. A.T. Kearny. (2019). *The 2019 Kearney Foreign Direct Investment Confidence Index*, Downloaded from https://www.kearney.com/foreign-direct-investment-confidence-index/2019-full-report#exec.
6. UNCTAD. (2020b). *Investment Trends Monitor*, March 8, 2020.
7. Du, J., Lu, Y., & Tao, Z. (2008). FDI location choice: agglomeration vs Institutions. *International Journal of Finance and Economics, 13*, 92–107.
8. Singh, D. S. (2019). Foreign Direct Investment (FDI) inflows in India: A review. *Journal of General Management Research, 6*(1), 41–53.
9. Dunning, J. H. (2001). The OLI paradigm of international production: past, present and future. *International Journal of the Economics of Business, 8*(2), 173–190.

10. *South China Morning Post, Could Robotic Automation Replace China's 100 Million Workers in its Manufacturing Industry?* February 14, 2019.
11. Campos, N. & Kinoshita, Y. (2002). *The Location Determinants of Foreign Direct Investment in Transition Economies, Working Paper Group 3–9 Kinoshita,* William Davidson Institute, Michigan.
12. Redding, S., & Venables, A. J. (2004). Economic geography and international inequality. *Journal of International Economics, 62,* 53–82.
13. Ma, A. C. (2006). Geographical location of foreign direct investment and wage inequality in China. *World Economy, 29,* 1031–1055.
14. Yin, W. (2018). An integration of different approaches for global value chains [in Korean]. *Review of International Area Studies, 27*(2), 37–54.
15. Yin, W. (2017). *Global value chain: theoretical integration, extension, and empirical analysis,* Unpublished Ph.D. dissertation, Seoul National University.
16. Moon, H. C., & Parc, J. (2014). Economic effects of foreign direct investment: A case study of samsung electronics' mobile phone [in Korean]. *Korea Business Review, 18*(3), 125–145.
17. *Wall Street Journal, Modi's First Two Years: Economic Report Card*, May 25, 2016.
18. Moon, H. C. (2016a). *Foreign Direct Investment: A Global Perspective.* Singapore: World Scientific.
19. Qian, G., Li, L., Li, J., & Qian, Z. (2008). Regional diversification and firm performance. *Journal of International Business Studies, 39*(2), 197–214.
20. Alcácer, J., & Chung, W. (2007). Location strategies and knowledge spillovers. *Management Science, 53*(5), 760–776.
21. Alcácer, J., & Chung, W. (2014). Location strategies for agglomeration economies. *Strategic Management Journal, 35*(12), 1749–1761.
22. Mudambi, R., Li, L., Ma, X., Makino, S., Qian, G., & Boschma, R. (2018). Zoom in, zoom out: Geographic scale and multinational activity. *Journal of International Business Studies, 49,* 929–941.
23. Moon, H. C. (2017). The strategy for Korea's economic success: Innovative growth and lessons from silicon valley [in Korean]. *Review of International Area Studies, 26*(3), 1–33.
24. Porter, M. E. (1990). *The Competitive Advantage of Nations.* New York: Free Press.
25. Li, P., & Bathelt, H. (2018). Location strategy in cluster networks. *Journal of International Business Studies, 49,* 967–989.
26. Moon, H. C., & Jung, J. S. (2010). Northeast Asian cluster through business and cultural cooperation. *Journal of Korea Trade, 14*(2), 29–53.
27. ASEAN Secretariat and UNCTAD. (2017). *ASEAN Investment Report 2017: Foreign Direct Investment and Economic Zones in ASEAN*, Jakarta, Indonesia: Association of Southeast Asian Nations (ASEAN).
28. HKTDC. (2017). *Vietnam Utilizes Preferential Zones as a Means of Offsetting Investment Costs,* Downloaded from https://hkmb.hktdc.com/en/1X0A9ID5/hktdc-research/Vietnam-Utilises-Preferential-Zones-as-a-Means-of-Offsetting-Investment-Costs, March 27, 2017.
29. Moon, H. C. (2013). The reasons for apple decisions of maintaining the Chinese factories in spite of Obama's request… [in Korean]. *Dong-A Business Review, 127,* 90–93.
30. New York Times. (2012). *How the US Lost Out on iPhone Work,* January 21, 2012.
31. KOTRA. (2015). *India to Act as a 'Factory in the World' through 'Make in India' Policy to Substitute China [in Korean]* March 3, 2015.
32. *The Hindu Business Line, Making 'Make in India' work,* March 19, 2020.
33. *Business Standard, Parliament Panel Expresses Concerns over Dip in Manufacturing Sector FDI,* March 11, 2020.
34. *Economist, Narendra Modi is a Fine Administrator, But Not Much of a Reformer*, June 24, 2017.
35. Millennium Post, Authorities Hint Negligence Caused UP Train Disaster, Downloaded from https://www.millenniumpost.in/big-stories/authorities-hint-negligence-caused-up-train-disaste-258250. August 20, 2017.

36. Moon, H. C. (2016b). *The Strategy for Korea's Economic Success.* New York: Oxford University Press.
37. *FDI Statistics, Development of Industry Policy & Promotion of India,* https://dipp.gov.in/publications/fdi-statistics.
38. Institute for Industrial Policy Studies (IPS). (2015). *IPS National Competitiveness Research 2014–2015,* Seoul: IPS.
39. International Institute for Management Development (IMD). (2017). *IMD World Competitiveness Yearbook 2017,* Geneva: IMD.
40. Lee, W, Song, Y. C., Cho, C., & Choi, Y. (2013). Changes in the Labor Market since India's Economic Reforms [in Korean], *Korea Institute for International Economic Policy* (KIEP) report.
41. World Bank, Ease of Doing Business. (2020). https://www.doingbusiness.org/rankings.
42. World Economic Forum (WEF). (2016). *The Global Competitiveness Report 2016–2017,* Geneva: WEF.
43. Moon, H. C., & Yin, W. (2019). *Strategic Direction for Promoting FDI in India,* Korea consulting report.

(Prof) Hwy-Chang Moon, Ph.D. is a Chair Professor of the Seoul Business School at aSSIST University and Professor Emeritus in the Graduate School of International Studies at Seoul National University, where he also served as the Dean. He has delivered special lectures at various institutions, including Helsinki School of Economics, Keio University, and Stanford University. He is currently the editor-in-chief of the Journal of International Business and Economy, and published numerous journal articles and books on topics covering International Business Strategy, Cross-Cultural Management, Foreign Direct Investment, and Economic Development in East Asia. He has also consulted several multinational companies, international organizations, and governments.

Wenyan Yin, Ph.D. is a Adjunct Professor of the Seoul Business School at aSSIST University and a lecturer in the Graduate School of International Studies at Seoul National University. She has published a number of articles including those in the journals indexed by SSCI, A&HCI, and SCOPUS. Her interest areas of research include national competitiveness, global value chain, foreign direct investment, international business strategy, and cross-cultural management. She has worked at the Institute for Industrial Policy Studies as a researcher for three years. She has also conducted many research projects related to the international competitiveness and foreign direct investment for firms and governments.

Chapter 14
Structured Approach to Build-in Design Robustness to Improve Product Reliability

Vic Nanda and Eric Maass

Abstract Robustness is defined as the degree to which a system or component can function correctly in the presence of invalid inputs or stressful environmental conditions. The objective of robustness is to deliver high reliability to customers. Robustness ensures that product design is immune to and can gracefully handle invalid inputs and stressful environmental conditions without any disruption or degradation of service to the end user. Robustness can be systematically built into any system, regardless of hardware or software, by following an end–to–end approach that encompasses product requirements, design, development, and testing. This chapter provides a structured approach to design in robustness by mapping baseline use case scenarios as 'sunny day' scenarios, identifying potential failures using P–diagrams and Design Failure Modes & Effects Analysis (or, "rainy day" scenarios), and proactively embedding design controls to strengthen product robustness and minimize field failures. The authors describe an innovative way to prioritize design improvements not just by traditional Risk Priority Number (RPN) of design failures but by considering the actual magnitude of risk reduction, as well as by factoring in cost of design improvements in prioritization decisions. Robustness once built–in product design must be validated through vigorous robustness testing to provide objective evidence of design robustness and support decision-making regarding product readiness for release. A comprehensive approach to robustness testing is described along with guidance on how to design a comprehensive suite of robust test cases for high reliability.

Keywords Robustness · Robust design · P-diagram · DFMEA · Robust testing

V. Nanda (✉)
NOKIA, Espoo, Finland
e-mail: vic_nanda@hotmail.com

E. Maass
Medtronic Restorative Therapy Group, Tempe, USA

K. B. Misra (ed.), *Handbook of Advanced Performability Engineering*,
https://doi.org/10.1007/978-3-030-55732-4_14

Fig. 14.1 Mars opportunity rover

14.1 Introduction to Robustness

Robustness is defined as the degree to which a system or component can function correctly in the presence of invalid inputs or stressful environmental conditions (IEEE [1]). The objective of robustness is to deliver products with high reliability to customers and end users.

One of the best examples of robustness is the Mars Opportunity rover that was launched on July 3, 2003 and landed on Mars on January 25, 2004, with a planned 90−sol duration of activity (slightly more than 90 earth days), yet it remained operational until June 10, 2018 and exceeded its operating plan by 14 years, 46 days in earth time, or 5111 sols, 55 times its designed lifespan! It withstood hard environmental conditions and stress including Martian dust storms far beyond its intended lifespan and performed remarkably well beaming stunning visuals of the Martian landscape and provided wealth of scientific data (Fig. 14.1).

Here on earth, there are several examples of robust products such as consumer products that work faithfully under all reasonable operating and environmental conditions such as mobile phones, personal computers, automobiles, and so on.

14.2 Why Robustness Matters?

When products are not robust, it can cause immense harm and inconvenience to end users and damage to company reputation. For example, in 2017, Amazon Web Services (AWS) experienced an 11-h outage due to a simple human error that crippled popular websites like Netflix and top 100 online retail websites. AWS was forced to issue a public apology and a detailed post-mortem that stated an authorized employee executed a command that was supposed to remove a small number of servers for one of the AWS sub-systems for maintenance but one of the parameters for the command was entered incorrectly and took down a large number of servers that supported critical services—the engineer intended to decommission 0.03% of the servers for maintenance but inadvertently entered 30%, that knocked out large part of the ASW network as shown in Fig. 14.2. Clearly, the system was not robust in that it had no designed-in control mechanism to preclude such a human error, and further it pointed to a deficiency in the AWS server test processes in that there was no test case that exercised this scenario. Consequently, AWS made changes to its internal tools and processes, so that servers were taken down more slowly and blocked operations that reduced server capacity below safety check levels.

Another robustness failure had even more catastrophic consequences—In 2015, an Airbus 400 M military transport aircraft as shown in Fig. 14.3 crashed due to a faulty software configuration. The crash investigation confirmed that there was no structural defect in the airplane but a pure software defect in the configuration settings programmed in the electronic control unit (ECU) in three of the aircraft's four ECUs—a file-storing torque calibration parameters for each engine were somehow 'accidentally wiped' when the software was being installed. As a result, three of the aircraft's engines automatically shut down in flight. Worse, as per the design of the software, the pilot of the A400M would not have gotten an alert about the missing data until the aircraft was already at an altitude of 400 feet. No cockpit alert about the data fault would appear while the aircraft was on the ground. The A400M, which was on a final test flight before delivery to the Turkish Air Force, reached an altitude of 1,725 feet after takeoff before three of the engines stalled and it crashed during an attempted emergency landing. There were no survivors.

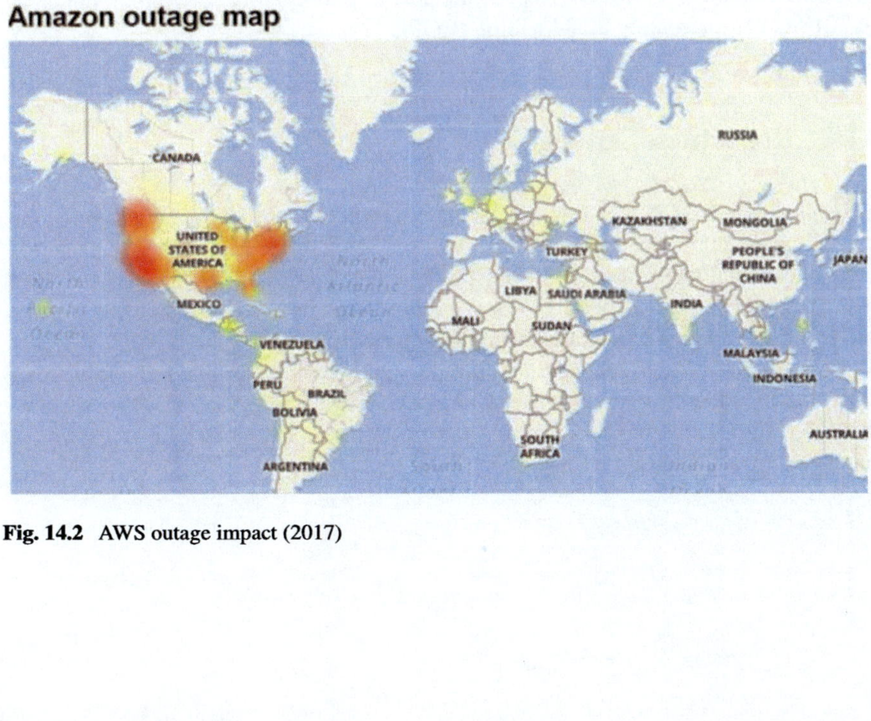

Fig. 14.2 AWS outage impact (2017)

Fig. 14.3 Airbus A400M military transport aircraft that crashed in 2015 due to software design defect

14.3 Benefits of Robustness

Robustness in products offers compelling benefits to companies producing those products as well as the customers and end users. It reduces Cost of Poor Quality (COPQ) from internal and external failures that directly contributes to reduced operating expenses and *improved bottom-line* for the producer. Reduction in customer-reported defects results in improved customer satisfaction and loyalty, improved customer retention, and sales growth from existing and new customers. Improved quality therefore directly contributes to *improved top-line*. These twin benefits of reduced operating expenses by virtue of reduced COPQ and improved top-line by virtue of sales growth from current and new customers *improve business profitability*.

As an example, in the software industry, according to The Cost of Poor Quality Software in the USA: A 2018 Report, the cost of poor quality software in the USA in 2018 is approximately $2.84 trillion (COPQ [2])!

14.4 Robustness Strategy

Robustness in products cannot be an afterthought and designed outside in—one cannot assure that a product is robust by merely testing for robustness. Robustness must be thought of proactively, it must be planned for all phases in a project. This is the fundamental concept of *shift left quality*—that is to prevent and find defects early in the product design and development process.

Robustness can be systematically built into any system, regardless of hardware or software, by following an end-to-end approach that encompasses product requirements, design, development, testing, and customer deliverables. In other words, robustness must be designed inside out.

14.5 End-to-end Robustness Lifecycle

The end-to-end lifecycle to plan, design-in and validate robustness in products covers four phases:

1. Robustness specifications
2. Robust architecture and design
3. Robust development
4. Robust testing.

We provide an overview of each of these phases before covering each phase in detail.

14.5.1 Robustness Specifications

For any product, the design and development lifecycle begins with requirements specifications, and planning for robustness starts with:

- Documenting robustness requirements that specify anticipated product behaviour in the event of:
  - Incorrect, incomplete, delayed, or missing inputs
  - Execution errors
  - *Edge cases*—errors associated with input(s) at maximum or minimum value
  - *Boundary cases* when one input is at or just beyond maximum or minimum limits
  - *Corner cases* that are outside normal operating parameters and indicate a stress scenario when multiple environmental variables or conditions are simultaneously at extreme levels even though each parameter is within the specified range for that parameter
- Specifying fault tolerance requirements, including failover requirement for a faulty system to 'failover' or gracefully handoff to a backup system in the event of a failure
- Documenting traceability of robustness requirements to design specifications and test cases to assure that the robustness requirements are designed in and validated during testing, and
- Requirement reviews including verification of robustness requirements and traceability.

14.5.2 Robust Architecture and Design

Robust architecture and design is about having a design that can gracefully handle invalid inputs, internal errors and failures without unrecoverable catastrophic product failure. The objectives of robust architecture and design are to:

- Conform to robustness specifications, so that subsequent product design and development can successfully be validated against requirements and designs for robustness, respectively
- Identify potential vulnerabilities or high-risk 'failure points'
  - 'Hot spots', or design elements with intense use and therefore higher risk of failure,
  - 'Weak spots', or design elements that are known to be fragile from historical defect data
  - 'Critical interfaces' between sub-systems and modules
- Gracefully handle invalid inputs, processing errors and failures
- Verify that the robustness requirements have been adequately met (verification is performed in design reviews).

14.5.3 Robust Development

Robust development entails developing the product in accordance with the robustness requirements and detailed designs, to:

- Ensure the design for robustness is fully implemented, including following development guidelines and best practices to minimize failures
- Include development reviews to verify that the documented robust design has been implemented.

14.5.4 Robust Testing

The purpose of robust testing is to inject invalid inputs, introduce error and failure scenarios and stressful environmental, and use conditions to verify whether the system behaves as accepted by the customer. Robust testing is during the entire product development lifecycle from individual modules, sub-systems, to the integrated system to ensure robustness at all levels. Robustness testing validates that all the robustness requirements have been implemented.

To summarize, from a robustness perspective, the primary goal during requirements specification, architecture, design and development is *fault prevention and fault tolerance*, while the goal during robust testing is *fault identification and removal.*

We will now look at each of these phases in detail, starting with robustness specifications.

14.5.5 *Creating Robustness Specifications*

The first step to defining robustness specifications is to map the baseline use case scenarios—understand the ideal use cases of the overall system with the help of a block diagram that depicts the end-user, inputs and interactions, to show what is the default scenario(s) assuming no exceptions and errors. Such a use cases is commonly referred to as the 'happy path' or 'sunny day scenario'. At this time, we assume no errors or failures. For example, for a customer using an A™ to withdraw money, we first assume that the customer correctly enters the PIN code and correctly makes the right user selections to withdraw the money.

Next, identify potential failures in the sunny day scenario—the 'what if' scenarios—what if the input is incorrect? What if the input is incomplete? While both of these are failure scenarios, but they are distinct and constitute individual *failure modes*—one when the input is incorrect or the other when it is incomplete. Likewise, there may be additional failure modes pertaining to the inputs. In the A™ example, what if the customer entered the PIN incorrectly? What should the A™ do? What does the customer reasonably expect it to do? Should it reject the transaction and return the card? Should it display an error message and allow the customer to make a second attempt?

Failure modes are not restricted to user or data inputs (from other systems) alone. What if there is an internal failure or execution error at a module or sub-system level? What should happen in these failure scenarios? 'What should happen' equates to the requirement specification for how the failure is expected to be handled (as expected by the customer and end user).

Figure 14.4 shows an example of a basic block diagram with some but not all potential points of failure. Such use cases that depict failure modes are also called 'rainy day scenario' or 'unhappy path'. Such block diagrams can be built at system, sub-system, sub-assembly and module or component level to fully map out potential failures at all levels of design.

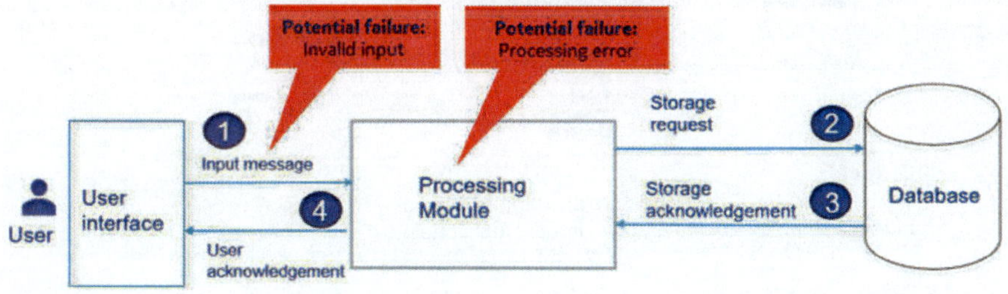

Fig. 14.4 Mapping of baseline use case with potential failure scenarios

14.6 Parameter diagram (P-diagram)

In his book 'Quality Planning & Analysis', Juran describes the purpose of P-diagrams as follows: "the most basic product feature is performance, i.e. the output—the colour density of a television set, the turning radius of an automobile. To create such output, engineers use principles to combine inputs of materials, parts, components, assemblies, liquids, etc. For each of these inputs, the engineer identifies parameters and specifies numerical values to achieve the required output of the final product" (Juran [3]).

The Parameter Diagram (P-Diagram) takes the inputs from a system/customer and relates those inputs to desired outputs of a design that the engineer is creating, also considering non-controllable outside influences (Fig. 14.5).

During requirement flow-down, potential problems can be anticipated at the sub-system, sub-assembly and the component levels. The system-level flow-down will involve a birds-eye view of failure modes and will involve a broader cross-section of expertise for this purpose—but the anticipation of failure modes and mechanisms at sub-system and component levels will involve a more focused set of experts to dissect the potential problems involved at that deeper, more detailed level. Essentially, at these subsequent iterations, the sub-system and component under consideration become 'the system' for the team. It is worth noting that many of the sub-systems for complex products could literally *be* the 'system' or product for the same or for other companies. For example, many cellular phones include digital cameras—and digital cameras are products for camera manufacturers.

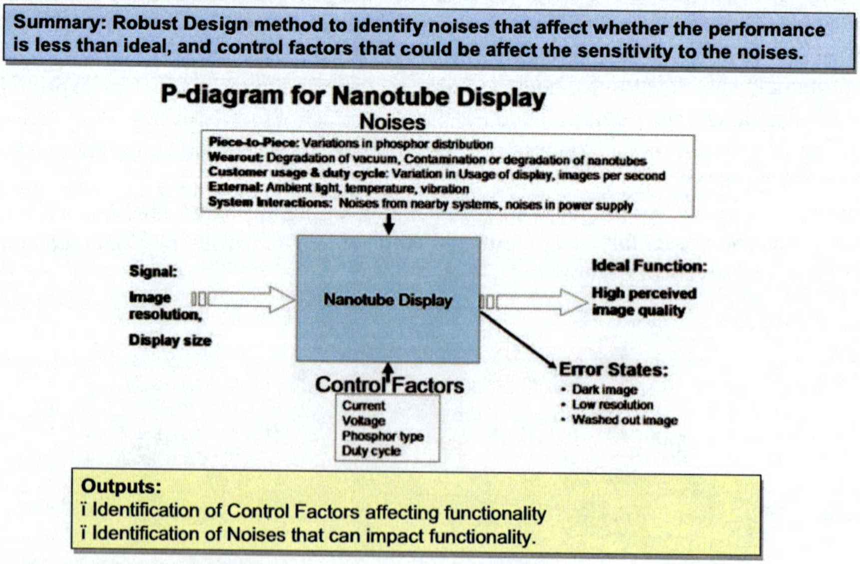

Fig. 14.5 Example of P-diagram

Either as an integrated sub-system, or as a separate product, anticipation of potential problems is a vital first step towards prevention of problems. The P-Diagram is valuable in anticipating and preventing problems if not adequately addressed could impact the success of the product. It does this by enabling engineering teams to document input signals, noise factors, control factors, error states, and ideal response:

- Input signals are a description of the inputs received by the system, which are processed to generate the intended output.
- Control factors are *parameters or design improvements that can potentially prevent, mitigate, and/or detect the failure modes.*
- Error states are any undesirable and unintended system outputs. These are referred to as *Failure Modes—'Real-life' failures that can happen when the system is deployed.*
- Noise factors are *environmental factors or internal factors that can potentially impact normal system operations and result in failure modes.* The design must be robust against the expected noise factors.
- Ideal response is the intended output of the system.

Steps to complete the P-diagram are:

1. Identify signal
2. Identify intended function or result
3. Identify noise factors
4. Identify 'real-life' failure modes from the noise factors
5. Identify control factors.

A P-Diagram can help with the development of the Design Failure Mode and Effects Analysis (DFMEA), in which the error states or deviations from the ideal function (at the lower right of P-diagrams) could suggest failure modes to be included in the DFMEA, and the noises (at the top of the P-Diagrams) could suggest potential causes for the failure modes.

The team approach for identifying control and noise factors used in developing the P-Diagram can be leveraged in the flowing down requirements to the next level. The P-Diagram can also prove useful in generation and subsequent evaluation of alternative concepts for the sub-system, module or component, particularly in terms of considering the noises that can affect performance when brainstorming potentially robust design approaches—and the relative insensitivity of the alternative concepts to those noises can and should be considered in selecting a superior concept for the sub-system, module, or component.

The P-Diagram can also prove valuable during transfer function determination and initializing the identification of control and noise factors to use in an experimental design approach. The P-Diagram will also prove valuable for evaluating and optimizing robustness against the noises and verification of reliability, and some of the noises from the P-Diagram can be used as stress factors for reliability evaluation (Maass [4]).

14.7 Identifying Detailed Failure Modes with D-FMEA

The primary objective of DFMEAs is to help analyze, improve, and control the risk of product or service or feature or functional area design failure in meeting customer needs or expectations. It is supposed to be a living document that is initially created during product design and then maintained after product release as fixes are made to the product and enhancements are made in future releases.

DFMEA is essentially a risk identification and mitigation tool and it progresses through the following phases: risk identification, risk characterization, risk aversion, and improvement actions prioritization.

Risk Identification: The failure modes from the P-diagrams are populated in the DFMEA table and these are further expanded to identify even more failure modes (from team brainstorming, past defects) which may not otherwise be possible to depict in the P-diagram (in order to minimize complexity of the P-diagram).

Risk Characterization: After listing each failure mode, the team lists the effect of each failure and scores:

- The severity of the risk on a 10-point scale with 10 indicating most severe,
- Likelihood of occurrence on a 10-point scale with 10 indicating most likely, and
- Detection mechanisms to detect or prevent the failure mode with 10 indicating no ability to detect or prevent and therefore risk of defect escape to the customer, and 1 indicating strong detection mechanism to prevent the defect escape.

The RPN (Risk Priority Number) score is then computed and it is the product of severity of impact, likelihood of occurrence and detection score, and it provides a risk score for each failure mode and helps assess the relative risk of each failure. Because each of the three relative rankings is on a scale that ranges from 1 to 10, the overall RPN will always be between 1 and 1000, with 1000 indicating the failure mode with the highest risk.

Risk Aversion: In order to avert design risks, one must completely understand all the root causes (Fig. 14.6, column 5) and assess each failure mode and its root causes individually with by assigning them separate RPN scores. Risk aversion strategies include:

1. Risk mitigation (acting to reduce the risk probability and impact),
2. Risk avoidance (choosing a course of action that eliminates the risk),
3. Risk transfer (transferring the risk to another more appropriate product team to own the risk), and
4. Risk acceptance (plan a contingency to contain the risk in case the risk is realized).

As a general rule, improvement action plans from DFMEA aim to minimize the RPN score, typically below a threshold of 100 by reducing likelihood of occurrence and improving detection and prevention (control), while reduction in severity of impact is possible only be altering the design. For example, if the brakes in a car fail, improvements will focus on reducing the likelihood of failure and improving detection of brake failure, but the severity of impact on the passengers (in this case,

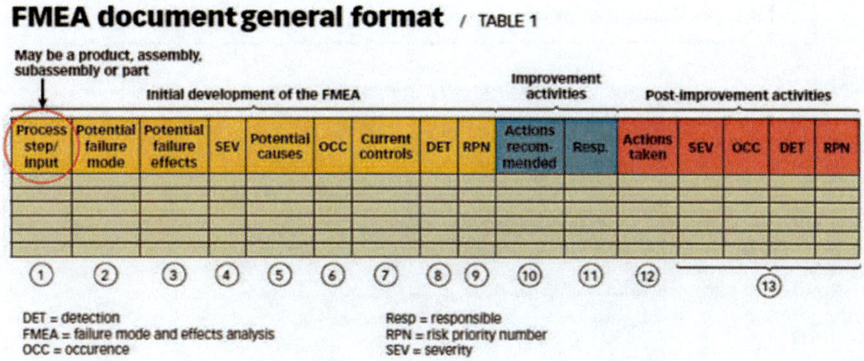

Fig. 14.6 FMEA template

risk of injury) in the event that they do fail would be the same unless design changes are made to lower the risk of injury, for example, adding airbags and crumble zones in the car to minimize injury to passengers (and thus lower the score for severity of impact).

After the improvement actions are identified, a new RPN is computed and compared with the original RPN to assess the % reduction in risk and to assess if the new RPN is below the acceptable threshold.

14.8 Improvement Actions Prioritization

The original RPNs are sorted in descending order in order to define prioritized improvement actions starting with the highest risk failure modes down to the threshold below which the risks are considered low risk (typically, failure modes with RPN score of 100 or less). This prioritized list of improvement actions is then translated into execution plans for 30–60–90 days (or other timeframes).

This approach does have some major limitations. It does not factor in the extent of business impact, or extent of reduction in risk exposure as a result of risk aversion actions, and it does not factor in cost, difficulty or effort to reduce the risk. Figure 14.7 shows risk prioritization approach that factors in reduction in risk exposure and cost to prioritize the risks. All failure modes that are above the threshold (old RPN) are reviewed for reduction in RPN after proposed improvement actions and the magnitude of reduction in RPN is plotted on the Y-axis. The scale may be numbered on the Y-axis from the least to the most RPN reduction seen in the project, for example, H on the Y-axis would be max RPN reduction and L would be least reduction in RPN. Likewise, the X-axis can be numbered according to the maximum cost of improvement actions to the least cost or may be used simply with low–medium–high cost variable. Alternatively, cost may be replaced with ‘difficulty’ in implementing the improvement actions.

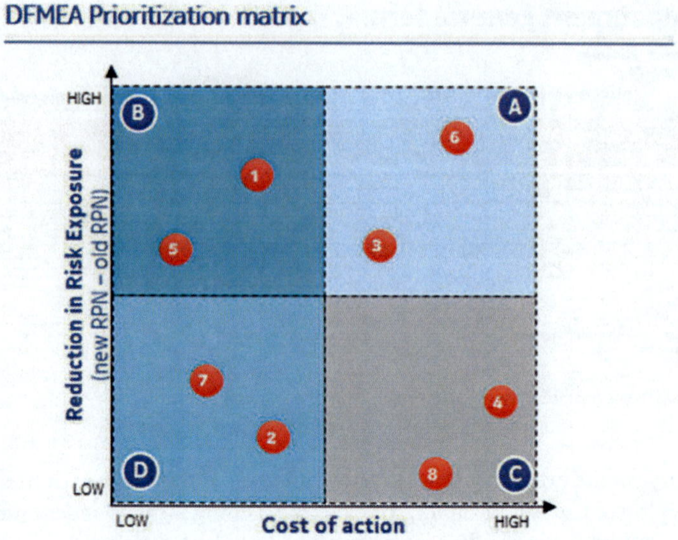

Fig. 14.7 Failure mode prioritization based on reduction in risk and cost of action

The failure modes that must be addressed first lie in quadrant B (maximum reduction in RPN with least to moderate cost), followed by quadrant A or D (depending on if the team prioritizes reduction in risk or cost of action). Finally, failure modes in quadrant C are those with the least reduction in RPN and require significant cost to reduce risk below the acceptable threshold. This sequence can also be used to accordingly define the improvement timeframe starting with immediate action on the failure modes that offer the greatest reduction in risk exposure.

14.8.1 Embedding Design Controls for Robustness

This step involves updating the original 'rainy day' block diagram to embed the design control actions from the DFMEA that mitigate the risks. It shows where the design vulnerabilities are and what control actions have been identified to avert those design risks. It therefore serves as powerful visual representation to view the landscape of design risks and actions identified to improve design robustness as shown in Fig. 14.8. Again, as previously mentioned, this analysis can be performed at sub-system and module level and all such block diagrams of lower levels of architecture and design would need to be updated to depict the control actions identified to avert risks in the entire product architecture and design.

While non-engineers tend to rely upon heuristic thinking for decisions, engineers traditionally use deterministic modelling in their tasks.

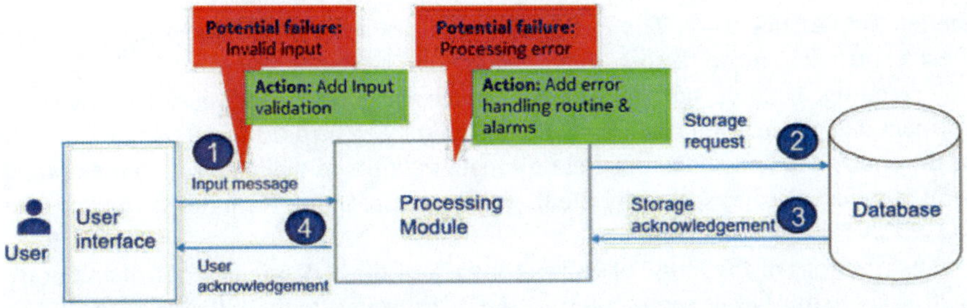

Fig. 14.8 Rainy day scenario with embedded design controls to reduce risk of failure modes

Design robustness drives beyond deterministic to probabilistic or stochastic modelling (Maass [4]). The melding of engineering modelling with probabilistic methods is referred to as Predictive Engineering. Probabilistic methods include Monte Carlo Simulation, the generation of system moments method, and Bayesian Networks to predict the probability that the product will meet expectations over the range of noise factors.

On the right side of Fig. 14.9, parallel line segments represent the set of requirements for the product. Critical requirements, two line segments with arrows represent a subset of the requirements that are prioritized for the intensity of robust design. The goal is to predict the distribution for each critical requirement over the range of noise factors—represented as the distributions to the far right of Fig. 14.9.

To predict the distributions for the critical requirements, the critical requirements are flowed down to control and noise factors as represented by the line segments on

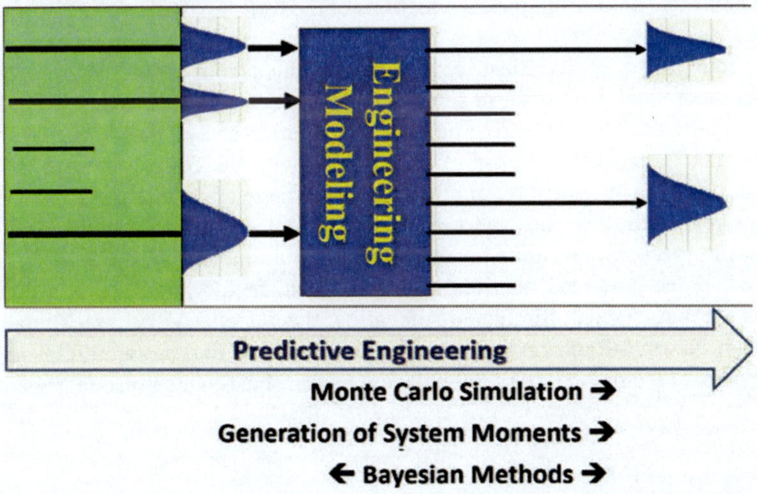

Fig. 14.9 Robust design/predictive engineering as the melding of deterministic engineering modelling with probabilistic methods

the left side of Fig. 14.9. The P-diagram provides a useful method for identifying these control and noise factors.

Screening methods such as fractional factorial experimental designs, Plackett–Burman designs, or Definitive Screening Designs can be used to determine a subset of the control and noise factors that dominate in terms of their impact on the critical requirements, as represented by the three arrows emerging from the shaded area to the left of Fig. 14.9.

Distributions of the range of values for each of these dominant control and noise factors are estimated as represented by the distributions to the left side of Fig. 14.9.

Engineering modelling is used to develop equations or transfer functions for how the subset of dominant control and noise factors affects each critical requirement. The engineering modelling can use theoretical methods, also called first principles modelling; examples include Ohm's Law, Voltage = Current x Resistance, or simple additive models like the total thickness of stacked layers or total delay from the sum of delays for each step.

Engineering modelling can also use designed experiments that vary the control and noise factors either through simulation or emulation. Mechanical simulation can use Finite Element Analysis; electronic simulation can use electronic circuit simulation tools. Emulation can use hardware that emulates the actual hardware and software system that may not yet exist in its final form.

Combinations of control and noise factors at varied settings can be defined according to an experimental plan such as Central Composite Design for Response Surface Modelling or Space-Filling Designs. These combinations of settings for control and noise factors are run through the simulation or emulation, and results for each critical requirement are obtained for each combination.

The combinations and the results can be analyzed using multiple regression to obtain empirical or semi-empirical equations of the form $y = f(x1, x2, \ldots, xp, noise1, noise2, \ldots, noiseq)$.

The equation for each critical requirement obtained from engineering modelling, whether theoretical, empirical, or semi-empirical is deterministic at this point. Probabilistic methods (Monte Carlo Simulation, Generation of System Moments method or Bayesian Networks) are used in conjunction with the equations to predict the distributions of each critical requirement over the range of control and noise factors.

If the predicted distributions for each critical requirement are satisfactory—that is, with the specification limits for that critical requirement, robust design has been achieved. If the predicted distributions are not satisfactory, optimization methods such as Steepest Ascent, Simulated Annealing, the Genetic Algorithm, Branch and Bound, or Newton–Raphson can be used to explore and find more optimal settings for the control factors that render the design more robust to the noise factors.

14.9 Robustness Testing

In robust testing, the primary objective is to validate that the system or component can function correctly in the presence of invalid inputs or stressful environmental conditions, including misuse and abuse test cases. Planning for robustness starts with robustness test strategy and includes robustness test planning and execution.

14.9.1 Robustness Test Strategy

Robust testing always starts from the inside, from the smallest component, module, sub-assembly, and it progresses outwards to greater aggregation of the product until the final finished product. Therefore, robust testing is the responsibility of the development and test organizations and not the test organization alone.

Robust testing begins with developing an overall robustness test strategy and plan, including identifying all phases of product development and test where robustness testing will be performed, identifying test resources, types of robustness testing, test cases, test procedures, test tools, problem management for defects reported from robust testing, defects database, and defect profiles from types of robust testing to inform future robust test strategy.

14.9.2 Robust Test Planning and Execution

How should product teams design a comprehensive suite of robust test cases for high reliability? There are four potential sources of robust test cases:

1. **Historical Defect Data**: One can review past customer-reported and internally founds defects to gain insights into which product sub-systems, components, and sub-assemblies have been most prone to robustness defects. This requires reviewing all customer-reported defect data and categorizing all defects, with techniques such as affinity analysis (clustering) of keywords in defect reports to categorize defects that indicate poor design robustness. These can then be used to design new robust test cases.
2. **Requirement Specifications**: This involves reviewing the requirement specifications to understand how the product is expected to gracefully handle invalid inputs and environmental stresses, when such requirements are specified, and identifying robust test cases based on requirements by considering 'what if' scenarios discussed earlier. In addition, one can identify what features, components, sub-assemblies, and interfaces are new, unique, and difficult (complex), collectively referred to as NUD design elements that pose greater risk of failure, and design test cases to test them

3. **Product Architecture and Design**: Review of the product architecture and design help identify the product hot spots, weak spots, critical interfaces, as previously described.

Examples of robust test cases include testing with invalid inputs, testing in unexpected environments, testing the product in stressful environmental conditions to predict how the product will perform while being exposed to expected stress levels or operating conditions above specification limits to create failure scenarios that would likely have occurred under normal stress over a period of time. Typically, this is done using one stress parameter at a time, such as vigorously shaking a mobile device at high intensity over a period of time, and this is referred to as Accelerated Life testing (ALT). Similarly, High Accelerated Life Testing (HALT) also tests a product for robustness to elevated stress beyond specification limits and may include multiple stress parameters such as shaking the mobile device while also raising the environmental temperature and continue to raise the stress up to the point of failure and discover the performance limit of the product (beyond the specification limit). This stress testing methodology is also referred to as 'test-to-fail' where a product is tested until its failure. Therefore, ALT helps answer the question when the product will fail and HALT helps define the difference between performance and specification limits to answer the question how much 'design margin' exists before eventual product failure.

14.10 Conclusion

Robust products can be designed by following a comprehensive end-to-end process as outlined in this chapter. To deliver reliable products, one must start with requirement specification, through design, development, and testing. Design robustness must be built inside out. Techniques described in this chapter can help proactively identify design vulnerabilities that can be systematically assessed, resolved through design improvements, and verified through robustness testing to assure customers that the product will perform reliably in the field.

References

1. IEEE standard glossary of software engineering terminology, IEEE Std 610.12–1990.
2. https://www.it-cisq.org/the-cost-of-poor-quality-software-in-the-us-a-2018-report/The-Cost-of-Poor-Quality-Software-in-the-US-2018-Report.pdf.
3. Juran, J. M., Gryna Frank, M. (1993). *Quality Planning and Analysis: From Product Development Through Use*, McGraw Hill.
4. Maass, E.C., McNair, P.D. (2009). *Applying Design for Six Sigma to Software and Hardware Systems*, Prentice Hall.

Vic Nanda is Head of Quality Capabilities Scaling & Nokia Quality Consulting at Nokia. He leads Nokia's Continuous improvement and Lean Six Sigma programs that have delivered over 1 Billion Euros of business impact from 2012–2019. His experience at major telecoms such as Motorola, Nortel, Ericsson and others spans operational excellence and quality management systems deployment using Lean Six Sigma, kaizen events, CMMi, ISO/TL 9000, and PMBOK practices. He has authored three books and several publications on Lean Six Sigma, quality management systems, and process improvement. He is a frequent industry speaker and has taught 1000+ Lean Six Sigma belts, coached 150+ Lean Six Sigma belts to certification, and delivered cumulative business impact in excess of 200M dollars. He is a Master Black Belt and holds 10 other quality certifications. He has consulted for the US Government, NASA (OSIRIS-REX mission to Asteroid Bennu), Arizona State University, and more.

He was awarded the American Society for Quality (ASQ) Golden Quill Award, and the ASQ Feigenbaum Medal by the ASQ for displaying outstanding characteristics of leadership, professionalism, and contributions to the field of quality. He has a Masters in Computer Science from McGill University, Bachelors in Computer Engineering from University of Pune, India, and executive education from Harvard Business School.

Dr. Eric Maass is Senior Director for DFSS/DRM for Medtronic Restorative Therapy Group. He is responsible for developing and leading the DRM strategic plan and focus for most of the company and has been the chief architect for Medtronic's DFSS/DRM BB and MBB programs. He was recognized with Medtronic's individual Star of Excellence award for 2012 and has been recognized as a Medtronic Technical Fellow. He joined Medtronic in October 2009, after 30 years with Motorola in roles ranging from Research and Development through Manufacturing, to Director of Operations for a $160 Million business and Director of Design and Systems Engineering for the Wireless group of Motorola SPS. He was a co-founder of the Six Sigma methods at Motorola, and had been the Lead Master Black Belt for DFSS at Motorola. His book, *Applying DFSS to Software and Hardware Systems*, provides clear step-by-step guidance on applying DFSS for developing innovative and compelling new products and technologies, while managing the business, schedule and technical risks. He received his Bachelor's degree in Biological Sciences from the University of Maryland Baltimore County, his Master's degree in Biomedical and Chemical Engineering from Arizona State University and his PhD in Industrial and Systems Engineering from Arizona State University. Dr Maass also currently serves as an Adjunct Professor at Arizona State University, and as chairman of the Industrial Advisory Board for the NSF-Sponsored B.R.A.I.N Industry/University collaborative research consortium.

Chapter 15
Time Series Modelling of Non-stationary Vibration Signals for Gearbox Fault Diagnosis

Yuejian Chen, Xihui Liang, and Ming J. Zuo

Abstract Gearboxes often operate under variable operating conditions, which lead to non-stationary vibration. Vibration signal analysis is a widely used condition monitoring technique. Time series model-based methods have been developed for the study of non-stationary vibration signals, and subsequently, for fault diagnosis of gearboxes under variable operating conditions. This chapter presented the latest methodologies for gearbox fault diagnosis using time series model-based methods. The main contents include widely used time-variant models, parameter estimation and model structure selection methods, model validation criteria, and fault diagnostic schemes based on either model residual signals or model parameters. Illustrative examples are provided to show the applications of model residual-based fault diagnosis methods on an experimental dataset collected from a laboratory gearbox test rig. Future research topics are pointed out at the end.

Keywords Gearboxes · Non-stationary · Time series models · Fault diagnosis · Vibration analysis

15.1 Introduction

Gearbox fault diagnosis refers to fault detection, fault mode identification, and severity assessment, which are critical for the prevention of sudden failures of gearboxes, enabling condition-based maintenance, and thus minimizing downtime and/or maintenance costs. Vibration analysis is the most widely used technique for gearbox fault diagnosis.

Y. Chen · M. J. Zuo (✉)
Department of Mechanical Engineering, University of Alberta, Edmonton, Alberta T6G 19, Canada
e-mail: ming.zuo@ualberta.ca; mzuo@ualberta.ca

X. Liang
Department of Mechanical Engineering, University of Manitoba, Winnipeg, MB R3T 5V6, Canada

K. B. Misra (ed.), *Handbook of Advanced Performability Engineering*,
https://doi.org/10.1007/978-3-030-55732-4_15

In many industrial applications, gearboxes operate under variable speed conditions. For instance, wind turbine gearboxes run under variable speed conditions due to the randomness of wind [1]. The gearbox that drives the fans of demand ventilation systems operates under variable speed conditions to reduce operating costs [2]. In railway systems, gearboxes experience run-up and coast-down conditions. The varying speeds modulate the amplitude and frequency of vibration signals. Therefore, the vibration signals become non-stationary. Effective non-stationary signal analysis tools are needed for gearbox fault diagnosis.

Time series model-based methods (TSMBMs) were initially employed in the structural health monitoring (SHM) field and have now drawn increased attention for gearbox fault diagnosis [3–5]. TSMBMs use time series models to model the vibration signals that are generated by gearbox systems. How to identify a time series model can be regarded as a response-only system identification problem.

Modelling non-stationary vibration signals need time-variant time series models that are realized by configuring the parameters of time-invariant models to be time variant. In this chapter, we will describe four widely used time-variant time series models. They are categorized based on how the parameters of time-invariant models are configured as time variant.

A time series model is generally composed of autoregressive (AR) and moving average (MA) terms. The MA terms are often ignored because (1) the AR terms can approximate the MA terms and (2) the consideration of MA terms makes the model identification more complex. Thus, in this chapter, we describe time-variant time series models that are composed of the AR terms only. Meanwhile, for simplicity, this chapter is limited to time series models for a single-channel vibration signal.

The materials in two of our earlier journal papers [3, 4] have been summarized and included in this chapter. Note that this chapter has also described other methods [5–10] to provide a comprehensive introduction of the latest TSMBMs.

The rest of this chapter is organized as follows: Sect. 15.2 introduces four time-variant time series models; Sect. 15.3 presents parameter estimation and model structure selection methods for the identification of time-variant time series models as well as the criteria for model validation; Sect. 15.4 describes two schemes (i.e. model residual-based scheme and model parameter-based scheme) for fault diagnosis; Sect. 15.5 presents the applications of the model residual-based fault diagnostic scheme on an experimental dataset collected from a laboratory gearbox test rig; conclusion remarks are drawn in Sect. 15.6.

15.2 Time Series Models for Non-stationary Vibration Signals

In this section, we present four time-variant AR models for representing non-stationary vibration signals. The first one is the periodic AR (PAR) model [6]. The PAR has AR parameters varying periodically with a specified period T. The PAR model has the following difference equation

$$y_t = \sum_{i=1}^{n_a} a_i(t) y_{t-i} + \varepsilon_t \tag{15.1}$$

where y_t and $y_{t\text{-}i}$ denote the vibration at time t and t-i, respectively; n_a is the AR model order; a_i stands for the AR parameters, which are periodic with the same period T; and ε_t is a zero-mean Gaussian white noise at time t. The PAR model is useful for representing non-stationary vibration signals with periodic time-varying characteristics. Wyłomańska et al. [6] used the PAR model for fault diagnosis of a gearbox in a bucket-wheel excavator that is a heavy-duty mining machine subjected to cyclic load/speed variation due to the digging/excavating process. However, the PAR model may not be applicable for representing non-stationary vibration signals collected under non-periodic variable speed conditions, such as the random variable speed condition that winds turbine gearbox experience.

The second one is the adaptive AR model with its model parameters adaptively (recursively) adjusted by recursive parameter estimation methods [7]. The adaptive evolution of model parameters enables the AR model time variant, and thus, it can track the non-stationary characteristics of vibration signals. Zhan et al. [7] and Shao et al. [8] used the adaptive AR model for fault diagnosis of fixed-axis gearboxes. The adaptive AR model requires a proper tuning of the convergence rate for its recursive parameter estimation algorithm. Too high of a convergence rate results in overfitting, and too low of a convergence rate causes underfitting.

The third one is the functional series time-dependent AR (FS-TAR) model [5, 10]. The FS-TAR model has a model difference equation the same as Eq. (15.1), but $a_i(t)$ is no longer periodic. Instead, $a_i(t)$ is represented by a function of time that is expanded in functional series (basis expansion). Therefore, the FS-TAR model is not limited to applications with cyclic load/speed variations. Reported basis functions include discrete cosine transform functions, Legendre polynomials, Harr functions, normalized B-spines, etc.[5]. Take the Legendre polynomials basis as an example. The dependency $a_i(t)$ is of the form:

$$a_i(t) = \sum_{j=1}^{p} a_{i,j} t^j \tag{15.2}$$

where $a_{i,j}$ stands for the AR parameters of projection and p specifies the order of functional spaces. The FS-TAR model has been widely used in the SHM field, such as a pick-and-place mechanism [10].

The last one is the functional pooled AR (FP-AR) model [3, 4]. The FP-AR model has the following model difference equation [3, 4, 11]

$$y_t = \sum_{i=1}^{n_a} a_i(k_t) y_{t-i} + \varepsilon_t \tag{15.3}$$

where k_t denotes the operating condition at time t and a_i is a function of k_t. We can see that the FP-AR model has the same model structure as the FS-TAR model, but with its AR parameters dependent on operating condition variable k_t. In the case when k_t is a vector, the FP-AR model is extended to the vector functional pooled autoregressive (VFP-AR) model in which the AR parameters are functions of a vector. Traditionally, the FP-AR models were identified to represent the vibrations under different levels of operating conditions [12–14]. It has recently been shown that the FP-AR model can be used to represent non-stationary signals which have a continuous time-varying spectrum [3, 4]. Chen et al. [3] presented an FP-AR model-based method for tooth crack fault detection of fixed-axis gearboxes under variable speed conditions. Chen et al. [4] presented a VFP-AR model-based method for tooth crack severity assessment of fixed-axis gearboxes under random variable speed conditions.

15.3 Model Identification and Validation

Model identification refers to the estimation of time series models based on the vibration data records y_t (for $t = 1, 2, \ldots, N$, where N is the number of data points). The identification of time series models includes parameter estimation and model structure selection. Model structure selection refers to the selection of lagged terms and/or functional basis. Once a time series model is identified, the model needs to be validated to ensure its modelling accuracy. In this section, we will describe the most widely employed methods for parameter estimation, model structure selection, and model validation.

15.3.1 Parameter Estimation Methods

Typical parameter estimation methods include the least squares (LS) and maximum likelihood (ML) [5]. The LS estimator can be used for PAR, FS-TAR, and FP-AR models. The LS estimator of the model parameter vector $\boldsymbol{\theta}$ is based on minimizing the squared summation of residuals $\mathbf{e}$

$$\hat{\boldsymbol{\theta}} = \text{argmin}\{\|\mathbf{e}\|\} = \text{argmin}\left\{\|\mathbf{y} - \boldsymbol{\Phi}^T\boldsymbol{\theta}\|\right\} \tag{15.4}$$

where $\|\bullet\|$ denotes the l_2 norm, $\mathbf{y} = [y_1, \ldots, y_N]^T$ denotes the observed time series, and $\boldsymbol{\Phi}$ is a hat matrix that is constructed from k_t and/or $y_{t-1}, \ldots, y_{t-n_a}$, depending on the time series model structure. The residual $\mathbf{e}$ means the one-step-ahead prediction error. The above minimization problem yields the solution expressed as

$$\hat{\boldsymbol{\theta}} = \left[\boldsymbol{\Phi}^T\boldsymbol{\Phi}\right]^{-1}\boldsymbol{\Phi}^T\mathbf{y} \tag{15.5}$$

The recursive least squares (RLS) estimator [5, 15] computes the model parameter vector $\boldsymbol{\theta}$ recursively by making use of the new data record at a given time instant t. When RLS is employed for estimating the AR model, we can realize an adaptive AR model (as introduced in Sect. 15.2) to represent non-stationary vibration signals. Readers may refer to Refs [5, 15] for more details about RLS estimators.

The maximum likelihood estimator can be used for PAR, FS-TAR, and FP-AR models. The ML estimator of the model parameter vector $\boldsymbol{\theta}$ is based on maximizing the log-likelihood function given as follows [5]

$$\widehat{\boldsymbol{\theta}} = \mathrm{argmax} L(\boldsymbol{\theta};\, \mathbf{e}|\mathbf{y}) \tag{15.6}$$

$$L(\boldsymbol{\theta};\, \mathbf{e}|\mathbf{y}) = -\frac{N}{2}\ln 2\pi - \frac{1}{2}\sum_{t=1}^{N}\left(\ln\sigma_t^2 + \frac{\varepsilon^2[t, \boldsymbol{\theta}]}{\sigma_t^2}\right) \tag{15.7}$$

where σ_t is the standard deviation of the residual, which is also dependent on the model parameter vector $\boldsymbol{\theta}$. The standard deviation σ_t is estimated directly from the residual sequence $\mathbf{e}$ using the sample standard deviation formula [5]. In Eq. (15.7), it is assumed that ε_t follows the zero-mean Gaussian distribution with a standard deviation σ_t.

Both LS and ML estimators are asymptotically Gaussian distributed with mean coinciding to the true value. The ML estimator, however, achieves lower model parameter estimation variance than the LS estimator [5]. On the other hand, the ML estimator has a higher computational cost than the LS method.

15.3.2 Model Structure Selection

Typical methods for model structure selection include various information criteria and regularization. The most widely used criteria are the Akaike Information Criterion (AIC) and Bayesian Information Criterion (BIC). The AIC is of the form

$$\mathrm{AIC} = N\ln(\mathrm{RSS}/N) + 2d \tag{15.8}$$

where d is the number of model parameters and RSS is the training residual sum of squares. The BIC is similar to the AIC, with a different penalty for the number of parameters as follows

$$\mathrm{BIC} = N\ln(\mathrm{RSS}/N) + d\ln(N) \tag{15.9}$$

Both AIC and BIC penalize the model structural complexity and thus avoid over-fitting. Based on minimizing these criteria, model structure selection becomes an integer optimization problem. Such an optimization problem can be solved via backward and/or forward regression, genetic algorithm, particle swarm algorithm, etc. It is important to note that these criteria-based methods require the assumption of consecutive AR set (and identical sets of functional spaces for FS-TAR and FP-AR) for the aforementioned four time-variant AR models to simplify the model structure selection procedure [14]. Without such an assumption, the integer optimization problem will have 2^d (d is usually greater than 100) different solutions, which is computationally impossible to find the global minimiser.

The regularization (e.g., l-1 norm)-based method has recently been adopted for model structure selection for the FP-AR model [3]. The regularization-based methods are free from the assumptions of consecutive AR set and identical sets of functional spaces, and, therefore, achieve higher modelling accuracy [3, 4]. With an initial (sufficient large) n_a and high dimension functional spaces, the least absolute shrinkage and selection operator (LASSO) estimator is given as follows,

$$\hat{\boldsymbol{\theta}} = \text{argmin}\left\{\|\mathbf{y} - \boldsymbol{\Phi}^T\boldsymbol{\theta}\|\right\} + \lambda|\boldsymbol{\theta}| \tag{15.10}$$

where $\lambda \geq 0$ is a tuning parameter and $|\bullet|$ denotes the l-1norm. The selection of λ is critical. When $\lambda = 0$, the LASSO will reduce to the LS estimator. Too large λ value will force too many coefficients to zero, whereas too small λ value will force a limited number of coefficients to zero. The λ can be selected by either the K-fold cross-validation [3] or validation set approach [4]. Other regularizations, such as l-2 norm and elastic net [15], are also options for time series model structure selection.

15.3.3 Model Validation

Model validation is mainly based on a validation signal. Upon the identification of a time series model, the inverse filter is constructed and then applied to process the validation signal. We refer to the residual obtained from the validation signal as 'residual-of-validation'. Model accuracy can be judged by the mean squared error (MSE) of the residual-of-validation, the randomness of the residual-of-validation, and the frozen-time spectrum [3, 5]. First, a model with a lower MSE of the residual-of-validation is more accurate in modelling the baseline vibration than those with a higher MSE [36, Sect. 7]. Second, the more random the residual-of-validation is, the more accurate the model is. Ljung–Box test [16] can be conducted to quantify the randomness. Last, the frozen-time spectrum $S(f, t)$ of time series models can be obtained and compared with the non-parametric spectrum (e.g. short-time Fourier transform) of non-stationary signals. An accurate time series model should give a parametric spectrum in good agreement with the non-parametric one.

15.4 Time Series Model-Based Fault Diagnosis

Fault diagnosis may be based on either model residuals or model parameters. In this section, we introduce both model residual-based methods and model parameter-based methods.

15.4.1 Model Residual-Based Method

For fault detection, the model residual-based method relies on the identification of a baseline time series model. Figure 15.1 shows the schematic of the model residual-based method for fault detection [3]. A time series model is identified to represent the baseline vibration signals. Then, the vibration signals collected under future unknown health state are processed by an inverse filter constructed from the baseline model. Any changes in the residual signals may indicate the occurrence of a fault. Researchers have examined the whiteness [8], variance [17, 18], Gaussionality [7], and impulsiveness [3, 8] to quantify the changes in residual signals.

For severity assessment or fault mode identification, the model residual-based method relies on the identification of time series models under each severity level or fault mode. Figure 15.2 shows the schematic of the model residual-based method for severity assessment [4, 19]. The presented scheme can also be used for fault mode identification by changing the fault severity states to fault modes. During the training phase, the training signals collected under each fault severity level and a wide range of the speed variation are used. Under each fault severity level, a time series model is identified to represent the vibration signals of that state. We refer these time series models as state models. During the testing phase, these trained state models are used for severity assessment. Vibration signal y_t, along with necessary operating condition variables is collected under an unknown health state of the gearbox. Afterwards, the inverse filters from each of the state models are applied to process the vibration signals collected under the unknown health state and to obtain residuals of the state models. The final health state is classified as the state with an inverse filter that gives minimal residual MSE. Note that in industrial applications, it is not easy to obtain the signals under known health states. This is the major challenge associated with this model residual-based method for severity assessment or fault mode identification.

15.4.2 Model Parameter-Based Method

The model parameter-based method requires the identification of a model during the testing stage. For fault detection, the model parameter-based method is based on comparing the parameters of the current model with the parameters of the baseline time series model. Figure 15.3 shows the schematic of the model parameter-based

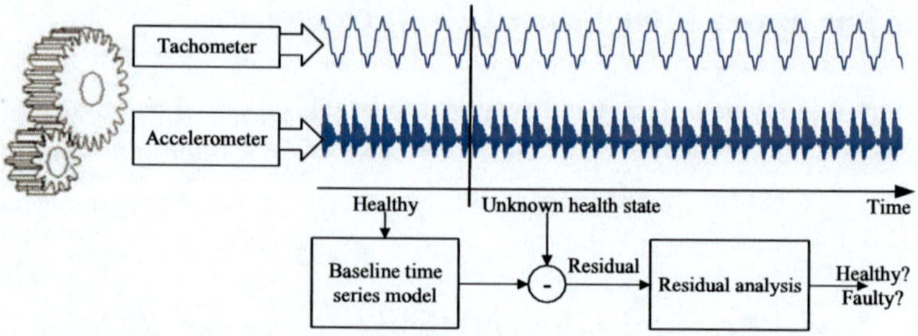

Fig. 15.1 Model residual-based fault detection method [3]

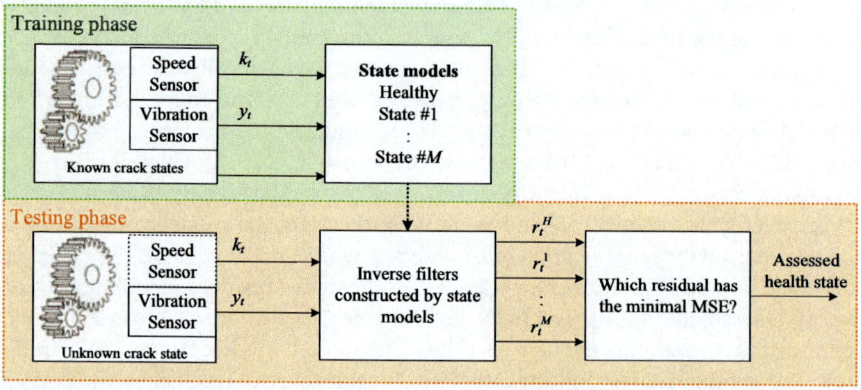

Fig. 15.2 Model residual-based severity assessment or mode identification method [4]

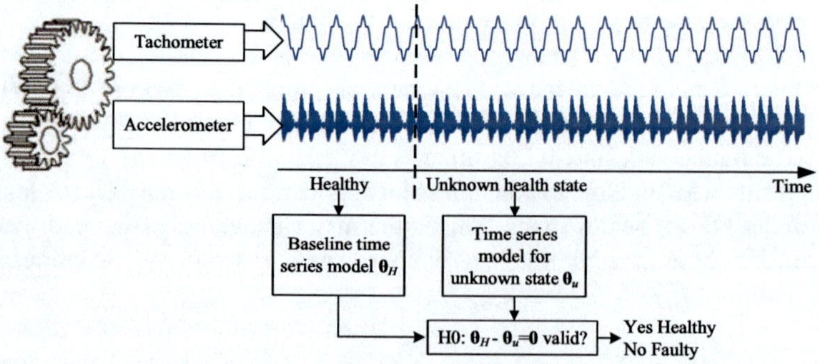

Fig. 15.3 Model parameter-based fault detection method [21]

method for fault detection [20, 21]. During the training stage, a time series model is identified to represent the baseline vibration signals. During the testing stage, another time series model is identified to represent the current vibration signals from an unknown health state. Fault detection is based on testing statistical differences between the model parameters under the healthy state $\boldsymbol{\theta}_H$ and the model parameters under the unknown health state $\boldsymbol{\theta}_u$ through the following hypotheses [20].

$$\text{H0} : \boldsymbol{\theta}_H - \boldsymbol{\theta}_u = 0; \text{H1}; \boldsymbol{\theta}_H - \boldsymbol{\theta}_u \neq 0 \tag{15.11}$$

where H0 denotes the null hypothesis and H1 denotes the alternative hypothesis. If H0 is valid, then the unknown health state is deemed healthy. Otherwise, the unknown health state is detected as faulty.

For severity assessment or fault mode identification, the model parameter-based method is based on comparing the parameters of the current model with the parameters of the trained state models [10]. Figure 15.4 shows the schematic of the model parameter-based method for severity assessment or fault mode identification. During the training stage, the state time series model is identified for each severity level or each fault mode. During the testing stage, another time series model is identified to represent the current vibration signals from an unknown health state. Severity assessment or fault mode identification is based on testing statistical differences between the model parameters obtained from the training stage ($\boldsymbol{\theta}_H$, $\boldsymbol{\theta}_1$, …, $\boldsymbol{\theta}_M$) and the model parameters under the unknown health state $\boldsymbol{\theta}_u$ through hypotheses similar to Eq. (15.11). The final health state is classified as the state with a valid null hypothesis. Since the model parameter-based method requires the identification of a model during the testing stage, it is generally not suitable for online applications.

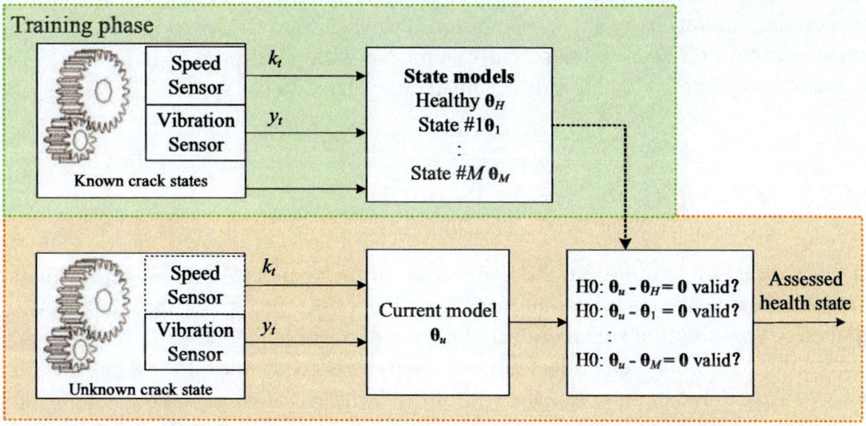

Fig. 15.4 Schematic of the model parameter-based severity assessment method [11]

15.5 Applications of Model Residual-Based Methods

This section presents the applications of the two model residual-based methods [3, 4] for fault detection and severity assessment of gear tooth crack. Section 15.5.1 will briefly describe the experimental dataset. Section 15.5.2 will present the application of the sparse FP-AR model residual-based fault detection method [3] on vibration signals different from the applications presented in ref.[3]. Section 15.5.3 will summarize the application of the sparse VFP-AR model residual-based fault detection method, as reported in Ref.[4].

15.5.1 Experimental Dataset

The experimental dataset was collected at the University of Pretoria, South Africa [22, 23]. Readers can refer to refs [22, 23]. for the detailed experimental setup. Two vibration sensors were equipped on this test rig. One vibration sensor is single axial and is labelled as #7. The other vibration sensor is a triaxial accelerometer. The experimental dataset contains 100 data files from a healthy gearbox and 1400 data files from a run-to-failure experiment with 50% initial crack and duration of around 21 days of continuous running. Each data file contains data collected within 20 s. The sampling frequency was $f_s = 25.6$ kHz. These vibration signals were further low passed using an FIR filter with a cut-off frequency of $f_c = 1.6$ kHz and then downsampled from $f_s = 25.6$ kHz to $f_s = 3.2$ kHz. When collecting each data file, an electrical motor drove the transmission train such that the rotating speed of the target gearbox followed a sinusoidal-like profile with a period equals to 10 s. The alternator generated a load torque positively correlated to the speed.

15.5.2 FP-AR Model-Based Fault Detection

In this subsection, we present the application of the sparse FP-AR model residual-based method [3] for gear tooth crack fault detection. In ref.[3], the method was applied to the vibration signal collected from the sensor labelled as #7. They did not analyze the vibration data collected from the triaxial accelerator. In this subsection, we are to apply the method reported in ref [3] to the vibration signal collected from the x-direction of the triaxial accelerometer (with a sensitivity of 100 mV/g). In other words, a one-dimensional vibration data series from a different sensor of the same test rig will be used to assess the effectiveness of the sparse FP-AR model residual-based method [3].

The following configurations are the same as used in Ref. [3]: The signals from a zebra-tape shaft encoder are used in this subsection to obtain the rotating speed information. Training and validation data are arbitrarily selected from the 100 baseline data files. The training data were 7.5 s in length, which was truncated from the length of 20 s. The validation data also have a length of 7.5 s. During the sparse FP-AR modelling, the initial set of functional spaces was configured as $\{1, \omega_t, \omega_t^2, \ldots, \omega_t^7\}$, where ω_t is the rotating speed. The candidate set for λ was configured as $[0, 1 \times 10^{-8}, 1 \times 10^{-7}, \ldots, 1 \times 10^{-1}]$.

The following results are obtained when we apply the sparse FP-AR model residual-based method to the vibration signal collected from the *x*-direction of the triaxial accelerometer: The initial n_a was determined as $\{1, 2, \ldots, 63\}$ by BIC. When $\lambda = 0$, we achieved the minimum CVMSE. By increasing λ, the CVMSE will get bigger. We need to use a larger λ value that does not have too big a CVMSE. When λ increases to 1×10^{-5}, the CVMSE is still within one standard deviation of the minimum CVMSE [3]. Therefore, $\lambda = 1 \times 10^{-5}$ is chosen. Table 15.1 lists the modelling performance of the identified sparse FP-AR. From the table, we can see that the residual-of-validation of the sparse FP-AR model has a *p*-value of 0.1509 from the Ljung–Box test. This *p*-value is significantly higher than the *p*-value of 0 of the validation signal. This means that the randomness of the residual-of-validation is much higher than the original validation signal.

Figure 15.5 shows the non-parametric spectrum (a) and the frozen-time spectrums (b) obtained from the sparse FP-AR model. The frozen-time spectrum obtained from the sparse FP-AR model aligns well with the non-parametric spectrum by tracking the time-varying spectral contents. Given the above model validation criteria,

Table 15.1 Modelling performance of the sparse FP-AR model. Algorithms were coded in MATLAB 2019a and implemented on a desktop with two Intel 2.4 GHz processors and 16 GB of RAM

Ljung–Box test, *p*-value, of validation signal	Ljung–Box test, *p*-value, of residual-of-validation	CPU time in training, (min)	CPU time in testing, (s)
0	0.1509	48	0.4

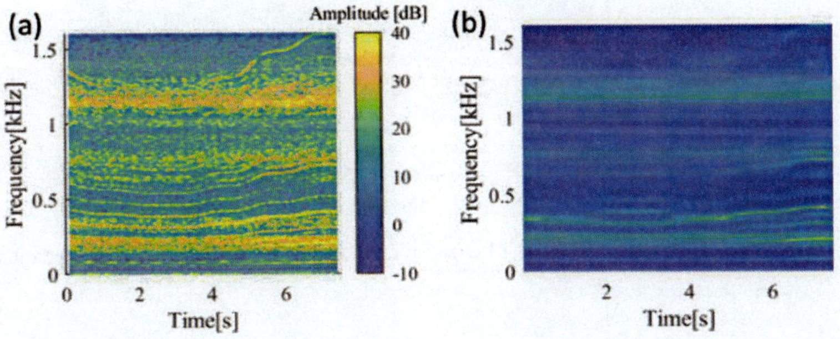

Fig. 15.5 Time–frequency spectrums: **a** non-parametric spectrum of the validation signal obtained by MATLAB spectrogram function; **b** Frozen-time spectrum obtained from the sparse FP-AR model. Z-axis scales are the same

we can conclude that the sparse FP-AR model has good modelling accuracy when representing the baseline vibration signal collected from the x-direction of the triaxial accelerometer. The sparse FP-AR model used about 48 min in the training stage and 0.4 s in the testing stage. Since this training process is completed offline, the length of time required is not very critical [3]. The computational time in the testing stage is more critical as it determines whether the TSMBM based on the sparse FP-AR model is practically useful or not [3]. In many applications where incipient faults do not immediately lead to a catastrophic failure of the gearbox system, updating the fault detection information every second is acceptable and thus requiring about 0.4 s in testing is acceptable [23].

The identified sparse FP-AR model was used for detecting the gear tooth crack faults. Each of the 99 baseline data files (the 100 baseline data files exclude the one used for training the sparse FP-AR model) and 1400 run-to-failure data files are truncated to have the speed profile the same as the validation signal. Figure 15.6 shows the normalized periodic modulation intensity (NPMI) [3] calculated from both the raw data and the residuals obtained from the identified sparse FP-AR model for each data file. The NPMI is the periodic modulation intensity (PMI) value of the residual divided by the PMI of the residual of baseline vibration, where the PMI represents the energy ratio between tooth crack-induced impulses and other components. Figure 15.6a shows the baseline case, whereas Fig. 15.6b shows the damaged case. It is clear that for the NPMI obtained from the raw signals, they are of similar magnitude for both the healthy data files and faulty data files. On the other hand, the NPMIs from the residuals of the sparse FP-AR model for the faulty data files are obviously higher than those for the healthy data files. This means that using the solid blue plots, we are able to detect the faults.

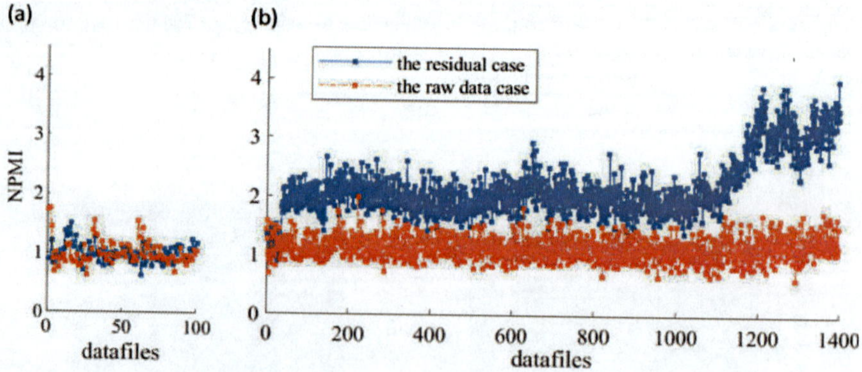

Fig. 15.6 Normalized PMI for detecting the tooth crack fault. **a** Healthy data files; **b** faulty data files

15.5.3 VFP-AR Model-Based Severity Assessment

In this subsection, we summarize the application of the sparse VFP-AR model-based method for gear tooth crack severity assessment, as reported in ref.[4].

Four discrete health states, namely, healthy (H), initial crack (F1), intermediate crack (F2), and missing tooth (F3) were considered. The F1 state corresponds to the gear with the initial 50% tooth crack. The F2 state corresponds to the gear that had run 17 days after the initial 50% tooth crack. The F3 state corresponds to the gear that ran right before the end of the run-to-failure experiment.

Under each health state, training, validation, and testing signals were prepared and preprocessed. In total, 43 data files under each health state were used in which one served for training (training signal), two for validation (one for model identification and the other for measuring modelling accuracy), and the rest 40 for testing severity assessment performance. The inverse filter constructed by the baseline sparse FP-AR model identified in Sect. 15.5.2 was used to preprocess these signals and to obtain residual signals.

For the training and validation signals, the first half (10 s) of the vibration signal in a data file was used. Such a vibration signal experienced a full cycle of the speed variation. On the other hand, the 40 segments of testing signals only lasted 5 s with a starting point p_s randomly sampled from [0, 0.25, 0.5, …, 10]s.

For the identification of the sparse VFP-AR models, Legendre polynomial basis functions were used for $\mathbf{G}(\omega_t)$ and the refined B-splines for $\mathbf{G}(\theta_t)$ [4] where θ_t is the rotating phase. The $\mathbf{G}(\omega_t)$ was configured as $\{1, \omega_t, \omega_t{}^2, \ldots, \omega_t{}^7\}$. As for the refined B-splines, r was configured as 3 and K as 40 (i.e. the number of teeth 37 plus $r = 3$). Two parameters k and n were further determined by estimating a sparse VFP-AR model with a small $n_a = 5$ and examining the occurrence of periodic B-spline bases. The k and n were determined as (24, 1) for F1 state; (24, 2) for F2 state; (24, 3) for F3 state. Since the vibration signal under the H state did not have crack induced impulses, its corresponding sparse VFP-AR model did not need to consider the phase. In other words, the sparse VFP-AR model for the H state reduced to a sparse FP-AR model. Afterwards, the n_a was to be determined after obtaining the k and n via the validation set approach. The n_a was determined to be (50, 40, 35, 35) for health states (H, F1, F2, F3), respectively.

Upon the identification of both the sparse VFP-AR models, the inverse filter was constructed and then applied to process the validation signals for measuring the modelling accuracy. Table 15.2 lists the MSE and the randomness of both residuals and the validation signals (i.e. the residual of the baseline sparse FP-AR model). The p-values from the Ljung–Box tests were reported, which means the probability of being random. From this table, we can see that the sparse VFP-AR models return a residual with reduced MSE and a higher probability of being random compared with the validation signals.

Computational costs were evaluated and listed in Table 15.2 as well. The training data points were 32,000. For four sparse VFP-AR models, the time required for training was around 17.9 h. Since this training process was completed offline, the

Table 15.2 Modelling accuracy and computational cost of sparse VFP-AR models [4]

Health state	H	F1	F2	F3
MSE of residual(normalized m/s^2)	0.881	1.172	1.114	1.191
MSE of validation signal (normalized m/s^2)	1.008	1.347	1.271	1.369
Randomness of residual (*p*-value)	1.55×10^{-5}	1.56×10^{-5}	0.0034	6.93×10^{-5}
Randomness of validation signal (*p*-value)	0	0	0	0
Time in training (h)	1.5	4.9	5.2	6.3
Time in testing (s)	2.2	2.4	2.4	2.6

length of time needed was not very critical [3, 4]. The computational time in the testing stage is more critical to determine whether the method is practical or not [3]. The inverse filter constructed by the four sparse VFP-AR models used less than 9.6 s. Again, in many applications where incipient faults do not immediately lead to a catastrophic failure of the gearbox system, updating the fault detection information every 10 s is acceptable.

Figure 15.7 shows the non-parametric spectrum (a, b, c, d) of the validation signals as well as the frozen-time spectrums of the sparse VFP-AR models (e, f, g, h). We can see the tooth crack-induced impulses as vertical lines in these spectrums. The vertical lines in the frozen-time spectrums of sparse VFP-AR models behave discretely, which are in good agreement with the discrete lines in non-parametric spectrums. The tooth crack-induced impulses can be represented using sparse VFP-AR models.

Figure 15.7 shows the non-parametric spectrum (a, b, c, d) of the validation signals as well as the frozen-time spectrums of the sparse VFP-AR models (e, f, g, h). We can see the tooth crack-induced impulses as vertical lines in these spectrums. The vertical lines in the frozen-time spectrums of sparse VFP-AR models behave discretely, which are in good agreement with the discrete lines in non-parametric spectrums. The tooth crack-induced impulses can be represented using sparse VFP-AR models.

The sparse VFP-AR models under known health state were applied to testing signals for the severity assessment. For each testing signal, four model residuals were obtained, and their MSE values were calculated. The health state was classified as the state with an inverse filter that gave minimal residual MSE. Figure 15.8 shows the classification results when processing the 40 testing signals under each health state. The classification accuracy was reported to be 93.8%. The results showed the effectiveness of the sparse VFP-AR model-based method.

15.6 Summary and Conclusion

This chapter presented the latest methodologies related to the time series model-based techniques for gearbox fault diagnosis. We described four most widely used time-variant time series models, typical parameter estimation and model structure

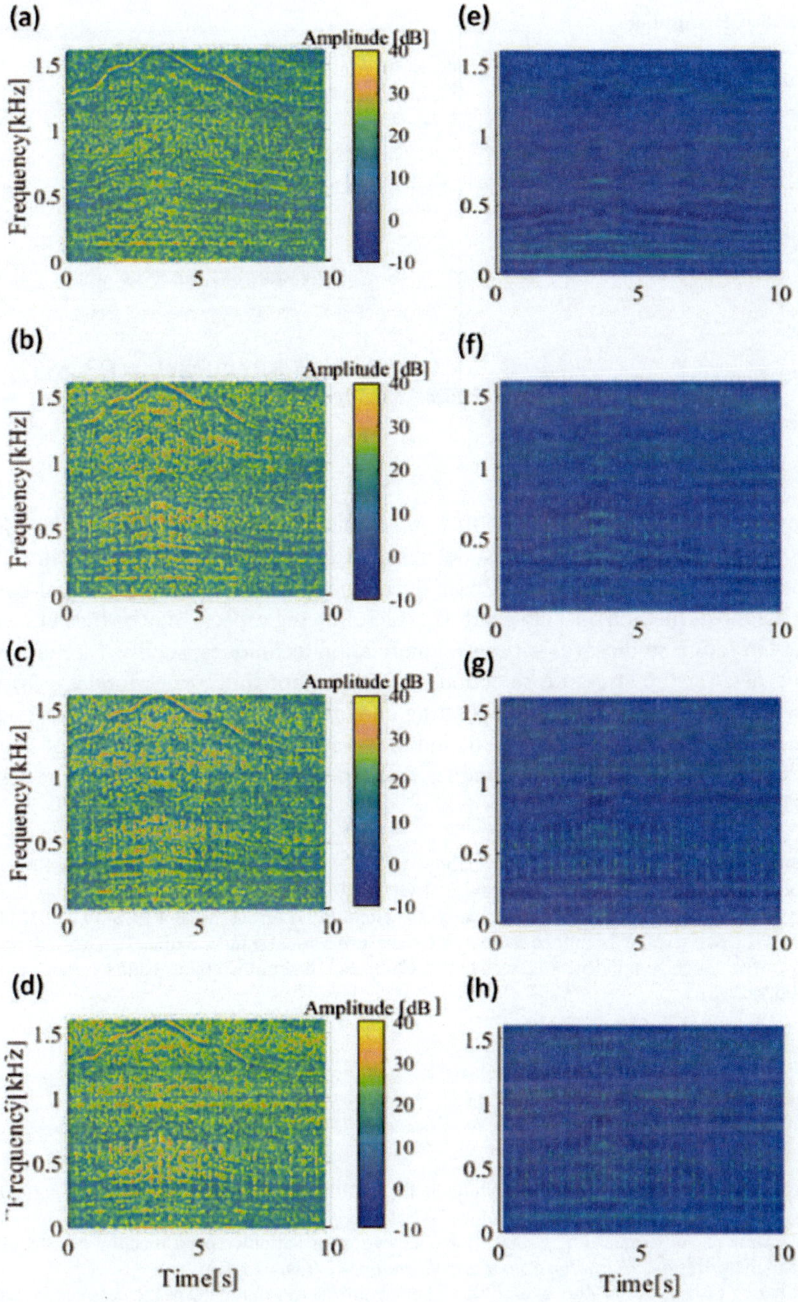

Fig. 15.7 Time–frequency spectrums: **a** ~**d** STFT spectrum of the validation signal; **e** ~ **h** Frozen-time spectrum of sparse VFP-AR models. Z-axis scales are the same for all spectrums [4]

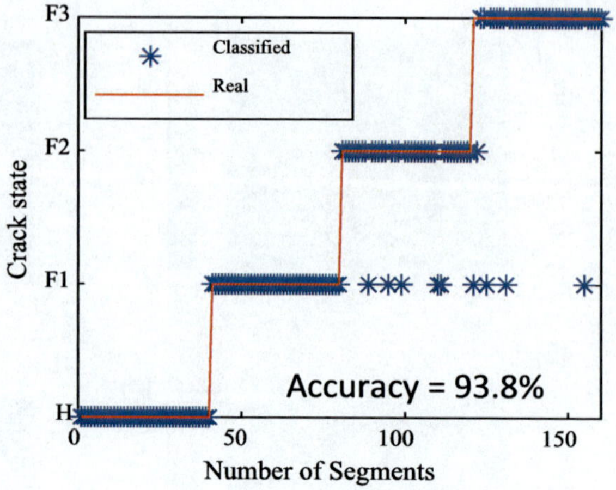

Fig. 15.8 Classification results [4]

selection methods for model identification, model validation criteria, and fault diagnosis schemes based on either model residual or model parameters. Finally, this chapter gave two examples to illustrate the applications of the model residual-based fault diagnosis method on a lab gearbox. The following aspects may be further investigated in future studies: (a) various regularization techniques, such as *l*-2 norm and elastic net, for the structure selection of time-variant time series models; (b) the consideration of more than two operating condition variables in a VFP-AR model, such as temperature, rotating speed, and load torque; and (c) the account of uncertainties of operating condition variables when identifying a time-variant time series model.

Acknowledgements This research is supported by the Natural Science and Engineering Research Council of Canada, Canada [grant number RGPIN-2015-04897, RGPIN-2019-05361]; Future Energy Systems under Canada First Research Excellent Fund [grant number FES-T11-P01, FES-T14-P02]; University of Manitoba Research Grants Program (URGP); Sadler Graduate Scholarship in Mechanical Engineering, Canada; and China Scholarship Council, China [grant number 201506840098].

References

1. Carlin, P. W., Laxson, A. S., & Muljadi, E. B. (2003). The history and state of the art of variable-speed wind turbine technology. *Wind Energy., 6*(2), 129–159.
2. McBain, J., & Timusk, M. (2009). Fault detection in variable speed machinery: Statistical parameterization. *Journal of Sound and Vibration, 327*(3), 623–646.
3. Chen, Y., Liang, X., & Zuo, M. J. (2019). Sparse time series modeling of the baseline vibration from a gearbox under time-varying speed condition. *Mech Syst Signal Process., 1*(134), 106342.
4. Chen, Y., Schmidt, S., Heyns, P. S., & Zuo, M. J. (2020). A time series model-based method for gear tooth crack detection and severity assessment under random speed variation. *Mechanical System and Signal Processing, 9*, 1–32.

5. Spiridonakos, M. D., & Fassois, S. D. (2014). Non-stationary random vibration modelling and analysis via functional series time-dependent ARMA (FS-TARMA) models—A critical survey. *Mechanical System and Signal Processing, 47*(1–2), 175–224.
6. Wyłomańska, A., Obuchowski, J., Zimroz, R., Hurd, H. (2017). Periodic autoregressive modeling of vibration time series from planetary gearbox used in bucket wheel excavator. In: Chaari, F., Leśkow, J., Napolitano, A., Sanchez-Ramirez, A., (Eds.), *Cyclostationarity: Theory and Methods* [Internet]. Springer International Publishing; 2014 [cited 2017 Mar 31]. pp. 171–86. (Lecture Notes in Mechanical Engineering). Available from: https://link.springer.com/chapter/10.1007/978-3-319-04187-2_12.
7. Zhan, Y., & Mechefske, C. K. (2007). Robust detection of gearbox deterioration using compromised autoregressive modeling and Kolmogorov-Smirnov test statistic—Part I: Compromised autoregressive modeling with the aid of hypothesis tests and simulation analysis. *Mechanical System and Signal Processing, 21*(5), 1953–1982.
8. Shao, Y., & Mechefske, C. K. (2009). Gearbox vibration monitoring using extended Kalman filters and hypothesis tests. *Journal of Sound and Vibration, 325*(3), 629–648.
9. Spiridonakos, M.D., Fassois, S.D. (2014). Adaptable functional series TARMA models for non-stationary signal representation and their application to mechanical random vibration modeling. *Signal Process, 96*, Part A, 63–79.
10. Spiridonakos, M. D., & Fassois, S. D. (2013). An FS-TAR based method for vibration-response-based fault diagnosis in stochastic time-varying structures: Experimental application to a pick-and-place mechanism. *Mechanical Systems and Signal Processing, 38*(1), 206–222.
11. Sakellariou, J.S., Fassois, S.D. (2007). A functional pooling framework for the identification of systems under multiple operating conditions. In: *2007 Mediterranean Conference on Control Automation*, pp. 1–6.
12. Kopsaftopoulos, F., Nardari, R., Li, Y.-H., & Chang, F.-K. (2018). A stochastic global identification framework for aerospace structures operating under varying flight states. *Mechanical Systems and Signal Processing, 1*(98), 425–447.
13. Sakellariou, J. S., & Fassois, S. D. (2016). Functionally Pooled models for the global identification of stochastic systems under different pseudo-static operating conditions. *Mechanical Systems and Signal Processing, 1*(72–73), 785–807.
14. Kopsaftopoulos, F. P., & Fassois, S. D. (2013). A functional model based statistical time series method for vibration based damage detection, localization, and magnitude estimation. *Mechanical Systems and Signal Processing, 39*(1), 143–161.
15. Hastie, T., Tibshirani, R., Friedman, J.H. (2009). *The Elements of Statistical Learning: data mining, inference, and prediction.* [Internet]. 2nd ed. 2009 [cited 2020 Feb 18]. Available from: https://web.stanford.edu/~hastie/ElemStatLearn/.
16. Ljung, G. M., & Box, G. E. P. (1978). On a measure of lack of fit in time series models. *Biometrika, 65*(2), 297–303.
17. Heyns, T., Godsill, S. J., de Villiers, J. P., & Heyns, P. S. (2012). Statistical gear health analysis which is robust to fluctuating loads and operating speeds. *Mechanical Systems and Signal Processing, 27,* 651–666.
18. Yang, M., & Makis, V. (2010). ARX model-based gearbox fault detection and localization under varying load conditions. *Journal of Sound and Vibration, 329*(24), 5209–5221.
19. Li, G., McDonald, G.L., Zhao, Q. (2017). Sinusoidal synthesis based adaptive tracking for rotating machinery fault detection. *Mechanical Systems and Signal Processing, 83*(Supplement C), 356–370.
20. Hios, J. D., & Fassois, S. D. (2014). A global statistical model based approach for vibration response-only damage detection under various temperatures: A proof-of-concept study. *Mechanical Systems and Signal Processing, 49*(1), 77–94.
21. Aravanis, T.-CI., Sakellariou, J.S., Fassois, S.D. (2019). A stochastic Functional Model based method for random vibration based robust fault detection under variable non–measurable operating conditions with application to railway vehicle suspensions. *Journal of Sound and Vibration*, 115006.

22. Schmidt, S., Heyns, P. S., & de Villiers, J. P. (2018a). A novelty detection diagnostic methodology for gearboxes operating under fluctuating operating conditions using probabilistic techniques. *Mechanical Systems and Signal Processing, 1*(100), 152–166.
23. Chen, Y., Liang, X., & Zuo, M. J. (2020). An improved singular value decomposition-based method for gear tooth crack detection and severity assessment. *Journal of Sound and Vibration, 3*(468), 115068.

Chapter 16
Risk-Informed Design Verification and Validation Planning Methods for Optimal Product Reliability Improvement

Zhaojun Steven Li and Gongyu Wu

Abstract This chapter proposes four types of mathematical optimization modeling approaches to optimize the product design Verification and Validation (V&V) planning during the New Product Development (NPD) process. These four optimization models provide four risk mitigation strategies from perspectives of cost efficiency to the optimal selection of a set of V&V activities for maximizing the overall system reliability improvement. The proposed approaches not only incorporate the critical product development constraints in V&V planning, such as the cost, time, reliability improvement, and sequencing and effectiveness of V&V activities, but also consider the decay of the improvement effectiveness when tackling the V&V activities' selecting and sequencing challenges. In addition, the concepts of set covering, set partition, and set packing are applied to assure that different levels of critical failure modes can be covered in different ways according to different risk mitigation requirements by the end of V&V execution. The application of the proposed optimization models and comparisons with existing methods for product V&V planning are illustrated through the product development of a power generation system within a diesel engine.

Keywords Verification and validation (V&V) planning · Reliability improvement · Risk mitigation · New product development · Set covering

---

Z. S. Li (✉)
Department of Industrial Engineering and Engineering Management, 1215 Wilbraham Rd, Springfield 01119, MA, USA
e-mail: zhaojun.li@wne.edu

G. Wu
School of Mechanical and Electrical Engineering, University of Electronic Science and Technology of China, Chengdu 611731, China

K. B. Misra (ed.), *Handbook of Advanced Performability Engineering*,
https://doi.org/10.1007/978-3-030-55732-4_16

16.1 Introduction

Developing and releasing new products is an important source of revenue for any organization because it brings higher sales, increased customer loyalty, and ultimately higher profits (Li et al. [1]). On the other hand, the growth in the number and variety of new products has brought tough global competition to these organizations. Severe competition forces organizations to shorten development cycles and reduce development budget so that these organizations can stay ahead of the competition through an effective and efficient New Product Development (NPD) process (Murthy et al. [2]). The NPD process starts with a new concept for a product or a system followed by identifying and defining the product requirements in the detail design stage. Then, a small number of pilot and prototype products are built and tested for performance and function verification. Mass production then proceeds following the verified product design objectives and requirements. Product design Verification and Validation (V&V) is an integral part of the NPD process to verify and validate that the newly developed product meets its engineering specifications and fulfills its intended functions (Maropoulos et al. [3]). In general words, verification is a quality control process that is used to evaluate whether or not the newly developed product complies with a regulation, requirement, specification, or imposed conditions (Babuska et al. [4]). Validation, on the other hand, often involves acceptance and suitability with external customers. During the V&V process, various V&V activities are planned and executed to, respectively, mitigate the risk of specific potential failure modes of the products. Such V&V activities are engineering tasks for design risk assessment and mitigation such as engineering analysis and calculations, design simulations, and physical tests. In summary, Product design V&V can confirm that the developed product conforms to its intended function and specifications through mitigating the risk of potential failure modes, and ultimately improve product reliability (Maropoulos et al. [3]). In addition, it is estimated that more than half of the NPD costs comes from V&V process (Belt et al. [5]). Therefore, an optimal V&V planning, which is a set of V&V activities, can ensure the effectiveness and efficiency of the NPD process.

Despite decades of industrial experiences, it is found that designing and developing increasingly complex products, e.g. aerospace products, still incurs significant cost overruns, schedule delays, and quality/reliability issues (Collopy et al. [6]). Such product development challenges can be seen from both industry and government projects (Reuters [7]). For instance, the United States Department of Defense (DoD) development programs are mostly plagued with cost overruns and schedule delays (Schwenn et al. [8]). A study of 96 major new weapon systems development programs in the United States DoD reported that almost 50% of the DoD's major defense acquisition programs do not meet projected cost goals. In addition, 80% of programs have experienced an increase in unit costs from initial estimates (Schwenn et al. [9]). Findings from a study in the construction industry also indicate the issues of cost overrun and schedule delays (Potty et al. [10]). The above discussed issues can be attributed to the lack of effective quantitative methods for providing the optimum plan for executing the product design V&V activities.

A simple example of a V&V planning is shown in Fig. 16.1. It is observed that each failure mode can be improved by one or more available V&V activities. These V&V activities can be obtained from the Design Failure Mode and Effect Analysis (DFMEA) process. Meanwhile, most V&V activities can be applicable for tacking multiple different failure modes. In general, the execution sequence requirements, time consumption, economic consumption, and improvement effectiveness in reducing risk of each V&V activity are also different. In addition, to ensure the safety of the product, while minimizing the overall risk of the product, some critical products have additional risk mitigation requirements; for example, the risk of each failure mode after improvement cannot exceed the pre-specified risk thresholds. Therefore, the product complexity, multiple and possible common and random failure modes, and various V&V options along with the demanding design requirements and objectives call for a cost- and time-effective V&V activity plan which optimally covers all critical failure modes of the products. From the reviewed literature, quantitative methodologies have not been well explored for optimal planning of V&V activities. The challenges in designing an optimal V&V planning include (1) how to assign a set of V&V activities to cover different failure modes in order to have maximum reliability improvements and meet additional requirements of the product, (2) how to optimally allocate the budget to each V&V activity, (3) how to schedule the V&V activities considering their execution sequencing requirements, and (4) how to consider the effectiveness and decay of different V&V activities in the risk mitigation.

This chapter proposed four types of mathematical modeling approaches to optimize the product design V&V planning during the NPD process. The ultimate goal of the proposed approaches is to mitigate the risks of critical failure modes for maximizing the overall system reliability improvement by optimally selecting a set

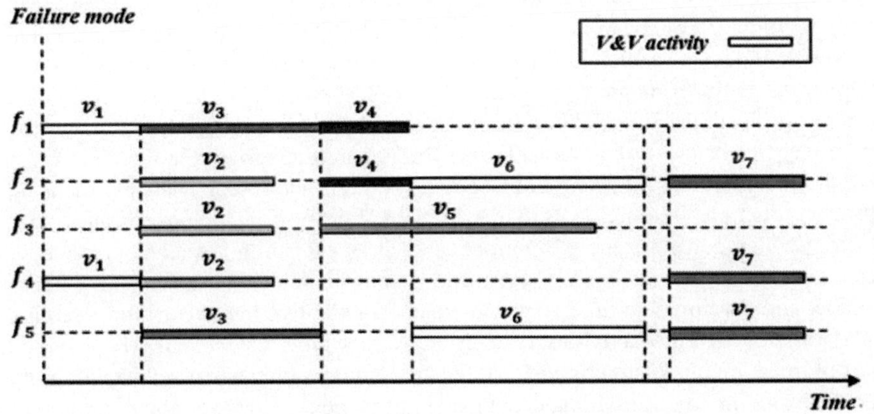

Fig. 16.1 A simple example of a sequence diagram of a verification and validation planning

of V&V activities. The remaining of the chapter is structured as follows. A literature review is presented in Sect. 16.2 which involves the novelties of this chapter. Section 16.3 formulates four mathematical V&V planning models by adopting the concepts of set covering, set partition, set packing, job shop scheduling, and reliability design risk. The application and results' comparison of these four types of mathematical optimization models in optimizing the product V&V planning are illustrated through the product development of a power generation system within a diesel engine in Sect. 16.4. In addition, the V&V planning results from the proposed approaches are also compared with three other existing product V&V planning approaches in the literature, such as Project Evaluation and Review Technique (PERT) (Bhattacharjee et al. [11]), cost-oriented, and time-oriented approaches, to show the advantages of the proposed V&V planning approaches. Section 16.5 concludes the chapter and discusses the future research directions.

16.2 Literature Review

Traditional V&V methods in modeling the NPD process such as Quality Function Deployment (QFD), Key Characteristics (KCs), new product functional decomposition, dimensional and shape verification, Design for X (DFX), PERT, and Design Structure Matrix (DSM) are mostly qualitative. With the improvement of customers' demand on function, product quality, and development cycle, the traditional product design V&V methods mentioned above can no longer meet the demand of customers. As a result, some researchers have improved the traditional V&V methods through different ways to adapt to the fierce market competition (Han et al. [12], Cho et al. [13], Estrada et al. [14]).

From another perspective, in order to provide guidance and strategies for the execution and optimization of V&V, the methods of simulation and modeling of V&V in the design of new complex engineering products/systems have also been widely studied. For example, Kukulies and Schmitt [15] proposed a conceptual approach of design verification planning based on uncertainty quantification to avoid unplanned engineering changes in the NPD process to improve the effectiveness of design verification activities. In addition, Chahin and Paetzold [16] have estimated the dependencies between requirements and product architecture in the model through the product maturity assessment. The above methods and research mainly focus on product functional requirements to optimize V&V process and do not consider other factors which constrain the NPD process, such as execution sequence and budget constraints. Mobin and Li [17] taken into account common constraints, including budget and development time, to propose a new qualitative framework for obtaining an optimal set of V&V activities. In addition, Ahmed and Chateauneuf [18] proposed an optimization model combining testing and design problems to meet the reliability goal considering the validation, design, and failure costs. However, the constraint of the sequencing of conducting V&V activities is lacked in above two models.

One of the very few studies in the literature that simultaneously considers product reliability, product development cycle and cost, and scheduling of conducting V&V activities is the model proposed by Mobin et al. [19]. In their research, they assumed that the risks of all identified failure modes need to be mitigated, and the set covering problem concept is applied to ensure that all failure modes are covered by the end of V&V execution. However, in the actual NPD process, low-risk failure modes may not be covered due to the limitations of development cycle and cost. In summary, the existing optimizing V&V planning methods are mostly qualitative, and the research that investigated the quantitative methods to model and optimize the V&V planning by considering the constraint of the sequencing of conducting V&V activities and the effectiveness of reducing risks is very limited. In addition, all of the existing studies in the literature do not consider different risk mitigation strategies under different requirements, the decay of the improvement effectiveness, as well as the special execution sequence requirements, such as the time gap during V&V process.

The contributions of this chapter are to formulate four types of mathematical optimization modeling approaches, including optimization models with set covering problem, set partition problem, and set packing problem, as well as an extended optimization model with set covering problem, to optimal design V&V planning. Different from the approaches in the literature, the proposed modeling approaches provide four risk mitigation strategies from perspectives of reliability and cost efficiency to cover different requirements of products. The proposed approaches not only incorporate all critical product development constraints in V&V planning, such as the cost, time, reliability improvement, as well as sequencing and effectiveness of V&V activities, but also incorporate the decay of the improvement effectiveness into the V&V activities' selecting and sequencing challenges. In addition, the concepts of set covering, set partition, and set packing are applied to assure that different levels of critical failure modes can be covered in different ways according to different risk mitigation requirements by the end of V&V execution.

16.3 Mathematical Modeling for V&V Planning

This section details the proposed four types of mathematical optimization modeling approaches which are formulated as integer programming problems to optimal design V&V planning, including optimization models with set covering problem, set partition problem, and set packing problem, as well as an extended optimization model with set covering problem. Most of the general additional risk mitigation requirements in addition to minimizing the overall risk of the product can be included in these four types of models.

16.3.1 Assumptions of the Proposed V&V Planning Optimization Formulations

The main assumptions of four proposed optimization modeling approaches for V&V planning, which are derived from the DFMEA handbook (Stamatis [20]) and the industry practices, are listed as follows:

(a) Each failure mode can be improved by one or more available V&V activities. Meanwhile, most V&V activities can be applicable for simultaneously improving multiple different failure modes without repeated economic and time consumption.

(b) V&V activities are divided into four categories, including design action, lab test, bench test, and performance test, of which only the design action can improve S. Each failure mode with S greater than seven has at least one corresponding design action.

(c) The improvement effectiveness of the j th V&V activity v_j on the i th failure mode f_i in *D*, *S*, and *O* are given and defined as θ_{ij}^D, θ_{ij}^S, and θ_{ij}^O, which represent the reduction percentage of failure mode f_i in *D*, *S*, and *O* by executing a V&V activity v_j, respectively.

(d) The risk reduction of all critical failure modes has additive effects. Meanwhile, it is assumed that multiple V&V activities can simultaneously improve *D*, *S*, or *O* for the same failure mode. The improvement effectiveness decay as the number of improvements increases. The overall improvement effectiveness of all executed V&V activity on the ith failure mode f_i in *D*, *S*, or *O* is subject to the exponential decay of base γ_D, γ_S, or γ_O, respectively ($0 < \gamma_D, \gamma_S, \gamma_O < 1$). More specifically, when a V&V activity v_j, which can improve the *D* of the failure mode f_i again, is executed, its improvement effectiveness in *D*, denoting as θ_{ij}^D, will decay to $\theta_{ij}^D * \gamma_D$. Therefore, when multiple such V&V activities are executed, the overall improvement effectiveness of all executed V&V activity on the i th failure mode f_i in D will decay to $\gamma_D^{N_i^D - 1}$ times the initial value, where N_i^D represents the number of times the *D* of failure mode f_i has been improved, which is similar for *S* and *O*.

(e) All V&V activities need to be executed following the given execution sequentially. Meanwhile, there is a necessary time gap between certain V&V activities. In addition, all V&V activities cannot be executed repeatedly.

16.3.2 Reliability Improvement Quantification

One goal of product design V&V is to increase the reliability of a product or a system by mitigating the critical failure modes' risks, which can be measured by the Risk Priority Number (RPN) of identified failure modes. The Reliability Improvement Index (RII) for each failure mode i, denoted as RII_i, which is defined as the ratio of the reduced value to the initial value of the RPN of the failure mode i, is adopted to quantify the overall improvement effectiveness of all executed V&V activities on the i th failure mode f_i (Barends et al. [21]). Note that a higher value of RII implies a larger reliability improvement. The mathematical formula of RII_i can be expressed as in Eq. (16.1):

$$\text{RII}_i = \frac{\text{RPN}_{i(\text{initial})} - \text{RPN}_{i(\text{new})}}{\text{RPN}_{i(\text{initial})}}, \tag{16.1}$$

where $\text{RPN}_{i(\text{initial})}$ and $\text{RPN}_{i(\text{new})}$ are the RPN of the failure mode f_i before and after executing selected V&V activities, respectively. RPN is the product of D, O, and S. Therefore, $\text{RPN}_{i(\text{initial})}$ and $\text{RPN}_{i(\text{new})}$ can be, respectively, presented as in Eqs. (16.2) and (16.3):

$$\text{RPN}_{i(\text{initial})} = D_{i(\text{initial})} * O_{i(\text{initial})} * S_{i(\text{initial})}, \tag{16.2}$$

$$\text{RPN}_{i(\text{new})} = D_{i(\text{new})} * O_{i(\text{new})} * S_{i(\text{new})}, \tag{16.3}$$

where $D_{i(\text{initial})}$, $D_{i(\text{new})}$, $O_{i(\text{initial})}$, $O_{i(\text{new})}$, $S_{i(\text{initial})}$, and $S_{i(\text{new})}$ are D, O, and S of the failure mode f_i before and after executing selected V&V activities, respectively. $D_{i(\text{new})}$, $O_{i(\text{new})}$, and $S_{i(\text{new})}$ can be, respectively, formulated as

$$D_{i(\text{new})} = D_{i(\text{initial})} * \prod_{j=1}^{m} \left[\left(1 - \theta_{ij}^{D} * u_j\right) * \gamma^{N_i^D - 1} \right], \tag{16.4}$$

$$O_{i(\text{new})} = O_{i(\text{initial})} * \prod_{j=1}^{m} \left[\left(1 - \theta_{ij}^{O} * u_j\right) * \gamma^{N_i^O - 1} \right], \tag{16.5}$$

$$S_{i(\text{new})} = S_{i(\text{initial})} * \prod_{j=1}^{m} \left[\left(1 - \theta_{ij}^{S} * u_j\right) * \gamma^{N_i^S - 1} \right], \tag{16.6}$$

where m is the number of V&V activities, i.e. $j = 1, 2, \cdots, m$. u_j is a binary decision variable, indicating whether the j th V&V activity v_j is selected, i.e. $u_j = 1$, if selected; $u_j = 0$, otherwise.N_i^D, N_i^O, and N_i^S represent the number of times the D, O, and S of the failure mode f_i has been improved, respectively. N_i^D, N_i^O, and N_i^S can be calculated using Eq. (16.7):

$$N_i^D = \sum_{j=1}^{m} a_{ij}^D * u_j,\ N_i^O = \sum_{j=1}^{m} a_{ij}^O * u_j,\ N_i^S = \sum_{j=1}^{m} a_{ij}^S * u_j, \tag{16.7}$$

Fig. 16.2 An example of the incidence matrix of D, i.e. a_{ij}^{D}, based on Data of Fig. 16.1

$$
\begin{array}{c}
 \\ f_1 \\ f_2 \\ f_3 \\ f_4 \\ f_5
\end{array}
\begin{array}{c}
\begin{matrix} v_1 & v_2 & v_3 & v_4 & v_5 & v_6 & v_7 \end{matrix} \\
\begin{pmatrix}
1 & 0 & 1 & 1 & 0 & 0 & 0 \\
0 & 1 & 0 & 1 & 0 & 1 & 1 \\
0 & 1 & 0 & 0 & 1 & 0 & 0 \\
1 & 1 & 0 & 0 & 0 & 0 & 1 \\
0 & 0 & 1 & 0 & 0 & 1 & 1
\end{pmatrix}
\end{array}
$$

where a_{ij}^{D}, a_{ij}^{O}, and a_{ij}^{s} are incidence matrices. The elements in the incidence matrix indicate whether the j th V&V activity v_j can improve the D, S, or O of the ith failure mode f_i, i.e. element is equal to one, if can improve; element is equal to zero, otherwise. Figure 16.2 illustrates an example of the incidence matrix of D, i.e. a_{ij}^{D}, using the data from Fig. 16.1, assuming that all these seven V&V activities can improve the D of the corresponding failure modes.

As mentioned, increasing the reliability of a product or a system by mitigating the crucial failure modes' risks is one goal of product design V&V. Based on Eqs. (16.1) to (16.6), the overall reliability improvement effectiveness of all failure modes after executing all selected V&V activities, denoted as $\mathrm{RII}_{\mathrm{Total}}$, can be calculated as follows:

$$
\begin{aligned}
\mathrm{RII}_{\mathrm{Total}} &= \sum_{i=1}^{n} \mathrm{RII}_i = \sum_{i=1}^{n} \frac{\mathrm{RPN}_{i(\mathrm{initial})} - \mathrm{RPN}_{i(\mathrm{new})}}{\mathrm{RPN}_{i(\mathrm{initial})}} \\
&= \sum_{i=1}^{n} \Bigg\{ 1 - \prod_{j=1}^{m} \big[\left(1 - \theta_{ij(D)} u_j\right) * \left(1 - \theta_{ij(O)} u_j\right) * \left(1 - \theta_{ij(S)} u_j\right) \big], \\
&\quad \gamma^{N_i^D - 1} \gamma^{N_i^O - 1} \gamma^{N_i^S - 1} \Bigg\},
\end{aligned}
\tag{16.8}
$$

where n is the total number of failure modes f_i.

16.3.3 V&V Sequencing Modeling Using Job Shop Scheduling Formulation

In this subsection, the job shop scheduling concept and its application in modeling the precedence constraints in planning V&V activities are presented. The goal of a job shop scheduling problem is to find an optimal schedule for a given collection of jobs (i) where each requires a known sequence of processors (j) that can accommodate one job at a time. Suppose that the processing times are given as t_{ij}, which represent the processing time of job i on the processor j. The typical decision variables for a job shop scheduling problem are s_{ij} representing the start time of job i on the processor j. The objective function can be to minimize the makespan, i.e. minimize the completion time of the last job. The precedence requirement that job i must complete processing on processor j before starting on processor j' can be expressed as $s_{ij} + t_{ij} < s_{ij'}$. To assure that jobs are not scheduled simultaneously on the same processor, this conflict constraint can be added to the model (Applegate et al. [22]).

In the V&V activity planning, the failure modes can be considered as jobs and V&V activities can be considered as processors. Each failure mode (job) can be mitigated by a sequence of V&V activities (processors). Since failure modes can be mitigated simultaneously when a certain V&V activity is executed, the precedence constraint can be relaxed to include equality such that multiple V&V activities can be executed simultaneously. The execution time of all V&V activities, i.e. the makespan of V&V process, should be minimized. The equivalent job shop scheduling objective function for the V&V activity planning can be mathematically modeled as in Eq. (16.9):

$$\text{Min:}[\text{Max}\{(s_j + t_j) * u_j\},\ \forall j = 1, 2, \cdots, m], \tag{16.9}$$

where s_j represents the start time of the jth V&V activity v_j. t_j is the duration of the jth V&V activity v_j. Equation (16.9) first finds the maximum completion time of V&V activities considering all failure modes ($i = 1, 2, \cdots, n$). The maximum completion time is also known as the makespan. Then, the makespan of V&V process is minimized for all V&V activities ($j = 1, 2, \cdots, m$).

Since the total time for V&V activities implementation (T_0) is limited, and all V&V activities should be executed under the time constraint, the objective function in Eq. (16.9) can be converted to a constraint presented as in Eq. (16.10):

$$\text{Max}\{(s_j + t_j)u_j\} \le T_0,\ \forall j = 1, 2, \cdots, m. \tag{16.10}$$

The start time of each V&V activity in the selected set of V&V activities, s_j, can be calculated by recursion in turn according to Eq. (16.11):

$$\begin{array}{c} \\ v_1 \\ v_2 \\ v_3 \\ v_4 \\ v_5 \\ v_6 \\ v_7 \end{array}\begin{array}{c} \begin{array}{ccccccc} v_1 & v_2 & v_3 & v_4 & v_5 & v_6 & v_7 \end{array} \\ \left(\begin{array}{ccccccc} - & 1 & 1 & 1 & 1 & 1 & 1 \\ - & - & 0 & 1 & 1 & 1 & 1 \\ - & - & - & 1 & 1 & 1 & 1 \\ - & - & - & - & 0 & 1 & 1 \\ - & - & - & - & - & 0 & 1 \\ - & - & - & - & - & - & 1 \\ - & - & - & - & - & - & - \end{array}\right) \end{array}$$

(a) The Upper Triangular Binary Matrix, $k_{l.j}$, supposing that only v_2 and v_3, v_4 and v_5, as well as v_5 and v_6 can be executed at the same time in this example.

$$\begin{array}{c} \\ v_1 \\ v_2 \\ v_3 \\ v_4 \\ v_5 \\ v_6 \\ v_7 \end{array}\begin{array}{c} \begin{array}{ccccccc} v_1 & v_2 & v_3 & v_4 & v_5 & v_6 & v_7 \end{array} \\ \left(\begin{array}{ccccccc} - & 0 & 0 & 0 & 0 & 0 & 0 \\ - & - & 0 & 0 & 0 & 0 & 0 \\ - & - & - & 0 & 0 & 0 & 0 \\ - & - & - & - & 0 & 0 & 0 \\ - & - & - & - & - & 0 & 0 \\ - & - & - & - & - & - & \Delta t \\ - & - & - & - & - & - & - \end{array}\right) \end{array}$$

(b) The Upper Triangular Matrix, $\Delta t_{l.j}$, supposing that only v_6 and v_7 cannot be executed continuously and supposing there is a time gap Δt in this example

Fig. 16.3 An example of the upper triangular binary matrix, $k_{l.j}$, and the upper triangular matrix, $\Delta t_{l.j}$, using the data of Fig. 16.1

$$s_j = \begin{cases} 0, & \text{if } j = 1 \\ \text{Max}\left\{(\Delta t_{j-l,j} + s_{j-l} + t_{j-l} k_{j-l,j}) u_{j-l}\right\},\ l = 1, 2, \cdots, j-1, & \text{if } j \neq 1 \end{cases} \tag{16.11}$$

where $k_{l.j}$ is an upper triangular binary matrix ($l < j$); the elements in this matrix indicate whether the lth V&V activity v_l and the jth V&V activity v_j can be executed at the same time, i.e. element is equal to one, if can be executed at the same time; element is equal to zero, otherwise. Figure 16.3a illustrates an example of the upper triangular binary matrix, $k_{l.j}$, using the data of Fig. 16.1, assuming that only v_2 and v_3, v_4 and v_5 as well as v_5 and v_6 can be executed at the same time in this example. $\Delta t_{l.j}$ is an upper triangular matrix ($l < j$); the elements in this matrix represent the time gap between the end of the lth V&V activity and the start of the jth V&V activity. Certain specific V&V activities cannot be executed continuously and need to wait for a specific time gap; such time gap may be the transfer of materials or devices, the cooling of devices, and the removal of internal stress in the materials, such as v_6 and v_7 in Fig. 16.1. Figure 16.3b illustrates an example of the upper triangular matrix, $\Delta t_{l.j}$, using the data of Fig. 16.1, assuming that only v_6 and v_7 cannot be executed continuously and supposing there is a time gap Δt in this example.

16.3.4 Failure Modes Coverage Modeling

In this subsection, the concepts of set covering, set partition, set packing and their applications in planning V&V activities are presented, respectively. The applications of these three concepts can cover different levels of critical failure modes in various ways from perspectives of reliability and cost efficiency.

16.3.4.1 Set Covering Formulation

A general set covering problem can be described as follows: given a universe U and a family G of subsets of U, a set covering is a subfamily $W \subseteq G$ of sets whose union is U (Vazirani [23]). In the set covering optimization problem, the input is a pair (U, G), and the task is to find a set covering with the minimum weight. The set covering optimization problem can be formulated as the integer linear program, which is shown as in Eqs. (16.12) and (16.13):

$$\text{Min:} \sum_{g \in G} \omega_g x_g \tag{16.12}$$

Subject to

$$\sum_{g: e \in G} x_g \geq 1, \ \forall e \in U, \tag{16.13}$$

where g represents one of the subsets in the family G $(g \in G)$. ω_g denotes the weight of each subset g. e represents one of the elements in the universe U. x_g is a binary decision variable, indicating whether each subset g is included in the subfamily W, i.e. $x_g = 1$, if included; $x_g = 0$, otherwise. The objective function, Eq. (16.12), is formulated to find a subfamily W with the minimum weight. Equation (16.13) ensures every element e in the universe U is included at least one time in the subfamily W.

In the V&V planning problem, all identified critical failure modes usually need to be improved at least one time. Meanwhile, the time and economic budget during the product development process are limited. Therefore, the objective function, Eq. (16.12), in the set covering optimization problem can be considered as a budget constraint which can find a set of V&V activities with the cost less than the budget in the V&V planning problem. This budget constraint is presented as in Eq. (16.14):

$$\sum_{j=1}^{m} c_j u_j \leq C_0, \tag{16.14}$$

where c_j represents the economic consumption of each V&V activity v_j. C_0 is the total economic budget constraint for V&V execution.

Then, the constraint, Eq. (16.13), in the set covering optimization problem can be considered as the constraint which can make each failure mode to be improved at least one time. This constraint of covering each failure mode is formulated as in Eq. (16.15):

$$\sum_{j=1}^{m} a_{ij}^{V} u_j \geq 1, \ \forall i = 1, 2, \cdots, n, \tag{16.15}$$

where a_{ij}^{V} is an incidence matrix, and the elements in this incidence matrix indicate whether the jth V&V activity v_j can improve the ith failure mode f_i, i.e. element is equal to one, if can improve; element is equal to zero, otherwise.

16.3.4.2 Set Partition Formulation

A general set partition problem can be described as follows: given a universe U and a family G of subsets of U, a set partition is a subfamily $W \subseteq G$ of sets whose union is U; meanwhile, all sets in W are pairwise disjoint, i.e. every element in the universe U is included in one and only one of the subsets of the subfamily W (Vazirani [23]). In the set partition optimization problem, the input is a pair (U, G), and the task is to find a set partition with the minimum weight. The set partition optimization problem can be formulated as the integer linear program, which is shown as in Eqs. (16.12) and (16.16):

$$\text{Min:} \sum_{g \in G} \omega_g x_g \tag{16.12}$$

Subject to

$$\sum_{g: e \in G} x_g = 1, \ \forall e \in U, \tag{16.16}$$

where the objective function, Eq. (16.12), is also formulated to find a subfamily W with the minimum weight. Equation (16.16) ensures every element e in the universe U is included one and only one time in the subfamily W.

In the set covering formulation, although the overall improvement effectiveness of all failure modes after executing all selected V&V activities, i.e. $\text{RII}_{\text{Total}}$, can be improved by multiple improvements to one failure mode through multiple V&V activities, as mentioned in assumption (d) in subsection 16.3.1, when multiple V&V activities improve the same failure mode, the improvement effectiveness decay as the number of improvements increases. In some cases, the increased economic and time consumption tends to be worth more than the increased overall improvement effectiveness, that is, the Cost Efficiency (CE) (Farrell [24]) of multiple improvements is lower than which of single improvement. CE can evaluate the producing ability of the current output at minimal cost, given its input cost. In the other words, CE is interpreted as a measure of evaluating whether production is executed at the lowest cost. According to the research proposed by Mirdehghan et al., [25] CE can be calculated as the ratio of average minimum cost, denoted as $\bar{C}_{\min}$, to average observed cost, denoted as $\bar{C}_{\text{observed}}$.

$$CE = \frac{\overline{C}_{\min}}{\overline{C}_{\text{observed}}}. \tag{16.17}$$

In the application in planning V&V activities, the output is represented by the average improvement effectiveness $\overline{\text{RII}}$ of each failure mode after executing all selected V&V activities.

$$\overline{\text{RII}} = \frac{\text{RII}_{\text{Total}}}{n}. \tag{16.18}$$

The input cost includes two aspects in planning V&V activities, including time and economic cost. In order to better evaluate input cost, a Time and Economic Metric, denoted as TE, is proposed to integrate time and economic cost, which is shown as in Eq. (16.19):

$$TE = T_{\text{Total}} + \alpha C_{\text{Total}}, \tag{16.19}$$

where T_{Total} is the total time consumption of a selected set of V&V activities, C_{Total} is the total economic consumption of a selected set of V&V activities, and α is a coefficient used to scale the ratio and unit of economic and time measurements. Then, the average observed cost, $\bar{C}_{\text{observed}}$, in the application in planning V&V activities can be calculated.

$$\overline{C}_{\text{observed}} = \frac{TE}{\overline{\text{RII}}} = \frac{n(T_{\text{Total}} + \alpha C_{\text{Total}})}{\text{RII}_{\text{Total}}}. \tag{16.20}$$

In order to avoid the decay of improvement effectiveness due to the multiple improvements, i.e. improve CE, each failure mode should be improved by one and only one V&V activity. Then, the constraint, Eq. (16.16), in the set partition optimization problem can be considered as the constraint which can make each failure mode to be improved by one and only one time.

$$\sum_{j=1}^{m} a_{ij}^{V} u_j = 1, \ \forall i = 1, 2, \cdots, n. \tag{16.21}$$

16.3.4.3 Set Packing Formulation

A general set packing problem can be described as follows: given a universe U and a family G of subsets of U, a set packing is a subfamily $W \subseteq G$ of sets such that all sets in W are pairwise disjoint; in other word, every element in the universe U is included in at most one subset of the subfamily W (Vazirani [23]). Therefore, set partition can be regarded as a special case of set packing. In the set packing optimization problem, the input is a pair (U, G), and the task is to find a set packing with the minimum weight. The set packing optimization problem can also be formulated as the integer linear program, which is shown as in Eqs. (16.12) and (16.22):

$$\text{Min: } \sum_{g \in G} \omega_g x_g \tag{16.12}$$

Subject to

$$\sum_{g: e \in G} x_g \leq 1, \ \forall e \in U, \tag{16.22}$$

where the objective function, Eq. (16.12), is also formulated to find a subfamily W with the minimum weight. Equation (16.22) ensures every element e in the universe U is included at most one time in the subfamily W.

In the set partition formulation, there is not always a set of V&V activities that can improve each failure mode just once. Sometimes, in pursuit of overall improvement effectiveness of the product development process under limited time and budget, some failure modes may not be improved. For instance, the initial RPN of some failure modes is low, and the improvement rate of RPN obtained by improving these failure modes is also usually low. In this case, the focus of improvement can be shifted to the failure modes with high RPN, so as to obtain a higher overall improvement effectiveness of the product development process. Therefore, each failure mode should be improved by at most one V&V activity, and the constraint, Eq. (16.22), in the set packing optimization problem can be considered as the constraint which can make each failure mode to be improved at most one time.

$$\sum_{j=1}^{m} a_{ij}^{V} u_j \leq 1, \ \forall i = 1, 2, \cdots, n. \tag{16.23}$$

16.3.5 *Four Types of Mathematical Models for the V&V Activity Planning*

Based on the concepts of set covering concepts in planning V&V activities, four types of modeling approaches are proposed to investigate four risk mitigation strategies to understand different product developing requirements.

16.3.5.1 Model I. Optimization Model with Set Covering Formulation

In the scenario where all identified failure modes need to be improved and each failure mode can be improved by multiple V&V activities, set covering formulation can be used, which is formulated as follows. The objective function, Eq. (16.24), is formulated to maximize the overall reliability improvement effectiveness of all failure modes after executing all selected V&V activities:

$$\text{Max}: \text{RII}_{\text{Total}} = \sum_{i=1}^{n} \left\{ 1 - \prod_{j=1}^{m} \left[\left(1 - \theta_{ij(D)} u_j\right) * \left(1 - \theta_{ij(O)} u_j\right) * \left(1 - \theta_{ij(S)} u_j\right) \right] \gamma^{N_i^D - 1} \gamma^{N_i^O - 1} \gamma^{N_i^S - 1} \right\} \tag{16.24}$$

Subject to

$$\text{Max}\{(s_j + t_j) u_j\} \le T_0, \ \forall j = 1, 2, \cdots, m, \tag{16.10}$$

$$\sum_{j=1}^{m} c_j u_j \le C_0, \tag{16.14}$$

$$\sum_{j=1}^{m} a_{ij}^{V} u_j \ge 1, \ \forall i = 1, 2, \cdots, n, \tag{16.15}$$

where $N_i^D = \sum_{j=1}^{m} a_{ij}^D u_j$, $N_i^O = \sum_{j=1}^{m} a_{ij}^O u_j$, and $N_i^S = \sum_{j=1}^{m} a_{ij}^S u_j$. $s_j = 0$, if $j = 1$; $s_j = \text{Max}\{(\Delta t_{j-l,j} + s_{j-l} + t_{j-l} k_{j-l,j}) u_{j-l}\}$, $\forall l = 1, 2, \cdots, j-1$, otherwise. u_j is a binary decision variable, indicating whether the jth V&V activity v_j is selected, i.e. $u_j = 1$, if selected; $u_j = 0$, otherwise. The first constraint, Eq. (16.10), guarantees that the total time consumption of the optimal set of V&V activities should be less than the maximum expected V&V execution time. The second constraint, Eq. (16.14), is that the total cost for executing the optimal set of V&V activities should be less than the budget for the product development process. The third constraint, Eq. (16.15), confirms that each failure mode is included by at least one V&V activity through executing the optimal set of V&V activities.

16.3.5.2 Model II. Extended Optimization Model with Set Covering Formulation

The goal of Model I is to maximize the overall reliability improvement effectiveness of all failure modes. However, in practice, even if the overall improvement effectiveness is maximized after executing all selected V&V activities, certain high-risk failure modes can also result in high frequency and hard-to-detect failures (e.g. failure modes with high O and D) of the product, or even loss of life (e.g. failure modes with high S). Hence, under the sufficient production cycle and cost, the improved RPN and S of each failure mode are usually required to meet the minimum improvement requirements to ensure that there is no reliability defect in the newly developed product. That is, the RPN and S of the failure mode f_i after executing selected V&V activities, i.e. $\text{RPN}_{i(\text{new})}$ and $S_{i(\text{new})}$, should not be greater than the pre-specified risk thresholds. $\text{RPN}_{i(\text{new})}$ and $S_{i(\text{new})}$ can be calculated by Eqs. (16.3) and (16.3), respectively. Then, the constraints of minimum improvement requirements of the improved RPN and S of each failure mode can be presented in Eqs. (16.25) and (16.26), respectively:

$$\text{RPN}_{i(\text{new})} \leq \text{RPN}_0, \ \forall i = 1, 2, \cdots, n, \tag{16.25}$$

$$S_{i(\text{new})} \leq S_0, \ \forall i = 1, 2, \cdots, n, \tag{16.26}$$

where RPN_0 and S_0 are the pre-specified risk thresholds of RPN and S, respectively.

The extended optimization model with set covering problem for optimizing V&V activity planning sacrifices some of the overall reliability improvement effectiveness to ensure that the risk of each failure mode of a product is within a safe range. Model II can be formulated as Eqs. (16.10), (16.14), (16.15), and (16.24), and add two constraints, Eqs. (16.25) and (16.26) to ensure that RPN and S of each failure mode after executing all selected V&V activities are not greater than the pre-specified risk thresholds.

16.3.5.3 Model III. Optimization Model with Set Partition Formulation

The optimization model with set partition problem not only ensures that the risk of each failure mode is mitigated, but also focuses on the cost efficiency (i.e. CE) of the NPD process. More specifically, this model limits each failure mode to be improved by one and only one V&V activity for avoiding the decay of improvement effectiveness due to the multiple improvements. Then, the overall reliability improvement effectiveness of all failure modes after executing all selected V&V activities, $\text{RII}_{\text{Total}}$, can be reduced to $\text{RII}^*_{\text{Total}}$.

$$\text{RII}^*_{\text{Total}} = \sum_{i=1}^{n} \left\{ 1 - \prod_{j=1}^{m} \left[\left(1 - \theta_{ij(D)} u_j\right) * \left(1 - \theta_{ij(O)} u_j\right) * \left(1 - \theta_{ij(S)} u_j\right) \right] \right\}. \tag{16.27}$$

The optimization model with set partition problem for optimizing V&V planning is presented in the following. The objective function, Eq. (16.28), is also formulated to maximize the overall reliability improvement effectiveness of all failure modes after executing all selected V&V activities:

$$\text{Max}: \text{RII}^*_{\text{Total}} = \sum_{i=1}^{n} \left\{ 1 - \prod_{j=1}^{m} \left[\left(1 - \theta_{ij(D)} u_j\right) * \left(1 - \theta_{ij(O)} u_j\right) * \left(1 - \theta_{ij(S)} u_j\right) \right] \right\} \tag{16.28}$$

Subject to

$$\text{Max}\{(s_j + t_j)u_j\} \le T_0,\ \forall j = 1, 2, \cdots, m, \tag{16.10}$$

$$\sum_{j=1}^{m} c_j u_j \le C_0, \tag{16.14}$$

$$\sum_{j=1}^{m} a_{ij}^{V} u_j = 1,\ \forall i = 1, 2, \cdots, n, \tag{16.21}$$

where the third constraint, Eq. (16.21), confirms that each failure mode is covered by one and only one V&V activity through executing the optimal set of V&V activities.

16.3.5.4 Model IV. Optimization Model with Set Packing Formulation

Compared with the optimization model with set packing problem (i.e. Model III), the optimization model with set packing problem (i.e. Model IV) focuses more on the cost efficiency (i.e. CE) of the NPD process. More specifically, this model gives up the improvement of some low-risk failure modes (described in subsection 16.3.4.3) to pursue a higher overall reliability improvement effectiveness, and the decay of improvement effectiveness due to the multiple improvements is also avoided in the V&V process. Then, Model IV can be presented in the following:

$$\text{Max}: \text{RII}^*_{\text{Total}} = \sum_{i=1}^{n} \left\{ 1 - \prod_{j=1}^{m} \left[\left(1 - \theta_{ij(D)} u_j\right) * \left(1 - \theta_{ij(O)} u_j\right) * \left(1 - \theta_{ij(S)} u_j\right) \right] \right\} \tag{16.28}$$

Subject to

$$\text{Max}\{(s_j + t_j)u_j\} \le T_0,\ \forall j = 1, 2, \cdots, m, \tag{16.10}$$

$$\sum_{j=1}^{m} c_j u_j \le C_0, \tag{16.14}$$

$$\sum_{j=1}^{m} a_{ij}^{V} u_j \leq 1, \ \forall i = 1, 2, \cdots, n, \tag{16.23}$$

where the third constraint, Eq. (16.23), ensures that each failure mode is included by at most one V&V activity through executing the optimal set of V&V activities.

In an NPD process, the large number of V&V activities results in the high dimension of decision variable u_j. Hence, the above four types of mathematical optimization models are the integer mixed optimization problem with high dimension. Genetic Algorithm (GA) (Mitchell [26]) which is commonly used to generate high-quality solutions to optimization and search problems by relying on bioinspired operators such as mutation, crossover, and selection is applied to solve above four optimization models.

16.4 Examples of V&V Planning for Optimal Reliability Improvement

In this section, the applications of proposed four types of mathematical optimization models are illustrated according to a case of a power generation system in a diesel engine; meanwhile, the results of these four types of mathematical optimization models as well as three other existing product V&V planning approaches in the literature are compared.

16.4.1 The Introduction of the Power Generation Unit

The diesel engine is a crucial part of the diesel generator, and the diesel engine is a device that converts the internal energy produced by burning diesel into mechanical energy.

Fifteen V&V activities (i.e. $m = 15$) from the product V&V process of the diesel engine are extracted as the numerical example of this chapter. These 15 V&V activities are divided into design action (i.e. v_3, v_8, and v_9), performance test (i.e. v_4, v_5, v_{10}, v_{11}, and v_{14}), lab test (i.e. v_6, v_{12}, and v_{13}), and bench test (i.e. v_1, v_2, v_7, and v_{15}). The combination of these 15 V&V activities can mitigate the design risks of 25 failure modes (i.e. $n = 25$). Recall that only the design action can improve S. In addition, the design actions usually consume the most economic resources and time, while the resource consumption of other types of V&V activities decreases in the order of performance test, lab test, and bench test. The effectiveness of various types of V&V activities are also different. The required input for the proposed optimization models, including the economy consumption (c_j) and time consumption (t_j) of each V&V activity, the upper triangular matrix $\Delta t_{l,j}$ ($l < j$) and the incidence matrices

a_{ij}^V, a_{ij}^D, a_{ij}^O, and a_{ij}^S of these 15 V&V activities, the improvement effectiveness of the j th V&V activity v_j on the i th failure mode f_i in D (θ_{ij}^D), O (θ_{ij}^O), and S (θ_{ij}^S), as well as the initial RPN $\left(\text{RPN}_{i(\text{initial})}\right)$ and S $\left(S_{i(\text{initial})}\right)$ of each failure mode f_i, can be found in our paper (Wu et al. [27]).

The total economic budget constraint for V&V execution is assumed to be \$600,000 (i.e. $C_0 = 600,000$), and the available time budget of executing the V&V process is constrained to be 600 days (i.e. $T_0 = 600$). The base of the exponential decay in D (γ_D), O (γ_O), and S (γ_S) are all assumed to be 0.9. For the extended optimization model with set covering problem, the improved RPN ($\text{RPN}_{i(\text{new})}$) of each potential failure mode f_i is assumed to be not greater than 60 (i.e. $\text{RPN}_0 = 60$), and the improved S ($S_{i(\text{new})}$) of each potential failure mode f_i is assumed to be not greater than 7 (i.e. $S_0 = 7$).

16.4.2 Results Using Model I to IV in Sect. 16.3.5

The schematic view of the optimal V&V activities plan using Model I is presented in Fig. 16.4, it can be seen that ten V&V activities, including v_1, v_3, v_4, v_5, v_7, v_9, v_{10}, v_{13}, v_{14}, and v_{15}, out of the proposed fifteen V&V activities are selected for execution to optimally mitigate all critical failure modes. The objective function value is obtained as $\text{RII}_{\text{Total}} = 13.77$. Total cost of executing the selected ten V&V activities is \$595,000, and the total execution time is obtained as 550 days.

Figure 16.5 shows that ten V&V activities, including v_1, v_2, v_3, v_4, v_7, v_8, v_9, v_{13}, v_{14}, and v_{15}, out of the proposed fifteen V&V activities are selected for execution to optimally mitigate all critical failure modes when Model II is used for optimizing V&V activities planning. The objective function value is obtained as $\text{RII}_{\text{Total}} = 13.37$. Total cost of executing the selected ten V&V activities is \$589,000, and the total execution time is obtained as 599 days.

The schematic views of the optimal V&V activities plan using Models III and IV are presented in Fig. 16.6a,b, respectively. It is observed through Fig. 16.6a that five V&V activities, including v_2, v_3, v_5, v_6, and v_{11}, out of the proposed fifteen V&V activities are selected for execution to optimally mitigate all critical failure modes, and each failure mode is improved exactly once. The objective function value is obtained as $\text{RII}_{\text{Total}} = 10.98$. Total cost of executing the selected ten V&V activities is \$321,000, and the total execution time is obtained as 325 days. In addition, the results in Fig. 16.6b show that four V&V activities, including v_3, v_4, v_9, and v_{12}, out of the proposed fifteen V&V activities are selected for execution to optimally mitigate all critical failure modes except f_{24} (whose initial RPN is equal to 18). The objective function value is obtained as $\text{RII}_{\text{Total}} = 12.10$. Total cost of executing the selected four V&V activities is \$301,000, and the total execution time is obtained as 314 days.

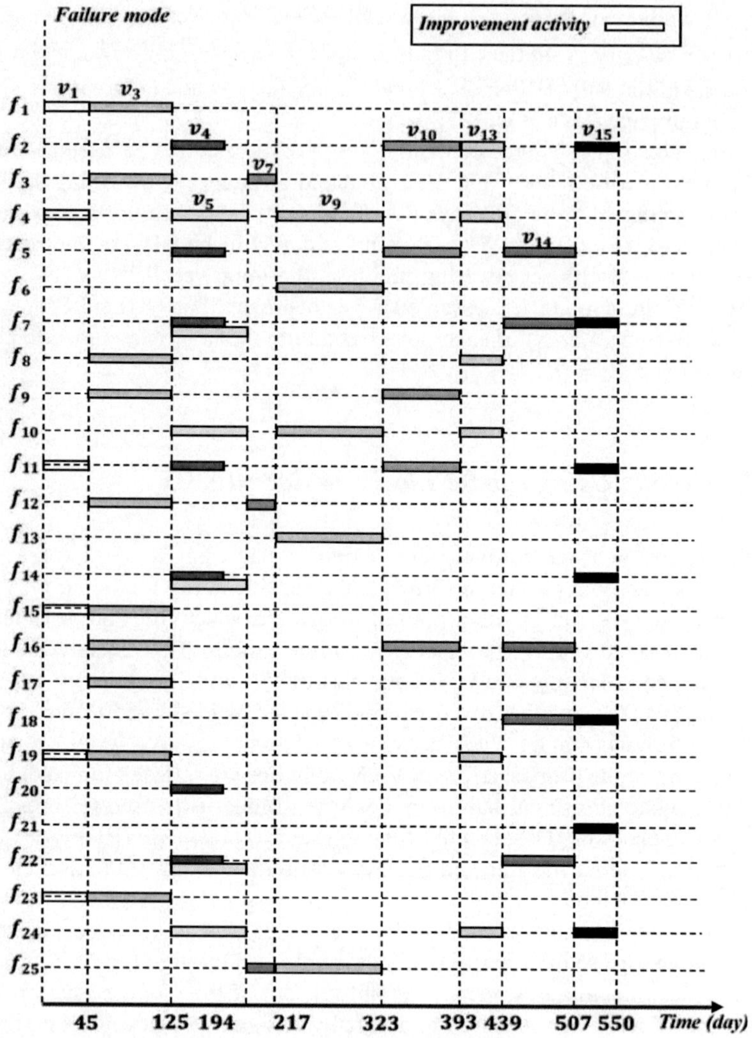

Fig. 16.4 The schematic view of V&V planning obtained from model I

16.4.3 Comparisons of Various V&V Modeling Approaches

The proposed four types of V&V activities planning model for maximum reliability improvement in NPD process are the first quantitative approaches considering the challenges of failure mode coverage, effectiveness of V&V activities, decay of effectiveness, scheduling, and budget and time constraints. Other quantitative approaches in the V&V planning literature, such as PERT, capture some aspects of V&V activities planning such as scheduling and budget, but these approaches lack in modeling

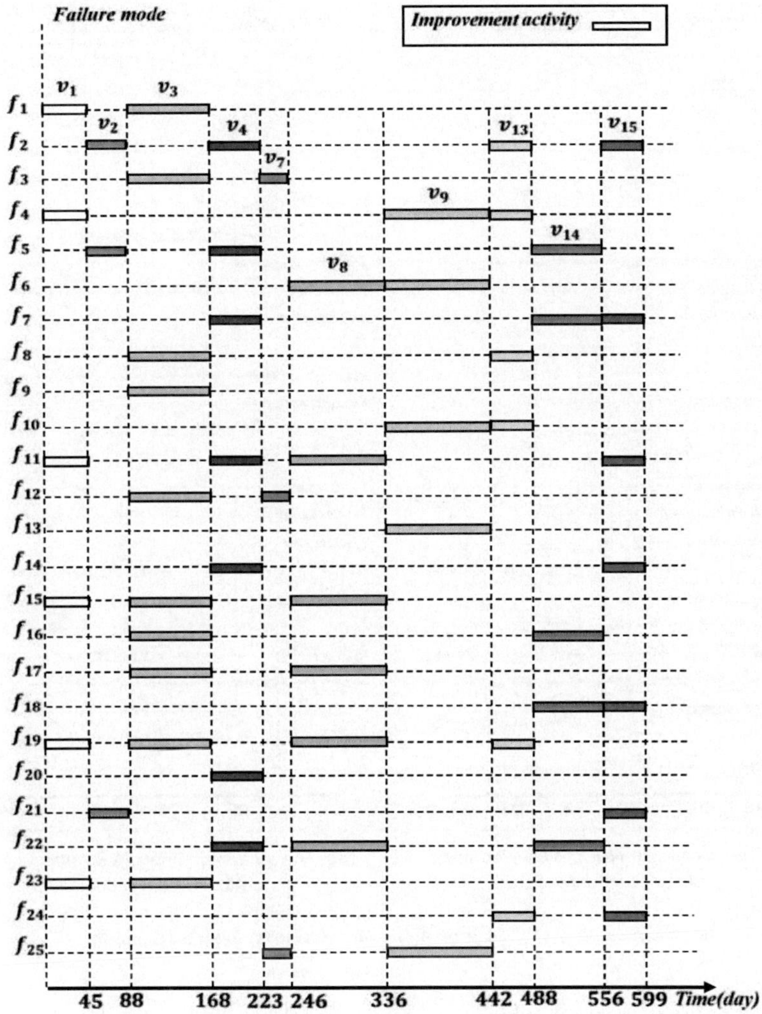

Fig. 16.5 The schematic view of V&V planning obtained from model II

the reliability improvement and prioritizing failure risks as well as covering critical failure modes. In this subsection, three approaches in V&V activities planning are applied to the case problem, and the results are compared with the proposed four types of V&V planning models in this chapter. These three approaches include (1) PERT approach in which only sequencing and scheduling of V&V activities are considered, (2) Cost-oriented V&V planning approach in which the total cost of V&V activities execution is used as the objective function, and (3) Time-oriented

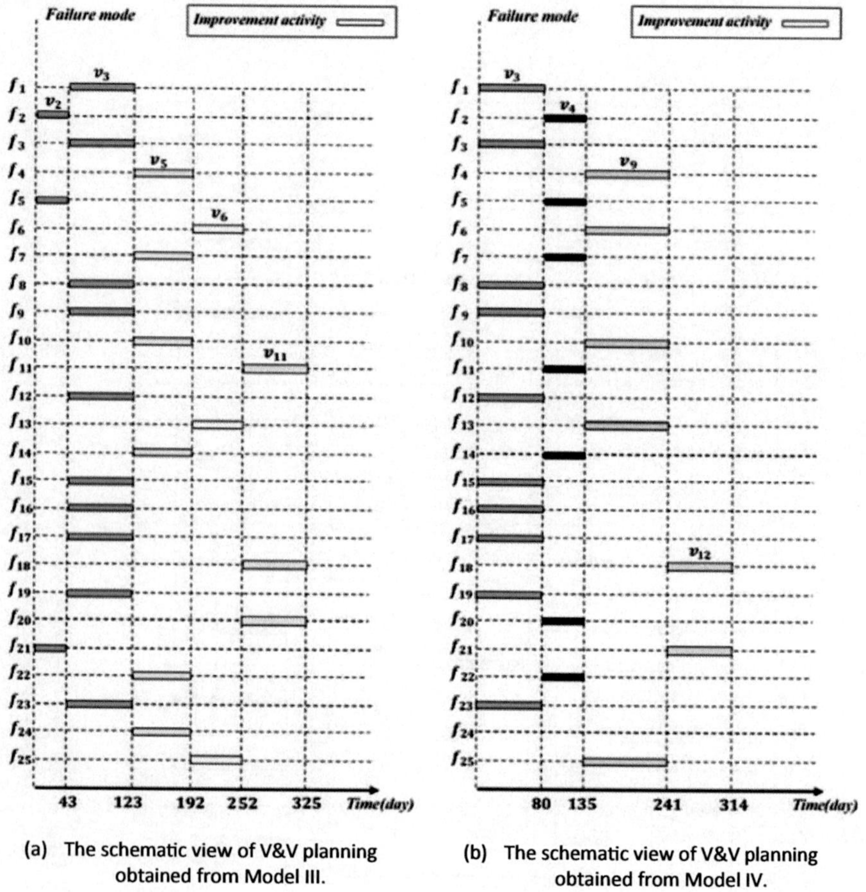

Fig. 16.6 The schematic views of V&V planning obtained from models III and IV

V&V planning approach in which the total time of V&V activities execution is minimized. Note that there is no reliability improvement maximization in the mentioned three approaches.

16.4.3.1 Planning V&V Activities Using PERT

In order to compare the results obtained from models proposed in this chapter with the traditional model in the literature, PERT approach is used to model the V&V activities planning problem in this subsection. Considering the sequencing, time gap, and duration of each V&V activity, the results of the V&V plan provided by PERT show that the overall reliability improvement effectiveness of all failure modes after executing all V&V activities is obtained as $RII_{Total} = 16.30$. Total cost of executing

V&V activities is equal to $870,000, while total time is obtained as 849 days. From the results, it can be seen that considering the sequencing of V&V activities, as well as the limited time and budget, PERT approach is only able to provide a plan for V&V activities that satisfies the sequencing constraints, while it is an infeasible solution in terms of satisfying time and cost constraints.

16.4.3.2 Cost-Oriented V&V Planning Approach

The goal of the cost-oriented V&V planning approach is to minimize the total cost of V&V activities execution while all failure modes are covered and improved. However, the reliability improvement effectiveness of each V&V activity is not included in this approach. Hence, the proposed V&V model with set covering problem (i.e. Model I) can be modified to show the cost-oriented V&V planning approach.

$$\text{Min: } \sum_{j=1}^{m} c_j u_j \tag{16.29}$$

Subject to

$$\text{Max}\left\{(s_j + t_j)u_j\right\} \leq T_0,\ \forall j = 1, 2, \cdots, m, \tag{16.10}$$

$$\sum_{j=1}^{m} a_{ij}^{V} u_j \geq 1,\ \forall i = 1, 2, \cdots, n. \tag{16.15}$$

The optimal V&V plan obtained through this cost-oriented formulation shows that four V&V activities, including v_3, v_4, v_9, and v_{15}, out of the proposed fifteen V&V activities are selected for execution to mitigate all critical failure modes. Total cost is minimized at $286,000 with the duration of V&V execution as 284 days. The overall reliability improvement effectiveness of all failure modes after executing all V&V activities is obtained as $\text{RII}_{\text{Total}} = 7.92$.

16.4.3.3 Time-Oriented V&V Planning Approach

The goal of the time-oriented V&V planning approach is to minimize the total time of V&V activities execution while all failure modes are covered and improved. Similar to the cost-oriented approach, the reliability improvement effectiveness of each V&V activity is not included. The proposed V&V model with set covering problem (i.e. Model I) can also be modified to show the time-oriented V&V planning approach.

$$\text{Min:}\left\{\text{Max}\left\{(s_j + t_j)u_j\right\}\right\} \tag{16.30}$$

Subject to

$$\sum_{j=1}^{m} c_j u_j \leq C_0, \tag{16.14}$$

$$\sum_{j=1}^{m} a_{ij}^{V} u_j \geq 1, \ \forall i = 1, 2, \cdots, n. \tag{16.15}$$

According to the time-oriented V&V planning approach, five V&V activities, including v_3, v_4, v_5, v_6, and v_{15}, out of the proposed fifteen V&V activities are selected for execution to mitigate all critical failure modes. Total time of V&V activities execution is minimized at 252 days with the cost as \$329,000. The overall reliability improvement effectiveness of all failure modes after executing all V&V activities is obtained as $\text{RII}_{\text{Total}} = 7.63$.

16.4.3.4 Summary of Comparative Analysis

The summary results of all seven approaches in planning V&V activities are provided in Table 16.1. The Cost Efficiency (CE) is calculated using Eqs. (16.17) to (16.20), in which the coefficient α is valued at 0.001 to equalize the magnitude of time and economy in this case. In addition, in order to compare the relative efficiency of seven types of models, the average minimum cost ($\bar{C}_{\text{min}}$) in the Eq. (16.17) is equal to the minimum average observed cost ($\bar{C}_{\text{observed}}$) of seven types of models in this case (i.e. the results obtained from Model IV).

Comparing the results obtained from the seven approaches, the maximum reliability improvement is obtained when the PERT approach is used to model and solve the case problem. This result makes sense since all V&V activities are executed in

Table 16.1 Summary of comparisons of seven approaches in planning V&V activities

Approach	Selected V&V activities	Total time (days)	Total cost (\$1000)	Reliability improvement ($\text{RII}_{\text{Total}}$)	Feasibility of solution	Cost efficiency (CE) (%)
Model I	1,3,4,5,7,9,10,13,14,15	550	595	13.77	Feasible	61.12
Model II	1,2,3,4,7,8,9,13,14,15	599	589	13.37	Feasible	56.77
Model III	2,3,5,6,11	325	321	10.98	Feasible	86.39
Model IV	3,4,9,12	314	301	12.10	Feasible	100
PERT	All	849	870	16.30	Infeasible	48.19
Cost oriented	3,4,9,15	284	286	7.92	Feasible	70.62
Time oriented	3,4,5,6,15	252	329	7.63	Feasible	66.75

this case; however, total time and cost of V&V execution are beyond the allowed time and budget for V&V execution. Hence, the solution provided by PERT is an infeasible solution in this sense. Comparing the six feasible solutions, total time of V&V execution has its minimum value in the time-oriented approach, and total cost has its minimum value in the cost-oriented approach. However, the overall reliability improvement effectiveness in these two approaches is significantly lower than that from the other four approaches. Comparing the four proposed approaches, the overall reliability improvement effectiveness in Models I and II are higher than those in Models III and IV, while the total cost and time in Models I and II are significantly larger than those in Models III and IV, and the cost efficiency in Models I and II are also significantly lower than those in Models III and IV. This phenomenon suggests that the overall reliability improvement effectiveness of all failure modes can be improved by multiple improvements to one failure mode through multiple available V&V activities, while the increased cost and time usually tend to be worth more than the increased improvement effectiveness due to the decay of improvement effectiveness, that is, the cost efficiency of multiple improvements is low. More specifically, comparing Model I with Model II, the overall reliability improvement effectiveness and cost efficiency of Model I are both higher than those in Model II. That is because Model II needs to ensure that the improved RPN and S of each failure mode reach the preset constraints at the expense of the overall improvement effectiveness under the same time and cost constraints. Comparing Model III with Model IV, the overall reliability improvement effectiveness and cost efficiency of Model III are higher than those in Model IV; it demonstrated that in pursuit of overall improvement effectiveness of the product development process, sometimes, some failure modes with low initial RPN or with low improvement effectiveness need not be improved.

In summary, comparing the four proposed approaches, Models I, II, and III can ensure that all crucial failure modes can be covered and improved, in which Model I shows the highest overall reliability improvement effectiveness, Model II can ensure that the improved RPN and S of each failure mode reach the preset constraints, and Model III shows the highest cost efficiency. In addition, when some non-critical failure modes do not need to be improved, Model IV shows the highest cost efficiency. Hence, V&V planning can be optimized and executed through four proposed optimizing strategies to meet the various demands of developers.

16.5 Discussions and Conclusions

To gain competitive advantages, companies that design and develop new complex products seek to increase the effectiveness and efficiency of their NPD processes. The product design V&V planning is one of the main processes in the early stages of the NPD, which includes a series of engineering activities defined to meet design objectives and performance requirements, such as a desired reliability level. This chapter proposed four novel mathematical models to optimize the V&V activities planning for improving the reliability of a new product while taking into account constraints in development time and budget, sequencing and effectiveness of V&V activities, as well as the decay of the improvement effectiveness.

The V&V planning of a new powertrain is used as an example, and numerical simulations are carried out to illustrate the applications of the proposed four V&V optimization models. Three types of optimization problem concepts are adopted to ensure that all critical failure modes are covered in different ways according to different risk mitigation requirements of organizations. The proposed modeling approaches provide four risk mitigation strategies from perspectives of reliability and cost efficiency to maximum overall reliability improvement effectiveness and cover different requirements of products. Taking the V&V planning of a new powertrain as an example, the numerical simulations and solutions are carried out to illustrate the applications of proposed four V&V optimization models. In future research, the iterations of V&V activities in a multistage development and the uncertainty of the input variables of the model will be investigated.

References

1. Li, Z., Mobin, M., & Keyser, T. (2016). Multi-objective and multi-stage reliability growth planning in early product-development stage. *IEEE Transactions on Reliability, 65*(2), 769–781.
2. Murthy, D. N. P., Rausand, M., & Virtanen, S. (2009). Investment in new product reliability. *Reliability Engineering and System Safety, 94*(10), 1593–1600.
3. Maropoulos, P. G., & Ceglarek, D. (2010). Design verifcation and validation in product lifecycle. *CIRP Annals, 59*(2), 740–759.
4. Babuska, I., & Oden, J. T. (2004). Verification and validation in computational engineering and science: basic concepts. *Computer Methods in Applied Mechanics and Engineering, 193*(36–38), 4057–4066.
5. Belt, P., Harkonen, J., Mottonen, M., et al. (2008). Improving the efficiency of verification and validation. *International Journal of Services and Standards, 4*(2), 150–166.
6. Collopy, P. D., & Hollingsworth, P. M. (2011). Value-driven design. *Journal of Aircraft, 48*(3), 749–759.
7. Reuters. *Boeing sees more narrowbody delivery delays in third quarter,* Downloaded from: https://www.reuters.com/article/us-boeing-production-delays/boeingsees-more-narrowbody-delivery-delays-in-third-quarter-idUSKBN1KT1QW, on Aug 8, 2018.
8. Schwenn, R. E., Chitikila, R. C., & Laufer, D. R. et al. (2011). *Defense acquisitions: Assessment of selected weapon programs,* DC, USA: United States Government Accountability Office, Report No. GAO-11-233SP.
9. Schwenn, R. E., Brink, H., & Mebane, C. T., et al. (2009). *Defense acquisitions: Assessment of selected weapon programs,* DC, USA: United States Government Accountability Office, Report No. GAO-09-326SP.
10. Potty, N. S., Irdus, A. B., & Ramanathan, C. (2011). Case study and survey on time and cost overrun of multiple D&B projects. *National Postgraduate Conference, Kuala Lumpur, Malaysia, Sep, 19–20,* 1–6.
11. Bhattacharjee, A. K., Dhodapkar, S. D., & Shyamasundar, R. K. (2001). PERTS: An environment for specification and verification of reactive systems. *Reliability Engineering and System Safety, 71*(3), 299–310.
12. Han, X., Li, R., Wang, J., et al. (2018). Identification of key design characteristics for complex product adaptive design. *The International Journal of Advanced Manufacturing Technology, 95*(1–4), 1215–1231.

13. Cho, H., & Park, J. (2019). Cost-effective concept development using functional modeling guidelines. *Robotics and Computer-Integrated Manufacturing, 55,* 234–149.
14. Estrada, G., Shunk, D. L., & Ju, F. (2018). Systematic continuous improvement model for variation management of key characteristics running with low capability. *International Journal of Production Research, 56*(6), 2370–2387.
15. Kukulies, J., & Schmitt, R. (2018). Stabilizing production ramp-up by modeling uncertainty for product design verification using Dempster-Shafer theory. *CIRP Journal of Manufacturing Science and Technology, 23,* 178–196.
16. Chahin, A. & Paetzold, K. (2018). *Planning validation and verification steps according to the dependency of requirements and product architecture,* IEEE International Conference on Engineering, Technology and Innovation (ICE/ITMC), Stuttgart, Germany, Jun.17–20, pp. 1–6.
17. Mobin, M., & Li, Z. (2018). An Integrated Approach to Plan the Design Verification and Validation Activities for the New Product Reliability Improvement. *Annual Reliability and Maintainability Symposium (RAMS) Reno, NV, USA, Jan., 22–25,* 1–7.
18. Ahmed, H., & Chateauneuf, A. (2014). Optimal number of tests to achieve and validate product reliability. *Reliability Engineering and System Safety, 131,* 242–250.
19. Mobin, M., Li, Z., & Cheraghi, S.H., et al., (2019) An approach for design verification and validation planning and optimization for new product reliability improvement. *Reliability Engineering and System Safety, 190.*
20. Stamatis, D.H. (2003). *Failure mode and effect analysis: FMEA from theory to execution,* 2nd ed, ASQ Quality Press: Milwaukee, WI, USA.
21. Barends, D. M., Oldenhof, M. T., Vredenbregt, M. J., & Nauta, M. J. (2012). Risk analysis of analytical validations by probabilistic modifcation of FMEA. *Journal of Pharmaceutical and Biomedical Analysis, 64,* 82–86.
22. Applegate, D., & Cook, W. (1991). A computational study of the job-shop scheduling problem. *ORSA Journal on Computing, 3*(2), 149–456.
23. Vazirani, V. V. (2001). *Approximation algorithms.* Berlin, Heidelberg: Springer.
24. Farrell, M. J. (1957). The Measurement of Productive Efficiency. *Journal of the Royal Statistical Society, 120*(3), 253–290.
25. Mirdehghan, S. M., & Shirzadi, A. (2012). Ranking decision making units based on the cost efficiency measure. *International Journal of Pure and Applied Mathematics, 81*(1), 55–63.
26. Mitchell, M. (1998). *An introduction to genetic algorithms.* Cambridge, MA, USA: MIT press.
27. Wu, G., Li, Z.S., & Liu, P. (2020). *Risk-informed reliability improvement optimization for verification and validation planning based on set covering modeling,* Proceedings of the Institution of Mechanical Engineers, Part O: Journal of Risk and Reliability.

Zhaojun Steven Li Associate Professor with the Department of Industrial Engineering at Western New England University in Springfield, MA, USA. Dr. Li's research interests include data analytics, applied statistics and operations research, reliability engineering, systems engineering and its applications in product design, diagnostics and prognostics of complex engineering systems. He received his Ph.D. in Industrial Engineering from the University of Washington. He is an ASQ Certified Reliability Engineer and Caterpillar Six Sigma Black Belt. Dr. Li's most recent industry position was a reliability team lead with Caterpillar to support the company's new engine development. He is serving on editorial boards for IEEE Transactions on Reliability and IEEE Access Reliability Society Section. He is also the Co-EIC of the International Journal of Performability Engineering journal. He is a senior member of IISE and IEEE. He has served as a board member of IISE Quality Control and Reliability Engineering (QCRE) Division and IEEE Reliability Society. He served as the VP for Publications of IEEE Reliability Society in 2019.

Gongyu Wu Currently pursuing the Ph.D. degree in the School of Mechanical and Electrical Engineering at the University of Electronic Science and Technology of China, Chengdu, China. His research interests include risk analysis of new product development processes, cascading failures modeling for cyber-physical power systems, and smart grid resilience.

Chapter 17
Efficient Use of Meta-Models for Reliability-Based Design Optimization of Systems Under Stochastic Excitations and Stochastic Deterioration

Gordon J. Savage and Young Kap Son

Abstract The main difficulty in the application of reliability-based design optimization (RBDO) to time-dependent systems is the continual interplay between calculating time-variant reliability (to ensure reliability policies are met) and moving the design point to minimize some objective function, such as cost, weight or size. In many cases, the reliability can be obtained readily using, for example, first-order reliability methods (FORM). However, this option is not available when certain stochastic processes are invoked to model, for example, gradual damage or deterioration. In this case, inefficient Monte Carlo simulation (MCS) must be used. The work herein provides a novel way to obviate this inefficiency. First, a meta-model is built to relate the system cumulative distribution function of time to failure (*cdf*) to the design space. A design of experiments paradigm helps determine a few training sets and then the mechanistic model and the uncertain characteristics of the variables, with MCS, help produce the corresponding *cdf* curves. The meta-model (using matrix methods) directly links an arbitrary sample from the design space to its *cdf*. The optimization process accesses the meta-model to quickly evaluate both objectives and failure constraints. A case study uses a electromechanical servo system. The meta-model approach is compared to the traditional MCS approach and found to be simple, accurate and very fast, suggesting an attractive means for RBDO of time-dependent systems under stochastic excitations and stochastic deterioration.

Keywords Design · Reliability · Optimization · Meta-models · Stochastic inputs

G. J. Savage (✉)
Department of Systems Design Engineering, University of Waterloo, Waterloo, ON, Canada
e-mail: gjsavage@uwaterloo.ca

Y. K. Son
Department of Mechanical and Automotive Engineering,
Andong National University, Andong, Korea

K. B. Misra (ed.), *Handbook of Advanced Performability Engineering*,
https://doi.org/10.1007/978-3-030-55732-4_17

17.1 Introduction

In engineering systems, extreme environmetal conditions and uncertain loads lead to component deterioration, and this in turn leads to poor performacne, or worse, failure. In mechanical and structural systems, wear is a critical source of failure since it effects the life span of hinges, bearings and coupling components. Examples include vehicle clutches, multi-bar linkages and servo systems that lose their ability to perform to specifications. In electrical systems, the parameters drift from their initial settings through both usage and tempertaure and hunidity variations. For example, the band frequencies in filters become altered and the attenuation effectiveness degrades.

The analysis of degradation started with Meeker and Escobar [1] who introduced statistical-based physics-based models. And after decades, degradation models [2–4] include (a) random variable models, (b) marginal distribution models and (c) cumulative damage models. The random variable (RV) models (also called degradation path models) randomize the parameters associated with some empirical deterioration law. For example, consider resistance R, then $R = R_0 \pm Ct$ where t is time, R_0 is the initial resistance and C is the random (or, perhaps deterministic) degradation rate.

The marginal distribution (MD) models (also referred to as degradation distribution models) provide a new distribution at any time t. A simple MD model has the form $R = R_0 \pm C(\mathbf{p}(t))$ where $\mathbf{p}$ are distribution parameters and $C(\mathbf{p}(\mathrm{t}))$ represents a particular distribution at time t. The cumulative damage (CD) models (also called shock models) assume that the degradation is caused by shocks or jumps and that damage accumulates additively [4]. These models are used when the temporal uncertainty associated with the deterioration cannot be ignored. In this model, $R = R_0 \pm C(t)$ where $C(t)$ is a stochastic process, such as Weiner, Gamma [5] and inverse Gaussian. It is apparent then that system deterioration leads to time-dependent reliability issues that may be mitigated by reliability-based design optimization (RBDO).

The RBDO problem has three issues. The first deals with how reliability is to be calculated over time; the second deals with parametric uncertainty in components and excitation uncertainty in loads. Finally, the third issue addresses the dynamical nature of the system; that is, the performance measures steady-state with algebraic equations, or transient' with differential equations. These areas are bridged by considering time-variant parametric uncertainties and stochastic processes.

Stochastic loads are one of the main sources of time-variant reliability. Kuschel and Rackwitz [6] employed the outcrossing rate to find time-variant reliability and solved the optimization problem under simple loads. Wang and Wang [7] developed a nested extreme response method to transform the time-variant RBDO problem into time-invariant RBDO problem. Hu and Du [8] devised the equivalent most likely failure point (MLFP) and extended the sequential reliability assessment algorithm (SORA) to solve time-variant RBDO problems containing stochastic loads. Therein FORM was invoked and design parameters comprised either deterministic variables or means of distributions. Jiang et al. [9] produced the time-invariant equivalent method (TIEM) to reduce the number of *cdf* calculations. FORM was used and design parameters were deterministic variables and means of distributions.

A few papers address RBDO and degradation. Savage and Son [10] applied the set theory method to find efficiently the *cdf* for multiple response systems with deterministic component degradation. Rathod et al. [11] treated probabilistic damage accumulation as a measure of degradation in material fatigue and modelled it as a stationary process that in turn became a constraint in the optimized solution. Singh et al. [12] considered deterministic degradation and introduced the composite limit state to convert a time-variant RBDO problem into a time-invariant problem and then invoked a genetic algorithm to search for the MLFP.

For an efficient design process, meta-models (often called surrogate models) have been introduced. In the past three decades, their impact in the design of systems has been significant. They are computationally efficient substitutes for the mechanistic model: they are both accurate and very fast. These two features allow for a variety of timely perform ability calculations in the optimization routines. The success of the meta-model depends on (a) the proper selection of the input variables (i.e. excitations and component parameters), (b) their ranges (e.g. design space), (c) the number of training samples, (d) the philosophy used to collect this data and finally (e) the form of the approximating function. Overviews of various meta-models are contained in Refs.[13, 14]. The popular kriging methods are detailed in Refs.[15–17], the moving least squares (sometimes called lazy learning) meta-models are described in Refs. [18, 19]. The Bayesian meta-models are illustrated in Ref. [20].

The use of meta-models to provide efficiencies in time-invariant reliability analysis is contained in references [21, 22]. Work using meta-models in RBDO with time-invariant reliability includes references [23–26]. There is some time-variant reliability analysis using meta-models. Savage et al. [27] predicted the reliability of degrading dynamic systems using various meta-models. Singh et al. [12] used a meta-model of the composite limit-state surface and then used it to determine time-invariant failure. Dregnei et al. [28] developed a random process meta-model that linked the left singular vectors of the responses of a system to the left singular vectors of an uncertain excitation matrix and augmented this with uncertain component dimensions. The meta-model was then used to help determine the lifetime reliability. Zhang et al. [29] established a meta-model based on response surface for time-variant limit-state function to estimate time-dependent reliability for nondeterministic structures under stochastic loads. Stochastic loads were discretized as static random variables in the model, and FORM was applied to estimate reliability.

Herein, we present a new method for RBDO of time-variant systems containing stochastic degradation and stochastic excitations. In the first stage, a meta-model is built that explicitly relates time-variant failure to the design space. In the second stage, the optimization process invokes the typical nested approach, but now the meta-model is used to quickly evaluate objective functions and failure constraints: the design time needed to conduct RBDO for time-variant systems is greatly reduced.

17.2 Time-Variant Reliability

For time-variant reliability, we denote the vector **V**, with elements V_j, $j = 1, \ldots n$, as the randomness in the problem. The probability density functions of **V** are assumed to exist and provide distribution parameters **p**. A conversion to standard normal (i.e. **U**-space) is usually possible through an iso-probabilistic transformation [30], denoted as $V(\mathbf{p}_V, \mathbf{U})$. Further, we let $\mathbf{W}(t)$ be a vector of stochastic processes for time t. These processes include excitations and loads (typically modelled by Gaussian processes) as well as cumulative damage degradation, often modelled by the Gamma process. These random effects come from dynamically varying environmental conditions and the temporal uncertainties of changes in material properties and structural dimensions.

We start with component level time-variant reliability in terms of the related *cdf*. For the *ith* component, let Z_i be the response and ξ_i an upper or lower specification. Then the limit-state function is $g_i(\mathbf{V}, \mathbf{W}(t), t) = \pm\{Z_i(\mathbf{V}, \mathbf{W}(t), t) - \xi_i\}$ where a positive value indicates success and a negative value indicates failure. For convenience, we write the failure event over lifetime span $[0, t_L]$ as

$$E_i(0, t_L) = \{g_i(\mathbf{V}, \mathbf{W}(t), t) \leq 0, \text{ for } \exists t \in [0, t_L]\} \tag{17.1}$$

and the *true cdf* for the *ith* component is

$$F^i(t_L) = P(E_i(0, t_L)) \tag{17.2}$$

The evaluation of Eq. (17.2) is generally intractable; however, discrete time is of help. Consider the planned time t_L with equally spaced time points obtained from a small, fixed, time step h (the length to be determined later). For a time index $l = 0$, $1, \ldots L$, where L is the number of time steps to the planned time, the time at the *lth* step is $t_l = l \times h$. We write a set that represents the *instantaneous failure* region of the *ith* limit-state function at any selected point-in-time t_l with reference to notation in Eq. (17.1) as

$$E_{l,i} = \{g_i(\mathbf{V}, \mathbf{W}(t_l), t_l) \leq 0\} \tag{17.3}$$

Note, we must find the stochastic processes $\mathbf{W}(t)$ at discrete times t_l.

Now, let us consider a system with say e components and extend Eq. (17.3) to *Parallel* and *Serial* connections. For parallel connections, the *system* instantaneous failure region at time t_l is defined to be the set $\mathbf{E}_l = \bigcap_{i=1}^{e} E_{l,i}$. For Series connections, the *system* instantaneous failure region at time t_l is defined to be the set $\mathbf{E}_l = \bigcup_{i=1}^{e} E_{l,i}$. Moreover, the *system cumulative failure set* A_l is defined as the set that represents the accumulation of all system instantaneous failure regions for all discrete times up to t_l, and is written as

$$A_l = \mathbf{E}_0 \cup \mathbf{E}_1 \cup \cdots \cup \mathbf{E}_l = \bigcup_{q=0}^{l} \mathbf{E}_q \tag{17.4}$$

The *system cumulative safe set* up to time $t_{l+1} = t_l + \Delta t$ is denoted as $\overline{A}_{l+1}$ and is simply

$$\overline{A}_{l+1} = \overline{\mathbf{E}}_0 \cap \overline{\mathbf{E}}_1 \cap \cdots \cap \overline{\mathbf{E}}_{l+1} = \bigcap_{q=0}^{l+1} \overline{\mathbf{E}}_q \tag{17.5}$$

We define the emergence of the incremental failure region from the system cumulative safe region, from time t_l during time interval Δt, as $\mathbf{B}_l = A_{l+1} \cap \overline{A}_l$. This term can be simplified by noting $\mathbf{E}_q \cap \left(\overline{\mathbf{E}}_0 \cap \cdots \cap \overline{\mathbf{E}}_q \cdots \cap \overline{\mathbf{E}}_l\right) = \emptyset$. Hence, for $q = 0, 1, \ldots, l$, we have more simply

$$\mathbf{B}_l = \mathbf{E}_{l+1} \cap \overline{A}_l \tag{17.6}$$

We write the *cdf* as the sum of increments, or

$$F(t_L) = P(\mathbf{E}_0) + P(\mathbf{B}_0) + \cdots + P(\mathbf{B}_l) + \cdots + P(\mathbf{B}_{L-1}) \tag{17.7}$$

The expression for $\boldsymbol{B}_l$ in Eq. (17.6) requires the time history of the system responses, and thus it is logistically difficult to determine the probability $P(\boldsymbol{B}_l)$. In MCS, a sample from the distributions of the design variables is chosen and then the sign of $\boldsymbol{E}_l$ is determined for time index $l = 0, 1, \ldots L$, stopping and recording the time of first failure [31]: as well, all future times for the sample are recorded as fail. For all MCS samples, a histogram that represents the terms in Eq. (17.7) is built, and then the *cdf* is found as the summation of all of the terms up to the time of interest.

17.3 Stochastic Processes

The approaches to modelling both excitations and degradation over discretized time are outlined next.

17.3.1 Excitations: Gaussian Stochastic Process

An excitation (i.e. a source or load) is denoted as $Y(t)$, and typically modelled by a nonstationary Gaussian stochastic process. There are many proposed modelling methods including Karhunen–Loeve [32], polynomial chaos expansion [33], proper orthogonal decomposition [34] and EOLE [35]. The EOLE model is easy to write in matrix form (thus simplifying computer programming) and hence is employed herein. Let the mean function be $\mu_Y(t_i)$, the standard deviation function $\sigma_Y(t_i)$ and the autocorrelation function $\rho_Y(t_i, t_j)$. Then, the Gaussian process takes the compact matrix form [36]

$$Y(t) = \mu_Y(t) + [\boldsymbol{\Sigma}(t)]^T (\boldsymbol{\Phi}\tilde{\Lambda})\mathbf{U} \tag{17.8}$$

where $(\boldsymbol{\Phi}\tilde{\Lambda})$ is a matrix of constants, $\boldsymbol{\Sigma}(t)$ is a time-related vector (containing standard deviation and correlation parameters) and $\mathbf{U}$ is vector of standard normal. For a simpler notation, let the distribution parameters in Eq. (17.8) be written as $\mathbf{q}(t) = [\mu(t), \boldsymbol{\Sigma}(t)]^T$, then more informatively

$$Y(t) = Y(\mathbf{q}(t), \mathbf{U}) \tag{17.9}$$

17.3.2 Component Degradation: Gamma Process C(t)

The Gamma process is suitable for modelling gradual damage or deterioration when it is monotonically accumulating over time: examples include wear, fatigue, corrosion, crack growth, erosion, consumption, creep, etc. [5]. The Gamma process is a continuous-time process with stationary, independent, non-negative Gamma increments, obtained from the Gamma distribution. Let $G(\alpha, \beta)$ denote the distribution and let its density function be

$$f(\gamma) = \frac{1}{\beta^{\alpha}\Gamma(\alpha)}\gamma^{\alpha-1}e^{-\gamma/\beta} \tag{17.10}$$

where $\Gamma(\alpha)$ is the so-called gamma function, and α and β are the shape and scale parameters, respectively. Finally, the Gamma process is denoted as $C(t|\mu, \sigma^2)$ with mean μ and variance σ^2. Then, for any time increment $\Delta t = t_L/L > 0$, the increments are [37]

$$C\big(t + \Delta t|\mu, \sigma^2\big) - C\big(t|\mu, \sigma^2\big) \sim G(\alpha\Delta t, \beta) \tag{17.11}$$

We note that the distribution of the increments depends on the length of Δt but not on the time t. Let us find now suitable distribution parameters for the Gamma distribution in Eq. (17.10). If the mean value of the process is linear, then we may write mean and variance of the process as

$$\begin{aligned} E[C(t = t_L)] &= \mu = (\alpha t_L)\beta, \\ Var[C(t = t_L)] &= \sigma^2 = (\alpha t_L)\beta^2 \end{aligned} \tag{17.12}$$

We have the new parameters for the desired Gamma distribution (in terms of mean and standard deviation) as

$$\alpha = \frac{1}{t_L} \frac{\mu^2}{\sigma^2} \tag{17.13}$$

and

$$\beta = \frac{\sigma^2}{\mu} \tag{17.14}$$

Let the parameters be written compactly as $\mathbf{r} = [\mu, \sigma, \Delta t]^T$, then equation Eq. (17.14) provides the series of random variables

$$C(\mathbf{r}, t_l) = C(t_0) + \sum_{i=1}^{l} G\left(\frac{1}{t_L} \frac{\mu^2}{\sigma^2} \Delta t, \frac{\sigma^2}{\mu}\right) l = 1, 2, \ldots L \tag{17.15}$$

It is a simple manner to generate the Gamma process over discrete time. For the kth manifestation of a Gamma process, say $C^{(k)}(t)$ with incremental samples $\gamma_i^{(k)}$ chosen according to the Gamma distribution in Eq. (17.10), we have the series of process values,$c^{(k)}(t_0) = 0, c^{(k)}(t_1) = c^{(k)}(t_0) + \gamma_1^{(k)}, \ldots, c^{(k)}(t_{l+1}) = c^{(k)}(t_l) + \gamma_l^{(k)}, \ldots,$ $c^{(k)}(t_L) = c^{(k)}(t_{L-1}) + \gamma_L^{(k)}$ where $k = 1, 2, \ldots N$. Hence, for simulation purposes, we generate N Gamma process paths to provide N jump values at each discrete time $t_0, t_1, t_2, \cdots, t_{L-1}$.

17.4 Meta-Model Development

The fundamentals and formulation of the meta-model to link the *cdf* to the design space are presented next.

17.4.1 Design Parameters and Training Data

A judicious selection of the design variables and their operating ranges is important to keep the meta-model manageable but effective. There are simple analyses that help.

(1) An importance analysis, that uses sensitivity information, can trim the number of design variables to a manageable few.
(2) A 'parameter design' to minimize $F(t = 0)$ can find nominal values of the design variables.

(3) A worst-case analysis to ensure $F(t = 0) \leq \delta$ helps find the variations of the variables

The design parameters **p** used to form the meta-model are herein means and tolerances written as

$$\mathbf{p}^T = [\boldsymbol{\mu}^T \mathbf{tol}^T] = [\mu_1\, \mu_2 \cdots \mu_m\, tol_1\, tol_2 \cdots tol_n] \tag{17.16}$$

The design space is set out by lower and upper limits whereby as $\mu_i \in [lsl_i, usl_i]$ for $i = 1, 2, \ldots m$ and $tol_i \in [LSL_i, USL_i]$ for $i = 1, 2, \ldots n$. Let the input or training samples be $\mathbf{p}_j, j = 1, 2, \ldots \delta$ selected from the design space using design of experiments (DOE) and Latin Hypercube sampling. Then, for the jth sample (i.e. $\mathbf{p}_j$), the corresponding input data vector, based on a selected polynomial fit, becomes

$$\mathbf{d}(\mathbf{p}_j)^T = \left[1\ \mathbf{p}_j^T\ f(\mathbf{p}_j^T)\right]_{1 \times q} \tag{17.17}$$

where we have allowed for a constant, linear terms and typically quadratic terms. The vector length q depends on the order of the polynomial and the sizes of m and n. The resulting input training matrix becomes

$$\mathbf{D} = \begin{bmatrix} [\mathbf{d}(\mathbf{p}_1)]^T \\ [\mathbf{d}(\mathbf{p}_2)]^T \\ \vdots \\ [\mathbf{d}(\mathbf{p}_\delta)]^T \end{bmatrix}_{\delta \times q} \tag{17.18}$$

To generate the output matrix, we invoke the mechanistic model along with the random and stochastic information about the variables to generate the *cdf* curves. Then, for discrete time,$\mathbf{t} = [t_1, t_2, \ldots, t_L]$, a corresponding vector $\mathbf{F}^T(\mathbf{p}_j)$ is obtained. For all δ experiments, the output matrix has the structure

$$\overline{\mathbf{F}} = \begin{bmatrix} [\mathbf{F}^T(\mathbf{p}_1)] \\ [\mathbf{F}^T(\mathbf{p}_2)] \\ \vdots \\ [\mathbf{F}^T(\mathbf{p}_\delta)] \end{bmatrix}_{\delta \times L} \tag{17.19}$$

17.4.2 A Moving Least Squares Meta-Model

The ubiquitous kriging meta-model can be found in a variety of places [13–17]; however, the moving least squares meta-model [18, 19] is simpler and is thus adapted herein. Consider the arbitrary input set of parameters $\tilde{\mathbf{p}}$, then a weight matrix $\mathbf{W}(\tilde{\mathbf{p}})$ is required that effectively selects the so-called nearby data sets in $\mathbf{D}$ and $\overline{\mathbf{F}}$. Some examples are presented in [18, 19, 22]. A new input matrix $\mathbf{W}(\tilde{\mathbf{p}})\mathbf{D}$ is formed and related to the new output matrix $\mathbf{W}(\tilde{\mathbf{p}})\overline{\mathbf{F}}$. For a least squares solution, the normal equations [38] (or orthogonal methods [39]) become

$$\left[\mathbf{D}^T\mathbf{W}^T(\tilde{\mathbf{p}})\mathbf{W}(\tilde{\mathbf{p}})\mathbf{D}\right]\boldsymbol{\Theta}(\tilde{\mathbf{p}}) = \left[\mathbf{D}^T\overline{\mathbf{W}}(\tilde{\mathbf{p}})\overline{\mathbf{F}}\right] \tag{17.20}$$

A solution to Eq. (17.20) produces the matrix

$$\boldsymbol{\Theta}(\tilde{\mathbf{p}}) = \left[\,\boldsymbol{\theta}_1(\tilde{\mathbf{p}})\ \boldsymbol{\theta}_2(\tilde{\mathbf{p}})\ \cdots\ \boldsymbol{\theta}_L(\tilde{\mathbf{p}})\,\right]_{q\times L} \tag{17.21}$$

Finally, an approximation of the *cdf* curve (i.e. row vector) for $\tilde{\mathbf{p}}$ is

$$\tilde{\mathbf{F}}(\tilde{\mathbf{p}})^T = \mathbf{d}(\tilde{\mathbf{p}})^T\boldsymbol{\Theta}(\tilde{\mathbf{p}}) \tag{17.22}$$

Note that the kth element of the *cdf* vector requires only the kth column of the weight matrix, hence

$$\tilde{F}_k = \mathbf{d}(\tilde{\mathbf{p}})^T\boldsymbol{\theta}_k(\tilde{\mathbf{p}}) \tag{17.23}$$

17.4.3 Error Analysis

Errors in the meta-models arise from the following sources: the first source is the number of time instances; that is, the size of Δt used in capturing the time histories of the output function. This number can be increased until a specified error metric has been met. The second source is the number of training excitation functions (i.e. δ) chosen. There are several ways to determine this number: the simplest is to use the rule-of-thumb that says multiply the number of parameters (or inputs) by a convenient factor (e.g. ten or twenty) and then add a small contingency factor. Also, the *leave-one–out* method is popular [25].

17.5 Case Study: Servo Actuator

The servo system of interest is shown in Fig. 17.1, and both the component models and interconnection model can be found in more detail in Savage and Carr [40].

The motor and tachogenerator pair are shown as $M_{7,9}$ and $G_{8,10}$, respectfully. Herein, the motor and the tachogenerator are identical devices, just interconnected

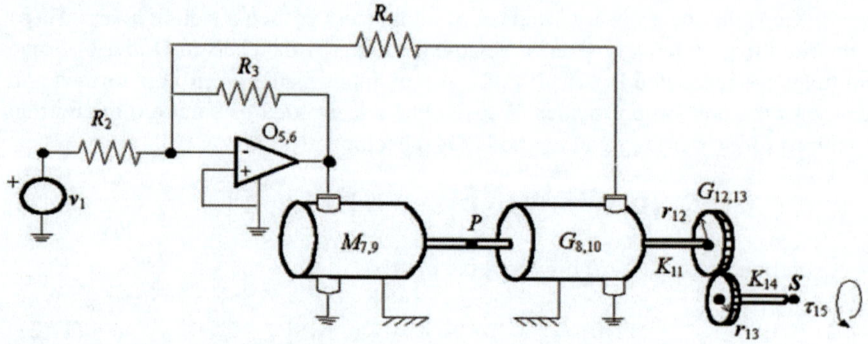

Fig. 17.1 Electromechanical servo system

differently. Their three parameters are the torque constant (denoted as κ), the rotational inertia (J) and the winding resistance R. The difference amplifier (denoted as $O_{5,6}$) comprises the three resistances R_2, R_3 and R_4 and an operational amplifier with open-loop gain A. The gear train, $G_{12,13}$ has gear ratio $r = r_{12}/r_{13}$. A voltage supply v_1 acts as the power source. An applied torque τ_{15} models the load arising from some connected subsystem at the shaft S. (The elastic shafts K_{11} and K_{14} are not factors in the steady-state performance.) The problem is to reduce the total cost (comprising production and loss of quality), and yet meet specific failure constraints, by adjusting a few important parameters.

17.5.1 Performance Measures

The three performance measures (responses) include the time constant t_c (indicates response time), the steady-state shaft speed ω_{SS} and the initial or starting torque τ_o. Their performance metrics and design specifications are given in Table 17.1 and all metrics must be satisfied for successful operation.

The mechanistic models for the performance measures, in terms of the electromechanical parameters, (with very large op-amp gain A) are, respectively,

$$t_c = \frac{4R(R + R_4)J}{\kappa^2(2R + R_4 + R_3)} \tag{17.24}$$

Table 17.1 Responses, performance metrics and specifications

Response	Metric	Specifications
Z_1 (t_c)	Smaller-is-best	$USL_1 = 0.051$ s
Z_2 (ω_{SS})	Target-is-best	$LSL_2 = 545$, T = 570 rad/sec, $USL_2 = 595$
Z_3 (τ_o)	Larger-is-best	$LSL_3 = 0.19$ N-m

Table 17.2 Specifications for the system variables and stochastic processes

Variable/Process	Distribution	Mean	Standard deviation	Standard normal variable
V_1 (κ_0)	Normal	μ_1	σ_1	U_1
V_2 (R_0)	Normal	2.9	0.02	U_2
V_3 (r)	Normal	0.461	0.005	U_3
V_4 (v_1)	Normal	12 V	0.04	U_4
$Y(t)$ (τ_{15})	Gaussian	0.01	0.0015	$U_5 \sim U_{22}$
$C_1(t)$ for $\kappa(t)$	Gamma	0.01 ($t = 10$)	0.005	N/A
$C_2(t)$ for R (t)	Gamma	2×10^{-7} ($t = 10$)	8×10^{-8}	N/A

$$\omega_{SS} = \frac{rR_3(R + R_4)}{\kappa R_2(2R + R_4 + R_3)} v_1 - \frac{r^2 R(R + R_4)}{\kappa^2(2R + R_4 + R_3)} \tau_{15} \tag{17.25}$$

$$\tau_o = \frac{\kappa R_3}{rRR_2} v_1 \tag{17.26}$$

17.5.2 Component Uncertainties

A sensitivity analysis tells us that the most important variable is the motor torque constant κ followed equally by the motor resistance R and the gear train gear ratio r. Thus, we let κ be the design variable and R and r be noise random variables with distributions given in Table 17.2. The supply voltage v_1 is obtained from a known power supply but may be uncertain owing to manufacturing abilities or the controller requirement: it becomes a noise variable with the distribution given in Table 17.2. The remaining variables are deterministic: for the rotor inertia, $J = 1/1{,}000{,}000$ kg-m^2. For the op-amp, $R_2 = 10$ k Ω, $R_3 = 40$ k Ω, $R_4 = 10$ k Ω and $A = 5 \times 10^6$.

The load torque τ_{15} is uncertain owing to the particular end-use of the servo and is modelled by the Gaussian stochastic process $Y(t)$ with parameters given in Table 17.2. The autocorrelation is $\rho(t_1, t_2) = \exp[-(t_2 - t_1)^2/\lambda^2]$ with $\lambda = 1$ year. The conversion to EOLE requires 17 singular values and provides profiles similar to those in Fig. 17.2.

17.5.3 Component Degradation Modelling

Many direct current motors use permanent magnets to provide the requisite magnetic flux. However, with overuse and extreme operating conditions, the magnetic field

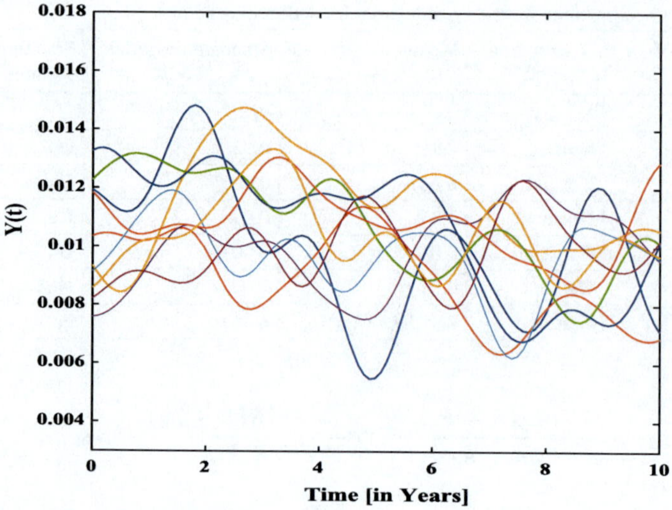

Fig. 17.2 Ten sampled torque load profiles of Y(t)

strength ebbs causing a corresponding reduction in the torque constant. The degradation is often written in the random path form, or $\kappa(t) = \kappa_0(1 - d \times t)$ where κ_0 is the initial torque constant value and d is a known degradation rate. For this case study, we let the degradation be modelled by a stochastic process. Thus, we have the torque constant over time t as

$$X_1(t) = V_1(1 - C_1(t)) \tag{17.27}$$

where V_1 is its initial uncertainty with parameters given in Table 17.2 and $C_1(t)$ is a Gamma process with $C_1(t = 0) = 0$, and parameters as given in Table 17.2. Some typical torque constant paths over lifetime are shown in Fig. 17.3.

The armature winding resistance R increases over time and the random path form is $R(t) = R_0 \exp(ct)$ where R_0 is the initial resistance and c is constant. With a stochastic degradation model, the armature winding resistance becomes

$$X_2(t) = V_2 \exp(C_2(t)) \tag{17.28}$$

where $X_2(t)$ is the resistance at time t, V_2 is the uncertain resistance at initial time and has the parameters given in Table 17.2. Finally, $C_2(t)$ is Gamma degradation process with the distribution parameters in Table 17.2.

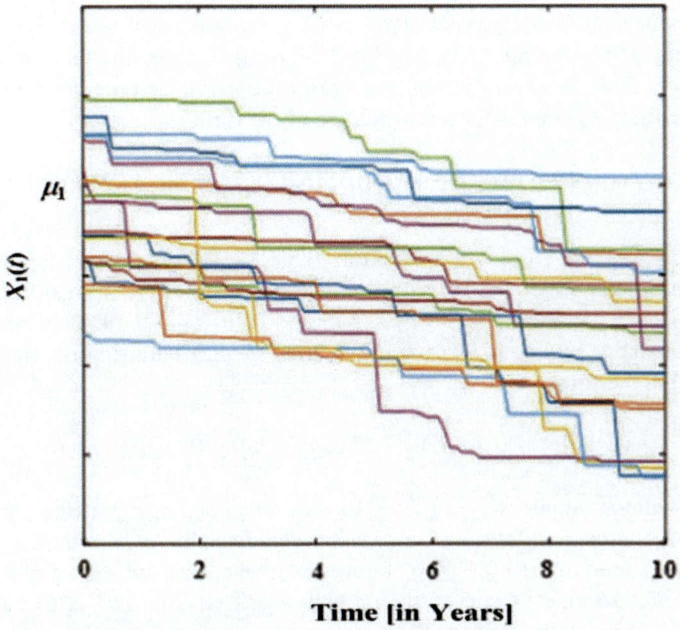

Fig. 17.3 Typical degradation profiles of $X_1(t)$

17.5.4 Design Space and the Meta-Model

The design parameters are the mean and tolerance of the torque constant, so $\mathbf{p}^{\mathrm{T}} = [\mu_1, tol_1]$ where tol_1 is the statistical tolerance $3\sigma_1$. To ensure a feasible design (such that the four limit specifications in Table 17.1 are satisfied), we find a nominal value for κ_o by setting all of the variables in the performance measures to their deterministic or mean values and then minimize the single, deterministic, loss function

$$L(\kappa) = \left(\frac{(t_c - 0.04)}{0.005}\right)^2 + \left(\frac{(\omega_{SS} - 570)}{19}\right)^2 + \left(\frac{(\tau_o - 0.24)}{.02}\right)^2 \tag{17.29}$$

where the performance measures, in terms of the variables, come from Eqs. (17.24) ~ (17.26). We get $\overline{\kappa}_o = 7.45 \times 10^{-3}$. A sensitivity analysis using initial failure $F(t = 0)$ and entries in $\mathbf{p}$ shows a very sensitivity system and thus to ensure a realistic initial failure minimum the design space is allotted as given in Table 17.3.

Table 17.3 Upper and lower specification limits for design parameters

Design parameter	*lsl*	*usl*
μ_1	7.300×10^{-3}	7.600×10^{-3}
tol_1	$0.5\%\mu_1$	$1.6\%\mu_1$

To form the meta-model, we let $\mathbf{d}(\mathbf{p})^T = \left[1\ \mu_1\ tol_1\ \mu_1^2\ tol_1^2\ \mu_1 tol_1\right]_{1\times 6}$; then 40 training sets are chosen appropriately from the design space, and the matrix $\mathbf{D}_{40\times 6}$ is built. To get the corresponding output matrix (i.e. *cdf*), we note the uncertainties in the three time-variant responses which are of the form

$$Z_i(t_l) = f((\mathbf{p}, U_1), (\mathbf{p}_V, \mathbf{U_V}), (\mathbf{q_Y}(t_l), \mathbf{U_Y}), \mathbf{r_C}, t_l), i = 1, 2\,and\,3 \quad (17.30)$$

where $\mathbf{p}_V, \mathbf{q_Y}(t_l), \mathbf{r_C}, \mathbf{U_V}, \mathbf{U_Y}$ are found in Table 17.2. Now, based on Table 17.1, the four limit-state functions for samples of the responses are symbolically $g_1 = 0.051 - z_1$, $g_2 = 595 - z_2$, $g_3 = z_2 - 545$ and $g_4 = z_3 - 0.19$. The incremental failure probability in Eq. (17.7) uses the series system failure event (needed for Eq. (17.6)) in the form

$$\mathbf{E}_l(\mathbf{p}) = \{g_1(\mathbf{p}, t_l) \le 0 \cup g_2(\mathbf{p}, t_l) \le 0 \cup g_3(\mathbf{p}, t_l) \le 0 \cup g_4(\mathbf{p}, t_l) \le 0\} \quad (17.31)$$

For the training samples $\mathbf{p}_j\, j = 1, 2, \ldots 40$, the corresponding *cdf* at time increments of $\Delta t = 0.1$ year over a lifetime of ten years is generated by MCS with $N = 100{,}000$ samples and stored in matrix $\overline{\mathbf{F}}_{40\times 11}$. Representative curves are shown in Fig. 17.4. The quite broad range of curves shows how sensitivity the *cdf* is to the design parameters. (The time to generate the training *cdf*'s is only 20 s.)

The meta-model links the two matrices through the weight matrix $\mathbf{W}(\mathbf{p})$. As a test of the efficacy of the meta-model, the *cdf* is obtained for arbitrary test values $[\mu_1, tol_1] = [7.3074 \times 10^{-3}, 1.4056]$. The results using both quadratic moving least squares (qMLS) meta-model and the ubiquitous kriging meta-model are compared to the traditional marching-out MCS method and shown in Fig. 17.5. The errors are acceptable with the average being about 2.5%. Other test values give similar errors.

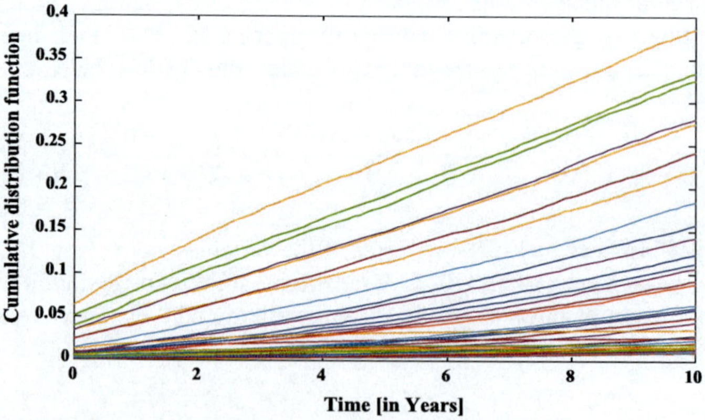

Fig. 17.4 *Cdf* Curves (i.e. output) from training data (40 samples)

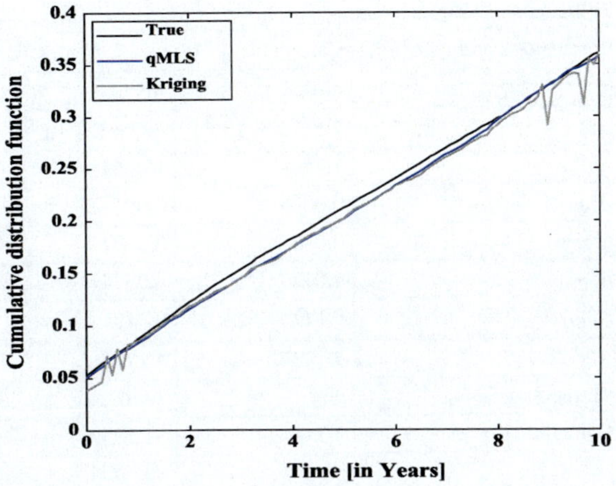

Fig. 17.5 Comparison of *cdf* curves for meta-models

17.5.5 Design Application

The meta-model is now used for design purposes invoking an optimization algorithm [41, 42].

The total cost is the objective to be minimized. The constraints comprise failure policies at initial time and some later time along with the design space. We write

$$\begin{aligned} &\text{Minimize } C_p(\mathbf{p}) + c_S(F(\mathbf{p},t_0)) + C_{LQ}(\mathbf{p}, \theta, c_F) \\ &\quad\text{subject to} \\ &\quad\quad F(\mathbf{p}, t_0) \le F_0, \\ &\quad\quad F(\mathbf{p}, t_M) \le F_M \\ &\quad\quad\quad \mathbf{p}_L \le \mathbf{p} \le \mathbf{p}_U \end{aligned} \tag{17.32}$$

where for production cost is $C_p(\mathbf{p}) = 3.5 + 0.57/tol_1$ and the loss-of-quality cost is $C_{LQ}(\mathbf{p}, \theta, c_F) = c_F \sum_{l=1}^{L} \left((F(\mathbf{p}, t_l) - F(\mathbf{p}, t_{l-1}))e^{-\theta t}\right)$. For two cases, the failure constraints and the scrap and loss-of-quality parameters are

Case (a): $F_0 = 0.001$, and $F_{\mathrm{M}} = 0.05$ at $t_{\mathrm{L}} = 10$ year; $c_s = \$20$, $c_F = \$15$ and $\theta = 3\%$

Case (b): $F_0 = 0.001$, and $F_{\mathrm{M}} = 0.005$ at $t_{\mathrm{M}} = 5$ year; $c_s = \$20$, $c_F = \$20$ and $\theta = 3\%$

The optimization results are shown in Table 17.4. The MCS approach used 20,000 samples to keep the elapsed time reasonable.

Table 17.4 Design results

Parameters and cost [$]	Case (a)		Case (b)	
	qMLS	MCS	qMLS	MCS
μ_1	7.5110×10^{-3}	7.4867×10^{-3}	7.5066×10^{-3}	No solution
tol_1	1.4539	1.4061	1.1414	–
F_O	0.0008	0.001	0.001	–
F_M	0.0137	0.0191	0.0044	–
C_P	4.0159	4.0334	4.1571	–
C_{LQ}^E	0.1778	0.2451	0.2028	–
C_T	4.1937	4.2785	4.3599	–
Iterations	944	3320	597	4596
Time [sec]	5.3	31,785.6	3.15	43,200.6

In Case (a) and Case (b), the meta-model approach found a solution in an extremely small time, met the constraints and produced costs comparable to the MCS solution for Case (a). Note that a single *cdf* meta-model has been used in both design scenarios. In essence, the investment of 40 MCS to train the meta-model has obviated the need to perform thousands of MCS for the optimization process.

17.6 Conclusions

Herein, we have presented an efficient, two-stage, methodology for RBDO of time-dependent engineering systems. The time dependence of interest is caused by stochastic degradations of dimensions and materials. To obviate the lengthy *cdf* computations by MCS at each optimization iteration, a meta-model that gives the *cdf* in terms of the design space has been built as a first step. Since the meta-model is essentially explicit, the *cdf* prediction becomes very fast. Sufficient training data ensures acceptable accuracy. The overhead to form the meta-model becomes trivial when compared to the time needed for optimization iterations with the traditional marching-out MCS. The meta-model adopted herein is based on the moving least squares paradigm and has been found to be much faster and as accurate as the ubiquitous kriging meta-model. The accuracy of the moving least squares meta-model arises from the use of a regularized formula for choosing the weights that determine the so-called nearby training samples of the *cdf*.

The case study has pointed out the efficacy of the meta-model approach. Herein, the meta-model captured accurately the nature of the *cdf* for an electromechanical servo system with multiple competing performance measures under both stochastic excitations and stochastic degradations. The approach has led to the shortening of the optimization time by several orders of magnitude with acceptable accuracy and thus presents a useful tool for RBDO of time-dependent systems.

References

1. Meeker, W. Q. & Escobar, L. A. (1993). *Statistical methods for reliability data*, Wiley, New York, USA, Chap. 7.
2. Ye, Z., & Xie, M. (2015). Stochastic modelling and analysis of degradation for highly reliable products. *Applied Stochastic Models in Business and Industry, 31*(1), 16–32.
3. Shahraki, A. F., Yadev, O. P., & Liao, H. (2017). A review on degradation modelling and its engineering applications. *International Journal Performability Engineering, 13*(3), 299–314.
4. Pandey, M. D., Yuan, X. X., & van Noortwijk, J. M. (2009). The influence of temporal uncertainty of deterioration in life-cycle management of structures. *Structure and Infrastructure Engineering, 5*(2), 145–156.
5. van Noortwijk, J. M., Kallen, M. J., & Pandey, M. D. (2005). Gamma processes for time-dependent reliability of structures, In K. Kolowrocki, (Ed.), *Advances in Safety and Reliability, Proceedings of ESREL 2005*, Tri City, Poland. 1457–1464.
6. Kuschel, N., & Rackwitz, R. (2000). Optimal design under time-variant reliability constraints. *Structural Safety, 22*(2), 113–127.
7. Wang, Z. & Wang, P. (2012). Reliability-based product design with time-dependent performance deterioration, *IEEE Conference on Prognostics and Health Management*, CO, USA 1–12.
8. Hu, Z., & Du, X. (2016). Reliability-based design optimization under stationary stochastic process loads. *Engineering Optimization, 48*(8), 1296–1312.
9. Jiang, C., Fang, T., Wang, Z. X., Wei, X. P., & Huang, Z. L. (2017). A general solution framework for time-variant reliability based design optimization. *Computer Methods in Applied Mechanics and Engineering, 323*, 330–352.
10. Savage, G. J., & Son, Y. K. (2009). Dependability-based design optimization of degrading engineering systems. *ASME Journal of Mechanical Design, 131*(1), 011002.
11. Rathod, V., Yadov, O. P., Rathore, A., & Jain, R. (2012). Reliability-based design optimization considering probabilistic degradation behaviour. *Quality and Reliability Engineering International, 28*(8), 911–923.
12. Singh, A., Mourelatos, Z. P., & Li, J. (2010). Design for life-cycle cost using time-dependent reliability. *ASME Journal of Mechanical Design, 132*(9), 091008.
13. Wang, G. G., & Shan, S. (2007). Review of metamodeling techniques in support of computer-based engineering design optimization. *ASME Journal of Mechanical Design, 129*(4), 370–380.
14. Simpson, T. W., Peplinski, J. D., Kock, P. N., & Allen, J. K. (2001). Metamodels for computer-based engineering design: Survey and recommendations. *Engineneing Computing, 17*(2), 129–150.
15. Sacks, J., Welch, W. J., Mitchel, T. J., & Wynn, H. P. (1989). Design and analysis of computer experiments. *Statistics Science, 4*(4), 409–423.
16. Martin, J. D., & Simpson, T. W. (2005). Use of kriging models to approximate deterministic computer models. *AIAA Journal, 43*(4), 853–863.
17. Nielsen, H. B., Lophaven, S. N., & Søndergaard, J. (2002). *DACE A matlab kriging toolbox*, Technical Report IMM-TR-2002-12, Technical University of Denmark.
18. Birattari, M., Bontempi, G., & Bersini, H. (1998). Lazy learning meets the recursive least squares algorithm, *Proceedings of the 1998 conference on Advances in neural information processing Systems* 11, MA, USA. pp. 375–381.
19. Most, T., & Bucher, C. (2005). A moving least squares weighting function for the element free Galerkin method which almost fulfils essential boundary conditions, *Structural Engineering and Mechanics, 21*(3):315–332.
20. Romero, D. A., Amon, C. H., Finger, S., & Verdinelli, I. (2004). Multi-stage Bayesian surrogates for the design of time-dependent systems, *ASME 16th International Conference on Design Theory and Methodology*, Utah, USA, Sept. 28–Oct. 2. pp. 405–414.
21. Youn, B. D., & Choi, K. K. (2004). A new response surface methodology for reliability-based design optimization. *Computers and Structure, 82*(2/3), 241–256.

22. Kang, S., Koh, H., & Choo, J. F. (2010). An efficient response surface method using moving least squares approximation for structural reliability analysis. *Probabilistic Engineering Mechanics, 25*(4), 365–371.
23. Song, C., & Lee, J. (2011). Reliability-based design optimization of knuckle components using conservative method of moving least squares models. *Probabilistic Engineering Mechanics, 26*(2), 364–379.
24. Kumar, A., & Chakraborty, A. (2017). Reliability-based performance optimization of TMD for vibration control of structures with uncertainty in parameters and excitation. *Structural Control and Health Monitoring, 24*(1), e1857.
25. Wehrwein, D., & Mourelatos, Z. P. (2009). Optimization of engine torque management under uncertainty for vehicle driveline clunk using time-dependent meta-models. *ASME Journal of Mechanical Design, 131*(5), 051001.
26. Son, Y. K., & Savage, G. J. (2018). A simple explicit meta-model for probabilistic design of dynamic systems with multiple mixed inputs. *International Journal of Reliability Quality and Safety Engineering, 25*(3), 1850011.
27. Savage, G. J., Son, Y. K., & Seecharan, T. S. (2013). Probability-based prediction of degrading dynamic systems. *ASME Journal of Mechanical Design, 135*(3), 031002.
28. Drignei, D., Baseski, I., Mourelatos, X. P., & Kosova, E. (2016). A random process metamodel approach for time-dependent reliability. *ASME Journal of Mechanical Design, 138*(1), 011403.
29. Zhang, D., Han, X., Jiang, C., Liu, J., & Li, Q. (2017). Time-dependent reliability analysis through response surface method. *ASME Journal of Mechanical Design, 139*(4), 041404.
30. Rosenblatt, M. (1952). Remarks on a multivariate transformation. *the Annals of Mathematical Statistics, 23,* 470–472.
31. Styblinski, M. A., & Huang, M. (1993). Drift reliability optimization in IC design: Generalized formulation and practical examples. *IEEE Transactions on Computer-Aided Design of Integrated Circuits and Systems, 12*(8), 1242–1252.
32. Loeve, M. (1977). *Probability theory I*. New York, USA: Springer.
33. Xiu, D. B., & Karniadakis, G. E. (2002). The Weiner-Askey polynomial chaos for stochastic differential equations. *SIAM Journal Science Computing, 24*(2), 619–644.
34. Zhang, J., & Ellington, B. (1994). Orthogonal series expansion of random fields in reliability analysis. *Journal of Engineering Mechanics, 120*(12), 2660–2677.
35. Li, C., & Kiureghan, A. D. (1993). Optimal discretization of random fields. *Journal of Engineering Mechanics, 119*(6), 1136–1154.
36. Savage, G. J., & Son, Y. K. (2019). Reliability-based design optimization of time-dependent systems with stochastic degradation. *Journal of Mechanical Science and Technology, 33*(12), 5963–5977.
37. Avramidis, A. N., L'Ecuyer, P., & Trembley, P. (2003). Efficient simulation of gamma and variance-gamma processes, *Proceedings of the 2003 Winter Simulation Conferences*, LA, USA. pp. 319–326.
38. Leon, S. J. (1998). *Linear algebra with application.* Upper Saddle River, N. J., USA: Prentice Hall.
39. Lee, D. Q. (2012). *Numerically efficient methods for solving least squares problems*, [Online].
40. Savage, G. J., & Carr, S. M. (2001). Interrelating quality and reliability in engineering systems. *Quality Engineering, 14*(1), 137–152.
41. Son, Y. K., & Savage, G. J. (2008). Economic-based design of engineering systems with degrading components. *International Journal of Production Economics, 111*(2), 648–663.
42. Chou, C. Y., & Chen, C. H. (2001). On the present worth of multivariate quality loss. *International Journal of Production Economics, 70,* 279–288.

Gordon J. Savage was born in Canada and received his Ph.D. from the University of Waterloo in 1977 in the area of Graph-Theoretic Models of Engineering Systems. He is presently a faculty member of, and a full professor in, Systems Design Engineering at the University of Waterloo, Waterloo, Ontario, Canada. His present research interests are in the areas of (a) modelling, formulation and computer implementation of linear graph models of engineering systems, and (b) design for quality, reliability and robustness in complex engineering systems. He has published over 100 refereed papers. Research is ongoing to integrate quality and reliability metrics using both mechanistic and empirical models. He has been involved in the study of the methodology of design for over 40 years and spent parts of seven years presenting the process of innovation in China. His practical experience in industry and his supervision of the Midnight sun solar car project give him a unique ability to link the methodology of design, the process of innovation and robust design.

Young Kap Son received his Ph.D. from the Department of Systems Design Engineering in 2006 at the University of Waterloo in Canada. Currently, he is a professor in Mechanical and Automotive Engineering at the Andong National University, Korea. His current research interests include economic-based design of dynamic systems for reliability improvement, and reliability estimation and design optimization of one-shot systems.

Chapter 18
Dynamic Asset Performance Management

Aditya Parida and Christer Stenström

Abstract Managing asset performance under prevailing dynamic business and industrial scenario is becoming critical and complex, due to technological advancements and changes like artificial intelligence, Industry 4.0, and advanced condition monitoring tools with predictive and prescriptive analytics. Under the dynamic asset management landscape, asset performance is an integral part of an industrial process to ensure performance assurance and acts as a key game changer. Therefore, managing the asset performance and data analytics throughout the asset life cycle is critical and complex for the long-term industrial and business viability, as it involves multiple stakeholders with dynamic inputs and outputs with conflicting expectations. Lack of linkage and integration between various stakeholders along the hierarchical levels of an organization with their changing requirements is still a major issue for industries. For integration within an organization, each asset needs predictive and prescriptive analytics, besides it needs to be linked and integrated for achieving the business goals. In this chapter, managing the various issues and challenges to dynamic asset performance is discussed.

Keywords Dynamic asset performance · Performance assurance · Artificial intelligence · Industry 4.0 · Predictive and prognostic analytics

18.1 Introduction

The manufacturing industry and asset managers are passing through a very complex and challenging time due to the dynamic global business scenario and the emerging disruptive technologies which compel them to transform, adopt, and manage their asset performance competitively to meet the business goals. The manufacturing industry is operating under a digital world where technology is the heart of all processes with a demand to meet challenges of Industry 4.0 with artificial intelligence [1], big data and industrial internet, predictive and prescriptive analytics, to

A. Parida (✉) · C. Stenström
Luleå University of Technology, 971 87 Luleå, Sweden
e-mail: Aditya.Parida@ltu.se

K. B. Misra (ed.), *Handbook of Advanced Performability Engineering*,
https://doi.org/10.1007/978-3-030-55732-4_18

fulfill the asset management strategies and goals as per ISO 55000 [2]. With 60–80% of asset's operation and maintenance cost, asset performance management plays a vital and critical role in asset life cycle management [3].

The landscape of asset management in 2020 is expected to touch 101.7 Trillion US $ [4]. As the world has become a global village, the global economies are increasingly integrated and interdependent. The asset under management of one part of the world may be influenced by the changes in the Gross Domestic Products (GDP) of another region/country. The unprecedented economic and political disturbances added by regulatory changes compel the asset managers to find time to look into the future. All these technological, economic, and regulatory changes are bringing in fundamental shifts which are going to dictate the future of the asset management for the industry. The asset managers need to plan for the future considering the likely changes for the asset management landscape and identified key game changers for the global and industrial competitive environment.

The regulatory bodies for asset management are focusing on risk-based framework under which the industry has to manage their aging assets. The ISO 55000 International Standard provides guidance and assurance for the industry and the asset managers to realize maximum value from the asset while balancing the risk, cost, and performance to meet the mission and objectives of their organization. Asset Management (AM) is defined as the "coordinated activity of an organization to realize value from assets" [2]. Performance cannot be managed and assured if it cannot be measured. It is only through performance measurement that assets can be managed to meet the challenging demands of the dynamic industrial objectives through increased productivity with better availability and utilization of assets. The asset managers and owners need to measure the performance of industrial and manufacturing process to understand the tangible and intangible contribution of assets toward business objectives. Asset performance assurance is required to evaluate, control, and improve various asset activities for ensuring achievement of organizational goals and objectives. Asset Performance Management (Asset PM) encompasses the capabilities of data capture, integration, visualization, and predictive and prescriptive analytics tied together for the explicit purpose of improving the reliability and availability of physical assets.

Under this context, in recent years, asset performance management is receiving a lot of attention from industry and academia. This chapter deals with the broad topic of asset performance assurance management and discusses various issues and challenges associated with dynamic asset performance under Industry 4.0 business scenario. The outline of the chapter is as follows: after introduction, the emerging trends of asset performance management for industry with its associated issues and challenges are presented in the next section. Managing data analytics for asset PM and digital twins for data analytics and industrial solutions are discussed in Sect. 18.3. The important issues associated with the development of a dynamic asset PM are discussed in Sect. 18.4. Section 18.5 presents a link and effect case study for asset performance management, and the final section concludes the chapter.

18.2 Emerging Trends of Asset Management

Managing asset is a critical function and activity for the asset owners and managers. Asset management is a structured and systematic approach adopted by the asset owners or managers to perform and deliver the expected goals over assets' entire life cycle. The asset management life cycle process includes acquiring, operating, maintaining, upgrading, and disposing of the asset for a cost-effective, least risk, and optimized performance. Managing infrastructure asset needs financial, economic, engineering, and other management practices cost effectively to meet the business goals. Managing infrastructure and industrial assets is a very important issue for the owners and infrastructure managers. The asset management's dynamic approach is based on asset health condition monitoring with predictive and prescriptive analytics to meet the conflicting demands of the various stakeholders and regulatory bodies through optimized asset performance and return on investment.

With the onset of the internet revolution, computing power and data networks provide an unprecedented flow of information and communication for managing asset. With more and more technological innovation and development, a transformational change is taking place for the industrial internet. This has brought us artificial intelligence enhanced machines dealing with big data and advance data analytics like predictive and prescriptive analytics [5] leading into a connection of human mind and machine. All these technological changes are bringing in a jump in global economy and asset management. The assets of all industrial sectors like infrastructure, transportation, healthcare, and other sectors need all these technological advancement and industrial internet for real-time monitoring, control, and right decision-making to meet the global competition and business goals. These are achieved through sensors and other condition monitoring tools, to see, smell, hear, and feel the asset health condition supporting the asset utilization and system optimization.

Sensors are there for some time now. But over time the sensors are developed to be less costly, with transmitting high volume data for longer duration. With these sensors and condition monitoring, the assets have become predictive, prognostics, and reactive. The assets are today equipped with artificial intelligence and communicate seamlessly with all stakeholders. Today, we are able to use embedded systems with multi-core processor technology, advanced cloud computing system, and a new software-defined machine infrastructure for real-time asset management by decoupling machine software from hardware. All these supports to monitor, manage, and upgrade industrial assets.

Preventive condition monitoring to zero-based failure with predictive and prescriptive analytics without any asset delays is core to asset PM. For example, today's airline industry global flight delays due to maintenance amount to 10% of flight delays which amount to 8 billion \$. The preventive aviation system allows the aircraft to communicate with the technician and reduces any delay as the technician already knows what actions to be taken on the aircraft by the time it lands. All these leads to reduce the turnaround time and delays for the industry sectors. In the energy sector today, the wind turbines communicate with each other to synchronize their blades producing more energy and reducing the cost of energy produced from 30 to 5 cents per kilowatt.

The paradigm shift in asset management starts with traditional financial perspectives prior to 1990. See Fig. 18.1. However, industry soon visualized that mere financial perspectives cannot help them from becoming uneconomical and bankrupt, unless they look into the quality of manufacturing, products and process inspection and control, i.e., taking care of both financial and nonfinancial perspectives. With this background, while ISO 9000 was introduced in 1987, which looks into all aspects of quality of the production and organization, Kaplan and Norton came out with the balanced score card [6] which looks into four perspectives of the business activities, both financial and nonfinancial. These four perspectives are customer, finance, internal business process, and learning and growth. However, industry was missing a standard which could take care of all aspects of asset management. While a lot of research works were going on, British Standards Institution (BSI) came out with PAS: 55 in 2004 to take care of asset management, while another ISO 14000 has come out for managing the environmental issues. During this period, a number of research works were undertaken to focus on performance measurement, linking it to the business objectives and strategy, KPIs, and Performance Indicators (PIs). While measuring the operational and maintenance performance, a holistic concept for asset performance measurement and asset management was developed during 2006. A group of industries from US, Europe, and Australia were simultaneously working to develop a standard on Asset management ISO 55000 [2].

The International Organization for Standardization published the asset management system standard ISO 55000 in 2014. The ISO 55000 series provides terminology, requirements, and guidance for implementing, maintaining, and improving an effective asset management system. This standard is increasingly used by the industry

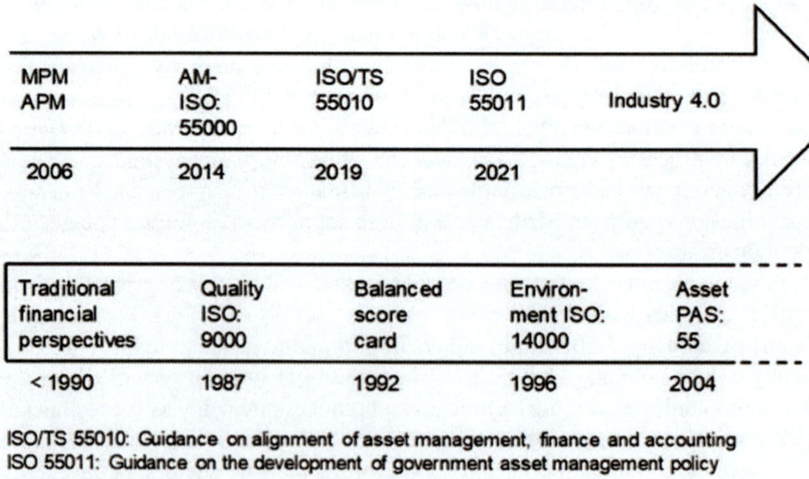

Fig. 18.1 Paradigm shift in asset management

and infrastructure sectors in order to achieve the optimization of costs, risks, performance, and sustainability. Standard ISO 55000 consists of three standards, where ISO 55000 provides a critical overview, concepts, and terminology for developing a long-term plan for the organization's objectives, business policies including stakeholders' requirements. ISO 55001 deals with the requirements of the organization, implementation, maintenance, and improvement of the asset management system. Asset management and asset management system relationship is provided in ISO 55001 [7] and grouped in consistent with fundamentals of AM, context of organization, leadership, planning, support, operation, performance evaluation, and improvement. Asset performance evaluation (clause 9 of ISO 55001) forms an integral part of the asset performance management for the industry. ISO 55002 offers interpretation and guidance for such a system to be implemented in accordance with the requirements [8]. As further improvement to ISO 55000, in 2016, the formal revision process of ISO 55002 was launched. In 2017, two new projects were launched: ISO/TS 55010: Guidance on alignment of asset management, finance, and accounting (published in September 2019), and ISO 55011: Guidance on the development of government asset management policy (targeted for publication in 2021).

An Asset Management System (AMS) needs to be developed and documented as per ISO 55001. AMS is a "set of interrelated or interacting elements to establish asset management policy, asset management objectives and processes to achieve those objectives." Strategic approach is followed in ISO 55001 adhering to ISO's new structure which follows the Deming's plan-do-check-act cycle which covers to organization, leadership, planning, support, operation, performance evaluation, and improvement. Engineering asset strategy is formulated from the corporate strategy considering the integrated and whole life cycle of the asset. An integrated approach is essential as an asset performance management is associated with various stakeholders with their conflicting needs with multiple inputs and outputs. From asset performance objectives, two set of activities are undertaken. One set of activity develops the key performance indicators for benchmarking performance with similar industry and the other set formulates the activity plan, implementation, and measurement.

18.2.1 Asset Performance Management for Industry

Managing asset is a complex issue involving various issues and challenges for the asset owners and managers. The asset management's dynamic approach is based on asset health condition monitoring to meet the conflicting demands of the various stakeholders and regulatory bodies through optimized asset performance and return on investment. Asset strategy is formulated from the corporate strategy considering the integrated and whole life cycle of the asset. An integrated approach is essential for the asset management as it is associated with various stakeholders with their conflicting needs besides multiple inputs and outputs. Assets cannot be managed if their performance cannot be measured and evaluated. The asset strategy is formulated from asset performance objectives based on the corporate objectives and strategy.

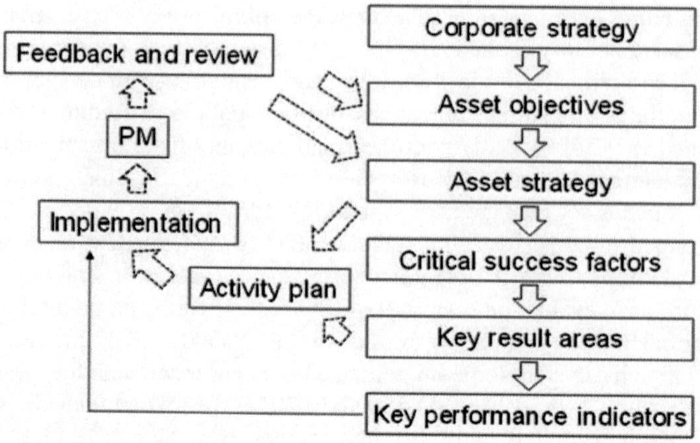

Fig. 18.2 Strategic asset performance measurement process. Adapted from Parida et al. [9]

From asset strategy follows the critical success factors, key result areas, KPIs, and PIs. The KPIs are used for benchmarking performance with similar industry and formulating the activity plan, implementation, measurement, and review as given in Fig. 18.2 as a continuous improvement process. As shown in the figure, asset performance objectives are formulated as per stakeholders' requirements and organization's integrated capability and capacity. KPIs are formulated from the objectives for the strategic and managerial levels for measuring and assessing the asset performance through various PIs and measures. The activity plans are made for undertaking implementations along with measurement and performance assessment, based on which feedback and reviewing action are undertaken to validate the asset performance objectives. The asset strategy and performance objectives are modified or updated based on the feedback and review as a continuous process.

The strategic asset performance requirements involve two activities [10]:

1. Cascading down the objectives from strategic to shop floor level.
2. Aggregation of performance measurements from shop floor to strategic level.

The business objectives are cascaded down through the corporate strategy, asset objectives, asset strategy, critical success factors, key result areas, KPIs, and PIs at the shop floor and individual asset level. Asset objectives and strategies are formulated based on both internal and external stakeholders needs. Based on the plant capacity, resources are allocated for implementation. The performance of the asset is monitored and controlled through compiled data from condition monitoring and is measured and aggregated through the PIs from functional level to the KPIs at tactical or managerial level to the strategic level in a bottom-up manner. This aggregation of PIs to KPIs are compared with the business and asset strategy so as to modify and improve the performance in a continuous manner.

The cascading down of the objectives from strategic to shop floor or operational level and aggregation of performance measurements from shop floor to strategic level can be seen under the APM concept of ISO 55000 as shown in the Fig. 18.3. Without a comprehensive description of strategy, executives cannot easily communicate the strategy among themselves or to their employees [6]. Therefore, it is essential that the corporate strategy and objectives of an organization is converted to the specific objectives integrating different hierarchical levels of the organization. With increasing technological advancements in condition monitoring tools and industrial internet for data collection, compilation, and data aggregation, the PIs and KPIs are calculated and compared. Under Industry 4.0 and industrial artificial intelligence, the asset capability and resources utilization become critical for implementing an appropriate Performance Measurement (PM) system. Without an integrated PM system, the assets cannot be managed and the desired objectives cannot be achieved. The PM system forms the foundation for making improvement decisions by modifying or changing objectives, production targets, priority areas, modifying resource allocations, and new or improved technology introduction amongst others. Thus, a PM system indicates how the overall organization, its collective, and individual assets are performing, through benchmarking internally and externally in all aspects of asset management, besides productivity improvement and optimization.

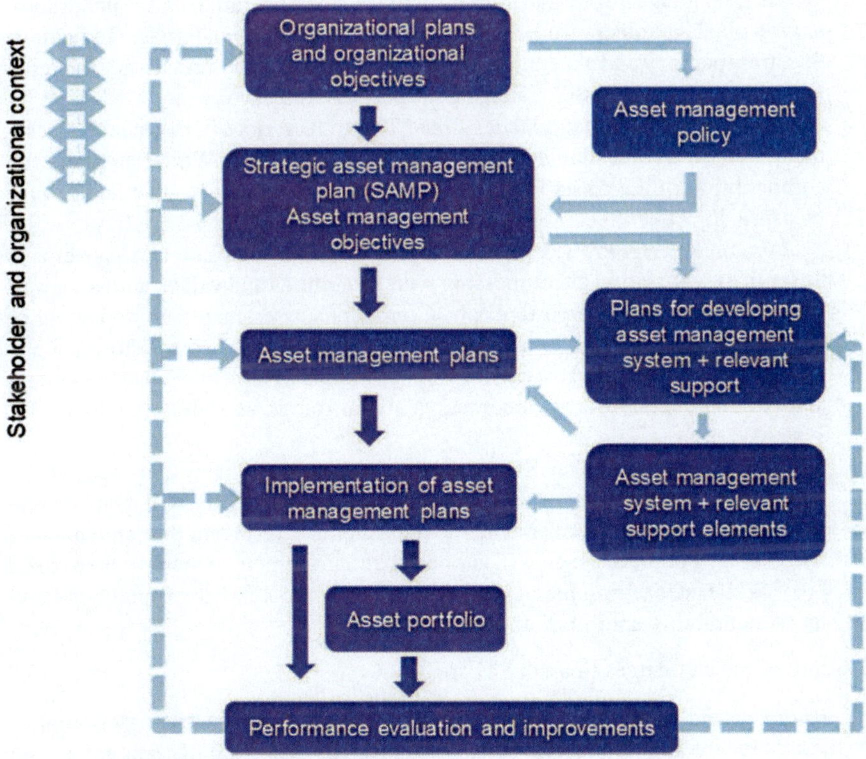

Fig. 18.3 APM Concept under ISO 55000. Adapted from ISO 55000 [2]

18.2.2 Issues and Challenges in Asset Performance Measurement

There are a number of issues and challenges in asset performance assessment. Some of the major issues related to asset management are

(a) *Developing asset management strategy*. Based on the internal and external stakeholders needs, the corporate objectives and strategy are formulated based on which the asset management objectives and strategy are made to translate into individual asset targets and goals at the operational level. Similarly, the integration of results and outcomes from shop floor level to managerial and strategic levels needs to be considered and formulated, so that the overall business targets can be compared with the achieved ones and necessary decision-making can be made. While doing so, following up the ISO 55000 can be a very useful and supporting tool.
(b) *Supporting organizational issue*. Alignment of asset management system with the corporate strategy is a complex issue in the entire organization. The asset owners and operators need to understand the asset management requirements in the same language by one and all. This will lead to cooperation and collaboration among all stakeholders for achieving the AM objectives and goals. To achieve this, transparency and accountability of information and communication related to various aspects of asset management plays a critical role.
(c) *Asset performance measurement system*. There are various performance assessment frameworks available [11] besides the ISO 55000. While choosing the compatibility of the Asset PM system, one needs to consider what and how to measure, how to collect the data, its storage and undertaking analytics.
(d) *Sustaining the asset PM system*. Developing an appropriate asset PM system for the entire organization encompassing with its cultures and values with a view to sustaining it plays an important role. Continuous assessment and reviewing of the system need to be undertaken for the improvement of asset performance at regular interval and build trust in the PM system. Involvement of all stakeholders and communication transparency plays a vital role in sustaining the asset PM system.
(e) *Asset PM indicators*. The KPIs and PIs of the asset PM system need to be specific, measurable, attainable, realistic, and timely. This will help all the stakeholders to clearly understand the measures and results in the same manner without any conflicts which will support the management to achieve their goals [10]. While identifying the asset PM indicators, the number of indicators and its accountability are important issues.

Some of the challenges in asset PM are

(a) *Choosing the right assets*. Within an organization, with smarter asset requirements for intelligent asset management solutions, selection of right asset is an important challenge. Today, the asset owners will like to procure assets which can undertake regular condition monitoring with embedded sensors and other tools to transmit their health data on real time to predict failures through diagnostics and prescribe solutions. While for a new organization the options are many with a number of high-tech assets, the challenge is for the aging assets with long residual life and less productivity.

(b) *Minimizing asset downtime*. Minimizing asset downtime is important both from production loss and downtime cost of the production process. For manufacturing, mining, or aviation, the downtime cost per hour is very high, and under the present competitive business scenario, it is very difficult for any company to withstand this loss. For this purpose, real-time asset health monitoring, data transmission, collection, storage for its prognostics and prescriptive analytics are challenging issues.

(c) *How to replace aging assets*. Replacing the aging assets which are not so high tech with sensors and data acquisition is a big problem for the asset owners at present. The owners need to consider the life cycle costs of the aging acquiring asset along with health condition and then to proceed with cost comparing and decision-making of replacement.

(d) *Optimizing cost not compromising risk*. While taking care of an aging asset or procuring a new asset, a study and analysis of cost-risk-profit from life cycle cost perspective is essential. With new national and international regulations on safety and risk, no one can overlook the challenge of productivity and profitability (cost) without sacrificing the risk.

(e) *Systematic asset PM approach*. Though asset PM system looks quite simple, a lot of activities are essential including training and skill improvement in PM system and a thorough understanding of the asset PM system. No two organizations nor assets are equal nor identical, for which the asset PM system for the organization has to be unique and develop their own PM system while following a similar PM system.

(f) *Organizational change and breaking the silos*. To bring in an organization change is a real challenge. Most of the existing and older organization gets used to their own structures, work cultures, and hierarchy. So, procuring a new asset with new technology or introducing a PM system or ISO 55000 is a challenging task which needs to be adopted and introduced in a planned and structured manner involving one and all within the organization and to break the existing working silos. The old and technically non-savvy personnel and conducting training for them becomes a difficult and challenging task.

(g) *Predictive and prescriptive analytical challenges*. The health condition monitoring of the assets with emerging predictive, prognostics, and prescriptive analytics are also challenging though these techniques allow organizations to predict failures, risks, and remaining useful life, besides the creation of value by the asset using historical and real-time data. These analytics provide support to asset managers in real-time decision-making not compromising the safety and cost.

18.3 Managing Data Analytics for Asset PM

For smart assets with predictive and prescriptive analytics, and industrial artificial intelligence, the amount of big data is enormous. The data consist of text, unstructured/semi-structured/structured and multimedia content (images/video/audio) with about 20 quintillion (10×18) bytes of data that are produced per day now. The fastest growing data are generated from physical observation, inspection, and measurements, real-time online data from health monitoring through embedded and wireless sensors, and Natural Language Processing (NLP) of inspection records. With the accelerated technological developments, it is estimated that, by 2020, more than 50 billion devices will be connected to the internet.

The real challenge in big data analytics is to arrange, collect, and analyze actionable data, which converts data into insight, information, and action, supporting faster decision-making for real-time operation and optimized business process.

The traits of actionable data are

(a) Accuracy: The data elements need to be right, legible, valid, and equivalent. For the "right data," it is important to understand, why data is collected and what we are trying to achieve with this data collection. Data should not be collected without a purpose and not from any data which can be collected. Even with the purpose of data collection, we need to determine whether the collected data will meet the purpose for which these are collected. Is there a need to modify the data collection method, technology to meet the specific objectives? All these considerations and questions will decide what right data is required to be collected.
(b) Data quantity: The quantity of the data required to be collected, stored and analyzed for the purpose these are collected to meet the business objectives, is an important issue? It is seen that organizations were collecting vast amounts of data without any meaning except they are stored without any data exploitation for results. It is mostly seen that a limited dataset is required for monitoring the health and operational condition of the assets. Though, it is not specifically possible to know which type of data will be required for future data exploitation, the data quantity needs to be specific to meet the designed purpose and objectives.
(c) Accessibility and data sharing: The data need to be accessible by all users, devices, and data modalities for whom these are designed and collected. The data sharing needs to support and enable personnel and organizations to share more data willingly and effectively. There may be conflicting and contradicting interests between two or more data users/groups or subunits which are required to be managed during data structure and designing stage.
(d) Quality: Quality of actionable data includes data security, completeness, duplication, and consistency. Data quality forms an important issue while data is collected, stored, shared, and analyzed. The data collection people may not know or interpret if the data collected is of the quality it is expected to interpret or deliver the expected results. The interpretation or exploitation may be

carried out by a different group or organization not in the same location. There is a scope for possible error if the people involved between data collection and decision-making chain are many in numbers.

(e) Data costs: Though the information technology costs are reducing with time, the number of users and devices has increased manifold along with very high volume of data flow; therefore, a significant cost is involved in data collection, its storage, and usage for exploitation. These costs need to be compared with the gains from the use of data, and a cost-benefit analysis could control the overall data costs and its usage.

(f) Ubiquity/liquidity: The actionable data need to possess interoperability, persistency, and virtual availability. These aspects are essential for real time as well as virtual data application like digital twins. Data availability in devices of all the real or virtual users needs to be compatible, so that the data analytics are performed as designed for right decision-making.

(g) Organization of the actionable data: The actionable data organization involves context of the data, data logic, data structure, and semantic consistency. Data parsing which is "the process of analyzing text made of a sequence of tokens to determine its grammatical structure with respect to a given formal grammar" is also an important issue of actionable data organization.

The analytics challenges in managing data for asset PM are

(a) Improving data collection and management: The existing data collection method needs to be critically examined for its validity for actionable data and its improvement from smart asset and asset PM perspectives. The new and emerging technological changes with regard to asset health monitoring and data analytics are also required to be considered.

(b) Collecting less, updating more: The data quantity needs to be in matching with the asset PM requirements, thus collecting less with more focus on up-dation. There is a tendency to collect more data without validation of the purpose for which they are collected and if they can be linked with the asset PM system.

(c) Constructing and evaluating alternative life cycle scenarios: With the aging assets, data collection and its analytics becomes important for evaluation of alternate life cycle costs for different assets alternatives. The competitive business scenario demands an optimized operation and maintenance costs where these alternatives with life cycle costs trade-offs play important roles.

(d) The maintenance/operation cost trade-offs: The operation and maintenance cost trade-offs need big data and its analytics in asset PM system. This is an area where all management wants this cost trade-offs to be minimized so that their business profitability will go up.

(e) The maintenance/service quality trade-offs: The maintenance and service quality costs are important from customer satisfaction and asset PM perspectives. The customer expects the best maintenance service quality, while the management wants these to be optimized and offered at minimum cost.

(f) Incorporate flexibility: Assets are becoming smart with advanced technology. The expectation of the owners and operators is changing with new asset models.

All these changes demand a change in mindset and flexibility for the asset owners, operators, and other stakeholders. To incorporate flexibility may look simple, but from an organization and management prospective, it is a complex issue to be taken care of.

(g) Institutional Challenges: Each organization has its structure and work culture which are developed over a period of time. Any organizational change is therefore not an easy task which needs cooperation, coordination, and collaboration among all the stakeholders, especially breaking the configuration of working in silos. The planning and implementation of any organizational change needs transparent information flow, involvement of related stakeholders, their training, and implementation with a PM system to evaluate the results till the organizational objectives are achieved.

18.3.1 Digital Twins for Data Analytics and Industrial Solutions

Digital twin is a digital and virtual representation of potential and actual physical assets that can be used for diagnostics, maintenance, and product innovation. Digital Twins serve as realistic models for fast testing to ensure that design issues of a product is sorted out before the product reaches the shop floor. The digital twin provides the dynamics of how an Internet of things device operates and lives throughout product's life cycle. Because model-based Digital Twins do not require physical performance data to predict behavior, they can be used for a greater range of engineering tasks like conceptual development and virtual commissioning. Besides conceptual development and virtual commissioning, digital twins can be used for

Online diagnostics: Simulating the Digital Twin in parallel with the real machine to provide valuable insight into where a problem might arise as the machine's response varies from the model as it ages. The online diagnostics of assets provide a warning with onset of degradation and from the rate of degradation with time can indicate the likely time of failure or the remaining useful life of the asset. With the warning from the asset, the operator and the owner have a possibility to react and take an appropriate decision to repair or replace with cost and risk analysis and the time plan for such action.

Virtual sensors: The virtual sensors with its dynamic response help the digital twin to provide inputs for the control system and point out the replacement or repair of the faulty sensors. There may be a scenario where sensors can be eliminated with advanced smart assets. These virtual sensors are going to support in reducing the cost with an increased productivity and effectiveness.

Predictive maintenance: Digital twin looks into big data analytics for not only prognostics and predictive analytics but also prescriptive analytics with real time and virtual data. The data are collected and analyzed under dynamic loads for the gears, bearings, and motors for confirming various maintenance and repair tasks besides replacement decision-making. All these analytics help to determine the loads to calculate the impact on the component's life.

There are many factors that determine the maintenance schedule for a machine, but one factor is frequently overlooked because it is difficult to predict without a Digital Twin. This factor is the impact of dynamic loading on bearings, gears, and motors, caused by changes in the duty cycle. Putting a Digital Twin through a proposed duty cycle can help to determine the loads on these components, which can be used to calculate the impact on the component's life.

Product optimization. The manufacturer of the product and the production process tries to find out the answers to the following questions through the digital twins.

- The safety limit of the production process, i.e., how much load or stress can the machine undertake?
- What is the warning or alarm limit of the operation before the occurrence of an actual failure?
- What data exploitation is required for answering these questions?

Customer support: The digital twin is also used to verify the customer specifications, accept customer feedback suggestion for product improvement, which are tried out on the virtual product and production and to improve the real product if found to be appropriate. While doing these, different operating conditions criteria are applied on the virtual product to see the performance effect.

Thus, it can be seen that digital twins use and integrate industrial internet of things, artificial intelligence, machine learning, and smart data analytics for creation of a real-time digital simulation models to update and change the physical product. A digital twin collects multiple inputs from different sources like historical data from users, operational condition and failure data from operators, repair and maintenance data from the maintenance personnel, etc., to undertake analytics in real time and decision-making in real time for the present and future asset management. These data and data analytics are utilized by all stakeholders not only for the single assets but also for multiple assets individually and under an integrated manner for operational and maintenance optimization of the assets, systems, and manufacturing processes. The digital twin is in its formative years, with advanced technology and smart assets, with IT–OT–ET (Information Technology–Operational Technology–Engineering Technology) has the potential to solve many major problems in real-time.

How the digital twin is used for continuous improvement through data exploitation can be seen from Fig. 18.4. The digital twin product and production represents the virtual product and virtual production. The specifications are taken from virtual product for virtual production and after validation used for commissioning of real production and after couple of variation and verification used for ideal delivery of real product. The collaborative platform is used for continuous improvement for variation of different parameters and verification of real production and product till the optimized and specified product is delivered. This process of continuous improvement of digital twin is also effectively used for optimized operation and maintenance of the production process.

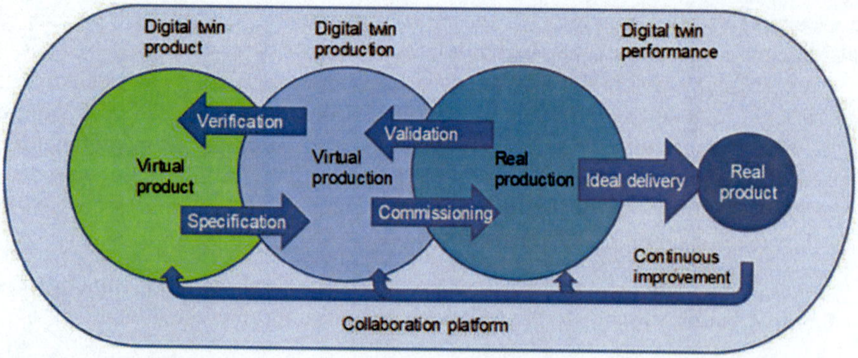

Fig. 18.4 Continuous improvement through data exploitation by digital twin. Adapted from Siemens [12]

18.3.2 Data Analysis and Asset PM

Since last couple of decades, various operational and organizational data are being collected through information technology; yet, an integration of these was a real challenge due to the silo working within an organization. Those organizations which broke these silos could perform better and could achieve optimized result. Yet, the Engineering Technology (ET) was not integrated into the IT–OT, thereby causing a lack of total solution to asset management. However, with onset of twenty-first century, the IT–OT–ET convergence emerged as the real solution to the Industrial Internet of Things (IIoT) which closes the asset life cycle management loop. The flow of data during an asset performance management is data collection, integration, and aggregation with quality management for data context visualization and modeling for data analytics after which data visualization and reporting. Analytics: The role of data under smarter assets for intelligent asset management solutions with digital transformation, use of machine learning and real-time data analytics through cloud computing and digital twin cannot be overlooked for continuous improvement and real-time asset management. See Fig. 18.5.

18.4 Integrated Dynamic Asset Performance Management Framework

Under the dynamic global business scenario, each successful organizations are aware to formulate the winning integrated strategy, implement, and manage it under a dynamic and competitive business environment. The word "Dynamic" is very relevant for asset performance management due to

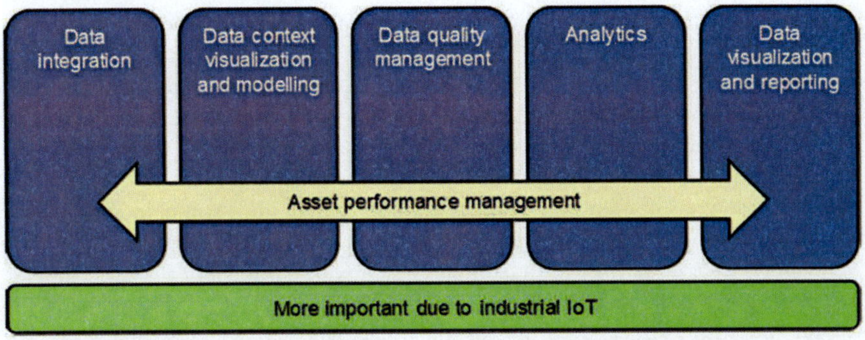

Fig. 18.5 Data analysis and asset performance management. Adapted from Hollywood [13]

(a) Dynamic business scenario and regulatory needs.
(b) Changing technological innovation and changes: Industry 4.0, artificial intelligence, digital twins, predictive and prescriptive analytics, and data analytics.
(c) Dynamic and conflicting requirements of various asset stakeholders (owners, managers, customers, suppliers, regulating authorities amongst others).

ISO 55000 and asset PM scorecard requirements needs the application of IT–OT–ET convergence and advanced technology like, asset health prognostics and digital twin. More and more organizations are applying asset PM as an integrated part of the strategic management. Asset PM system specifies and translates the business objectives and strategy across the hierarchical levels of the organization to one and all. The asset PM strategy is broken down to strategic, aggregate, and short-term plan with set targets and aligns various strategic initiatives enhancing the feedback for improvement and learning. The integrated issues in engineering asset PA are discussed as under Parida [10].

(a) *Stake holder's requirement*. Stakeholders' requirements are considered from both the external and internal stakeholders needs matching with the internal assets, their capability and capacity and other resources, based on which the corporate objectives and strategies are formulated. The external stakeholders' needs are considered from competitors and futuristic business scenario while the internal needs are considered from employees, owners and organizational perspectives. These corporate and strategies are translated into the targets at managerial and operational level.
(b) *Organizational issues*. The asset PM system needs to be aligned and form integral part of the corporate strategy. This will require commitments from the top management and all employees to be aware of the asset PM system through effective communication and training, so that they all speak the same language and are fully involved. The involvement of the employees in the asset PM system at every stage, like the planning, implementation, monitoring, and control, and at each hierarchical level can ensure the success of achieving the asset performance and business strategies. Besides, all functional processes and areas like

logistics, IT, human resources, marketing, and finance need to be integrated with engineering assets.

(c) *Engineering asset requirements*. From the stakeholders' need, the demand analysis of engineering asset is perceived and designed. After concept development, validation and engineering asset specifications are worked out. Besides, competitive product, cost of maintenance, risk management, correct product design, asset configuration and integration are considered from strategic and organizational perspective. From operation and maintenance, the engineering asset's maintenance tasks may be outsourced partially or completely.

(d) *How to measure?* It is essential to select the right asset PM system and KPIs/PIs for measuring asset PM from an integrated whole life cycle perspective for benchmarking besides collecting the relevant data and analysis for appropriate decision-making. The asset PM reports developed after the data analysis are used for subsequent preventive and/or predictive decisions though support of data analytics. The asset PM needs to be holistic, integrated, and balanced [14]

(e) *Sustainability*. Sustainability development is the development that is consistent while contributing for a better quality of life for the stakeholders. This concept integrates and balances the social, economic, and environmental factors with risk issues amongst others.

(f) *Linking strategy with integrated asset performance assessing criteria*. The linkage between integrated Enterprise Asset Management (EAM) measuring criteria with condition monitoring, IT, and hierarchical level for decision-making at different hierarchical level is given at Fig. 18.6. This figure describes

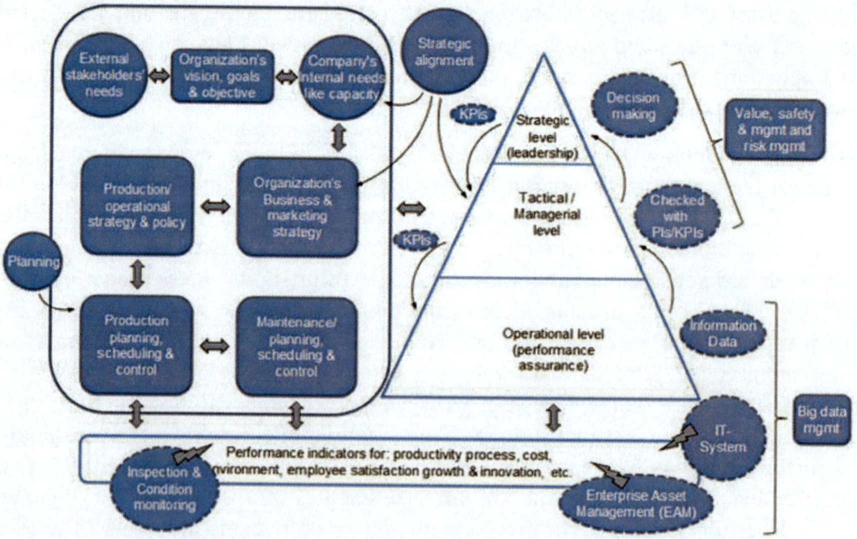

Fig. 18.6 Integrated dynamic asset performance management framework, big data, and predictive analytics under a link and effect concept. Adapted from Parida and Kumar [16]

the linkage between the external and internal stakeholders' needs and considers the concept of integrated enterprise asset management from the different hierarchical needs, while linking the performance measurement and assessment from engineering asset's operational level to decision-making at strategic level [15]. The external effectiveness is highlighted by stakeholders' need like return on investment and customer satisfaction. The internal effectiveness is highlighted through the desired organizational performance reflected by optimized and integrated resources utilization for EAM. For example, availability and performance speed of the equipment and the machineries forms part of the internal effectiveness or back end process. Quality is the most important aspect, which is not only related to the products' quality of the back end process, but also with customer satisfaction of external effectiveness. From external stakeholders, the quantity of annual production level is decided, considering the customer's requirements, return on investment and internal plant capacity and plant availability, etc. From internal stakeholders, the organization considers department's integration, employee requirements, organizational climate, and skill enhancement. After formulation of the asset PM system, the multi-criteria PIs are placed under the multi-hierarchical levels of the organization.

While considering the Asset PM from a holistic and balanced perspective, it is essential that the organization should be in a state of readiness to adopt the asset PM. The state of readiness of the organization will envelope and consider the issues related to asset PM to take care of the challenges. Considering the criticality of asset's KPIs for the complex system, an Integrated and Dynamic Asset Management (IDAM) framework considers the goals and levels with PM hierarchy through a link and effect relationship, with a real-time eMaintenance solution for "real time remote operation" through digital twins and Big Data analytics.

Thus, the asset management framework needs to consider

- Analyzing organization's business goals as per stakeholders needs
- Issues and challenges of asset management system related to business goals
- Associated regulatory compliance for safety and environmental issues
- Reviewing of operation and maintenance strategy with asset management strategy
- Return On Asset (ROA) investment and Return On Capital Employed (ROCE)
- Reliability centered design, Dynamic Reliability, Availability and Maintainability (RAMS) and risk analysis
- Spare parts and inventory management and optimization
- Multi-criteria and hierarchical asset PM and integrating condition monitoring tools and Big Data analytics
- Identifying KPIs and benchmarking for each unit/department's performance
- Integrating IT with operation and logistics, like CMMS/EAM and eMaintenance
- Life Cycle Costing (LCC) for dynamic business scenarios and decision-making.

Specifying the end results, i.e., developing the asset performance management framework with software, hardware, and demonstrator is required to be incorporated for verification of the framework. The success and failure of the asset PM of the organization will depend on the asset management and asset management system concept complemented with IDAM and link and effect concept considering the relevant issues and challenges. Those organizations, who consider, integrate, and implement asset PM concept, taking care of issues and challenges are going to succeed and survive under the competitive business scenario. The gap between the success and failure needs to be assessed by the organization and all measures to achieve business goals are required to be accepted strategically for implementation.

The success and failure of the asset PM of the organization will depend on the asset management and asset management system concept complemented with IDAM and link and effect concept considering the relevant issues and challenges. Those organizations, who consider, integrate and implement asset PM concept, taking care of issues and challenges are going to succeed and survive under the competitive business scenario. The gap between the success and failure needs to be assessed by the organization and all measures to achieve business goals are required to be accepted strategically for implementation.

18.5 Link and Effect Model Case Study for Asset Performance Management

The link and effect model concept is defined as "a methodology for developing performance measurements systems, by combining performance measurement and engineering principles for proactive asset management." Infrastructure Managers (IMs) have grown with the expansion of railway networks, and consequently the operation and maintenance practices have grown with the specific needs of each IM and country. However, harmonization and increased use of standards have come with the globalization, especially in the European Union (EU), considering increasing interoperability and building of a trans-European railway network. Therefore, performance measurement needs to be dynamic and versatile. Another important element in performance measurement of railways is the fast development of new technologies, including computers (hardware and software) and condition monitoring. Changes in the Enterprise Resource Planning (ERP) system or the Computerized Maintenance Management System (CMMS) within an organization can alter the performance measurement practices and monitoring of historical asset condition data. Besides globalization and technological changes, organizational changes can also affect the success of measuring performance.

The improvement methods applied by the industry is usually based on a continuous improvement process, like the Plan-Do-Study-Act (PDSA) cycle. Also, it is common practice to use the key elements of strategic planning, like vision, mission, goals, objectives, etc. The link and effect model is therefore based on the PDSA cycle along with emphasis on the key elements of strategic planning.

The link and effect model has two main components: a four-step continuous improvement process, and a top-down and bottom-up approach (Fig. 18.7).

Step 1: The first step of the link and effect model concentrates on the strategic planning, which also includes gathering stakeholders' objectives (usually conflicting) and assembling them into a common framework. For railways in the EU, aligning and harmonization start at the European level and are broken down to national governmental and infrastructure manager levels, i.e., from strategic to operational planning.

Step 2: The performance measurement system of organizations is under constant pressure from strategic planning, organizational changes, new technologies and changes in physical asset structure. Therefore, Step 2 in the link and effect model concerns updating the measurement system according to new stakeholder demands and objectives. See Fig. 18.8.

Step 3: Organizations collect a large amount of data, but turning the data into information is often lacking. Accordingly, analysis methodologies are developed in Step 3 by use of various statistical methods, for construction of performance indicators and identification of performance killer and drivers. Since data collection costs resources, another important aspect in Step 3 is to identify what data is required and what data is superfluous. Aggregation of data is a weakness of traditional performance measurement systems since it can make the indicators abstract as the underlying factors can be unknown, e.g., total train delay or total number of failures. Therefore, the link and

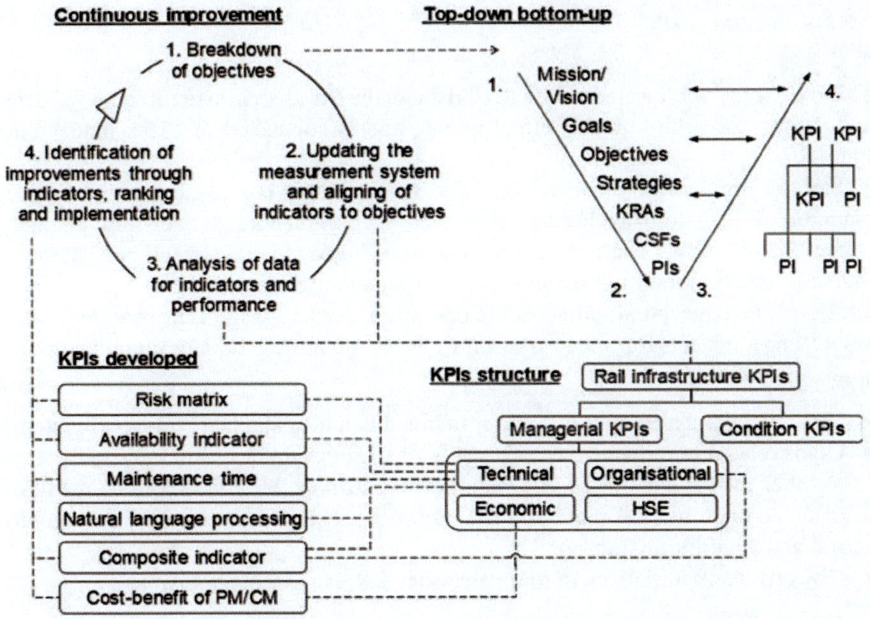

Fig. 18.7 The link and effect model, based on **a** A four-step continuous improvement. Process and **b** A top-down and bottom-up process [15]

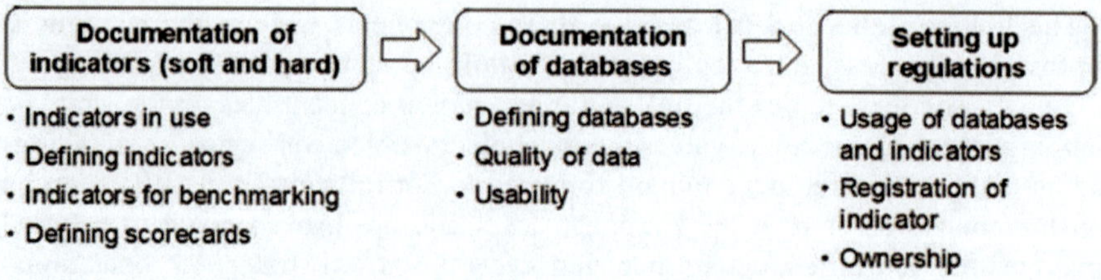

Fig. 18.8 Key requirements of strategic planning

effect model complements thresholds with the underlying factors responsible for the performance. Indicators with thresholds are commonly only given attention when some limit has been passed, making them reactive in nature. In contrast, the link and effect model gives the underlying performance drivers and killers, providing a starting point for improvements.

Step 4: The link and effect model utilizes continuous improvement with the ultimate goal of facilitating decision-making, by providing an up-to-date performance measurement system. Step 4 includes simulation, ranking, reengineering of physical assets and processes, implementing prognostic techniques and further defining indicators and databases.

18.5.1 Case Study

This case study was carried out on Malmbanan, the Swedish iron ore line, to validate and demonstrate the link and effect model, and is connected to "Risk matrix" of Fig. 18.7.

Step 1: Breakdown of objectives. The goal of Step 1 is to align the strategic planning of different stakeholders at the various organizational levels into a single frame. There are two challenges: firstly, identifying key elements and putting them into the same terminology; secondly, translating the high-level initiatives and goals, which can be conceptual, into specific operational tasks. At the European level, the White Paper on the European transport system identifies the key components of strategic planning as

- Vision: Toward a competitive and resource efficient/sustainable transport system.
- Goals related to railways: by 2030, 30% of road freight over 300 km should shift to other modes such as rail or waterborne transport; by 2050, 50% of medium distance intercity passenger and freight journeys should be shifted from road to rail and waterborne transport.
- Objectives: 40 initiatives in four categories KRAs (Key Result Areas).

Key elements of the strategic planning of transportation in Sweden are

- Overall goal: to ensure the economically efficient and sustainable provision of transport services for people and businesses throughout the country.
- Objectives: Railway operation and maintenance related objectives can be found in Trafikverket's (Swedish Transport Administration) Quality of Service (QoS) scorecard.

By studying the QoS scorecard, two indicators are of interest to this case study: train delay due to infrastructure problems and punctuality. Once the goals and objectives are identified and put into a common framework, it is easy to align to operational measures. By studying the objectives, it is found that service quality is a key facilitator at both the international and the national level. Availability is a vital component of service quality. The focus in this case study is on availability, more specifically, on failures and downtime in railway infrastructure as shown in Fig. 18.9.

Step 2: Updating the measurement system and aligning indictors. Indicators need to be set up and aligned to measure the results. Indicators related to failures and downtime specific to railways include

- Failures or work orders (in total, per item, per track-km, or per train-km)
- Train delay (in total, per item, per track-km, or per train-km)
- Punctuality (per line, line class, or area).

Punctuality, failures, and train delay are included as indicators on Trafikverket's QoS scorecard, i.e., failures, work orders, and downtime will directly affect the strategic objectives. However, indicators need to be further defined within an organization after analysis has been carried out. Thus, an objective of the link and effect model is to present an indicator along with its underlying factors, not just as an aggregated measure.

Step 3: Analysis of data for indicators, performance killers, and cost drivers. Operation and maintenance data of the Swedish railway section 111 have been collected,

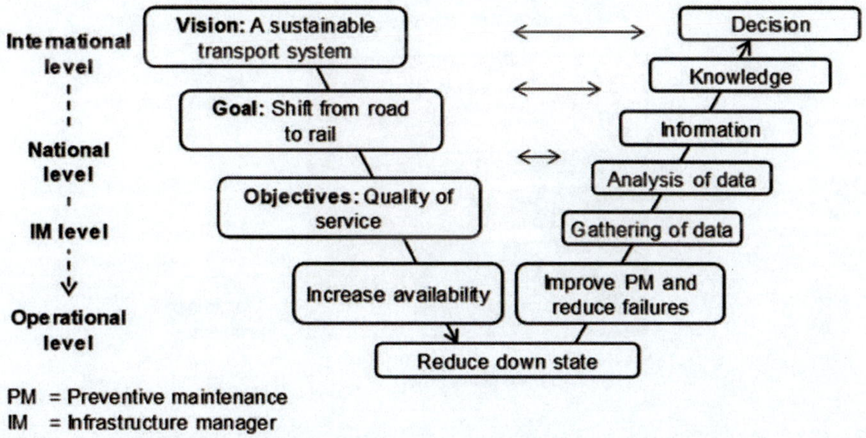

Fig. 18.9 Breakdown of strategy into failures and downtime

verified, and analyzed. Section 111 is a 128 km 30 tons mixed traffic section of Malmbanan. The data consist of corrective maintenance Work Orders (WOs) from 2001.01.01 to 2009.12.01. Out of 7476 WOs in total, 1966 mentioned train delays, i.e., 26%. This analysis is based on the 1966 WOs connected to delays. In addition, the two percent longest delays were considered as outliers, reducing the number of WOs from 1966 to 1926.

The outliers are preferably analyzed before decision-making, but this is out of scope of this study. With reference to Failure Mode Effect Analysis (FMEA) and risk matrices, also known as probability–consequence diagrams, a composite indicator, or index, of operational risk was defined as

$$R = \sqrt{(a\alpha)^2 + (b\beta)^2} \tag{1}$$

where α equals the infrastructure failures and β the train delay; and a and b are constants for weighting. Figure 18.10 shows the resulting operational risk at the system level. The poorest performing systems are the Switches and Crossings (S&C) and the track. These two systems are further analyzed.

Table 18.1 summarizes the results where the risk ranks equal to the length of the hypotenuse after normalizing the x-values to the y-values by dividing the delays by 100.

Step 4: Identification of improvements through indicators and implement: by redesigning or applying preventive maintenance to the identified items giving poor performance, the overall delay can be reduced. However, it is preferable to simulate improvements before moving to action. Figure 18.11 gives an example of simulation. In the figure, (a) shows the result on S&C when all the WOs of the switch controller system are removed from the dataset, i.e., the controller system never fails. Such a

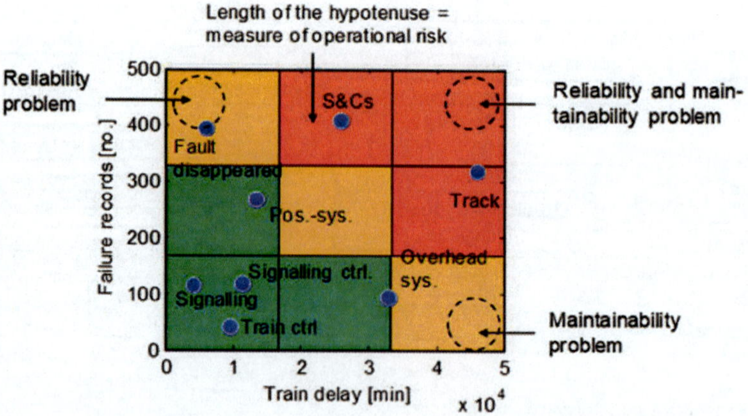

Fig. 18.10 Probability–consequence diagram at the system level displaying failure work orders and the corresponding train delay

Table 18.1 Work Orders (WOs) and train delays of performance killers

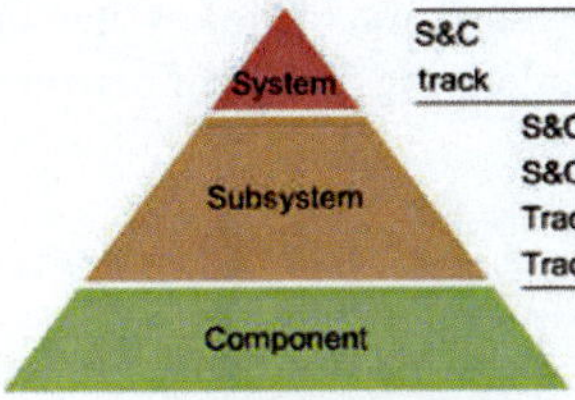

	WOs [No.]	Delay [Min]	Risk rank
S&C	404 (21%)	16880 (15%)	496
track	308 (16%)	28590 (25%)	575
S&C: Ctrl sys.	91 (4.7%)	3069 (2.7%)	105
S&C: Motor sys.	78 (4.0%)	2724 (2.4%)	91
Track: Joints	127 (6.6%)	4325 (3.8%)	147
Track: Rail	98 (5.1%)	18470 (16%)	329
S&C: Connector	37 (1.9%)	989 (0.9%)	41
S&C: Point drive	53 (2.8%)	1898 (1.7%)	62

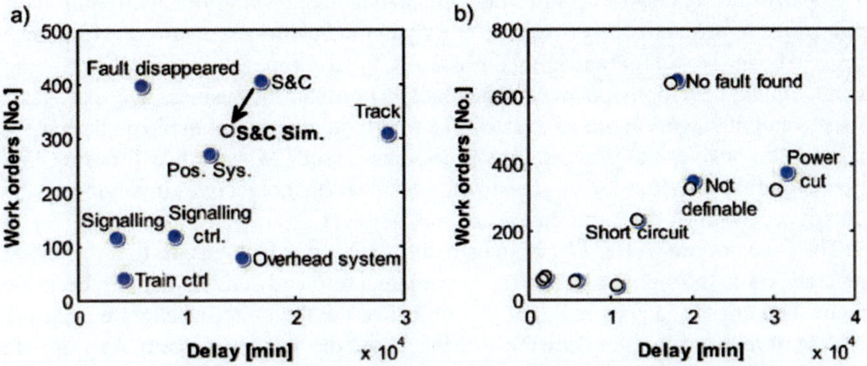

Fig. 18.11 Simulation at system level of the railway section. **a** Impact on the system level when all the failures of the switch controller subsystem are removed from the dataset. **b** All failures sorted according to the registered actual fault. The circles show the result when all the WOs of the switch controller system are removed from the dataset; power outage maintenance work is less

change in the dataset affects other factors at the system level. In (b), all WOs of the railway section are sorted by the actual faults found by the repair team. The black circles show the result from (a) when all the WOs of the switch controller system are removed from the dataset. It can be seen that power outage faults in the railway reduce most.

The link and effect model has been developed for improving performance measurement systems, by combining performance measurement and engineering principles. The performance measurement system needs to be able to handle implementation issues and challenges. The link and effect model was developed with emphasis on

- Continuous improvement
- The key elements of strategic planning
- The underlying factors responsible for the performance.

The link and effect model differs from other performance measurement systems and focuses on providing a breakdown process with description of the key elements of strategic planning, but especially, it focuses on the underlying factors of performance indicators.

18.6 Conclusions

A properly developed APM framework with management commitment is required for companies to remain competitive under a dynamic business scenario. Performance needs to be measured for managing the assets. Global competitiveness and increasing technological developments makes the asset PM critical for business success. Asset owners and managers are keen to know the return on investment made on their asset to meet the business objectives. An appropriate Asset PM system will ensure that all operational activities are aligned to the organizational objectives involving all the employees to fulfill the requirements of stakeholders.

The focus of any Asset PM Management system needs to ensure that the assets generate value throughout its life for the organization and stakeholders in the value chain. The critical factors enabling this objective are the effective capture, sharing, and use of relevant data for decision-making across the business system. Appropriate PIs and KPIs need to be identified and developed to improve the performance of assets as well as the asset management system.

Organizations must evolve to enable better decision-making and share knowledge and skills, breaking down silos and boundaries resulting from functional specialism and multiple cost centers, data capture, sharing, and standards. To improve the quality and availability of the information available for decision-making. The key to success is to gain the commitment from top management to drive the changes in organizational culture to improve the understanding of how good asset PM contributes to organizational goals.

An asset cannot be managed without considering the integrated strategic issues for an appropriate PM system. This is because of the various stakeholders conflicting interest needs, associated multiple inputs and outputs including the tangible and intangible gains from the asset. For engineering asset PM, strategic issues are essential to be considered. Under prevailing dynamic business scenario, asset PM is extensively used by the business units and industries to assess the progress against the set goals and objectives in a quantifiable way for its effectiveness and efficiency. An integrated asset PM with a link and effect model concept provides the required information to the management for effective decision-making. Research results demonstrate that companies using integrated balanced performance systems perform better than the one who do not manage measurement.

References

1. Baur, C., & Wee, D. (2015, June). Manufacturing next act. Web site: https://www.mckinsey.com/business-functions/operations/our-insights/manufacturings-next-act.
2. ISO, ISO 55000. (2014). *Asset management, overview, principles and terminology*. Geneva: International Organizations for Standardization (ISO).
3. Barringer (2017, August 15) *Maintenance cost is 60–80% of the life cycle cost*. Downloaded from https://www.barringer1.com/pdf/lcc_rel_for_pe.pdf.
4. PwC, *Worldwide asset cost will be $101.7 trillion by 2020*. Downloaded from https://www.pwc.co.za/en/press-room/asset-manage.html. Dated 15 Aug 2017.
5. Shi-Nash, A., & Hardoon, D. R. (2016). Data analytics and predictive analytics in the era of big data. *Chapter 19 of Internet of things and data analytics handbook*. Wiley on line library.
6. Kaplan, R. S., & Norton, D. P. (2004). *Strategy maps, converting intangible assets into tangible outcomes*. USA: Harvard Business School Press.
7. ISO, ISO 55001. (2014). *Asset management, requirements*. Geneva: International Organizations for Standardization (ISO).
8. ISO, ISO 55002. (2014). *Asset management, requirements*. Geneva: International Organizations for Standardization (ISO)
9. Parida, A., Åhren, T., & Kumar, U. (2003). *Integrating maintenance performance with corporate balanced scorecard*. In *Proceedings of the 16th International Congress*, 27–29 August 2003 (pp. 53–59). Växjö, Sweden.
10. Parida, A. (2006). *Development of a multi-criteria hierarchical framework for maintenance measurement*. Ph.D. thesis, Luleå University of Technology, Sweden. https://epubl.ltu.se/1402-1544/2006/37/LTU-DT-0637-SE.pdf.
11. Parida, A., Galar, D., Kumar, U., & Stenström, C. (2015). Performance measurement and management for maintenance: A literature review. *Journal of Quality in Maintenance Engineering, 21*(1), 2–33.
12. Siemens, *Maximize productivity from operation through maintenance*. Downloaded from https://new.siemens.com/global/en/markets/automotive-manufacturing/digital-twin-performance.html. Dated 20 January 2020.
13. Hollywood, P. (2017). IT/OT/ET convergence. ARC Insights, ARC Advisory Group.
14. Parida, A., & Chattopadhyay, G. (2007). Development of multi-criteria hierarchical framework for maintenance performance measurement (MPM). *Journal of Quality in Maintenance Engineering, 13*(3), 241–258.
15. Stenström, C., Parida, A, Galar, D., & Kumar, U. (2013). Link and effect model for performance improvement of railway infrastructure. *Institution of Mechanical Engineers, Proceedings Part. F, Journal of Rail and Rapids Transit, 27*(4), 392–402.
16. Parida, A., & Kumar, U. (2009). *Integrated strategic asset performance assessment*. In *Proceedings of World Congress of Engineering and Asset Management* (pp. 369–371). Athens, Greece, 28–30 September 2009.

Aditya Parida, Ph.D. was a Professor at the Luleå University of Technology, Sweden. He was involved in teaching Masters and Ph.D. courses, and associated with several industries, Swedish National and European Union research projects. He has worked for the Indian Army and retired as a Director Management Studies at Army HQ New Delhi. He was a visiting Professor with a number of Universities in Sweden, Finland, Iran, and India. He is the author of two books, three book chapters, and over a hundred peer-reviewed papers on asset engineering and management, performance measurement management, and eMaintenance. He has been International Chair and key-note speakers, and Editors/Guest-editors to several Journals. He is external examiner for a Master's Program of Manchester University 2018–22. He has been awarded with Emerald Outstanding Paper Award-2007 and Eminent Alumni Award from VSSUT, India, 2017.

Christer Stenström, Ph.D. has an M.Sc. in Engineering Physics from Luleå University of Technology (LTU), Sweden and a M.Sc. in Engineering Mechanics from University of Nebraska Lincoln, USA. After his Ph.D. in Maintenance Performance Measurement in the railway industry, from the Division of Operation and Maintenance Engineering in 2014, he worked as an Assoc. lecturer at LTU till 2019. He has a large number of peer-reviewed journal publications and technical reports to his credit. His Ph.D. thesis was awarded with the best Ph.D. thesis award by the European Federation of National Maintenance Societies (EFNMS). He is now working at LKAB mining company of Sweden.

Chapter 19
Asset Management Journey for Realising Value from Assets

Gopinath Chattopadhyay

Abstract Assets in line with ISO55000 standard for asset management are items, things and entities which have value or potential value to the organisation. Asset management is for what we do with those assets. The journey begins with understanding the needs of the organisation in line with business objectives to deliver goods and services in a reliable, safe, timely and cost-effective manner. Realising value from assets is a holistic approach addressing complexities of expectations of stakeholder and providing competitive advantage to the business. It starts from the concept of the asset and continues to the design, manufacturing/construction, operations, maintenance and disposal of the asset known as asset life cycle. Focus is on reduced risks, enhanced performance including safety of the operation, environment and the wider communities and achieving reduced Life Cycle Costs. Systematic approach in asset management helps in improving reliability, availability, maintainability, safety and security. Leadership, good organisation culture, alignment with other systems and assurance that assets will perform when needed contributes significantly to the success of any organisation. This chapter covers how to balance cost, risk and performance in informed decision-making for maintaining value of and realising value from assets.

Keywords Asset management · ISO55000 · Life cycle cost · Risks · Performance

19.1 Introduction

The history of asset management goes long back to the days of terotechnology covering installation, commissioning, maintenance, replacement and removal of plants and equipment. It helped in better management of physical assets for reducing life cycle costs through reliability, availability and maintainability. In the past, major focus was on maintenance and managing the assets. In this asset management journey, the focus is now shifted more on what we do with these assets. In addition, there are

G. Chattopadhyay (✉)
Federation University Australia, Gippsland, Australia
e-mail: g.chattopadhyay@federation.edu.au

K. B. Misra (ed.), *Handbook of Advanced Performability Engineering*,
https://doi.org/10.1007/978-3-030-55732-4_19

other types of assets including financial assets, information asset, human asset and intangible assets including knowledge and goodwill. This journey from maintenance to asset management has therefore taken a holistic approach for balancing costs, risks and enhancing performance (Chattopadhyay [1, 2]).

Assets in line with ISO55000 for asset management are any items, things and entity which has value or potential value to any organisation. Fundamental of asset management is focused on value, leadership and culture, alignment with corporate objectives and other systems and assurance that assets will perform as and when they are needed (ISO [3]). This chapter is mainly on physical assets. However, there are other important assets such as finance, information technology, human assets and nontangible assets such as intellectual property, goodwill, tacit knowledge and know-hows.

Asset management linked to physical assets was first used by Dr. Penny Burns in 1980s (in Asset Management History Project, 1984) (Wikipedia [4]). Infrastructure Asset Management Manual, published in 1996 in New Zealand, on asset management for infrastructure sector became international infrastructure management manual (IIMM) in 2000 (IPWEA [5]).

The professional societies: The Asset Management Council (AMC) in Australia, Institute of Public Works Engineers Australasia (IPWEA) and the Institute of Asset Management (IAM) in the UK along with various professional bodies around the world contributed significantly to the development of body of knowledge in the area of asset management (AMC [6], IAM [7]). Global Forum of Maintenance and Asset Management (GFMAM) provided a platform for better understanding of needs of various countries around the world in asset management and defining and interpreting technical terms in a consistent manner. This helped in developing guidelines for addressing issues and challenges in asset management from global perspective in a coordinated and consistent manner (GFMAM [8]).

Asset management, as per Peterson, covers the following concepts:

- Business goals driving decisions for the use and care of assets,
- Asset strategy determined by operational considerations,
- Maintenance and reliability for a defined goal (not an end in itself),
- Intent for optimising the application of all resources (not just maintenance) (Peterson [9]).

Moore suggested a view of asset management covering

- Incorporation of an understanding and alignment between the business expectations for the assets both currently and into the future;
- An understanding of the assets' current condition and capability today and into the future;
- The centrality of the consideration of how and why the assets are operated;
- A consideration of asset life cycle, e.g. design considerations in terms of capability, reliability and ease of asset management at both initial and rehabilitation phases of an assets' life;
- How asset management needs to be implemented (Moore [10]).

Asset management is expected to provide a strategic platform to connect the physical assets of the business, their utilisation and maintenance along with all the other assets. Woodhouse proposed it as a set of disciplines, methods, procedures and tools aimed at optimising the Whole of Life Business Impact of costs, performance and risk exposures associated with the availability, efficiency, quality, longevity and regulatory, safety and environmental compliance of the company's physical assets (Woodhouse [11]).

International Infrastructure Management Manual (IIMM) proposed how to develop

- Asset Management (AM) policy,
- Organisational structure to deliver AM functions and
- Quality management processes that support the AM functions.

Publicly available specifications on asset management PAS55 (1 and 2 of British Standard Institute, 2008) were developed by industries to cover holistic asset management and paved the way for international standard on asset management ISO55000:2014 series for risk-based and informed decision-making with an aim for reducing cost and risks and enhancing performance over entire life covering various stages of asset life such as acquisition, utilisation and disposal (BSI [12, 13]).

Capital-intensive industries around the world have been facing an ever-increasing pressure of demand growth, geographical locations and ageing assets for doing more with less. There are credit constraints and scarcity of capital. However, showing the board members what is the risk of doing nothing and the actual cost of risk for that option, then one will be surprised to see that there is money available for preventions and continual improvements. What is needed is to show the value of the proposed initiatives and not just limiting the proposition limited to costs and benefits. Asset management journey begins with understanding the needs of the organisation in line with business objectives. The concept of the asset is developed and continues with the design, manufacturing/construction, operations, maintenance and finally, disposal of the asset at the end of the asset life cycle in a cost-effective, reliable, safe, secured and timely manner.

19.2 Overview of Asset Management

ISO standard for asset management which is practically a management standard for asset management consists of three parts:

- ISO 55,000 Asset management—Overview, principles and terminology;
- ISO 55,001 Asset management—Management systems—Requirements;
- ISO 55,002 Asset management—Management systems—Guidelines on the application of ISO 55,001 (ISO [14, 15]).

Asset management in line with this ISO standard is defined as 'coordinated activities of an organisation to realise value from assets' covering the following principles:

- Assets exist to provide value to the organisation and its stakeholders.
- People are key determiners of asset value realisation.
- An asset management organisation is a learning organisation covering.
- Strategic asset management plan.
- AM system.
- Asset management plans.
- Asset management requires understanding of the organisation's operating context and opportunities.
- Asset management decisions consider both short-term and long-term economic, environmental and social impacts.
- Asset management transforms strategic intent into technical, economic and financial decisions and actions (ISO55000) (Fig. 19.1).

ISO55000 series of standards tell what needs to be done and do not tell how it can be done. How the requirements can be addressed by individual organisations needs to be addressed according to the context and expectations of the organisation covering the following:

- Normative reference
- Terms and definitions
- Context of the organisation
  - Understanding the organisation and its context
  - Understanding the needs and expectations of stakeholders
  - Determining the scope of the asset management system
  - Asset management system
- Leadership

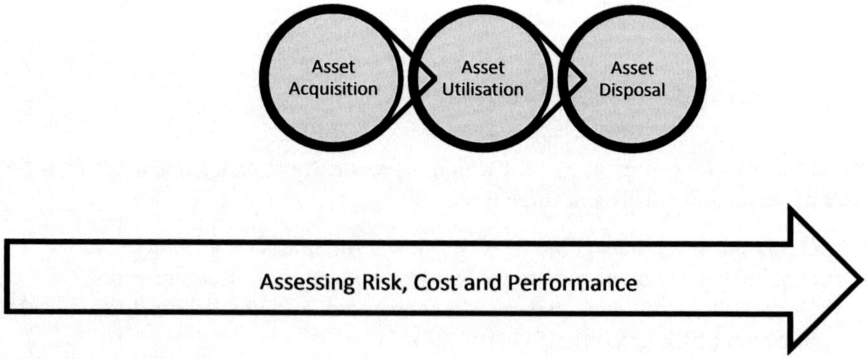

Fig. 19.1 Asset life cycle

  - Leadership and commitment
  - Policy
  - Organisational roles, responsibilities and authorities
  - Planning

- Actions to address risks and opportunities
  - Planning for the asset management system
  - Planning for assets
  - Asset management objectives and planning to achieve them
  - Asset management objectives
  - Asset management planning

- Support
  - Resources
  - Competence
  - Awareness
  - Communication
  - Information system support
  - Documented Information

    General
    Creating and updating
    Control of documented Information

- Operation
  - Operational planning and control
  - Management of change
  - Outsourcing of asset management activities

- Performance evaluation
  - Monitoring, measurement, analysis and evaluation
  - Internal audit
  - Management review

- Improvement
  - Nonconformity and corrective action
  - Continual improvement
  - Preventive and predictive action

The Global Forum on Maintenance and Asset Management (GFMAM) published the Asset Management Landscape, which covers the subject areas for the asset management required to address the knowledge and skills needed for good asset management. These are as follows:

Asset Management Strategy and Planning

- Asset Management Policy

- Asset Management Strategy
- Demand Analysis
- Strategic Planning
- Asset Management Plan

Asset Management Decision-Making

- Whole-life Cost and Value Optimisation
- Operations and Maintenance Decision-Making
- Capital Investment Decision-Making
- Resourcing Strategy and Optimisation
- Shutdowns and Outage Strategy and Optimisation
- Ageing Assets Strategy

Life cycle Delivery Activities

- Technical Standards and Legislation
- Asset Acquisition and Commissioning
- Systems Engineering
- Configuration Management
- Maintenance Delivery
- Reliability Engineering
- Asset Operations
- Resource Management
- Shutdown and Outage Management
- Fault and Incident Response
- Asset Rationalisation and Disposal

Asset Knowledge Enablers

- Asset Information Strategy
- Asset Knowledge Standards
- Asset Information Systems
- Asset Data and Knowledge

Organisation and People Enabler

- Contract and Supplier Management
- Asset Management Leadership
- Organisational Structure & Culture
- Competence and Behaviour

Risk and Review

- Criticality, Risk Assessment and Management
- Contingency Planning and Resilience Analysis
- Sustainable Development
- Weather and Climate Change
- Asset and Systems Change Management
- Assets and Systems Performance and Health Monitoring

- Management Review, Audit and Assurance
- Accounting Practices
- Stakeholder Relations

There is a need to develop tools and techniques along with artefacts for further enhancing capabilities of personnel engaged in asset management and related activities for better managing value from assets in various stages of asset life cycle including procurement, operation and maintenance and disposal of assets for as minimum as possible but as far as practicable life cycle costs (LCC). Organisations need to know about their assets, their conditions, maintenance history, costs and informed risk-based decision for inspections, maintenance and replacements including options for overhaul, major repairs and life extension.

Understanding the condition of assets from failure and maintenance history and estimating the remaining life and option engineering for life enhancement are key steps in life cycle management of capital-intensive assets. Asset management is therefore not treated as a destination. It is like a journey for realising value from assets though appropriate allocation of funds for maintenance and upgrades known as Operational expenditure (opex) and replacements covered in Capital expenditure (capex).

19.3 Understanding the Asset and Its Remaining Life

Estimation of remaining life is a comprehensive and multidisciplinary activity that takes into account a range of factors such as asset life cycle asset management principles, needs of the users of the asset, competing demands of stakeholders, current and future policy and legislative environment, the entity's corporate governance and planning framework, technical adequacy and commercial viability, external or market factors (commercial, technological, environmental or industry implications), the need to rationalise operations to improve service delivery and cost-effectiveness of any life extension. It helps in sound decisions that are appropriate to address the identified risks and the associated impacts on value, carrying out appropriate tasks at the 'right' time and at the right level of expenditure, achieving the right balance between competing factors, such as performance, cost and risk. The starting point of this is understanding the failure mechanism.

Failure is not an easy term to explain to different stakeholders in a consistent way. It is generally accepted as the inability of an item to perform its required function. Causes of failures are mainly the limitations of the system, subsystem or components to perform due to design, manufacture, user and maintenance-related issues resulting in failures. Modes of failures are the resulting effects of failure causes. Mechanisms of failures are physical, chemical or other process causing failures. Analysis of failure needs logical, systematic examination to identify and analyse the probability, causes and consequences of failures and/or potential failures including near hits. In addition, failures can also be due to misuse and/or overloading.

When failure occurs directly and without any influence of any failure of another item, it is termed as primary failure. If failure occurs either direct or indirect failure of another item, then it is termed as secondary failure. Where failure occurs with probability of failure increasing with time such as age and/or usage, it is termed as wear-out failure. If failures do not give any indication or not detected by inspection or monitoring, it is termed as sudden failure. Where failures give some indication or can be detected by prior inspection or monitoring, those are termed as gradual failures. Where loss of functional ability is up to level where it does not stop the item to perform some of the required functions, it is called a partial failure. If the loss of functionality resulting from deviations in characteristic(s) is beyond specified limits causing complete lack ability for required function, it is called a complete failure. When failures are sudden and complete then termed as catastrophic failure. When failures are gradual and partial, then those are termed as degradation.

If failures are likely to cause injury to persons or significant damage to material, then those are considered as critical failures. If failures are other than a critical failure, which is likely to reduce the ability of a more complex item to perform its required function, then those are considered as major failures. Failures not reducing the ability of a more complex item to perform its required function are considered as minor failures.

In the life cycle of any asset, failures can occur due to design, manufacturing, testing and installation-related problems in the early stage, operations and usage-related wear and tears in the middle phase of the life and faster rate in the last phase of the life due to ageing, operation and maintenance-related problems at the end of the life of the asset. This is captured in the bathtub failure curve comprising of decreasing, constant and increasing rates of failures such as.

- Early Failure Period
- Constant Failure Rate Period
- Wear-Out Failure Period

Failure analysis considers life data from maintenance history and mathematical and or statistical modelling for using those life data (age, usage, number of times usage and many other) in prediction and intervention of failures through appropriate inspection, maintenance, repairs and replacements. In the following analysis, life data is taken as time and item is in operation before failure. Let T denote the time to failure, t denote age and $F(t)$ denote the failure distribution function. Then,

$$F(t) = \text{Probability}\,(T \leq t) \tag{19.1}$$

The reliability function corresponds to the probability that an item survives to any given age.

For an item which starts to operate at age $t = 0$, the reliability function, $R(t)$, is the probability that failure does not occur in the interval 0–t. Then,

$$R(t) = \text{Probability}(T > t) \tag{19.2}$$

$$R(t) = 1 - F(t) \tag{19.3}$$

The probability density function (pdf) of the time to failure is a function of age, such that the area under the curve between any two age values gives the probability that a new item will fail in that age interval. The probability density function, $f(t)$, is the differential coefficient of the distribution function $F(t)$. We have the following equations:

$$f(t) = \mathrm{d}F(t)/\mathrm{d}t \tag{19.4}$$

Probability of failure in t to $t + \partial t = f(t)\partial t$

$$F(t) = \int_0^t f(u)\mathrm{d}u \tag{19.5}$$

Note

$$\int_0^\infty f(t)\mathrm{d}t = 1 \tag{19.6}$$

The hazard function $h(t)$ is a function such that the probability that an item which has survived to age t fails in the small interval t–$t + \partial t$ is $h(t)\partial t$.

The hazard function can be related to the reliability function $R(t)$ and the probability density function $f(t)$ as follows. The probability of failure in the interval t–$t + \partial t$ is $R(t)\ h(t)\ \partial t$. Then,

$$f(t)\partial t = (t)h(t)\partial t \tag{19.7}$$

$$f(t) = R(t)h(t) \tag{19.8}$$

and

$$h(t) = f(t)/R(t) = f(t)/(1 - F(t)) \tag{19.9}$$

Probability of failure in the interval

$$t_1 \text{ to } t_2 = \int_{t_1}^{t_1} f(t)\mathrm{d}t \tag{19.10}$$

where

$$\int_0^{\infty} f(t)\,\mathrm{d}t = 1 \quad (19.11)$$

In industries, some simple terms are used for analysis. These are as follows:

Mean Time to Failure (MTTF), which is average of the observed ages at failure.

Mean Time Between Failures (MTBF), which is the ratio of the component hours of service to the number of failures. In some organisations, this is known as uptime and used as a measure of reliability. The higher the MTBF, the better is the reliability.

Mean Time to Repair (MTTR), which is average of the observed times between failures and back to operation through maintenance actions. In some organisation, this is known as downtime and used as a measure of maintainability. The lower the MTTR, the better is the maintainability.

Availability, which is measured as a ratio of uptime to uptime plus downtime.

Analysis of remaining life is critical to decision-making for future operation and maintenance of plant. It includes how much longer the plant can operate safely in its current condition, what components should be replaced to keep the plant operating, what design life to be considered for replacement components, the cost of future replacements and the cost of planned operating modes. Remaining life can be estimated using life consumed to date, and future operating modes and maintenance plans.

Understanding the asset and its remaining life helps in analysing costs for life enhancements based on degradation, target performance and residual risks. Cost-effectiveness of any capacity and capability gains through upgrade is analysed using revised life cycle costs. Any capital injection and/or reducing inspection and maintenance intervals are worthwhile if the value realised through these activities for reducing cost of operations, risks and associated safety outweighs the total cost for life enhancement activities over the revised remaining life.

19.4 Life Cycle Costing

Life cycle cost considers total cost for the asset over the entire life of the asset. Life cycle costs (LCC) considers all expenses for

- deciding what is needed
- acquisition
- installation
- utilisation (operation) and maintenance
- refurbishment or replacement
- discarding and disposal costs (ISO [16]).

ISO 15,686–5: 2017 (ISO, 2017) suggests whole of life costs and LCC and is given by

$$\begin{aligned}\text{Life cycle costs (LCC)} = {} & \text{Capital cost } (C) + \text{lifetime operating costs } (O) \\ & + \text{lifetime maintenance costs} \\ & + \text{lifetime maintenance costs (M)} \\ & + \text{lifetime plant losses } (L) + \text{plant disposal cost } (D)\end{aligned} \tag{19.12}$$

Analysis of life cycle costing (LCC) considers

- Service life, life cycle and design life
- period of analysis
- costs covering
- acquisition
- maintenance, operation and management
- residual values/disposal
- discounting
- inflation
- taxes
- utility costs including energy
- risks.

Some of the costs in different phases of life of any asset need to be considered covering:

Cost of planning and acquisition covering:

- need study,
- design and development,
- construction,
- installation,
- testing and commissioning,
- modification and fixing teething problems,
- spare parts,
- training of people, and
- operations and maintenance manuals and relevant drawings.

Operating costs covering:

- labour,
- power,
- consumables,
- equipment and
- overhead charges.

Maintenance costs covering:

- labour,
- parts,
- materials,

- consumables,
- equipment and
- overhead charges.

Life Cycle Cost

- Inflation rate constant at $i \times 100\%$ pa.
- Discount rate $r \times 100\%$ pa.
- Annual operating costs, maintenance costs and plant losses are incurred at the end of the year; we have

$$\mathrm{LCC} = C + \sum_{1}^{N} \mathrm{On}\frac{(1+i)^n}{(1+r)^n} + \sum_{1}^{N} \mathrm{Mn}\frac{(1+i)^n}{(1+r)^n} + \sum_{1}^{N} \mathrm{Ln}\frac{(1+i)^n}{(1+r)^n} + \mathrm{Dn}\frac{(1+i)^n}{(1+r)^n} \tag{19.13}$$

In any real-life capital-intensive assets, life consumption and maintenance costs add complexity to LCC modelling. For example, for rail network, rail life ends due to two major failure modes. First one is Rolling Contact Fatigue (RCF)-initiated cracks and undetected propagations resulting in rail breaks and derailments. Second one is rail–wheel friction-initiated wear resulting in early replacement decisions when it reaches wear limit sooner. Failure to replace might lead to wheel rollover and derailment. There are decision variables such as inspection intervals and grinding intervals for rail surface for controlling crack propagation. In the same manner, there is decision variable for placement of lubricators and choice of lubricants for providing lubricants at the gauge face for reducing wear and therefore further enhancing asset life. All these have an impact on replacement intervals of rails and are used for reducing risk cost associated with derailments, early replacement and unplanned maintenance actions (Chattopadhyay et al. [17, 18].).

19.5 Balancing Cost, Risk and Performance Through Asset Management

Balancing cost, risk and performance is both art and science. There are regulatory requirements to comply with and discretionary decisions by organisations over and above the regulatory requirements. Any capital investment in this process needs systematic approach using the following steps:

- Defining the objective/s
- Defining the alternative options
- Estimating the lifetime
- Estimating the benefits and costs

- Specifying the time value for money (discounting rates)
- Developing/defining the performance measures for effectiveness
- Comparing apples to apples for ranking the alternatives
- Analysing sensitivity using what-if scenarios
- Recommending the option based on cost, risk and performance (Parida et. al [19]).

Alternative capital investment options are analysed using several techniques including the following:

- The payback method: the period when return from the investment covers the capital investment. Any investment is ok if payback period is below acceptable limit (say, asset life).
- Present Worth (PW): an amount at some beginning or base time that is equivalent to a schedule of receipts and/or disbursements for any investment option. Any investment is ok if present worth of benefits is more than investment.
- Annual Worth (AW): a uniform series of money for a certain period equivalent in amount to a schedule of receipts and/or disbursements for any investment option.
- Future Worth (FW): an amount at some ending or termination time which is equivalent to a schedule of receipts and/or disbursements for any investment option. Any investment is ok if future worth of benefits is more than the future worth of investment.
- Rate Of Return (ROR): the acceptability of individual investment option. Any option is acceptable if its internal rate of return (IRR) is not less than a predetermined minimum attractive rate of return (MARR). The higher the IRR, the better is the option.
- Benefit–Cost Analysis (BCA): ratio of the equivalent worth of benefits to the equivalent worth of costs and options are accepted for this ratio more than one. The higher the ratio, the better is the option (Canada et al. [20]).

Example of Payback Analysis

A new asset costing $20,000 will cost $1000 to install and $4000 per year to operate, with a useful life estimated at 15 years. The resale value of this existing asset is $5000 and is now costing $8000 per year to operate. Both assets have the same output capacity.

Solution

Purchase price	$20,000
+Installation cost	$ 1,000
−Sale of existing machine	−$ 6,000
Net cost of equipment	$15,000
Old asset operational costs/yr	$ 8,000
New asset operational costs/yr	$ 5,000

(continued)

(continued)

Net additional profit/year	$ 3,000
Payback Period = 15,000/3000 = 5 years	
If assets life [say 10 years] is greater than payback of 5 years, investment option is acceptable	

Example of Present Worth Analysis:

A new production unit is being considered for purchase. The following facts are available:

(a) Installed cost of the equipment = $240,000.
(b) Estimated additional earnings per year = $80,000 (compared to the present process).
(c) Useful life of the equipment = 8 years.
(d) Estimated resale value of new unit in 8 years' time is $15,000. Resale value of the old machine is included in value at (a).
(e) Assume depreciation is straight line over 8 years and that tax on income is at 50% (Table 19.1).

Annual cash inflow		
Additional earnings/year	$80,000	$80,000
Less depreciation 24,000/8	−$30,000	
Taxable income	$50,000	
Less tax at 50%	−$25,000	−$25,000
Annual profit after tax	$25,000	
Annual cash inflow after tax		$65,000
Total cash outflow		
Installed cost	$240,000	
Resale value in 8 years	$15,000	

Example of selecting from alternative options:

MARR = 15%, Life = 5 years, Salvage value is realised at the end of the life (Tables 19.2, 19.3 and 19.4).

From the above analysis, option B is the preferred option.

The true rate of return is the discount value at which the present value outflows equal the present value inflows and can be calculated using Excel for extrapolation.

When internal rate of return is more than MARR, the investment option is acceptable.

Maintainability

It is the ability of the system to be back to operation when maintenance is performed using standard procedure, right spared and trained people. This is shown using 20 data from maintenance history (Table 19.5).

Table 19.1 PW calculation

Discount rate (%)	Present worth (PW) factor for single payment (P/F) (resale value)	Present worth (PW) outflow	Present worth (PW) factor for uniform series (P/F) (annual cash inflow)	Present worth (PW) inflow
16	0.305	$235,425	4.344	$238,920
14	0.351	$234,735	4.639	$255,145
18	0.266	$236,010	4.078	$224,290

Table 19.2 Options A and B

Option	Initial investment(P)	Salvage value (S)	Annual receipts
A	– 7000	1000	2000
B	–10,000	2000	3000

Table 19.3 PW and FW analysis

Option A	Year 0	Year 1	Year 2	Year 3	Year 4	Year 5	Totals
Initial investment	–7000						
Annual net receipts		2000	2000	2000	2000	2000	
Salvage value						1000	
Net receipts	–7000	2000	2000	2000	2000	3000	
Present worth (PW)	–7000	1739	1512	1315	1144	1492	201
Future worth (FW)	–14,080	3498	3042	2645	2300	3000	405
Option B							
Initial investment	–10,000						
Annual net receipts		3000	3000	3000	3000	3000	
Salvage Value						2000	
Net Value	–10,000	3000	3000	3000	3000	5000	
Present Worth (PW)	–10,000	2609	2268	1973	1715	2486	1051
Future Worth (FW)	–20,114	5247	4563	3968	3450	5000	2114

Table 19.4 Summary of options

Option	Present worth (PW)	Future worth (FW)
A	201	405
B	1051	2114

If the maintainability test fails, then there is need for further enhancing the process, maintenance strategy and/or design of the system.

All decisions need to be prioritised based on risks. One of the tools used in risk assessment is Risk Priority Number (RPN).

Table 19.5 Maintainability test

Data	Observed maintenance time	Deviation from mean	Square of the deviation from mean
1	39	−17.05	290.70
2	57	0.95	0.90
3	70	13.95	194.60
4	51	−5.05	25.50
5	74	17.95	322.20
6	63	6.95	48.30
7	66	9.95	99.00
8	42	−14.05	197.40
9	85	28.95	838.10
10	75	18.95	359.10
11	42	−14.05	197.40
12	43	−13.05	170.30
13	54	−2.05	4.20
14	65	8.95	80.10
15	47	−9.05	81.90
16	40	−16.05	257.60
17	53	−3.05	9.30
18	32	−24.05	578.40
19	50	−6.05	36.60
20	73	16.95	287.30
Total	1121		4,078.95
Mean time	56.05		
Std Dev			14.65
Risk factor	0.1		
Z from table	1.28		
Upper Limit = Mean time + z*Std Deviation/Sqrt of Number of data	60.24	Less than contracted time, 65 min	
Maintainability is performing			

Risk Prioritisation Number (RPN)

It is given as

RPN = Severity x Likelihood X Detectability

where the severity is ranked (commonly from 1–5) using metrics such as

- Negligible: minor treatment (1).

- Marginal: injury requiring < 10 days hospitalisation/medical leave (2).
- Serious: injury requiring > 10 days hospitalisation/medical leave (3).
- Very Serious: injury requiring > 30 days hospitalisation/medical leave (4).
- Critical: fatality/permanent body injury (5).

The severity ranking can relate to environmental, plant damage and downtime metrics.

Where the detectability is ranked (commonly from 1–5) using metrics such as

- No or Very low detectability: inevitable, potential failure not detectable (5)
- Low detectability: unlikely to detect a potential failure (4).
- Moderate detectability: may be able to detect a potential failure (3).
- High detectability: a good chance to detect a potential failure (2).
- Very high detectability: it is almost certain to detect a potential failure (1).

The likelihood is similarly ranked (say 1–5), e.g.

- Unlikely: might occur once in 10 years (1).
- Remote: might occur once in 5 years (2).
- Occasional: might occur once in 3 years (3).
- Moderate: likely to occur once per year (4).
- Frequency: likely to occur many times per year (5).

Priority is allocated based on RPN. The higher the RPN, the higher the rank in selecting any items for risk mitigation.

In majority of infrastructure sector, a traffic light type approach of green, yellow, orange and red signal is used for flagging actions to be taken for risk mitigation. Red means the highest priority, orange is flagged to be monitored closely or inspections to be tightened and actions to be taken in the nearest future in line with corporate guideline and/or regulatory requirements. And Green means no action is required other than normal inspections and monitoring. Risk matrix in line with ISO31000 can be analysed similar to Table 19.6 (ISO [21]).

Overall Equipment Effectiveness (OEE)

The overall equipment effectiveness (OEE) is used to better understand the performance of the maintenance. It evaluates how effectively a manufacturing operation is utilised and is expressed well in terms of Performance, Availability and Quality. It is measured in terms of whether plant is operated as per expected speed, reduced

Table 19.6 Risk matrix

Likelihood	Consequence				
	Insignificant	Minor	Moderate	Major	Catastrophic
Almost certain	High risk	High risk	Extreme risk	Extreme risk	Extreme risk
Likely	Medium risk	High risk	High risk	Extreme risk	Extreme risk
Possible	Low risk	Medium risk	High risk	Extreme risk	Extreme risk
Unlikely	Low risk	Low risk	Medium risk	High risk	Extreme risk
Rare	Low risk	Low risk	Medium risk	High risk	High risk

speed or with minor stops. Availability is analysed in terms of breakdowns and product changeovers. Quality is analysed in terms of acceptance and rejects in start-up, during production runs and customer returns. Therefore, OEE indicates the health and performance of assets and productivity and considers.

- Breakdowns
- Setup and Adjustment
- Small stops
- Slow running
- Start-up defects
- Production defects

Effectiveness (OEE) is widely expressed as

$$\text{OEE} = A \times P \times Q \tag{19.14}$$

where
A: Availability
P: Performance and
Q: Quality.

Good asset management helps in reducing losses, enhancing availability, performance of the assets and assuring quality of products and services using OEE (Chundhoo et. al [22, 23]).

19.6 Realising Value from Assets

Asset management, if practised well, retains value of assets and realises value from assets. Some of the important factors including value judgement may not be fully quantifiable and are generally analysed by industries using experience (by resolving conflict of brain vs. heart). Decisions are taken based on risks and not just based on costs and benefits. Risk is the 'effect of uncertainty on objectives' where uncertainty is the 'state, even partial, of deficiency of information related to, understanding or knowledge of an event, its consequence, or likelihood' (ISO [21]). Risk is 'susceptible' to measurement (e.g. we might know the distribution of likelihood or the possible consequences). However, uncertainty reflects that the exact outcome is unpredictable. Global warming and rare events including cycle, tornados, tsunamis, earthquakes and many other similar challenges put additional difficulties in accurately assessing impacts of those events on asset management (Komljenovic [24]).

Risk management is generally limited to what we know about events, probabilities and outcomes (Knight [25]). As per ISO 31,000, risk management—principles and guidelines—the basic steps need to be used are

- Establishing the context
- Identifying the risks

- Analysing the risks
- Assessing the risks
- Treating the risks
- Monitoring and reviewing progress and performance.

Balancing act of cost, risk and performance is a complex process. At a business level, balancing needs to consider context of the organisation in line with ISO55001. Decision-makers need to better understand the needs and expectations of stakeholders.

There are financial, legal, image/reputation, safety, environmental, service delivery and many other risks for any asset management-related decisions. Mandatory levels of performance and risk are generally regulatory driven. Discretionary levels beyond that require a clear understanding of what customers are willing to pay, what competitors are charging and costs associated with providing expected performance and managing risks (Aven [26]).

Historically, cost used to be based on what level of service the customer should have. Organisations used to be conservative and risk-averse. Therefore, the recovery of cost was the criteria for pricing. In today's competitive market, the balancing of discretionary levels requires an iterative step-by-step approach over a period of time. It would be worth looking at what customers are prepared to pay for different levels of service and determine the life cycle costs of the assets for providing the agreed level of service along with the risks associated with each of the options. Balancing at a facility/asset level is dictated by the business requirements covering capex and opex. This means to balance the risk and cost to achieve the specified performance.

Asset management journey for realising value from assets' needs to be an iterative process from time to time over the life cycle of the assets. Options are generated based on asset condition, remaining physical, technical and economic life, operational costs, costs for upgrade and replacements. Intervention actions are justified based on comparing value realisation from assets for 'Doing nothing', 'Minimal repairs', 'Overhauls' and Replacements by 'as is' or capacity and/or capability improved options.

19.7 Conclusions

Asset management decisions are generally taken based on risk appetite of the board. Options include avoiding, treating, transferring, terminating or retaining risk based on decisions from balancing act. A 'desired' option is recommended based on stakeholders' perception of 'value for cost' in line with AS4183 for 'value' in general and 'value for money' in particular. Balancing is proposed in this chapter for distributing weights to important areas of the decision model for enabling someone or something to remain upright and steady in the perspective of the business. It is required to be reviewed from time to time for a long-term sustainability of the business. Asset owner/s can retain and grow the business considering a 'desired balance' in line with

ISO55000 and proposed by Asset Management Council (AMC) for the concept of an ‘accepted level’ of trade-off in cost, risk and performance (SA [27]).

International standard on asset management provides consistency in the interpretation of principles and the application of asset management across the industries. There is a need for further developing tools and techniques on how to implement and correctly measure success of good asset management supported by continual improvements.

There is huge opportunity for future work for various industry sectors as follows:

- Alignment of ISO55001 with other systems such as ISO9001, 31000, 14001, 45001, Information Technology (IT) and financial standards (ISO [28–30]).
- Further developing and applying asset management standards for other assets including natural, environmental and social assets.
- Assuring that asset management and audit teams have depth and breadth in line with asset management landscape and provide opportunity for building capability for required competency.

Good asset management helps in the journey of any organisation towards excellence for their business through a balancing act for costs, risks and performance for maintaining value of and realising value from assets. It is a journey which requires leadership and a long-term view along with commitment to financial, human and information system-related resources. Good in no good in today’s world. What matters is the aspiration from the whole organisation for leading towards excellence.

Acknowledgements Sincere acknowledgement to colleagues and researchers from asset management professional bodies, universities and industries including members of GFMAM, Asset Management Council (AM Council), Power Generation Programmes in Central Qld University and Maintenance and Reliability Engineering programmes in Federation University Australia.

References

1. Chattopadhyay, G. (2016). Issues and challenges of balancing cost, performance and risk in heavy-haul rail asset management. *In International Conference on Engineering Management, IEEM2016, IEEE Xplore*, 521-525, 978-1–5090-3665–3/16, 4–7, Bali, Dec 2016.
2. Chattopadhyay, G. (2020). Balancing cost, performance and risk in maintenance and capital expenditure. *IAPQR Transactions, 44*(2).
3. International Organization for Standardization, ISO 55000: 2014 “Asset Management—Overview, Principles and Terminology”, 2014.
4. Wikipedia, https://en.wikipedia.org/wiki/Infrastructure_asset_management, 26 May 2020.
5. International Infrastructure Management Manual (IIMM), IPWEA, 2015.
6. Asset Management Council (AMC) Australia: https://www.amcouncil.com.au/AMBoK_glossary_detail.aspx?glId=1905, 26 May 2020.
7. Global Forum on Maintenance and Asset Management, 2015, Asset Management Landscape, GFMAM: https://www.gfmam.org/, 26 May 2020
8. The Institute of Asset Management “Asset Management - an anatomy V3”, 84, 2015.
9. Peterson, S. B. (2002). Defining Asset Management. Strategic Asset Management Inc (SAMI), Unionville, Connecticut, USA. p.1, 2002.

10. Moore, R. (2007). Selecting the right manufacturing improvement tools—What Tool? When? Butterworth-Heinemann Burlington, Massachusetts, USA. pp. 32–34.
11. Woodhouse, J. (2001). *Asset management processes & tools.* p. 2. (London: The Woodhouse Partnership Ltd).
12. British Standard Institute, PAS55:2008–1:2008. Specification for the optimized management of physical assets, 2008.
13. British Standard Institute, PAS55:2008–2:2008. Guidelines for the application of PAS 55–1, 2008.
14. International Organization for Standardization, ISO 55001 Asset management—Management systems—Requirements, 2014.
15. International Organization for Standardization, ISO 55002 Asset management—Management systems—Guidelines on the application of ISO 55001, 2014.
16. International Organization for Standardization, ISO 15686–5:2017, Buildings and constructed assets—Service life planning—Part 5: Life-cycle costing, 2017.
17. Chattopadhyay, G., Webster, C., Phillips, H., & Leinster, M. (2011). *Asset management systems: Learning guide.* Australia: CQUniversity.
18. Chattopadhyay, G., Reddy, V., & Larsson, P. O. (2005). Decision on economical rail grinding interval for controlling rolling contact fatigue (RCF). *International Transactions in Operational Research, 12*(6), 545–558.
19. Parida, A., & Chattopadhyay, G. (2007). Development of a multi criteria hierarchical maintenance performance measurement (MPM) model. *Journal of Quality in Maintenance Engineering, 13*(3), 241–258.
20. Canada, J. R., Sullivan, W. G., White, J. A., & Kuponda, D. (2005). *Capital investment analysis for engineering and management*, Prentice-Hall.
21. International Organization for Standardization, ISO31000:2018, Risk management—Guidelines, 2018.
22. Chundhoo, V., Chattopadhyay, G., Gunawan, I., & Ibrahim, Y. M. (2017). OEE improvement of thermoforming machines through application of TPM at Tibaldi Australasia. In *International Conference on Engineering Management, IEEM2017*, 10–13, Singapore, Dec, 2017.
23. Chundhoo, V., Chattopadhyay, G., & Parida, A. (2019). Productivity Improvement through OEE measurement: A TPM case study for meat processing plant in Australia, eMaintenance2019, Stockholm, Sweden, 14–15 May, 2019.
24. Komljenovic, D. M., Gaha, G., Abdul-Nour, C., Langheit, & Bourgeois, M. (2016). Risks of extreme and rare events in asset management. *Safety Science, 88*, 129–145.
25. Knight, F. H. (1921). *Risk, uncertainty, and profit.* New York: The Riverside Press.
26. Aven, T. (2016). Risk assessment and risk management: Review of recent advances on their foundation. *European Journal of Operational Research, 253*, 1–13.
27. Standards Australia, Sai Global: AS 4183–2007 (R2018), Value Management, 2017.
28. International Standards Organisation, ISO 9001:2015 Quality management systems—Requirements, 2015.
29. International Standards Organisation, ISO 14001:2015, Environmental management systems—Requirements with guidance for use, 2015.
30. International Standards Organisation, ISO 45001:2018 Occupational health and safety management systems—Requirements with guidance for use, 2018.

Gopinath Chattopadhyay has Ph.D. in Mechanical Engineering (1999, from University of Queensland), Bachelor in Mechanical Engineering (1979, first-class first, Gold Medallist, from the University of Calcutta), Master of Industrial Engineering (1984, first-class first, Gold Medallist, from the University of Calcutta), MBA (1987, first-class second, Silver Medallist, from the University of Calcutta), Graduate certificate in education for higher education (1999 from Queensland University of Technology) and certificate in asset management from Institute of Asset Management, UK (2015). He is Post Graduate Programme Coordinator for Maintenance and

Reliability Engineering in Federation University, Australia. He is Adjunct Professor of Manipal University; Indian Institute of Engineering, Science and Technology and Indian Institute of Social Welfare and Business Management. He has worked in Qld University of Technology as Senior Lecturer and Postgraduate Programme coordinator for Engineering Management, Professor and Head of Post Graduate Programmes in Central Qld University and Visiting Professor in Indian Institute of Technology, Kharagpur, Lulea University of Technology, Sweden and University of Natural Resources and Environment, PNG. He has supervised 18 research higher degree students to successful completion. He has secured over 2.5 Million Australian dollars of funding for research projects and contributed to more than 170 international journal and conference papers including one prizewinning publication in 2008. He is currently Chair of Gippsland and past Chairs of Brisbane and Gladstone of Asset Management Council, Peak Professional body in asset management and a technical society of Engineers Australia. He was President of Australian Society for Operations Research (ASOR. Qld) and Vice President of Maintenance Engineering Society of Australia (MESA, Qld). He was industry reviewer of ISO55000 series of standards for Asset Management and Total Asset Management Plan (TAMP) for Queensland Government.

Chapter 20
Reliability-Based Performance Evaluation of Nonlinear Dynamic Systems Excited in Time Domain

Achintya Haldar and Francisco J. Villegas-Mercado

Abstract A novel concept of reliability-based performance evaluation using multiple deterministic analyses of nonlinear dynamic systems excited in time domain is presented. The dynamic excitations can be natural events like earthquakes or wave loadings. It can be thermo-mechanical loading caused by the use of computers. Unpredictability of the dynamic loadings, modeling the structural systems under uncertainty, and predicting the response behavior considering dynamic amplification and the different energy dissipation mechanisms can be very challenging. A transformational theoretical concept is presented to address this knowledge gap. The research objectives are achieved by using several advanced mathematical concepts including sensitivity analysis, model reduction techniques, intelligent sampling schemes, several advanced factorial schemes producing a compounding beneficial effect, and surrogate meta-modeling techniques to obtain efficiency without sacrificing accuracy. They are implemented in a multi-scale environment exploiting state-of-the-art computational power. The formulation extracts stochastic dynamical behavior using only hundreds of intelligent analyses instead of thousands of simulation-based analyses. This is a new design paradigm using intelligent multiple deterministic analyses. It will provide an alternative to simulation and the classical random vibration concept.

Keywords Reliability tools · Risk estimation · Nonlinear dynamic systems · Time-domain excitation · Performance-based design · Response surface method · First-order reliability method · Kriging method

---

A. Haldar (✉) · F. J. Villegas-Mercado
Department of Civil and Architectural Engineering and Mechanics, University of Arizona, 85721 Tucson, AZ, USA
e-mail: haldar@u.arizona.edu

K. B. Misra (ed.), *Handbook of Advanced Performability Engineering*,
https://doi.org/10.1007/978-3-030-55732-4_20

20.1 Introduction

Performance evaluation of engineered structures excited by dynamic loadings is expected to be very challenging. Dynamic loadings can be natural events like earthquakes or wave loadings during high seas. It can be thermo-mechanical loading caused by the use of computers. The cost of designing structures against such dynamic excitation is enormous, and there is considerable room for improvement in the current state of knowledge. Unpredictability of the dynamic loading, modeling the structural systems under uncertainty, and predicting the response behavior considering dynamic amplification and the different energy dissipation mechanisms can be very challenging. The presence of a large amount of uncertainty in every phase of the evaluation process increases the level of complexity in several orders of magnitude. Even in the period of very advanced information in the related areas, the knowledge gap is enormous. The most sophisticated dynamic analysis requires that the loading must be applied in time domain. Most major sources of nonlinearity need to be incorporated in the formulation explicitly. The amplified dynamic response information needs to be tracked, and the performance requirements suggested by the owners or users need to be met. The performance objectives cannot be assured with certainty since the presence of uncertainty in the formulation is enormous. There will be some amount of underlying risk, and the information on it needs to be quantified. If the risk is not acceptable, alternative designs need to be considered and compared. The final design must indicate the underlying risk under different operating or loading environments.

Issues related to the dynamic loadings in the presence of uncertainty are generally addressed using the classical random vibration concept. It is a very sophisticated mathematical technique. In spite of many sophistications, its applications to estimate underlying risk of the dynamic systems are very limited. Observing some of the shortcomings, the authors and their team members started developing the concept of the stochastic finite element method (SFEM) in mid-eighties. They incorporated all major sources of uncertainty in a stress-based nonlinear finite element (FE) algorithm. In the stress-based FE formulation, the tangent stiffness can be expressed in explicit form, fewer elements are required in describing large deformation configurations, integration is not required to obtain the tangent stiffness, and the stresses of an element can be obtained directly. It is specifically appropriate for frames. However, the displacement-based FE method, where shape functions are used to describe displacements at the nodes of the elements, is commonly used in the profession. The works of the research team are widely referred; however, the lack of availability of the program they developed is a major issue.

The research team considered developing a completely new nonlinear FE-based concept for the reliability estimation. This will enable users to use any program available to them capable of conducting nonlinear FE analysis of structural systems excited by dynamic loading applied in time domain. This effort was encouraged by a prestigious research grant from the United States National Science Foundation. This new and novel concept is briefly discussed in this chapter.

In developing the new concept, the authors considered the following related issues. Risk is always estimated for a specific limit state function (LSF) or performance function (PF). For explicit PFs, the classical first-order or second-order reliability method (FORM/SORM) will be very efficient [1]. For the nonlinear problems excited by dynamic loadings in time domain, PFs are expected to be implicit in nature. For implicit PFs, Haldar and Mahadevan [2] suggested a Monte Carlo simulation (MCS) approach in place of SFEM. It requires numerous deterministic analyses, sometimes of the order of millions for low probability events, making it very inefficient. To improve its efficiency, several space-reduction techniques, parallel computing, and advanced mathematical concepts can be used. However, they may be problem-dependent and will require advanced expertise which is not expected from the everyday users for routine applications. In addition, the results may not be acceptable to all concerned parties.

The research team observed that a typical nonlinear time-domain FE analysis of real structural systems may require from few minutes to several hours of continuous running of a computer. If one wants to simulate about 10,000 times, a very small number to estimate risk of the order of about 10^{-6}, it may require several years continuous running of a computer. The team realized that an alternative to the classical MCS is necessary. Using the concept, the underlying risk can be extracted using only few hundreds of simulations using any computer program instead of millions. The team proposed such a concept in this chapter.

20.2 Proposed Reliability Evaluation Concept for Dynamic Systems Excited in Time Domain

Following the common practice, a real structure will be represented by any FE algorithm capable of conducting nonlinear time-domain analysis. All major sources of nonlinearity will be incorporated in the formulation by following the practices used by the deterministic community. All major sources of uncertainty in the resistance-related structural design variables will be incorporated in the formulation similar to the procedure used in the SFEM concept. For routine applications, serviceability and strength PFs will be used for risk estimation. Since the PFs are implicit, the first major challenge will be how to develop an alternative.

As mentioned earlier, for the efficient implantation of FORM the PFs need to be explicit. Implicit PFs can be approximately represented by response surfaces (RSs) in the failure region considering all random variables (RVs) present in the formulation. They will be mathematical expressions, capable of incorporating the uncertainty information as realistically as practicable, in an acceptable way to the reliability community. The response surface method (RSM) can be used to generate an expression for a required RS explicitly. The RSM concept was initially developed to study chemical reactions [3]. The original concept needs to be modified for the structural reliability estimation. Some of the basic requirements of generating an acceptable

RS are 1) its mathematical form, 2) the realistic incorporation of uncertainty in all RVs, 3) it should be generated in the failure region, and 4) the information required to fit a polynomial through them to make an RS should be in explicit form. They are discussed next.

20.2.1 Mathematical Form of an RS

For nonlinear problems, a linear function of an RS will not be appropriate. To increase efficiency and considering many alternatives, the team decided to use a second-order polynomial without and with cross terms. Mathematically, they can be expressed as

$$\hat{g}(\mathbf{X}) = b_0 + \sum_{i=1}^{k} b_i X_i + \sum_{i=1}^{k} b_{ii} X_i^2 \tag{20.1}$$

and

$$\hat{g}(\mathbf{X}) = b_0 + \sum_{i=1}^{k} b_i X_i + \sum_{i=1}^{k} b_{ii} X_i^2 + \sum_{i=1}^{k-1} \sum_{j>i}^{k} b_{ij} X_i X_j \tag{20.2}$$

where k is the number of RVs; X_i ($i = 1,2,\ldots,k$) is the *ith* RV; b_0, b_i, b_{ii}, and b_{ij}, are the unknown coefficients; and $\hat{g}(\mathbf{X})$ is the expression for an RS. The total number of coefficients needed to generate Eqs. (20.1) and (20.2) can be shown to be ($2\,k + 1$) and ($k + 1$) ($k + 2$)/2, respectively. For large values of k, the cross terms will improve accuracy but will require significant large numbers of coefficients to be evaluated. This will require further attention and will be discussed later.

20.2.2 Realistic Incorporation of Uncertainty in All RVs

RSM concept when initially proposed used the coded variable space to incorporate uncertainty. This will not satisfy the reliability community. The uncertainty in an RV was expressed as

$$X_i = X_i^C \pm h_i x_i \sigma_{X_i} \quad \text{where} \quad i = 1, 2, \ldots, k \tag{20.3}$$

where X_i is the *ith* RV region; X_i^C is the coordinate of the center point of RV X_i; σ_{X_i} is the standard deviation of RV X_i; h_i is an arbitrary factor controlling the experimental region [4]; x_i is the coded variable that assumes values of 0, ± 1, or $\pm\sqrt[4]{2^k}$ depending on the coordinates of the sampling point with respect to the center point and sampling schemes, and k is the number of RVs. It is clear that Eq. (20.3) does not use distribution information explicitly.

FORM is implemented in the standard normal variable space. It is not realistic to transform a non-normal RV to a normal RV over the space of the RV. However, all non-normal RVs can be expressed in terms of equivalent normal mean ($\mu_{X_i}^N$) and equivalent normal standard deviation ($\sigma_{X_i}^N$) at the checking point (x_i^*) as in [1]:

$$\mu_{X_i}^N = x_i^* - \Phi^{-1}\left[F_{X_i}\left(x_i^*\right)\right]\sigma_{X_i}^N \tag{20.4}$$

and

$$\sigma_{X_i}^N = \frac{\phi\left\{\Phi^{-1}\left[F_{X_i}\left(x_i^*\right)\right]\right\}}{f_{X_i}\left(x_i^*\right)} \tag{20.5}$$

where $F_{X_i}\left(x_i^*\right)$ and $f_{X_i}\left(x_i^*\right)$ are the cumulative distribution function (CDF) and the probability density function (PDF) of the original non-normal X_i RV at the checking point (x_i^*), respectively; and $\Phi()$ and $\phi()$ are the CDF and PDF of the standard normal variable, respectively.

Therefore, the equivalent normal mean ($\mu_{X_i}^N$) and standard deviation ($\sigma_{X_i}^N$) values can be used in Eq. (20.3) to incorporate the distribution information of non-normally distributed RVs.

20.2.3 Response Surface to Be Generated in the Failure Region

The failure region of a realistic structure will be unknown in most cases. FORM identifies the coordinates of the most probable failure point in an iterative way. Since FORM is a major building block of the new method, the integration will assure generating the RS in the failure region.

20.2.4 Information Required to Fit a Polynomial to Generate an RS

To generate an expression of an RS, response data are required to fit a polynomial through them. The response data can be generated by conducting deterministic nonlinear FE analyses at sampling points following specific schemes around sampling points. Following the iterative strategy of FORM, the coordinates of the initial sampling point will be the mean values of all RVs. To generate coordinates of the sampling points, two schemes are commonly used. They are Saturated Design (SD) [4] and the Central Composite Design (CCD) [3]. It can be shown that the total number of unknown coefficients will be $2k + 1$ and $(k + 1)(k + 2)/2$ for Eqs. (20.1) and (20.2), respectively. Cross terms are not required to generate an RS using SD, and the total number of FE analyses required will be the total number of unknown coefficients. It is very efficient but lacks few statistical properties and may not be very accurate. CCD scheme is very accurate but inefficient. It requires cross terms to generate an RS and will require a total of $2^k + 2k + 1$ sampling points or deterministic FE analyses. Since the new method is iterative in nature, in order to increase the efficiency without compromising the accuracy, the team decided to use the SD scheme without cross terms in the intermediate iterations and CCD in the final iteration. To differentiate between an RS generated using the classical RSM concept and the integrated approach proposed here, it will be denoted hereafter as all-inclusive RS or AIRS.

20.3 Generation of a Mathematical Expression of an RS

At this stage, using the mathematical schemes discussed in Sect. 20.2, sufficient response data will be available to fit a second-order polynomial. However, the response data generated using the scheme discussed in Sect. 20.2.4, CCD will be used in the final iteration. To fit a polynomial to the data, generally regression analysis is used. Regression analysis fits the data on an average sense. To improve the accuracy, the team decided to use the Kriging approach. It was initially developed to improve the accuracy for tracing gold in ores. It is a geostatistical approach and uses weight factors. Weight factors decay as the distance between the sampling points and the RS increases. Some of the advantages of Kriging method (KM) are as follows: RS generate will pass through the sampling points, it will predict responses more accurately between two sample points, it is uniformly unbiased, and prediction errors are less than all other forms. It can be considered as the best linear unbiased surrogate for an RS. Universal Kriging (UK) method is used in developing the procedure.

Estimation of weight factors is complicated. Interested readers are referred to [5–7]. Conceptually, it is very briefly discussed here. The predicted value for an RS, at a point in space with coordinates $\mathbf{x}_0$, can be expressed as:

$$\hat{g}(\mathbf{x}_0) = \boldsymbol{\omega}^T \mathbf{Z} \quad (20.6)$$

where $\boldsymbol{\omega}$ is a vector for the unknown weights calculated based on the distance between the sample points and the unknown point, and $\mathbf{Z}$ is a vector containing the estimated values by FE analyses. $Z(\mathbf{X})$, can be decomposed into two components:

$$Z(\mathbf{X}) = \mu(\mathbf{X}) + Y(\mathbf{X}) \quad (20.7)$$

where $\mathbf{X}$ is a vector indicating the coordinates of the point in the space, $\mu(\mathbf{X})$ is a second-order polynomial with cross terms, and $Y(\mathbf{X})$ is an intrinsically stationary function with zero mean and underlying variogram function $\gamma_Y(\mathbf{h})$. A dissimilarity function and the variogram cloud are required to generate the variogram $\gamma_Y(\mathbf{h})$. The dissimilarity function is defined as:

$$\gamma^*(h_i) = \frac{1}{2}[Z(\boldsymbol{x}_i + h_i) - Z(\boldsymbol{x}_i)]^2 \tag{20.8}$$

where h_i is the distance between the sample points $Z(\boldsymbol{x}_i + h_i)$ and $Z(\boldsymbol{x}_i)$ in dimension of the ith RV. Then, the following mathematical model is used to fit the experimental variogram:

$$\gamma_Y(\mathbf{h}) = b\left\{1 - \exp\left[\sum_{i=1}^{n} -\left(\frac{h_i}{a_i}\right)^2\right]\right\} \tag{20.9}$$

where $\mathbf{h}$ is a vector of h_i components, a_i and b are unknown to be estimated, and called range and sill parameters, respectively, and n is the number of RVs. Finally, $\boldsymbol{\omega}$ in Eq. (20.6) can be estimated as [8]:

$$\boldsymbol{\omega} = \boldsymbol{\Gamma}_Y^{-1}\left[\boldsymbol{\gamma}_{Y,0} - \mathbf{F}\left(\mathbf{F}^{\mathrm{T}}\boldsymbol{\Gamma}_Y^{-1}\mathbf{F}\right)^{-1}\left(\mathbf{F}^{\mathrm{T}}\boldsymbol{\Gamma}_Y^{-1}\boldsymbol{\gamma}_{Y,0} - \mathbf{f}_0\right)\right] \tag{20.10}$$

where $\boldsymbol{\gamma}_{Y,0} = (\gamma_Y(\mathbf{X}_1 - \mathbf{X}_0), \ldots, \gamma_Y(\mathbf{X}_n - \mathbf{X}_0))^T$, $\Gamma_{Yi,j} = \gamma_Y\left(\mathbf{X}_i - \mathbf{X}_j\right)$, and $\mathbf{F}$ and $\mathbf{f}_0$ are the ordinary regression design matrices for the determinic points and the required point, respectively. The predicted value of an RS at the coordinate $\mathbf{x}_0$ can be shown to be:

$$\hat{g}(\mathbf{x}_0) = \left[\boldsymbol{\gamma}_{Y,0} - \mathbf{F}\left(\mathbf{F}^{\mathrm{T}}\boldsymbol{\Gamma}_Y^{-1}\mathbf{F}\right)^{-1}\left(\mathbf{F}^{\mathrm{T}}\boldsymbol{\Gamma}_Y^{-1}\boldsymbol{\gamma}_{Y,0} - \mathbf{f}_0\right)\right]^T \boldsymbol{\Gamma}_Y^{-1}\mathbf{Z} \tag{20.11}$$

20.4 Performance Functions

A specific PF will be mathematically represented hereafter as g(**X**). As discussed earlier, the serviceability and strength PFs are generally used in structural engineering applications. The serviceability PF is considered to be:

$$g(\mathbf{X}) = \delta_{\text{allow}} - \hat{g}(\mathbf{X}) \tag{20.12}$$

where δ_{allow} is the allowable drift, generally defined in design guidelines or specified by the owners or users and $\hat{g}(\mathbf{X})$ is the surrogate model for the global drift for an RS using Kriging. Strength PFs for steel structures for both the axial load and bending moment can be defined as [5]:

When

$$\frac{P_u}{\phi P_n} \geq 0.2; \; g(X) = 1.0 - \left(\frac{P_u}{P_n} + \frac{8}{9}\frac{M_u}{M_n}\right) = 1.0 - \left[\hat{g}_P(\mathbf{X}) + \hat{g}_M(\mathbf{X})\right] \quad (20.13)$$

When

$$\frac{P_u}{\phi P_n} < 0.2; \; g(X) = 1.0 - \left(\frac{P_u}{2P_n} + \frac{M_u}{M_n}\right) = 1.0 - \left[\hat{g}_P(\mathbf{X}) + \hat{g}_M(\mathbf{X})\right] \quad (20.14)$$

where P_n and P_u are the nominal and required tensile/compressive strength, respectively, M_n and M_u are the nominal and required flexural strengths, respectively, ϕ is the resistance factor, and $\hat{g}_P(\mathbf{X})$and $\hat{g}_M(\mathbf{X})$ are the surrogate models for axial force and bending moment, respectively, using Kriging and AIRS.

20.5 Reliability Estimation Using the Proposed Method

With the availability of an explicit expression for the required PFs, it will be straightforward to extract the reliability information using FORM. The first iteration will be initiated at the mean values of all RVs, and a required PF will be generated using SD without cross terms in the intermediate iterations. In the final iteration, the PF will be generated by the UK. Once the coordinates of the most probable failure point $\mathbf{x}^*$ are available, the reliability index β can be computed as [1]:

$$\beta = \sqrt{(\mathbf{x}^*)^t(\mathbf{x}^*)} \quad (20.15)$$

and the corresponding probability of failure *pf* can be estimated as:

$$p_f = \Phi(-\beta) = 1.0 - \Phi(\beta) \quad (20.16)$$

where $\Phi(\;)$ is the CDF of the standard normal.

After generating an AIRS at the mean values of all RVs in the first iteration, the coordinates of the center point will be updated using the FORM strategy. The iterative process of FORM will stop when both the reliability index and the coordinates of the center point converge.

20.6 Implementation of the Proposed Method

The proposed method as discussed above cannot be implemented as this stage for large realistic structural systems. Suppose the total number of RVs present in the formulation is relatively small, say $k = 8$. The total number of dynamic analysis (TNDA) required to implement the procedure will be over 290. Suppose $k = 20$, it will require over 1.05 million analyses, making it unimplementable. The team used several techniques to increase the implementation potential of the proposed method as discussed next.

20.6.1 Reduction of the Total Number of RVs

To increase the efficiency without compromising the accuracy, the total number of RVs present in the problem needs to be reduced at the earliest opportunity. In the first iteration, FORM will be initiated at the mean values of all k RVs. The required RS will be generated using SD without cross terms. Haldar and Mahadevan [2] observed that all RVs are not equally important in the risk or reliability estimation. The importance of an RV can be estimated from the information of the sensitivity indexes, the information generated to implement FORM. All RVs with smaller sensitivity indexes can be considered as deterministic at their respective mean values. Considering this, the total number of RVs is reduced from k to k_R. To reduce the size of the problem, all the discussions made so far need to be carried out in terms of k_R instead of k.

20.6.2 Improvements of Kriging Method

As discussed earlier, CCD will be used in the final iteration. CCD consists of one center point, $2k_R$ axial points, and 2^{k_R} factorial points. The efficiency can be improved only by reducing 2^{k_R} factorial points. After attempting many alternatives, the team decided to use the cross terms and the necessary sampling points only for the most significant RVs, k_m, ($k_m \le k_R$) in sequence in order of their sensitivity indexes until the reliability index converges with a pre-determined tolerance level [5–7]. The reduction of factorial points less than the number of coefficients may cause ill-conditioning of the regression analysis. To avoid ill-conditioning, only the cross terms for k_m most significant variables are considered in the polynomial expression. Since k_m is lower than $k_{R,}$, it can be shown that TNDA required to extract the reliability information will be reduced from $2^{k_R} + 2k_R + 1$ to $2^{k_m} + 2k_R + 1$. Reducing the size of the problem this way will be denoted hereafter as the modified Universal Kriging or MUK method. Both the original and the modified schemes will be denoted hereafter as Advanced Factorial Design (AFD).

In summary, to increase the accuracy of the estimated risk, in the last iteration, the PF will be generated using CCD and the MUK scheme. The total number of dynamic analysis required to implement the proposed reliability estimation procedure can be shown to be:

$$TNDA = (1 + 2k) + n(1 + 2k_R) + \left(2^{k_m} + 2k_R + 1\right) \tag{20.17}$$

where k is the total number RVs, k_R is the number of important RVs after the first iteration, n is the number of intermediate iterations, and k_m is the number of the most significant RVs for factorial points in the last iteration. The proposed reliability estimation method for nonlinear dynamic system excited in time domain and LSF generated by the MUK method will be denoted hereafter as AIRS-MUK-FORM. It is now necessary to verify the proposed method.

20.7 Applications of AIRS-MUK-FORM in Performance-Based Seismic Design

After several major earthquakes all over the world in the mid-nineties, it was observed that the economic consequences could be catastrophic. Before those incidences, the structures were designed to protect human life, at least in the U.S. The Northridge earthquake of 1994 caused over 40 billion US dollars of damage [9]. This prompted the profession to find an alternative design approach that could minimize the level of damage/economic losses. It led to the development of the Performance-Based Seismic Design (PBSD) concept. PBSD was proposed by the Federal Emergency Management Agency (FEMA) supported by the background work conducted by SAC [a joint venture consisting of the Structural Engineers Association of California (SEAC), Applied Technology Council (ATC), and the California Universities for Research in Earthquake Engineering (CUREE)]. It is essentially a sophisticated risk-based concept. Structures can be designed to satisfy a performance requirement if the owner or the users are willing to accept the corresponding risk. The basic concept of performance level as stated by FEMA 350 [10] is: "The intended post-earthquake condition of a building; a well-defined point on a scale measuring how much loss is caused by earthquake damage." The American Society of Civil Engineers (ASCE) 41-13 [11] recommended four performance levels. They are Operational (OP), Immediate Occupancy (IO), Life Safety (LS), and Collapse Prevention (CP). The damage level ranges from minimal or no damage to both structural and nonstructural elements to extensive damage to all components (prior to collapse). Figure 20.1 qualitatively indicates the performance levels. The performance levels suggested by FEMA are summarized in Table 20.1.

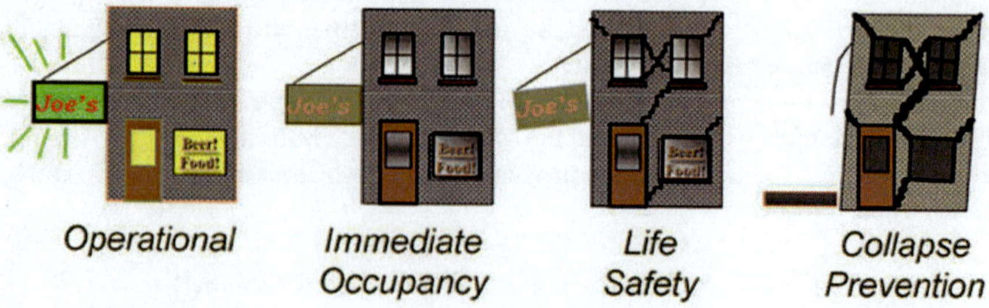

Fig. 20.1 Different performance levels defined in ASCE 41-13

Table 20.1 Structural performance levels reported in FEMA-350 [10]

Performance level	Return period	Probability of exceedance	Allowable drift
IO	72-year	50% in 50 years	0.007*H
LS	475-year	10% in 50 years	0.025*H
CP	2475-year	2% in 50 years	0.050*H

FEMA does not specifically suggest a specific reliability evaluation method to estimate the underlying risk, and it is a major obstacle in implementing PBSD. The research team of the authors believes that no reliability method is currently available to estimate risk corresponding to a performance level by applying the earthquake loading in time domain considering all major sources of nonlinearity satisfying the deterministic community. The authors believe that the method proposed here can be used to implement the PBSD guidelines. Referring to Eq. (20.12) to generate a PF, the response surface $\hat{g}(\mathbf{X})$ will be generated using AIRS-MUK-FORM and the δ_{allow} value will be selected from Table 20.1 according to a specific performance requirement. It will be showcased later with the help of examples.

20.8 Applications of AIRS-MUK-FORM Incorporating Post-Northridge Improved Structural Features

In addition to proposing PBSD, to improve structural behavior during earthquake excitations, several novel design features were introduced to dissipate the input energy. During the Northridge earthquake of 1994, welds in beam–column connections failed in a brittle manner in several steel frame structures. A typical connection that failed during the 1994 Northridge earthquake is shown in Fig. 20.2a. Following

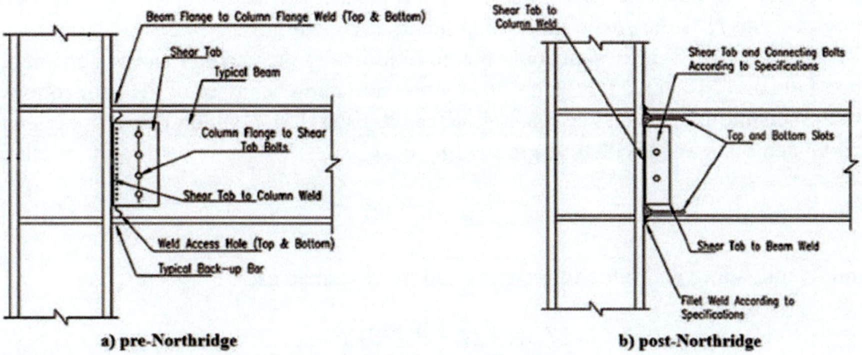

Fig. 20.2 Typical steel connection configurations as in Mehrabian et al. [12]

the earthquake, the Structural Engineering Association of California recommended not to use them. It will be referred to hereafter as the pre-Northridge connections.

Several alternative improved connections were proposed to increase ductility and the energy absorption capacity. They include cover plated, spliced beam, connections with Reduced Beam Sections (RBS) or Dog-Boned [10, 12], and slotted-web beam–column connections [13]. Seismic Structural Design Associates, Inc. (SSDA) proposed a unique proprietary slotted-web (SlottedWebTM) moment connection by cutting slots in the web of the beam as shown in Fig. 20.2b. SSDA tested several full-scale beam–column connection models using ATC-24 test protocol [13] and shared the test data with the research team. The test results clearly indicated that the slots in the web of the beam introduced the desirable behavior without reducing the initial stiffness of the connection. However, the presence of two slots in the web raised concern to some scholars.

Before the Northridge earthquake of 1994, beam–column connections in steel frames were generally considered as fully restrained (FR) type. However, both the experimental and analytical studies indicate that they are partially restrained (PR)-type with different rigidities. They introduce another major source of nonlinearity even when the load is very small. In addition, the structural dynamic properties including stiffness, damping, frequency, mode shape, etc. are affected [14]. They will change the response surface required for a PF represented by Eq. (20.12).

PR connections are commonly described in terms of M-θ curves; M represents the moment transmitted by the connection, and θ is the relative rotation angle of connecting members. Among many alternatives, the Richard four-parameter M-θ model is selected to represent the PR connections in steel frames. Using the four-parameter Richard model, a typical M-θ curve can be expressed as:

$$M(\theta) = \frac{(k - k_P)\theta}{\left(1 + \left|\frac{(k-k_P)\theta}{M_0}\right|^N\right)^{\frac{1}{N}}} + k_P\theta \tag{20.18}$$

where M is the connection moment, θ is the relative rotation between the connecting elements, k is the initial stiffness, k_P is the plastic stiffness, M_0 is the reference moment, and N is the curve shape parameter, as shown in Fig. 20.3a.

Equation (20.18) represents only the monotonically increasing loading portion of the M-θ curve. To consider the unloading and reloading behavior of PR connections for the seismic excitation, the Masing rule can be used [15]. A general class of Masing model can be defined with a virgin loading curve as:

$$f(M, \theta) = 0 \tag{20.19}$$

and its unloading and reloading curves can be described as:

$$f\left(\frac{M - M_a}{2}, \frac{\theta - \theta_a}{2}\right) = 0 \tag{20.20}$$

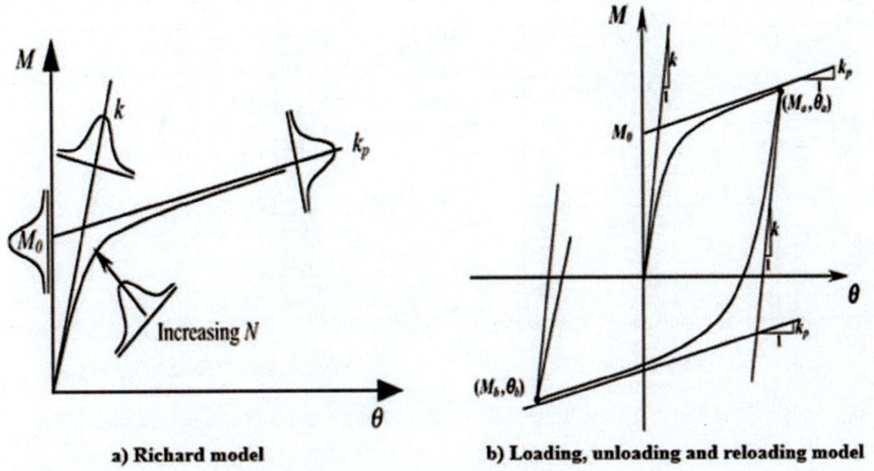

Fig. 20.3 M-θ curves for PR connections as in Mehrabian et al. [12]

where (M_a, θ_a) is the load reversal point as shown in Fig. 20.3b. The unloading and reloading behavior of a PR connection considered in the study is:

$$M(\theta) = M_a - \frac{(k - k_P)(\theta_a - \theta)}{\left(1 + \left|\frac{(k-k_P)(\theta_a-\theta)}{2M_0}\right|^N\right)^{\frac{1}{N}}} - k_P(\theta_a - \theta) \tag{20.21}$$

Thus, Eq. (20.18) is used when the connection is loading, and Eq. (20.21) is used when the connection is unloading and reloading. It represents the hysteretic behavior of a PR connection. The above formulations are incorporated in the proposed reliability approach to implement PBSD, as discussed next.

20.9 Example to Showcase PBSD and Flexible Connection Behavior in Steel Frames

A relatively small two-story steel frame as shown in Fig. 20.4a is considered to facilitate the comparison of the reliability estimations by the proposed AIRS-MUK-FORM and the standard MCS methods. The structure was excited for 20 s using the Northridge earthquake time history recorded at Canoga Park station shown in Fig. 20.4b. Both PR connection configurations are considered. Table 20.2 summarizes the Richard model parameters. Table 20.3 lists uncertainty information of all RVs. Reliability analysis results are summarized in Table 20.4. The reliability information estimated by AIRS-MUK-FORM and 50,000 MCS matches very well, clearly indicating the validity of the proposed method. It also indicates the weaknesses of

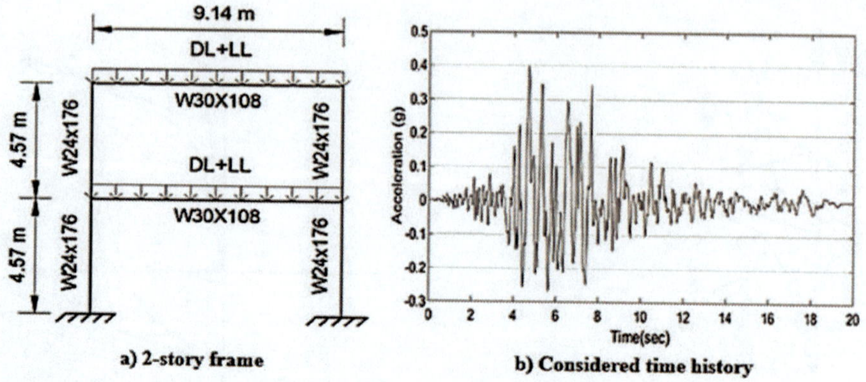

Fig. 20.4 **a** Steel frame. **b** Time history recorded at 1994 Northridge Canoga park station

Table 20.2 Pre- and Post-Northridge four Richard parameters of two-story steel frame

Parameter	PR Connection	
	Pre-Northridge	Post-Northridge
k (kN-m/rad)	6.6663E + 05	1.9546E + 07
kp(kN-m/rad)	1.1113E + 04	4.5194E + 03
Mo (kN-m)	7.7835E + 02	2.0145E + 03
N	1.10	1.00

Table 20.3 Uncertainty of RVs used for two-story steel frame

Random variable	Distribution	Mean ($\bar{X}$)	COV
E (kN/m^2)	Lognormal	1.9995E + 08	0.06
Fy (kN/m^2)	Lognormal	3.6197E + 05	0.10
A_C (m^2)	Lognormal	3.3355E-02[a]	0.05
A_G (m^2)	Lognormal	2.0452E-02[a]	0.05
I_{xC} (m^4)	Lognormal	1.8606E-03[a]	0.05
I_{xG} (m^4)	Lognormal	2.3642E-03[a]	0.05
DL (kN/m^2)	Normal	4.0219	0.10
LL (kN/m^2)	Type 1	1.1970	0.25
g_e	Type 1	1.00	0.20

[a]Mean values of A and Ix can be found in AISC manual

pre-Northridge PR connections. For additional information, interested readers are encouraged to refer to [16].

Table 20.4 Reliability results for two-story steel frame

LSF	Method	FR		PR (pre-northridge)		PR (post-northridge)	
		β (TNDA)	p_f	B (TNDA)	p_f	β (TNDA)	p_f
Overall drift	Proposed	3.5232 (83)	2.13E-04	3.4063 (94)	3.29E-04	3.6208 (94)	1.47E-04
	MCS	3.5149 (50,000)	2.20E-04	3.4141 (50,000)	3.20E-04	3.6331 (50,000)	1.40E-04
Inter-story drift	Proposed	3.2429 (105)	5.92E-04	2.9825 (94)	1.40E-03	3.3357 (83)	4.25E-04
	MCS	3.2389 (50,000)	6.00E-04	2.9845 (50,000)	1.42E-03	3.3139 (50,000)	4.60E-04

20.10 Consideration of Uncertainty in Dynamic Excitations in Time Domain

In the previous section, the reliability of a structure is estimated for one known earthquake time history, ignoring the uncertainty in selecting earthquake time histories. The seismic loading may be the most unpredictable natural loading. It depends on many features including the seismicity of the region, local soil conditions [17, 18], types of structures to be built, etc. Because of the unpredictability of seismic excitation, the design guidelines changed, both conceptually and analytically very frequently [19–21]. The current design guidelines in the U.S. require to consider at least 11 earthquake time histories fitting the design or target spectrum. A suite of earthquake time histories fitting the spectrum can be generated in several ways. Two most common practices in the U.S. are scaling past recorded time histories available at Pacific Earthquake Engineering Research Center (PEER) database [5, 6] and using the Broadband Platform (BBP) [22] if required information is available. The first alternative is reasonable for most practical applications.

In this approach, initially, the ground motions in the database are scaled to match the design spectrum at the fundamental period of the structure to be built at the site, site conditions, etc. The most desirable scale factor (SF) will be 1.0, but it is very rarely obtained. In this study, ground motions with SF of more than 4 and less than 0.25 are ignored [6, 17, 18]. Following the above-ground motion selection process, the database can be reduced from several thousands to several hundreds.

To select the most suitable site-specific 11 time histories, a ranking process is used. The team develops a suitability factor concept to rank them. A response spectrum is generally developed for a wide frequency range. They considered the range of the period to be 0.2 and 2 times the fundamental period of the structure. This range is then subdivided into at least 40 equally spaced intervals in the log scale. At each of these periods, the differences between the selected ground motion spectral acceleration spectrum and the target spectrum are estimated. The total error for all intervals, denoted as Square Root of the Sum of the Squares (SRSS), is estimated for each of

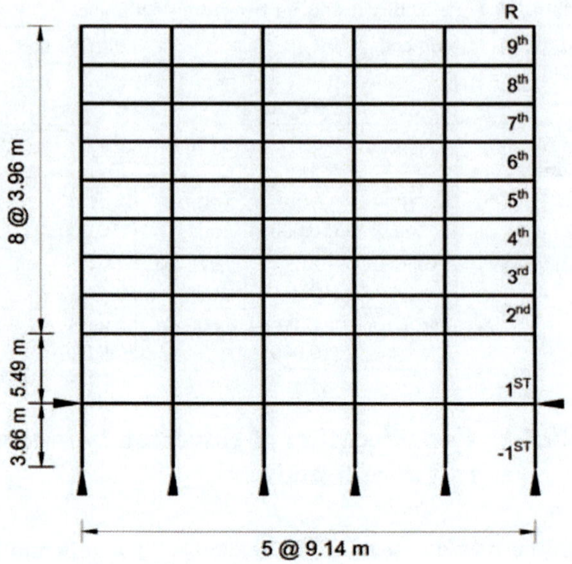

Fig. 20.5 SAC-2000 nine-story steel building, FEMA 355F [24]

the scaled ground motions. Eleven earthquake time histories with lower SRSS values are considered for further study. They are expected to contain all major sources of uncertainty in the earthquake excitations and will also satisfy the current seismic design guidelines in the U.S. Due to severe space limitations, the whole process cannot be discussed in detail. The reader is referred to [6, 17, 18, 23].

After the validation, the research team members considered a nine-story steel frame, as shown in Fig. 20.5, designed by SAC experts [24]. It is supposed to be considered as the benchmark design for further consideration. Cross-sectional properties of the members and the corresponding Richard model parameters are given in Table 20.5. Uncertainty information of all RVs is listed in Table 20.6. Time histories considered are given in Table 20.7. Reliability results are summarized in Table 20.8 [23].

20.11 Design of Offshore Structures for Wave Loading Applied in Time Domain

A typical offshore structure (OFS) is shown in Fig. 20.6a. OFSs are increasingly used to address energy-related issues. In reliability-based design of onshore structures (ONSs), it has become very common to address major sources of uncertainty. Engineering design of OFSs has not kept up with similar improvements as in ONSs. This is very important since the failure of OFSs will have not only disastrous economic consequences but also huge environmental impacts. OFSs are needed to be designed

Table 20.5 Size of steel cross sections and four Richard parameters for the nine-story building

Story/Floor	Columns		Girder	Four richard parameters			
	Exterior	Interior		*k*(kN-m/rad)	*kp*(kN-m/rad)	*Mo*(kN-m)	*N*
9/Roof	W14 × 233	W14 × 257	W24 × 62	8.6433E + 06	4.5194E + 03	8.1519E + 02	1.00
8/9	W14 × 257	W14 × 283	W27 × 94	1.5705E + 07	4.5194E + 03	1.5920E + 03	1.00
7/8	W14 × 257	W14 × 283	W27 × 102	1.7230E + 07	4.5194E + 03	1.7597E + 03	1.00
6/7	W14 × 283	W14 × 370	W33 × 130	2.6382E + 07	4.5194E + 03	2.7664E + 03	1.00
5/6	W14 × 283	W14 × 370	W33 × 141	2.9037E + 07	4.5194E + 03	3.0585E + 03	1.00
4/5	W14 × 370	W14 × 455	W33 × 141	2.9037E + 07	4.5194E + 03	3.0585E + 03	1.00
3/4	W14 × 370	W14 × 455	W33 × 141	2.9037E + 07	4.5194E + 03	3.0585E + 03	1.00
2/3	W14 × 370	W14 × 500	W36 × 150	3.2822E + 07	4.5194E + 03	3.4748E + 03	1.00
1/2	W14 × 370	W14 × 500	W36 × 150	3.2822E + 07	4.5194E + 03	3.4748E + 03	1.00
−1/1	W14 × 370	W14 × 500	W36 × 150	3.2822E + 07	4.5194E + 03	3.4748E + 03	1.00

Table 20.6 Uncertainty of RVs used for the nine-story steel frame

Random variable	Distribution	Mean ($\bar{X}$)	COV
E (kN/m^2)	Lognormal	1.9995E + 08	0.06
Fy_G (kN/m^2)[a]	Lognormal	3.3509E + 05	0.10
Fy_C (kN/m^2)[a]	Lognormal	3.9645E + 05	0.10
A (m^2)	Lognormal	[b]	0.05
I_x (m^4)	Lognormal	[b]	0.05
DL_R(kN/m^2)	Normal	4.1727	0.10
DL_F(kN/m^2)	Normal	4.8263	0.10
LL_R(kN/m^2)	Type 1	0.9576	0.25
LL_F(kN/m^2)	Type 1	0.9576	0.25
g_e	Type 1	1.00	0.20

[a]Yield stress of girder or column cross section reported in FEMA-355C [24]

[b]Mean values of A and Ix can be found in AISC manual

Table 20.7 Ground motions sets for CP, LS, and IO performance levels

Set 1: 2% PE in 50 years; CP Performance level			Set 2: 10% PE in 50 years; LS Performance level			Set 3: 50% PE in 50 years; IO Performance level		
EQ	Name	SF	EQ	Name	SF	EQ	Name	SF
1	1995 Kobe	1.2	21	Imperial Valley, 1940	2.0	41	Coyote Lake, 1979	2.3
2	1995 Kobe	1.2	22	Imperial Valley, 1940	2.0	42	Coyote Lake, 1979	2.3
3	1989 Loma Prieta	0.8	23	Imperial Valley, 1979	1.0	43	Imperial Valley, 1979	0.4
4	1989 Loma Prieta	0.8	24	Imperial Valley, 1979	1.0	44	Imperial Valley, 1979	0.4
5	1994 Northridge	1.3	25	Imperial Valley, 1979	0.8	45	Kern, 1952	2.9
6	1994 Northridge	1.3	26	Imperial Valley, 1979	0.8	46	Kern, 1952	2.9
7	1994 Northridge	1.6	27	Landers, 1992	3.2	47	Landers, 1992	2.6
8	1994 Northridge	1.6	28	Landers, 1992	3.2	48	Landers, 1992	2.6
9	1974 Tabas	1.1	29	Landers, 1992	2.2	49	Morgan Hill, 1984	2.4
10	1974 Tabas	1.1	30	Landers, 1992	2.2	50	Morgan Hill, 1984	2.4
11	Elysian Park (sim.)	1.4	31	Loma Prieta, 1989	1.8	51	Park., 1966, Cholame	1.8
12	Elysian Park (sim.)	1.4	32	Loma Prieta, 1989	1.8	52	Park., 1966, Cholame	1.8
13	Elysian Park (sim.)	1.0	33	North., 1994, Newhall	1.0	53	Park., 1966, Cholame	2.9
14	Elysian Park (sim.)	1.0	34	North., 1994, Newhall	1.0	54	Park., 1966, Cholame	2.9
15	Elysian Park (sim.)	1.1	35	North., 1994, Rinaldi	0.8	55	N. Palm Springs, 1986	2.8
16	Elysian Park (sim.)	1.1	36	North., 1994, Rinaldi	0.8	56	N. Palm Springs, 1986	2.8
17	Palos Verdes (sim.)	0.9	37	North., 1994, Sylmar	1.0	57	San Fernando, 1971	1.3
18	Palos Verdes (sim.)	0.9	38	North., 1994, Sylmar	1.0	58	San Fernando, 1971	1.3

(continued)

Table 20.7 (continued)

Set 1: 2% PE in 50 years; CP Performance level			Set 2: 10% PE in 50 years; LS Performance level			Set 3: 50% PE in 50 years; IO Performance level		
EQ	Name	SF	EQ	Name	SF	EQ	Name	SF
19	Palos Verdes (sim.)	0.9	39	N. Palm Springs, 1986	3.0	59	Whittier, 1987	1.3
20	Palos Verdes (sim.)	0.9	40	N. Palm Springs, 1986	3.0	60	Whittier, 1987	1.3

using the reliability-based concept considering all major sources of uncertainty not only considering the wave but also the seismic loading [16]. The authors believe that design method for seismic loading can also be used for the wave loading.

The basic governing equation of motion for any dynamic system in matrix notation can be expressed as:

$$\mathbf{M}\ddot{\mathbf{X}} + \mathbf{C}\dot{\mathbf{X}} + \mathbf{K}\mathbf{X} = \mathbf{F} \tag{20.22}$$

where **M**, **C**, and **K** are the mass, damping, and stiffness matrixes, respectively; **F** is the external force vector; and $\ddot{\mathbf{X}}$, $\dot{\mathbf{X}}$, and **X** are acceleration, velocity, and displacement vectors, respectively. For ONSs excited by the seismic loading, the information on **M, C, K,** and **F** can be generated by following widely used procedures available in the literature.

For OFSs vibrating in water, generating information on dynamic parameters can be more involved. Since they are not widely used, the information on them are not readily available. The mass matrix of an OFS consists of the mass of the structure plus the added mass produced by the displacement of water caused by a structural element. The added mass is generally calculated as $\rho(C_m - 1)V$, where V is the effective volume of the member in water, ρ is the mass density of water, and C_m is the inertia coefficient. For jacket-type OFSs, as shown in Fig. 20.6a, most of the mass is concentrated at the top because of the deck. The damping matrix also includes structural damping and the damping caused by the motion of members in water attenuating the velocity of the structure. The hydrodynamic damping is calculated as ρAUC_d, where A is the effective projected area, U is the velocity of water particles, and C_d is the drag coefficient. The external force vector **F** includes the weight of the structural elements, deck weight, and the forces caused by the environment, including the wave loading. Since it is virtually impossible that the critical wave and earthquake loadings act simultaneously, OFSs are designed separately for them.

The wave and earthquake loadings do not act on structures in a similar way. Usually, members located close to the free water surface are subjected to the wave loading. The wave forces also attenuate with depth, known as the Wheeler effect. The most important structural parameter in the wave loading is the surface area perpendicular of the wave direction. The seismic loading acts at the base of a structure. It is then

Table 20.8 Structural reliability in terms of β for the steel frame using FR and PR connections [16]

CP (2% PE in 50 years)					LS (10% PE in 50 years)					IO (50% PE in 50 years)				
EQ	Overall drift		Inter-story drift		EQ	Overall drift		Inter-story drift		EQ	Overall drift		Inter-story drift	
	FR	PR	FR	PR		FR	PR	FR	PR		FR	PR	FR	PR
	β^a	β^a	β^a	β^a		β^a	β^a	β^a	β^a		β^a	β^a	β^a	β^a
1	4.30 (341)	4.44 (341)	3.83 (341)	3.95 (341)	21	6.69 (341)	6.97 (341)	4.30 (356)	4.66 (341)	41	2.68 (356)	2.72 (356)	2.28 (341)	2.19 (341)
2	6.32 (341)	5.96 (341)	6.00 (341)	5.55 (356)	22	6.70 (341)	6.75 (341)	6.43 (341)	6.43 (356)	42	3.62 (356)	3.77 (341)	3.12 (371)	3.30 (371)
3	7.46 (356)	8.84 (341)	6.21 (341)	7.06 (341)	23	6.80 (341)	7.02 (356)	6.33 (341)	6.58 (356)	43	4.56 (356)	4.68 (356)	4.29 (356)	4.37 (341)
4	4.09 (356)	4.81 (341)	4.69 (356)	4.42 (371)	24	2.74 (356)	3.92 (341)	10.08 (341)	9.09 (356)	44	7.41 (341)	7.41 (341)	6.89 (356)	6.92 (356)
5	5.23 (356)	5.32 (341)	4.72 (341)	4.80 (341)	25	7.37 (341)	7.49 (341)	6.87 (356)	7.04 (341)	45	4.88 (356)	4.95 (341)	4.35 (341)	4.50 (356)
6	4.55 (356)	4.62 (341)	4.18 (356)	4.22 (356)	26	6.85 (356)	6.78 (341)	6.99 (341)	7.12 (356)	46	5.11 (341)	5.12 (341)	4.61 (356)	4.63 (341)
7	6.33 (341)	6.83 (341)	5.97 (341)	6.34 (341)	27	7.17 (341)	7.18 (371)	6.65 (371)	6.66 (356)	47	3.60 (341)	3.81 (356)	3.20 (371)	3.42 (341)
8	4.67 (356)	4.80 (356)	4.06 (341)	4.17 (356)	28	6.78 (356)	6.83 (341)	6.37 (341)	6.43 (356)	48	7.13 (356)	7.39 (341)	6.67 (341)	7.55 (356)
9	9.00 (341)	9.64 (356)	8.30 (356)	8.56 (341)	29	3.47 (356)	3.63 (341)	3.15 (356)	3.32 (356)	49	3.64 (341)	3.85 (356)	3.19 (341)	3.41 (356)
10	8.04 (341)	8.97 (341)	7.59 (341)	8.16 (356)	30	5.42 (341)	5.47 (356)	5.10 (341)	5.04 (356)	50	4.50 (341)	5.21 (341)	4.50 (356)	4.39 (341)

(continued)

Table 20.8 (continued)

CP (2% PE in 50 years)					LS (10% PE in 50 years)					IO (50% PE in 50 years)				
EQ	Overall drift		Inter-story drift		EQ	Overall drift		Inter-story drift		EQ	Overall drift		Inter-story drift	
	FR	PR	FR	PR		FR	PR	FR	PR		FR	PR	FR	PR
	β^a	β^a	β^a	β^a		β^a	β^a	β^a	β^a		β^a	β^a	β^a	β^a
11	4.63 (341)	5.73 (341)	4.14 (341)	4.26 (341)	31	4.31 (341)	4.43 (341)	3.94 (356)	4.08 (356)	51	4.33 (356)	4.42 (371)	4.10 (356)	4.15 (371)
12	4.29 (341)	4.31 (341)	3.94 (356)	3.92 (371)	32	8.30 (356)	8.55 (371)	7.97 (371)	8.86 (356)	52	7.05 (341)	7.02 (356)	6.20 (341)	7.50 (356)
13	6.99 (341)	6.71 (341)	6.94 (341)	7.33 (341)	33	5.10 (341)	4.97 (356)	4.38 (341)	4.44 (356)	53	4.24 (356)	4.27 (356)	3.67 (356)	3.79 (356)
14	6.46 (341)	6.75 (341)	4.42 (341)	4.59 (341)	34	3.93 (356)	4.01 (341)	3.47 (356)	3.54 (341)	54	3.98 (356)	4.47 (356)	3.86 (341)	4.40 (356)
15	3.82 (341)	4.03 (341)	3.47 (356)	3.67 (356)	35	4.31 (356)	4.37 (341)	3.93 (371)	4.01 (356)	55	3.12 (341)	3.19 (356)	2.73 (356)	2.78 (341)
16	3.73 (341)	3.89 (356)	3.33 (341)	3.50 (341)	36	3.77 (341)	3.83 (356)	3.39 (356)	3.44 (356)	56	3.15 (356)	3.12 (341)	2.70 (341)	2.69 (356)
17	5.54 (356)	5.66 (341)	5.30 (341)	5.39 (356)	37	5.45 (356)	5.77 (356)	5.11 (356)	5.35 (341)	57	5.85 (341)	5.88 (341)	5.76 (356)	5.65 (341)

(continued)

Table 20.8 (continued)

CP (2% PE in 50 years)					LS (10% PE in 50 years)					IO (50% PE in 50 years)				
EQ	Overall drift		Inter-story drift		EQ	Overall drift		Inter-story drift		EQ	Overall drift		Inter-story drift	
	FR	PR	FR	PR		FR	PR	FR	PR		FR	PR	FR	PR
	β^a	β^a	β^a	β^a		β^a	β^a	β^a	β^a		β^a	β^a	β^a	β^a
18	4.05 (356)	4.18 (341)	3.64 (356)	3.78 (341)	38	3.90 (356)	4.03 (356)	3.29 (356)	3.41 (356)	58	3.35 (341)	3.50 (356)	3.00 (341)	3.14 (356)
19	7.13 (341)	7.42 (356)	6.64 (341)	7.01 (341)	39	6.85 (341)	6.93 (356)	6.64 (341)	6.85 (341)	59	1.45 (341)	1.50 (356)	1.41 (356)	1.44 (341)
20	4.55 (356)	4.75 (341)	4.14 (356)	4.32 (341)	40	4.05 (341)	4.12 (356)	3.60 (341)	3.63 (341)	60	1.45 (341)	1.36 (341)	2.02 (341)	1.95 (356)
Mean	5.56	5.88	5.08	5.25	Mean	5.50	5.65	5.40	5.50	Mean	4.25	4.38	3.93	4.11

[a]The numbers below the reliability index inside the parentheses indicate the TNDA for each case

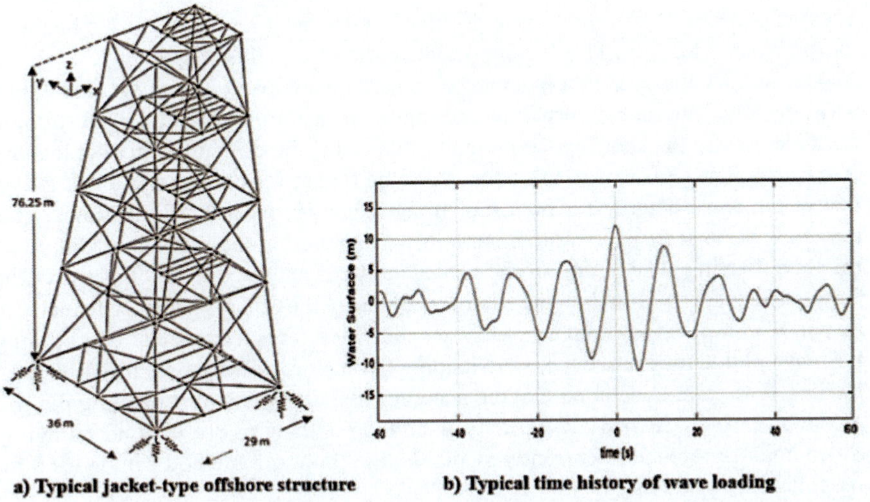

Fig. 20.6 **a** Offshore structure configuration; **b** Conceptual wave loading time history

distributed according to the mass distributions along the height of a structure. Thus, the points of application for both loadings are very different. The dynamic response characteristics and the underlying risks are also expected to be very different. Uncertainty associated with the resistance-related variables can be assumed to be similar to that of ONSs. However, it is essential at this stage to quantify uncertainties in the wave loading.

20.11.1 Uncertainty Quantifications in Wave Loading

A typical time history of wave loading is similar to earthquake time histories but not very irregular. Time history of wave loading is conceptually shown in Fig. 20.6b. The frequency contents of the wave and earthquake loadings are very different. The most important factors in modeling wave loading are the profile of the water level surface and the wave height. Initially, New Wave (NW) theory was proposed to model sea surface fluctuations as a function of time. It is deterministic in nature and accounts for the spectral content of the sea [25]. To address randomness in it, the Constraint New Wave (CNW) [26] theory can be used. Using the CNW theory, many hours of random wave loading in time domain can be simulated in a computationally efficient manner. Figure 20.6b indicates a sample water surface level in time domain using CNW. Vazirizade et al. [27] discussed in detail how to generate water level surface levels for a particular sea state. In other words, all of generated water level surfaces have the same maximum wave height and frequency content but different profiles.

They considered 11 different wave profiles similar to 11 earthquake time histories for the seismic loading. This is not possible to discuss in this chapter.

The authors observed that in some cases earthquake loading is more critical than wave loading. The uncertainty in predicting the probability of failure p_f in strength for a member in the splash zone indicates that it may be difficult to predict loading in the area. The COVs of p_f are observed to be higher for the seismic loading than that of the wave loading for both the strength and serviceability limit states. This may indicate that there is more uncertainty in predicting the seismic loading than the wave loading. The period of the wave loading is expected to be higher than the earthquake loading. However, the submerged state of the OFSs is expected to increase its period compared to when they are not submerged. This tendency of approaching the wave period may cause the wave loading to be more critical than the earthquake loading. The team concluded that the accuracy and efficiency of estimating the risk of the proposed AIRS-MUK-FORM for offshore structures are very encouraging. Since no such method is currently available, the proposed method can be used for the reliability of OFSs.

20.12 Multidisciplinary Applications of the Proposed Method

The research team members believe that if a concept is very advanced, it would have multidisciplinary application potential. The concept behind AIRS-MUK-FORM has been applied to estimate underlying risk of both onshore and offshore structures in the previous sections. They are excited by seismic and wave loadings in time domain.

Solders are used in electronic packaging (EP), and they are subjected to cyclic thermo-mechanical loading. The team explored the possibility of extracting reliability information using the proposed concept, and the FE representation of a typical solder ball system is shown in Fig. 20.7a. The subject is difficult, and interested readers are referred to Azizsoltani and Haldar [7]. It is only conceptually presented in this chapter.

Both thermal displacement and thermal loading are applied in the cyclic form as shown in Fig. 20.7b with the ramp time, T_1, of 18 min, the dwell time, T_2, of 30 min in high temperature and the idle time, T_3, of 10 min at low temperature. Thermal loading starts at 20 °C [7]. Then, the temperature reduced to −55 °C and cycled between 125 °C and −55 °C. To incorporate the variation or uncertainty in the intensity of the thermal displacement and thermal loading, two magnification factors denoted as MF_{TD} and MF_T, respectively, are introduced, as shown in Fig. 20.7b. MF_{TD}, MF_T, T_1, T_2, and T_3 are all considered to be random variables with different statistical characteristics. Figure 20.7b shows the time-dependent displacement and temperature variations for the first two cycles. The results match very well with the MCS results. One reviewer commented that it was a groundbreaking work.

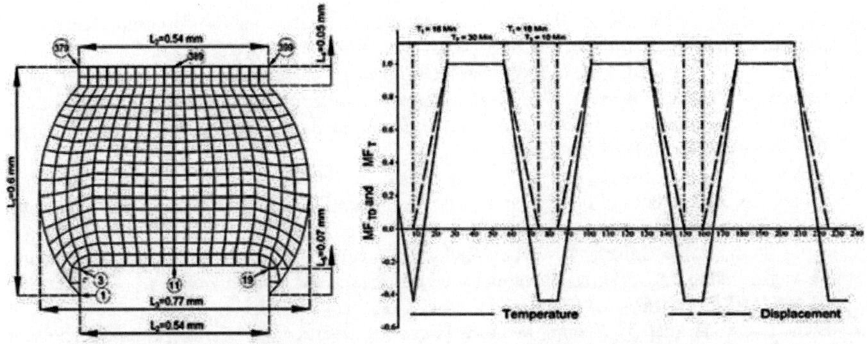

a) Solder ball geometry and FE mesh

b) Variation of MF_{TD}, MF_T.

Fig. 20.7 **a** Considered solder ball. **b** Thermal loading applied in time domain

20.13 Summary

The novel methodology discussed in this chapter is found to be very efficient and accurate. The proposed reliability approach for the design of more earthquake-resistant structures can consider dynamic structural systems excited by seismic loading applied in time domain. The algorithm can also incorporate all major sources of nonlinearity and uncertainty. The concept is also used to estimate the reliability of OFSs and solder balls used in electronic packaging. The results are verified using MCS. Using this approach, the underlying risk can be evaluated by performing a reduced number of deterministic analyses, in the order of hundreds instead of multiple thousands. The methodology represents an alternative to both the classical random vibration and the simulation approaches.

Acknowledgements The study is also partially supported by the National Science Foundation under Grant No. CMMI-1403844. Additional funding was provided by the government of Mexico through the *Consejo Nacional de Ciencia y Tecnologia* (CONACyT) and the *Universidad Autónoma de Sinaloa* (UAS). Any opinions, findings, or recommendations expressed in this chapter are those of the authors and do not necessarily reflect the views of the sponsors. The research team members who participated in developing the concept presented in this chapter are Dr. Jungwon Huh, Dr. Seung Yeol Lee, Dr. Hamoon Azizsoltani, Dr. J. Ramon Gaxiola-Camacho, and Dr. Sayyed Mohsen Vazirizade.

References

1. Haldar, A., & Mahadevan, S. (2000). *Probability, reliability, and statistical methods in engineering design*. New York, NY, USA: Wiley.
2. Haldar, A., & Mahadevan, S. (2000). *Reliability assessment using stochastic finite element analysis*. New York, NY: Wiley.

3. Box, G. E., & Wilson, K. B. (1951). On the experimental attainment of optimum conditions. *Journal of the Royal Statistical Society. Series B. Methodological, 13*, 1–45.
4. Khuri, A. I., & Cornell, J. A. (1996). *Response surfaces: designs and analyses*. CRC press.
5. Azizsoltani, H., & Haldar, A. (2017). Intelligent computational schemes for designing more seismic damage-tolerant structures. *Journal of Earthquake Engineering*, 1–28. https://doi.org/10.1080/13632469.2017.1401566.
6. Azizsoltani, H., Gaxiola-Camacho, J. R., & Haldar, A. (2018). Site-specific seismic design of damage tolerant structural systems using a novel concept. *Bulletin of Earthquake Engineering, 16*(9), 3819–3843. https://doi.org/10.1007/s10518-018-0329-5.
7. Azizsoltani, H., & Haldar, A. (2018). Reliability analysis of lead-free solders in electronic packaging using a novel surrogate model and kriging concept. *Journal of Electronic Packaging, ASME, 140*(4), 041003-1–11. https://doi.org/10.1115/1.4040924.
8. Lichtenstern, A. (2013). *Kriging methods in spatial statistics*. Ph.D. dissertation, Technische Universität München, Germany.
9. Eguchi, R., Goltz, J., Taylor, C., Chang, S., Flores, P., Johnson, L., et al. (1998). Direct economic losses in the northridge earthquake: A three-year post-event perspective. *Earthquake Spectra, 14*(2), 245–264. https://doi.org/10.1193/1.1585998.
10. FEMA-350. (2000). *Recommended seismic design criteria for new steel moment-frame buildings*. Federal Emergency Management Agency (FEMA).
11. ASCE/SEI 41-13. (2014). *Seismic evaluation and retrofit of existing buildings*. American Society of Civil Engineers., Reston, Virginia 20191: ASCE.
12. Mehrabian, A., Haldar, A., & Reyes-Salazar, A. (2005). Seismic response analysis of steel frames with post-northridge connection. *Steel and Composite Structures, 5*(4), 271–287. https://doi.org/10.12989/scs.2005.5.4.271.
13. Richard, R. M., Allen, C. J., & Partridge, J. E. (1997). *Proprietary slotted beam connection designs*. Modern steel construction, Chicago, pp. 28–33.
14. Reyes-Salazar, A., & Haldar, A. (2001). Energy dissipation at PR frames under seismic loading. *Journal of Structural Engineering, 127*(5), 588–592. https://doi.org/10.1061/(ASCE)0733-9445(2001)127:5(588).
15. Colson, A. (1991). Theoretical modeling of semirigid connections behavior. *Journal of Constructional Steel Research, 19*(3), 213–224.
16. Gaxiola-Camacho, J. R., Haldar, A., Azizsoltani, H., Valenzuela-Beltran, F., & Reyes-Salazar, A. (2017). Performance-based seismic design of steel buildings using rigidities of connections. *ASME Journal Risk Uncertain Part A, 4*(1), 04017036. https://doi.org/10.1061/AJRUA6.0000943.
17. Haldar, A., Gaxiola-Camacho, J. R., Azizsoltani, H., Villegas-Mercado, F. J., & Vazirizade, S. M. (2020). *A novel geomechanics concept for earthquake excitations applied in time domain*. International Journal of Geomechanics, ASCE. Accepted for publication.
18. Villegas-Mercado, F. J., Azizsoltani, H., Gaxiola-Camacho, J. R., & Haldar, A. (2017). Seismic reliability evaluation of structural systems for different soil conditions. *IJGEE, 8*(2), 23–38. https://doi.org/10.4018/IJGEE.2017070102.
19. FEMA-222A & 223A. (1995). *NEHRP recommended provisions for seismic regulations for new buildings*. Washington, DC.
20. FEMA-302 & 303. (1997). *NEHRP recommended provisions for seismic regulations for new buildings and other structures, part 1—provisions, prepared by the building seismic safety council for the federal emergency management agency*. Washington, DC.
21. ISO-19901-2. (2017). Petroleum and natural gas industries—Specific requirements for offshore structures—Part 2: Seismic design procedures and criteria.
22. SCEC. (2016). *Broadband platform; Southern California earthquake center (SCEC)*. USA.
23. Gaxiola-Camacho, J. R., Azizsoltani, H., Villegas-Mercado, F. J., & Haldar, A. (2017). A novel reliability technique for implementation of performance-based seismic design of structures. Engineering Struct, *142*, 137–147. https://doi.org/10.1016/j.engstruct.2017.03.076.
24. FEMA-355C & F. (2000). *State of the art report on performance prediction and evaluation of steel moment-frame buildings*. Washington, DC.

25. Tromans, P. S., Anaturk, A. R., & Hagemeijer, P. (1991). *A new model for the kinematics of large ocean waves-application as a design wave*. The first international offshore and polar engineering conference, 11–16 August, Edinburgh, The United Kingdom.
26. Mirzadeh, J. (2015). *Reliability of dynamically sensitive offshore platforms exposed to extreme waves*. Ph.D. dissertation, The University of Western Australia.
27. Vazirizade, S. M., Haldar, A., & Gaxiola-Camacho, J. R. (2019). Uncertainty quantification of sea waves—An improved approach. *Oceanography and Fisheries, 9*(5). https://juniperpublishers.com/ofoaj/pdf/OFOAJ.MS.ID.555775.pdf.

Achintya Haldar completed his Ph.D. from University of Illinois. He worked for Bechtel Power Corporation after graduation. After returning to academic career, he taught at Illinois Institute of Technology, Georgia Tech, and now at the University of Arizona. He is a Distinguished member of ASCE and a Fellow of SEI. He developed the Stochastic Finite Element Method and several reliability evaluation concepts applicable to many engineering disciplines. He received numerous research and teaching awards listed at haldar.faculty.arizona.edu. He authored over 615 technical articles including several well accepted books.

Francisco J. Villegas-Mercado is a Ph.D. Candidate of the Civil and Architectural Engineering and Mechanics Department at the University of Arizona. In 2014, he joined Professor A. Haldar's research team and he started working under his guidance and supervision. Since then, he has been author and coauthor of several research papers and a book chapter. His research interests are related to reliability analysis and probabilistic methods, performance-based seismic design, and earthquake engineering.

Chapter 21
Probabilistic Physics-of-Failure Approach in Reliability Engineering

Mohammad Modarres

Abstract This chapter describes an overview of the probabilistic physics-of-failure for applications to reliability engineering problems. As reliability engineering experts face situations where system and component reliability failure data are lacking or with the poor quality, a powerful modeling approach is to relay on the underlying processes and phenomena that lead to failures. Originally derived from chemistry, mechanics, and metallurgy, the processes that lead to failures are called failure mechanisms that include phenomena such as fatigue, creep, and corrosion. Physics-of-failure is an empirically based mathematical and analytical approach to modeling these underlying processes of failures. Due to limitations of information and test data available for the understanding of these processes, the PoF-based reliability should include formal accounting of the uncertainties. The physics-of-failure methods in reliability engineering that consider uncertainties lead us to the probabilistic physics-of-failure. This chapter covers some important analytical and practical aspects of the probabilistic physics-of-failure modeling, including some examples.

Keywords Reliability prediction · Physics-of-failure · Probabilistic physics-of-failure · Uncertainty analysis · Accelerated life testing · Accelerated degradation testing

21.1 Introduction

Reliability methods have been progressively relying on the modeling of failure phenomena rather than historical data observed. Reliability modeling has evolved from the constant hazard rate assumption to more representative life distributions (such as the Weibull and lognormal) to address wear-out and aging mechanisms better. However, consideration of the physics and mechanistic principles that govern the occurrence of failure data is critical to better model a more realistic prediction of failures. Engineering models that describe and trace degradation over time and

M. Modarres (✉)
University of Maryland, College Park, Maryland, USA
e-mail: modarres@umd.edu

K. B. Misra (ed.), *Handbook of Advanced Performability Engineering*,
https://doi.org/10.1007/978-3-030-55732-4_21

the ultimate failure of a component are referred to as the physics-of-failure (PoF) models. These models can be used to predict failures of engineering units that undergo specific operational and user experience, whereby making such models specific to a component conditioned on its design and operational experience.

The formal consideration of physics and mechanistic methods in reliability engineering is referred to as the physics-of-failure (PoF) approach. The PoF approach is an empirically based approach to reliability engineering and prediction as well as prognosis and health management (PHM), in contrast to the traditional statistical approach that solely relies on historical data. It uses physics-based principles and simulations to assess design and reliability. The approach can evaluate and predict system performance while reducing subjectivities by directly relying on failure mechanisms such as fatigue, fracture, wear, and corrosion. The PoF approach is a comprehensive representation of the wear-out and aging, and is capable of bringing relevant physical factors into the life assessment and reliability models of the structures, components, and systems. Unlike the reliability models developed based on field data that suffer from the wide variation in operating conditions and practices, reliability models based on PoF, developed using accelerated life and degradation tests, take into account operational conditions (applied stresses) that permit flexibility in applied stresses, leading to more relevant and component-specific life models.

The development of PoF models has typically relied on limited empirical information. The uncertainties associated with this limitation make the PoF models and their parameters uncertain. The need to formally characterize this uncertainty in the PoF model has led to the probabilistic physics-of-failure (PPoF) approach [1]. A more fundamental extension of the empirically based PoF model has more recently gained much steam by making PoF science-based models that rely on the physical laws such as the thermodynamic entropy and the 2nd law of thermodynamics or the statistical mechanics Boltzmann entropy [2].

Mechanistic-based failure models can be categorized into three core groups: stress strength, damage endurance, and performance requirements. In all these models, metrics representing one or more failure-inducing agents such as applied loads and environmental attack variables, for instance moisture, is related to the lifetime or amount of damage as the component operates. The operating stresses are either directly applied such a cyclic load due to rotations or vibrations in machinery, or indirectly through existing mechanical, thermal, electrical, chemical, and radiation-induced forces that lead to stresses on an item. Both stress and time in the PoF models may either be analyzed deterministically (e.g., identifying and studying the sources of stresses) or probabilistically (e.g., treating stress variation as a random variable). Substantial uncertainties associated with failure-inducing agents can originate from environmental and operational conditions, and from the emergence of failure mechanisms that were not considered or well understood at the time of design. These uncertainties should be fully understood and accounted for a PPoF analysis.

Accelerated life testing (ALT) has traditionally been used as a leading approach to obtain the empirical data for mechanistic modeling of wear-out, damage process, and failure in PoF and PPoF modeling practice. Before performing ALT, the direct or indirect stress agent, which could be an aggregate effect of a single or multiple physical and operational conditions, should be identified. The next step involves accelerating this stress agent and applying it to samples of the structure, system or component and monitor the degradation, manner of failures, and times of failure. The PoF and PPoF models of failure, damage, and degradation developed by using accelerated test data provide a more accurate- and component-specific representation of the damage, failure phenomena, performance, and life as compared to the traditional reliability prediction techniques.

The interdependence of components and parts in a system can also be a critical factor in the reliability modeling of system and component reliability. In the study of system behavior, there are situations in which progressive failure of one component may activate or accelerate other failure mechanisms or the failure of other components. There are usually many links between different components through their properties and common environmental conditions. The PoF approach properly incorporates these interdependencies in complex structures, systems, and components.

21.1.1 Physics-of-Failure Modeling Process

The PoF concept initially evolved from fracture mechanics. For example, Paris et al. (Paris, Gomez and Anderson 1961) introduced methods for predicting the rate of fatigue crack growth to the point of fracture. Also, other researchers related the stress and strain in materials to the life of materials [3].

Given this background, Rome Air Development Center (RADC—the predecessor to the U.S. Air Force Rome Laboratory) introduced a PoF program in 1961 to address the growing complexity of military equipment and the resulting increase in the number of failures observed. In 1962, researchers from Bell Labs [4] justified using the kinetic theory's interpretation of the Arrhenius equation: a simple yet accurate formula for the temperature dependence of the chemical reaction rate constant as a basis for assessment of temperature-induced aging of semiconductor devices. Later, the RADC and Armor Research Foundation of the Illinois Institute of Technology (now IIT Research Institute) organized the first PoF symposium in electronics in Chicago in September 1962. This symposium laid the groundwork for future research and development activities related to PoF by RADC and several other organizations. Numerous original papers and ideas, introducing and explaining the PoF concepts and methods, were presented in these symposia.

The PoF approach to reliability relies on the empirical knowledge of damage and degradation processes and the operational and environmental stresses applied to a component, including its geometry, material properties, and potential failure mechanisms that individually or in combination lead to its failure. The PoF mathematical models would then be used to assess the amount of degradation or performance reduction, expended, and remaining life. Using PoF reduces the need and over-dependence on a huge amount of historical and field data to reach the same level of confidence over reliability predictions. Further, the PoF models show how and why a component fails under a given failure mode.

The most critical step in a PoF approach is to understand and assign representative failure mechanisms (such as corrosion or fatigue) that cause the dominant failure modes of a component. Once failure mechanisms are assigned, accelerated life testing would be the choice method to decide the proper mathematical form of the PoF model and to estimate its parameters. Accelerated tests used to develop the PoF models can reduce long and costly life testing. In the mathematical form of the PoF, one seeks to relate the fundamental physical and chemical properties of materials parametrically along with applied stresses to reliability metrics (such as degradation, life, or cycles-to-failure).

Sometimes it is impossible to build a limited number of identical units or prototypes for reliability testing. Cases in point include large-scale systems (like buildings and space vehicles), one-of-a-kind or highly expensive systems, and units that must work properly for the first time. In these cases, performance and field data are not available, and the PoF approach to degradation and life assessment is the leading choice for reliability assessment. As such, the PoF approach is particularly useful in the design stage when there are limited prototypes or test facilities. Finally, the PoF approach has great utility when dealing with highly reliable units that don't produce much, if any failure data.

The PoF techniques can also be used to interpret and extrapolate field data for failure prediction for in-service components. Sometimes the field data might include features that are related to physical measures and degradation of the unit. A good example of this is the vibration of a bearing. The vibration is indirect suggestive of a flaw, but since the flaw itself cannot be tracked, the vibration can be used to estimate failure as such PoF models can be used to relate these indirect features as well as direct variables that related to component life. This is the reason that, more recently, the PoF models have formed the basis for diagnostic and remaining useful life estimation in the currently flourishing prognosis and health management (PHM) field. Coupled with machine learning methods, PoF models learn and become the knowledge base for diagnostics and prognostics in components. This is useful for maintenance practitioners, as it provides a means of predictive maintenance and condition monitoring.

There is no single unique methodology for performing PoF-based reliability analysis. If an item involves multiple subassemblies (parts and components), each subject to different failure mechanisms, then the combined effect of applicable failure mechanisms should be modeled. Figure 21.1 depicts the structural and dynamic hierarchy of PoF analysis elements for a multi-component system. The lowest level in this hierarchy is inter- and intra-environmental factors that affect failure mechanisms. The intra-environmental factors refer to conditions resultant from the unit operation itself. This includes, for example, heat dissipation or vibration caused by an imbalanced rotating shaft. The inter-environmental factors are those imposed externally from its design boundary. Examples include relative humidity and the prevalence of dust particles. There may be a causal chain among inter- and intra-environmental

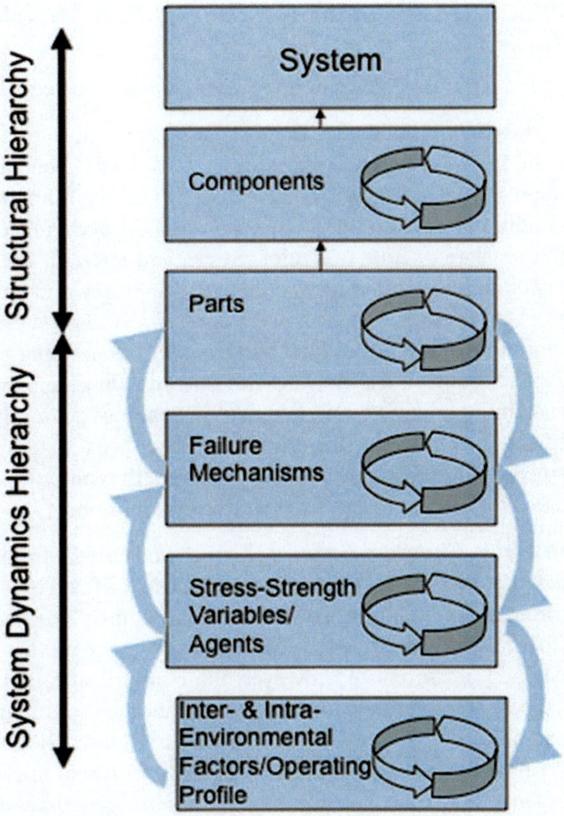

Fig. 21.1 System hierarchy used in PoF analysis

factors such that one may lead to another or vice versa synergistically. For example, low temperatures may cause condensation, leading to accelerated corrosion.

All environmental factors potentially lead to various forms of stress. For example, high temperature (as either an inter- or intra-environmental factor) leads to thermal expansion, and (if the unit is confined) can cause mechanical stresses. Such stress agents are key actors in activating or accelerating degradation through corresponding failure mechanisms. While one failure mechanism may also accelerate another (such as accelerating fatigue when corrosion exists), failure mechanisms can also produce new stresses. For example, wear in a journal bearing can cause vibration-induced fatigue. The top part of the hierarchy in Fig. 21.1, known as the structural hierarchy, depicts the formal organization and topology of the system showing the functional and support relationships among parts, components, and the whole systems. On the other hand, the lower part of the figure, the system's dynamics hierarchy, shows the underlying processes (failure mechanisms) and conditions that lead to the occurrence and acceleration of such mechanisms.

21.1.2 Mathematical Forms of PoF Models

Three possible PoF modeling methods described earlier are discussed briefly below.

Stress–Strength Model. In this model, the item (e.g., a structure, system, or component) fails if the applied stresses caused by design, operation, and the external environment exceed its strength (Fig. 21.2). This failure model may depend on environmental conditions, applied operating loads, and the occurrence of critical events, rather than the passage of time or cycles. Stress and strength are treated as a random variable encompassing variability in all conditions. Two examples of this model include a steel bar under the mean tensile stress lower than its yielding point but which will be randomly subjected to load that exceeds the yielding point over time.

The second is a transistor with a mean voltage applied across the emitter–collector remaining below a failed level but which may randomly exceed the limit. In the case of the steel bar, the likelihood of failure is estimated from the probability that the stress random variable exceeds the strength random variable, which is obtained from a convolution of the two respective distributions.

Damage–Endurance Model. This model differs from the stress–strength model in that the *stress* (load) causes degradation in the form of irreversible cumulative damage through, for example, corrosion, wear, embrittlement, creep, or fatigue. The stress (load) aggregate drives the cumulative damage metric. Cumulative damage may not degrade performance; however, the component fails when the cumulative damage exceeds its endurance limit (i.e., endurance to the amount of cumulative damage). For example, a crack grows on a structure until it reaches a critical length beyond which the growth will be prompt and catastrophic. Accumulated damage does not disappear when the stresses are removed, although sometimes treatments such as annealing can repair cumulative damage. Variables representing damage and endurance may be treated as random and represented by probability density functions to capture distributions of initial damage, model parameter uncertainties, and model errors. Therefore, at any time or cycle (Fig. 21.3), the likelihood of failure may be represented by the exceedance of the damage distribution from the endurance probability density functions. If endurance is not a random variable and remains constant, then the distribution of the time-to-failure may be obtained when cumulative damage values

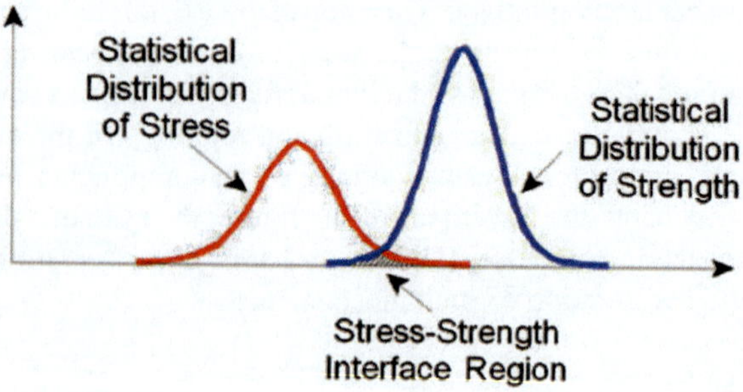

Fig. 21.2 Stress–strength modeling

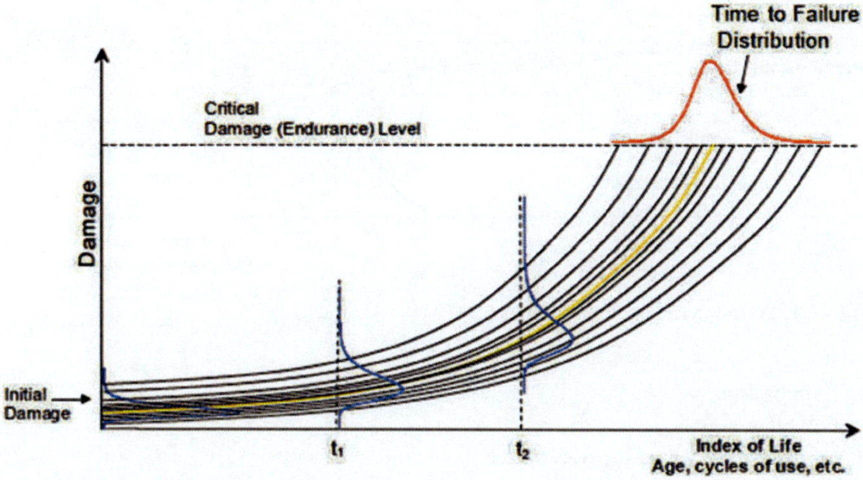

Fig. 21.3 Damage–endurance model

randomly exceed the constant value of the endurance (Fig. 21.3). The distribution of the time-to-failure shown in Fig. 21.3 is based on the assumption of a constant endurance limit around the median of the distribution of the endurance. Clearly, at a given time or cycle, N, the probability that the damage distribution exceeds endurance level (or distribution of endurance) would be equal to the probability that the random variable, time-to-failure (as represented by the time-to-failure distribution) is lower than N.

Performance Requirements Model. In this modeling approach, a system performance characteristic (such as system output capability, efficiency, or availability) is satisfactory if it remains within acceptable tolerance limits. Examples include rotating machinery efficiency, or reduction of resistivity in a resistor as a function of time and higher temperatures, and printer print quality (such as one that is based on a level of efficiency or output at the pump head). Systems start with a positive margin of performance that cumulatively and irreversibly degrades due to the underlying failure mechanisms. These mechanisms cause internal degradation and damage until performance falls below the minimum requirement level (i.e., fails). As the stress applied to the unit increases the rate of performance degradation, the time-to-failure (the point at which the system reaches the minimum or acceptable performance limit) is reduced. The concept is depicted in Fig. 21.4.

Simpler mathematical forms to describe a degradation increase, life reduction, and similar variables in PoF are preferable. This is because the amount of testing to generate empirical data are often limited, whereby models with multiple terms and parameters often lead to the overfitting of the PoF model and perform badly during life prediction and prognosis. Simple PoF mathematical form includes

- Linear: $y = ax + b$,

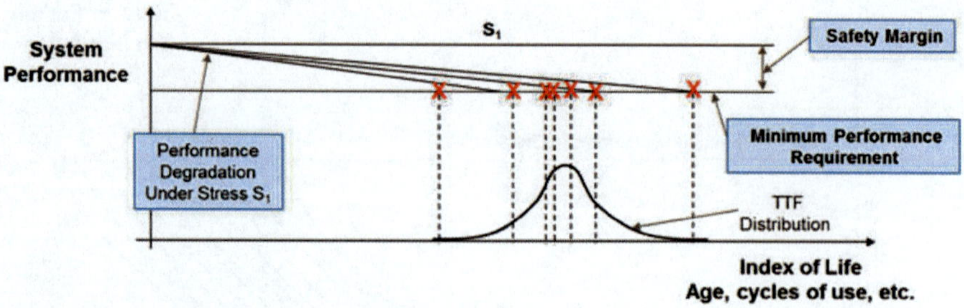

Fig. 21.4 Performance requirement model

- Exponential: $y = be^{ax}$
- Power: $y = bx^{a}$
- Logarithmic: $y = a\ln(x) + b$.

For example, the empirical form of the Arrhenius model used as a PoF model of life is an exponential form described by the expression $t = Ae^{\frac{E_a}{KT}}$ where t is "life", and in physics, K is the Boltzmann constant and E_a is the activation energy constant, but when used as a PoF model, A and $\frac{E_a}{K}$ can be viewed as constants, and temperature T (or $x = \frac{1}{T}$) is considered the "stress".

Combinations of the above equation forms may also be used as the PoF models. An example of this is the so-called Eyring relationship in chemistry in which an inverse power and exponential forms are combined. The PoF representation of the Eyring relationship in reliability may be expressed in the form of $t = A\frac{1}{T}e^{\frac{B}{T}}$, where parameters A and B are constants and temperature T is the "stress".

21.2 PPoF Approach to Life Assessment

Due to the inevitable stochastic variations of the many factors involved in the degradation and failure processes described by the PoF models, probabilistic physics-of-failure (PPoF) models can be necessary to formally account for the uncertainties in model parameters and model errors. The earliest effort in PPoF modeling was by Haggag et al. [5] who presented a PPoF approach to reliability assurance of high-performance chips by considering common defect activation energy distribution. Hall and Strutt [6] have presented PPoF models for component reliabilities by considering parameter and model uncertainties. Azarkhail and Modarres [7] have presented a Bayesian framework for uncertainty management in PPoF reliability models. Matik and Sruk [8] highlighted the need for PoF to be probabilistic to include inevitable variations of variables involved in processes contributing to the occurrence of failures in the analysis.

The fundamental elements of building a PPoF model is illustrated in Fig. 21.5. The lowest element in this figure shows the inter- and intra-environmental factors

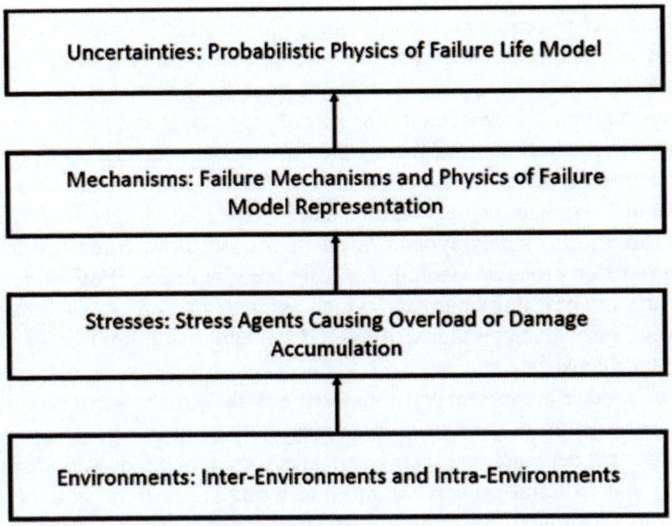

Fig. 21.5 System hierarchy in probabilistic–mechanistic reliability life model

that produce the stresses that cause degradation and failure. The next level addresses how these factors translate into stresses that trigger damage accumulating failure mechanisms or cause a failure. The relationship between the stress and the progression of damage or time of failure will be addressed in the next level by consideration of the underlying mechanisms of failure triggered. The final element (top-level) is the probabilistic life assessment that formally accounts for the PoF parameter and model uncertainties, leading to the PPoF models. Typically, a probabilistic approach (such as Bayesian inference) is shown to characterize the corresponding PoF model uncertainties. The arrows in Fig. 21.5 show the direction of influences, such as how external ambient temperature may affect viscosity, thereby triggering a wear mechanism. Usually, the direction of influences is upward (i.e., sequential causal relationships), but it is possible to have some influences going downward, causing a circular synergy among variables, for example, certain operating conditions, such as high internal temperature generated by poor lubrication during operation, lower lubricant viscosity, which in turn can increase the friction that further exacerbates the high internal temperature (intra-environment).

There are two basic types of uncertainties that can be associated with a PPoF model of failure mechanism: aleatory and epistemic uncertainty. Aleatory uncertainty is the inherent randomness in the PPoF model. This type of uncertainty is intrinsic and cannot be reduced. Examples of aleatory uncertainty include random environmental variations such as the level of humidity or temperature, random vibration in stress amplitude, and certain material properties such as size and density of existing flaws. Epistemic uncertainty is about lack of knowledge and information that consists of an incomplete description of the phenomena modeled (e.g., the failure mechanism), measurement errors, and a lack of sufficiently accurate measurements to fully capture the failure phenomena. Incorporating additional PoF model data and information reduces this type of uncertainty: as such this uncertainty is reducible, whereas aleatory uncertainty is not. Since there can be uncertainties associated with failure-inducing agents (i.e., stresses), including model parameters and the form of the PoF model itself, the prediction of failures by the PoF models is inherently a probabilistic problem requiring reliance on the PPoF models for reliability predictions.

Each failure, damage, or degradation mechanism should have its PPoF model. All applicable PPoF models applied to an item need to be combined to find the overall degradation process. Methods for combining multiple PPoF models include the use of the weakest link approach, which assumes that one of such degradation mechanisms causes damage that will exceed the endurance limit before the other applicable mechanisms.

The PPoF models are formulated considering all the variables that can initiate and propagate degradation in the item under study. As part of this process, one should identify important degradation causing variables such as applied loads, displacement amplitudes, and material properties, and contacting surfaces in an adhesive wear process. In this example, the amount of degradation may be measured in terms of the volume of materials lost and then correlated with the applied load. Experimental degradation data from accelerated testing would be needed to determine the PoF-based correlation between degradation and the applied loads. The next step is to characterize all forms of uncertainties associated with the PoF models and data, and estimates model parameters including their uncertainties (such a confidence or credibility interval). This step converts the PoF models into PPoF models.

A suitable regression approach should be developed to characterize all uncertainties formally. Bayesian regression is a powerful technique for estimating probability distributions of model parameters. For this purpose, one requires experimental degradation data under prevailing environments experiencing operational conditions corresponding to each degradation mechanism [9]. Other factors that can lead to uncertainties in the time of failure or amount of damage (such as manufacturing methods and material properties) should also be considered. Each failure mechanism has specific stress agents that cause degradation. Stress variables that trigger and promote the failure mechanisms may be obtained using a finite element analysis considering the component's geometry and material properties, including the prevailing operating conditions applied. The input parameters for the finite element analysis (e.g., geometry, material properties) need to be entered probabilistically—not deterministically—with the corresponding stresses estimated as probability distributions.

When used to predict reliability characteristics of components, a Monte Carlo simulation approach may complement the PPoF models by propagating all the associated uncertainties (such as those associated with the model, its parameters, and initial material flaws) to estimate the probability distribution of the unit failure or amount of damage as a function of time under the prevailing stresses. Monte Carlo simulation is the leading method for simulating systems, especially in the presence of coupled input variables.

21.2.1 Accelerated Life Testing for PPoF Model Development

To develop the PoF models and estimate their parameters and model uncertainties, it is imperative to rely on evidence and data describing events of failure or amount of degradation (damage) versus time data. These data can be obtained from life and degradation testing or valid field data. Many of today's structures, systems, and components are capable of operating under benign environmental stresses for an extended period. This makes normal life (non-accelerated) testing of such equipment difficult and costly, if not impossible. Field data, in many cases, are scarce, and even when they are available, it is hard to judge their uniformity and precision. Alternatively, accelerated life testing (ALT) provides a quicker way to understand the component life and degradation processes better and generates data for the development of PoF and PPoF models. As such, generating reliability data in the shortest possible time can be achieved by relying on formal ALT methods.

Accelerated life testing can effectively gather more reliability and life information in a shorter time by utilizing a more severe test environment than what would be otherwise experienced under normal use conditions. Accelerated life testing is performed by increasing the stress variables and loads that are known to trigger failure mechanisms that cause accumulation of damage and failure, thereby reducing the time needed for a failure to occur. The concept is conceptually illustrated in Figs. 21.6 and 21.7, where the trajectory of a cumulative degradation shown. The trajectories shift to faster damage accumulation and earlier times of failure as the applied stresses (loads or inter- and intra-environmental factors) increase.

Accelerated life tests are the prime method of generating data needed to develop the PPoF models, which in turn can be used to estimate and predict the equipment life or degradation and damage under normal operating conditions. This step in PPoF analysis underlines the importance of formally characterizing all the uncertainties in the PPoF models to reflect such uncertainties in the predicted life from such models.

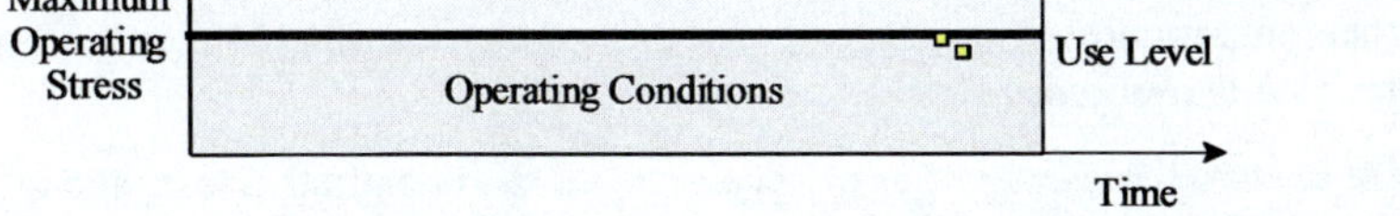

Fig. 21.6 Conceptual acceleration of stress agents at two overstress conditions and corresponding data points generated from ALT

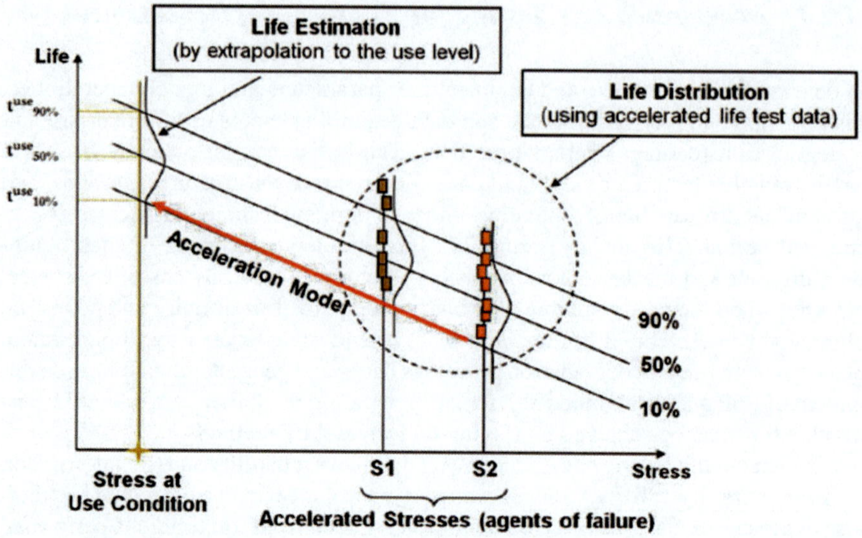

Fig. 21.7 Conceptual PPoF model development and extrapolation in ALT

Figure 21.7 illustrates the stress regions of a conceptual accelerated test that generate several failure data points at two stress levels in the "overstress" region. These data are then used to develop the PoF model that best describes them, including the associated uncertainties to extrapolate the resulting PPoF models (associated with each quantile of the life) to the "use" stress level to estimate the corresponding life distribution (see Fig. 21.7). A closely related measure in ALT is the acceleration factor (AF) associated with an ALT, defined as the fraction of life under normal ("use") operating or environmental conditions divided by the reduced amount of life when the same component is tested at higher stresses. This means that $\mathrm{AF} = \frac{t_\gamma^{\mathrm{use}}}{t_\gamma^{\mathrm{Accelerated}}}$, where t is life and γ is the quantile of the life.

Acceleration of the stress variable is achieved thoroughly applying loads either singly or in combination. Examples include

- More frequent power cycling
- Higher vibration levels
- Higher humidity
- More severe temperature cycling
- Higher temperatures
- Higher load amplitudes.

There are two basic categories of accelerated tests: quantitative tests and qualitative tests. The former commonly refers to Accelerated Life Tests (ALT) and Accelerated Degradation Tests (ADT). The latter is characterized by tests that aim to enhance the reliability of the item during design and operation.

Quantitative tests are conducted on structures, systems, or components. They can take a few weeks to a few months to complete. ALT is fundamentally based on the assumption that the unit under test will exhibit the same behavior under a shorter time frame (at a high stress level) as it would in a longer time frame at use stress conditions. Hence, there are several important planning considerations when conducting ALT tests so that this assumption remains valid.

Qualitative accelerated tests are designed to find failures linked to design or manufacturing without providing any life or damage characteristics associated with the items. Qualitative accelerated tests are not useful for developing PPoF models. However, they have many uses, especially during design, manufacturing, and production. By accelerating failures of structures, components, or systems, these tests can determine the robustness of the unit in its useful life. When a failure occurs during a qualitative accelerated test, one needs to determine the root cause of the failure and judge whether the failure mode and mechanism observed would occur under normal use conditions. The most common type of qualitative test is Highly Accelerated Life Testing (HALT). HALT is not a life test: its purpose is not to determine life characteristics. Rather, it is a test designed to promote the occurrence of failure modes (mechanical or electronic) that will occur during the life of the product under normal use conditions. HALT provides valuable information to determine design weaknesses as well as the product's upper and lower destruct limits. Another example of a qualitative accelerated test is known as Highly Accelerated Stress Screening (HASS). HASS tests are applied during the manufacturing phase and are used to screen marginal and defective units. HASS can expose infant mortality failures and other latent defects that would otherwise occur when the unit is being used.

21.3 An Example of Developing a PPoF Model: Proportional Hazard (PH) Model

A useful PoF life–stress relationship consisting of either single or multiple applied stresses can be addressed through the family of the Proportional Hazards (PH) models. This family is often of great utility in terms of modeling component life. Several commonly used and well-understood life stress models belong to this class, including the Arrhenius model, the inverse power law (IPL) model, temperature–humidity model, and the generalized Eyring model, as discussed earlier.

If we let S be the covariate that represents stress, and if $h_0(t;\boldsymbol{\gamma})$ is the baseline hazard rate with the parameters vector $\boldsymbol{\gamma}$, then Eq. (21.1) which is a measure of the effect that the stress has on the hazard rate expressed as

$$h(t;\gamma|S;\boldsymbol{\theta}) = h_0(t;\boldsymbol{\gamma})g(S;\boldsymbol{\theta}) \tag{21.1}$$

where $\boldsymbol{\theta}$ is a row vector of parameter and $g(S, \boldsymbol{\theta})$ is a modifier function describing the effect of the stress *S*. This model is readily expandable to more than one stress and parameter represented by the row vectors $\boldsymbol{S}$ and $\boldsymbol{\theta}$ as

$$h(t; \boldsymbol{\gamma}|\boldsymbol{S}; \boldsymbol{\theta}) = h_0(t; \boldsymbol{\gamma})g(\boldsymbol{S}; \boldsymbol{\theta}) \tag{21.2}$$

The modifier function can be a combination of different independent models such as exponential, inverse power law, etc. Equation 21.3 is an example of this kind of modifier function.

$$g(\boldsymbol{S}; \boldsymbol{\theta}) = \mathrm{e}^{\sum_{i=1}^{n} \theta_i S_i} \tag{21.3}$$

Therefore, the full PH model, commonly known as the Cox PH model, can be written as

$$h(t; \boldsymbol{\gamma}|\boldsymbol{S}; \boldsymbol{\theta}) = h_0(t; \boldsymbol{\gamma})\mathrm{e}^{\sum_{i=1}^{n} \theta_i S_i} \tag{21.4}$$

One important point to be made is that the values of the (S_i) can be the raw data themselves, or some useful transformation (logarithms, reciprocals, etc.) of them.

Suppose we know (or assume) that the failure times for a particular component operating under a constant (but arbitrary) stress *S* are distributed according to the Weibull probability density function (PDF) having the shape parameter β and characteristic life α. Therefore, the baseline hazard rate would be

$$h_{\text{weibull}} = \frac{\beta}{\alpha^{\beta}} t^{\beta-1} \tag{21.4}$$

This can be used as the hazard rate,

$$h(t; \alpha, \beta|\boldsymbol{S}; \boldsymbol{\theta}) = \frac{\beta}{\alpha^{\beta}} t^{\beta-1} \mathrm{e}^{\sum_{j=1}^{m} \theta_j S_j} \tag{21.5}$$

As such, the reliability function of the PH-Weibull model is

$$R(t|\boldsymbol{S}; \boldsymbol{\theta}) = \exp\left[-t^{\beta} \mathrm{e}^{\sum_{j=0}^{m} \theta_j S_j}\right] \tag{21.6}$$

And the PDF for the PH-Weibull distribution will be

$$f(t|\boldsymbol{S}; \boldsymbol{\theta}) = \beta t^{\beta-1} \exp\left[\sum_{j=0}^{m} \theta_j S_j - t^{\beta} \mathrm{e}^{\sum_{j=0}^{m} \theta_j S_j}\right] \tag{21.7}$$

21.4 Application of Cox PH Model to Shock-Type Stresses

A shock model includes two random variables: the time between shock occurrences and the magnitude of the shock (e.g., stress shocks). A shock, for example, would be the mechanical impact due to a drop from a given elevation or an abusive operation of a device. A single shock of small size may not cause the failure of the device, but a certain number of consecutive small shock stresses or a large shock stress may fail the device.

In case of multiple socks (e.g., drops) and other stress loads applied, the lifetime of the product, t, would be $t = \sum_i T_i$, where the sequence $T_1, \ldots, T_N$ represents the times between shock arrivals (i.e., interarrivals). The random variable N represents the number of shocks to a complete failure. Typically, one can assume that the random variable T_i follows a Poisson distribution with a constant known rate (i.e., frequency of drops), whereby the random variable of the time t can also be described as a function of the number shocks. The random variable N may be defined as a function of the sequence of multiple shocks of size S_i representing the magnitudes of the shock (e.g., drop heights and other high stress uses). Note that an *extreme shock* may exist in which the occurrence of n shocks (ordered from least to the highest) ends with a single (last) shock of the size d_h that guarantees a failure, where $\{N = n\}$ if $\{S_1 < d_h, \ldots, S_{n-1} < d_h, S_n \geq d_h\}$. So, d_h is the stress (drop level, for example) beyond which the unit would certainly fail.

Alternatively, the *run shock* model would normally apply, where the device fails when k consecutive shocks of any size of at least d_l occur. In this case, the number of shocks to failure would be $N = \min\{n : S_{n-k+1} \geq d_l, \ldots, S_n \geq d_l\}$, where shocks below d_l don't accumulate any damage. Assuming T_i is independent of S_i for all device i's, then $T_i \cdot s$ would be independent of N. The dependence between S_i and T_i needs a more involved method to obtain the lifetime distribution.

The PPoF model, in this case, is best described by the family of the Proportional Hazards (PH) models. The PH PPoF model has a covariate hazard rate that measures the effect that the stress has on the expected life, and is expressed by Eq. 21.2 with $h(t; \boldsymbol{\gamma}|S < d_l; \boldsymbol{\theta}) = h_0(t; \boldsymbol{\gamma})$. Since the parameters described by vectors $\boldsymbol{\gamma}$ and $\boldsymbol{\theta}$ are uncertain due to limitation of drop data, then these parameters should be represented by PDFs, $f(\boldsymbol{\gamma})$ and $g(\boldsymbol{\theta})$ in a Bayesian estimation or by the corresponding classical confidence intervals. As such, in case of the Bayesian estimation, the expected PPoF hazard function would be

$$h(t|S) = \iint_{\boldsymbol{\theta},\boldsymbol{\gamma}} h(t; \boldsymbol{\gamma}|S, \boldsymbol{\theta}) f(\boldsymbol{\gamma}) g(\boldsymbol{\theta}) d\boldsymbol{\theta} d\boldsymbol{\gamma} \tag{21.8}$$

The PH model is readily expandable to more than one stress level, including multiple stress types. The modifier function can be a combination of different independent models such as exponential (e.g., Arrhenius model for temperature) and

Table 21.1 Stress levels and failure times

Temp (°K)		Relative humidity (%)		Time-to-failures (h)
T	$1/T$	RH	1/RH	
393	2.54×10^{-3}	0.60	1.68	102, 115, 151, 196, 210
353	2.83×10^{-3}	0.85	1.18	184, 193, 244, 265, 281
353	2.83×10^{-3}	0.60	1.67	298, 311, 337, 347, 405

inverse power law (e.g., when fatigue loading, vibration, voltage, or thermal cycling such as the Coffin–Manson model are appropriate). PH can model both independent and dependent stresses. A commonly used linear form of the modifier function for shocks is the Cox model expressed by Eq. 21.4. However, when non-shock models are also involved besides the shock models, then the hazard rate would take the form

$$h(t;\gamma|S;\boldsymbol{\theta},\boldsymbol{\delta}) = h_0(t;\gamma)e^{\sum_{i=1}^{n}\theta_i S_i}(g(S_{\text{non}},\boldsymbol{\delta}))^{-1} \tag{21.9}$$

where $g(S_{\text{non}},\boldsymbol{\delta})$ is the non-shock stress acceleration term with a vector of parameters, $\boldsymbol{\delta}$. For example, for temperature, T, the normalized Arrhenius model $g(T,\alpha,\beta) = \alpha e^{\frac{\beta}{T}}$ is used; and for vibration having a root-mean-square (RMS), G, $g(G,\rho,\sigma) = \frac{1}{\rho G^{\sigma}}$ is used; and for thermal cycling, ΔT, the normalized Coffin–Manson, $g(G,\mu,\varphi) = \frac{\mu}{\Delta T^{\varphi}}$ is used. The uncertainties associated with the parameters in Eq. 21.9 would be needed in a PPoF analysis. A common and powerful way to describe these uncertainties is to find their PDFs through the Bayesian inference. For more discussions on Bayesian estimation in regression equations such as the PPoF mathematical models, see [9].

As an example, consider a component whose reliability may be affected by both the ambient temperature and relative humidity under which it operates. Five samples at each of three (T, H) combinations were run until failure. The stress levels and failure times are given in Table 21.1.

Analyzing the data using the reciprocals of temperature and humidity as the covariates x_1 and x_2 provides the following MLE parameter point estimates: $\hat{\theta}_0 = 12.15$, $\hat{\theta}_1 = -14{,}474$, $\hat{\theta}_2 = -4.38$, $\hat{\beta} = 6.17$. The 95% confidence bounds based on the MLE estimates using the Fisher information matrix [9] are computed as

$$2.89 < \hat{\theta}_0 < 21.4;\ -15,277 < \hat{\theta}_1 < -13,671;$$
$$-6.56 < \hat{\theta}_2 < -2.20;\ 4.60 < \hat{\beta} < 7.74$$

Suppose that the actual anticipated usage conditions for this component were 20 °C (293 K, or $1/T = 3.41 \times 10^{-3}$) and 40% relative humidity (RH = 0.40, or 1/RH = 2.5). Then, the Weibull parameters for the usage conditions would be $\beta = 6.17$ and

$$\alpha = \left(\exp\left[\sum_{j=0}^{m}\theta_j S_j\right]\right)^{-1/\beta} = \left[e^{(12.15-14474\times3.41\times10^{-3}-4.38\times2.5)}\right]^{-1/6.17} = 2452.6\,\mathrm{h}$$

If the mission time for the component was 1500 h, then the estimated mission reliability would be

$$R(1500\,h) = e^{-\left(\frac{1500}{2452.6}\right)^{6.17}} = 0.96$$

21.5 Degradation Models

Unlike the life versus stress models, the general degradation path model (Meeker [10]) may be used to show the accumulated amount of damage or degradation over time until the damage exceeds the endurance to stand damage and a failure follows. As such degradation data (for now assume no measurement errors and detection probability) at a fixed level of stress at a given time t would be needed to estimate the path to failure and the time-to-failure distribution. In the degradation PPoF models, $D(t)$ denotes the degradation path of a particular component. Values of $D(t)$ can be monitored continuously, but in practical applications, they are often sampled at discrete points in time. Suppose the observed sample degradation path for some unit i at time t_{ij} is a unit's actual degradation path $D(t)$ plus a model error, as given by

$$y_{ij} = D_{ij} + \varepsilon_{ij}, i = 1, \ldots, n;\ j = 1, \ldots, m_i \tag{21.10}$$

where $D_{ij} = D\left(t_{ij}|\boldsymbol{\theta}\right)$ is the degradation path given the vector of parameters $\boldsymbol{\theta} = \{\theta_1, \ldots \theta_k\}$ and t is the actual time (operation time), cycles, expended life, or an index of age. The actual degradation path of unit i at sampling number j is $t_{ij} \cdot \varepsilon_{ij} \sim N(0, \sigma_\varepsilon)$ and is a residual deviation (model error) for unit i at t_{ij} (corresponding to the sampling number j).

The total number of inspections to measure the cumulative damage on unit i is denoted by m_i. Note that time t could be represented as real time, operating time, or some other appropriate quantitative measures such as miles for automobile tires or the number of loading cycles for fatigue applications. Figure 21.8 conceptually shows the data points (cumulative degradation or damage) for unit i, at each measurement time t_{ij}. In reality, there will only be one measurement of the degradation variable at each time, even though for each unit, the measurement time need not be the same. For instance, the first measurement time for unit 1, t_{11}, does not necessarily need to be the same as the first measurement time for unit 2, t_{21}.

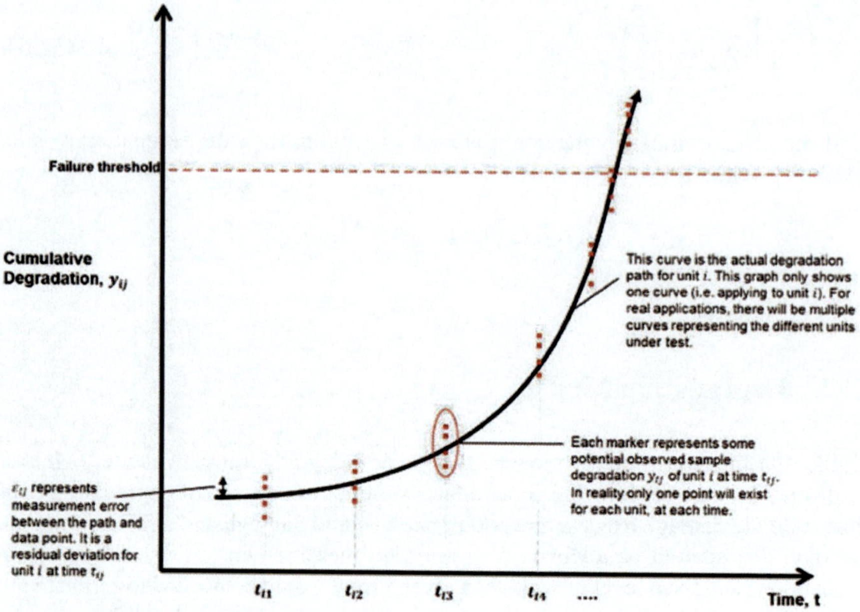

Fig. 21.8 General degradation path model for unit i

The scales of y and t can be chosen to simplify the form of $D(t_{ij}|\theta_{1i}, \ldots, \theta_{ki})$. The choice of a degradation PPoF model requires not only the determination of the mathematical form of $D(t_{ij}|\theta_{1i}, \ldots, \theta_{ki})$, but also estimation of the parameters in $\theta_{1i}, \ldots, \theta_{ki}$. Especially since the elements of the vector $\boldsymbol{\theta}$ may be correlated, the parameters' covariance or the joint distribution of the parameters of this vector (in a Bayesian estimation) should be estimated. Meeker and Escobar, in Statistical Methods for Reliability Data (1998), describe the use of a general family of transformations to a multivariate normal distribution with mean vector $\boldsymbol{\mu}_{\boldsymbol{\theta}}$ and covariance matrix $\boldsymbol{\Sigma}_{\boldsymbol{\theta}}$.

In PPoF, it is assumed that the parameters $\theta_{1i}, \ldots, \theta_{ki}$ are random and independent of the measurement error ε_{ij}. It is also possible to assume that ε_{ij} are independently and identically distributed (iid). Since each degradation observationy_{ij} is taken sequentially, there is potential for autocorrelation between the ε_{ij}'s, especially for closely spaced readings. However, in many practical applications involving the modeling of degradation of units from a population or process, provided that the model fit is adequate and measurement processes are in control, this autocorrelation is typically weak. Moreover, variability is dominated by unit-to-unit variability in the $\boldsymbol{\theta}$ values, and point estimates of regression models are not seriously affected by autocorrelation. Although, in some cases, ignoring autocorrelation can result in standard errors that are seriously biased, this is not as much of a problem when confidence intervals are employed.

Consider the accelerated degradation data given by NIST/SEMATECH [11]. The degradation data are from a component that degrades linearly by time at a different rate under the stress caused by the operating temperature. Assume the endurance level is when the cumulative damage changes by 30% or more. Fifteen components were tested under three different temperature conditions (five at 65 °C, five at 85 °C, and the last five at 105 °C). Degradation percent values were read out at 200, 500, and 1000 h. The readings are given by unit in the following three temperature cell tables.

Percent degradations at 65 °C

200 h		500 h	1000 h
Unit 1	0.87	1.48	2.81
Unit 2	0.33	0.96	2.13
Unit 3	0.94	2.91	5.67
Unit 4	0.72	1.98	4.28
Unit 5	0.66	0.99	2.14

Percent degradations at 85 °C

200 h		500 h	1000 h
Unit 1	1.41	2.47	5.71
Unit 2	3.61	8.99	17.69
Unit 3	2.13	5.72	11.54
Unit 4	4.36	9.82	19.55
Unit 5	6.91	17.37	34.84

Percent degradations at 105 °C

200 h		500 h	1000 h
Unit 1	24.58	62.02	124.10
Unit 2	9.73	24.07	48.06
Unit 3	4.74	11.53	23.72
Unit 4	23.61	58.21	117.20
Unit 5	10.90	27.85	54.97

Note that one unit failed in the 85 °C cell, and four units failed in the 105 °C cell. Because there were so few failures, it would be impossible to fit a life distribution model in any cell but the 105 °C cell, and therefore no PPoF model can fit with reasonable confidence using the failure events of these data. Therefore, we can rely on the PPoF models using degradation as the dependent variable. For this purpose, we propose the following power law model

$$D_{i,j}(t, T; k, \theta, \eta) = kT_i^{\theta} t_j^{\eta} \quad (21.11)$$

where D_{ij} is the unit degradation percentages, and $k, \theta,$ and η are parameters and t is time in hours, and T is the temperature in °C. We further account for uncertainties assuming a lognormal distribution for damage. Therefore, the PPoF for this problem would be expressed as

$$D(t, T; k, \theta, \eta, \sigma_t) = \text{LOGNORM}\left(kT^{\theta}t^{\eta}; \sigma_t\right) \quad (21.12)$$

where σ_t is the standard deviation of the lognormal model.

Using a Bayesian estimation of the PPoF model parameters (See [1]), at the use temperature of 20 °C assuming priors of $\sigma_t \sim \text{UNIF}\left(10^{-6}, 100\right), \theta \sim \text{UNIF}(0.1, 10)$, $\eta \sim \text{NORM}(0.99, 0, 0057)$, and $\ln(k) \sim \text{UNIF}(-100, -1)$ provides posterior distribution of the amount damage at the use temperature of 20 °C and the posterior time-to-failure distribution described in terms of logarithmic mean and standard deviation values as $L_{\text{use } 20°\text{C}} \sim \text{LOGNORM}(16.57; 0.58)$. The distributions of the damage and time-to-failure at the temperature of 20 °C are shown in the plots of Figs. 21.9 and 21.10, respectively. Further, the mean time-to-failure at the use temperature of 20 °C is calculated as 1.86×10^6 h.

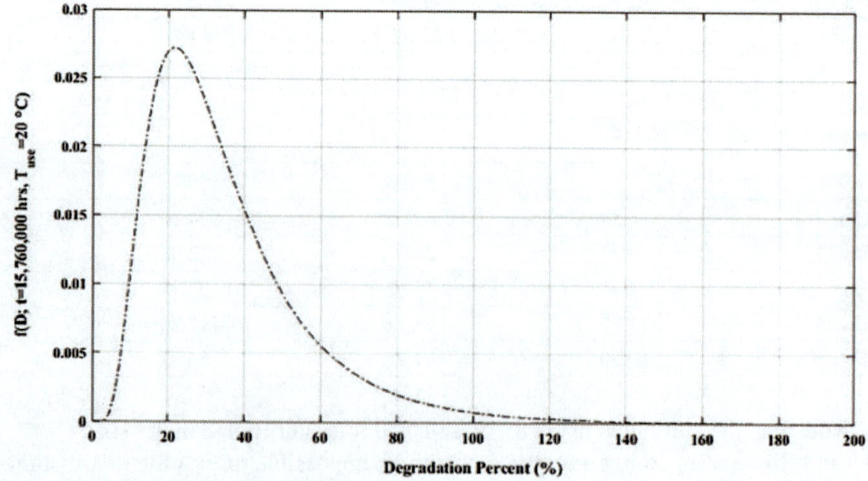

Fig. 21.9 Damage percent distribution at 20 °C

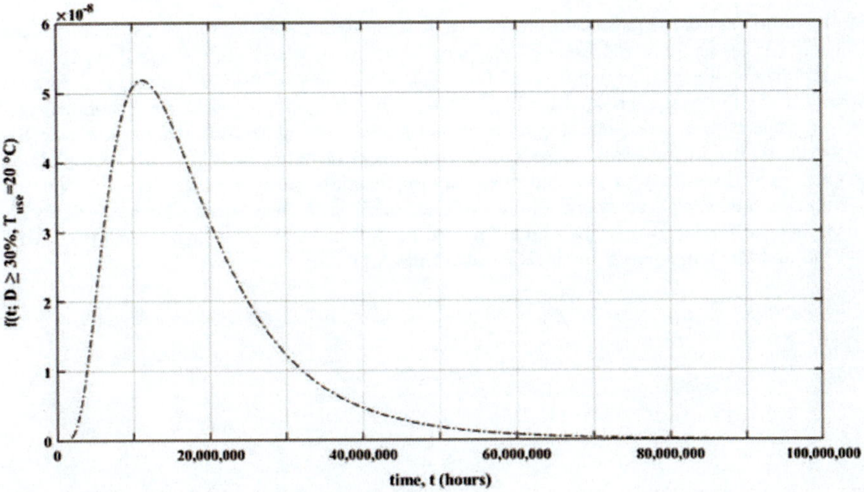

Fig. 21.10 Time-to-failure distribution at 20 °C

References

1. Modarres, M., Amiri, M., & Jackson, C. (2017). *Probabilistic physics of failure approach to reliability: Modeling, accelerated testing, prognosis and reliability* (p. 288). Wiley-Scrivener.
2. Jaynes, E. T. (1965). Gibbs vs. Boltzmann Entropies. *American Journal of Physics, 33*, 391–398.
3. Dowling, N. (2012). *Mechanical behavior of materials: Engineering methods for deformation, fracture, and fatigue* (4th ed., p. 960). Boston: Pearson.
4. Dodson, G., & Howard, B. (1961). High stress aging to failure of semiconductor devices. In *Proceedings of the Seventh National Symposium on Reliability and Quality Control*. Philadelphia, PA.
5. Haggag, A., McMahon, W., Hess, K., Cheng, K., Lee, J., & Lyding, J. (2000). A probabilistic-physics-of-failure/short-time-test approach to reliability assurance for high-performance chips: Models for deep-submicron transistors and optical interconnects. In *Proceedings of IEEE Integrated Reliability Workshop* (pp. 179–182).
6. Hall, P., & Strutt, J. (2003). Probabilistic physics-of-failure models for component reliabilities using Monte Carlo simulation and Weibull analysis: A parametric study. *Reliability Engineering & System Safety, 80*(3), 233–242.
7. Azarkhail, M., & Modarres, M. (2007). A novel Bayesian framework for uncertainty management in physics-based reliability models. In Proceedings of ASME International Mechanical Engineering Congress and Exposition. Seattle, WA.
8. Matik, Z., & Sruk, V. (2008). The physics-of-failure approach in reliability engineering. In *Proceedings of IEEE International Conference on Information Technology Interfaces* (pp. 745–750). Dubrovnik, Croatia.
9. O'Connor, A. N., Modarres, M., & Mosleh, A. (2019). *Probability distributions used in reliability engineering* (p. 214). College Park: Center for Risk and Reliability.
10. Meeker, W., & Escobar, L. (1998). *Statistical methods for reliability data* (p. 712). Hoboken: Wiley.
11. *NIST/SEMATECH e-Handbook of Statistical Methods*. Downloaded from https://www.itl.nist.gov/div898/handbook/, on May 12, 2020.

Mohammad Modarres, Ph.D. is the Nicole J. Kim Eminent Professor of Engineering and Director, Center for Risk and Reliability at the University of Maryland, College Park. He is an expert in reliability engineering, probabilistic risk assessment, physics-of-failure, and fracture mechanics. His interests in risk, reliability, structural integrity and prognosis, and health management include both experimental and probabilistic model development efforts. He has over 400 papers in archival journals and proceedings of conferences including multiple textbooks and book chapters in various areas of risk and reliability engineering. He is a University of Maryl and Distinguished Scholar-Teacher and a fellow of the American Nuclear Society. He received his BS in Mechanical Engineering from Tehran Polytechnic, MS in Mechanical Engineering from MIT, and MS and Ph.D. in Nuclear Engineering also from MIT.

Chapter 22
Reliability and Availability Analysis in Practice

Kishor Trivedi and Andrea Bobbio

Abstract Reliability and availability are key attributes of technical systems. Methods of quantifying these attributes are thus essential during all phases of system lifecycle. Data (measurement)-driven methods are suitable for components or subsystems but, for the system as a whole, model-driven methods are more desirable. Simulative solution or analytic–numeric solution of the models are two major alternatives for the model-driven approach. In this chapter, we explore model-driven methods with analytic–numeric solution. Non-state-space, state-space, hierarchical, and fixed-point iterative methods are explored using real-world examples. Challenges faced by such modeling endeavors and potential solutions are described. Software package SHARPE is used for such modeling exercises.

Keywords Availability · Reliability · Fault tree · Markov model · Non-state-space model · State-space model · Hierarchical model · Fixed-point iteration technique

22.1 Introduction

This chapter discusses techniques that are found to be effective for reliability and availability assessment of real systems. Modern life heavily relies on man-made systems that are expected to be reliable. Many high-tech cybersystems are found wanting since their failures are not so uncommon. Such failures and consequent downtimes lead to economic losses, to a loss of reputation, and to even loss of lives. To ameliorate the situation, methods have been developed that reduce failure occurrences and resultant downtimes. In order to gauge the effectiveness of these improvement methods, scalable and high-fidelity techniques of reliability and availability assessment are needed.

K. Trivedi (✉)
Electrical and Computer Engineering, Duke University, Durham, NC 27708, USA
e-mail: ktrivedi@duke.edu

A. Bobbio
DiSit, Università del Piemonte Orientale, 15131 Alessandria, Italy

K. B. Misra (ed.), *Handbook of Advanced Performability Engineering*,
https://doi.org/10.1007/978-3-030-55732-4_22

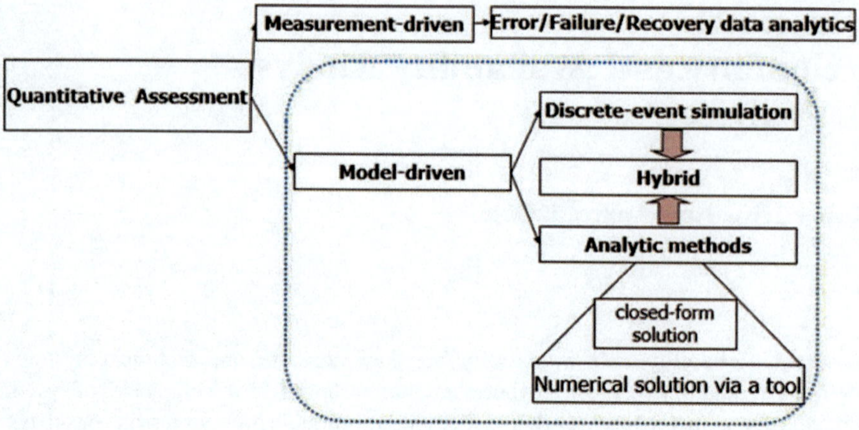

Fig. 22.1 Reliability/availability assessment methods

This chapter discusses techniques that are found to be effective for reliability and availability assessment in practice. Assessment methods can be divided into measurement-driven (or data-driven) versus model-driven methods (Fig. 22.1). Data-driven methods are suitable for small subsystems, while model-driven methods are appropriate for large systems. Using model-driven methods, we can derive the dynamic behavior of a system consisting of many components using first principles (of probability theory) rather than from measurements.

In practice, these two approaches are combined together so that subsystem or component behavior is derived using data-driven methods, while the system behavior is derived using model-driven methods.

This chapter focuses on model-driven methods. Models can be solved using discrete-event simulation or using analytic–numeric techniques. Some simple models can be solved analytically to yield a closed-form formula while a much larger set of models can be dealt with by a numerical solution of their underlying equations. The latter approach is known as analytic–numeric solution. Distinction between analytic–numeric solution versus discrete-event simulation-based solution ought to be noted. We believe that simulative solution and analytic–numeric solutions should be judiciously combined in order to solve complex system models. This chapter, however, is on analytic–numeric methods, providing an overview of a recently published book by the authors of this chapter [1].

Our approach to exposing the methods is example-based. Chosen examples are of real systems that we have ourselves analyzed for some companies. Overall modeling process is depicted in Fig. 22.2.

Non-state-space (or combinatorial) models can deal with large systems if based on the drastic assumption of statistical independence among components. State-space model types, specifically continuous-time Markov chains and Markov reward models, are commonly utilized for higher fidelity. Multi-level models that judiciously combine non-state-space and state-space methods will be seen to have the scalability

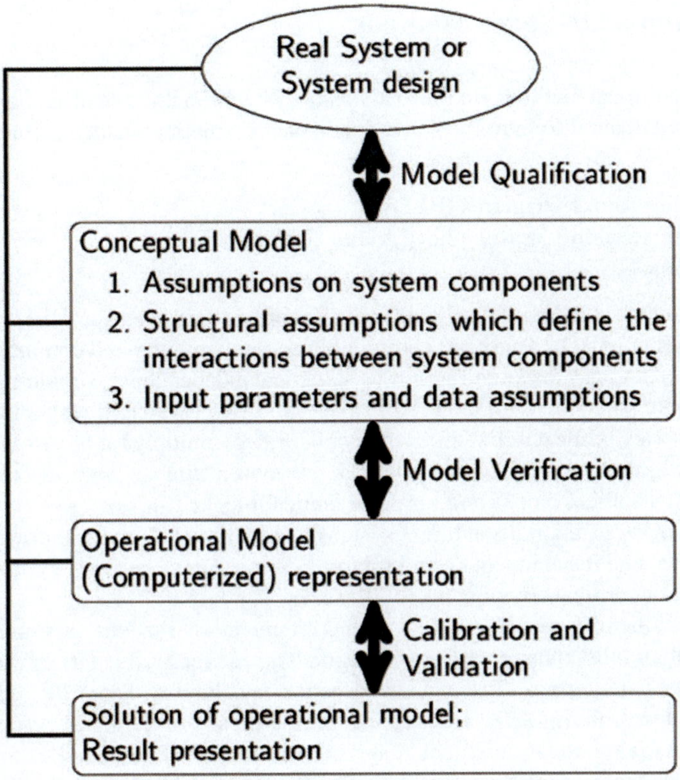

Fig. 22.2 The overall modeling process

and fidelity needed for capturing the dynamic behavior of real systems. Depending on the application, a model may be solved for its long-term (steady-state) behavior or its time-dependent (or transient) behavior. Solution types for such models are classified in Fig. 22.3 [1, 2]. Software packages that are used in solving the examples of this chapter are SHARPE [2, 3] and SPNP [4, 5].

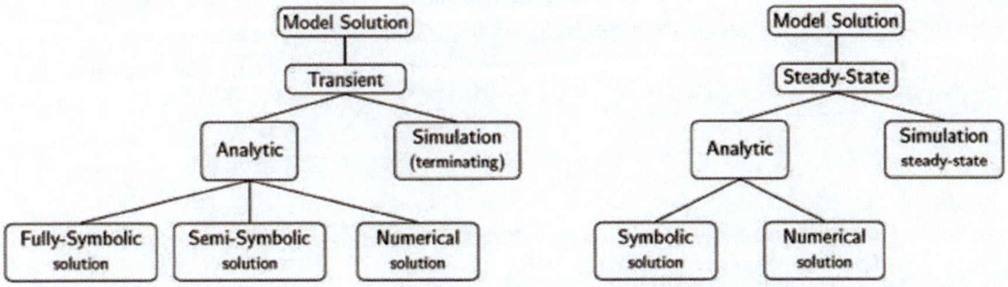

Fig. 22.3 Solution techniques

22.2 Non-state-Space Methods

Several traditional methods for the analysis of system reliability and availability can be classified under the umbrella of non-state-space (sometimes called combinatorial) methods:

- Reliability Block Diagrams (RBD)
- Network reliability or Reliability graphs (RelGraph)
- Fault Trees.

The simplest paradigm for reliability/availability is the (series–parallel) reliability block diagram (RBD). These are commonly used in computer and communications industry and are easy to use and assuming statistical independence, simple algorithms are available to solve very large RBDs. Reliabilities (availabilities) multiply for blocks in series, while unreliabilities (unavailabilities) multiply for blocks in parallel. Efficient algorithms for *k-out-of-n* blocks are also available, both in the case of statistically identical blocks and for non-identical blocks [1].

Besides system reliability at time *t*, system mean time to failure, system availability (steady-state and instantaneous), and importance measures can also be computed so as to point out critical components (bottlenecks) [1].

High availability requirement in telecommunication systems is usually more stringent than most other sectors of industry. The carrier-grade platform from Sun Microsystems requires a "five nines and better" availability. From the availability point of view, the top-level architecture of a typical carrier-grade platform was modeled in [6] as a reliability block diagram consisting of series, parallel, and *k-out-of-n* subsystems, as shown in Fig. 22.4. The SCSI series block is further expanded as in the inset of Fig. 22.4.

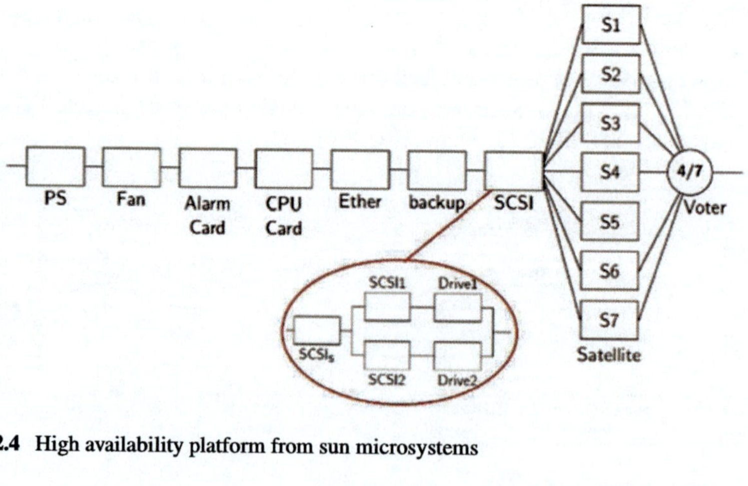

Fig. 22.4 High availability platform from sun microsystems

Series–parallel structure is often violated in practice. Non-series–parallel block diagrams are often cast as s-t connectedness problems, also known as network reliability problems or just relgraph in SHARPE. The price to be paid for this additional modeling power is the increased complexity of solution methods. Known solution methods are factoring (or conditioning), finding all minpaths followed by the use of one of many sum-of-disjoint-product (SDP) algorithms, the use of binary decision diagrams (BDD), or the use of Monte Carlo simulation. SDP- and BDD-based algorithms have been implemented in the SHARPE software package [2, 3]. Nevertheless, real systems pose a challenge to these algorithms. For instance, the reliability of the current return network subsystem of Boeing 787 was modeled as a relgraph shown in Fig. 22.5. However, the number of minpaths was estimated to be over 4.2 trillion.

To solve the model, for the purpose of FAA certification, a new bounding algorithm was developed, patented, and published [7]. Table 22.1 reports the results showing that the upper and lower bounds to the s-t reliability were close enough, with a very small number of minpaths and mincuts selected for the computation. The computation time was very short for this otherwise intractable problem. This new bounding algorithm is implemented in the SHARPE software package and continues to be

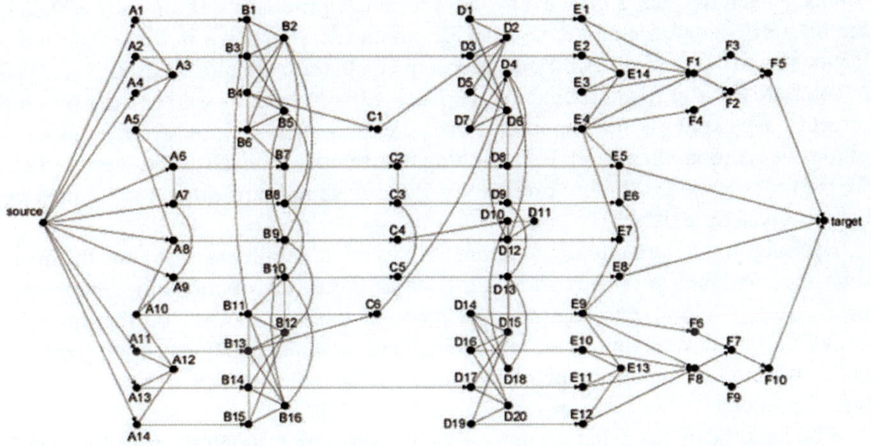

Fig. 22.5 Boeing relgraph example

Table 22.1 Unreliability upper/lower bounds

Runtime	20 s	120 s	900 s
Up bound	1.146036×10^{-8}	1.081432×10^{-8}	1.025519×10^{-8}
Low bound	1.019995×10^{-8}	1.019995×10^{-8}	1.019995×10^{-8}
#minpaths	28	29	33
#mincuts	113	113	113

used by Boeing (via their IRAP software package [8]) for the reliability assessment of current return network of all Boeing commercial airplanes.

Table 22.2 shows a comparison of SDP and BDD methods for various benchmark networks of increasing complexity. The different BDD columns show the effect of node ordering on the computational time [9]. The used benchmark networks are shown in Fig. 22.6 and were inspired by the literature [10]. Note also that the bounding method is not utilized in the comparison table.

In the aerospace, chemical, and nuclear industries, engineers use fault trees (FT) to capture the conditions under which system fails. These Boolean conditions are encoded into a tree with AND gates, OR gates, and *k-out-of-n* gates as internal nodes, while leaf nodes represent component failures and the top or root node indicates system failure.

Fault trees without repeated events are equivalent to series–parallel RBDs, while those with repeated events are more powerful [1, 2, 11]. Solution techniques for fault trees with repeated events are the same as those for the network reliability problem discussed in the previous paragraph [1]. Fault trees with several thousand components can be solved with relative ease.

Figure 22.7 shows an FT for a GE Equipment Ventilation System. Notice that leaves drawn as circles are basic events, while inverted triangles represent repeated events. Assuming that all the events have a failure probability equal to $q = 0.001$, the SHARPE input file and the SHARPE output file are shown in Fig. 22.8 on the left-hand and on the right-hand side, respectively. In this example, SHARPE is asked to compute the Top Event probability (QTE = 1.0945e−02) as well as the list of the mincuts. We could ask for importance measures as well as a closed-form expression of top event probability [1, 3]. By assigning failure rates for each event, we could ask for the time-dependent failure probability of the system. Many other possibilities for output measures exist.

By assigning failure rates to components, system reliability at time t and the mean time to system failure can be computed. Time-to-failure distribution other than exponential (e.g., Weibull) can be used in such non-state-space models. Furthermore, by assigning failure rate and repair rate to each component, steady-state and instantaneous availability can be computed (assuming independence in repair besides failure independence).

FTs have been extended to non-coherent gates such as NOT gates, to multi-state components [12], phased-mission systems [13], and with dynamic gates [14]. SHARPE fault trees allow NOT gate, multi-state components, and phased-mission systems. Dynamic gates are not explicitly included in SHARPE but can easily be implemented since (static) fault trees, Markov chains, and their combination via hierarchical modeling are provided [1].

Table 22.2 Comparison of SDP and BDD with various orderings

Example	SDP		BDD(O1)		BDD(O2)		BDD(O3)		BDD(O4)	
	#DP	Time(s)	Size	Time(s)	Size	Time(s)	Size	Time(s)	Size	Time(s)
Relex1	7	0.03	15	0.04	15	0.04	19	0.04	17	0.04
Relex2	11	0.04	27	0.05	19	0.05	27	0.05	27	0.05
Relex3	16	0.04	21	0.05	28	0.05	32	0.05	28	0.05
Relex4	40	0.05	28	0.05	57	0.05	54	0.05	33	0.05
Relex5	78	0.06	64	0.06	65	0.06	85	0.06	67	0.06
Relex6	150	0.10	291	0.08	347	0.08	277	0.08	277	0.08
Relex7	402	0.31	120	0.06	187	0.07	1178	0.12	444	0.09
Relex8	2294	6.46	966	0.16	505	0.19	4865	0.38	743	0.15
Relex9	47,312	148.45	2083	0.41	14,821	2.26	12,277	1.25	3360	0.46

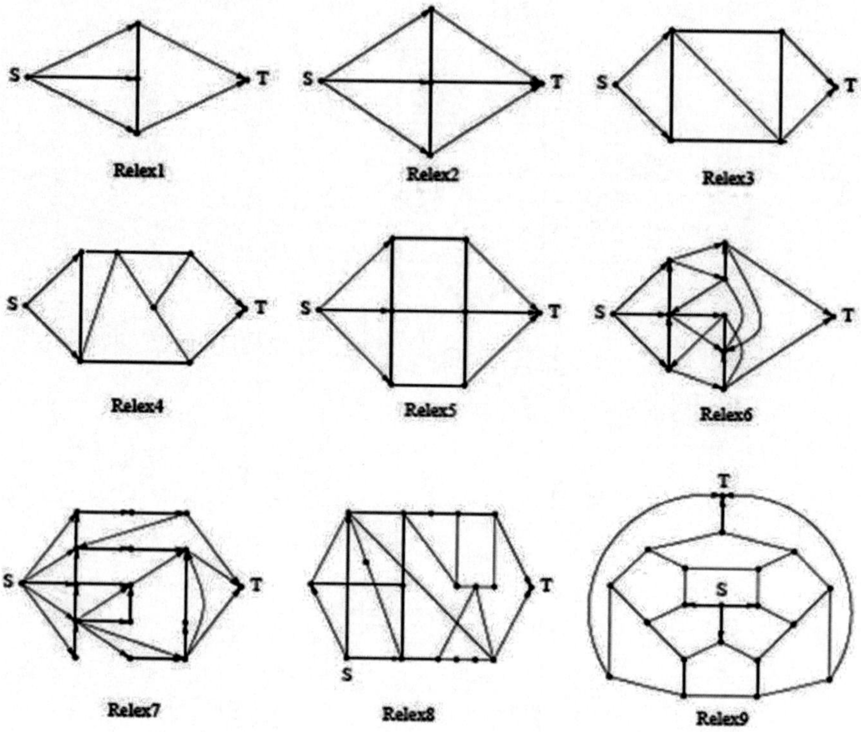

Fig. 22.6 Benchmark networks

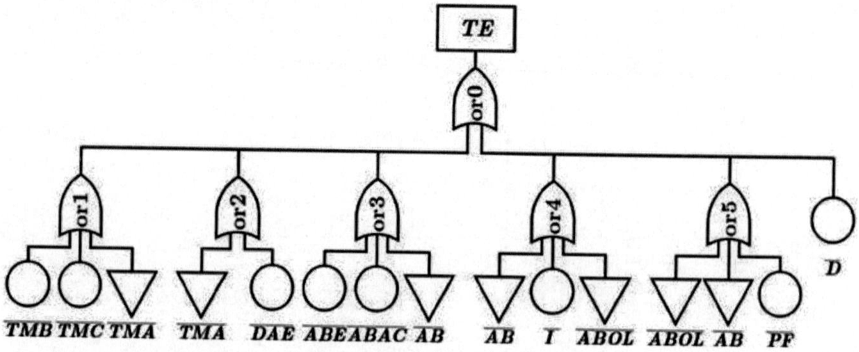

Fig. 22.7 Fault tree model equipment ventilation system

```
echo Ventilation System

ftree vent
basic TMB prob(lp)
basic TMC prob(lp)
repeat TMA prob(lp)
basic DAE prob(lp)
basic ABE prob(lp)
basic ABAC prob(lp)
repeat AB prob(lp)
basic I prob(lp)
basic PF prob(lp)
basic D prob(lp)
repeat ABOL prob(lp)
or orr1 TMB TMC TMA
or orr2 TMA DAE
or orr3 ABE ABAC AB
or orr4 AB I ABOL
or orr5 AB PF ABOL
or te orr1 orr2 orr3 orr4 orr5 D
end
bind
lp 0.001
end
var sysunrel sysprob(vent)
expr sysunrel

mincuts(vent)
end
```

```
Ventilation System
sysunrel:    1.0945e-002

Mincuts for system vent:

[(TMC)],
[(TMB)],
[(PF)],
[(ABOL)],
[(I)],
[(AB)],
[(ABAC)],
[(ABE)],
[(TMA))],
[(DAE)],
[(D)]
```

Fig. 22.8 SHARPE input/output files for ventilation system

22.3 State-Space Methods

As stated in the last section, non-state-space models with thousands of components can be solved without generating their underlying state space by making the independence assumption. But in practice, dependencies do exist among components. We then need to resort to state-space models such as (homogeneous) continuous-time Markov chains (CTMC).

Markov models have been used to capture dynamic redundancy, .imperfect coverage (e.g., failure to failover or failure to detect, etc.), escalated levels

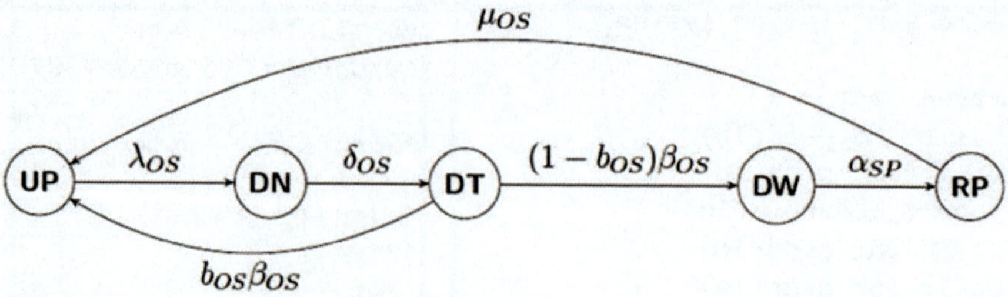

Fig. 22.9 CTMC availability model of Linux OS

of recovery, concurrency, contention for resources, combined performance and reliability/availability, and survivability [1, 15]. Markov availability model will have no absorbing states (Fig. 22.9), while Markov reliability models will have one or more absorbing states (Fig. 22.11). Markov models can be solved for steady-state, transient, and cumulative transient behavior according to the following equations [1, 15]:

Steady-state	$\boldsymbol{\pi} \boldsymbol{Q} = 0$ with $\sum \pi = 1$
Transient	$\mathrm{d}\boldsymbol{\pi}(t)/\mathrm{d}t = \boldsymbol{\pi}(t)\boldsymbol{Q}$ given $\boldsymbol{\pi}(0)$
Cumulative transient	$\mathrm{d}\boldsymbol{b}(t)/\mathrm{d}t = \boldsymbol{b}(t)\ \boldsymbol{Q} + \boldsymbol{\pi}(0)$

In the above formulas, $\boldsymbol{Q}$ is the infinitesimal generator matrix of the CTMC, $\boldsymbol{\pi}(t)$ is the state probability vector at time t, $\boldsymbol{\pi}(0)$ is the initial state probability vector, $\boldsymbol{\pi} = \lim_{t\to\infty} \boldsymbol{\pi}(t)$ is the steady-state probability vector, and $\boldsymbol{b}(t) = \int_0^t \boldsymbol{\pi}(u)\mathrm{d}u$ is the vector of the expected state occupancy times in the interval from 0 to t. Derivatives of these measures with respect to the input parameters can also be computed numerically [1].

22.3.1 *CTMC Availability Models*

The system availability (or instantaneous, point, or transient availability) is defined as the probability that at time t the system is in an up state:

$$A(t) = P\{\text{system working at } t\}$$

Steady-state availability (A_{ss}) or just availability is the long-term probability that the system is up:

$$A_{ss} = \lim_{t\to\infty} A(t) = \frac{\text{MTTF}}{\text{MTTF} + \text{MTTR}}$$

where MTTF is the system mean time to failure and MTTR is the system mean time to recovery. When applied to a single component, the above equation holds without any distributional assumptions. For a complex system with redundancy, the equation holds if we use "equivalent" MTTF and "equivalent" MTTR [1].

The availability model of the Linux operating system used in IBM's SIP implementation on WebSphere was presented in [16] and is shown in Fig. 22.9. From the up state, the model enters the down state DN with failure rate λ_{OS}. After failure detection, with a mean time of $1/\delta_{OS}$, the system enters the failure-detected state DT.

The OS is then rebooted with the mean time to reboot given by $1/\beta_{OS}$. With probability b_{OS} the reboot is successful, and system returns to the UP state. However, with probability $1 - b_{OS}$, the reboot is unsuccessful, and the system enters the DW state where a repairperson is summoned. The travel time of the repairperson is assumed to be exponentially distributed with rate α_{SP}. The system then moves to the state RP. The repair takes a mean time of $1/\mu_{OS}$, and after its completion, the system returns to the UP state.

Solving the steady-state balance equations, a closed-form solution for the steady-state availability of the OS is easily obtained in this case due to the simplicity of the Markov chain.

$$A_{ss} = \pi_{UP} = \frac{1}{\lambda_{OS}}\left[\frac{1}{\lambda_{OS}} + \frac{1}{\delta_{OS}} + \frac{1}{\beta_{OS}} + (1 - b_{OS})\left(\frac{1}{\alpha_{SP}} + \frac{1}{\mu_{OS}}\right)\right]^{-1}$$

We can alternatively obtain a numerical solution of the underlying equations by using a software package such as SHARPE. Either graphical or textual input can be employed. The SHARPE textual input file modeling the CTMC of Fig. 22.9 is shown in Fig. 22.10. Noting that UP (labeled 1) is the only upstate, the steady-state availability is computed using the command *expr prob (LinuxOS,1)*. With the assigned numerical values for parameters (see Fig. 22.10), the result is $A_{ss} = 0.99963$.

```
echo Linux OS in IBM WebSphere Model
markov LinuxOS
1 2 los
2 3 dos
3 1 bos*beta
3 4 (1-bos)*beta
4 5 asp
5 1 mos
 Parameter values
```

```
bind
los 1/4000
dos 1
beta 6
bos 0.9
asp 1/2
mos 1
end
echo  Steady-state availability equal
echo  to probability of state 1
expr prob (LinuxOS,1)
end
```

Fig. 22.10 SHARPE input file for the CTMC of Fig. 22.9

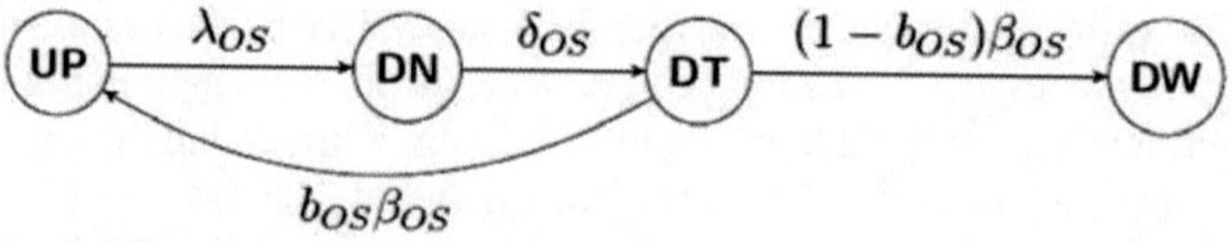

Fig. 22.11 CTMC reliability model of Linux OS

22.3.2 CTMC Reliability Models

While CTMC availability models have no absorbing states, CTMC reliability models have one or more absorbing states and the reliability at time t is defined as the probability that the system is continuously working during the interval $(0 - t]$. Further, since in a reliability model the system down state is an absorbing state, the MTTF can be calculated as the mean time to absorption in the corresponding CTMC model [1, 2, 15].

The reliability model extracted from the availability model of the Linux operating system used for IBM's SIP application is shown in Fig. 22.11. The repair transition from state RP to state UP and the transition from state DW are removed, that is, the down state reached starting from the UP state is made an absorbing state. Note that states DN and DT are down states but the sojourns in these states are likely to be short enough to be considered as glitches that can be ignored while computing system reliability and MTTF.

In this case, the model is simple enough so that a closed-form solution can be obtained by hand (or using Mathematica) by setting up and solving the underlying Kolmogorov differential equations. Alternatively, a numerical solution of the underlying equations can be obtained using SHARPE. The SHARPE textual input file for the reliability model of Fig. 22.11 is shown in Fig. 22.12. Note that in this case, since the CTMC is not irreducible, an initial probability vector must be specified.

The system reliability at time t is defined in this case as $R(t) = \pi_{\mathrm{UP}}(t)$ and, in the SHARPE input file of Fig. 22.11, is computed from $t = 0$ to $t = 10{,}000$ in steps of 2000. As noted earlier, the MTTF is defined as the mean time to absorption and is computed using the SHARPE command *expr mean (LinuxOS)*. With the assigned numerical values, the result is MTTF = 40,012 h.

The CTMC of a reliability model can be, but need not be, acyclic, as in the case of Fig. 22.11. If there is no component level repair (recovery), then the CTMC will be acyclic but if there is component level repair (but no repair after system failure) then the CTMC will have cycles. However, the model will always have one or more absorbing states.

Reliability modeling techniques have wide applications in different technical fields and have been proposed to provide new frontiers in predicting healthcare outcomes. With the rise in quantifiable approaches to health care, lessons from reliability modeling may well provide new ways of improving patient healthcare. Describing the development of conditions leading to organ system failure provides motivation for quantifying disease progression. As an example, a simple model for

```
markov LinuxOS
1 2 los
2 3 dos
3 1 bos*beta
3 4 (1-bos)*beta
end

* initial state probabilities
1 1.0
2 0.0
3 0.0
4 0.0
end

* Parameter values
bind
los 1/4000
dos 1
beta 6
bos 0.9
end
```

```
echo Reliability vs time equal to probability
echo of state 1 at time t

echo System reliability at times 0 thru 10000
echo in steps of 2000
func rel(t) tvalue(t;LinuxOS,1)
loop t,0, 10000, 2000
expr rel(t)
end

expr mean (LinuxOS)
end
```

Fig. 22.12 SHARPE input file for CTMC of Fig. 22.11

progressive kidney disease leading to renal failure is reported in Fig. 22.13 [17] where five discrete conditions are enumerated in keeping with clinical classification of kidney function.

The parameter values, used in solving the model of Fig. 22.13, are reported in Table 22.3. These values are estimated for a 65-year-old Medicare patient and are

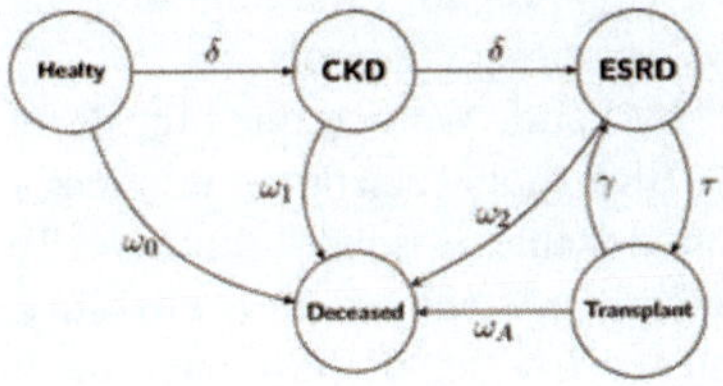

State meaning:
Healthy: Normal renal function,
CKD: Chronic Kidney Disease without renal failure,
ESRD: End-Stage Renal Disease
Transplant: Patients who have received a transplant,
Deceased.

Fig. 22.13 Markov model of renal disease

Table 22.3 Parameter values for a 65-year-old medicare patient

Description	Symbol	Value (event/year)
Decline	δ	0.1887
Transplant	τ	0.1786
Graft rejection	γ	0.0050
Prognosis-healthy	ω_0	0.0645
Prognosis-CKD	ω_1	0.1013
Prognosis-ESRD	ω_2	0.2174
Prognosis-Transplant	ω_A	0.0775

based on the latest available statistics from the United States Renal Data System (USRDS) annual report [18].

The model of Fig. 22.13 is solved for the survival rate and expected cost incurred by a patient in a 1-year interval [17].

Efficient algorithms are available for solving Markov chains with several million states [19–21] both in the steady-state and in the transient regime. Furthermore, measures of interest such as reliability, availability, performability, survivability, etc. can be computed by means of reward rate assignments to the states of the CTMC [1, 15]. Derivatives (sensitivity functions) of the measures of interest with respect to input parameters can also be computed to help detect bottlenecks [22–24]. Nevertheless, the generation, storage, and solution of real-life-system Markov models still pose challenges. Higher level formalisms such as those based on stochastic Petri nets (SPNs) and their variants [4, 15, 25–27] have been used to automate the generation, storage, and the solution of large state-space Markov models [26]. Our own version of SPN is known as stochastic reward nets (SRN). SRNs extend SPN formalism in several useful ways besides allowing specification of reward rates at the net level. This enables more concise description of real-world problems and an easier way to get the output measures [4].

An example of the use of stochastic reward nets to model the availability of an Infrastructure-as-a-Service (IaaS) cloud is shown in Fig. 22.14 [28]. To reduce power usage costs, physical machines (PMs) are divided into three pools: Hot pool (high performance and high power usage), warm pool (medium performance and medium power usage), and cold pool (lowest performance and lowest power usage). PMs may fail and get repaired. A minimum number of operational hot PMs are required for the system to function but PMs in other pools may temporarily be assigned to the hot pool in order to maintain system operation. Upon repair, PMs migrate back to their original pool. Migration creates dependencies among the pools.

A monolithic CTMC is too large to construct by hand. We use our high-level formalism of Stochastic Reward Net (SRN) [26]. An SRN model can be automatically converted into an underlying Markov (reward) model that is solved numerically for the measures of interest such as expected downtime, steady-state availability, reliability, power consumption, performability, and sensitivities of these measures.

In Fig. 22.14, place P_{h} initially contains n_{h} tokens (PMs of the hot pool), P_{w} contains n_{w} tokens (PMs of the warm pool), and P_{c} contains n_{c} tokens (PMs of the cold pool). Assuming the number of PMs in each pool is identical and equal to n, the number of states for the monolithic model of Fig. 22.14, is reported in the second column of Table 22.4. From this table, it is clear that this approach based merely on a higher formalism such as SRN, which we call largeness tolerance, soon reaches its limits as the time needed for the generation and storage of the state space becomes prohibitively large for real systems.

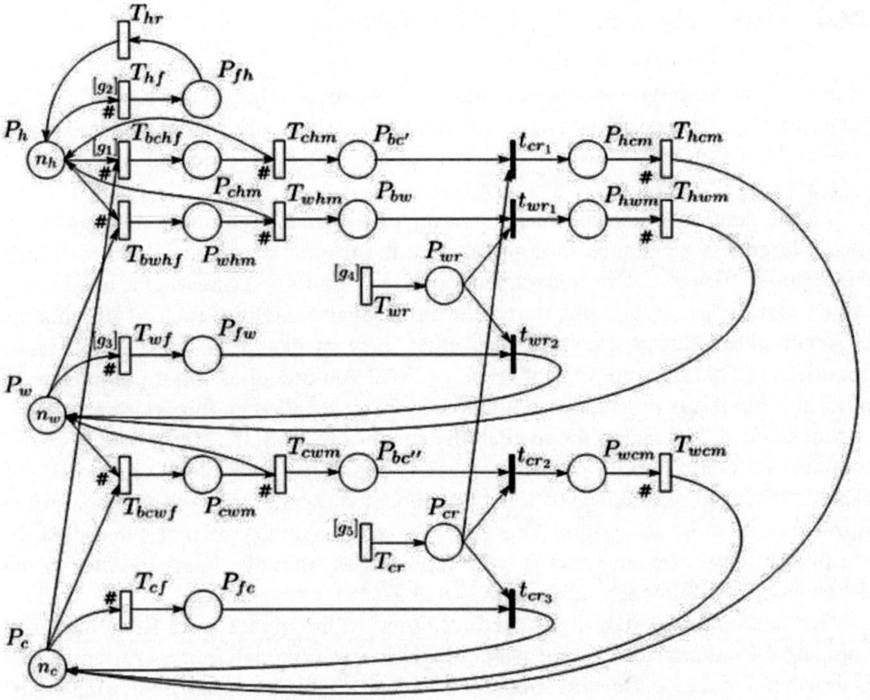

Fig. 22.14 SRN availability model of IaaS cloud

Table 22.4 Comparison of monolithic versus decomposed model

n	Monolithic model	Interacting sub-model
	#states	#states
3	10,272	196
4	67,075	491
5	334,948	1,100
6	1,371,436	2,262
7	4,816,252	3,770
8	Memory overflow	6,939
10	–	20,460
20	–	21,273
40	–	271,543
60	–	1,270,813
80	–	3,859,083
100	–	9,196,353

n is the initial #PM in each pool

22.4 Hierarchy and Fixed-Point Iteration

In order to avoid large models as is the case in a monolithic Markov (or generally state space) model, we advocate the use of multi-level models in which the modeling power of state-space models and efficiency of non-state-space models are combined together (Fig. 22.15).

Since a single monolithic model is never generated and stored in this approach, this is largeness avoidance in contrast with the use of largeness tolerance (recall stochastic Petri nets, SRNs, and related modeling paradigms) wherein the underlying large model is generated and stored. In multi-level modeling, each of the models is solved and results are conveyed to other relevant models to use as their input parameters. This transmission of results of one sub-model as input parameters to other sub-models is depicted as a graph that we have called an import graph [29].

Consider, for instance, the availability model of the SUN Microsystem whose top-level RBD availability model is shown in Fig. 22.4. Each block of the RBD of Fig. 22.4 is a complex subsystem that was modeled separately using the appropriate formalism in order to compute the steady-state availability of that subsystem. In the present case, the subsystems were modeled as Markov chains to cater to the dependencies within each subsystem.

The subsystem availability is then rolled up to the higher level RBD model to compute the system steady-state availability. The import graph for this system model is shown in Fig. 22.16. Specification, solution, and passing parameters for such multi-level models are facilitated by the SHARPE software package [2, 3]. The import graph in this case is acyclic. We can then carry out a topological sort of the graph resulting in a linear order specifying the order in which the sub-models are to be solved and the results rolled up in the hierarchy.

As the next example, we return to the IaaS cloud availability model and improve its scalability. The monolithic SRN model of Fig. 22.14 is decomposed into three sub-models to describe separately the behavior of the three pools [28, 29] while taking into account their mutual dependencies by means of parameter passing. The three sub-models are shown in Fig. 22.17.

Its import graph is shown in Fig. 22.18, indicating the output measures and input parameters that are exchanged among sub-models to obtain the overall model solution. Import graphs such as the one shown in Fig. 22.18 are not acyclic, and hence the

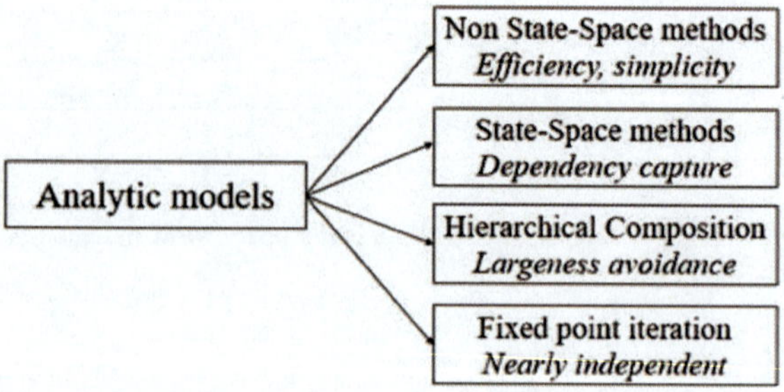

Fig. 22.15 Analytic modeling taxonomy

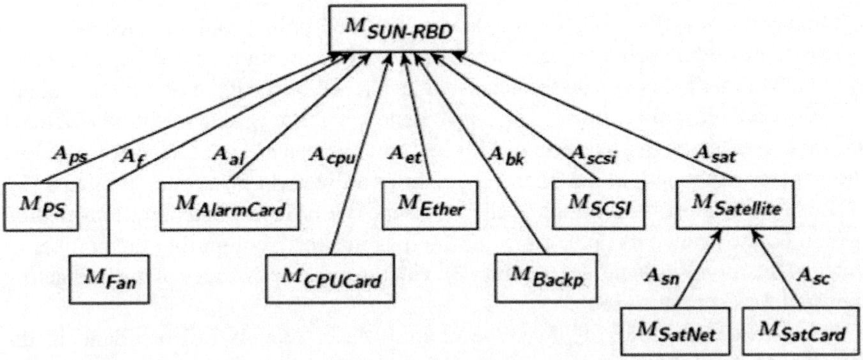

Fig. 22.16 Import graph for high availability platform from Sun Microsystems [6]

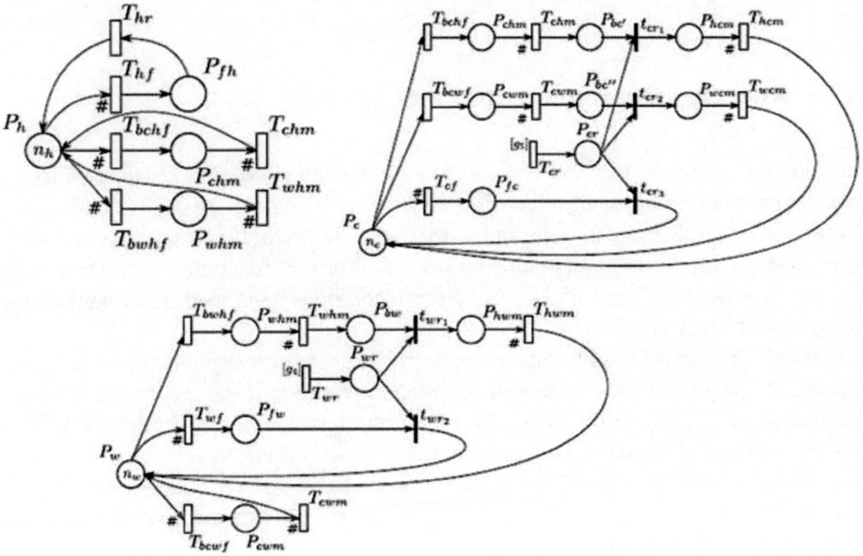

Fig. 22.17 Decomposed SRN availability model of IaaS cloud

Fig. 22.18 Import graph describing sub-model interactions

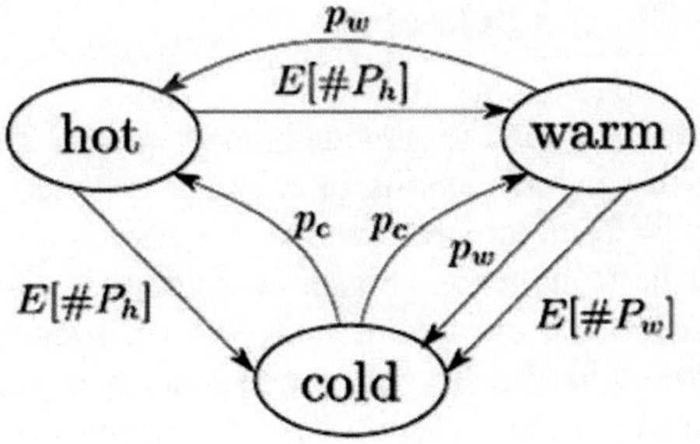

solution to the overall problem can be set up as a fixed-point problem. Such problems can be solved iteratively by successive substitution with some initial starting point. Many mathematical issues arise such as the existence of the fixed point, the uniqueness of the fixed-point, the rate of convergence, accuracy, and scalability. Except for the existence of the fixed point [30], all other issues are open for investigation. Nevertheless, the method has been successfully utilized on many real problems [1].

Table 22.4 shows the effect of the decomposition/fixed-point iteration method (which is also known as interacting sub-models method), comparing the number of states of the monolithic model (column 2) with the number of states of the interacting sub-model case (column 3).

Many more examples of this type of multi-level models can be found in the literature [1, 2, 16, 29–35]. A particular example is the implementation of the Session Initiation Protocol (SIP) by IBM on its WebSphere. A hierarchical availability model of that system is described in detail in [16].

22.5 Relaxing the Exponential Assumption

One standard complaint about the use of homogeneous continuous-time Markov chains is the ubiquitous assumption of all event times being exponentially distributed. There are several known paradigms that can remove this assumption: non-homogeneous Markov chain, semi-Markov and Markov regenerative process, and the use of phase-type expansions. All these techniques have been used, and many examples are illustrated in [1].

Nevertheless, there is additional complexity in using non-exponential techniques in practice, partly because the analytical–numeric solution is more complex but also because of additional information about the non-exponential distributions which is then needed and is often hard to come by.

A flowchart comparing the modeling power of the different state-space model types is shown in Fig. 22.19 [1], and in Fig. 22.20, we provide a classification of the modeling formalisms considered in [1].

22.6 Conclusions

We have tried to provide an overview of known modeling techniques for the reliability and availability of complex systems. We believe that techniques and tools do exist to capture the behavior of current-day systems of moderate complexity. Nevertheless, higher and higher complexity is being designed into systems, and hence the techniques must continue to evolve. Together with the largeness problem, the need for higher fidelity will require increasing use of non-exponential distributions, the need to properly combine performance, power, and other measures of system effectiveness together with failure and recovery. Parameterization and validation of the

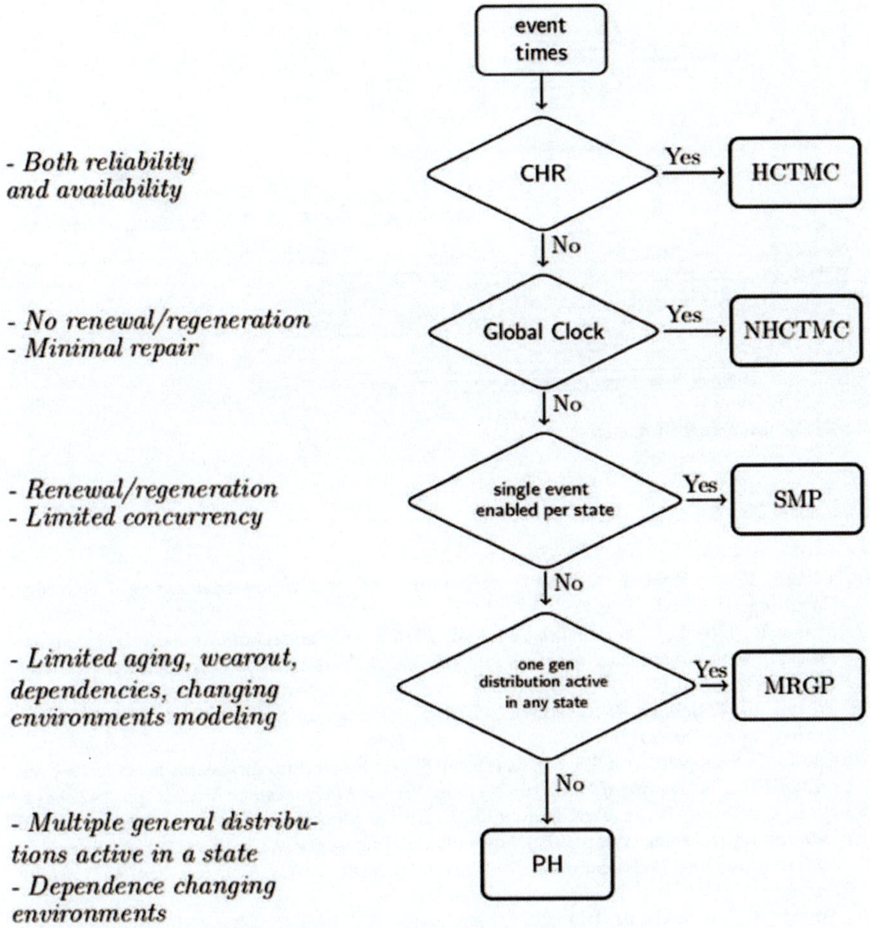

Fig. 22.19 Flow chart comparing the modeling power of the different state space model types [1]

models need to be further emphasized and aided. Tighter connection between data-driven and model-driven methods on the one hand, and combining simulative solution with analytic–numeric solution on the other hand, is desired. Validated models need to be maintained throughout the life of a system so that they can be used for tuning at operational time as well. Besides system-oriented measures such as reliability and availability, user-perceived measures need to be explored [34–36]. Uncertainty in model parameters, so-called epistemic uncertainty, as opposed to aleatory uncertainty already incorporated in the models discussed here, needs to be accounted for in a high-fidelity assessment of reliability and availability [37]. For further discussion on these topics, see [1].

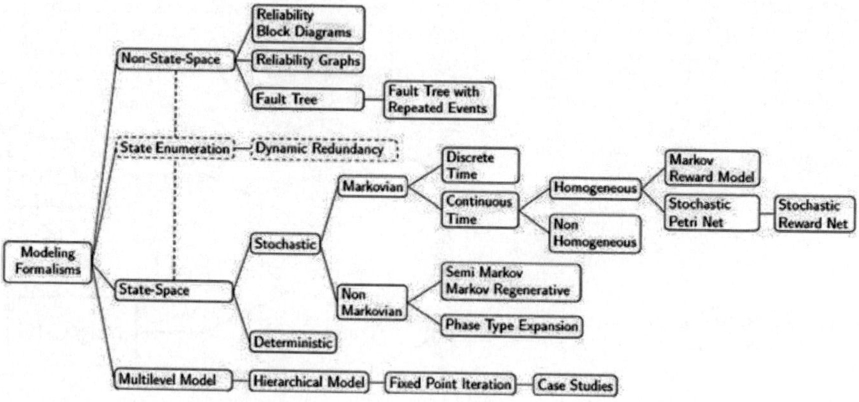

Fig. 22.20 Modeling formalisms

References

1. Trivedi, K., & Bobbio, A. (2017). *Reliability and availability engineering*. Cambridge: Cambridge University Press.
2. Sahner, R., Trivedi, K., & Puliafito, A. (1996). *Performance and reliability analysis of computer systems: An example-based approach using the SHARPE software package*. Kluwer Academic Publishers.
3. Trivedi, K., & Sahner, R. A. (2009). SHARPE at the age of twenty two. *ACM Performance Evaluation Review, 36*(4).
4. Ciardo, G., Muppala, J., & Trivedi, K. (1989). SPNP: Stochastic petri net package. In *Proceedings of Third International Workshop on Petri Nets and Performance Models* (pp. 142–151).
5. Hirel, C., Tuffin, B., & Trivedi, K. (2000). SPNP: Stochastic petri nets. Version 6. In B. Haverkort & H. Bohnenkamp (Eds.), *International Conference on Computer Performance Evaluation: Modelling Techniques and Tools (TOOLS 2000), LNCS 1786* (pp. 354–357). Berlin: Springer.
6. Trivedi, K., Vasireddy, R., Trindade, D., Nathan, S., & Castro, R. (2006). Modeling high availability systems. In *Proceedings of IEEE Pacific Rim International Symposium on Dependable Computing (PRDC)*.
7. Sebastio, S., Trivedi, K., Wang, D., & Yin, X. (2014). Fast computation of bounds for two-terminal network reliability. *European Journal of Operational Research, 238*(3), 810–823.
8. Ramesh, V., Twigg, D., Sandadi, U., Sharma, T., Trivedi, K., & Somani, A. (1999). An integrated reliability modeling environment. *Reliability Engineering and System Safety, 65*, 65–75.
9. Zang, X., Sun, H., & Trivedi, K. (2000). *A BDD-based algorithm for reliability graph analysis*. Department of Electrical & Computer Engineering: Duke University, Technical Report.
10. Soh, S., & Rai, S. (2005). An efficient cutset approach for evaluating communication-network reliability with heterogeneous link-capacities. *IEEE Transactions on Reliability, 54*(1), 133–144.
11. Malhotra, M., & Trivedi, K. (1994). Power-hierarchy among dependability model types. *IEEE Transactions on Reliability, R-43*, 493–502.
12. Zang, X., Wang, D., Sun, H., & Trivedi, K. (2003). A BDD-based algorithm for analysis of multistate systems with multistate components. *IEEE Transactions on Computers, 52*(12), 1608–1618.
13. Zang, X., Sun, H., & Trivedi, K. (1999). A BDD-based algorithm for reliability analysis of phased mission systems. *IEEE Transactions On Reliability, 48*(1), 50–60.

14. Merle, G., Roussel, J., Lesage, J., & Bobbio, A. (2010). Probabilistic algebraic analysis of fault trees with priority dynamic gates and repeated events. *IEEE Transactions on Reliability, 59*(1), 250–261.
15. Trivedi, K. (2001). *Probability & statistics with reliability, queueing & computer science applications* (2nd ed.). Wiley.
16. Trivedi, K., Wang, D., Hunt, J., Rindos, A., Smith, W. E., & Vashaw, B. (2008). Availability modeling of SIP protocol on IBM © Websphere ©. In *Proceedings of Pacific Rim International Symposium on Dependable Computing (PRDC)* (pp. 323–330).
17. Fricks, R., Bobbio, A., & Trivedi, K. (2016). Reliability models of chronic kidney disease. In *Proceedings IEEE Annual Reliability and Maintainability Symposium* (pp. 1–6).
18. United States Renal Data System. (2014)."2014 annual data report: An overview of the epidemiology of kidney disease in the United States. National Institutes of Health—National Institute of Diabetes and Digestive and Kidney Diseases, Tech. Rep., 2014.
19. Stewart, W. (1994). *Introduction to the numerical solution of markov chains*. Princeton University Press.
20. Reibman, A., & Trivedi, K. (1988). Numerical transient analysis of Markov models. *Computers and Operations Research, 15*:19–36.
21. Reibman, A., Smith, R., & Trivedi, K. (1989). Markov and Markov reward model transient analysis: An overview of numerical approaches. *European Journal of Operational Research, 40*, 257–267.
22. Bobbio, A., & Premoli, A. (1982). Fast algorithm for unavailability and sensitivity analysis of series-parallel systems. *IEEE Transaction on Reliability, R-31*, 359–361.
23. Blake, J., Reibman, A., & Trivedi, K. (1988). Sensitivity analysis of reliability and performability measures for multiprocessor systems. *ACM SIGMETRICS Performance Evaluation Review, 16*(1), 177–186.
24. Matos, R., Maciel, P., Machida, F., Kim, D. S., & Trivedi, K. (2012). Sensitivity analysis of server virtualized system availability. *IEEE Transactions on Reliability, 61*, 994–1006.
25. Bobbio, A. (1990). System modelling with petri nets. In A. Colombo & A. de Bustamante (Eds.), *System reliability assessment* (pp. 103–143). Kluwer Academic P.G.
26. Ciardo, G., Muppala, J., & Trivedi, K. (1991). On the solution of GSPN reward models. *Performance Evaluation, 12*, 237–253.
27. Ciardo, G., Blakemore, A., Chimento, P., Muppala, J., & Trivedi, K. (1993). Automated generation and analysis of Markov reward models using stochastic reward nets. In C. Meyer & R. Plemmons (Eds.), *Linear algebra, markov chains, and queueing models, The IMA Vol in mathematics and its applications* (Vol. 48, pp. 145–191). Berlin: Springer.
28. Ghosh, R., Longo, F., Frattini, L., Russo, S., & Trivedi, K. (2014). Scalable analytics for IaaS cloud availability. *IEEE Transactions on Cloud Computing*.
29. Ciardo, G., & Trivedi, K. (1993). A decomposition approach for stochastic reward net models. *Performance Evaluation, 18*, 37–59.
30. Mainkar, V., & Trivedi, K. (1996). Sufficient conditions for existence of a fixed point in stochastic reward net-based iterative models. *IEEE Transactions on Software Engineering, 22*(9), 640–653.
31. Sukhwani, H., Bobbio, A., & Trivedi, K. (2015). Largeness avoidance in availability modeling using hierarchical and fixed-point iterative techniques. *International Journal of Performability Engineering, 11*(4), 305–319.
32. Ghosh, R., Longo, F., Naik, V., & Trivedi, K. (2013). Modeling and performance analysis of large scale IaaS clouds. *Future Generation Computer Systems, 29*(5), 1216–1234.
33. Ghosh, R., Longo, F., Xia, R., Naik, V., & Trivedi, K. (2014). Stochastic model driven capacity planning for an infrastructure-as-a-service cloud. *IEEE Transactions Services Computing, 7*(4), 667–680.
34. Trivedi, K., Wang, D., & Hunt, J. (2010). Computing the number of calls dropped due to failures. *ISSRE*, 11–20.
35. Mondal, S., Yin, X., Muppala, J., Alonso Lopez, J., & Trivedi, K. (2015). Defects per million computation in service-oriented environments. *IEEE Transactions Services Computing, 8*(1), 32–46.

36. Wang, D., & Trivedi, K. (2009). Modeling user-perceived service reliability based on user-behavior graphs. *International Journal of Reliability, Quality and Safety Engineering, 16*(4), 1–27.
37. Mishra, K., & Trivedi, K. (2013). Closed-form approach for epistemic uncertainty propagation in analytic models. In *Stochastic reliability and maintenance modeling* (Vol. 9, pp. 315–332). Springer Series in Reliability Engineering.

Dr. Kishor Trivedi is a Professor of Electrical and Computer Engineering and Computer Science at Duke University and a Life Fellow of IEEE. He is the author of a well-known text entitled, *Probability and Statistics with Reliability, Queuing and Computer Science Applications.* His latest book, co-authored with Andrea Bobbio, *Reliability and Availability Engineering,* is published by Cambridge University Press in 2017. He has supervised 48 Ph.D. dissertations and has an h-index 100 as per Google scholar. As a recipient of IEEE Computer Society's Technical Achievement Award for research on Software Aging and Rejuvenation, he has worked closely with industry in carrying out reliability/availability analysis and in the development and dissemination of modeling software packages such as HARP (with NASA), SAVE (with IBM), SHARPE, SPNP, and Boeing's IRAP.

Dr. Andrea Bobbio is a Professor of Computer Science at Università del Piemonte Orientale in Italy and Senior Member of IEEE. His academic and professional activity has been mainly in the area of reliability engineering and system reliability. He contributed to the study of heterogeneous modeling techniques for dependable systems, ranging from non-state-space techniques to Bayesian belief networks, to state-space-based techniques, and fluid models. He has visited several important institutions and is the author of 200 papers in international journals and conferences. He is co-author of the book, *Reliability and Availability Engineering*, published by Cambridge University Press in 2017.

Chapter 23
WIB (Which-Is-Better) Problems in Maintenance Reliability Policies

Satoshi Mizutani, Xufeng Zhao, and Toshio Nakagawa

Abstract There have been many studies of maintenance policies in reliability theory, so that we have to select better policies that are suitable for objective systems in actual fields, such as age replacement, periodic replacement, replacement first and last, replacement overtime, and standby or parallel system, appeared in research areas. This chapter compares systematically maintenance policies and shows how to select one from the point of cost theoretically. The expected cost rates of maintenance policies and optimal solutions to minimize them are given, and their optimal policies such as replacement time T^*, number N^* of working cycle, and number K^* of failures are obtained. Furthermore, we discuss comparisons of optimal policies to show which is better analytical and numerically. These techniques and tools used in this chapter would be useful for reliability engineers who are worried about how to adopt better maintenance policies.

Keywords Which-is-better problems · Age replacement · Replacement first and last · Replacement overtime · Standby or parallel system

23.1 Introduction

Most units deteriorate with age and use, and eventually fail; however, we cannot predict the exact time of failures as the units are used in random environments, so that it becomes an important problem to make maintenance and replacement plans for such units preventively. There have been many studies of maintenance policies in reliability theory [1–4]. A very classical replacement policy is based on its age, which

S. Mizutani (✉) · T. Nakagawa
Aichi Institute of Technology, Toyota 470-0392, Japan
e-mail: mztn@aitech.ac.jp

T. Nakagawa
e-mail: toshi-nakagawa@aitech.ac.jp

X. Zhao
Nanjing University of Aeronautics and Astronautics, Nanjing 211106, China

K. B. Misra (ed.), *Handbook of Advanced Performability Engineering*,
https://doi.org/10.1007/978-3-030-55732-4_23

is called *age replacement*, and has been extended theoretically by many researchers. Another classical replacement policy is with minimal repairs at failures, which is called *periodic replacement*. That is, units are replaced periodically at planned times kT $(k = 1, 2, \ldots)$, and minimal repairs are done at failures. However, it becomes a very difficult problem to select the right maintenance policy according to real phenomena, and there have not been research works done for this yet. In recent years, we have been studying several comparisons of maintenance policies with analytical and numerical approaches. In this chapter, we summarize concisely our results titled on *Which-Is-Better* (WIB) problems for maintenance policies.

When the unit is replaced at random times, the maintenance is called as *random maintenance* [5–8]. For example, the unit is replaced at a completion time of working cycles because replacement done during the work may cause of wastefulness or delay [9–13]. Nakagawa and Zhao considered replacement first and last policies when there are two triggers of replacement [8, 14–16]. *Replacement first* means that the unit is replaced at events such as its failure or maintenance times, whichever occur first, and *replacement last* means that it is replaced at the events before failure, whichever occur last. In addition, *replacement overtime* means the unit repeats random cycles such as working times and is replaced at a completion time of the first cycle over a planned time [15, 17, 18].

For the above replacement models, it has been shown that it is an interesting problem to determine which policies are better than others. In our research, we have compared several replacement policies such as replacement first, last, and overtime from the points of cost and performance [19–25]. From the above studies, Sect. 23.2 gives comparisons of random age replacement policies such as replacement first, last, and overtime [5, 20]. Section 23.3 gives comparisons of random periodic replacement policies, when the unit undergoes minimal repair at failures [5, 20]. Section 23.4 gives comparisons of cycle N or failure K [3, 5], when the unit is replaced at the Nth $(N = 1, 2, \ldots)$ working cycle and at the Kth $(K = 1, 2, \ldots)$ failure. Numerical examples are given for optimal N^* and K^* that minimize the expected cost rates. Section 23.5 gives comparisons of replacement first, last, or overtime with the Nth working cycle and the Kth failure. Numerical examples are given for optimal (K_F^*, N_F^*) and (K_L^*, N_L^*) [5, 25]. In Sect. 23.6, we obtain reliabilities, mean times to failure and failure rates of standby and parallel systems, and compare them analytically and numerically [21].

For our *Which-Is-Better* (WIB) problems, the following assumptions are given:

(i) Let X be a random variable of the failure time of an operating unit that has a general distribution $F(t) \equiv \Pr\{X \le t\}$ with finite mean $\mu \equiv \int_0^\infty \bar{F}(t)\mathrm{d}t$, density function $f(t) \equiv \mathrm{d}F(t)/\mathrm{d}t$, and failure rate $h(t) \equiv f(t)/\bar{F}(t)$, where $\bar{F}(t) \equiv 1 - F(t)$. It is assumed that $h(t)$ increases strictly with t from $h(0) = 0$ to $h(\infty) \equiv \lim_{t\to\infty} h(t) = \infty$.

(ii) The unit operates for random working cycles Y_i $(i = 1, 2, \ldots)$. It is assumed that Y_i are independent random variables and have an identical distribution $G(t) \equiv \Pr\{Y_i \le t\}$ with finite mean $1/\theta \equiv \int_0^\infty \bar{G}(t)\mathrm{d}t$, where $\bar{G}(t) \equiv 1 - G(t)$. The

j-fold Stieltjes convolution of $G(t)$ is $G^{(j)}(t) \equiv \Pr\{Y_1 + Y_2 + \cdots + Y_j \le t\}$ $(j = 1, 2, \ldots)$, $G^{(0)}(t) \equiv 1$ for $t \ge 0$, and $M(t) \equiv \sum_{j=1}^{\infty} G^{(j)}(t)$.

(iii) When the unit undergoes minimal repairs at failures, the time for repair is negligible and failure rate remains undisturbed after repairs. The probability that j failure occurs exactly in the time interval $[0, t]$ is $p_j(t) \equiv \left[H(t)^j / j!\right] e^{-H(t)}$ $(j = 0, 1, 2, \ldots)$, where $H(t) \equiv \int_0^t h(u)du$. Let $P_j(t) \equiv \sum_{i=j}^{\infty} p_i(t)$, and $\bar{P}_j(t) \equiv 1 - P_j(t) = \sum_{i=0}^{j-1} p_i(t)$, where note that $\sum_{j=0}^{-1} \equiv 0$. Then, $P_0(t) = P_j(\infty) = \bar{P}_\infty(t) = \bar{P}_j(0) = 1$ and $P_\infty(t) = P_j(0) = \bar{P}_0(t) = \bar{P}_j(\infty) = 0$.

23.2 Random Age Replacement

23.2.1 Replacement First

Suppose that the unit is replaced at time T $(0 < T \le \infty)$ or at a random working time Y, whichever occurs first. Then, the probability that the unit is replaced at time T is $\bar{G}(T)\bar{F}(T)$, the probability that it is replaced at time Y is $\int_0^T \bar{F}(t)dG(t)$, and the probability that it is replaced at failure is $\int_0^T \bar{G}(t)dF(t)$. The mean time to replacement is

$$T\bar{G}(T)\bar{F}(T) + \int_0^T t\bar{F}(t)\,dG(t) + \int_0^T t\bar{G}(t)\,dF(t) = \int_0^T \bar{G}(t)\bar{F}(t)\,dt.$$

Therefore, the expected cost rate is

$$C_{AF}(T) = \frac{c_P + (c_F - c_P)\int_0^T \bar{G}(t)dF(t)}{\int_0^T \bar{G}(t)\bar{F}(t)dt}, \tag{23.1}$$

where c_F = replacement cost at failure, and c_P = replacement cost at T or at time Y, i.e., c_P is a cost for preventive replacement with $c_P < c_F$. We find optimal T_{AF}^* to minimize $C_{AF}(T)$. Differentiating $C_{AF}(T)$ with respect to T and putting it equal to zero,

$$h(T)\int_0^T \bar{G}(t)\bar{F}(t)\,dt - \int_0^T \bar{G}(t)\,dF(t) = \frac{c_P}{c_F - c_P} \tag{23.2}$$

Letting $L_{AF}(T)$ be the left-hand side of (23.2), $L_{AF}(T)$ increases strictly with T from 0 to ∞. Thus, there exists an optimal T_{AF}^* $\left(0 < T_{AF}^* < \infty\right)$ that satisfies (23.2), and the resulting cost rate is

$$C_{AF}\left(T_{AF}^*\right) = (c_F - c_P)h\left(T_{AF}^*\right) \tag{23.3}$$

23.2.2 Replacement Last

Suppose that the unit is replaced at time T $(0 \le T < \infty)$ or at a random working time Y, whichever occurs last. Then, the probability that the unit is replaced at time T is $\bar{F}(T)G(T)$, the probability that it is replaced at completion of a work is $\int_T^\infty \bar{F}(t)\,\mathrm{d}G(t)$, and the probability that it is replaced at failure is $F(T) + \int_T^\infty \bar{G}(t)\,\mathrm{d}F(t)$. The mean time to replacement is

$$T\bar{F}(T)G(T)+\int_T^\infty t\bar{F}(t)\,\mathrm{d}G(t)+\int_0^T t\mathrm{d}F(t)+\int_T^\infty t\bar{G}(t)\,\mathrm{d}F(t)$$
$$=\int_0^T \bar{F}(t)\,\mathrm{d}t+\int_T^\infty \bar{G}(t)\bar{F}(t)\,\mathrm{d}t.$$

Therefore, the expected cost rate is

$$C_{\mathrm{AL}}(T)=\frac{c_{\mathrm{F}}-(c_{\mathrm{F}}-c_{\mathrm{P}})\int_T^\infty G(t)\,\mathrm{d}F(t)}{\int_0^T \bar{F}(t)\,\mathrm{d}t+\int_T^\infty \bar{G}(t)\bar{F}(t)\mathrm{d}t} \tag{23.4}$$

We find optimal T_{AL}^* to minimize $C_{\mathrm{AL}}(T)$. Differentiating $C_{\mathrm{AL}}(T)$ with respect to T and putting it equal to zero,

$$h(T)\left[\int_0^T \bar{F}(t)\,\mathrm{d}t+\int_T^\infty \bar{G}(t)\bar{F}(t)\,\mathrm{d}t\right]-\left[1-\int_T^\infty G(t)\,\mathrm{d}F(t)\right]=\frac{c_{\mathrm{P}}}{c_{\mathrm{F}}-c_{\mathrm{P}}} \tag{23.5}$$

Letting $L_{\mathrm{AL}}(T)$ be the left-hand side of (23.5), $L_{\mathrm{AL}}(T)$ increases strictly with T from $-\int_0^\infty \bar{G}(t)\,\mathrm{d}F(t) < 0$ to ∞. Thus, there exists an optimal T_{AL}^* $\left(0 \le T_{\mathrm{AL}}^* < \infty\right)$ that satisfies (23.5), and the resulting cost rate is

$$C_{\mathrm{AL}}\left(T_{\mathrm{AL}}^*\right)=(c_{\mathrm{F}}-c_{\mathrm{P}})h\left(T_{\mathrm{AL}}^*\right) \tag{23.6}$$

23.2.3 Replacement Overtime

Suppose that the unit is replaced at the first completion of working cycles over time T $(0 < T \le \infty)$. Then, the probability that the unit is replaced at the first completion of working cycles over time T is

$$\sum_{j=0}^{\infty}\int_{0}^{T}\left[\int_{T-t}^{\infty}\bar{F}(t+u)\,\mathrm{d}G(u)\right]\mathrm{d}G^{(j)}(t),$$

the probability that it is replaced at failure before time T is

$$\sum_{j=0}^{\infty}\int_{0}^{T}[G^{(j)}(t)-G^{(j+1)}(t)]\mathrm{d}F(t)=F(T),$$

and the probability that it is replaced at failure after time T is

$$\sum_{j=0}^{\infty}\int_{0}^{T}\left\{\int_{T-t}^{\infty}[F(t+u)-F(T)\mathrm{d}G(u)]\right\}\mathrm{d}G^{(j)}(t).$$

The mean time to replacement is

$$\sum_{j=0}^{\infty}\int_{0}^{T}\left[\int_{T-t}^{\infty}(t+u)\bar{F}(t+u)\,\mathrm{d}G(u)\right]\mathrm{d}G^{(j)}(t)+\sum_{j=0}^{\infty}\int_{0}^{T}\left\{\int_{T-t}^{\infty}\left[\int_{T}^{t+u}v\,\mathrm{d}F(v)\right]\mathrm{d}G(u)\right\}$$

$$\mathrm{d}G^{(j)}(t)+\int_{0}^{T}t\,\mathrm{d}F(t)=\int_{0}^{T}\bar{F}(t)\,\mathrm{d}t+\sum_{j=0}^{\infty}\int_{0}^{T}\int_{T}^{\infty}[\bar{G}(t-u)\bar{F}(u)\mathrm{d}u]\mathrm{d}G^{(j)}(t).$$

Therefore, the expected cost rate is

$$C_{\mathrm{AO}}(T)=\frac{c_{\mathrm{F}}-(c_{\mathrm{F}}-c_{\mathrm{P}})\sum_{j=0}^{\infty}\int_{0}^{T}\left[\int_{T-t}^{\infty}\bar{F}(t+u)\,\mathrm{d}G(u)\right]\mathrm{d}G^{(j)}(t)}{\int_{0}^{T}\bar{F}(t)\,\mathrm{d}t+\sum_{j=0}^{\infty}\int_{T}^{\infty}\left[\bar{G}(t-u)\bar{F}(u)\mathrm{d}u\right]\mathrm{d}G^{(j)}(t)}\qquad(23.7)$$

We find optimal T^{*}_{AO} to minimize $C_{\mathrm{AO}}(T)$ when $G(t)=1-\mathrm{e}^{-\theta t}$. Differentiating $C_{\mathrm{AO}}(T)$ with respect to T and putting it equal to zero,

$$Q_1(T;\theta)\int_{0}^{T}\bar{F}(t)\,\mathrm{d}t-F(T)=\frac{c_{\mathrm{P}}}{c_{\mathrm{F}}-c_{\mathrm{P}}},$$

where

$$Q_1(T;\theta)\equiv\frac{\int_{T}^{\infty}\mathrm{e}^{-\theta t}\mathrm{d}F(t)}{\int_{T}^{\infty}\mathrm{e}^{-\theta t}\bar{F}(t)\,\mathrm{d}t}\geq h(T)\qquad(23.8)$$

Note that $Q_1(T;\theta)$ is greater than $h(T)$ and increases strictly with T to $h(\infty)$. Thus, there exists an optimal T^{*}_{AO} $\left(0\leq T^{*}_{\mathrm{AO}}<\infty\right)$ that satisfies (23.8), and the resulting cost rate is

$$C_{AO}\left(T_{AO}^*\right) = (c_F - c_R)Q_1\left(T_{AO}^*; \theta\right) \tag{23.9}$$

23.2.4 Comparisons of T_{AF}^*, T_{AL}^*, and T_{AO}^*

Compare the left hands of (23.2) and (23.5). Letting

$$D_A(T) \equiv L_{AL}(T) - L_{AF}(T) = \int_0^T G(t)\bar{F}(t)[h(T) - h(t)]\,dt - \int_T^\infty \bar{G}(t)\bar{F}(t)[h(t) - h(T)]\,dt,$$

we easily have

$$D_A(0) \equiv \lim_{T\to 0} D_A(T) = -\int_0^\infty \bar{G}(t)\,dF(t) < 0, \quad D_A(\infty) \equiv \lim_{T\to\infty} D_A(T) = \infty,$$

$$D_A'(T) = h'(T)\left[\int_0^T G(t)\bar{F}(t)dt + \int_T^\infty \bar{G}(t)\bar{F}(t)dt\right] > 0.$$

Thus, there exists a finite and unique T_A^* $\left(0 < T_A^* < \infty\right)$ that satisfies $D_A(T) = 0$. Therefore, from (23.3) to (23.6), we have the following results:

(i) If $L_{AF}\left(T_A^*\right) \geq c_P/(c_F - c_P)$, then $T_{AF}^* \leq T_{AL}^*$ and replacement first is better than replacement last.
(ii) If $L_{AF}\left(T_A^*\right) < c_P/(c_F - c_P)$, then $T_{AF}^* > T_{AL}^*$ and replacement last is better than replacement first.

Next, we compare the left hands of (23.2) and (23.8) when $G(t) = 1 - e^{-\theta t}$. Then, we have

$$Q_1(T;\theta)\int_0^T \bar{F}(t)dt - F(T) - h(T)\int_0^T e^{-\theta t}\bar{F}(t)\,dt + \int_0^T e^{-\theta t}dF(t)$$

$$= \left[Q_1(T;\theta) - h(T)\right]\int_0^T \bar{F}(t)\,dt + \int_0^T \left(1 - e^{-\theta t}\right)\bar{F}(t)[h(T) - h(t)]\,dt > 0,$$

and hence, $T_{AO}^* < T_{AF}^*$. Therefore, from (23.3) to (23.9), we have the following results:

(i) If $Q_1(T^*_{\mathrm{AO}};\theta) \geq h(T^*_{\mathrm{AF}})$, then replacement first is better than replacement overtime.
(ii) If $Q_1(T^*_{\mathrm{AO}};\theta) < h(T^*_{\mathrm{AF}})$, then replacement overtime is better than replacement first.

Furthermore, we compare the left hands of (23.5) and (23.8) when $G(t) = 1-\mathrm{e}^{-\theta t}$. Then, we have

$$Q_1(T;\theta)\int_0^T \bar{F}(t)\mathrm{d}t - F(T) - h(T)\left[\int_0^T \bar{F}(t)\,\mathrm{d}t + \int_T^\infty \mathrm{e}^{-\lambda t}\bar{F}(t)\,\mathrm{d}t\right]$$
$$+1-\int_T^\infty \left(1-\mathrm{e}^{-\theta t}\right)\mathrm{d}F(t) = [Q_1(T;\theta)-h(T)]\int_0^T \bar{F}(t)\,\mathrm{d}t + \int_T^\infty \mathrm{e}^{-\theta t}\bar{F}(t)[h(t)-h(T)]\,\mathrm{d}t > 0,$$

and hence, $T^*_{\mathrm{AO}} < T^*_{\mathrm{AL}}$. Therefore, from (23.6) to (23.9), we have the following results:

(i) If $Q_1(T^*_{\mathrm{AO}};\theta) \geq h(T^*_{\mathrm{AL}})$, then replacement first is better than replacement overtime.
(ii) If $Q_1(T^*_{\mathrm{AO}};\theta) < h(T^*_{\mathrm{AL}})$, then replacement overtime is better than replacement first.

23.3 Random Periodic Replacement

23.3.1 Replacement First

Suppose that the unit operates for a random working time Y with $G(t) \equiv \Pr\{Y \leq t\}$ and undergoes minimal repair at each failure. The unit is replaced at time T $(0 \leq T \leq \infty)$ or at time Y, whichever occurs first. Then, the expected number of failures until replacement is

$$H(T)\bar{G}(T) + \int_0^T H(t)\,\mathrm{d}G(t) = \int_0^T \bar{G}(t)h(t)\,\mathrm{d}t,$$

and the mean time to replacement is

$$T\bar{G}(T) + \int_0^T t\,\mathrm{d}G(t) = \int_0^T \bar{G}(t)\,\mathrm{d}t.$$

Therefore, the expected cost rate is

$$C_{\mathrm{RF}}(T)=\frac{c_{\mathrm{M}}\int_0^T \bar{G}(t)h(t)\,\mathrm{d}t+c_{\mathrm{P}}}{\int_0^T \bar{G}(t)\,\mathrm{d}t},\tag{23.10}$$

where c_{M} = minimal repair cost at each failure. We find optimal T^*_{RF} to minimize $C_{\mathrm{RF}}(T)$. Differentiating $C_{\mathrm{RF}}(T)$ with respect to T and putting it equal to zero,

$$\int_0^T \bar{G}(t)[h(T)-h(t)]\mathrm{d}t=\frac{c_{\mathrm{P}}}{c_{\mathrm{M}}}.\tag{23.11}$$

Letting $L_{\mathrm{RF}}(T)$ be the left-hand side of (23.11), $L_{\mathrm{RF}}(T)$ increases strictly with T from 0 to ∞. Thus, there exists an optimal T^*_{RF} $(0<T^*_{\mathrm{RF}}<\infty)$ that satisfies (23.11), and the resulting cost rate is

$$C_{\mathrm{RF}}(T^*_{\mathrm{RF}})=c_{\mathrm{M}}h(T^*_{\mathrm{F}}).\tag{23.12}$$

23.3.2 Replacement Last

Suppose that the unit is replaced at time T $(0\le T\le\infty)$ or at time Y, whichever occurs last. Then, the expected number of failures until replacement is

$$H(T)G(T)+\int_T^\infty H(t)\,\mathrm{d}G(t)=H(T)+\int_T^\infty \bar{G}(t)h(t)\,\mathrm{d}t,$$

and the mean time to replacement is

$$TG(T)+\int_T^\infty t\,\mathrm{d}G(t)=T+\int_T^\infty \bar{G}(t)\,\mathrm{d}t.$$

Therefore, the expected cost rate is

$$C_{\mathrm{RL}}(T)=\frac{c_M\left[H(T)+\int_T^\infty \bar{G}(t)h(t)\,\mathrm{d}t\right]+c_P}{T+\int_T^\infty \bar{G}(t)\,\mathrm{d}t}.\tag{23.13}$$

We find optimal T^*_{RL} to minimize $C_{\mathrm{RL}}(T)$. Differentiating $C_{\mathrm{RL}}(T)$ with respect to T and putting it equal to zero,

$$\int_0^T[h(T)-h(t)]\mathrm{d}t-\int_T^\infty \bar{G}(t)[h(T)-h(t)]\mathrm{d}t=\frac{c_{\mathrm{P}}}{c_{\mathrm{M}}}.\tag{23.14}$$

Letting $L_{\mathrm{RL}}(T)$ be the left-hand side of (23.14), $L_{\mathrm{RL}}(T)$ increases strictly with T from $L_{\mathrm{RL}}(0) = -\int_T^\infty \bar{G}(t)h(t)\,\mathrm{d}t < 0$ to ∞. Thus, there exists an optimal T^*_{RL} $\left(0 < T^*_{\mathrm{RL}} < \infty\right)$ that satisfies (23.14), and the resulting cost rate is

$$C_{\mathrm{RL}}\left(T^*_{\mathrm{RL}}\right) = c_{\mathrm{M}} h\left(T^*_{\mathrm{RL}}\right). \tag{23.15}$$

23.3.3 Replacement Overtime

Suppose that the unit is replaced at the first completion of working cycles over time T $(0 < T \le \infty)$. Then, the expected number of failures until replacement is

$$\sum_{j=0}^{\infty} \int_0^T \left[\int_T^\infty H(u)\,\mathrm{d}G(u-t)\right] \mathrm{d}G^{(j)}(t) H(T) + \int_T^\infty \bar{G}(t)h(t)\,\mathrm{d}t$$
$$+ \int_0^T \left[\int_T^\infty \bar{G}(u-t)h(u)\,\mathrm{d}u\right] \mathrm{d}M(t),$$

and the mean time to replacement is

$$\sum_{j=0}^{\infty} \int_0^T \left[\int_T^\infty u\,\mathrm{d}G(u-t)\right] \mathrm{d}G^{(j)}(t) = T + \int_T^\infty \bar{G}(t)\,\mathrm{d}t + \int_0^T \left[\int_T^\infty \bar{G}(u-t)\,\mathrm{d}u\right] \mathrm{d}M(t).$$

Therefore, the expected cost rate is

$$C_{\mathrm{RO}}(T) = \frac{c_{\mathrm{M}}\left\{H(T) + \int_T^\infty \bar{G}(t)h(t)\,\mathrm{d}t + \int_0^T \left[\int_T^\infty \bar{G}(u-t)h(u)\,\mathrm{d}u\right]\mathrm{d}M(t)\right\} + c_{\mathrm{P}}}{T + \int_T^\infty \bar{G}(t)\,\mathrm{d}t + \int_0^T \left[\int_T^\infty \bar{G}(u-t)\,\mathrm{d}u\right]\mathrm{d}M(t)}. \tag{23.16}$$

We find optimal T^*_{RO} to minimize $C_{\mathrm{RO}}(T)$. Differentiating $C_{\mathrm{RO}}(T)$ with respect to T and putting it equal to zero,

$$\int_0^\infty \theta\bar{G}(t)\left(Th(T+t) - H(T) + \int_T^\infty \bar{G}(u)[h(T+t) - h(u)]\,\mathrm{d}u\right.$$
$$\left. + \int_0^T \left\{\int_T^\infty \bar{G}(u-x)[h(T+t) - h(u)]\,\mathrm{d}u\right\}\mathrm{d}M(x)\right)\mathrm{d}t = \frac{c_{\mathrm{P}}}{c_{\mathrm{M}}}, \tag{23.17}$$

whose left-hand side increases strictly with T from 0 to ∞. Thus, there exists an optimal T_{RO}^* $(0 < T_{RO}^* < \infty)$ that satisfies (23.17), and the resulting cost rate is

$$C_{RO}(T_{RO}^*) = c_M \int_0^\infty \theta \bar{G}(t) h(T_{RO}^* + t)\, dt. \tag{23.18}$$

23.3.4 Comparisons of T_{RF}^*, T_{RL}^*, and T_{RO}^*

Comparing with the left hand of (23.11) and (23.14),

$$\begin{aligned} D_R(T) \equiv L_{RL}(T) - L_{RF}(T) &= \int_0^T G(t)[h(T) - h(t)]\, dt \\ &- \int_T^\infty \bar{G}(t)[h(t) - h(T)]\, dt, \end{aligned} \tag{23.19}$$

which increases strictly from $-\int_0^\infty \bar{G}(t)h(t)\,dt < 0$ to ∞. Thus, there exists a finite and unique T_P^* $(0 < T_P^* < \infty)$ that satisfies $D_R(T) = 0$. Therefore, from (23.12) to (23.15), we have the following results:

(i) If $L_{RF}(T_P^*) \geq c_P/c_M$, then $T_{RF}^* \leq T_{RL}^*$ and replacement first is better than replacement last.
(ii) If $L_{RF}(T_A^*) < c_P/c_M$, then $T_{RF}^* > T_{RL}^*$ and replacement last is better than replacement first.

Next, we compare the left hands of (23.11) and (23.17) when $G(t) = 1 - e^{-\theta t}$. Then, we have

$$\begin{aligned} &T \int_0^\infty \theta e^{-\theta t} h(t+T)\, dt - H(T) - \int_0^T e^{-\theta t}[h(T) - h(t)]\, dt \\ &= T + \int_0^\infty \theta e^{-\theta t}[h(t+T) - h(T)]\, dt + \int_0^T (1 - e^{-\theta t})[h(T) - h(t)]\, dt > 0, \end{aligned}$$

and hence, $T_{RO}^* < T_{RF}^*$. Thus, we can compare $C_{RO}(T_{RO}^*)$ in (23.18) with $C_{RF}(T_{RF}^*)$ in (23.12), and determine which policy is better. For example, when $H(t) = (\lambda t)^2$, i.e., $h(t) = 2\lambda^2 t$, from (23.18),

$$C_{RO}(T_{RO}^*) = 2c_M\lambda^2\left(T_{RO}^* + \frac{1}{\theta}\right),$$

and from (23.12),

$$C_{\mathrm{RF}}\left(T_{\mathrm{RF}}^*\right) = 2c_{\mathrm{M}}\lambda^2 T_{\mathrm{RF}}^*.$$

Thus, we have the following results:

(i) If $T_{\mathrm{RO}}^* + 1/\theta \geq T_{\mathrm{RF}}^*$, then replacement first is better than replacement overtime.
(ii) If $T_{\mathrm{RO}}^* + 1/\theta < T_{\mathrm{RF}}^*$, then replacement overtime is better than replacement first.

Furthermore, we compare the left hands of (23.14) and (23.17) when $G(t) = 1 - \mathrm{e}^{-\theta t}$. Then, we have

$$T\int_0^\infty \theta \mathrm{e}^{-\theta t} h(t+T)\,\mathrm{d}t - H(T) - \int_0^T [h(T) - h(t)]\mathrm{d}t + \int_T^\infty \mathrm{e}^{-\theta t}[h(t) - h(T)]\,\mathrm{d}t$$
$$= T + \int_0^\infty \theta \mathrm{e}^{-\theta t}[h(t+T) - h(T)]\,\mathrm{d}t + \int_T^\infty \mathrm{e}^{-\theta t}[h(t) - h(T)]\,\mathrm{d}t > 0,$$

and hence, $T_{\mathrm{RO}}^* < T_{\mathrm{RL}}^*$. Thus, we can compare $C_{\mathrm{RO}}\left(T_{\mathrm{RO}}^*\right)$ in (23.18) with $C_{\mathrm{RL}}\left(T_{\mathrm{RL}}^*\right)$ in (23.15), and determine which policy is better. For example, when $h(t) = 2\lambda^2 t$, from (23.15),

$$C_{\mathrm{RL}}\left(T_{\mathrm{RL}}^*\right) = 2c_{\mathrm{M}}\lambda^2 T_{\mathrm{RL}}^*.$$

Thus, we have the following results:

(i) If $T_{\mathrm{RO}}^* + 1/\theta \geq T_{\mathrm{RL}}^*$, then replacement last is better than replacement overtime,
(ii) If $T_{\mathrm{RO}}^* + 1/\theta < T_{\mathrm{RL}}^*$, then replacement overtime is better than replacement last.

23.4 Replacement with Cycle *N* or Failure *K*

23.4.1 Replacement with Cycle N

We consider that the unit operates for a job with random cycles Y_j $(j = 1, 2, \ldots)$ and undergoes minimal repair at each failure, and the unit is replaced at cycle N $(N = 1, 2, \ldots)$ [5]. Then, the expected number of failures until replacement is

$$\int_0^\infty H(t)\,\mathrm{d}G^{(N)}(t) = \int_0^\infty \left[1 - G^{(N)}(t)\right] h(t)\,\mathrm{d}t,$$

and the mean time to replacement is N/θ.

Therefore, the expected cost rate is

$$C(N) = \frac{c_N + c_M \int_0^\infty [1 - G^{(N)}(t)] h(t)\,dt}{N/\theta}, \tag{23.20}$$

where c_N = replacement cost at cycle N. We find optimal N^* to minimize $C(N)$. Forming the inequality $C(N+1) - C(N) \geq 0$,

$$\int_0^\infty [1 - G^{(N)}(t)][Q_2(N) - h(t)]\mathrm{d}t \geq \frac{c_N}{c_M}, \tag{23.21}$$

where for $0 < T \leq \infty$ and $N = 0, 1, 2, \ldots$,

$$Q_2(N;T) \equiv \frac{\int_0^T [G^{(N)}(t) - G^{(N+1)}(t)]h(t)\mathrm{d}t}{\int_0^T [G^{(N)}(t) - G^{(N+1)}(t)]\mathrm{d}t} \leq h(T),$$

$$Q_2(N) \equiv \lim_{T \to \infty} Q_2(N;T) = \theta \int_0^\infty [G^{(N)}(t) - G^{(N+1)}(t)]h(t)\mathrm{d}t.$$

In particular, when $G(t) = 1 - \mathrm{e}^{-\theta t}$,

$$Q_2(N) = \int_0^\infty \frac{\theta(\theta t)^N}{N!}\mathrm{e}^{-\theta t}h(t)\mathrm{d}t,$$

which increases strictly with N to $h(\infty)$. Thus, there exists optimal N^* $(1 \leq N^* < \infty)$ that satisfies (23.21), and the resulting cost rate is

$$c_M Q_2(N^* - 1) < C(N^*) \leq c_M Q_2(N^*). \tag{23.22}$$

23.4.2 Replacement with Failure K

Suppose that the unit is replaced at failure K $(K = 1, 2, \ldots)$ [3]. Then, the expected cost rate is

$$C(K) = \frac{c_K + c_M K}{\int_0^\infty \overline{P}_K(t)\mathrm{d}t}, \tag{23.23}$$

where c_K = replacement cost at failure K. We find optimal K^* to minimize $C(K)$. Forming the inequality $C(K+1) - C(K) \geq 0$,

$$Q_3(K)\int_0^\infty \bar{P}_K(t)\mathrm{d}t - K \geq \tfrac{c_K}{c_M}, \tag{23.24}$$

where for $0 < T \leq \infty$ and $K = 0, 1, 2, \ldots$,

$$Q_3(K;T) \equiv \frac{\int_0^T p_K(t)h(t)\mathrm{d}t}{\int_0^T p_K(t)\mathrm{d}t}, \quad Q_3(K) \equiv \lim_{T\to\infty} Q_3(K;T) = \frac{1}{\int_0^\infty p_K(t)\mathrm{d}t},$$

which increases strictly with K to $h(\infty)$. Thus, because the left-hand side of (23.24) increases strictly with K to ∞, there exists an optimal K^* $(1 \leq K^* < \infty)$ that satisfies (23.24), and the resulting cost rate is

$$c_M Q_3(K^* - 1) < C(K^*) \leq c_M Q_3(K^*). \tag{23.25}$$

23.4.3 Numerical Comparison

It is very difficult to discuss analytically comparisons of optimal N^* and K^*. Thus, we give numerical examples of optimal N^* and K^* when $G(t) = 1 - \mathrm{e}^{-t}$ and $H(t) = (\lambda t)^2$, i.e., $h(t) = 2\lambda^2 t$. In this case,

$$Q_2(N) = \int_0^\infty \frac{t^N}{N!}\mathrm{e}^{-t} 2\lambda^2 t\mathrm{d}t = 2\lambda^2(N+1),$$

$$\int_0^\infty [1 - G^{(N)}(t)]h(t)\,\mathrm{d}t = \sum_{j=0}^{N-1}\int_0^\infty \frac{t^j}{j!}\mathrm{e}^{-t}2\lambda^2 t\mathrm{d}t = \lambda^2 N(N+1),$$

and from (23.21), optimal N^* is given by

$$\lambda^2 N(N+1) \geq \frac{c_N}{c_M},$$

and from (23.20),

$$\frac{C(N^*)}{c_M} = \frac{c_N/c_M + \lambda^2 N^*(N^*+1)}{N^*}.$$

Furthermore, noting that

$$\int_0^\infty p_K(t)\,\mathrm{d}t = \int_0^\infty \frac{(\lambda t)^{2K}}{K!}\mathrm{e}^{-(\lambda t)^2}\mathrm{d}t = \frac{1}{2\lambda}\frac{\Gamma(K+1/2)}{\Gamma(K+1)},$$

$$\sum_{j=0}^{K-1}\int_0^\infty p_j(t)\mathrm{d}t = \frac{1}{\lambda}\frac{\Gamma(K+1/2)}{\Gamma(K)},$$

optimal K^* is, from (23.24),

$$\frac{2\Gamma(K+1/2)/\Gamma(K)}{\Gamma(K+1/2)/\Gamma(K+1)} - K = K \geq \frac{c_K}{c_M},$$

where $\Gamma(\alpha) \equiv \int_0^\infty x^{\alpha-1}\mathrm{e}^{-x}\mathrm{d}x$ for $\alpha > 0$. Thus, if c_K/c_M is an integer, then $K^* = c_K/c_M$, and from (23.23),

$$\frac{C(K^*)}{c_M} = \frac{c_K/c_M + K^*}{\Gamma(K^*+1/2)/[\lambda\Gamma(K^*)]}.$$

Table 23.1 gives optimal K^*, N^*, $C(K^*)/c_M$ and $C(N^*)/c_M$ for $\lambda = 0.1, 1.0$, and $c_N/c_M = 1, 2, \ldots, 10$. We can see that for $\lambda = 1.0$, $C(K^*)/c_M < C(N^*)/c_M$, that is, replacement with K^* is better than replacement with N^*. On the other hand, for $\lambda = 0.1$, $C(K^*)/c_M > C(N^*)/c_M$ for $K^* = c_N/c_M \leq 5$, and $C(K^*)/c_M < C(N^*)/c_M$ for $K^* \geq 6$. Optimal N^* decreases with λ. The reason would be that when λ is large, interval times of failures become small and we should replace early to avoid the cost of failures; however, λN^* are almost the same for λ.

Table 23.1 Optimal N^*, K^*, $C(K^*)/c_M$ and $C(N^*)/c_M$ when $G(t) = 1 - \mathrm{e}^{-t}$, $H(t) = (\lambda t)^2$, and $c_N = c_K$

$\frac{c_N}{c_M}$	K^*	$\lambda = 0.1$			$\lambda = 1.0$		
		N^*	$\frac{C(K^*)}{c_M}$	$\frac{C(N^*)}{c_M}$	N^*	$\frac{C(K^*)}{c_M}$	$\frac{C(N^*)}{c_M}$
1	1	10	0.226	0.210	1	2.257	3.000
2	2	14	0.301	0.293	1	3.009	4.000
3	3	17	0.361	0.357	2	3.611	4.500
4	4	20	0.413	0.410	2	4.127	5.000
5	5	22	0.459	0.457	2	4.585	5.500
6	6	22	0.500	0.503	2	5.002	6.000
7	7	24	0.539	0.542	3	5.387	6.333
8	8	26	0.575	0.578	3	5.746	6.667
9	9	28	0.608	0.611	3	6.084	7.000
10	10	32	0.640	0.643	3	6.404	7.333

23.5 Replacement First, Last, or Overtime with Cycle *N* and Failure *K*

23.5.1 *Replacement First*

Suppose that the unit is replaced at cycle N $(N = 1, 2, \ldots)$ or at failure K $(K = 1, 2, \ldots)$, whichever occurs first. The probability that the unit is replaced at cycle N is $\int_0^\infty \bar{P}_K(t)\,\mathrm{d}G^{(N)}(t)$, and the probability that it is replaced at failure K is $\int_0^\infty [1 - G^{(N)}]\,\mathrm{d}P_K(t)$. The expected number of failures until replacement is

$$\int_0^\infty H(t)\bar{P}_K(t)\,\mathrm{d}G^{(N)}(t) + \int_0^\infty H(t)[1 - G^{(N)}(t)]\,\mathrm{d}P_K(t)$$
$$= \int_0^\infty [1 - G^{(N)}(t)]\bar{P}_K(t)h(t)\,\mathrm{d}t,$$

and the mean time to replacement is

$$\int_0^\infty t\bar{P}_K(t)\,\mathrm{d}G^{(N)}(t) + \int_0^\infty t[1 - G^{(N)}(t)]\,\mathrm{d}P_K(t)$$
$$= \int_0^\infty [1 - G^{(N)}(t)]\bar{P}_K(t)\,\mathrm{d}t.$$

Therefore, the expected cost rate is

$$C_F(N, K) = \frac{c_K - (c_K - c_N)\int_0^\infty \bar{P}_K(t)\,\mathrm{d}G^{(N)}(t) + c_M\int_0^\infty [1 - G^{(N)}(t)]\bar{P}_K(t)h(t)\,\mathrm{d}t}{\int_0^\infty [1 - G^{(N)}(t)]\bar{P}_K(t)\,\mathrm{d}t}. \quad (23.26)$$

We find optimal N_F^* and K_F^* to minimize $C_F(N, K)$ when $c_K = c_N$ and $G(t) = 1 - \mathrm{e}^{-\theta t}$. At first, we find optimal N_F^* for a fixed K. Forming the inequality $C_F(N + 1, K) - C_F(N, K) \geq 0$,

$$Q_4(N, K)\int_0^\infty [1 - G^{(N)}(t)]\bar{P}_K(t)\,\mathrm{d}t - \int_0^\infty [1 - G^{(N)}(t)]\bar{P}_K(t)h(t)\,\mathrm{d}t \geq \frac{c_N}{c_M}, \quad (23.27)$$

where

$$Q_4(N,K) \equiv \frac{\sum_{j=0}^{K-1} \int_0^\infty (\theta t)^N \mathrm{e}^{-\theta t} p_j(t) h(t)\,\mathrm{d}t}{\sum_{j=0}^{K-1} \int_0^\infty (\theta t)^N \mathrm{e}^{-\theta t} p_j(t)\,\mathrm{d}t},$$

which increases strictly with N from $Q_4(0, K)$ to $h(\infty)$ and increases strictly with K from $Q_4(N, 1)$ to

$$Q_4(N,\infty) = \int_0^\infty \frac{\theta(\theta t)^N}{N!} \mathrm{e}^{-\theta t} h(t)\,\mathrm{d}t = Q_2(N).$$

Thus, because the left-hand side of (23.27) increases strictly with N to ∞, there exists optimal N_F^* $(1 \le N_F^* < \infty)$ that satisfies (23.27), and the resulting cost rate is

$$c_M Q_4(N_F^* - 1, K) < C_F(N_F^*, K) \le c_M Q_4(N_F^*, K) \tag{23.28}$$

Next, we find optimal K_F^* for a fixed N. Forming the inequality $C_F(N, K+1) - C_F(N, K) \ge 0$,

$$Q_5(K,N) \int_0^\infty [1 - G^{(N)}(t)]\bar{P}_K(t)\,\mathrm{d}t - \int_0^\infty [1 - G^{(N)}(t)]\bar{P}_K(t) h(t)\,\mathrm{d}t \ge \frac{c_K}{c_M}, \tag{23.29}$$

where

$$Q_5(K,N) \equiv \frac{\sum_{j=0}^{N-1} \int_0^\infty [(\theta t)^j/j!] \mathrm{e}^{-\theta t} p_K(t) h(t)\,\mathrm{d}t}{\sum_{j=0}^{N-1} \int_0^\infty [(\theta t)^j/j!] \mathrm{e}^{-\theta t} p_K(t)\,\mathrm{d}t},$$

which increases strictly with N from $Q_5(K, 1)$ to $Q_5(K, \infty) = Q_3(K)$, and increases strictly with K from $Q_5(0, N)$ to $h(\infty)$. Thus, because the left-hand side of (23.29) increases strictly with K to ∞, there exists optimal K_F^* $(1 \le K_F^* < \infty)$ that satisfies (23.29), and the resulting cost rate is

$$c_M Q_5(K_F^* - 1, N) < C_F(N, K_F^*) \le c_M Q_5(K_F^*, N). \tag{23.30}$$

23.5.2 Replacement Last

Suppose that the unit is replaced at cycle N $(N = 0, 1, 2, \ldots)$ or at failure K $(K = 0, 1, 2, \ldots)$, whichever occurs last. The probability that the unit is replaced at cycle N is $\int_0^\infty P_K(t)\,\mathrm{d}G^{(N)}(t)$, and the probability that it is replaced at failure K is $\int_0^\infty G^{(N)}(t)\,\mathrm{d}P_K(t)$. The expected number of failures until replacement is

$$\int_0^\infty H(t)P_K(t)\,dG^{(N)}(t) + \int_0^\infty H(t)G^{(N)}(t)\,dP_K(t) = \int_0^\infty \left[1 - G^{(N)}(t)P_K(t)\right]h(t)\,dt,$$

and the mean time to replacement is

$$\int_0^\infty tP_K(t)\,dG^{(N)}(t) + \int_0^\infty t\,G^{(N)}(t)\,dP_K(t) = \int_0^\infty \left[1 - G^{(N)}(t)P_K(t)\right]dt.$$

Therefore, the expected cost rate is

$$C_L(N,K) = \frac{c_K - (c_N - c_K)\int_0^\infty P_K(t)\,dG^{(N)}(t) + c_M\int_0^\infty \left[1 - G^{(N)}(t)P_K(t)\right]h(t)\,dt}{\int_0^\infty \left[1 - G^{(N)}(t)P_K(t)\right]dt}. \quad (23.31)$$

We find optimal N_L^* and K_L^* to minimize $C_L(N,K)$ when $c_K = c_N$ and $G(t) = 1 - e^{-\theta t}$. At first, we find optimal N_L^* for a fixed K. Forming the inequality $C_L(N+1,K) - C_L(N,K) \geq 0$,

$$\tilde{Q}_4(N,K)\int_0^\infty \left[1 - G^{(N)}(t)P_K(t)\right]dt - \int_0^\infty \left[1 - G^{(N)}(t)P_K(t)\right]h(t)\,dt \geq \frac{c_N}{c_M}, \quad (23.32)$$

where

$$\tilde{Q}_4(N,K) \equiv \frac{\sum_{j=K}^\infty \int_0^\infty (\theta t)^N e^{-\theta t} p_j(t)h(t)\,dt}{\sum_{j=K}^\infty \int_0^\infty (\theta t)^N e^{-\theta t} p_j(t)\,dt},$$

which increases strictly with N from $\tilde{Q}_4(0,K)$ to $h(\infty)$, and increases strictly with K from

$$\tilde{Q}_4(N,0) = \int_0^\infty \frac{\theta(\theta t)^N}{N!} e^{-\theta t} h(t)\,dt = Q_2(N)$$

to $h(\infty)$, and $\tilde{Q}_4(N,K) \geq Q_4(N,K)$. Thus, because the left-hand side of (23.32) increases strictly with N to ∞, there exists optimal N_L^* $\left(0 \leq N_L^* < \infty\right)$ that satisfies (23.32), and the resulting cost rate is

$$c_M\tilde{Q}_4(N_L^* - 1, K - 1) < C_L\left(N_L^*, K\right) \leq c_M\tilde{Q}_4\left(N_L^*, K - 1\right). \quad (23.33)$$

Next, we find optimal K_L^* for a fixed N. Forming the inequality $C_L(N, K+1) - C_L(N, K) \geq 0$,

$$\tilde{Q}_5(K, N) \int_0^\infty [1 - G^{(N)}(t) P_K(t)] \mathrm{d}t - \int_0^\infty [1 - G^{(N)}(t) P_K(t)] h(t) \, \mathrm{d}t \geq \frac{c_K}{c_M}, \quad (23.34)$$

where

$$\tilde{Q}_5(K, N) \equiv \frac{\int_0^\infty G^{(N)}(t) p_K(t) h(t) \, \mathrm{d}t}{\int_0^\infty G^{(N)}(t) p_K(t) \, \mathrm{d}t},$$

which increases strictly with K from $\tilde{Q}_5(0, N)$ to $h(\infty)$. Thus, because the left-hand side of (23.34) increases strictly with K to ∞, there exists optimal $K_L^* (0 \leq K_L^* < \infty)$ that satisfies (23.34), and the resulting cost rate is

$$c_M \tilde{Q}_5(K_L^* - 1, N) < C_L(N, K_L^*) \leq c_M \tilde{Q}_5(K_L^*, N). \quad (23.35)$$

23.5.3 Numerical Comparison

We show numerically optimal (K_F^*, N_F^*) and (K_L^*, N_L^*) when $c_N = c_K$, $G(t) = 1 - \mathrm{e}^{-t}$ and $H(t) = (\lambda t)^2$. They are computed by enumeration and comparison of the expected cost rates. Tables 23.2 and 23.3 give (K_F^*, N_F^*), $C_F(K_F^*, N_F^*)/c_M$, (K_L^*, N_L^*), and $C_L(K_L^*, N_L^*)/c_M$ for $\frac{c_N}{c_M} = 1, 2, \ldots, 10$ when $\lambda = 0.1$ and $\lambda = 1.0$, respectively.

Table 23.2 Optimal (K_F^*, N_F^*), $C_F(K_F^*, N_F^*)/c_M$, (K_L^*, N_L^*) and $C_L(K_L^*, N_L^*)/c_M$ when $G(t) = 1 - \mathrm{e}^{-t}$, $H(t) = (\lambda t)^2$, $c_N = c_K$, and $\lambda = 0.1$

$\frac{c_N}{c_M}$	(K_F^*, N_F^*)	(K_L^*, N_L^*)	$\frac{C_F(K_F^*, N_F^*)}{c_M}$	$\frac{C_L(K_L^*, N_L^*)}{c_M}$
1	(3, 11)	(0, 9)	0.208	0.210
2	(4, 15)	(1, 13)	0.291	0.292
3	(5, 19)	(2, 16)	0.354	0.355
4	(6, 22)	(3, 18)	0.407	0.408
5	(7, 25)	(4, 20)	0.454	0.455
6	(8, 27)	(5, 22)	0.496	0.497
7	(9, 30)	(6, 24)	0.535	0.536
8	(10, 32)	(7, 25)	0.572	0.572
9	(11, 34)	(8, 26)	0.606	0.607
10	(12, 36)	(9, 28)	0.638	0.639

Table 23.3 Optimal(K_F^*, N_F^*), $C_F(K_F^*, N_F^*)/c_M$, (K_L^*, N_L^*) and $C_L(K_L^*, N_L^*)/c_M$ when $G(t) = 1 - e^{-t}$, $H(t) = (\lambda t)^2$, $c_N = c_K$, and $\lambda = 1.0$

$\frac{c_N}{c_M}$	(K_F^*, N_F^*)	(K_L^*, N_L^*)	$\frac{C_F(K_F^*, N_F^*)}{c_M}$	$\frac{C_L(K_L^*, N_L^*)}{c_M}$
1	(2, 3)	(2, 1)	2.221	2.617
2	(3, 4)	(3, 1)	2.995	3.222
3	(4, 5)	(4, 1)	3.604	3.750
4	(5, 5)	(5, 1)	4.123	4.224
5	(6, 6)	(6, 1)	4.583	4.657
6	(7, 7)	(7, 1)	5.001	5.056
7	(8, 7)	(8, 1)	5.386	5.429
8	(9, 8)	(9, 1)	5.745	5.779
9	(10, 8)	(9, 1)	6.083	6.110
10	(11, 9)	(10, 1)	6.404	6.426

We can see from these tables that $C_F(K_F^*, N_F^*)/c_M < C_L(K_L^*, N_L^*)/c_M$; however, two costs are almost the same. Optimal (K_F^*, N_F^*) and (K_L^*, N_L^*) for $\lambda = 1.0$ are smaller than those for $\lambda = 0.1$. This indicates that if λ is large, i.e., possibility of failure is high, we should replace early the unit to avoid the cost of failures. We could make similar discussions for the case of $c_N \neq c_K$.

23.6 Standby and Parallel Systems

Finally, we compare reliability measures and replacement policies of a standby system and a parallel system that consist of n ($n = 1, 2, \ldots$) identical units. Each unit has a failure distribution $F(t)$ with finite mean μ, density function $f(t) \equiv \mathrm{d}F(t)/\mathrm{d}t$, and failure rate $h(t) \equiv f(t)/\bar{F}(t)$.

23.6.1 Reliability Measures

For a standby system: When an operating unit has failed, the next unit begins to operate immediately. The system fails when all units have failed. Then, reliability, MTTF (mean time to failure), and failure rate are, respectively,

$$R_S(t) = 1 - F^{(n)}(t), \quad \mu_S = \int_0^\infty [1 - F^{(n)}(t)]\mathrm{d}t,$$

$$h_S(t) = \frac{f^{(n)}(t)}{1 - F^{(n)}(t)} \quad (n = 1, 2, \ldots). \tag{23.36}$$

For a parallel system: All units begin to operate a time 0, and the system fails when they have failed. Then, reliability, MTTF, and failure rate are, respectively,

$$R_P(t) = 1 - F(t)^n, \quad \mu_P = \int_0^\infty [1 - Ft^n]\mathrm{d}t, \quad h_P(t) = \frac{nh(t)\bar{F}(t)F(t)^{n-1}}{1 - F(t)^n}. \tag{23.37}$$

When $F(t) = 1 - \mathrm{e}^{-\lambda t}$, (23.36) and (23.37) are, respectively,

$$R_S(t) = \sum_{j=0}^{n-1} \frac{(\lambda t)^j}{j!}\mathrm{e}^{-\lambda t}, \quad \mu_S = \frac{n}{\lambda}, \quad h_S(t) = \frac{\lambda(\lambda t)^{n-1}/(n-1)!}{\sum_{j=0}^{n-1}[(\lambda t)^j/j!]}, \tag{23.38}$$

and

$$R_P(t) = 1 - \left(1 - \mathrm{e}^{-\lambda t}\right)^n, \quad \mu_P = \frac{1}{\lambda}\sum_{j=1}^{n}\frac{1}{j}, \quad h_P(t) = \frac{n\mathrm{e}^{-\lambda t}\left(1 - \mathrm{e}^{-\lambda t}\right)^{n-1}}{1 - \left(1 - \mathrm{e}^{-\lambda t}\right)^n}. \tag{23.39}$$

Comparing the above two results, we easily have $R_S(t) \geq R_P(t)$ and $\mu_S \geq \mu_P$. Furthermore, using a mathematical induction, we have

$$h_P(t) = \frac{n\left(1 - \mathrm{e}^{-\lambda t}\right)^{n-1}}{\sum_{j=0}^{n-1}\left(1 - \mathrm{e}^{-\lambda t}\right)^j} \geq \frac{(\lambda t)^{n-1}/(n-1)!}{\sum_{j=0}^{n-1}[(\lambda t)^j/j!]}.$$

Table 23.4 presents values of $R_i(t)$, $\mu_i(t)$, and $h_i(t)$ $(i = S, P)$ for n at time 1.0 when $\lambda = 1.0$. When n becomes large, we can see the obvious differences between the two systems. For MTTF, μ_S increases strictly with n to ∞, but μ_P increases slowly and will be $\gamma + \log n$ when n is large enough, where γ is Euler's constant and $\gamma = 0.577215\ldots$. For reliability and failure rate, it is much easier for a standby system to keep it reliable by increasing the number of units, e.g., when n increases to 7, $R_S(t) = 1.000 > R_P(t) = 0.960$, and $h_S(t) = 0.001 < h_P(t) = 0.171$.

Table 23.4 Values of $R_S(t)$, $R_P(t)$, μ_S, μ_P, $h_S(t)$, and $h_P(t)$ when $F(t) = 1 - e^{-t}$

n	$R_S(t)$	$R_P(t)$	μ_S	μ_P	$h_S(t)$	$h_P(t)$
1	0.368	0.368	1.000	1.000	1.000	1.000
2	0.736	0.600	2.000	1.500	0.500	0.775
3	0.920	0.747	3.000	1.833	0.200	0.590
4	0.981	0.840	4.000	2.083	0.062	0.442
5	0.996	0.899	5.000	2.283	0.015	0.327
6	0.999	0.936	6.000	2.450	0.003	0.238
7	1.000	0.960	7.000	2.590	0.001	0.171
8	1.000	0.975	8.000	2.718	0.000	0.121
9	1.000	0.984	9.000	2.829	0.000	0.086
10	1.000	0.990	10.000	2.929	0.000	0.060

23.6.2 *Replacement Policies*

Suppose that both systems work for the same job with processing times and fail when all of units have failed. Let c_A be the acquisition cost for one unit and c_R be the replacement cost at each failure. Then, the total acquisition and replacement cost for a standby system until failure is $n(c_A + c_R)$, and the expected cost rate is

$$C_S(n) = \frac{n(c_A+c_R)}{\int_0^\infty [1-F^{(n)}(t)]\,\mathrm{d}t}. \tag{23.40}$$

The total acquisition and replacement cost for a parallel system until failure is $nc_A + c_R$, and the expected cost rate is

$$C_P(n) = \frac{nc_A+c_R}{\int_0^\infty [1-Ft^n]\,\mathrm{d}t}. \tag{23.41}$$

When $F(t) = 1 - e^{-\lambda t}$, (23.40) and (23.41) become, respectively,

$$\frac{C_S(n)}{\lambda} = c_A + c_R, \qquad \frac{C_P(n)}{\lambda} = \frac{nc_A + c_R}{\sum_{j=1}^n (1/j)}. \tag{23.42}$$

We find optimal n_P^* to minimize $C_P(n)$ in (23.42). From the inequality $C_P(n+1) - C_P(n) \geq 0$ and $C_P(n) - C_P(n-1) < 0$,

$$(n+1)\sum_{j=2}^{n+1}\frac{1}{j} \geq \frac{c_A}{c_R} > n\sum_{j=1}^{n}\frac{1}{j}. \tag{23.43}$$

Thus, optimal number n_P^* $(1 \leq n_P^* < \infty)$ is given by a finite and unique integer that satisfies (23.43). Therefore, from (23.42) to (23.43), if

$$\frac{1}{n_P^*-1}\sum_{j=2}^{n_P^*}\frac{1}{j} \geq \frac{c_A}{c_R+c_A}, \tag{23.44}$$

then a parallel system would save more cost than a standby system. However, (23.44) is rewritten as

$$\frac{\sum_{j=2}^{n_P^*}(1/j)}{n_P^*-1-\sum_{j=2}^{n_P^*}(1/j)} \geq \frac{c_A}{c_R}, \tag{23.45}$$

and from (23.43),

$$n\sum_{j=2}^{n}\frac{1}{j} - \frac{\sum_{j=2}^{n}(1/j)}{n-1-\sum_{j=2}^{n}(1/j)} = \frac{\sum_{j=2}^{n}(1/j)}{n-1-\sum_{j=2}^{n}(1/j)}\left[n\sum_{j=2}^{n}\left(1-\frac{1}{j}\right)-1\right] \geq 0,$$

which implies that there does not exist n_P^* that satisfies (23.45). This shows actually that under the above conditions, a standby system saves more cost than a parallel system.

We next compute a modified cost $\hat{c}_A \geq c_A$ and optimal number $\hat{n}_P$ for a parallel system, by setting $C_P(\hat{n}_P)/\lambda = C_S(n)/\lambda = c_A + c_R$ when the original c_A is still used for a standby system. This might be the case when a parallel system is better than a standby one.

We solve the following simultaneous equations for given $c_A + c_R$:

$$(\hat{n}_P+1)\sum_{j=2}^{\hat{n}_P+1}\frac{1}{j} \geq \frac{\hat{c}_A}{c_R},$$

and

$$\frac{n\hat{c}_A + c_R}{\sum_{j=1}^{\hat{n}_P}(1/j)} = c_A + c_R.$$

That is, we compute $\hat{n}_P$ that satisfies

$$\hat{n}_P(\hat{n}_P+1)\sum_{j=2}^{\hat{n}_P+1}\frac{1}{j} \geq \left(\frac{c_A}{c_R}+1\right)\sum_{j=1}^{\hat{n}_P}\frac{1}{j} - 1. \tag{23.46}$$

Using $\hat{n}_P$, we compute $\hat{c}_A/c_R$ which is given by

$$\frac{\hat{c}_A}{c_R} = \frac{1}{\hat{n}_P}\left[\left(\frac{c_A}{c_R}+1\right)\sum_{j=1}^{\hat{n}_P}\frac{1}{j} - 1\right]. \tag{23.47}$$

Table 23.5 Optimal n_P^*, $C_P(n_P^*)/\lambda c_R$, $\hat{n}_P$, and $\hat{c}_A/c_R$ when $F(t) = 1 - e^{-t}$

$\frac{c_A}{c_R}$	$\frac{C_S(n)}{\lambda c_R}$	n_P^*	$\frac{C_P(n_P^*)}{\lambda c_R}$	$\hat{n}_P$	$\frac{\hat{c}_A}{c_R}$
1.0	2.000	1	2.000	1	1.000
2.0	3.000	2	3.333	2	1.750
3.0	4.000	3	5.455	2	2.500
4.0	5.000	3	7.091	3	3.250
5.0	6.000	4	10.080	3	3.333
6.0	7.000	4	12.000	3	3.944
7.0	8.000	5	15.766	4	4.556
8.0	9.000	5	17.956	4	4.437
9.0	10.000	6	22.449	4	4.958
10.0	11.000	6	24.898	4	5.479

Table 23.5 presents optimal n_P^* and its $C_P(n_P^*)/\lambda c_R$, and modified $\hat{n}_P$ and $\hat{c}_A/c_R$. It is shown that $\hat{c}_A \le c_A$ can be found, and we can provide less unit for a parallel system under $\hat{c}_A$, as shown in Table 23.5 that $\hat{n}_P \le n_P^*$. That is, if the unit acquisition cost $\hat{c}_A$ for a parallel system is lower than that c_A for a standby system, then we could adopt a parallel system to save the expected cost rate; otherwise, a standby system should be used.

23.7 Conclusion

We have summarized systematically and shortly various comparisons of replacement first, last, and overtime policies in random age and periodic replacements, periodic replacement with cycle number and failure number, and reliability properties of standby and parallel systems, based on our original research works. We have derived optimal policies theoretically and decided which policy is better than the other by comparing them. These results would be applied to some real systems such as industrial equipment, and network and computer systems. The comparisons given in this chapter would be greatly useful for researchers in reliability to search for new topics of studies and for engineers and managers who are worried about which policy should be adopted for objective systems. Furthermore, we believe that "Which-is-Better (WIB) Problems" is the first name in reliability fields and would have great impact among the readers in reliability.

References

1. Barlow, R. E., & Proschan, F. (1965). *Mathematical theory of reliability*. New York: Wiley.
2. Osaki, S. (2002). *Stochastic models in reliability and maintenance*. Berlin: Springer.
3. Nakagawa, T. (2005). *Maintenance theory of reliability*. London: Springer.
4. Nakagawa, T. (2008). *Advanced reliability models and maintenance policies*. London: Springer.
5. Nakagawa, T. (2014). *Random Maintenance Policies*. London: Springer Verlag.
6. Sugiura, T., Mizutani, S., & Nakagawa, T. (2006). Optimal random and periodic inspection policies. In H. Pham (Ed.), *Reliability modeling, analysis and optimization* (pp. 393–403). Singapore: World Scientific.
7. Chen, M., Mizutani, S., & Nakagawa, T. (2010). Random and age replacement policies. *International Journal of Reliability, Quality and Safety Engineering, 17*(1), 27–39.
8. Nakagawa, T., Zhao, X., & Yun, W. (2011). Optimal age replacement and inspection policies with random failure and replacement times. *International Journal of Reliability, Quality and Safety Engineering, 18*(5), 405–416.
9. Chen, M., Nakamura, S., Nakagawa, T. (2010). Replacement and preventive maintenance models with random working times. *IEICE Transactions on Fundamentals of Electronics, Communications and Computer Sciences, E93-A*(2), 500–507.
10. Chang, C. C. (2014). Optimum preventive maintenance policies for systems subject to random working times, replacement, and minimal repair. *Computers and Industrial Engineering, 67*, 185–194.
11. Zhao, X., & Nakagawa, T. (2016). Over-time and over-level replacement policies with random working cycles. *Annals of Operations Research, 244*, 103–116.
12. Berrade, M., Scarf, P., & Cavalcante, C. (2017). A study of postponed replacement in a delay time model. *Reliability Engineering and System Safety, 168*, 70–79.
13. Sheu, S., Liu, T., Zhang, Z., & Tsai, H. (2018). The general age maintenance policies with random working times. *Reliability Engineering and System Safety, 169*, 503–514.
14. Zhao, X., & Nakagawa, T. (2012). Optimization problems of replacement first or last in reliability theory. *European Journal of Operational Research, 223*(2), 141–149.
15. Zhao, X., & Nakagawa, T. (2013). Optimal periodic and random inspection with first, last, and overtime policies. *International Journal of Systems Science*. https://doi.org/10.1080/00207721.2013.827263.
16. Chang, C. C., & Chen, Y. L. (2015). Optimization of preventive maintenance policies with replacement first and last for system subject to shocks. *International Journal of Systems Science: Operations and Logistics, 2*(1), 35–48.
17. Nakagawa, T., & Zhao, X. (2015). *Maintenance overtime policies in reliability theory*. London: Springer.
18. Mizutani, S., Zhao, X., & Nakagawa, T. (2015). Overtime replacement policies with finite operating interval and number. *IEICE Transaction, Fundamentals, E98-A*(10), 2069–2075.
19. Zhao, X., & Nakagawa, T. (2014). Comparisons of periodic and random replacement policies. In I. B. Frenkel, A. Karagrigoriou, A. Lisnianski, & A. Kleyner (Eds.), *Applied reliability engineering and risk analysis: Probabilistic models and statistical inference*. New York: Wiley.
20. Zhao, X., Mizutani, S., & Nakagawa, T. (2015). Which is better for replacement policies with continuous or discrete scheduled times? *European Journal of Operational Research, 242*(2), 477–486.
21. Zhao, X., Chen, M., & Nakagawa, T. (2015). Comparisons of standby and parallel systems in reliability, replacement, scheduling and application. *Journal of Risk and Reliability, 230*(1), 101–108. https://doi.org/10.1177/1748006X15593831.
22. Zhao, X., Al-Khalifa, K. N., Hamouda, A. M., & Nakagawa, T. (2016). First and last triggering event approaches for replacement with minimal repairs. *IEEE Transactions on Reliability, 65*(1), 197–207.
23. Zhao, X., Qian, C., & Nakagawa, T. (2017). Comparisons of replacement policies with periodic times and repair numbers. *Reliability Engineering and System Safety, 168*, 161–170.

24. Zhao, X., Qian, C., & Nakagawa, T. (2019). Periodic replacement policies and comparisons with their extended policies. In Q. L. Li, J. Wang, & H. B. Yu (Eds.), *Stochastic models in reliability, network security and system safety*. Berlin: Springer. https://doi.org/10.1007/978-981-15-0864-6.
25. Mizutani, S., Zhao, X., & Nakagawa, T. (2020). Which replacement is better at working cycles or numbers of failures. *IEICE Transaction, Fundamentals, E103-A*(2), 523–532.

Satoshi Mizutani received Ph.D. degree from Aichi Institute of Technology in 2004. He was a Visiting Researcher at Kinjo Gakuin University in Nagoya City, from 2004 to 2007. He worked as Assistant Professor from 2007 to 2013. and as Associate Professor from 2013 to 2018 at Aichi University of Technology. He is now Associate Professor at Aichi Institute of Technology, Japan. His research interests are extended optimal replacement and inspection policy in reliability theory. He has got IEEE Reliability Society Japan Chapter 2010, Outstanding Young Scientist Award, and. APIEMS 2017, Best Paper Award.

Xufeng Zhao is a Professor at Nanjing University of Aeronautics and Astronautics, China. He received his bachelor's degree in information management and information system in 2006, and master's degree in system engineering in 2009, both from Nanjing Tech University, China; and his doctoral degree in business administration and computer science in 2013 from Aichi Institute of Technology, Japan. Dr. ZhaoHe has worked as Postdoctoral Researcher from 2013 to 2017 at Aichi Institute of Technology and Qatar University, respectively. Dr. ZhaoHe is interested in probability theory, stochastic process, reliability and maintenance theory, and applications in computer and industrial systems. He has published two 2 books in maintenance theory from Springer and more than forty 40 research papers in peer- reviewed journals; and he is the author or co-author of eight book chapters from Springer, Wiley, and World Scientific. He has gotten one best paper award from IEEE Reliability Society and five best paper awards from International conferences in reliability, maintainability, and Qquality.

Toshio Nakagawa received B.S.E. and M.S. degrees from Nagoya Institute of Technology in 1965 and 1967, respectively; and a Doctor degree from Kyoto University in 1977. He worked as a Research Associate at Syracuse University for two 2 years from 1972 to 1973. He is now an Honorary Professor at Aichi Institute of Technology, Japan. He has published 6 books from Springer, and about 200 journal papers. His research interests are optimization problems in operations research and management science, and analysis for stochastic and computer systems in reliability and maintenance theory.

Chapter 24
A Simple and Accurate Approximation to Renewal Function of Gamma Distribution

R. Jiang

Abstract Renewal function (RF) of a life distribution has many applications, including reliability and maintenance-related decision optimizations. Such optimization problems need a simple and accurate approximation of RF so as to facilitate the solving process. Several such approximations have been developed for the Weibull distribution, but it seems that no such approximation is available for the gamma distribution. This may result from the fact that the convolutions of the gamma distribution are known so that its RF can be evaluated by a partial sum of gamma distribution series. However, the partial sum usually involves a number of terms and hence is not simple. Thus, a simple and accurate RF approximation for the gamma distribution is still desired. This chapter proposes such an approximation. The proposed approximation uses a weight function to smoothly link two known asymptotical relations. The parameters of the weight function are given in the form of empirical functions of the shape parameter. The maximum relative error of the proposed approximation is smaller than 1% for most of typical range of the shape parameter. The approximation is particularly useful for solving optimization problems that need to iteratively evaluate a gamma RF.

Keywords Gamma distribution · Renewal function · Approximation · Asymptotical relation · Weight function

24.1 Introduction

The gamma distribution has many applications in reliability and maintenance-related areas. Typical applications include

- Modeling the lifetime of a unit [1–4]
- Modeling the demand amount of a spare part [5, 6]

R. Jiang (✉)
Changsha University of Science and Technology, Changsha, Hunan 410114, People's Republic of China
e-mail: jiang@csust.edu.cn

K. B. Misra (ed.), *Handbook of Advanced Performability Engineering*,
https://doi.org/10.1007/978-3-030-55732-4_24

- Modeling a random degradation quantity, i.e., the gamma process model [7]
- Modeling heterogeneity in lifetime data as a frailty model [8], and
- Modeling inter-arrival times in a queuing system [9].

Similarly, the renewal function (RF) of a cumulative distribution function (cdf) has wide applications such as in reliability theory; continuous sampling plans; maintenance optimization; warranty cost estimation; spare part inventory planning, control, and management; and queuing systems [10–13]. The applications can be roughly divided into two categories:

(a) Calculating the RF for a known t and
(b) Decision optimization for the problems that need to iteratively evaluate the RF for different values of t.

For case (a), many numerical methods are available (e.g., see [14]). Jin and Gonigunta [12] propose the generalized exponential function to approximate the gamma distributions and then solve for the RF using the Laplace transform. It is concluded that the proposed model can achieve good approximations when the gamma shape parameter is in the range of 1–10. However, using a partial sum of the gamma cdf series to evaluate the gamma RF appears more straightforward, simpler, and more accurate than their approximations.

For case (b), when the shape parameter is a positive integer, the gamma distribution becomes the Erlang distribution, for which a closed-form expression exists and the expression becomes complicated as the shape parameter increases [15]. Thus, a simple and accurate approximation to the gamma RF is desired when the shape parameter is not a positive integer or is a large integer. It seems that no such approximation is available currently.

This chapter presents such an approximation to fill this gap. The proposed approximation uses a weight function to smoothly link two asymptotical relations. The first asymptotical relation is the first two terms of the gamma cdf series, which is accurate for small to moderate t; and the second is the well-known asymptotical relation that is applicable for large t. The weight function is the reliability function associated with the normal distribution and its parameters are functions of the shape parameter of the gamma distribution. An accuracy analysis is carried out and the results show that the proposed approximation is very accurate besides being simple.

This chapter is organized as follows. Some important results of the gamma RF are given in Sect. 24.2. The proposed approximation is presented and its accuracy is analyzed in Sect. 24.3. A real-world example is given in Sect. 24.4 to confirm its usefulness and accuracy. The chapter is concluded in Sect. 24.5.

24.2 Some Important Results of the Gamma RF

24.2.1 Gamma Distribution and Its Moments

The gamma probability density function (pdf) is given by

$$f(t) = \frac{1}{\eta^{\alpha}\Gamma(\alpha)} t^{\alpha-1} e^{-t/\eta} \tag{24.1}$$

where α is the shape parameter, η is the scale parameter, and $\Gamma(.)$ is the gamma function. Generally, the gamma cdf does not have a closed-form analytic expression and many software packages (e.g., MS Excel) have a standard function to evaluate it. However, when α is a positive integer, the gamma cdf becomes the Erlang distribution and its cdf is given by Nakagawa [16]

$$F(t) = 1 - e^{-\lambda t} \sum_{k=0}^{\alpha-1} \frac{(\lambda t)^k}{k!} \tag{24.2}$$

where $\lambda = 1/\eta$.

The kth raw moment ($k = 1, 2, \ldots$) is given by

$$m_k = \int_0^{\infty} t^k f(x) \mathrm{d}x = \frac{\eta^k \Gamma(k+\alpha)}{\Gamma(\alpha)} \tag{24.3}$$

Letting $k = 1$ yields its mean.

$$\mu = m_1 = \alpha\eta. \tag{24.4}$$

Letting $k = 2$ yields its the second raw moment.

$$m_2 = \mu(\mu + \eta). \tag{24.5}$$

Its variance (denoted as σ^2) and coefficient of variation (CV, denoted as ρ) are given respectively by

$$\sigma^2 = \alpha\eta^2; \rho = 1/\sqrt{\alpha}. \tag{24.6}$$

Jin and Gonigunta [12] mention that the typical range of α is 1–10 when modeling the product reliability with increasing failure rate in many practical applications. Jiang [17] mentions that the Weibull shape parameter value found in practice rarely exceeds 4. That is, the CV of time to failure is usually larger than 0.2805, which corresponds to $\alpha < 12.71$. Therefore, the upper bound of typical range of α can be taken as 13.

24.2.2 Convolution of Two Gamma Distributions with a Common Scale Parameter

Let X and Y be two independent gamma random variables with a common scale parameter η and different shape parameters α_1 and α_2, respectively. The pdf of $Z = X + Y$ is given by Blischke and Murthy [18]

$$f(t) = \int_0^t \frac{x^{\alpha_1-1}(t-x)^{\alpha_2-1}}{\eta^{\alpha_1+\alpha_2}\Gamma(\alpha_1)\Gamma(\alpha_2)} e^{-x/\eta-(t-x)/\eta} dx$$

$$= \frac{t^{\alpha_1+\alpha_2-1}e^{-t/\eta}}{\eta^{\alpha_1+\alpha_2}\Gamma(\alpha_1+\alpha_2)} \int_0^1 \beta(u;\alpha_1,\alpha_2)du \quad (24.7)$$

where $u = x/t$ and $\beta(u;\alpha_1,\alpha_2)$ is the standard beta distribution with shape parameters α_1 and α_2. Since

$$\int_0^1 \beta(u;\alpha_1,\alpha_2)du = 1, \quad (24.8)$$

Equation (24.7) becomes

$$f(t) = \frac{t^{\alpha_1+\alpha_2-1}e^{-t/\eta}}{\eta^{\alpha_1+\alpha_2}\Gamma(\alpha_1+\alpha_2)}. \quad (24.9)$$

This implies that the sum of two independent gamma random variables with a *common scale parameter* is still a gamma random variable with the same scale parameter and its shape parameter is the sum of two individual shape parameters. This is an important property of the gamma distribution. The middle part of Eq. (24.7) clearly shows that the "*common scale parameter*" is the key condition for the property to hold.

24.2.3 Renewal Function and Its Asymptotical Relations

Consider a sequence of independent and identically distributed random variables $\{X_k; k = 1, 2, \ldots\}$. Let $F(t)$ denote the cdf of X_k. The cdf of $T_k = X_1 + X_2 + \cdots +$

X_k is given by the k-fold convolution of $F(t)$, denoted as $F^{(k)}(t)$ with $F^{(1)}(t) = F(t)$. Let $N(t)$ denote the number of renewals occurring in the interval (0, t]. The renewal function is the expected value of $N(t)$ and given by Nakagawa [16] and Jiang [19]

$$M(t) = \sum_{k=1}^{\infty} F^{(k)}(t) \tag{24.10}$$

or

$$M(t) = F(t) + \int_0^t M(t-x)dF(x) \tag{24.11}$$

When t is small, $F^{(k)}(t) \approx 0$ for $k > 1$ so that

$$M(t) \approx F(t). \tag{24.12}$$

When t is large, the well-known asymptotical relation derived by Smith [20] is given by

$$M(t) \approx \frac{t}{\mu} - 0.5\left(1 - \rho^2\right). \tag{24.13}$$

24.2.4 Some Special Cases of the Gamma RF

When α is a positive integer, the gamma distribution becomes into the Erlang distribution; and when $\alpha = 1$, the Erlang distribution reduces into the exponential distribution.

For the Erlang distribution with small shape parameter, the analytical solution of the integral equation given by Eq. (24.11) can be obtained using the Laplace transform approach [15]. Some such solutions are shown in Table 24.1, where $\lambda =$

Table 24.1 RFs of the Erlang distribution

α	RF
1	$M(t) = \lambda t$
2	$M(t) = \left(e^{-2\lambda t} + 2\lambda t - 1\right)/4$
3	$M(t) = \left[e^{-1.5\lambda t}\left(\cos\left(\frac{\sqrt{3}}{2}\lambda t\right) + \frac{1}{\sqrt{3}}\sin\left(\frac{\sqrt{3}}{2}\lambda t\right)\right) + \lambda t - 1\right]/3$
4	$M(t) = \left[e^{-2\lambda t} + 2e^{-\lambda t}(\cos(\lambda t) + \sin(\lambda t)) + 2\lambda t - 3\right]/8$

$1/\eta$. As seen from the table, the expression of the solution becomes complicated as α increases so that the analytical solution for $\alpha > 4$ is rarely useful.

24.2.5 Series Representation of the Gamma RF

According to Eq. (24.9), the k-fold convolution of the gamma cdf is a gamma cdf with shape parameter $k\alpha$ and scale parameter η. That is,

$$F^{(k)}(t) = F(t; k\alpha, \eta). \tag{24.14}$$

Thus, the gamma RF can be evaluated by the following partial sum of gamma cdf series

$$M(t) = \sum_{k=1}^{\infty} F^{(k)}(t) \approx \sum_{k=1}^{N} F(t; k\alpha, \eta). \tag{24.15}$$

Here, N is called the stop point and its value is determined according to the following condition

$$\varepsilon(t, N) = \sum_{k=N+1}^{\infty} F(t; k\alpha, \eta) \leq \varepsilon \tag{24.16}$$

where ε is a specified tolerance level, e.g., 10^{-6}. The value of N is determined through solving the following inequality [16]

$$\varepsilon(t, N) \leq F(t; N\alpha, \eta)F(t)/[1 - F(t)]. \tag{24.17}$$

For $\varepsilon = 10^{-6}$, Table 24.2 shows the value of N as a function of α and t/μ. As seen, N increases as t increases and α decreases. For $t/\mu \geq 3$, $N \geq 8$. This implies that the

Table 24.2 Values of N

$\alpha \backslash t/\mu$	2.5	3	4	5	$\alpha \backslash t/\mu$	2.5	3	4	5
1	15	17	20	23	7	6	9	11	14
1.5	13	14	17	20	8	6	8	11	13
2	11	13	15	18	9	6	8	11	13
3	10	11	13	16	10	6	8	10	13
4	9	10	12	15	11	6	8	10	13
5	7	9	12	14	12	5	8	10	13
6	7	9	11	14	13	5	8	10	12

partial sum of gamma cdf series as a gamma RF approximation is not simple enough. Therefore, it is necessary to develop a simple and accurate RF approximation for the gamma distribution.

24.3 Proposed Gamma RF Approximation

Several attempts have been made to approximate the RF using a twofold sectional model, which smoothly links two asymptotical relations [17, 21–23]. In this section, we introduce a time-varying weight function to smoothly link the two asymptotical relations. Specific details are outlined as follows.

24.3.1 Definition of the Proposed RF Approximation

Let $M_1(t)$ denote the asymptotical relation for small and moderate t, and $M_2(t)$ denote the asymptotical relation for large t. The phrase "small t" ["large t"] implies the time range where Eq. (24.12) [Eq. (24.13)] is relatively accurate. Let τ_1 and τ_2 [which are defined in Eq. (24.24)] denote the applicable ranges of Eqs. (24.12) and (24.13), respectively. Thus, the phrase "moderate t" implies the time range between τ_1 and τ_2.

The proposed RF approximation is given by

$$M_a(t) = p(t)M_1(t) + [1 - p(t)]M_2(t) \tag{24.18}$$

where $p(t)$ is the weight function and it monotonically decreases from 1 to 0 as t increases. According to this property of $p(t)$, the proposed approximation meets the following relations:

$$M_a(t) \to M_1(t) \text{ for small } t, \text{ and } M_a(t) \to M_2(t) \text{ for large } t. \tag{24.19}$$

The approximation is characterized by $M_1(t)$, $M_2(t)$ and $p(t)$, which are specified as follows.

To extend the applicable range of $F(t)$ as an RF approximation for small t, $M_1(t)$ is taken as the sum of the first two terms of Eq. (24.15), i.e.,

$$M_1(t) = F(t; \alpha, \eta) + F(t; 2\alpha, \eta). \tag{24.20}$$

$M_2(t)$ is given by Eq. (24.13), i.e.,

$$M_2(t) = \frac{t}{\mu} - 0.5\left(1 - \rho^2\right) = \frac{t}{\mu} - 0.5(1 - 1/\alpha). \tag{24.21}$$

The weight function is specified as a reliability function of the normal distribution, given by

$$p(t) = 1 - \Phi(t; \mu_\tau, \sigma_\tau) \tag{24.22}$$

where Φ(.) is the normal cdf, μ_τ and σ_τ are the model parameters to be specified in the next subsection.

24.3.2 Determination of μ_τ and σ_τ

We first analyze the applicable range of $M_1(t)$ and $M_2(t)$ by examining their relative errors with the exact values obtained from Eq. (24.15) with $N = 15$, which is sufficiently large since $N \leq 15$ for $t/\mu < 2.5$ (see Table 24.2).

Referring to Fig. 24.1, we define two relative error curves

$$\varepsilon_1(t) = 1 - M_1(t)/M(t), \varepsilon_2(t) = 1 - M_2(t)/M(t). \tag{24.23}$$

According to these two curves, we define two time points

$$\tau_1 = \sup\{t : |\varepsilon_1(t)| \leq 0.01\}, \tau_2 = \inf\{t : |\varepsilon_2(t)| \leq 0.01\}. \tag{24.24}$$

Clearly, τ_1 [τ_2] is a measure of the applicable range of $M_1(t)$ [$M_2(t)$].

Table 24.3 shows the values of τ_1 and τ_2 as functions of α. From the table, we have the following observations:

- Max(τ_1, τ_2) < 2.5μ, implying $N = 15$ is appropriate.
- $\tau_1 < \tau_2$ in the range of $\alpha = (1.0147, 2.2295)$; otherwise, $\tau_1 \geq \tau_2$.
- For case of $\tau_1 < \tau_2$, the maximum relative error of the approximation can be larger than 1%; for case of $\tau_1 \geq \tau_2$, the maximum relative error is smaller than 1%.

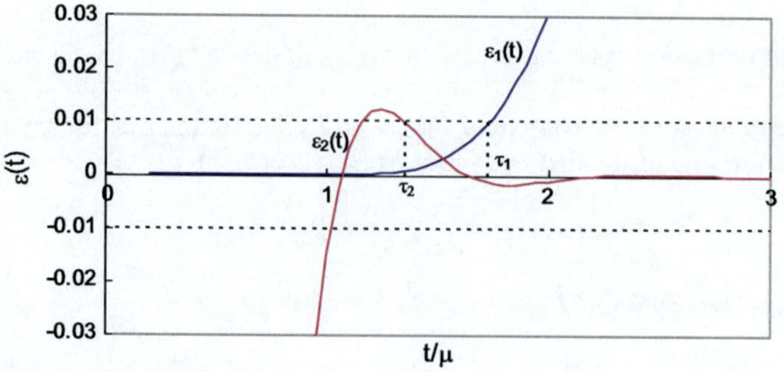

Fig. 24.1 Relative error curves ($\alpha = 7$)

Table 24.3 Values of τ_1 and τ_2

α	τ_1/μ	τ_2/μ	$\tau_1 < \tau_2$?	α	τ_1/μ	τ_2/μ	$\tau_1 < \tau_2$?
1	0.2613	0.0000	No	4	1.3489	1.0061	No
1.0147	0.2695	0.2695	$\tau_1 = \tau_2$	5	1.5131	1.0366	No
1.25	0.4018	0.8632	Yes	6	1.6366	1.0294	No
1.5	0.5365	0.9451	Yes	7	1.7336	1.3521	No
1.75	0.6605	0.9407	Yes	8	1.8125	1.4191	No
2	0.7726	0.9072	Yes	9	1.8783	1.4457	No
2.2295	0.8654	0.8654	$\tau_1 = \tau_2$	10	1.9345	1.4589	No
2.25	0.8732	0.8614	No	11	1.9832	1.8161	No
2.5	0.9635	0.8106	No	12	2.0260	1.8713	No
3	1.1178	0.7114	No	13	2.0640	1.8997	No

Table 24.4 Correlation coefficients

	μ_τ	Ln(μ_τ)	r_τ	Ln(r_τ)
α	0.9584	0.7745	0.0420	0.1781
Ln(α)	0.9846	0.8705	0.1115	0.1506

It is reasonable to take μ_τ as close to the average of τ_1 and τ_2 as possible. To find the function relation between μ_τ and α, we examine the correlation coefficients (CCs) between $\{\alpha, \ln(\alpha)\}$ and $\{\mu_\tau, \ln(\mu_\tau)\}$, and the results are shown in the second and third columns of Table 24.4. As seen, μ_τ and $\ln(\alpha)$ are highly correlated. After some trials, the following model provides a good fitting to the data in Table 24.3:

$$\mu_\tau = (\tau_1 + \tau_2)/2 \approx a/\alpha + b\ln(\alpha + c). \tag{24.25}$$

The parameters are estimated using the least square method, which minimizes the following

$$\mathrm{SSE} = \sum_{\alpha=1}^{13} \left[(\mu_{\tau,\alpha} - \tau_{1,\alpha})^2 + (\mu_{\tau,\alpha} - \tau_{2,\alpha})^2\right]. \tag{24.26}$$

This yields $(a, b, c) = (-0.5048\mu, 0.7207\mu, 2.2965)$.

The fitted model is displayed in Fig. 24.2. As seen, μ_τ is generally in-between τ_1 and τ_2, as expected.

Let r_τ denote the range of τ_1 and τ_2, i.e.,

$$r_\tau = |\tau_1 - \tau_2|. \tag{24.27}$$

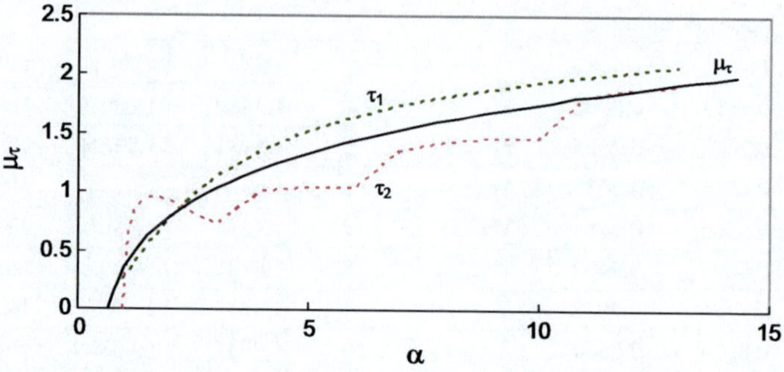

Fig. 24.2 Plot of μ_τ versus α

Similarly, we examine the CCs between $\{\alpha, \ln(\alpha)\}$ and $\{r_\tau, \ln(r_\tau)\}$ and the results are shown in the last two columns of Table 24.4. As seen, all the four values of CCs are small, implying that r_τ can be thought to be independent of α. Thus, the value of r_τ is taken as the average of those values of r_τ calculated from Table 24.3, i.e.,

$$r_\tau = 0.3030\mu. \tag{24.28}$$

The plot of r_τ versus α is shown in Fig. 24.3, which confirms the appropriateness of Eq. (24.28). According to the three-sigma rule of thumb, the value of σ_τ can be determined as

$$\sigma_\tau = r_\tau/6 = 0.0505\mu. \tag{24.29}$$

As a result, the proposed RF approximation is fully specified by four parameters: (a, b, c) in Eq. (24.25) and r_τ in Eq. (24.28).

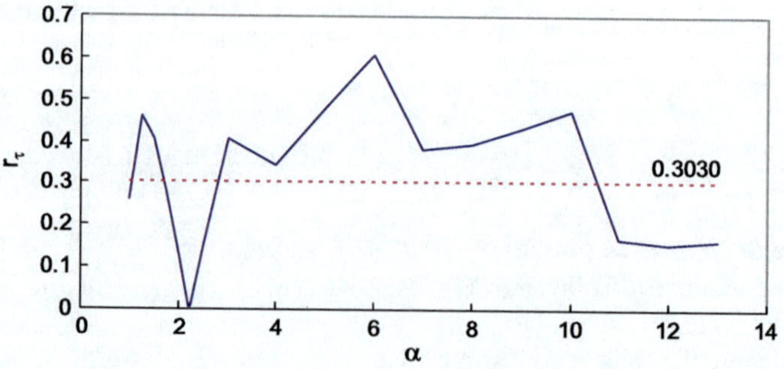

Fig. 24.3 Plot of r_τ versus α

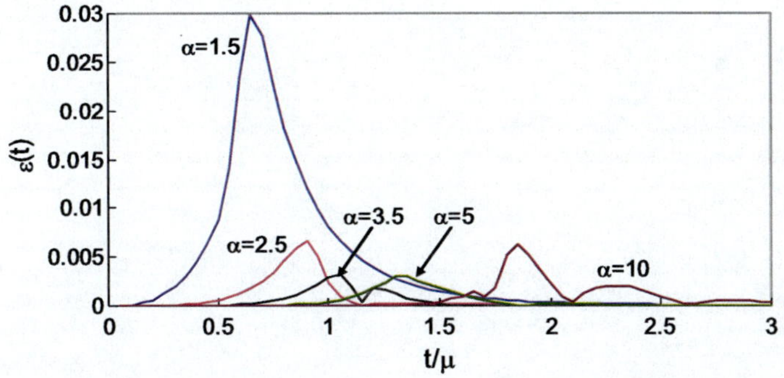

Fig. 24.4 Relative error curves

24.3.3 *Accuracy of the Proposed RF Approximation*

According to Table 24.3, $\max(\tau_1/\mu,\ \tau_2/\mu) = 2.0640$. Therefore, we evaluate the accuracy in the range of $t/\mu = (0, 3)$. For a given α, the relative errors (denoted as ε_i) at 60 time points (i.e., $t_i/\mu = 0.05(0.05)3.0$, $1 \le i \le 60$) are computed as

$$\varepsilon_i = |1 - M_a(t_i)/M(t_i)|. \tag{24.30}$$

Figure 24.4 shows the plots of the relative error curves for several values of α. As seen, the shape is complex.

To be simple, we use the following two measures to evaluate the accuracy

$$\varepsilon_a = \frac{1}{60}\sum_{i=1}^{60} \varepsilon_i,\ \varepsilon_M = \max_i(\varepsilon_i). \tag{24.31}$$

The values of the measures associated with a set of α's values are shown in Table 24.5 and displayed in Fig. 24.5. From the table and figure, we have the following observations:

- The maximum of relative errors is 2.976%, the average of maximum relative errors is 0.682% and the average of relative errors is 0.093%. More specifically,

Table 24.5 Accuracy measures as functions of α

α	ε_a (%)	ε_M (%)	α	ε_a (%)	ε_M (%)	α	ε_a (%)	ε_M (%)
1.5	0.416	2.976	4	0.028	0.261	9	0.068	0.511
2	0.186	1.482	5	0.042	0.307	10	0.079	0.623
2.5	0.066	0.664	6	0.042	0.390	11	0.086	0.607
3	0.052	0.413	7	0.042	0.296	12	0.093	0.456
3.5	0.037	0.317	8	0.053	0.323	13	0.104	0.598

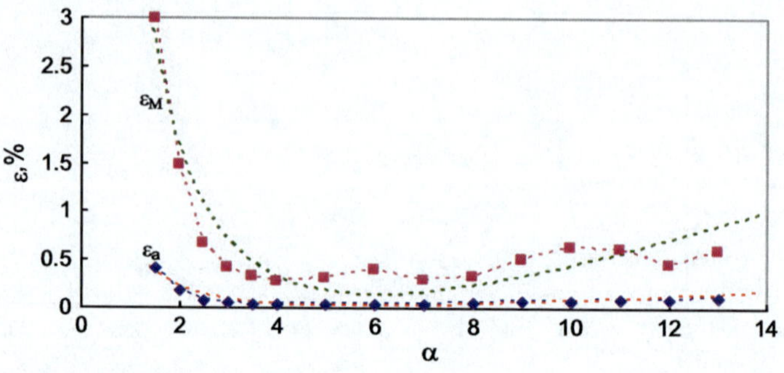

Fig. 24.5 Plots of maximum and average of relative errors

the maximum relative errors are between 1 and 3% for $\alpha < 2.23$ and smaller than 1% for $\alpha > 2.23$. As a whole, the approximation is accurate.

- Trend analysis shows that both ε_a and ε_M first decrease and then increase as α increases. This is because the facts that $M_1(t)$ to approximate the RF for small t is not very well when α is small and the RF fluctuates when α is large. As a result, the approximation is very accurate in the range of $\alpha = (3.5, 8.0)$.

24.4 An Application to Maintenance Policy Optimization

The data shown in Table 24.6 are from Lawless [24] and deal with the number of million revolutions before failure for 23 deep groove ball bearings in the life tests.

Using the maximum likelihood method, the data are fitted to four optional models: Weibull, Normal, Lognormal, and Gamma distributions. The estimated parameters and corresponding log-likelihood values are shown in Table 24.7. As seen, the gamma distribution is the most appropriate model in terms of the log-likelihood value.

Table 24.6 Data of time to failure for bearings

17.88	28.92	33.00	41.52	42.12	45.60	48.80	51.84
51.96	54.12	55.56	67.80	68.44	68.64	68.88	84.12
93.12	98.64	105.12	105.84	127.92	128.04	173.40	

Table 24.7 Maximum likelihood estimates and log-likelihood values

	Weibull	Normal	Lognormal	Gamma
β, μ or α	2.103	72.23	4.151	4.028
η or σ	81.88	36.66	0.5215	17.93
ln(L)	−113.688	−115.472	−113.125	−113.025

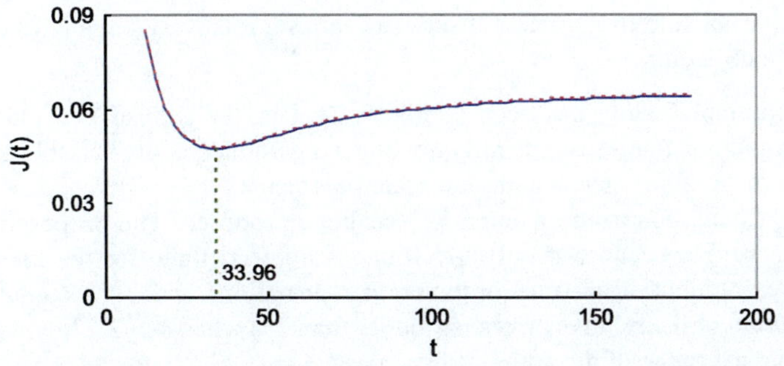

Fig. 24.6 Cost rate curves obtained from the exact RF and proposed RF approximation

Table 24.8 Optimal solutions and corresponding cost rate and RF values

RF	T^*	$M(T^*)$	$J(T^*)$
Exact	33.9555	0.121646	0.047363
Approximate	33.9560	0.121651	0.047363

Suppose that a block replacement policy is implemented to preventively replace the bearing. The optimal replacement time can be determined by minimizing the cost rate function given by Jiang and Murthy [25]

$$J(t) = \frac{c_p + c_f M(t)}{t} \tag{24.32}$$

where c_p is the cost of a preventive replacement and c_f is the cost of a failure replacement.

Without loss of generality, assume that $c_p = 1$ and $c_f = 5$. Figure 24.6 shows the cost rate curves obtained from the exact RF (i.e., the solid line) and proposed RF approximation (i.e., the dotted line). It is clear that the two curves almost overlap. This is because α (= 4.028) is within the range where the approximation is very accurate.

The optimal solutions, the corresponding cost rate and RF values are shown in Table 24.8. As seen, the two solutions are almost the same. This confirms the accuracy and usefulness of the proposed RF approximation.

24.5 Conclusions

In this chapter, we have discussed the necessity to develop an approximation to the gamma RF. The main reasons for such an approximation have been

(a) Using the Laplace transform approach to solve the renewal integral equation can obtain a relatively simple analytic solution only for $\alpha = 1, 2, 3$, and 4 and

(b) The partial sum of gamma cdf series requires a relatively large N to obtain an adequate accuracy.

An important finding has been obtained. That is, the sum of two independent gamma random variables is generally no longer a gamma random variable unless the two random variables have a common scale parameter.

An approximation to the gamma RF has been proposed. The proposed approximation is simple because it only has four parameters (three for the mean of the normal weight function and one for its standard deviation); and it is accurate because the maximum of the relative errors is smaller than 1% when $\alpha > 2.23$, which covers most of typical range of the gamma shape parameter.

If $M_1(t)$ is taken as the first three terms of Eq. (24.15), the accuracy can be further increased and the corresponding RF approximation is still relatively simple. In this case, the values of τ_1 will get larger and the parameters of the weight function should be slightly adjusted.

Similar to the gamma distribution, the convolution of the normal distribution is known. A topic for future research is to examine the possibility to develop a similar approximation for the normal distribution.

Acknowledgements The research was supported by the National Natural Science Foundation of China (No. 71771029).

References

1. Pham-Gia, T. (1999). System availability in a gamma alternating renewal process. *Naval Research Logistics, 46*(7), 822–844.
2. Sarkar, J., & Chaudhuri, G. (1999). Availability of a system with gamma life and exponential repair time under a perfect repair policy. *Statistics and Probability Letters, 43*(2), 189–196.
3. Das, R. N., & Park, J. S. (2012). Discrepancy in regression estimates between log-normal and gamma: Some case studies. *Journal of Applied Statistics, 39*(1), 97–111.
4. Kempa, W. M., Paprocka, I., Kalinowski, K., & Grabowik, C. (2014). Estimation of reliability characteristics in a production scheduling model with failures and time-changing parameters described by gamma and exponential distributions. *Advanced Materials Research, 837,* 116–121.
5. Moors, J. J. A., & Strijbosch, L. W. G. (2002). Exact fill rates for (R, s, S) inventory control with gamma distributed demand. *Journal of the Operational Research Society, 53*(11), 1268–1274.
6. Khaniyev, T., Turksen, I. B., Gokpinar, F., & Gever, B. (2013). Ergodic distribution for a fuzzy inventory model of type (s, S) with gamma distributed demands. *Expert Systems with Applications, Vo., 40*(3), 958–963.
7. van Noortwijk, J. M. (2009). A survey of the application of gamma processes in maintenance. *Reliability Engineering and System Safety, 94*(1), 2–21.
8. Economou, P., & Caroni, C. (2005). Graphical tests for the assumption of gamma and inverse Gaussian frailty distributions. *Lifetime Data Analysis, 11,* 565–582.
9. Miller, G. K., & Bhat, V. N. (1997). Estimation for renewal processes with unobservable gamma or Erlang inerarrival times. *Journal of Statistical Planning and Inference, 61*(2), 355–372.

10. Yang, G. L. (1985). Application of renewal theory to continuous sampling plans. *Naval Research Logistics Quarterly, 32,* 45–51.
11. Frees, E. W. (1986). Warranty analysis and renewal function estimation. *Naval Research Logistics Quarterly, 33,* 361–372.
12. Jin, T., & Gonigunta, l. (2008). Weibull and gamma renewal approximation using generalized exponential functions. *Communications in Statistics: Simulation and Computation, 38*(1), 154–171.
13. Brezavšèek, A. (2013). A simple discrete approximation for the renewal function. *Business Systems Research, 4*(1), 65–75.
14. Dohi, T., Kaio, N., & Osaki, S. (2002). Renewal processes and their computational aspects. In *Stochastic models in reliability and maintenance* (pp. 1–30). Berlin: Springer.
15. Helvaci, D. (2013). *Generating renewal functions of uniform, gamma, normal and Weibull distributions for minimal and non negligible repair by using convolutions and approximation methods,* Ph.D. Dissertation, Auburn University, USA.
16. Nakagawa, T. (2011). Renewal processes. In *Stochastic processes* (pp. 47–93). London: Springer.
17. Jiang, R. (2020). A novel two-fold sectional approximation of renewal function and its applications. *Reliability Engineering and System Safety, 193.*
18. Blischke, W. R., & Murthy, D. N. P. (2000). *Reliability: Modeling, prediction, and optimization* (p. 730). New York: Wiley.
19. Jiang, R. (2008). A gamma–normal series truncation approximation for computing the Weibull renewal function. *Reliability Engineering and System Safety, 93*(4), 616–626.
20. Smith, W. L. (1954). Asymptotic renewal theorems. *Proceedings of the Royal Society of Edinburgh: Section A Mathematics, 64,* 9–48.
21. Spearman, M. L. (1989). A simple approximation for IFR Weibull renewal functions. *Microelectronics Reliability, 29*(1), 73–80.
22. Garg, A., & Kalagnanam, J. R. (1998). Approximations for the renewal function. *IEEE Transactions on Reliability, 47*(1), 66–72.
23. Jiang, R. (2020). Two approximations of renewal function for any arbitrary lifetime distribution. *Annals of Operations Research.*
24. Lawless, J. F. (2003). *Statistical models and methods for lifetime data.* New York: Wiley.
25. Jiang, R., & Murthy, D. N. P. (2008). *Maintenance: Decision models for management.* Beijing: Science Press.

Renyan Jiang is a Professor of Changsha University of Science and Technology, China. He got his Ph.D. at the University of Queensland, Australia. His research interests are in various aspects of quality, reliability, and maintenance. He is the author or co-author of five reliability/maintenance related books and has published more than 200 papers.

Chapter 25
Transformative Maintenance Technologies and Business Solutions for the Railway Assets

Uday Kumar and Diego Galar

Abstract In the past, railway systems were overdesigned and underutilized making the need for effective, coordinated, and optimized maintenance planning non-existence. With passing years, these assets are getting old and at the same time, their utilization has increased manifold mainly due to societal consciousness about climate and cost. With steeply increasing utilization of railway systems, the major challenge is to find the time slot to perform maintenance on the infrastructure and rolling stocks to maintain its functionality and ensure safe train operation. This has led the sector to look for new and emerging technologies that will facilitate effective and efficient railway maintenance and ensure reliable, punctual, and safe train operation. This chapter presents the current status and the state-of-the-art of maintenance in railway sector transformative maintenance technologies and business solutions for the railway assets. It discusses the digital transformation of railway maintenance, application of artificial intelligence (AI), machine learning, big data analytics, digital twins, robots, and drones as part of the digital railway maintenance solutions. The chapter presents a conceptual road map for developing transformative maintenance solutions for railway using new and enabling technologies which are founded on data-driven decisions.

Keywords Railway assets · Maintenance · Big data · Maintenance analytics · Digital twin · Augmented reliability · Transformative technologies · AI · IIoT · Machine learning

---

U. Kumar (✉) · D. Galar
Division of Operation and Maintenance Engineering, Luleå Railway Research Center, Luleå University of Technology, 97187 Luleå, Sweden
e-mail: Uday.Kumar@ltu.se

D. Galar
e-mail: Diego.Galar@ltu.se

K. B. Misra (ed.), *Handbook of Advanced Performability Engineering*,
https://doi.org/10.1007/978-3-030-55732-4_25

25.1 Introduction

Railway is an attractive mode of transport because of its environmental friendliness and its sustainable way of transporting large amounts of freight and passengers in a cost-effective and comfortable way. Other advantages are its low energy consumption, owing to its high transport capacity, and its high level of safety. Since the railway system runs on rails and wheels, it has an inherent advantage of lesser frictional resistance which helps carry more load or wagons. However, the railway system has one degree of freedom, which implies poor flexibility and low redundancy. High reliability is a key factor for a stable railway system. If an unexpected incident occurs, it can take a much longer time (than failure recovery time) due to additional factor of train queues and related safety regulations (traffic recovery time) to restore the normal traffic [1]. In this scenario, maintenance is perceived as a key business function for achieving punctuality, reliability, and cost-effectiveness of railway assets and rail transport.

Hence, a key challenge for the modern-day railway sector is to ensure a reliable, punctual, and cost-effective mode of the transport system for passengers and goods. This necessitates a high level of reliability, availability (and capacity), and safety of railway infrastructure and rolling stocks which can only be ensured through effective, coordinated and efficient maintenance planning and executions.

So far, railway has been using the standard technologies and tools required to run the railway in a safe, effective, and efficient way. These technologies can be broadly classified as supporting and optimizing technologies and collectively provides the foundation for the predictive technologies, and are used for the estimation of the remaining useful life (RUL) popularly using condition monitoring tools and technologies, RAMS (reliability, and maintainability and safety) modeling, Life Cycle Costing (LCC) analysis, etc., to arrive at the correct maintenance decision. These challenges have led to the search for innovative maintenance solutions and deployment of new and emerging technologies such as AI, Machine learning, Big data analytics, IIoT, Virtual reality, etc., when it is economically and technologically viable.

This real-time data-driven approach to operate railway is expected to transform the way railway asset is operated and maintained ensuring increased reliability and quality of service, increased capacity, and reduced life cycle costs for the asset. To get useful information out of the high volume of data generated by railway assets, advanced tools, are developed and implemented so that data can be systematically processed into information and facilitate decision-making. Such solutions are expected to support railway's digital transformation journey and operations goals.

The discussions in the chapter will be centered on the capability of enabling technologies that will facilitate the development of transformative technologies for the effective maintenance of railway asset using the power of predictive and prescriptive analytics.

25.2 Railway Systems

Railway systems have complex technologies, with a wide range of standard engineering and business solutions and organization forms. Railway systems can be broadly divided into three groups, namely linear distributed assets, point assets, and mobile assets as shown in Fig. 25.1.

25.2.1 Linear Distributed Railway Assets

Linear assets can be defined as engineering structures or infrastructures that often cross a long distance and can be divided into different segments that perform the same function but are subjected to different loads and conditions [2]. Linear assets often form networks that consist of a number of 'lines.' These lines are functionally similar but can have different characteristics due to various construction materials, operational environments, or geometric sizes. In the railway context, the overhead contact wires and the tracks could be considered such assets. For a linear asset, it is necessary to define the location of a point or a section along with the asset for maintenance purposes. If any single section of a linear asset malfunctions, the

Fig. 25.1 An illustrative example of railway system

entire asset will not function properly. These characteristics of linear assets mean that their registry (categorization), reliability and cost analysis, and maintenance decision modeling methods are different from those of non-linear assets [2].

25.2.2 Point Assets

An asset which is geographically located at one place and crossings, etc., for example, a railway switch or a camera. With a point asset, maintenance planning is more or less planned similar to any other mechanical equipment.

25.2.3 Mobile Assets

Assets that are not at fixed locations but are mobile in nature, for example, locomotives, railway wagons, coaches, inspection trains, etc., maintenance actions on such assets are planned in locations/workshops as per convenience considering costs and availability of resources.

25.3 Current Status of Railway Maintenance: The Preventive Culture

The area of operation and maintenance of infrastructures and rolling stock is multidisciplinary in nature transcending boundaries s of several disciplines of engineering and management. In general, the maintenance intervals of railway assets are determined "statistically," that is, by operating time and/or distance traveled or by the number of actions performed by the system. These intervals are based on previous experiences or on operational load and the average life of the components involved. However, some components may deteriorate more rapidly than expected due to changes in the operating environment. Maintenance actions on railway can be broadly divided into two broad categories:

Corrective Maintenance (CM): The aim of corrective maintenance is to "fix a failed item after fault recognition to bring its normal function. CM actions are performed when the asset has a failure (in the case of railway equipment) or has degraded sufficiently (in the case of infrastructure). The most common form of CM is "minimal repair" where the state of the asset after repair is nearly the same as that just before failure. The other extreme is "as good as new" repair and this is seldom possible unless one replaces the failed asset with a new one [3].

Preventive Maintenance (PM): The aim of PM is to take actions to prevent the occurrence of failures as per the prescribed criteria of time usage, or condition. For examples, these are the actions carried out at to fix minor or major problems in case of infrastructure (*e.g.*, small potholes in a section of a road) or components that have degraded in the case of rolling stocks and other equipment due to age and/or usage.

There are different kinds of preventive maintenance (PM) policies. Some are presented here:

- *Age-based maintenance*: PM actions are taken based on the calendar age of the items or components.
- *Clock-based maintenance*: PM actions are taken based on the usage of the items or components. For example, railway wheels are taken out for PM actions (grinding of wheels) every 40, 000 km.
- *Opportunistic maintenance*: Opportunistic PM deals with carrying out PM actions as a result of opportunities that occur in an uncertain manner [3]. For example, in a locomotive, there are several components. Failure of one component will provide an opportunity to carry out inspections or PM actions on one or more non-failed items. This implies that planned maintenance actions are moved forward to exploit opportunities. Such PM actions are taken considering several factors and such as cost, criticality of the component, and functional requirements of the locomotive, etc.
- *Condition-based maintenance*: It includes a combination of condition monitoring and/or inspection and/or testing, analysis, and the ensuing better operation. When machine health deteriorates below predefined thresh hold limit, the machine is taken out of service for maintenance actions.

The EN-13306 standard describes 14 different types of maintenance policy, one of which is the condition-based maintenance policy and involves condition monitoring of the items or components. Condition monitoring is defined as the application of the appropriate sensors and sensors technologies to estimate the health and track the degradation of railway assets and its components.

The general concept of condition monitoring of an item is shown in Fig. 25.2. Here, the P–F interval is the warning period, the period between the point at which the onset of failure is detectable and the point of functional failure. If the condition monitoring is performed at intervals longer than the P–F interval, the potential failure may not

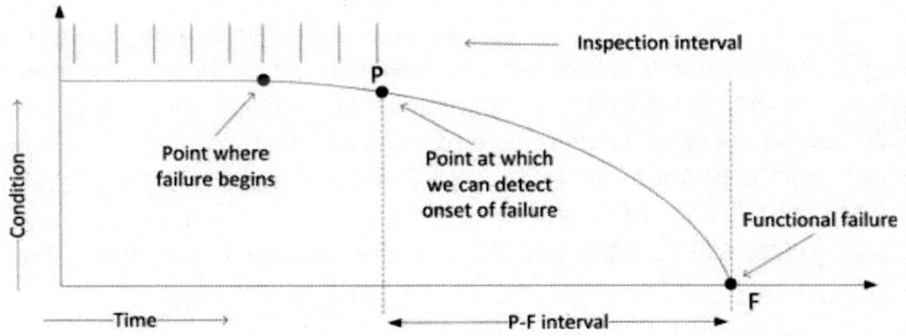

Fig. 25.2 The concept of condition monitoring [4]

be detected, and if the condition monitoring is performed at too high frequency compared to the P–F interval, it will be expensive. The monitoring interval must be selected in consideration of the cost and risk; the cost often increases with a higher monitoring frequency and the risk increases with a lower monitoring frequency [4].

Online condition monitoring is defined as the continuous monitoring of machines or production processes with the support of appropriate sensors. Currently, the railway sector all over the world has a mixed approach to maintenance strategy characterized by corrective, preventive, and condition-based maintenance.

- Tools to monitor the railway infrastructure (off Board wayside monitoring systems): These are installed on the infrastructure to observe its status. Some of these tools are able, starting from the monitoring of railway infrastructure status, to trace back the deterioration phenomenon of rolling stock. Monitoring the condition of the infrastructure can also facilitate the understanding of influences of infrastructures condition on the health of rolling stocks wheels, *i.e.*, the infrastructure condition influences the status of the rolling stock.
- Tools to monitor the railway infrastructure (On-board): These tools are installed on rolling stock and measure the state of infrastructure. For example, lindometer installed on locomotives is used to identify missing fasteners on the track. Dedicated diagnostic or service trains are equipped with sensors to measure the track geometry and other track parameters.
- Tools to monitor rolling stock status: These tools are generally installed on the wayside to monitor the condition of wheels and bogie, etc.

A *condition-based maintenance* strategy that integrates these tools will be a powerful instrument to reduce operating and maintenance costs. The information obtained by the various measurements, although disaggregated and heterogeneous, will be integrated using predictive models and innovative expert systems to support management decisions.

During the last few years, the focus has been to find transformative maintenance technology and business solutions for these assets, which will ensure safe and failure-free train operations at the lowest possible maintenance cost. Such solutions should make the use of railway assets (majority of these assets are old) possible with almost less risks.

From a quick review of the global technology trend in railway, it is evident that leaders in railway sectors have aligned their strategic thinking toward assets wide application of digital technologies and transformative business solutions for achieving excellence in their operations. Today, many railway operators have developed their operation-specific road map for the digital transformation of maintenance processes using the power of new and emerging technologies. For the digital transformation of maintenance work processes in maintenance, we propose a framework (see Fig. 25.3) that will facilitate transforming maintenance in the railway sector.

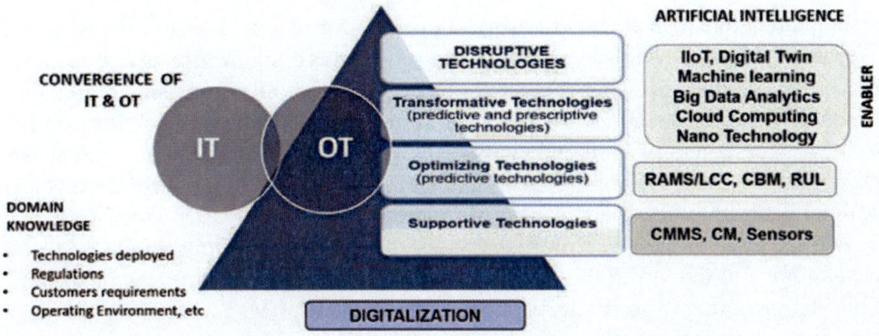

Fig. 25.3 Transformative maintenance technology framework for railway system

25.4 Digital Transformation of Railway and Transformative Maintenance Technologies

Technologies such as AI, ML, Big data analytics (predictive and prescriptive analytics), Virtual Reality (VR), Augmented Reality (AR), Industrial Internet of Things (IIoT), 5G communication technologies, etc., that offer near-perfect solutions (even in real time) for the maintenance of the aging railway assets, are collectively termed as transformative technologies.

The framework consists of four different stages of technological development relevant for maintenance.

- Supportive Technologies
- Optimizing technologies
- Transformative Technologies
- Disruptive technologies.

It also emphasizes the importance of domain knowledge. To operationalize the framework, digitalization of railway assets and maintenance processes is necessary.

25.4.1 Digitization, Digitalization and Digital Transformation

Digitization: It is the process of converting analog data and information into digital form. Not so long ago, maintenance data was analog as it was recorded manually on paper. However, with the advent of computers, engineers, and managers started recording and storing data and information into digital computer files. This process of collecting data and information digitally is called the process of digitization. In short, it is the process of converting information from analog to digital.

Digitalization: The process of using digitized information to make a decision and execute maintenance and other actions are referred to as digitalization of maintenance processes. Digitalization in the context of maintenance is all about using digital data to simplify decision-making processes and enhance the capability of condition monitoring processes and systems and facilitate optimal decision-making in maintenance. Digitalization makes it possible that data and information essential for correct and optimal decision-making are is instantly accessible.

Digital transformation: It is the process of using digital information and new technologies to transform the way maintenance actions are planned and executed. This reengineering of maintenance business in the digital age is called digital transformation of maintenance. A key element of digital transformation understands the underlying value-adding processes and potential of technologies under consideration. This will facilitate the selection of Best Available Technology (BAT).

Digital transformation of railway transport systems opens new opportunities allowing improving efficiency of the rail system in general including maintenance processes. Digital transformation of maintenance work processes and functions can help developing algorithms and models, which, in turn, may facilitate the implementation of digital tools for transforming maintenance management practices in the railway sector.

Without engineering and business domain knowledge related to the railway system, transforming maintenance processes will be difficult if not impossible [5].

25.4.2 Domain Knowledge

As evident from the Transformative Maintenance Technologies (TMT) framework, domain knowledge essentially consists of

- Understanding of railway systems and technology deployed,
- Operating environment,
  - Physical environment such as temperature, gradient,
  - Logistics and supportability organization,
  - Supportability & logistics, etc.
- Business-associated operational risks,
- Acts and regulations governing the railway operation,
- Customer's requirements, etc.

Prior to the use and implementation of transformative technologies in railway, knowledge of the system and its function, customers' requirements regulations, and interface with other assets is a must for the success of digital transformation.

In the following, different stages of technologies deployed for railway and future technology need is discussed.

25.4.3 *Supportive Technologies*

Supportive technologies relevant for maintenance are either provided by the manufacturers or are installed by the users as per need. These technologies are often used to monitor the health of the railway systems and facilitate decision-making in maintenance. Examples are condition monitoring modules, sensors designed and installed by manufacturers on railway systems, or condition monitoring systems installed by the railway operators.

In general, today most of the railway maintenance organizations are equipped with the smart Computerized Maintenance Management System (CMMS). Reliable railway maintenance enabled by CMMS technology is required to improve critical issues like safety, delays, and overall system capacity. An interconnected CMMS can help maintain, manage and connect tracks, terminals, rolling stock, and communications infrastructure. It can identify maintenance issues before these influence safety, operations, or revenue. It can collect, store, and analyze data to prevent breakdowns and issue predictive maintenance algorithms to extend equipment life [6].

25.4.3.1 Condition Monitoring Tools/Modules

One of the challenges in the implementation of condition monitoring as supportive tools for maintenance process is finding the measurement technologies suited to monitor the status of a specific system and able to provide valid and reliable measures. Physical sensors monitor the operating environment conditions. Monitoring critical parameters are necessary to guarantee safe execution of critical processes, avoid hazardous controls, and prevent possible critical failures or damage to the infrastructure (in case of rolling stocks).

25.4.3.2 Sensors

Several instruments and sensor devices are available to monitor the condition of rolling stocks and infrastructure and operating environmental parameters. These include smart sensors. Smart sensors have introduced the possibility of processing the data on the sensor boards itself, sending alarm messages in the event of suspect environmental conditions or event detection. Other kinds of smart devices, such as sensor networks, provide protection mechanisms able to isolate faulty and misbehaving nodes.

25.4.3.3 Radio-Frequency Identification (RFID) Technology

RFID is a technology for retrieving data on an object from a distance. The RFID technology is useful for tracking objects such as vehicles or spare parts. It makes the planning of maintenance activities more efficient by providing valuable and timely information on the state of the object. RFID has already proved effective in military, security, healthcare, and real-time object tracking and is widely being used by railway engineers. It can improve railway processes in various ways, such as automatic vehicle tracking and identification, operation and maintenance, asset management, and others [7].

25.4.4 Optimizing Technologies

The purpose of optimizing technologies is to facilitate optimization of the maintenance actions and essentially these include RAMS technologies, predictive technologies, etc., to optimize maintenance actions depending on the estimated remaining useful life of components or items. Often the focus is to get an estimate of the remaining useful life of the railway assets.

25.4.4.1 RAMS and LCC Technologies

Railway RAMS is an engineering discipline that integrates reliability, availability, maintainability, and safety characteristics appropriate to the operational objectives of a railway system into the inherent product design through railway systems engineering. It has the potential to improve the competitiveness of railway compare to other transport, especially road transport. Therefore, RAMS management is a significant issue in today's global railway projects and many of the leading railway companies have adopted it as a significant performance parameter both during design and operation phase. RAMS parameters are driver for the maintenance management and its effectiveness. There exists a vast literature on RAMS modeling, optimization, and technology [3, 8] and some of these are dedicated to RAMS in railway [9].

RAMS methodology together with LCC and risk analysis facilitates optimizing the failure prevention efforts during design or operation of the railway systems. RAMS standards for railway system (EN 501 26) define the responsibilities within the RAMS process throughout the life cycle of the railway system. An integrated methodology of RAMS, LCC, and RM can facilitate maintenance decision-making [10].

25.4.4.2 Predictive Technologies

Predictive maintenance is described as further development of condition-based maintenance strategy. In predictive maintenance, contextual information (such as operating environment, availability of support and location, etc.) is also integrated to estimate the best possible maintenance time or location (in case of rolling stocks) that will reduce the total business risks considering several engineering technical and business parameters.

The integration of various data, information, and processes is essential to the success of a predictive maintenance program. Predictive technologies facilitate the analysis of the trend of measured physical parameters against known engineering limits to detect, analyze, and correct a problem before a failure occurs either periodically or real time. Predictive technologies facilitate predictive analytics and in turn facilitates correct decision-making for remaining useful life considering some contextual information [11].

25.4.5 Transformative Maintenance Technologies and Business Solutions: The Role of Prescription

During the last few years, the focus has been to find transformative maintenance technologies and business solutions for the railway assets which will ensure safe and failure-free train operations at the lowest possible maintenance cost and time. Such solutions should make the use of railway assets (majority of these assets are old) with almost no risks. This real-time data-driven approach to operate railway is expected to transform the way railway asset is operated and maintained ensuring increased reliability and quality of service, increased capacity, and reduced life cycle costs for the asset. These transformative technologies such as artificial intelligence (AI), machine learning (ML), Big data analytics, IIoT, virtual reality, augmented reality, which can transform the railway maintenance practices and will enhance the maintenance effectiveness and efficiency, are discussed in brief in the following with examples from the railway sector.

25.4.5.1 Artificial Intelligence and Machine Learning

AI is defined as a technology able to perform tasks normally requiring human intelligence, such as visual perception, speech recognition, decision-making, and translation between languages, etc. AI refers to an artificial creation of human-like intelligence that can learn, reason, plan, perceive, and execute tasks. Artificial Intelligence in Industrial domain is often referred to as Industrial AI [12, 13].

Industrial AI capacity to analyze very large amounts of high-dimensional data can facilitate arriving at the most appropriate time for maintenance actions by integrating conventional data such vibration, current, or temperature with unconventional additional data, such as audio and image data and business data. AI can extend the life of an asset beyond what is possible using traditional analytics technique by combining model data information from designer and manufacturer, maintenance history, and Internet of Things (IoT) sensor data such as anomaly detection on the engine vibration data, and images and video of engine condition, from end user.

The railway sector is gradually adopting AI in its operation due to digitization of its various business units and work processes. In maintenance, AI is used for estimation of remaining useful life and for finding the best possible actions that will minimize the total business risks using the power of big data analytics and machine learning.

25.4.5.2 Machine Learning (ML)

ML is defined as the study of computer algorithms that improves automatically through experiences. It is seen as a subset of AI and is closely related to computational statistics [14]. In ML, a model typically learns from the training set and then performs the learned task, for example, classification or prediction, on new data. In this scenario, the model does not automatically learn from newly arriving data but instead carries out the already learned task on new data. To accommodate the knowledge embedded in new data, these models must be retrained.

Without retraining, they may become outdated and cease to reflect the current state of the system. ML uses data-independent variables, based on particular datasets. ML thrives on efficient learning algorithms, large datasets, and substantial computational performances to discover information and knowledge from raw data.

The ML algorithms can broadly be classified into three groups [15]:

- **Supervised Learning**: Supervised learning infers the relationships between a set of independent variables and a known dependent variable by mapping the independent variables with the dependent variable.
- **Unsupervised learning**: The goal of unsupervised learning is to find patterns or hidden structures from datasets consisting of a collection of input variables with the unknown output variable.
- **Reinforced learning**: In reinforcement learning, the system learns by means of the feedback given through rewards and punishments associated with actions. Similar to unsupervised learning, a reinforcement learning system is not provided with datasets containing pairs of known input–output variables.

Recent progress in machine learning has been driven both by the development of new learning algorithms and theory and by the ongoing explosion in the availability of online data and low-cost computation. Because of its ability to make predictions and reveal hidden insights from massive datasets, ML is one of the critical elements of big data analytics and provides a foundation for AI technology. The main role of ML in predictive maintenance is to automate the analysis of railway inspection and condition monitoring data and to reduce the subjectivity inherent in condition assessment [15].

As was mentioned above, in machine learning, a model typically learns from the training set and then performs the learned task. Therefore, to adapt to new information, algorithms must support incremental learning [16], sometimes referred to as sequential learning, which is defined as an algorithm's ability to adapt its learning based on the arrival of new data without the need to retrain on the complete dataset. This approach does not assume that the entire training set is available before learning begins but processes new data as they arrive. Although incremental learning is a relatively old concept, it is still an active research area due to the difficulty of adapting some algorithms to continuously arriving data.

Machine learning and data mining often employ the same methods and overlap significantly, but while machine learning focuses on prediction, based on known properties learned from the training data, data mining focuses on the discovery of (previously) unknown properties in the data.

25.4.5.3 Big Data

The on-going digitalization in industry and railway provides enormous capabilities for the railway sector to collect a vast amount of data and information (i.e., Industrial Big Data), from various processes and data sources such as operation, maintenance, and business processes. Accurate data and information available are one the prerequisites in maintenance knowledge discovery. Beside the collecting data and information, another challenge is to understand the patterns and relationships of these data useful and relevant for maintenance decisions [17].

The use of Big Data and the Internet of Things will allow transportation modes to communicate with each other and with the wider environment, paving the way for truly integrated and inter-modal transport solutions. In general, Big Data is associated with three unique characteristics, namely [18]:

- Volume
- Velocity
- Variety.

Volume: The increased use of an extensive network of sensors in rolling stocks and point assets are generating a huge amount of data every second.

Velocity: Velocity reflects the speed of data being collected and processed in short speed of data in and out.

Variety: Big data is usually collected from different sources and in different formats. These sources can be text, videos, images, or readings from sensors. The collected data can be structured and unstructured. Variety indicates

Apart from these three, two more terms are used to describe the characteristics of Big data and these are

Veracity: Some maintenance-related data are structured while some are not, such as free text comments for performed maintenance actions or failure reports. Those data have potential value when properly employed in asset management, but in order to achieve this, there is a need to assess and manage the veracity of the data, i.e., the data uncertainty.

Value of data: Collected data must have value for the purpose of analysis, i.e., how data can enable efficiency and effectiveness in maintenance management, for instance, for improved decision-making, and to choose the most cost-effective means to process the data is important [19].

25.4.5.4 Maintenance Data Management

In general, a lot of information and data needs to be captured and analyzed to assess the overall condition of railway assets, maintenance actions taken or planned, inspection data of railway tracks, rolling stocks, etc. Examples of information that are collected include track availability, use of track time, track condition, performance history, and work performed. Measurements of the condition of the track typically include continuous and spot measurements from automatic inspection vehicles, visual inspections from daily walking inspections, and records of in-service failures. Examples of conditions measured by automatic inspection vehicles are geometry car measurements (deviation from design curves, geometry exceptions to railroad standards, vehicle ride quality exceptions), rail measurements, gage restraint measurements, track deflection and stiffness measurements, clearance measurements, and substructure measurements.

Despite the plethora of information, decision-making can be difficult. Even with an accurate map of the corridor, rail, ties, and other corridor assets have no physical characteristics that lead to easy identification. Furthermore, problem areas for targeted maintenance often do not have discrete physical boundaries, such as the beginning and ending of a rail section. In addition, the transport administration collects large amounts of data on the railroad and rail traffic. This information is divided into a variety of databases/systems, and it is not easy to get an overall picture of what information is available. These challenges can be met by effectively using Big Data [20].

25.4.5.5 Maintenance Analytics

The concept for Maintenance Analytics (MA) is based on three interconnected time-lined phases, which aim to facilitate maintenance actions through an enhanced understanding of data and information. The MA phases are (see Fig. 25.4) [21]: (i) Maintenance Descriptive Analytics, (ii) Maintenance Predictive Analytics, and (iii) Maintenance Prescriptive Analytics.

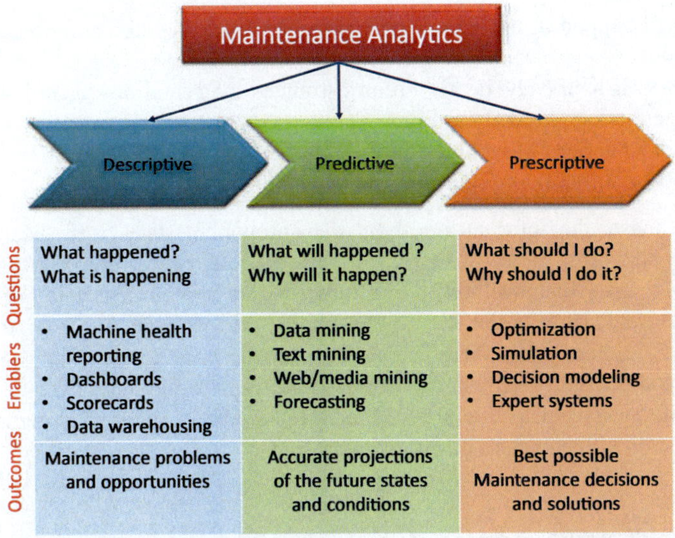

Fig. 25.4 The phases of maintenance analytics [22]

Maintenance Descriptive Analytics

The maintenance descriptive analytics aims to answer:

- What has happened? The algorithm provides information about the state and condition of the component. It identifies the failed items.
- Why has it happened? It explains the reason for failure or explain the current state of the item,
- What is happening? Algorithm visualizes the current state,
- Why is it happening? Algorithm explains the reason why the event is taking place.

Maintenance descriptive analytics provides "now casting" of the physical state or health of infrastructure or rolling stocks.

Here, access to data related to system operation, system condition, and expected performance and threshold are important. Such solutions are expected to support railway's digital transformation journey and operations goals. Another important aspect in order to understand the relationship of events and states during the descriptive analytics is time and time frame associated with each specific time log. For this phase, the availability of reliability data is necessary besides the data used in the descriptive phase.

Maintenance Predictive Analytics

The maintenance predictive analytics aims to answer:

- What will happen in the future? The algorithm identifies the components likely to fail.
- When will it happen? The algorithm estimates the remaining useful life of the component which is likely to fail.
- Why will it happen? Most of the time predictive analytics also identifies the reason for failure if enough and relevant data is available.

In this phase, the information about the component state from 'Maintenance descriptive analytics' is used to arrive at the remaining useful life. For the success of predictive analytics, availability of reliability data and context data is necessary besides the data used in the descriptive phase. Furthermore, in order to predict upcoming failure and fault, there is a need to provide business data such as planned operation and planned maintenance of different items (rolling stocks) and different section (infrastructure) to this phase. Predictive analytics gives information about the future state of the system through forecasting.

Maintenance Prescriptive Analytics

The maintenance prescriptive analytics aims to answer the following:

- What needs to be maintained? The algorithm identifies the maintenance actions needed for the identified nonperforming items that should be maintained or replaced.
- When should it be maintained? Once an item is identified for maintenance, when should it be taken out of operation for maintenance?
- How should it be maintained? What is the best possible means of maintenance? It identifies and recommends different maintenance scenario or actions with associated risks.
- Who should maintain it? In house or contractor? Maintenance prescriptive algorithm can also identify whether it is cost-effective to perform the maintenance action in the house or outsource it to contractors with relevant data and information is available.
- Where should it be maintained (specifically for rolling stocks)? The algorithm should be able to identify the best possible workshop which can perform the maintenance action. This is useful for maintenance of rolling stocks.
- What other components should be maintained together (e.g., opportunistic maintenance for infrastructure and rolling stocks)? In this phase, outcomes from 'Maintenance Diagnostic Analytics' and 'Maintenance Predictive Analytics' are used.

In addition, in order to predict upcoming failure and fault, there is a need to provide resource planning data and business data also. As described above, the different phases of maintenance analytics are highly dependent on the availability of a vast amount of data from various data sources. Maintenance analytics provides the

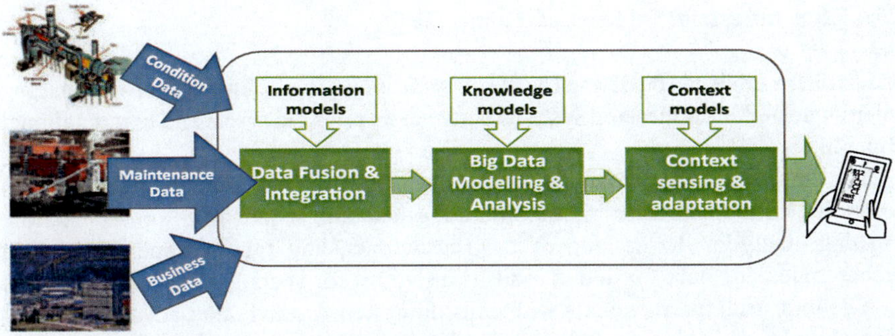

Fig. 25.5 Maintenance overarching approach [21]

foundation for eMaintenance platform for railway maintenance used in Sweden for maintenance optimization.

25.4.5.6 eMaintenance

There is no standard definition of eMaintenance. We define it as a concept which connects all the stakeholders, integrates their requirements, and facilitates optimal decision-making on demand or in real time to deliver the planned and expected function, capacity and services from the assets while minimizing the total business risks [21].

It can also be defined as materialization of information logistics aimed to support maintenance decision-making. eMaintenance solutions essentially combine information, knowledge, and context model, as illustrated in Fig. 25.5. However, eMaintenance implies a wide range of tools, technologies, and methodologies aimed for maintenance decision-making, including data analytics.

25.4.5.7 5G Communication

Development in 5G capability has the capability to facilitate automation of maintenance decision and actions as a huge amount of data can be transferred and analyzed in a fraction of time. The area of Industrial AI and more specifically transformative maintenance technologies will see an accelerated development as 5G will facilitate real-time data transfer making it possible to make decisions in real time the for maintenance of railway systems.

25.4.5.8 Industrial Internet of Things (IoT)

Industrial internet can be defined as the new and emerging technologies for managing interconnected machines and systems between its physical assets and computational capabilities [23].

The Industrial Internet of Things (IIoT) is the use of Internet of Things (IoT) technologies in manufacturing incorporating machine to machine communication, big data analytics, harnessing of the sensor data, and robotics and automation technologies that have existed in industrial settings for years.

The Industrial Internet starts with embedding sensors and other advanced instrumentation in an array of machines from the simple to the highly complex. This allows the collection and analysis of an enormous amount of data, which can be used to improve machine performance, and inevitably the efficiency of the systems and networks that link them. Even the data itself can become "intelligent," instantly knowing which users it needs to reach.

The three main components of this concept, namely intelligent devices, intelligent systems, and digital instrumentation to industrial machines are the first step in the Industrial Internet Revolution.

Cyber-Physical Architecture

Cyber-Physical Systems (CPS) is defined as transformative technologies for managing interconnected systems between its physical assets and computational capabilities. The five-level CPS structure, namely the 5C architecture, provides a step-by-step guideline for developing and deploying a CPS for manufacturing application [23].

Connection: Acquiring accurate and reliable data from machines and their components is the first step in developing a Cyber-Physical System application. The data might be directly measured by sensors or obtained from controller or enterprise resource planning (ERP) systems.

Conversion: Data-to-information conversion: Meaningful information to be inferred from the data. In recent years, the extensive focus has been applied to develop these algorithms specifically for prognostics and health management applications. By calculating health value, estimated remaining useful life, etc., the second level of CPS architecture brings context awareness to machines.

Cyber: The cyber level acts as a central information hub in IoT architecture. Information is being pushed to it from every connected machine to form the machines' network.

Cognition: Implementing CPS upon this level generates a thorough knowledge of the monitored system. Presentation of the acquired knowledge to expert users supports the correct decision.

Configuration: The configuration level is the feedback from cyberspace to physical space and acts as supervisory control to make machines self-configure and self-adaptive. This stage acts to apply the corrective and preventive decisions, which has been made in the cognition level, to the monitored system.

IoT technology has made predictive maintenance more affordable and available to railway maintenance engineers. By using the data that connected machines provide to measure damage, wear and tear, and other indicators of operational success, operators are gaining unprecedented insight into machine health and this, in turn, is changing popular methods of maintenance. Equipment can be monitored and fixed remotely, for example, with the potential capability to even repair itself. The combined use of sensors, embedded electronics and analytics services, and cloud-based systems result in increased productivity and a significant savings in maintenance costs [24].

25.4.5.9 Cloud and Edge Computing as Digital Enablers

Cloud computing is the delivery of computing services—including servers, storage, databases, networking, software, analytics, and intelligence—over the Internet ("the cloud") to offer faster innovation, flexible resources, and economies of scale. The term is generally used to describe data centers available to many users over the Internet.

However, because of various limitations, i.e., computational complexity, cloud computing has been overtaken by edge computing in IoT-based industrial applications. Intelligent and accurate resource management by artificial intelligence (AI) is also of increasing interest. The coordination of AI with edge computing will remarkably enhance the range and computational speed of IoT-based devices in industry [25].

An edge computing architecture that operates without the cloud is not to be confused with local scenarios in which all data are processed on individual devices. While such on-board computing can support critical decision-making in real time, the device hardware is costly. Moreover, the ability of local configurations to support operations such as machine learning is often limited. Conversely, an AI-enabled edge computing system in a factory could contextualize data from multiple machines to detect and ultimately predict problems causing downtime. That doesn't mean deploying machine learning at the edge is necessarily easy, however. It "requires more mature machine learning models and new ways to manage their deployments" [26].

Most of the time, the flow of data will be bidirectional between the edge and the cloud. While the cloud can foster the tracking of broad trends and second-order effects such as changes in energy consumption or air quality, edge computing gives local answers to local questions.

To give an example, Talgo (Spanish Train Manufacturer) has deployed telematics and remote diagnostic systems in its newly manufactured train sets with edge computing platforms. The system works, in part, by using the on-board computer that detects abnormal parameters and triggers trouble codes. From there, its communication system streams troubling operational data to train the health-monitoring center, which can coordinate responses from relevant parties, such as repair shops, dealers, and customer service agents.

25.4.5.10 Virtual Reality and Augmented Reality

Virtual Reality in Maintenance

Virtual reality (VR) is generally considered a natural extension to 3D computer graphics and is now mature enough to be used in industrial applications. VR, together with up-to-date software systems, supports such industrial applications as design, engineering, manufacturing, operations, and maintenance. The integration of virtual reality and industrial operations will help develop a cost-efficient production system with sophisticated maintenance management [27].

The functionality and acceptance of VM depend on the following issues:

- How the virtual maintenance activities help the maintenance engineer get a cohesive view of maintenance issues;
- How the VM activities support the maintenance engineer in decision-making processes;
- How the relevant supporting technologies can be applied to real mechanical maintenance needs;
- How efficiently and comfortably the maintenance engineer can perform the relevant tasks using the developed system.

Maintenance professionals have a lot to gain by capitalizing on the benefits of virtual reality. It starts with training. Technicians can immerse themselves in their future work environment to explore the site topology and become familiar with the environment in which they will be working. They can practice the activities they will need to perform, repeat them any number of times, and learn from their mistakes with no risk of customer repercussions or danger. In a virtual world, errors can't damage equipment.

With virtual reality, technicians have a far better ability to understand the layout and equipment at the customer site than they do by reviewing technical manuals or topography of the railway network. Virtual reality can be used to simulate any number of scenarios at the site as well as specific events and weather conditions such as darkness, rain, and snow associated it difficult network landscape [28].

Augmented Reality (AR)

Augmented reality (AR) is an innovative technology used to supplement a real-world environment with computer-generated sensory input. These virtual components seem to coexist with real ones in the same space, enhancing the user's perception of reality and enriching the information provided. There are many application areas for AR, ranging from the engineering field to various aspects of everyday life. AR has been defined as a human–machine interaction tool that overlays computer-generated information on the real-world environment. It has three key features [29]:

- Combination of real and virtual objects in a real environment;
- Real-time interaction with the system, able to react to the user's inputs;
- Geometrical alignment of virtual objects to those in the real world.

The strengths of AR are the following:

- Immersive system: Information is directly integrated into the real world;
- Immediate interpretation of information;
- Paperless ability to provide a large amount of knowledge;
- Possibility of integrating the system with other computer-aided devices;
- Faster procedures: The operator's attention is not taken away from the real environment when s/he is consulting procedural instructions.

With digitalization, application of AR is gradually becoming popular.

AR for Maintenance and Repair

Maintenance and repair activities represent a great number of AR applications; these use various overlay methods and hardware. AR techniques have great potential in remote maintenance applications, as they are capable of providing a mixed image (virtual and real) to the worker in the field and remotely to the expert assistant. In this context, work on an AR strategy for aiding the remote collimator exchange in an energy particle accelerator explores whether and how virtual co-location based on AR can be used to remotely support maintenance during space missions, wearing a Head Mounted Display (HMD). AR remote maintenance sessions have also been tested in the railway sector [29].

Augmented Reality for Training and Remote Maintenance

With its ability to provide remote instructions, AR is enabling new paradigms for maintenance, including remote maintenance and maintenance customized to the workers' understanding and skills. AR technology can also facilitate the training of maintenance workers by an equipment vendor, or even by other more experienced workers. The training concept involves presenting a cyber-representation that demonstrates how to perform the maintenance.

Virtual reality (VR) can be also used for training purposes in a similar context, based on pre-recorded visual presentations of the task at hand. The use of AR for training and remote maintenance is likely to expand in the coming years. The rising complexity of industrial equipment and machinery makes it increasingly difficult and more expensive to detect, troubleshoot, and repair failures. Maintenance workers may struggle in an era of numerous product variants and configurations, particularly given the technologically advanced capabilities that accompany most equipment. There are already several vendors that offer enterprise-scale AR solutions for maintenance tasks. In addition, some Industrial IoT solution providers (such as Thingworx.com) bundle and offer AR as a unique selling proposition. The AR solution suggests multiple applications, particularly to facilitate maintenance processes [30].

25.5 Transformative Technologies: Some Applications and Examples from Railways

25.5.1 Digital Twins

A digital twin is a dynamic digital model of a physical asset or system. It uses real-time data from sensors to continuously represent the physical reality of that asset or system. A digital twin can be maintained throughout the life cycle and is easily accessible at any time.

A digital twin can be defined as "a digital model capable of rendering state and behavior of a unique real asset in (close to) real time." It has five core characteristics [31]:

- It connects to a single, real, and unique physical asset. When we observe a state in the digital twin, it corresponds one-to-one with a potential observation on a particular physical asset.
- It captures essential physical manifestation of the real asset in a digital format, such as CAD or engineering models with corresponding metadata.
- It has the capability to render quantifiable measures of the asset´s state in (close to) real time.
- It reflects basic responses to external stimuli (forces, temperatures, chemical processes, etc.) in the present context.
- It describes the external operating context, such as wind, temperature, etc., within which the asset exists or operates.

Apart from providing an estimate of remaining useful life of an asset or components dynamically, digital twin can be useful for the prediction of future state of an asset and related risk scenario.

These can be specific end-user applications for monitoring and control or legacy applications for maintenance and asset management, or the data stream might feed into data analytics and machine learning stacks for pattern recognition and decision support [31].

25.5.1.1 Digital Twin in Railways

A digital twin is a virtual replica of a physical asset, making it a vital element of the digital rail solution. Because it is continuously updated, engineers can test detailed what-if scenarios that help in the planning of enhancement and maintenance programs.

Optimization of operational availability can reduce operating costs due to maintenance. A digital twin can display the state of the asset while it is running. Later, it can be linked to IT systems to help streamline and optimize maintenance processes and operational availability.

For each asset, engineers compile thousands of data points, specifically during the design and manufacturing phases. These are used to build a digital model that tracks and monitors an asset in real time, providing essential information throughout the life cycle. It can provide early warnings, predictions, and even a plan of action by simulating what-if conditions, thus keeping an asset in service longer [6].

25.5.1.2 Convergence of IT and OT to Create Railway Digital Twins

The information technology (IT) and operational technology (OT/operations) departments within a railway company have traditionally functioned somewhat independently. OT has kept the trains running smoothly and the infrastructure in good condition, while IT has managed business applications.

Railway leaders recognize that the operational data they use to support real-time decision-making could create additional value for the company. But these data must be merged with IT data in a meaningful way and made accessible across the organization. At the same time, IT needs to achieve the vision of a connected railway by driving innovation and minimizing downtime. But to get there, IT needs the knowledge and support of OT, as operations departments understand and control the assets.

The technology and operation of railway assets are complex, but the adoption of the Internet of Things (IoT) and its use with OT platforms enable the use of digital twins to manage, monitor, and maintain assets. Digital twins connect complex assets and their OT systems to an IT environment by capturing data to monitor performance, deterioration, and failure, as well as location, and safety compliance, for scheduling and asset utilization.

Through data fusion, railway digital twins become virtual and digital representations of physical entities or systems. However, a railway clone created with IT and OT convergence to forecast failures, demand, customer behavior, or degradation of assets is not complete since it lacks engineering knowledge. This happens because the digital engineering models developed during the engineering phase of railway projects do not typically play a role in the operational phase.

Therefore, railway digital transformation demands that engineering technology (ET) be included in the IT/OT convergence process as the importance of integrating product design increases. For that purpose, railway digital twins must be complemented by other information to assess the overall condition of the whole fleet/system, including information from design and manufacturing, as this obviously contains the physical knowledge of assets.

The integration of asset information throughout the entire life cycle is required to make accurate health assessments, determine the probability of a shutdown or slow down, and avoid black swans and other unexpected or unknown asset behaviors. Moreover, the lack of data on advanced degradation makes the data-driven approach, where IT and OT are only actors, vulnerable to such situations and ET is slowly gaining entry to the convergence conversation, even though engineering models often remain stranded in information silos, inhibiting the ability to leverage this information to optimize railway operations.

Despite these challenges, hybrid models comprising engineering knowledge and data collected from the field will soon be part of digitization all over the world. In short, the engineering technology (ET) of a railway asset, together with IT and OT, will help operation and management departments forecast problems, conduct better planning, and improve performance. Fortunately, it is now possible for companies to merge their IT, OT, and ET to enable asset performance modeling to deliver actionable intelligence for decision support [32].

25.5.1.3 Digital Transformation of Railways

The creation of smart, environment, and user-friendly mobility systems is a high priority in the evolution of transport worldwide. Rail transport is recognized as a vital part of this process. Meanwhile, radical advancement in the business environment, facilitated by ICT technologies, requires the existing business models and strategies adopted by rail operators to be brought up to date. The thorough understanding of the concept of digital transformation is paramount in the development of rail transport in the new economy.

Digitalization, as an ongoing process of convergence of the physical and virtual worlds, is bound toward Cyber-Physical Systems (CPS) and is responsible for the innovation and change in multiple sectors of the economy.

The main technologies and solutions which have accelerated digital transformation in the railway sector in recent years are

- Internet of Things (IoT),
- Cloud computing,
- Big Data analytics (BDA),
- Automation and robotics.

25.5.2 Self-Maintenance

Self-maintenance is a new concept. Self-maintenance machines are expected to be able to monitor, diagnose, and repair themselves to increase their uptime. One approach to self-maintenance is based on the concept of functional maintenance. Functional maintenance aims to recover the required function of a degrading machine by trading off functions, whereas traditional repair (physical maintenance) aims to recover the initial physical state by replacing faulty components, cleaning, etc. The way to fulfil the self-maintenance function is by adding intelligence to a machine, making it clever enough for functional maintenance so that the machine can monitor and diagnose itself. In other words, self-maintainability is appended to an existing machine as an additional embedded reasoning system.

The required capabilities of a self-maintenance machine (SMM) are the following:

- Monitoring capability: SMMs must have the ability to perform online condition monitoring using sensor fusion.
- Fault judging capability: From the sensory data, the SMM can judge whether the machine condition is in a normal or abnormal state.
- Diagnosing capability: If the machine condition is in an abnormal state, the causes of faults must be diagnosed and identified to allow repair planning actions.
- Repair planning capability: The machine must be able to propose repair actions based on the result of diagnosis and functional maintenance.
- Repair executing capability: The maintenance is carried out by the machine itself without human intervention.
- Self-learning and improvement: When faced with unfamiliar problems, the machine must be able to repair itself. If such problems recur, the machine will take a shorter time to repair itself, and the outcome of maintenance will be more effective and efficient.

Efforts to realize self-maintenance have been mainly in the form of intelligent adaptive control, where investigation of control has been achieved using fuzzy logic control. Self-maintenance requires the development and implementation of an adaptive artificial neuron–fuzzy inference system which allows the fuzzy logic controller to learn from the data it is modeling and automatically produce appropriate membership functions and the required rules. The controller must be able to handle sensor degradation, and this leads to self-learning and improvement capabilities.

Another system approach to self-maintenance is to add a self-service trigger function to a machine. The machine self-monitors, self-prognoses, and self-triggers a service request before a failure occurs. The maintenance task may still be conducted by a maintenance crew, but the no gap integration of machine, maintenance schedule, dispatch system, and inventory management system will minimize maintenance costs and raise customer satisfaction [33].

25.5.3 *Unmanned Aircraft System Technology in Railway Applications*

UAS technology is having a powerful and transformative impact on the rail industry. In railroad environments, UASs are particularly suitable for the following [34]:

- Structural monitoring, especially for critical assets like bridges and tunnels, and for fault detection (i.e., diagnostics/prognostics).
- Environmental security monitoring, such as assessments of fire, explosions, earthquakes, floods, and landslides along the track.
- Physical security monitoring, including detection of intrusions, objects stolen or moved, graffiti, etc.
- Safety monitoring, for example, to early detect failures on track elements/devices or obstacles on the track.
- Situation assessment and emergency/crisis management to monitor accident scenarios and coordinate the intervention of first responders.

The use of UAS technology offers the following direct benefits for routine inspection activities [34]:

- Reduction of risk to staff and people and infrastructure in the project area,
- Reduced planning cycle,
- More efficient work processes,
- More flexible, affordable verification tools,
- Higher quality data available in larger quantities at lower costs.

When natural disasters strike, many railway assets can be at risk. In such situations, it is critical to determine which part of the railway needs repair prior to the movement of trains. UASs can gather information on the condition of the track or bridges, as well as the presence of debris on the right of way.

The aging of rail infrastructure poses challenges. Visual condition assessment remains the predominant input to the decision-making process. Many railways use machine-vision technology installed on rail-bound vehicles, but there are situations in which inspectors on foot or in hi-rail vehicles assess the track's surroundings. In the case of high or steep slope embankments, UASs can collect detailed information that could be missed by inspectors [34].

25.5.3.1 UAVs Suitable for Railway Applications

Two primary UAV types are available for railway operations: "rotary wing" aircraft, and "fixed-wing" aircraft, shown in the lower portion.

Rotary wing UAVs share many characteristics with manned helicopters. Rather than a continuous forward movement to generate airflow, these units rely on a lift from the constant rotation of the rotor blades. There is no limit on how many blades an aircraft has, but the average is between four and eight. Unlike fixed-wing units, rotary-wing units have the ability to vertically take off and land, so they can be deployed virtually anywhere. This enables the aircraft to lift vertically and hover at a specific location. These UAVs can move in any direction, hovering over important areas, collecting the most intricate data. This ability makes them well suited for inspections where precision manoeuvring is critical to the operation.

Fixed-wing UAVs are designed for higher speeds and longer flight distances. This type of UAV is ideal for coverage of large areas, such as aerial mapping and surveillance applications. It can often carry heavier payloads than a rotary UAV. Fixed-wing UAVs glide efficiently, and the single fixed-wing drastically reduces the risk of mechanical failure. The maintenance and repair requirements for these units are often minimal, saving time and money. However, the current Beyond Visual Line of Sight (BVLOS) regulations limit the utility of fixed-wing UAVs. Several railways are using multi-rotor or hybrid vehicles that employ multiple rotors along with fixed wings to facilitate short take-offs. Among the various types of UAVs, the one with the highest number of units worldwide is the rotary-wing followed by the fixed-wing [34].

The nano-type UAV is becoming prevalent in the UAS market space. It is a palm-sized platform with a maximum take-off mass of less than 30 g. It has advanced navigation systems, full-authority autopilot technology, digital data links, and multi-sensor payloads. The operational radius for this type of platform is more than 1.5 km, and it can be flown safely in strong wind. Future development is anticipated to yield even smaller and more advanced nano-UAVs with high levels of autonomy. Cameras are still the most common sensor used on a UAV.

25.5.4 Disruptive Technology

There is no standard definition of disruptive technologies, but usually, it is defined as the technology that changes and disrupts and challenges the established business models and practices. Examples are Uber car rental, etc. We define disruptive technology as the transformative technologies that are adopted by the industry because of its ease of use, cost-effectiveness by use of new and emerging technologies.

25.6 Transformative Business Solution

25.6.1 Maintenance as a Service (MaaS)

Rail transportation is also experiencing another emerging trend raised in the last years, the so-called servitization, a process of creating value by adding services which involves a transition from product-centric where end user acquires the asset and responsibility toward offerings to progressively increase the content of services.

This implies the adoption of a more customer-centric approach, addressing concrete customer needs with more tailored solutions instead of just products. These services are closely related to maintenance whenever equipment manufacturers offer the solution to their clients and they can be provided throughout the whole equipment life cycle. Hence, the purpose of this chapter is to show the new role of maintenance and new actors powered by the digitization and knowledge discovery with a focus on Industry 4.0. This concept in transportation could be seen as a smart system consisting of physical complex assets such as vehicles and infrastructure that are highly connected and integrated.

The ability to collect and process large volumes of maintenance data in a cloud can enhance maintenance capabilities and services. Such services are increasingly provided for specific machinery or equipment and include:

- Predicting the lifetime of a product or providing insights on the optimal time for maintenance;
- Providing context-aware information about service maintenance, including manuals, videos, VR representations, and interactive support;
- Configuring IT and business information systems (e.g., ERP and asset management systems) based on the results of the analysis;
- Providing in-depth statistics and reports about the operation of the equipment.

All these services can be delivered on-demand, when and where they are needed. This gives rise to an entirely new paradigm for industrial maintenance, "Maintenance-as-a-Service." In this paradigm, the equipment vendor is able to charge the plant operator based on the actual use of maintenance services, rather than a flat service fee associated with the equipment.

Maintenance-as-a-Service (MaaS) could become a game-changer in industrial maintenance. It can motivate machine vendors to provide the best service while providing versatile, reliable, and functional equipment. In the future, we will likely see maintenance service revenues increasing.

Note that MaaS implies a bundling of machine maintenance in the broader pool of customer services. In this direction, IT vendors (such as Microsoft) are integrating maintenance functions with their customer relationship management (CRM) and service platforms.

Early MaaS features are already provided by some equipment vendors. For example, ThyssenKrupp Elevators come with a proactive maintenance program, which predicts maintenance problems before they occur and notifies maintenance engineers accordingly. MaaS is likely to extend to consumer equipment and goods as well. For example, BMW, the German automotive manufacturer, is planning to offer MaaS programs, which will let car owners know the best possible time for maintenance, repair, and service activities [30].

25.7 Concluding Remarks

A quick review of development in the railway sector shows that the railway sector has embraced and aligned their strategic thinking toward assets wide application of digital technologies and solutions for achieving excellence in their operations. In that respect, application of Big Data analytics, machine learning, applications of drone technology, robotics, etc., to facilitate the implementation of state-of-the-art predictive maintenance technologies has been a dominant theme for future investments to for ageing infrastructure and rolling stocks maintenance. The railway sector has boarded the digital train and exploring means for the implementation and exploitation of the new transformative technologies to integrate digital and physical infrastructures to ensure robust and reliable railway assets.

References

1. Patra, A. P., Kumar, U., Larsson-Kråik, P. O. (2010). Availability target of the railway infrastructure: An analysis. In *2010 Proceedings: Annual Reliability and Maintainability Symposium*, San Jose, California, USA, January 28, 2010.
2. Seneviratne, D., Ciani, L., Catelani, M., & Galar D. (2018). Smart maintenance and inspection of linear assets: An Industry 4.0 approach. *ACTA IMEKO, 7*(1), 50–56. ISSN: 2221-870X.
3. Ben-Daya, M., Kumar, U., & Murthy, D. P. (2016). *Introduction to maintenance engineering: modelling, optimization and management*. USA: Wiley.
4. Asplund, M., Famurewa, S., & Rantatalo, M. (2014). Condition monitoring and e-Maintenance solution of railway wheels. *Journal of Quality in Maintenance Engineering, 20*(3), 216–232.
5. Salesforce. (2020). *What is digital transformation?* https://www.salesforce.com/products/platform/what-is-digital-transformation/. Accessed 21-03-20.
6. Shukla, D. (2019). *Industry 4.0 solutions for new-age railways and airways.* June 18, 2019.
7. Kans, M., Galar, D., & Thaduri, A. (2016). Maintenance 4.0 in railway transportation industry. In *Proceedings of the 10th World Congress on Engineering Asset Management (WCEAM 2015)* (pp. 317–331). https://doi.org/10.1007/978-3-319-27064-7-30.
8. Misra, K. B. (2011). *Principles of reliability engineering*. LTU Press.
9. Mahboob, Q., & Zio, E. (2018). Handbook of RAMS in railway systems. In *Theory and practice* (1st Ed.). Published March 29, 2018 by CRC Press. ISBN 9781138035126.
10. Thaduri, A. & Kumar, U. (2020). Integrated RAMS, LCC and risk assessment for maintenance planning for railways. In *Advances in RAMS engineering* (pp. 261–292). Cham: Springer.
11. Thaduri, A., Kumar, U., & Verma, A. K. (2017). Computational intelligence framework for context-aware decision making. *International Journal of System Assurance Engineering and Management, 8*(4), 2146–2157.
12. Lee, J. (2020). Industrial AI. In *Applications with sustainable performance.* © 2020 Springer Nature Switzerland AG.
13. Kumar, U., Galar, D., & Karim, R. (2020). *Industrial AI in maintenance: False hopes or real achievements?* Maintenance World, March 2020.
14. Wikipedia (2020). *Machine learning*. https://en.wikipedia.org/wiki/Machine_learning. Accessed 16-03-20.
15. Jordan, M. I., & Mitchell, T. M. (2015). Machine learning: Trends, perspectives, and prospects. *Science, 349*(6245), 255–260.
16. Ross, D. A., Lim, J., Lin, R. S., & Yang, M. H. (2008). Incremental learning for robust visual tracking. *International Journal of Computer Vision, 77*(1–3), 125–141.

17. Kohli, S., Kumar, S. A. V., Easton, J. M., & Clive, R. (2017). *Innovative applications of big data in the railway industry. IGI Global*. November 30, 2017.
18. Galar, D., & Kumar, U. (2017). *eMaintenance: Essential electronic tools for efficiency*. Academic Press.
19. Galar, D., Thaduri, A., Catelani, M., & Ciani, L. (2015). Context awareness for maintenance decision making: A diagnosis and prognosis approach. *Measurement, 67,* 137–150.
20. Thaduri, A., Galar, D., & Kumar, U. (2015). *Railway assets: A potential domain for big data analytics*. In *2015 INNS Conference on Big Data* (Vol. 53, pp. 457–467). Lulea, Sweden: Lulea University of Technology.
21. Karim, R., Westerberg, J., Galar, D., & Kumar, U. (2016). Maintenance analytics—The new know in maintenance. *IFFAC on Line Paper, 49*(28), 214–219.
22. Kumar, U., & Galar, D. (2018). Maintenance in the era of industry 4.0: issues and challenges. In *Quality, it and business operations* (pp. 231–250). Berlin: Springer.
23. Lee, J., Bagheri, B., & Kao, H.-A. (2014). *A cyber-physical systems architecture for Industry 4.0-based manufacturing systems*. NSF Industry/University Cooperative Research Center on Intelligent Maintenance Systems (IMS), University of Cincinnati, Cincinnati, OH, United States.
24. Yoskovitz, S. (2016). Predictive maintenance will change the future. In *The shift to tool-assisted predictive maintenance is coming soon*. March 28, 2016. Accessed 15-03-2020.
25. Sodhro, A. H., Pirbhulal, S., & Albuquerque, V. H. C. (2019). Artificial intelligence driven mechanism for edge computing based industrial applications. *IEEE Transactions on Industrial Informatics, PP*(99), 1–1. https://doi.org/10.1109/tii.2019.2902878.
26. Buntz, B. (2020). *Edge computing benefits for ai crystallizing*. January 30th, 2020. Copyright © 2020 Informa PLC. https://www.iotworldtoday.com/2020/01/30/edge-computing-benefits-for-ai-crystallizing/. Accessed 10-04-2020.
27. Shamsuzzoha, A., Helo, P., Kankaanpää, T., Toshev, R., & Vu Tuan, V. (2018). Applications of virtual reality in industrial repair and maintenance. In *Proceedings of the International Conference on Industrial Engineering and Operations Management*, Washington DC, USA, September 27–29, 2018.
28. Biseul, X. (2019). *The benefits of virtual reality and augmented reality for maintenance*. https://en.praxedo.com/blog/benefits-virtual-reality-augmented-reality-maintenance/. Accessed 15-04-2020.
29. Dini, G., & Dalle Mura, M. (2015). Application of augmented reality techniques in through-life engineering services. In *The Fourth International Conference on Through-life Engineering Services*. Department of Civil and Industrial Engineering, University of Pisa, Via Diotisalvi, 2, Pisa 56122, Italy.
30. Soldatos, J. (2019). 5 transformational trends reshaping industrial maintenance. In *Category: Digital transformation*, July 25, 2019. https://www.prometheusgroup.com/posts/5-transformational-trends-reshaping-industrial-maintenance. Accessed 20-02-2020.
31. Erikstad, S. O. (2017). *Merging physics, big data analytics and simulation for the next-generation digital twins*. NTNU and SAP, Trondheim/Norway. September 2017.
32. Galar, D. (2018). *The creation of railway digital twins through the convergence of IT and OT*. August 2018. https://www.globalrailwayreview.com/article/72072/digital-twins-it-ot/. Accessed 10-04-2020.
33. Lee, J., & Wang, H. (2008). *New technologies for maintenance*. NSF Center for Intelligent Maintenance Systems. PO Box 0072, Univ. of Cincinnati, OH 45221, USA. January 2008.
34. U.S. Department of Transportation/Federal Railroad Administration. (2018). *Unmanned aircraft system applications in international railroads*. Final Report, Office of Research, Development and Technology Washington, DC, USA, February 2018.

Prof. Uday Kumar is a Professor of Operation and Maintenance Engineering at Luleå University of Technology, Luleå Sweden. His research interests are RAMS, LCC, Risk analysis, predictive maintenance, etc. He has co-authored five books and published more than 300 papers in peer-reviewed journals and conference proceedings. He is also a member of the Royal Swedish Academy of Engineering Sciences, Sweden.

Prof. Diego Galar is a Professor of Condition Monitoring in the Division of Operation and Maintenance Engineering at LTU, Luleå University of Technology where he is coordinating several H2020 projects related to different aspects of cyber-physical systems, Industry 4.0, IoT, or industrial Big Data, etc. He is also a principal researcher in Tecnalia (Spain), heading the Maintenance and Reliability research group within the Division of Industry and Transport. He has authored more than 500 journal and conference papers, books, and technical reports in the field of maintenance, working also as a member of editorial boards, scientific committees, and chairing international journals and conferences and actively participating in national and international committees for standardization and R&D in the topics of reliability and maintenance. In the international arena, he has been a visiting Professor in the Polytechnic of Braganza (Portugal), University of Valencia and NIU (USA) and the Universidad Pontificia Católica de Chile. Currently, he is a visiting professor in the University of Sunderland (UK), the University of Maryland (USA), the University of Stavanger (NOR), and Chongqing University in China.

Chapter 26
AI-Supported Image Analysis for the Inspection of Railway Infrastructure

Joel Forsmoo, Peder Lundkvist, Birre Nyström, and Peter Rosendahl

Abstract The focus in this chapter is on the use of object detection and image segmentation for railway maintenance using complex, real-world image-based data. Image-based data offer the ability to collect data across large spatial areas in a user-friendly manner, as stakeholders often have a basic and intuitive understanding of image-based content. By using already existing videos shot from track measurement vehicles traversing the railway network, it was possible to inspect the reindeer fence lining the railway in northern Sweden. The chapter suggests a framework for the costs and benefits of this type of analysis and adds some other possible applications of the image analysis of these videos.

Keywords Image analysis · Machine learning · AI · Railway · Reindeer fence · Cost-benefit analysis

26.1 Introduction

Infrastructure failures in the railway incur repair costs, as well as other costs, for example, the costs of unpunctuality. Fencing is an infrastructure designed to keep wildlife (and humans) off the track. In 2018 in Sweden, more than 7000 failures labeled "animal on track" were reported by the railway. This includes animals that died, were injured, or escaped unharmed. In addition to the harm done to animals, the train can be damaged. This is of special concern in northern Sweden, where large herds of semi-tame reindeer wander to find food and may be hit by trains transporting either passengers or iron ore. For example, the train's air cable can be damaged causing the train to brake; the result may be delays or damaged wheels due to heavy braking. Large costs could be reduced by faster detection of deviations on the reindeer fence and better knowledge of how the fence degrades.

J. Forsmoo · P. Lundkvist · B. Nyström (✉) · P. Rosendahl
Sweco, Box 50120, 97324 Luleå, Sweden
e-mail: birre_nystrom@hotmail.com

K. B. Misra (ed.), *Handbook of Advanced Performability Engineering*,
https://doi.org/10.1007/978-3-030-55732-4_26

The railway infrastructure, including the reindeer fence, is largely measured and inspected using manual, time-consuming approaches. Inspections are typically carried out in pairs/groups by certified individuals because of the high-risk nature of the work.

26.1.1 Deep Learning

Deep learning is a sub-category of machine learning and is often built on neural network architecture [3]. A neural network aims to mimic human perception and consists of one or more hierarchical layers extracting features in the data, with increasing order of abstraction. One deep learning algorithm is YOLO.

26.1.2 YOLO (You Only Look Once) Object Detection

The YOLO algorithm is inspired by the human visual system, whereby complex features and patterns can be identified and deduced at a glance [4]. YOLO predicts what features of interest are in a given scene and where they are by predicting bounding boxes using probabilities. The result is one or more bounding box for the objects in a scene that an algorithm is trained to identify. In this case, the objects were poles in the reindeer fence. YOLO was chosen for sake of its speed and its detection of one object per box, although its real-time performance is not needed per se. Traditional YOLO, with 24 convolutional layers, v3, was used.

26.1.3 Selecting Images

Rain, snow, mist, and reflected light can negatively impact the quality of the collected video- and image-based data. Moreover, light conditions vary over the course of the day and year. Hence, to meet the quality requirements of the project, i.e., the extent of blurriness or contrast, images were manually selected.

26.1.4 Calculating Absolute Angle of Segmented Objects

To determine the absolute angle in degrees of a given fence pole, we applied a color threshold. Depending on the time of the year and day, the reindeer fence is darker than the background. Hence, we used a simple color-based threshold to segment out the darker pixels in the given bounding box given by YOLO. Once the darker pixels were segmented, a line was fitted to these pixels. The line could then be used to determine the absolute angle in degrees of a given reindeer fence pole by knowing the line's equation.

26.2 Image Analysis of Reindeer Fence

The image analysis of reindeer fence along the railway line used existing videos (with 2–3 m between images), shot from track measurement vehicles, often traveling at the track section's maximum allowed speed (up to 160 km/h). Track measurement vehicles traverse the railway network, at intervals of 2 months or longer, depending on the type of railway. The videos had to be converted and processed using the following three steps:

1. Acquire data, i.e., video datasets (automatic).
2. Extract images from each video (automatic).
3. Discard superfluous and blurry images (manual).

The image-data extracted from the videos were then processed using a deep learning and image-based workflow following four steps:

1. Manually label features of interest as bounding boxes to train the AI model.
2. Augment the labeled images using various augmentation workflows, such as brighten, darken, contrast, and soften, to artificially produce a large training database.
3. Use trained AI model to automatically detect features of interest.
4. Match features of interest across images taken at the same location at different events in time and use the relationship between matched features to determine change over time. Use a color-based threshold to segment out the pole from the background in each bounding box. Fit a line to the segmented pole and calculate the absolute angle in degrees.

The original version of the extracted images was used to label features of interest using LabelIMG [6]. A rectangular bounding box was drawn around each visible reindeer fence pole for each image, as shown in Fig. 26.1.

This process created a separate.xml-file (one for each image; a few hundred were used) specifying where the labeled features were located for each image; the file comprised the training data. When training data were created for each image and object of interest, each image was augmented in terms of, for example, contrast and brightness, as shown in Fig. 26.2. In this way, the training data created for each input image could be used for each augmented version of that image—as the location of objects of interest (i.e., poles) in the image did not change. Thus, the amount of training data generated per unit time could increase significantly.

Once images were augmented in terms of brightness and contrast, and there were training data for each version of the augmented images, the deep learning/AI model was ready to be trained. We used a workflow based on TensorFlow and the YOLO object detection framework to automatically draw a rectangular bounding box around

Fig. 26.1 Rectangular bounding boxes drawn around each visible reindeer fence pole

probable features of interest, as shown in Fig. 26.3 [4]. First, the bounding boxes in Fig. 26.3a were automatically created across the images. These bounding boxes were combined with a probability map, as shown in Fig. 26.3b [4]. The bounding boxes across the image and the probability map together formed the final location of the detected objects, as shown in Fig. 26.3c. The two farthest away poles in Fig. 26.3c cannot be detected. As these poles soon get closer to the camera in another image, it was best to ignore the far most part of the images.

By using original images containing individual poles and artificially changing the angle of a smaller part of the images, we created examples of poles at different angles (absolute degrees compared to the original image). Figure 26.4 shows an example of a single pole rotated to four different extents. The estimated differences in angle (relative angle in degrees) between the single pole in the original image and the four augmented smaller images are shown in Table 26.1. This scenario is arguably comparable to the analysis of images taken at the same location at different points in time.

In the next step, we segmented the identified objects of interest from the background. Once this was done, it was possible to fit a line to the segmented area. From the fitted line, we could calculate the absolute angle of the line and, in turn, the segmented object through the line's equation. This is shown in Fig. 26.5.

26.3 Costs and Benefits

If benefits (the traffic) are considered static, we should strive to minimize the total cost as follows:

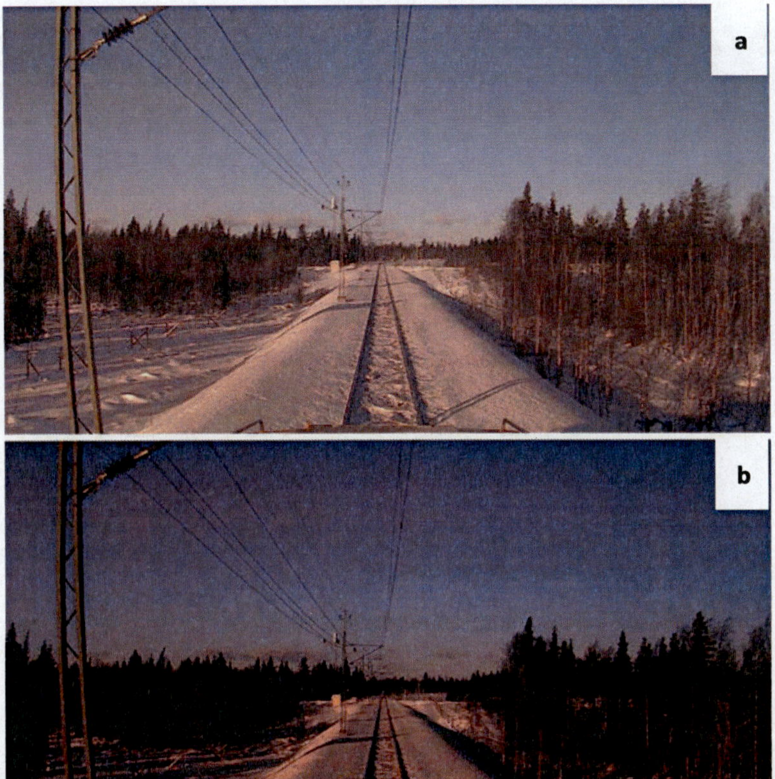

Fig. 26.2 Image augmentation methods used for all images; **a** shows the original image as captured by the camera; **b** shows a darkened version of the same image; **c** shows a version where the brightness and contrast have been increased

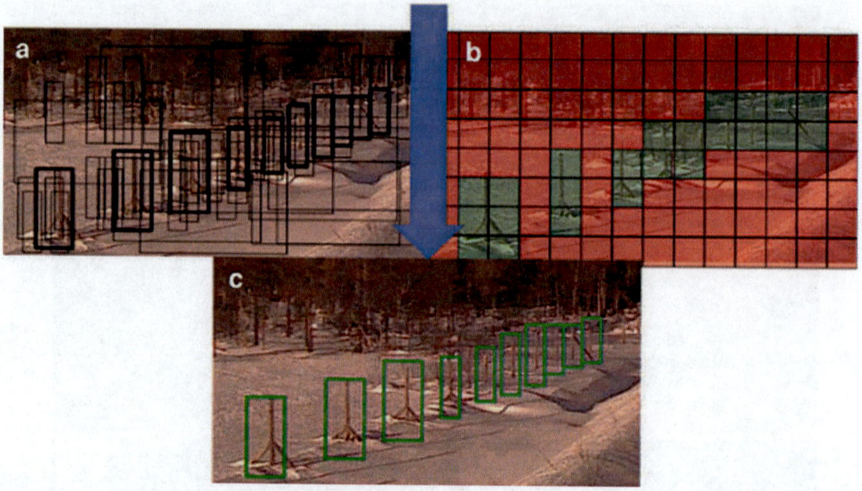

Fig. 26.3 YOLO object detection technique automatically creates bounding boxes throughout the image in (**a**) alongside a probability map for the image in (**b**). The bounding boxes in (**a**) and the probability map in (**b**) determine the final (likely) location of detected objects in (**c**)

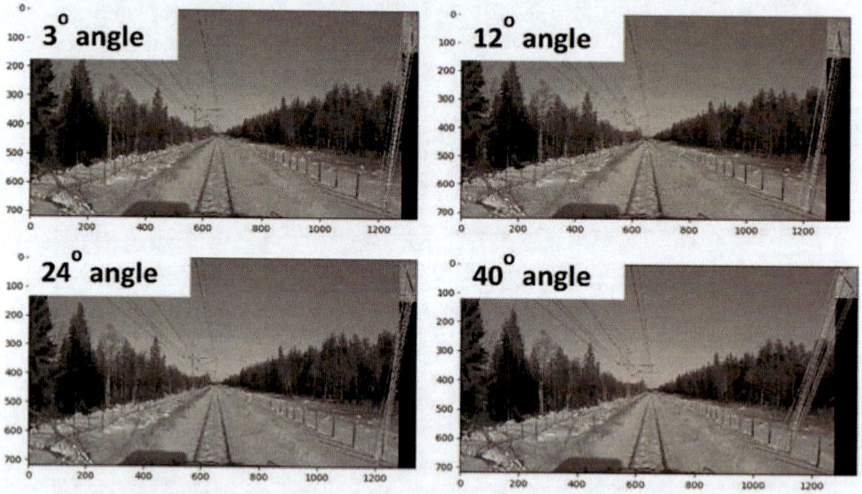

Fig. 26.4 During winter, each individual pole is identified (smaller images in the upper right corner) together with the original image. The degrees in the pictures show the actual difference in the angle of the slope between the pole in smaller images and the pole in the original image

Table 26.1 The known angle (the amount an image has been artificially rotated) compared to the estimated angle. The proposed method can be used with uncalibrated cameras or with cameras of unknown specifications, such as focal length and sensor size

Method	Actual rotation/angle (°)	Estimated rotation/angle (°)
Uncalibrated camera, feature matching	3	3.6
	12	12.2
	24	24.8
	40	39.3

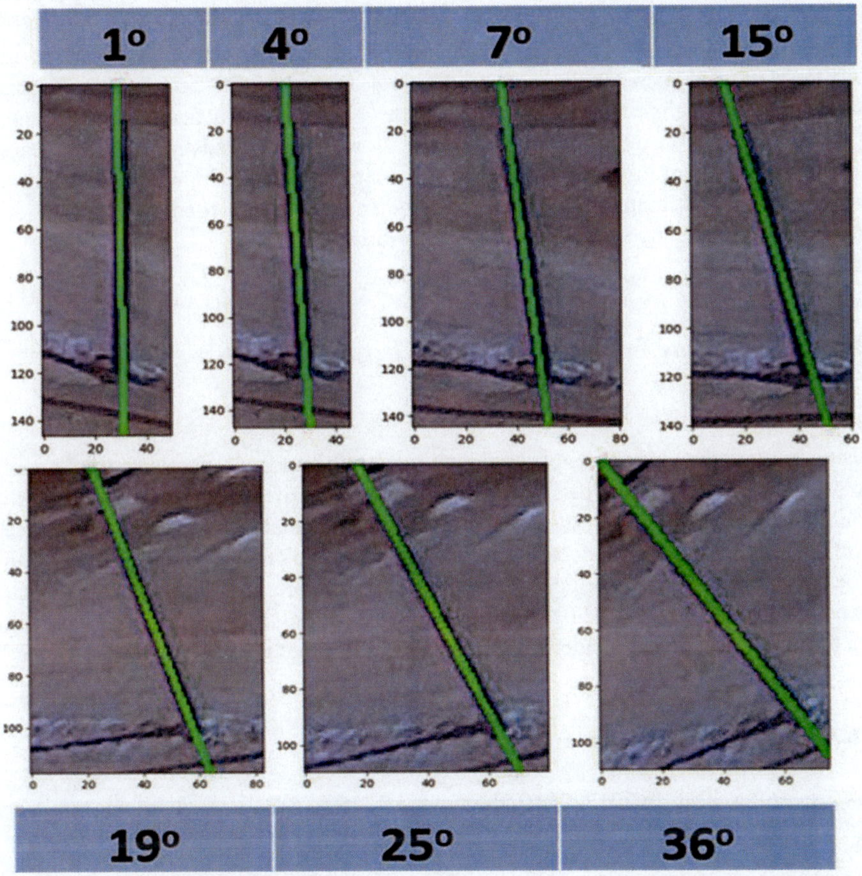

Fig. 26.5 During winter, an individual pole is identified, together with its slope (green line). The numbers represent the calculated angle of the slope

$$\text{Total cost} = \text{Unpunctuality cost} + \text{Other operations cost} + \text{Inspection cost} + \text{Preventive maintenance cost (other than inspection cost)} + \text{Corrective maintenance cost}$$

The decision on whether to use a manual inspection or image analysis inspection is not just a question of the cost of the inspection itself. For example, we might save money on inspections by performing the image analysis inspection at the same intervals as the more expensive manual inspection. Alternatively, we may shorten the inspection interval, allowing earlier detection of failures and thus sooner repair. Further, on densely trafficked lines, we must consider that manual inspection takes train traffic capacity away from the track, and this makes delays more likely. So there is a trade-off between maintenance and operation, as well as between preventive and corrective maintenance. For example, the high cost of a fault may advocate a shorter inspection interval. Of course, the slower the degradation and the more likely the method of inspection is to detect the degradation, the longer the inspection interval can be. Ultimately, data on degradation and costs are needed for optimization decisions (see Lyngby et al. [2], who studied railway track).

26.3.1 Manual Inspections Versus Inspections Using Image Analysis

Using image analysis and identifying leaning poles in reindeer fence as a complement to ordinary inspections yields various benefits, the chief of which is cost saving. With the help of image analysis, the parts of an inspection that aim to inspect the fence and fence poles can be conducted at a faster pace, as the person doing the inspection does not have to focus as much on the poles in the fence. Cost savings can also be expected because there will be fewer accidents involving hit animals, i.e., reindeer, on the track. The cost considerations include the following.

26.3.1.1 Cost of Accidents

As Fig. 26.6 shows, in 2012–2019, there were 3776 accidents involving hit reindeer (in the northern district, where most reindeer live). On average, 1106 reindeer are killed every year, with an average of 2.3 reindeer per accident.

Accidents involving hit and killed reindeer incur several costs. When an accident has occurred, maintenance workers are called to the scene to clear the tracks and make sure nothing else has been affected. When the accident happens at night, two workers go out together, at the cost of 1000 SEK/h/worker. They use a vehicle at the cost of 2500 SEK/h. Depending on the location of the accident, the time it takes to get there varies; we can assume it takes about 6 h to get to the location and take care of the animal/s. The compensation paid to the reindeer owner is 3000 SEK/reindeer.

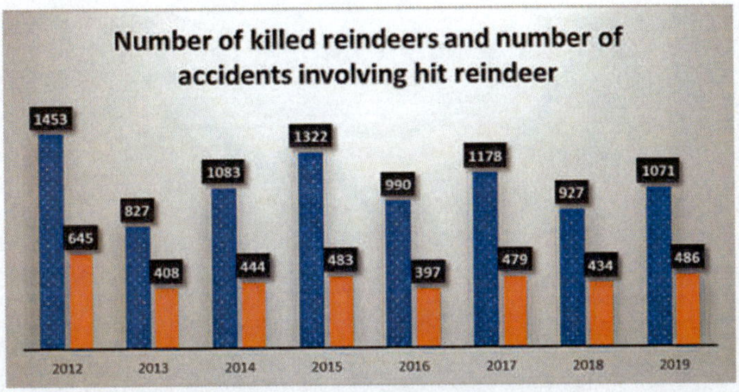

Fig. 26.6 Number of reindeer killed per year (dotted blue) and number of incidents involving hit reindeer (solid orange)

26.3.1.2 Cost of Delays

Between 2012 and 2019, there were 256 accidents when reindeer were hit and traffic was delayed. This resulted in 8207 min of delay in train traffic. The socio-economic cost of a minute's delay for a passenger train is 666 SEK/min and for a freight train is 66 SEK/min [5]. Assuming affected trains are 50% passenger and 50% freight, the cost of one minute's delay is 366 SEK. On average, there is a delay of 32 min for every accident involving reindeer on or around the track, whether they are hit or not. Therefore, the cost of delay connected to one accident can be calculated as 11,712 SEK/accident. The average cost of the accident itself and its incurred delay is shown in Fig. 26.7.

At times, reindeer are not hit, but their presence around the track still causes delays. These delays can occur for several reasons; for example, speed limitations may be set because of previous reindeer accidents or sightings. Collisions with moose also happen. They cause, on average, longer delays, as a moose is a bigger animal and causes more damage to the train. In some cases, the reindeer fence might hinder a moose from crossing the track, although it is not designed to do so.

On the one hand, the benefit (lower cost) of having a well-maintained reindeer fence might be greater than the cost of one accident with hit reindeer, multiplied by their number. On the other hand, the benefit might be less, as even a well-maintained reindeer fence is not a 100% guarantee that reindeer do not pass, and not all tracks have a reindeer fence, although there has been a long-standing effort to install reindeer fence in areas where reindeer accidents happen more frequently. In addition, there may be other operations costs, such as outdoor life being hampered.

Another concern is unpunctuality, taken here as the delay along each section. However, train traffic is planned with slack, so the train may reach the destination on time anyway; i.e., there is no unpunctuality cost. Yet slack might mean the railway's capacity is not used to its full extent, and this is also a cost, albeit a hidden one.

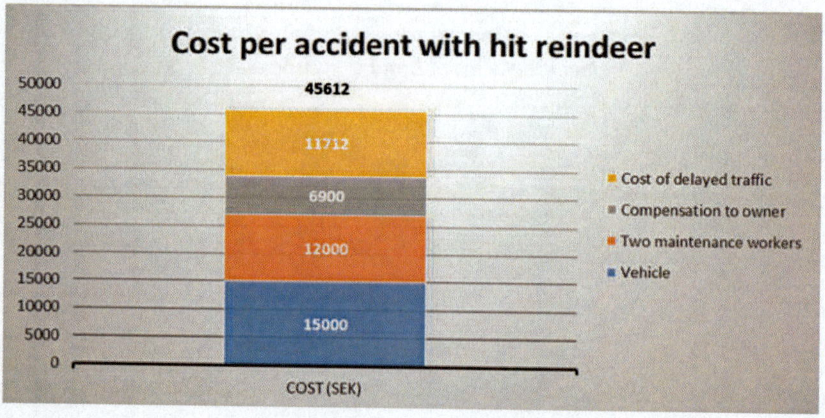

Fig. 26.7 Total cost per accident, calculated from a nighttime scenario with 6 h of corrective maintenance and 2.3 reindeer killed. Vehicle cost is 2500 SEK/h, two maintenance workers cost 1000 SEK/h, and compensation to the reindeer owner is 3000 SEK/reindeer. The cost of delayed traffic with an assumed 50/50 mix of passenger and freight trains at 32 min delay/accident is 366 SEK/min

26.3.1.3 Cost of Inspection

To keep reindeer away from the track, in northern Sweden, the railway has installed 850 km of reindeer fence on both the right and left sides of the track for 1600 km. Inspection is performed once a year. Since the fence is not the only part of the infrastructure inspected, the specific cost is difficult to quantify. However, it takes approximately 150 min to inspect 10 km of track, including fence (3–4 km/h), at a cost of 500 SEK/km. Therefore, it takes approximately 212 h to inspect all fencing, not including the set time between inspections, the time it takes for the maintenance worker to get to the fences, etc. The pace at which the inspector moves along the track should be able to be increased if the fences were not included among the objects subject to inspection. Sometimes, it is not known how often, the fence is inspected from vehicles, lowering personnel cost but also reducing the quality of inspection.

Inspection by image analysis should only require the inspector to verify image analysis results from the office desk. No performance measures, typically including shares of misses (i.e., a leaning pole is not detected) and false alarms (a leaning pole is incorrectly detected), were calculated in this study. However, to establish whether the share of misses and false alarms are at acceptable levels, inspectors should be tested. Specifically, they should look at images with poles and then asked to judge whether the number of times the algorithm was mistaken is acceptable. One of the factors complicating the design and evaluation of such a test is that a spurious missed leaning pole may have no relevance if the adjacent leaning poles are detected, as the missed pole will probably be detected by the inspector or, at the latest, by maintenance workers during repair.

However, it is not clear whether the manual inspection can be entirely substituted, as the stability of the poles should be tested by gently pushing them. Experience also says that old poles tend to have loose cramps, and this is difficult to see in an image. To more precisely calculate the cost, we should know how long it takes a skilled inspector to judge from the pictures of poles using image analysis which poles should be repaired (This likely requires judging the combined effect of several sloping poles; such judgement tests might themselves be used to enhance future image analysis.) An estimation of this work time is one minute per inspection remark. Assuming the same number of remarks as today results in a cost of 8 SEK/km. Admittedly, this only includes the inspection of reindeer fence, not all inspections. Yet inspection quality might increase if inspectors focus on one kind of asset at a time.

26.3.1.4 Cost of Maintenance

The cost assessment of maintaining reindeer fence is 5500 SEK for isolated parts (approximately 10 m). This includes the cost of personnel and transport to the right position. For longer parts of the fence, the assessment is 230,000 SEK/km. Hence, unplanned maintenance is roughly two times more expensive than planned maintenance.

26.3.1.5 Cost and Benefits Summarized

From the numbers above, it is hard to quantify the differences between manual and image analysis inspection because reindeer fences are often inspected simultaneously with other assets. Nor is it clear whether all aspects of manual inspection can be handled by image analysis, although the quality of image analysis inspection might be higher than manual inspection, in some respects. Key parameters to consider when making an informed decision on inspection methods are shown in Table 26.2, with reindeer fence costs in the first row.

26.3.2 Avenues for Future Work

Future work should include training the deep learning model on more images of higher quality and in more conditions, such as during summer or misty weather. It would also be interesting to use cameras on ordinary trains as often as several times a day, in order to better monitor the state of the infrastructure. This would require a company to invest in cameras and administer them and the videos. Vehicle suspension, quality of track, camera mounting, and the camera itself would all affect image quality. A dedicated group of freight trains with similar characteristics and a small number of personnel would likely be the best option if these trains travel large parts of the railway network. At the moment, it is not possible to eliminate

Table 26.2 Possible applications of image analysis of track measurement vehicle videos, as suggested by a group of railway experts

Asset inspection need	Unpunctuality cost	Other operations cost	Inspection cost	Preventive maintenance cost (other than inspection)	Corrective maintenance cost	Periodicity	Limitations of image analysis of track measurement vehicle videos	Examples of data to complement image analysis
Reindeer fence	Yes, might be calculated from the delay of an average incident involving reindeer, which is 32 min	Yes, e.g., owners of dead reindeer are paid 3000 SEK/reindeer	Up to 500 SEK/km fence for manual inspection, roughly 8 SEK/km for semi-automated inspection by image analysis	230,000 SEK/km fence	550,000 SEK/km fence	Regularly, before rein-deer movements in spring	Does not see the net	Asset register data, including the position of gates
Sighting distance from road at level crossings	Yes, train speed restrictions might be imposed	Too short sighting might cause accidents	Manual inspection is about 10 min of work			Regularly in high summer	Does not see the road from the railway	Sighting data from manual inspection
Swing bar height	Yes, road vehicles might tear down the contact wire, causing delays	Accidents to road traffic	Inspected together with the height of the contact wire	Adjust road, swing bar or contact wire	Clear place of accident and repair contact wire	Regularly	Difficult to identify the ground level in the images	

(continued)

Table 26.2 (continued)

Asset inspection need	Unpunctuality cost	Other operations cost	Inspection cost	Preventive maintenance cost (other than inspection)	Corrective maintenance cost	Periodicity	Limitations of image analysis of track measurement vehicle videos	Examples of data to complement image analysis
Switch heating	Yes, as a frozen switch might cause delays	Train traffic control tests the switch prior to train movements	An infrared camera is needed		Clear snow and ice. Change heating elements	In the beginning of the winter season	An IR camera is needed to do this	Time when switch heating is active
Switch snow covers	Yes, a switch without snow cover might be impossible to actuate	Train traffic control tests the switch prior to train movements	Inspected together with other track components		Clear snow and ice	Beginnings of winter and summer	To check every two months is too long an interval	Dates when snow covers are to be put in place and removed, according to contract
Gauge. Different kinds of objects might make the gauge narrower	Yes, caused by contact wire power outage and/or blocked track	Accidents to rolling stock. Land-owner discussions and compensation	Included in track in general	Clear wood before there is a risk	Clear trees that have fallen or might fall	Regularly and ad hoc (private fences too close to the tracks, after storms that might cause trees to fall)	Does not discriminate between hard and soft objects	Asset registers with data from vehicles measuring clearance gauge. Satellite photos

(continued)

Table 26.2 (continued)

Asset inspection need	Unpunctuality cost	Other operations cost	Inspection cost	Preventive maintenance cost (other than inspection)	Corrective maintenance cost	Periodicity	Limitations of image analysis of track measurement vehicle videos	Examples of data to complement image analysis
Noise walls	Components, some of them electrically conductive, fly away from the noise wall. However, no delays have been reported due to noise walls	Increased noise to the public. Accidents to trespassers	Similar and simultaneous with fences	Fasten loose components, paint	Clear debris. A large portion of failures is caused by accidents and sabotage, so preventive mainten-ance is of small value	Ad hoc inventory is needed now, as there have been no inspections of noise walls in the last years	Sometimes fences and noise walls run in parallel, which may make it difficult to recognize them	
Signs	Yes, missing signs for, e.g., speed increase and maximum traction current may cause delays		Manual inspection means walking the distance	Put up new signs	Put up new signs. Only a few failures a year	Regularly, ad hoc (e.g., outdated signs should be removed)	The track measurement vehicles do not traverse all tracks in shunting yards	Asset register data, data on performed maintenance actions (e.g., that a sign has been reported as removed by the maintenance contractor)

in situ inspection of reindeer fences. However, since a reindeer fence is more varied in its appearance than many other types of fences and other assets in the railway, it is likely that image analysis would work well. To get a better understanding of costs and benefits, future work should include estimations of how the downtime of the fence (caused by not knowing the faults) relates to the number of accidents involving hit reindeer. In addition, more of the assets inspected simultaneously with the reindeer fence could be included in the estimations (see Table 26.2). Another interesting possibility is to let the results of image analysis generate work orders for maintenance personnel.

26.3.2.1 More Possible Applications of Image Analysis of Existing Videos

A group of railway experts was asked to suggest more applications of image analysis of the available videos beyond the inspection of reindeer fence. Table 26.2 presents some possible applications, together with mostly qualitative descriptions of the parameters.

As noted previously, it is difficult to determine the cost of inspection of one type of asset, so the numbers for reindeer fence are approximations. Nevertheless, they give an idea of the different costs and what can be gained by decreasing costs of unpunctuality and other operation costs such as accidents, as well as how much might be gained by inspecting the fence often, allowing operators to better plan maintenance, thus lowering maintenance costs.

Table 26.2 suggests some of the proposed applications would fulfill a regular and large need, while others are ad hoc or have small benefits. Interesting possibilities include signs and switch snow covers; using the experiences from reindeer fence they should be easy to implement.

26.4 Discussion and Conclusions

By using remotely analyzed video- and image-based material along with the railway network, it is possible to identify railway infrastructure in need of maintenance automatically and in a timely manner, thus reducing the number of in situ inspections by trained personnel to identify actual and potential faults and risks. Early identification of maintenance needs—including probable faults—should lead to a reduction in urgent corrective maintenance and, in terms of reindeer fence, fewer animals on the track, which, in turn, will lead to a reduction in the number of injured animals and damaged trains.

Furthermore, image analysis of reindeer fence using existing video shots taken from track measurement vehicles allows the poles of the reindeer fence to be distinguished and the angle of their slope to be compared over time, making image analysis of interest to maintainers. However, the resolution of the videos is too poor to distinguish the net between the poles, so the technique cannot yet be expanded to include other details of the fence.

While manual inspection is regulated and widely used, it is prone to mistakes because of fatigue, lack of transparency, and quality assurance measures. Until now, the status of key railway infrastructure, such as reindeer fence, has been assessed using resource-intensive manual approaches; inspectors often drive or walk along a stretch of railway. With automated data collection and analysis, decisions on the maintenance of critical railway infrastructure can be made remotely, and in situ inspection and maintenance can be scheduled when and where needed, based on actual, up to date information on the status of a stretch of railway [1]. This would reduce corrective maintenance; maintenance tasks could be scheduled well in advance, improving working conditions and reducing costs.

That said, a full cost-benefit analysis is difficult to carry out, as the results of implementing image analysis inspections are not known beforehand. It is difficult to determine how much manual inspection might be eliminated, how inspection intervals might be changed, or how much downtime might be reduced.

Acknowledgements The authors thank JVTC for funding, Infranord for videos taken by the track measurement vehicles and knowledgeable help to improve the images, and Strukton. The authors are also grateful to their colleagues for their ideas and sharing their opinions on drafts of the manuscript.

References

1. Barke, D., & Chiu, W. K. (2005). Structural health monitoring in the railway industry: A review. *Structural Health Monitoring, 4*(1), 81–93.
2. Lyngby, N., Hokstad, P., & Vatn, J. (2008). RAMS management of railway tracks. In K. B. Misra (Ed.), *Handbook of performability engineering* (pp. 1123–1145). London: Springer. https://doi.org/10.1007/978-1-84800-131-2.
3. Rampasek, L., & Goldenberg, A. (2016) TensorFlow: Biology's gateway to deep learning? *Cell Systems, 5*(1), 12–14. https://doi.org/10.1016/j.cels.2016.01.009.
4. Redmon, J., Divvala, S., Girshick, R., & Farhadi, A. (2016). You only look once: Unified, real-time object detection. In *Proceedings of the IEEE Computer Society Conference on Computer Vision and Pattern Recognition* 2016-December (pp. 779–788). https://doi.org/10.1109/CVPR.2016.91.
5. Trafikverket, Översiktlig beräkning av den samhällsekonomiska kostnaden för förseningar i samband med evakuering – del 1, 2017. *Downloaded from* https://www.trafikverket.se/TrvSeFiler/Samhallsekonomiskt_beslutsunderlag/Regionoverskridande/Regions%C3%B6verskridande/7.%20%C3%96vrigt/Samh%C3%A4llsekonomiska%20f%C3%B6rseningskostnader%20vid%20t%C3%A5gevakuering/PM_Samh%C3%A4llsekonomisk%20kostnad%20evakuering%20-%20del1.pdf. 8 May 2020.
6. Tzutalin, *LabelImg, Git code*, 2015, *Downloaded from* https://github.com/tzutalin/labelImg. 8 May 2020.

Joel Forsmoo will defend his Ph.D. thesis in drone-based remote sensing from the University of Exeter in 2020. He is currently working as a consultant at Sweco, Sweden.

Peder Lundkvist received his Ph.D. in quality engineering from Luleå University of Technology in 2015. He is currently working as a consultant at Sweco, Sweden.

Birre Nyström received his Ph.D. in maintenance engineering from Luleå University of Technology in 2008. He is currently working as a consultant at Sweco, Sweden.

Peter Rosendahl holds an M.Sc. in industrial engineering. He is currently working as a consultant at Sweco, Sweden.

Chapter 27
User Personalized Performance Improvements of Compute Devices

Nikhil Vichare

Abstract Over the last decade, personalization has been used widely in products and services such as web search, product recommendations, education, and medicine. However, the use of personalization for managing performance on personal computer devices such as notebooks, tablets, and workstations is rare. Performance optimization on computing devices includes all aspects of the device capability such as power and battery management, application performance, audio and video, network management, and system updates. In each case, personalization first involves learning how the user uses the system along with the context of that experience. This is followed by tuning the hardware and software settings to improve that experience by providing individualized performance gains. This chapter discusses the need, complexities, and methods used for performance personalization. A method and case study of improving application performance using utilization data and a Deep Neural Network is presented.

Keywords Performance personalization · Machine learning · Application performance · Artificial intelligence · Edge inference · End-user computing

27.1 Introduction

Personalization is the process of tailoring a product, service, experience, or communication to the specific needs of the end user. Most consumers are familiar with the personalization on the web, which is about delivering content (news, entertainment, products, etc.) that is relevant to the user based on their past preferences and/or other demographics data. Personalization in medicine provides an opportunity for patients and healthcare providers to benefit from more targeted and effective treatments, potentially delivering more healthcare gain and improved efficiency [1]. There is a growing interest in personalized education and learning where recommender systems can automatically adapt to the interests and knowledge levels of learners.

N. Vichare (✉)
Dell Technologies, Austin, TX, USA
e-mail: Nikhil.Vichare@dell.com

K. B. Misra (ed.), *Handbook of Advanced Performability Engineering*,
https://doi.org/10.1007/978-3-030-55732-4_27

These systems recognize the different patterns of learning style and learners' habits through testing and mining their learning activity [2]. The research firm Gartner predicts [3] a 15% profit boost for those who successfully handle personalization in eCommerce.

Personal computer devices such as notebooks (laptop), desktops, tablets, and workstations allow a lot of customization at point of sale. Customers can select between a wide variety of hardware and software configurations to build their own personalized device. These customizations may include installing additional hardware and/or accessories. Modern Operating Systems have several personalization options such as language packs [4]. These options provide interfaces, menus, help topics, and dialog boxes in the language of user's choice. Other experiences and themes available for personalization include backgrounds, settings, and menus.

Although there are many ways a hardware or software can be configured at point of sale, there are very few methods to personalize the computing device for a user's specific workload. Workload is the amount of CPU, memory, storage, network, GPU, and other resources required for execution. Workload may be comprised of processes generated from a single application or multiple applications. This chapter will introduce a method to personalize a computing device for user-specific workloads by learning and characterizing the workload and using a Machine Learning model to map the workload to system settings that provide a performance boost.

27.2 Personalized Performance Improvements

A computing device generally processes, compiles, stores, and/or communicates information or data for business or personal use. Because technology and computing needs vary between different users and applications, computing devices are designed to serve a wide variety of applications and user personas. Furthermore, at point of sale, computing devices can be configured to be general-purpose or for a specific purpose such as gaming, high-performance computing, financial transaction processing, airline reservations kiosks, enterprise data storage, global communications, etc.

The same computing device can be customized to provide different levels of performance by configuring the hardware settings or Operating System (OS) settings. Figure 27.1 shows the methodology for performance personalization. A computing device may serve one or more experiences for the user. These user experiences include optimal audio/video quality for conferencing and entertainment, video/graphics performance (for games and graphics-based computations), battery life while travelling, CPU and memory for faster computation while running applications, and faster charging of batteries for users who need a runtime for next meetings or service appointments. Often, the user may switch between these experiences or may multitask with two or more experiences.

The first step for personalization is to understand and rank which experiences are most valuable to the customer and can be optimized by changing the hardware/software settings. The detection and ranking can be a complex task. In some

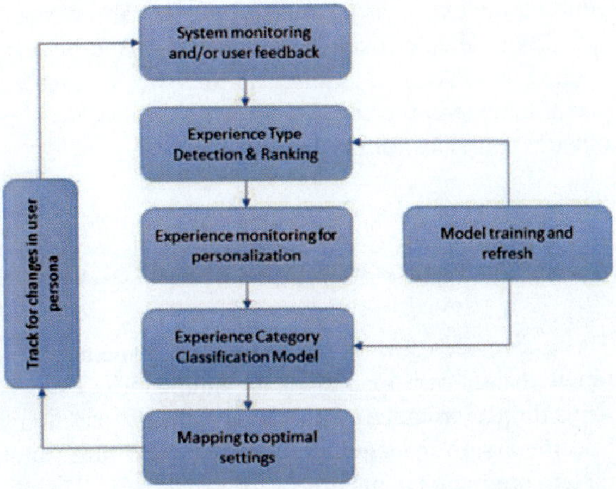

Fig. 27.1 Method for performance personalization

scenarios, a software application or survey may get direct feedback from the customer. In commercial environments where the devices are managed by IT organizations, the IT team may be aware of the device experiences valued by their users. In the absence of direct feedback, data can be collected on system and application usage to detect and rank the most frequent patterns. This may be accomplished with a pre-trained supervised Machine Learning model that takes the system and application data as input and detects and ranks the experiences.

Once the experience to be optimized has been identified, the next step is to collect detailed data on the specific category of experience. This is needed because users use computing devices for the same experience in different ways. For example, an office user may use an audio subsystem for a conference call in a noisy environment (which requires noise suppression), while a gamer will need a fully immersive gaming experience. Each experience category has its own set of best audio settings. Learning and tuning to these differences is the key to personalization.

Learning the experience category requires data about how the experience is utilized. The data along with labels of the category can be used to train a supervised ML (Machine Learning) model that can classify the experience categories. Such a model can be deployed on the computing device by integrating it as part of a performance application. Depending on the cost and complexity, the actual model may reside on the device or may be hosted on the cloud. In either case, the ML model needs to be managed, retrained, and updated as new data is available, which may need significant changes to the data to ML pipeline [5]. In the absence of data or labels to train a model, a rule-based system can be developed where the rules are generated by experts in the domain.

A critical component of performance personalization is the mapping of the experience category with optimal hardware, software, and OS settings. This mapping can be developed during the device development phase and updated as user experiences evolve. Later in the chapter, a case study is presented that provides an example of building the experience category to optimal settings mapping. Once the experience category is detected using an ML model or rule-based system, the optimal settings are applied on the device to tune and improve the user's personal experience. A mature performance monitoring system will keep track of changes in user patterns to select other experiences for optimization.

27.3 Personalized Application Performance Improvements

There have been several studies and implementations of workload characterizations to understand workload intensity and patterns, especially for server-side web workloads for performance evaluations, capacity planning, and resource provisioning [6]. Specifically, to manage the cloud infrastructure, studies have focused on workload pattern prediction using time-series analysis [7]. In other efforts, ML techniques have been applied to tune the performance of large data centers by optimizing the memory using prefetch [8, 9]. Similarly, characterization methods and case studies are available on understanding the characteristics of Big Data workloads for designing efficient configurations and improving throughput [10]. Most of the work listed above is focused on high-end servers and data centers. There has been alimited effort on automatically (either using ML or other means) providing performance improvements on devices used by individual end-users for their specific tasks and applications.

Computing devices are used with a range of software applications that performs tasks such as word processing, business communication and email, video processing, 3D modelling and simulations, etc. Different applications leverage system resources including hardware resources differently. Some applications may be multi-threaded (allows multiples threads or sub-processes within the context of single process), and some applications may be single-threaded. Some applications can benefit from a faster CPU speed and others from faster Input–Output (IO) performance (e.g., due to transactional access to storage). Certain applications may require more power and hence need to be run when devices are connected to AC power. Others may need strong network speeds to transfer data and computations back and forth from the cloud.

An application may run in different modes, resulting in different workloads. For example, a 3D modelling application working on a simple mechanical part versus a complex aircraft design will result in different levels of utilization of the CPU, graphics, memory, etc. Hence, learning and tuning the system settings to the user's specific application workload is important for getting the best performance. Figure 27.2 shows the workflow to improve the performance of user's application-specific workload. As discussed in the previous section, the key components of providing personalized improvements include (1) creating a model that can detect the various workload types and (2) developing a mapping between the workload and best settings to provide the performance improvements [11–13].

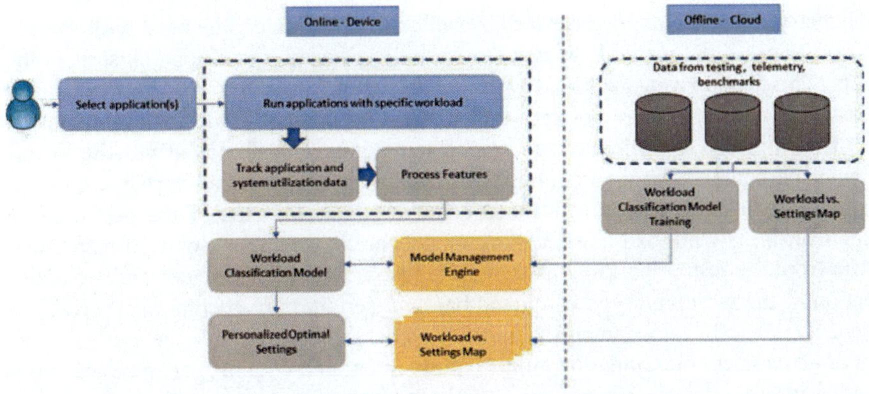

Fig. 27.2 Application workload performance optimization workflow

27.3.1 Workload Classification Model Training

Each workload generates different levels of utilization stress on the computing devices. This stress can be measured by monitoring the utilization of CPU, memory, storage, graphics, network, etc. The utilization can be tracked at the system level (across all applications and process running concurrently) and for the individual processes spun by the application. Along with the utilization data, various thermal and power sensors can be monitored to understand the correlating between utilization and stress levels. If the device is powered by battery, the battery sensors related to charge-level, current, voltage, and temperature can be included in the dataset. On a typical notebook computer with a Windows OS, 1000+ variables can be tracked to characterize the workload and device. These variables and the features extracted from this data are the predictors' variables (X) for the workload classification model. Most computing devices go through extensive validation testing under different application stresses. Devices in the gaming and workstation class are also tested using the industry-standard benchmark tests. The workload features can be collected and labelled during either of these test runs.

In some implementations of the workload classification, the solution developer may be aware of the different types of workloads used by their users. For example, a storage workload may be characterized by metrics such as read/write ratio, type of operations (random vs. sequential), distribution of block sizes, depth of IO operations, and use of cache. A grouping of these metrics can be used as a workload type and hence as a label or response variable (Y) for a supervised ML problem [14].

In other implementations, the solution developer may not have the knowledge of the different workload types. In that situation, data can be collected from systems in the lab where software application generating the workload is run on the computing devices and the device and applications are instrumented to collect the utilization data discussed earlier in this section. This data can be collected over a wide variety of device configurations and SW applications. It can be further augmented with telemetry data from fielded devices if available. The data can be used with various Unsupervised ML methods to find the naturally occurring workload clusters in the data. This data-driven workload type identification can lead to discoveries of new workload types that were not known to domain experts and solution developers.

The workload classification model is then integrated and deployed with the performance personalization application. The deployment methods can vary based on the overall software architecture utilized and the target audience of the performance application. If workload classification models are hosted on a cloud infrastructure, data from the user's workload is transferred from individual devices and a classification of the workload type is retuned back. In certain implementations, privacy or security concerns may demand minimal or no data transfer outside the user's device. In those case, the classification model is fully integrated with the performance application running the computing device. In both implementation scenarios, the models must be managed using a model management engine that connects to the backend ML pipeline for re-training the model offline and updating it for continuous use.

The entire process of tracking workload and device instrumentation data, feature creation, and execution of ML models for scoring must be carefully architected and developed to minimize the computing resources consumed on the user's device. When ML models are fully integrated with client applications, the actual inference can be performed on the CPU or GPU (if available). The overall goal must be to have none or minimal perceptible performance impact to the user for personalizing and improving the performance.

To meet that goal, models that perform inference on the edge must be thoroughly regularized to reduce the numbers of features used by the model. Reduced number of features also results in less data collection and hence further reduces the resources consumed on the user's device. ML models are usually selected based on their performance metrics such as accuracy or F1 scores for classification tasks and mean absolute error for regression tasks. However, for edge inference, overall resource impact of the model is also a key essential metric. The overall resource impact includes the resources needed for data collection, model execution, and any processing of the data before or after model execution. Software developers and data scientist need to carefully balance both accuracy and performance-related metrics for model selection.

27.3.2 Workload Versus Best Settings Map

The next step involves creating a mapping of workload type and a set of hardware, OS, and firmware settings that provide the maximum performance improvement for that workload. Across computing devices, the HW and SW settings available for tuning vary by the type of hardware, type of OS, firmware version, etc. For example, certain CPUs may not have the capability to enable or disable hyperthreading which allows parallelization of computing tasks. Similarly, some operating systems may not have the Superfetch capability that allows the preloading of applications in memory to run faster. This leads to variations in the performance improvements that can be achieved across configurations.

Across the different HW sub-systems like CPU, graphics, storage, memory and network—over hundred different setting options can be set in two or more ways. This creates a large number of combinations of system settings that can influence the performance of the device and application. Several of these combinations can be eliminated based on the domain expertise of conflicts between the setting themselves. The rest have to be evaluated using actual testing and data. The mapping between best settings and workload type can be developed by augmenting several sources of information such a performance testing and industry benchmark studies.

One approach is to develop this map during the labelled data collection phase for workload classification. When a specific workload type is executed on the computing device, it runs with a set of system settings, during each run performance metrics cans be tracked for evaluations. For instance, the metrics could quantify the time to completion of the workload, the volume of workload executed, and metrics for responsiveness. The tests are run iteratively by changing each (or a group of) system setting and measuring the performance over a test run. It is recommended to run multiple test runs for each setting to quantify and address the run to run variability on the metrics. Once the mapping of workload versus most optimal settings is available, it must be integrated with the performance application. As testing progresses on newer device configurations and with the availability of new system settings, the map can be updated either by updating the performance application or by client-cloud architecture where the mapping is hosted on a cloud server.

27.3.3 Solution Workflow

In the last two sections, we discussed the methods to develop a workload classification model and a mapping between workload type and optimal settings. These two components are the essential features of the performance personalization application that can be pre-installed on a computing device or downloaded and installed by individual users or IT administrators. The first step is to select the application(s) that need a performance boost. For general-purpose office users, these could be the most frequently used applications. For users using special high-performance applications, these applications could be selected by the user.

Once the application to be optimized is selected, the next step is to learn the type of workload generated by that application. This is done while the user is using the application. As the user uses the application to perform the intended tasks, the monitoring service of the performance application tracks the system-level and process-level utilization during this time. The learning period may continue for a few minutes to several hours depending on the variations in the workload patterns over time. The data collected during the application learning period is analyzed and transformed in a manner consistent with the input to the workload classification model. The model scores this data to output a workload type. The workload type is then matched with the best system settings from the map. The system settings changes are applied to provide a performance improvement.

27.4 Application Performance Optimization Case Study

In this section, we discuss the implementation details of an application performance optimization software. The implementation follows the methodology and workflow presented in Sect. 27.3. The performance optimization software, in this case is mainly targeted to workstation-class computing devices. These devices are high-powered system designed to execute CPU and Graphics heavy applications. Prior to the development of the ML-based performance optimization, end-users and HW/SW OEMs would run various applications on the computing device and individually change the system settings to get better performance. This association between best HW settings and an application was typically created manually in a lab using benchmarks and application analysis by engineers. A set of optimal system settings (also known as profiles) were developed per application. However, this approach is not scalable to support the wide variety of applications and a variety of workloads generated by the same application (Fig. 27.3).

As explained in the previous section, the same application can generate various type of workloads. Hence, what is needed is the ability to learn the applications workload profile without being explicitly programmed (i.e., using Machine Learning). The profile is learned using the features of the applications workload related CPU, memory, storage, etc. Over 1000+ variables can be tracked at a system and individual process level in time-series. These variables are transformed to create features that characterize the workload [11, 12] (Fig. 27.4).

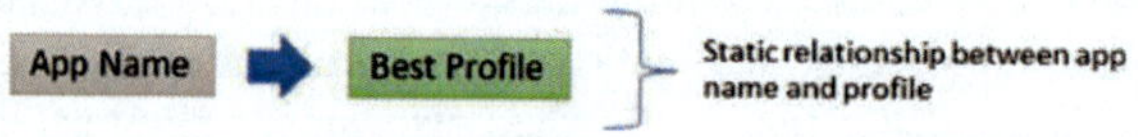

Fig. 27.3 Association between application name and best system settings (Profile)

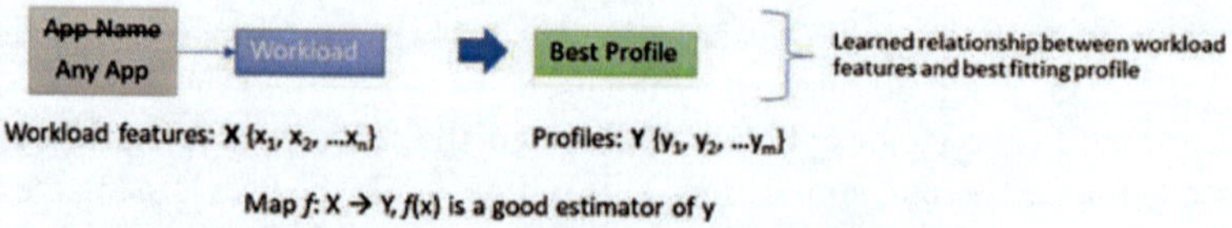

Fig. 27.4 Learned relationship between workload features and best profile

27.4.1 Collection of Labelled Data and Optimal Profiles Mapping

Data for training the workload classification model was collected on 50+ different HW/SW configurations belonging to workstation-class machines. The configurations have a mix of CPUs, storage, memory, and GPU technologies. Figure 27.5 shows the process of data collection under different workload types. In this case study, the number of workload types (*Y*) was known prior to the project. There are 27 workload types which means the ML model has to classify among 27 classes. Each data collection run consists of running a workload under a known set of system settings (*Z*). While the workload is getting executed on the device, over 300+ system and process variables are monitored in time-series. These time-series variables are processed to extract features from the data (*X*) which are used as predictor variables for the ML model.

Example of variables tracked in time-series include Utilization by core, Utilization per thread, Processor queue length, # of logical cores on the system, Current Hyperthreading setting, normalized CPU utilization by a thread—for all threads, IO operations by the process (read bytes, write bytes, idle bytes), Memory utilization by the process, Cache operations of the process, Time elapsed per process, Page file utilization by the process, Memory-committed bytes, Process virtual bytes, IO operations at physical and logical disk levels, etc.

The test run also tracks the performance metrics (*P*) for the workload. These metrics are the direct measure of how fast and/or responsive the workload runs on the device under those settings. Multiple runs are conducted for the same group of workload and settings on a given configuration to capture the variability between runs. More than ~200,000 h of data are collected over the various combinations of configurations, settings, and workloads. The response variables collected over these runs are used for training the workload classification model (*X*, *Y*). The settings and performance metrics are used to develop the mapping set resulting in the best performance improvements (*Z*, *P*) [11].

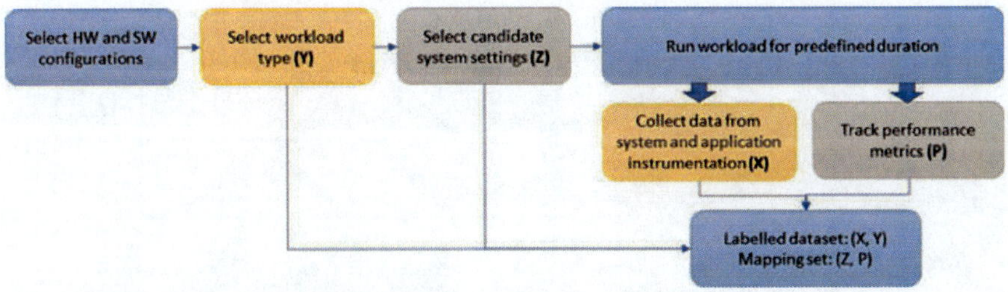

Fig. 27.5 Data collection for model training and settings mapping

27.4.2 Feature Extraction and Selection

Over 300 variables are collected in time-series. Each time variable is summarized using various statistical descriptors such as mean, median, max, min, range, percent of time at 0, and unique counts. Finally, the total number of features = each variable X number of relevant descriptors. One of the goals is to build a parsimonious model to reduce the resource consumption on the computing device. The data has predominantly numerical features but there were several categorial features for configuration options and few dichotomous features for binary setting options. The categorical features were encoded using one-hot encoding. The distribution and correlation of all features was studied to understand the relationships. Figure 27.6 shows the correlation matrix over selected variables.

Using the logistic regression framework, Variance Inflation Factor (VIF) is calculated to evaluate features with high VIF. For each feature x_i, R_i is computed by regressing the i-th feature on remaining features

$$\mathrm{VIF}_i = \frac{1}{\left(1 - R_i^2\right)} \tag{27.1}$$

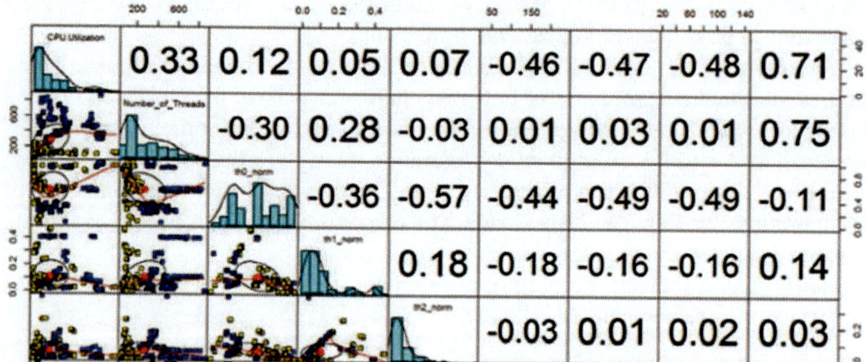

Fig. 27.6 Correlation matrix of select workload variables

The above expression estimates how much variance of a coefficient is inflated because of linear dependency with other predictors. VIF has a lower bound of 1. Features with very high VIF > 20 were evaluated and some were removed from the dataset.

27.4.3 Model Development

A Multilayer Perceptron (Deep Neural Network—DNN) was used to learn the relationship between the predictor features and the 27 storage workload patterns. Model was trained using Tensor Flow and Keras [15] with Python as the scripting language. A fully connected NN with dropout regularization was used to fit the data [16].

$$z_i^L = \sum_{j=1}^{d^L} W_{ij}^L a_j^{L-1} + b_i^L \tag{27.2}$$

$$a_i^L = \sigma^L(z_i^L) \tag{27.3}$$

where,

L	is number of Layer of hidden units $L = 1, 2,\ldots, M + 1$;
$x_j = a_j^0$	are input features into the network;
a_i^L	is the output of the Lth layer;
$f_i = a_i^{M+1}$	are the output values of the final layer;
W	weights;
b	bias;
z_i^L	is the o/p of the neuron before activation; and
σ^L	is the activation function of the Lth layer.

In this case, 2 activation functions were used. For the final layer that provides the workload type, a Softmax layer is used:

$$t = \mathrm{e}^{z^L} \tag{27.4}$$

$$a_L = \frac{\mathrm{e}^{z^L}}{\sum_{j=1}^{C} t_i} \tag{27.5}$$

where $C =$ number of classes. In this case, $C = 27$ workloads. The above equation normalizes the values of a_i. Note $\sum a_i = 1$ (all probabilities should add up to 1).

For rest of the layers, Sigmoid and Rectified Linear Unit (ReLu) were used in hyperparameter selection. The final model uses a ReLu:

$$\sigma^x = \max(0, x) \tag{27.6}$$

The whole network is trained by minimizing the supervised loss function:

$$\sum_{i=1}^{C} L(y_i, f_i(x)) \tag{27.7}$$

$$L(y_i, f_i) = -y_i \log f_i - (1 - y_i) \log(1 - f_i) \tag{27.8}$$

where y_i is the labels (actual workloads) and *fi* is the network output.

The loss function was minimized using the Adam optimizer [17] implemented in Keras [15].

27.4.4 Hyperparameter Evaluation

Hyperparameter selection was done in two stages. In the first stage, a candidate model was selected that provides reasonably good accuracy on the test set. Using this model, several parameters were evaluated individually, holding all other variables constant. The performance metric for selecting the hyperparameter values is either model loss or F1 score on the test set.

$$\text{F1} = 2 \times \frac{\text{Precision} * \text{Recall}}{\text{Precision} + \text{Recall}} \tag{27.9}$$

where Precision is the ratio of true positive to the total predicted positive and Recall is the ratio of true positive to total actual positive. Training time was not used as a metric for any evaluation. This section provides some insights into the hyperparameter evaluation effort on this dataset (Table 27.1).

Based on the results of the first stage, the values of the 5 parameters above were selected. These values were then used in stage 2 of hyperparameter search which included a full grid evaluation.

Three methods were evaluated for initialization. There was no significant difference in performance among the three (all other variables held constant). He initializer was selected for the final model. In the case of He, the weights are initialized keeping considering the size of the previous layer. The weights are random, but the range is different depending on the size of the previous layer of neurons. This provides a controlled initialization which has been reported to provide faster and efficient gradient descent [18]. He initializer also works better with the Relu activation used in the model.

In the case of ReLu, He initializer draws samples from a truncated normal distribution centered on 0 with a variance of

Table 27.1 Hyperparameters and values used in each stage

Stage	Hyperparameter	Values
1	Type of scaling	Standard, Min–max, Robust
1	Type of initializer	Uniform, He Uniform, Golrot Normal
1	Type of optimizer	Gradient Descent, Adam
1	Activation function for hidden layers	ReLu, Sigmoid
1	Batch size	64, 128, 256, 512, 1024
2	# of layers (includes o/p layer)	2, 3
2	Neurons per layer	10, 20, 30, 35, 40, 45, 50, 100
2	Dropout rate	0.1, 0.2, 0.3, 0.4, 0.5
2	Learning rate	0.1, 0.01, 0.001

$$\mathrm{Var}(W_i) = \frac{2}{n_i} \tag{27.10}$$

where n_i is the # of inputs to the tensor (Fig. 27.7).

Evaluating optimizers in detailed is a long time/compute resource-consuming task. A high-level evaluation was performed using the reference model. The default parameters from Keras were used to evaluate the F1 metric on Gradient. Among the three different scaling methods used, min–max scaling offered the best performance in terms of F1 score on the test dataset. For each variable, min-max scaling simply scales using the following equation (Fig. 27.8):

$$X_{\mathrm{scaled}} = \frac{X - X_{\mathrm{min}}}{X_{\mathrm{max}} - X_{\mathrm{min}}} \tag{27.11}$$

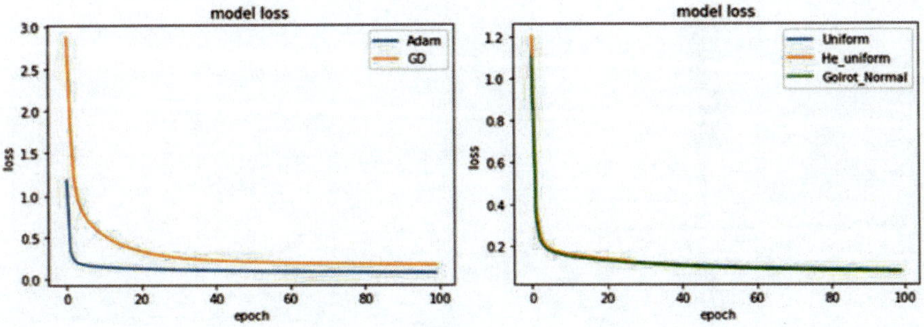

Fig. 27.7 Evaluation of initializers and optimizers

Fig. 27.8 Evaluation of feature scaling methods

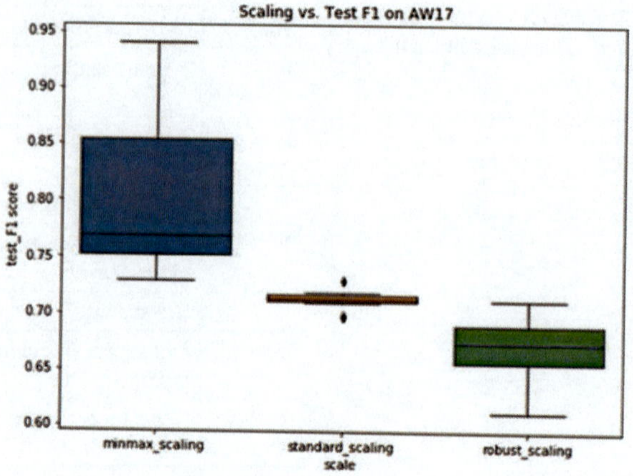

The block size and type of activation function did not provide any significant difference in terms of F1 score. In terms of the full grid search, four hyperparameters (# of layers, neurons per layer, dropout rate, and learning rate for Adam optimizer) were combined to create an exhaustive grid of values. Learning rate of 0.001 was selected for the best F1 score and time to learn. The Fig. 27.9 below shows the variations in the F1 scores after tenfold cross validation with various NN configs and dropout rates.

Based on the figure the smaller networks with large drop-out rates (30 neurons per layer + 0.5 dropouts) has the highest misclassifications. The figure does not show smaller networks (10, 20 neurons per layer) evaluated that had the worst F1 scores. Similarly, larger networks with 100 neurons did not improve the F1 over NN with 40 or 50 neurons per layer.

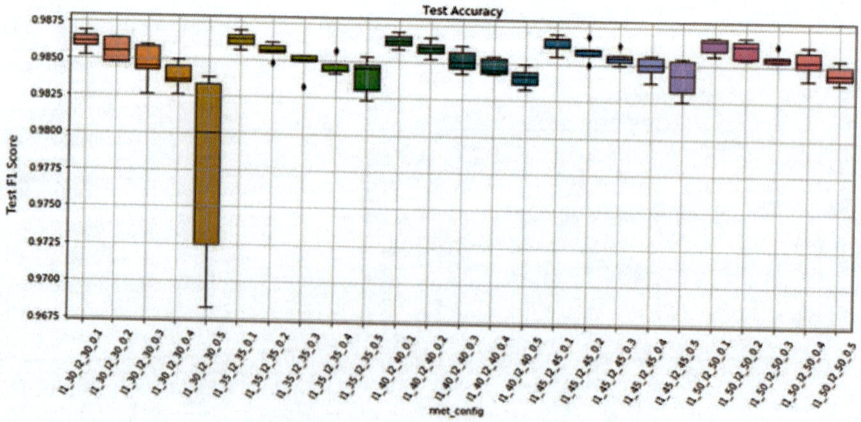

Fig. 27.9 Evaluation of network configs and dropout rates

27.4.5 *Regularization and Model Selection*

Two methods were used for regularization—Dropout and Early Stopping. The dropout rates of 0.1–0.5 in increments of 0.1 were evaluated in hyperparameter search. Based on the dropout rate, the method randomly drops out neurons from the network. This means the contribution to the activation of downstream neurons is temporally removed on the forward pass and any weight updates are not applied to the neuron on the backward pass. The theory is if neurons are randomly dropped out of the network during training, that other neurons will have to step in and handle the representation required to make predictions for the missing neurons. The effect is that the network becomes less sensitive to the specific weights of neurons and results in a network that is capable of better generalization and is less likely to overfit the training data [19].

Dropout was implemented using the Dropout layer in Keras [15]. The same dropout rate was applied to each layer.

$$a_i^L = \text{drop}\left(\sigma^L\left(\sum_{j=1}^{d^L} W_{ij}^L a_j^{L-1} + b_i^L\right)\right) \tag{27.12}$$

where $\text{drop}(x) = 0$ with a probability of dropout rate, else $\text{drop}(x) = x$.

For all runs during training, early stopping was used along with dropout for regularization. Early stopping was implemented using the Early Stopping callback in Keras [15]. The callback was set to monitor the loss over a patience of three values. Patience is the number of epochs with no additional improvement after which the training is terminated.

The training data for learning storage workloads was collected on ~50+ systems with varying configuration. Multiple samples of data were collected from each system. Such data is likely dependent on that particular configuration. Although several variables related to the configuration of the machine are used as inputs to the ML model, there are many more that cannot be accounted. In this work, machine id was treated as a group identifier. In this case, we would like to know if a model trained on a particular set of groups generalizes well to the unseen groups. To measure this, we need to ensure that all the samples in the validation fold come from groups that are not represented at all in the paired training fold.

This is implemented using the Group K-Fold feature in Scikit Learn [20]. It is a K-fold iterator variant with non-overlapping groups. The same group will not appear in two different folds. The number of distinct groups has to be at least equal to the number of folds. Final model selection was done based on the results of hyperparameter search and Group K-fold CV. The final model provided an F1 score of 0.96 on a test set. The model has 2 hidden layers and an output Softmax layer and has a total of 4267 trained parameters.

27.4.6 Application Learning Windows

Characterizing the application workload on a user's device can be a difficult process. One of the main sources of the uncertainty is user behavior. For example, a user may start the application learning process and may head out for lunch. In this scenario, the application is open and not running and the utilization variables of the application are not useful for characterization. Smart thresholds are implemented to account for lack of activity. Also, application workloads change over time, an application make start by reading a large amount of data which generates storage and memory activity which may be followed by several intense computations using CPU and GPU.

To account for these variabilities over time, the learning process is divided into multiple phases. The application is learned over several fixed windows of time. For each window, the data is collected, and the workload is classified along with the probability of classification. The final classification is based on voting across the windows [21]. Thus, using the methodology presented in Sect. 27.3 and a detailed example in this section, a solution to provide performance improvements to user-specific workloads can be developed on existing HW configuration.

27.5 Summary

The chapter provides an overview of the various personalization opportunities that exist on computing devices to improve the performance for tasks and experiences that are most important for the user. This type of personalization can be achieved by using system usage data and Machine Learning methods to learn the experiences to be personalized. A detailed methodology is presented to improve the performance of applications selected by the end-user. The method works by further classifying and characterizing the application workloads and tuning the hardware, OS, and firmware settings in order to provide the best improvement possible on the given hardware configuration. A detailed case study on workstation devices presents the various steps involved including a collection of labelled data, ML model training for classification of workloads, mapping workloads to system settings, and deployment notes.

References

1. Stratified, personalized or P4 medicine: a new direction for placing the patient at the center of healthcare and health education, Summary of a joint FORUM meeting. Academy of Medical Sciences, 2015. Downloaded from https://acmedsci.ac.uk/viewFile/564091e072d41.pdf.
2. Aleksandra, K., Vesin, B., Ivanovic, M., & Budimac, Z. (2011). E-Learning personalization based on hybrid recommendation strategy and learning style identification. *Computers and Education, 56*(3), 885–899.
3. Gartner Report. (2009, April) *Maximize the Impact of Personalization*. Gartner Inc. https://www.gartner.com/en/confirmation/executive-guidance/personalization.

4. Personalize your PC. https://support.microsoft.com/en-us/help/14165/windows-personalize-your-pc. Last Updated: Dec 3, 2018.
5. Schelter, S., Biessmann, F., Januschowski, T., Salinas, D., Seufert, S., & Szarvas, G. (2018, December). On challenges in machine learning model management. In *IEEE Data Engineering* (Vol. 41, No. 4).
6. Calzarossa, M., Massari, L., & Tessera, D. (2016, February). Workload characterization: A survey revisited. *ACM Computing Surveys (CSUR)*, Article No.: 48.
7. Khan, A., Yan, X., Tao, S., & Anerousis N. (2012, April). Workload characterization and prediction in the cloud: A multiple time series approach. In *IEEE Network Operations and Management Symposium*. Maui, HI.
8. Rahman, S., Burtscher, M., Zong, Z., & Qasem, A. (2015, August). Maximizing hardware prefetch effectiveness with machine learning. In *2015 IEEE 17th International Conference on High Performance Computing and Communications*. New York, NY.
9. Liao, S., Hung, T., Nguyen, D., Chou, C., Tu, C., & Zhou, H. (2009, November) *Machine learning-based prefetch optimization for data center applications*. In *Proceedings of the Conference on High Performance Computing Networking, Storage and Analysis*. Portland, Oregon.
10. Ren, Z., Xu, X., Wan, J., Shi, W., & Zhou, M. (2012, November). Workload characterization on a production Hadoop cluster: A case study on Taobao. In *IEEE International Symposium on Workload Characterization (IISWC)*. La Jolla, CA.
11. Vichare, N., & Khosrowpour, F. (2017, May). *Methods to associate workloads to optimal system settings*. United States Patent Application, 15/583078.
12. Vichare, N., & Khosrowpour, F. (2017, April). *Methods for system performance measurement of stochastic workloads*. United States Patent Application, 15/499050.
13. Vichare, N., & Khosrowpour, F. (2017, September). *Application profiling via loopback method*. United States Patent Application, 15/719789.
14. Vichare, N., & Khosrowpour, F. (2019, March). *Method to increase an applications performance by dynamic storage optimization*. United States Patent Application, 16/353153.
15. Chollet, F. (2015). Keras. https://keras.io.
16. Goodfellow, I., Bengio, Y., & Courville, A. (2016). *Deep learning*. MIT Press. https://www.deeplearningbook.org.
17. Kingma, D. P., & Ba, J. (2015). *Adam: A method for stochastic optimization*. In *3rd International Conference for Learning Representations*. San Diego.
18. He, L., Zhang, X., Ren, S., & Sun, J., *Delving deep into rectifiers: Surpassing human-level performance on ImageNet classification*. https://arxiv.org/abs/1502.01852.
19. Srivastava, N., Hinton, G., Krizhevsky, A., Sutskever, I., & Salakhutdinov, R. (2014). Dropout: A simple way to prevent neural networks from overfitting. *Journal of Machine Learning Research, 15*, 1929–1958.
20. Pedregosa, F., Varoquaux, G., Gramfort, A., Michel, V., & Thirion, B. (2011). Scikit-learn: Machine learning in Python. *JMLR, 12*, 2825–2830.
21. Vichare, N., & Khosrowpour, F. (2019, February). *Method for profile learning window optimization*. United States Patent Application, 16/265464.

Nikhil Vichare is a Distinguished Engineer at Dell Technologies. His primary focus is applied to Machine Learning and Data Science. He was the lead data scientist in developing the industry-first predictive support service for PCs and industry-first AI-based performance optimization solution. In his role, Nikhil has trained Machine Learning and Deep Learning models that have been deployed on millions of edge devices. He has 21 US patents granted and 20 additional patents pending for developing innovative solutions using instrumentation and algorithms. Nikhil received his Ph.D. in Mechanical Engineering from the University of Maryland, College Park. He is the author and co-author of over 25 publications in archival Journals and Proceedings and is a contributing author for 4 books.

Chapter 28
The Neglected Pillar of Science: Risk and Uncertainty Analysis

Terje Aven

Abstract Science, in general, is built on two pillars: on the one hand, *confidence*, obtained through research and development, analysis, argumentation, testing, data and information, and on the other *humbleness*, acknowledging that the knowledge—the justified beliefs—generated can be more or less strong and even erroneous. The main thesis of the present conceptual work is that the latter pillar—humbleness—has not been given the scientific attention it deserves. This pillar is founded on risk and uncertainty analysis, but the fields of this type of analysis are weak, lacking authority. The volume of research on risk and uncertainty analysis is small and the quality of current approaches and methods is not satisfactory. A strengthening of the fields of risk and uncertainty analysis is urgently and strongly needed. Several suggestions for how to meet these challenges are presented, including measures to stimulate further research on the fundamentals of these fields—and crossing established study borders, and initiatives to be taken by relevant societies to increase the awareness of the issue and deriving suitable strategies for how to develop risk and uncertainty analysis as a distinct science.

Keywords Science · Risk analysis · Uncertainty analysis · Knowledge · Black Swans

28.1 Introduction

Science tells us that smoking is dangerous. The evidence is strong; there is basically no discussion about it. The scientific method has been used to prove that smoking has severe negative health effects. A number of statistical models have been established, linking lung cancer and smoking [e.g. 2, 3]. We also have strong phenomenological knowledge about why smoking is having these effects.

The chapter is to a large extent based on Aven [1], with permission from the publishers.

T. Aven (✉)
University of Stavanger, Stavanger, Norway
e-mail: terje.aven@uis.no

K. B. Misra (ed.), *Handbook of Advanced Performability Engineering*,
https://doi.org/10.1007/978-3-030-55732-4_28

It is, however, not that many years since the statement that smoking is dangerous was very much contested. In 1960, a survey by the American Cancer Society found that not more than a third of all US doctors agreed that cigarette smoking was to be considered "a major cause of lung cancer" [4]. As late as 2011, a research work conducted by the International Tobacco Control Policy Evaluation Project in The Netherlands showed that only 61% of Dutch adults agreed that cigarette smoke endangered non-smokers [4, 5].

Science provides knowledge about the health effects of smoking, in the form of statements such as "Smoking is dangerous" and "Smoking causes lung cancer", supported by statistical analysis. This statistical analysis is concerned about two main issues:

(a) What does it mean that smoking is dangerous? And that smoking causes lung cancer?
(b) Uncertainty related to the correctness of these statements. How sure can we be that these statements are correct?

Issue (a) is commonly answered by referring to a suitable statistical and risk analysis framework. For example, a frequentist probability p may be introduced, expressing the fraction of persons belonging to a special population (e.g. women of a specific age group) that get lung cancer. By comparing estimates of this probability for non-smokers and for smokers, and considering variations, for example, related to the number of cigarettes per day and the duration of smoking, significant differences can be revealed, justifying the statements.

Hence, the statements can be interpreted as saying, for example, that smoking significantly increases the chances of getting lung cancer, where chance is understood in a frequency manner. In this framework, uncertainty is dealt with using concepts like variance and confidence intervals. Other frameworks exist, for example, the Bayesian one, in which epistemic uncertainties of unknown quantities—such as p—are represented by subjective probabilities expressing degrees of beliefs.

The smoking example demonstrates that communication of scientific findings is challenging. The problems relate to both (a) and (b). Two main concerns have to be balanced: the need to show confidence by drawing some clear conclusion (expressing that smoking is dangerous) and to be humble by reflecting uncertainties. Standard statistical frameworks as outlined above provide guidance on both (a) and (b) and their interactions, but they have limitations; they cannot provide strict proof. They can demonstrate correlation, but not causality. This has of course been used by the cigarette manufacturers, who have disputed any evidence supporting the conclusion that smoking is dangerous. The tobacco industry is powerful, and it has taken a long time to convince people that smoking kills. In some countries, the severe consequences of smoking have still not been acknowledged.

The present work is based on the conviction that there is also a problem in the way the humbleness concern is understood and communicated; it is not only about the propaganda from the tobacco industry. What does it really mean that we do not know for sure that smoking or passive smoking is dangerous? If John smokes a few cigarettes every day, will this actually cause him harm, given the fact that smoking makes him relax and feel good? How do we formulate and communicate this? The risk and uncertainty analysis fields should provide authoritative answers but they do not.

The risk and uncertainty analysis fields lack authority on fundamental concepts, principles and terms. This can be traced back to the fact that these fields are not broadly acknowledged as sciences. The volume of research and funding, as well as the number of academic positions and programmes, is rather small. If, for example, we compare the number of university professors in statistics with that in the risk field, the result is astounding. There are in fact very few professors worldwide specifically in the risk analysis field.

The consequences are rather limited scientific work directed at fundamental risk and uncertainty analysis research, which, in its turn, has serious implications for the quality of the humbleness dimension of science as described above.

A full risk and uncertainty analysis needs to capture at least these four types of elements:

(1) Modelling of variation and other phenomena of the real world (this is often done using probabilistic models)
(2) Representing and/or expressing epistemic uncertainties using probability, probability intervals, or another measurement tool
(3) Representing and/or expressing the strength of knowledge supporting the judgements in (2)
(4) The potential for surprise relative to the available knowledge and judgements made.

Current frameworks are basically limited to (1) and (2), and even for (2) there is a lack of clarity on the fundamentals. To illustrate this, consider the use of probabilities to represent/express epistemic uncertainties. In the literature these probabilities are often used without interpretation or, if an interpretation is provided, it is not really meaningful, as it mixes uncertainty analysis and value judgements; see Sect. 28.3.2. Similar problems exist for probability intervals, reflecting imprecision; see Sect. 28.3.2. The risk and uncertainty elements (3) and (4) have been given little attention in the scientific literature, but they are critical for the proper understanding and treatment of risk and uncertainty. These elements do not allow the same elegancy as (1) and (2), in terms of mathematical and technical formulation and precision, but are equally important. Statisticians and operations researchers usually work within a quantitative mathematical framework, and (3) and (4) are outside the scope for these categories of researchers.

For the smoking example, and seeing it from a historical perspective, all these four types of elements are important and they have all been addressed to some extent. However, most studies are founded on (1), although some also use (2). Today, the knowledge strength is very strong, but, if we go back some years, it was much weaker and (3) and (4) were highly relevant aspects to consider.

The scientific work related to climate change is another current topic illustrating this discussion. The international authority evaluating climate risk is the Intergovernmental Panel on Climate Change (IPCC). The panel has devoted considerable effort and competence to characterising climate risk and uncertainties [6–8]. However, the conceptualisation and treatment of risk and uncertainties lack a proper foundation, as demonstrated by Aven and Renn [9]. The IPCC evaluations cover, to a varying degree and quality, the four elements (1)–(4). The panel seems to have developed their approach from scratch, without really consulting the scientific community and literature on risk and uncertainty analysis. For IPCC, this community and the literature have clearly not provided the authoritative guidance that could support it to form its approach to risk and uncertainty. This demonstrates the point made above: the fields and science of risk and uncertainty analysis are too weak to have an impact on important scientific work such as climate change research. The result is a poor IPCC conceptualisation and treatment of risk and uncertainties.

Another example is the ISO 31000 Guideline on risk management [10], which has a strong impact on risk knowledge generation and decision-making. It suffers from similar problems to those of the IPCC reports. The presentation of the fundamental concepts of probability and risk lacks rigour and is not understandable [11, 12].

These are just examples, showing that the risk and uncertainty analyses are not sufficiently developed to be able to provide authoritative scientific guidance. The present work aims at pointing to this situation, clarifying what it is about and discussing measures to improve it. Section 28.2 first reflects briefly on what science means in general and for risk and uncertainty analysis in particular, to frame its further discussion. Section 28.3 examines specific challenges in risk and uncertainty analysis, based on the elements (1)–(4) presented above. Section 28.4 investigates ways of improving the current situation and strengthening the risk and uncertainty analyses and the related fields and sciences. Finally, Sect. 28.5 provides some conclusions.

28.2 What is Science?

If we consult a dictionary, the first definition of science that is presented is of the type: "The intellectual and practical activity encompassing the systematic study of the structure and behaviour of the physical and natural world through observation and experiment" [13]. This captures what is commonly referred to as natural sciences (physics, chemistry, etc.). The idea can be extended to society, to basically cover the systematic study of the world, as captured by this definition, "The observation, identification, description, experimental investigation, and theoretical explanation of phenomena" [14]. These dictionaries also link science to systematic knowledge generation, and, combining the two types of definitions, we are led to systematic studies of the world to gain knowledge.

However, such an understanding of science is narrow and would exclude many fields, for example mathematics and statistics, and many broader science concepts exist. The present work is based on Hansson [15], who argues that science (in the broad sense) should be seen as the practice that provides us with the most epistemically warranted statements that can be made, at the time being, on subject matters covered by the community of knowledge disciplines, i.e., on nature (natural science), ourselves (e.g. psychology and medicine), our societies (social sciences), our own physical constructions (e.g. technology and engineering), and our own mental constructions (e.g. linguistics, mathematics and philosophy) [16].

A knowledge discipline generates knowledge in the form of warranted or justified statements or beliefs. We can refer to the most warranted or justified statements or beliefs as scientific knowledge. The justified belief that smoking is dangerous for human health is a result derived from the knowledge discipline of medicine. It is based on science.

There is not much discussion today about the validity of this belief that smoking is dangerous. However, as discussed in Sect. 28.1, it has not always been like that. In general, there is a battle regarding what are the *most* justified beliefs, a battle that to a large extent is about power and institutions. There are different perspectives, ideas and schools of thought arguing for their statements and beliefs. Deciding which are the most justified beliefs is always contentious. Nonetheless, practice has shown that the process of reaching these most warranted statements and beliefs works well. In a longer perspective, the different knowledge disciplines move forward and make developments and progress [17, 18].

Today, the climate change issue is a good example of this battle. The climate knowledge disciplines have concluded that global warming takes place and is extremely likely (greater than 95% probability) to be the result of human activity. There is opposition to these views, it is indeed a battle, but these beliefs are those supported by the majority of climate scientists, and they are strongly influencing the political decision-making.

The statistical science provides knowledge on how to do the statistical analysis in cases like these (smoking, climate change), but there are issues of importance for the understanding and follow-up of the beliefs on smoking and climate change that extend beyond this science, as discussed in relation to items (1)–(4) mentioned in Sect. 28.1. These issues are addressed by the fields and sciences of uncertainty and risk analysis. As discussed in Sect. 28.1 and will be further argued for in the coming section, the knowledge produced by the uncertainty and risk sciences is too weak to influence the climate sciences. The meaning of the above statement—"greater than 95% percent" is not made clear in the IPCC documents, as explained in detail by Aven and Renn [9]. There is a lack of precision on a fundamental concept for reporting the scientific climate-change knowledge, which has serious implications for the communication and trustworthiness of the findings. Ideas about risk are used which risk researchers argue are unsuitable [9]. A serious lack of quality in the risk and uncertainty treatment is observed. The thesis is that this is due to weak risk and uncertainty analysis sciences.

Today, there is no broad acknowledgement of risk analysis (in the broad Society for Risk Analysis (SRA) sense, covering risk assessment, characterisation, communication, management and policy) being a science itself, a distinct science. However, it can be argued that the pillars for such a science exist [19] by including two different types of knowledge generation for risk analysis [20]:

(A) Risk knowledge related to an activity (interpreted in a broad sense, also covering natural phenomena) in the real world, for example, the use of a medical drug, the design of an offshore installation, or the climate.

(B) Knowledge of concepts, theories, frameworks, approaches, principles, methods and models to understand, assess, characterise, communicate and (in a broad sense) manage risk.

The (B) part is the specific one for risk analysis; no other science has (B) as its subject matter. The (A) part is driven by other sciences, like medicine and physics, and is supported by risk analysis and the (B)-generated knowledge.

Similar types of knowledge production can be defined by the field/science of statistics [19]:

(A1) Knowledge related to an activity in the real world using statistical analysis;

(B1) Knowledge on concepts, theories, frameworks, approaches, principles, methods and models for collecting, analysing, presenting, and interpreting data.

Risk analysis uses statistics but covers many topics which are not addressed by statistics, for example the role of the precautionary principle in risk management [19]. For uncertainty analysis, we can make a similar knowledge production to that for risk analysis, but as uncertainties always need to be seen in relation to what one is uncertain about, uncertainty analysis will have basically the same elements as risk analysis as defined by (A) and (B). It is common to focus on future consequences and quantities in risk analysis, but the analysis also covers other types of quantities, and uncertainty analysis can thus be considered a part of risk analysis in this sense. In the following, we refer to the science of 'risk and uncertainty analysis' to highlight that we cover this broad interpretation of (A) and (B).

28.3 Fundamental Challenges in Risk and Uncertainty Analysis

In this section, the four elements referred to in Sect. 28.1 to be covered by a full risk and uncertainty analysis are discussed. The purpose is to point to issues that are essential for science, but for which risk and uncertainty analysis struggles to provide appropriate support.

28.3.1 Modelling of Variation

Probability models are commonly used to model variation [21]. These models allow for sophisticated probabilistic and statistical analysis. There is a broad acknowledgement of the use of such models for these purposes. However, a probability model is a model and needs justification. In many cases, it cannot be meaningfully defined and interpreted. Models are often simply assumed to exist in applications, and no discussion about validity is performed.

To illustrate the problem raised, consider the important task of analysing rare events with extreme consequences. To this end, a probabilistic framework is often used, founded on the use of probability models. Reference is made to concepts like heavy and fat distribution tails. However, we seldom see that this framework is justified or questioned: is it, in fact, suitable for studying extreme event phenomena?

A probability model is a representation of variation in a population of similar situations or units. The probability model reflects the fraction of situations or units which have a specific property, also referred to as frequentist probabilities [22, 23]. The population considered is in practice always finite, but the model presumes an infinite number of situations. The probability is a thought-construction obtained by going from the finite state to the infinite.

A probability model is established based on reasoning, as for the binomial or Poisson distributions, or by estimations based on observations. Both approaches introduce uncertainties, as explained in the following.

If the probability model is based on reasoning, there will be a set of assumptions on which the modelling is founded. For example, in the homogenous Poisson case, the probability of an event occurring in a small interval $(t, t + h)$ is approximately equal to λh, for a fixed number λ, independent of the history up to t. Verifying such assumptions is, however, in general difficult, as there may be little relevant data that can be used to check them, in particular in the case of rare events. Estimations and model validation using observations are applicable when huge data sets are available but not when studying extreme events. The consequence is that the analysis simply needs to presume the existence of the model and the results interpreted as conditional on these assumptions. Thus, care has to be shown in making conclusions based on the analysis, as the assumptions could cover or conceal important aspects of uncertainties and risks.

To introduce a probability model, it needs to serve a purpose. The common argument used is that it allows for statistical inference, to apply the strong machinery of statistics and Bayesian analysis, updating our knowledge when new information becomes available [21]. For situations where variation is the key quantity of interest, such models surely have a role to play, but, in cases of extreme events (events with extreme consequences), variation is not really the interesting concept, as there is no natural family of situations to which these events belong. Take major nuclear accidents. For the industry, historical data are informative on what has happened and how frequently. But will the development and use of a probability model to represent the variation in the occurrences of such accidents lead to new and important insights? To provide an answer to this question, let us review the potential purposes for developing such a model:

(a) To predict the occurrence of coming events
(b) To show trends
(c) To present ‘true’ risk levels

(d) To facilitate continuous updating of information about the risk levels.

Clearly, the use of such models does not allow for accurate prediction of occurrences, as the data are so few and the future is not necessarily reflected well by these data. Hence (a) is not valid. We have to make the same conclusion when it comes to (b) for the same reasons: meaningful trends cannot be established when the data basis is weak. According to some risk perspectives, risk as a concept is closely linked to the existence of probability models [24]. Consider the risk quantity defined by the frequentist probability that a major nuclear accident occurs in a country in the next year. As discussed above, giving this probability an interpretation is challenging, as it requires the definition of an infinite population of similar situations to the one studied. Anyway, it is unknown and needs to be estimated. With a weak data base, this estimate could deviate strongly from the 'true' risk. Hence, (c) is also problematic. Probability modelling is an essential pillar for using Bayesian analysis to systematically update the knowledge when new information becomes available. However, the modelling needs to be justified for the results to be useful. As discussed above, the problem is that it is often difficult to establish in a meaningful way an infinite population of similar situations or units. There is always a need to formulate a hypothesis, as knowledge generation is built on theory [25–27], but, in cases of rare events, broader frameworks than high-level probabilistic modelling are required. Judgements of risk for such events cannot be based on macro statistical data and analysis. More in-depth analysis of risk sources, threats, barriers, consequences is needed, in other words, more in-depth risk assessments.

Instead of considering probability models as a tool for modelling variation, it is also common to think of them as representations of characteristics of the activity or system, using terms like 'propensity interpretation of probability' and 'logical probability'. For the propensity interpretation, suppose we have a special coin; its characteristics (centre of mass, weight, shape, etc.) are such that, when tossing the coin over and over again, the head fraction will reach a number: the head propensity of the coin. However, accepting the framework of the frequentist probability, i.e. that an infinite sequence of similar situations can be generated, is practically the same as accepting the idea of the propensity interpretation, as it basically states that such a framework exists [22]. The propensity can be seen as a repeatable experimental set-up, which produces outcomes with a limiting relative frequency, which is equal to the frequentist probability [23].

The idea of the logical probability is that it expresses the objective degree of logical support that some evidence gives to the event (a hypothesis being true). It is believed that there is a direct link between evidence and the probability. However, this idea has never received a satisfactory interpretation [28]. Using logical probabilities, we are not able to interpret what a probability of say 0.1 means compared to 0.2. It should, therefore, be rejected.

28.3.2 Representing and/or Expressing Epistemic Uncertainties

The future is unknown and many quantities are unknown—related to the past, present or the future; thus, there are uncertainties, epistemic uncertainties. We lack knowledge.

Let X be such a quantity. It can be a quantity in real life, for example, the time to failure of a specific system, or it could be a model quantity like the occurrence rate λ in the above Poisson model, the true quantity defined as the average number of events occurring for the period considered if we could hypothetically repeat the situation over and over again infinitely. Or it could be the model error $M_e = F - h$, where h is the true variation in a population being studied and F the probability model used to model h.

Next, we would like to represent or express our uncertainty about the true value of X. 'We' refers here to the analyst or any other person who conducts the judgements. Let Q be such a representation or expression of uncertainty. Basically, there are two ways of thinking in specifying Q, giving it an interpretation and determining its value in a concrete case:

(i) Use Q to express our uncertainties and degrees of beliefs about X, using the available knowledge K. Thus, Q is subjective (or inter-subjective in the sense that people can agree on the same Q value).
(ii) Seek to obtain an objective representation/transformation of the knowledge K available, to Q.

Approach (i) is commonly implemented using subjective probabilities; hence $Q = P$. The scientific literature on subjective probabilities is, however, rather chaotic, in the sense that the earlier and historical interpretations of this probability are still referred to, despite being based on unfortunate mixtures of uncertainty judgements and value judgements (Aven 2013). If the science of uncertainty analysis offers this type of interpretation, it is not surprising at all that it is not very much used in practice. Consider the following example. We are to assign a subjective probability for the event A that a specific hypothesis is true, for example, that global warming is the result of human activity. Following common schools of thought in uncertainty analysis, this probability $P(A)$ is to be understood as expressing that 0.95 is "the price at which the person assigning the probability is neutral between buying and selling a ticket that is worth one unit of payment if the event occurs, and worthless if not" (see e.g. SEP 2011, Aven 2013). Such an interpretation cannot and should not be used for expressing uncertainty, as it reflects the assigner's attitude to money [22, 29, 30]. If we are to be informed by the IPCC's uncertainty judgements, we would not like them to be influenced by these experts' attitude to dollars. It is irrelevant for the uncertainty judgement. Note that IPCC does not refer to this type of interpretation of probability.

There are many other perspectives on subjective probabilities and one often referred to is the Savage interpretation, based on *preferences* between acts; see

Bedford and Cooke [31]. The idea is that the subjective probability can be determined based on observations of choices in preferences. However, as these preferences relate to money or other value attributes, the same problem occurs as above; we do not produce pure uncertainty judgements but a mixture of uncertainty and value judgements, which makes, for example, a statement like $P = 0.95$ impossible to meaningfully interpret.

Fortunately, there exist a theory and meaningful operational procedures that can be used to specify subjective probabilities as a pure measure of uncertainty [21, 30, 32]. A subjective probability of 0.95 is here interpreted as expressing that the assigner has the same uncertainty and degree of belief in the event A occurring as randomly drawing a red ball out of an urn which comprises 100 balls, of which 95 are red. This way of understanding a probability was referred to by Kaplan and Garrick [33] in their celebrated paper about risk quantification, but there are relatively few examples of researchers and probabilists adopting this way of interpreting probability [22]. This is unfortunate, as it provides a simple, elegant and easily understandable basis and theory for subjective probability. It is also common to refer to these probabilities as 'knowledge-based probabilities' [22].

A subjective probability is personal, depending on the assigner and the knowledge supporting the assignment. This fact has led scholars to look for alternative approaches for representing or expressing the uncertainties, as the probability number produced in many cases has a weak basis. The probability assigned seems rather arbitrary and too dependent on the assigner. Scientific knowledge generation requires more objective results, is a common way of reasoning. It motivates the alternative approach (ii), an objective representation/transformation of the knowledge K available, to Q. There are different ways of obtaining such a representation/transformation, but the most common is the use of probability intervals—also referred to as imprecise probabilities. In the IPCC case, an interval probability of (0.95,1] is specified. This does not mean that the probability is uncertain, as there is no reference to a 'true' probability; it simply means that the assigner is not willing to be more precise, given the knowledge available. Hence the assigner expresses that his/her degree of belief for the event to occur or the statement to be true, is higher than an urn chance of 0.95. His/her uncertainty or degree of belief is comparable with randomly drawing a red ball out of an urn comprising 100 balls for which more than 95 are red. Betting type of interpretations are also commonly used for interpreting interval probabilities, but they should be rejected of the same reasons as given above for the subjective probabilities.

Studying the literature related to the challenge of (ii), one soon realises that this is indeed a rather confusing area of analysis and research. There are different theories: possibility theory, evidence theory, fuzzy set theory, etc., with fancy mathematics, but the essential points motivating these theories are often difficult to reveal. Interpretations of basic concepts are often missing.

The previous paragraphs are an attempt to clarify some of the issues discussed. The aim of the alternative approaches is to obtain a more objective representation of uncertainty, given the knowledge available. This is often misinterpreted as saying that the representation is objective. Clearly, the objectivity here just refers to the transformation from K to Q. Using P alone, it is acknowledged that there is a leap from K to Q, which is subjective. With a probability interval, this leap is reduced or eliminated. The knowledge K can, however, be strongly subjective, more or less strong and even erroneous, for example, if it represents the judgement by one expert.

In practice it can be attractive to use both (i) and (ii). The former approach ensures that the analysts' and experts' judgements are reported and communicated, whereas the latter approach restricts its results to a representation of documented knowledge.

28.3.3 *The Strength of the Knowledge K*

Any judgement of uncertainty is based on some knowledge K and this knowledge can be more or less strong. How should this be reported? In the IPCC work, a qualitative scale of confidence is used with five qualifiers: very low, low, medium, high and very high, reflecting the strength of evidence and degree of agreement [6, 7]. The strength of evidence is based on judgements of "the type, amount, quality, and consistency of evidence (e.g., mechanistic understanding, theory, data, models, expert judgment)" [7]. Consider the following statements from the IPCC [7]:

> Ocean acidification will increase for centuries if CO_2 emissions continue, and will strongly affect marine ecosystems (with high confidence). ([7], p. 16) (28.1)
>
> The threshold for the loss of the Greenland ice sheet over a millennium or more, and an associated sea level rise of up to 7 m, is greater than about 1 °C (low confidence) but less than about 4 °C (medium confidence) of global warming with respect to pre-industrial temperatures. ([7], p. 16) (28.2)

There are no explicit uncertainty judgements of the form Q used in these cases. But could not the first example (28.1) be interpreted as expressing that "Ocean acidification will increase for centuries if CO_2 emissions continue, and will strongly affect marine ecosystems" is true with very high probability? Yes, such an interpretation is reasonable, but, according to IPCC [6, p. 3], "Confidence is not to be interpreted probabilistically". How should we then interpret (28.1)? It is then even more difficult to understand (28.2). For example, if "...is greater than about 1 °C..." is expressed with low confidence, what does this statement really express? According to Aven and Renn [9], the IPCC framework lacks a proper uncertainty and risk analysis foundation, as the link between the strength of knowledge (level of confidence) and Q is not clarified. In IPCC documents, Q is sometimes used (as in the 95% case mentioned in Sect. 28.2), sometimes not [as in (28.1) and (28.2)].

The IPCC concept of confidence is based on the two dimensions, evidence and agreements. The latter criterion needs to be implemented with care; if agreement is among experts within the same 'school of thought', its contribution to confidence is much less than if the agreement is built on experts representing different areas, disciplines, etc. [12, 34].

Yet we find this criterion in most systems for assessing the strength of knowledge and confidence, see for example Flage and Aven [35] and Aven and Flage [36] who establish a qualitative strength of knowledge scheme based on judgements of the reasonability of assumptions made, the amount and relevance of supporting data and information, agreement between experts, the understanding of the phenomena studied, the degree of model accuracy, and to what degree this knowledge has been examined (with respect to , e.g. signals and warnings, unknown knowns, etc.).

For a related qualitative scoring scheme for assessing the knowledge strength, see the so-called NUSAP system (NUSAP: Numeral, Unit, Spread, Assessment, and Pedigree) [37–43]. Also in this system, agreement is identified as a criterion, in fact among both peers and stakeholders. Other criteria include the influence of situational limitations, plausibility, choice space, sensitivity to views of analysts, and influence on results.

As the IPCC case demonstrates, the scientific findings of climate change are strongly intertwined with judgements of the strength of the knowledge supporting these findings. Although there are weaknesses in the IPCC framework for uncertainty and risk treatment, the use of confidence statements in the IPCC setting is a step in the right direction. A lot of scientific work lacks this type of consideration. Results have been and still are produced without stressing that these are conditional on some knowledge and this knowledge could be more or less strong, and even erroneous. Critical assumptions are commonly reported as an integrated feature of the results, but more comprehensive knowledge considerations, as discussed in this section, are seldom carried out. If we also include potential surprises relative to this knowledge, as will be discussed in the coming section, they are even more seldom conducted. The scientific literature on uncertainty and risk analysis has devoted little attention to this type of issue, and there is no established practice on how to deal with them.

28.3.4 The Potential for Surprises

As discussed in Sect. 28.1, knowledge can be considered as justified beliefs. Hence knowledge can be more or less strong and also erroneous. Experts can agree and the data available can generate beliefs as formulated above in the IPCC case. Yet there is a potential for surprise; the knowledge can be wrong.

Dealing with this type of risk is challenging, as it extends beyond the knowledge available. Nonetheless, it is an essential component of science, of a type that forces scientists to balance confidence with humbleness, as discussed in Sect. 28.1.

There are different types of surprises. One of the most important ones is unknown knowns, as reflected by the origin of the black swan metaphor. Before the discovery of Australia, people in the Old World believed all swans were white, but then in 1697, a Dutch expedition to Western Australia discovered black swans [44], a surprise for us but not for people living there. The September 11 event is an example of an unknown known. It came as a surprise to most of us, although, of course, not to those planning the attack. By proper analysis, many unknown knowns can be revealed, but in practice, there will always be limitations, and surprises of this type can occur. Unknown unknowns—events not known to anybody—are more challenging to identify, but fortunately, such events are rarer. Testing and research are generic measures to meet this type of event, as well as a focus on resilience, signals and warnings [45].

There is also a third category of surprises, it is of a different type. In this case, the event is known but not believed to occur because of low-judged probability [45]. To illustrate the idea, think about an event A, for which a probability of 0.000001 is assigned, given the knowledge K, that is $P(A|K) = 0.000001$, or we could think about a situation where an imprecision interval is instead specified: $P(A|K) < 0.000001$. The point is that the probability is judged so low that the occurrence of the event is ignored for all practical purposes. Now suppose the probability assignment is based on a specific assumption, for example, that some potential attackers do not have the capacity to carry out a type of attack. Given this assumption, the probability is found to be negligible. Hence, if the event occurs it will come as a surprise, given the knowledge available. However, the assumption could be wrong and clearly, with a different knowledge base, the probability could be judged high, and the occurrence of the event would not be seen as surprising.

An illustrating example is provided by Gross [46, p. 39]: In a power plant in the USA in 2002, experts did not even think about the possibility of a special type of leak (nozzle leak deposits could eat into carbon steel of the reactor vessel), before workers discovered that boric acid had eaten almost all the way through the reactor pressure vessel head [46, p. 39]. Similar surprises are reported in the oil and gas industry [47].

This discussion relates to the fundamentals of risk assessments. Current practice has to a large extent been based on a frequentist understanding of probabilities, seeing probability judgements as reflecting states of the world. In this view, it is believed that an event with an estimated probability will occur sooner or later: it is like a physical law. However, this 'destiny perspective' on probability and risk is not very meaningful or fruitful for assessing and managing risk in cases with a potential for extreme outcomes and large uncertainties. Yet this type of thinking is largely prevalent in university programmes, in particular in engineering and business. The risk and uncertainty analysis sciences have not yet been able to challenge this thinking in a way that has changed common practices.

28.4 How to Improve Science by Strengthening Risk and Uncertainty Analysis

The topics discussed in Sect. 28.3 are all key ones in scientific work. Yet no authoritative guidance exists on how to deal with them. The present work is based on the conviction that this is a result of weak risk and uncertainty analysis fields and sciences. The topics are essential for understanding, for example, climate change, but, as the IPCC reports document, current practice suffers. Unfortunately, the situation is not expected to change in the near future, as the problem is a fundamental one. It takes time to build the necessary knowledge and a research community with institutions that can change the current state. What is needed is—as for all sciences—a broad recognition of being a science, as this in its turn creates a base for project funding, training programmes at all levels and academic positions. Today, the fields of risk and uncertainty analysis have some journals addressing the topics but few academic positions and study programmes. Funding schemes typically lack categories for risk and uncertainty analysis [16].

Authoritative guidance does not mean that there is no need for research challenging existing ideas. On the contrary, a basic feature of science is the continuous scrutiny of current thinking, methods and models, to further improve these. The essential pillar for such developments is that there is a strong fundamental research on "Knowledge on concepts, theories, frameworks, approaches, principles, methods and models to understand, assess, characterise, communicate and (in a broad sense) manage risk", as was referred to in Sect. 28.2 as the (B) part of the risk analysis science. This is, however, lacking today. The volume of research on the generic topics of this type is too small. There are scholars specifically working on (B) but not many on a world basis, compared to, for example, statistics.

Fortunately, there is substantial research in areas which are close to (B). Probability theory and statistics have already been mentioned. In addition, there is a considerable body of work on sensitivity analysis and uncertainty importance analysis, where the challenge is to identify the most critical and essential contributors to output uncertainties and risk; see e.g. Borgonovo and Plischke [48]. This type of analysis is useful in identifying critical assumptions and parameters, which is an important task of a risk and uncertainty analysis. However, current research in this area to a little extent covers issues related to knowledge strength and surprises (items (3) and (4) listed in Sect. 28.1).

Another area to which considerable attention has been devoted is the use of alternative approaches to probability to represent or express uncertainties. Some scholars argue that probability is the only tool needed to express uncertainties; it is acknowledged that there could be elicitation and imprecision issues, but in principle, the probability is perfect [30, 49]. However, this view is easily refuted for scientific applications; see for example Flage et al. [50]. Two situations can be characterised by the same probabilities: in one case the knowledge supporting the probability could be weak, whereas in the other it could be strong. Should not this difference in knowledge base be considered an integrated part of the uncertainty description? Yes, it should. For the decision-makers—who are not normally the same as the assessors—this difference is of course critical. The IPCC has acknowledged this need by the use of the confidence scale.

There seems to be a growing number of people recognising the need to see beyond probability to represent or express uncertainties. However, the rationale for this and the solutions presented to replace probability are in many cases poor. Imprecision is mixed with uncertainty, and concepts are introduced without a meaningful interpretation [50].

As discussed in Sect. 28.3.2, a probability interval (imprecision interval) is also based on some knowledge, which comprises justified beliefs, which can be more or less strong, or even erroneous. In much of the literature, it seems that authors believe that the use of such intervals, relying either on possibility theory or more generally evidence theory, represents an objective characterisation of uncertainties. The transformation from K to Q is more objective, yes, but it does not eliminate the subjectivity of K.

There is a considerable body of literature addressing uncertainties related to vague statements like 'few events'. A concept of fuzzy probability is introduced, with detailed mathematical formalism. However, a proper interpretation of this concept does not exist, and it is rejected by many authors [51, 31]: We cannot build a theory on a concept for which we cannot explain what the difference in meaning is between, say, 0.2 and 0.3. Yet a high number of papers are published using such probabilities. The present author considers that these works mess up the fields and sciences of risk and uncertainty analysis. Vague information can always be taken into account, but the risk and uncertainty analysis should focus on quantities for which underlying 'true' values exist.

28.5 Conclusions

Climate change is important as it relates to our future life on our planet. If climate change is mainly man-made, it has serious implications for how to confront this change. Climate change knowledge and science are to provide us with the necessary confidence for taking actions; our policies can be knowledge- and science-based.

Unfortunately, knowledge and science of this type are subject to uncertainties. We are not able to accurately predict what will happen in the future; we face risks. The best we can do is to characterise what will happen and related phenomena, using risk and uncertainty characterisations. The quality of these characterisations is thus essential. However, the fields and sciences producing the knowledge on how to perform such characterisations and use them in a context of management, governance and decision-making are relatively poor, as discussed in previous sections. This calls for measures. This work considers the following to be the most urgent:

- More research on foundational issues for risk and uncertainty analysis, providing the necessary pillars for risk and uncertainty analysis sciences. This can be stimulated by initiatives, for example, taken by editors and board members of relevant journals.
- More interactions between individual researchers, societies, associations, etc., dealing with different aspects of risk and uncertainty analysis, as well as other related areas of performability (quality, reliability, security, safety, and maintainability). For example, risk analysis scientists can benefit from working more closely with uncertainty analysis scientists, and vice versa.
- Relevant societies, associations, etc., to take responsibility for addressing foundational issues for risk and uncertainty analysis, as for example the Society for Risk

Analysis has done in developing a Glossary, and documents on Core Subjects and Fundamental Principles of Risk Analysis [52–54].

- Societies, associations and renowned experts to raise their voice regarding the need for and importance of strong and distinct sciences of risk and uncertainty analysis.
- Relevant societies, associations, etc., to derive suitable strategies for how to develop risk and uncertainty analysis as distinct sciences or a unified distinct science.

The basic message from the present discussion is that there is a strong need for the sciences of risk and uncertainty analysis, but they are not yet in a satisfactory state, and substantial improvements are required. The author's vision is that risk and uncertainty analysis are broadly acknowledged as sciences per se, or an integrated distinct science, and can provide strong support for the various application areas like climate change, medicine and health, etc., on issues related to risk and uncertainty. If this vision can be realised, the right balance between confidence and humbleness can be achieved.

References

1. Aven, T. (2020). *The science of risk analysis*. New York: Routledge.
2. Flanders, W. D., Lally, C. A., Zhu, B.-P., Henley, S. J., & Thun, M. J. (2003). Lung cancer mortality in relation to age, duration of smoking, and daily cigarette consumption. *Cancer Research, 63*, 6556–6562.
3. Yamaguchi, N., Kobayashi, Y. M., & Utsunomiya, O. (2000). Quantitative relationship between cumulative cigarette consumption and lung cancer mortality in Japan. *International Journal of Epidemiology, 29*(6), 963–968.
4. Proctor, R. N. (2011). The history of the discovery of the cigarette–lung cancer link: Evidentiary traditions, corporate denial, global toll. *Tobacco Control, 21*, 87–91.
5. ITC Netherlands Survey. (2011). *Report on smokers' awareness of the health risks of smoking and exposure to second-hand smoke*. Ontario, Canada: University of Waterloo.
6. IPCC Guidance Notes for Lead Authors of the IPCC Fifth Assessment Report on Consistent Treatment of Uncertainties: IPCC Cross Working Group Meeting on Consistent Treatment of Uncertainties, 2010.
7. IPCC Climate Change 2014 Synthesis Report Summary for Policymakers. https://www.ipcc.ch/pdf/assessment-report/ar5/syr/AR5_SYR_FINAL_SPM.pdf. 2014. Accessed October 29, 2019.
8. IPCC Fifth Assessment Report—Webpage (2018). https://www.ipcc.ch/report/ar5/index.shtml. Assessed October 29, 2019.
9. Aven, T., & Renn, O. (2015). *An evaluation of the treatment of risk and uncertainties in the IPCC Reports on Climate Change. Risk Analysis, 35*(4), 701–712 (Open access).
10. ISO 31000 Risk Management. https://www.iso.org/iso-31000-risk-management.html. Accessed October 29, 2019.
11. Aven, T. (2017). *The flaws of the ISO 31000 conceptualisation of risk. Journal of Risk and Reliability, Editorial, 231*(5), 467–468 (Open access).
12. Aven, T., & Ylönen, M. (2019). *The strong power of standards in the safety and risk fields: A threat to proper developments of these fields? Reliability Engineering and System Safety, 189*, 279–286 (Open access).

13. Oxford English Dictionary. https://www.oed.com. Accessed October 29, 2019.
14. Free Dictionary. https://www.thefreedictionary.com/. Accessed February 24, 2020.
15. Hansson, S. O. (2013). *Defining pseudoscience and science.* In M. Pigliucci & M. Boudry (Eds.), *Philosophy of pseudoscience.* Chicago, IL: University of Chicago Press.
16. Hansson, S. O., & Aven, T. (2014). Is risk analysis scientific? *Risk Analysis, 34*(7), 1173–1183.
17. Aven, T., & Ylönen, M. (2018). *The enigma of knowledge in the risk field.* In T. Aven & E. Zio (Eds.), *Knowledge in risk assessments.* NY: Wiley.
18. Bourdieu, P., & Wacquant, L. J. D. (1992). *An invitation to reflexive sociology.* Chicago: University of Chicago Press.
19. Aven, T. *An emerging new risk analysis science: Foundations and implications. Risk Analysis, 38*(5), 876–888 (Open access).
20. Aven, T., & Zio, E. (2014). Foundational issues in risk analysis. *Risk Analysis, 34*(7), 1164–1172.
21. Lindley, D. V. (2000). *The philosophy of statistics. The Statistician, 49,* 293–337 (With discussions).
22. Aven, T. (2013). *How to define and interpret a probability in a risk and safety setting.* Discussion paper Safety Science, with general introduction by Associate Editor. *Genserik Reniers, 51*(1), 223–231.
23. Stanford Encyclopedia Philosophy Interpretations of Probability (2011). https://plato.stanford.edu/entries/probability-interpret/. Accessed October 29, 2019.
24. Aven, T. (2011a). *Quantitative risk assessment: The scientific platform.* Cambridge University Press.
25. Bergman, B. (2009). Conceptualistic pragmatism: A framework for Bayesian analysis? *IIE Transactions, 41,* 86–93.
26. Deming, W. E. (2000). *The new economics* (2nd ed.). Cambridge, MA: MIT CAES.
27. Lewis, C. I. (1929). *Mind and the world order: Outline of a theory of knowledge.* New York, NY: Dover Publications.
28. Cowell, R. G., Dawid, A. P., Lauritzen, S. L., & Spiegelhalter, D. J. (1999). *Probabilistic networks and expert systems.* New York: Springer.
29. Cooke, R. M. (1986). Conceptual fallacies in subjective probability. *Topoi, 5,* 21–27.
30. Lindley, D. V. (2006). *Understanding uncertainty.* Hoboken, NJ: Wiley.
31. Bedford, T., & Cooke, R. (2001). *Probabilistic risk analysis.* Cambridge: Cambridge University Press.
32. Lindley, D. V. (1985). *Making decisions.* New York: Wiley.
33. Kaplan, S., & Garrick, B. J. (1981). On the quantitative definition of risk. *Risk Analysis, 1,* 11–27.
34. Miller, B. (2013). When is consensus knowledge-based? Distinguishing shared knowledge from mere agreement. *Synthese, 190,* 1293–1316.
35. Flage, R., & Aven, T. (2009). Expressing and communicating uncertainty in relation to quantitative risk analysis (QRA). *Reliability and Risk Analysis: Theory and Applications, 2*(13), 9–18.
36. Aven, T., & Flage, R. (2018). *Risk assessment with broad uncertainty and knowledge characterisations: An illustrating case study.* In T. Aven & E. Zio (Eds.), *Knowledge in risk assessments.* NY: Wiley.
37. Funtowicz, S. O., & Ravetz, J. R. (1990). *Uncertainty and quality in science for policy.* Dordrecht: Kluwer, Academic Publishers.
38. Funtowicz, S. O., & Ravetz, J. R. (1993). Science for the post-normal age. *Futures, 25,* 735–755.
39. Kloprogge, P., van der Sluijs, J., & Petersen, A. (2005). *A method for the analysis of assumptions in assessments.* Bilthoven, The Netherlands: Netherlands Environmental Assessment Agency (MNP).
40. Kloprogge, P., van der Sluijs, J. P., & Petersen, A. C. (2011). A method for the analysis of assumptions in model-based environmental assessments. *Environmental Modelling and Software, 26,* 289–301.

41. Laes, E., Meskens, G., & van der Sluijs, J. P. (2011). On the contribution of external cost calculations to energy system governance: The case of a potential large-scale nuclear accident. *Energy Policy, 39,* 5664–5673.
42. van der Sluijs, J., Craye, M., Futowicz, S., Kloprogge, P., Ravetz, J., & Risbey, J. (2005). Combining quantitative and qualitative measures of uncertainty in model-based environmental assessment. *Risk Analysis, 25*(2), 481–492.
43. van der Sluijs, J., Craye, M., Funtowicz, S., Kloprogge, P., Ravetz, J., & Risbey, J. (2005). Experiences with the NUSAP system for multidimensional uncertainty assessment in model based foresight studies. *Water Science and Technology, 52*(6), 133–144.
44. Taleb, N. N. (2007). *The Black Swan: The impact of the highly improbable.* London: Penguin.
45. Aven, T. (2015). *Implications of black swans to the foundations and practice of risk assessment and management. Reliability Engineering & System Safety, 134,* 83–91 (Open access).
46. Gross, M. (2010). *Ignorance and surprises.* London: MIT Press.
47. Black swan. *Norwegian Oil and Gas Association.* https://www.norskoljeoggass.no/en/search/?q=black+swan. Accessed April 23, 2020.
48. Borgonovo, E., & Plischke, E. (2015). Sensitivity analysis: A review of recent advances. *European Journal of Operational Research, 248,* 869–887.
49. O'Hagan, A., & Oakley, J. E. (2004). Probability is perfect, but we can't elicit it perfectly. *Reliability Engineering and System Safety, 85,* 239–248.
50. Flage, R., Aven, T., Baraldi, P., & Zio, E. (2014). Concerns, challenges and directions of development for the issue of representing uncertainty in risk assessment. *Risk Analysis, 34*(7), 1196–1207.
51. Aven, T. (2011b). On the interpretations of alternative uncertainty representations in a reliability and risk analysis context. *Reliability Engineering and System Safety, 96,* 353–360.
52. Glossary Society for Risk Analysis. (2015). www.sra.org/resources. Accessed April 23, 2020.
53. Core Subjects of Risk Analysis. (2017). www.sra.org/resources. Accessed April 23, 2020.
54. Risk Analysis: Fundamental Principles. (2017). www.sra.org/resources. Accessed April 23, 2020.

Terje Aven, Ph.D. is the Professor of Risk Analysis and Risk Management at the University of Stavanger, Norway. He has many years of experience as a risk analyst and consultant in industry, and is the author of many books and papers covering a broad range of risk science topics. He has served as the Chair of the European Safety and Reliability Association (ESRA) and as the President of the Society for Risk Analysis (SRA) worldwide. He is the Editor-in-Chief of the Journal of Risk and Reliability, and Associate editor for Risk Analysis.

Chapter 29
Simplified Analysis of Incomplete Data on Risk

Bernhard Reer

Abstract A framework for simplified analysis of incomplete data on risk is presented and illustrated in five feasibility case studies on road traffic and occupational safety. The Bayesian theorem is utilized for both structuring cases of incomplete input data and providing options for dealing with incompleteness. The application of the framework requires the availability of an interval scale of an index of prevention in a situation exposed to failure. A key parameter of the framework is bounded in the range from 0 to 1 and represents the average degree of prevention (v) in failure exposure situations for a given type of risk. The Bayesian structure of the framework allows to verify an expert judgement for v by a quantitative evaluation of failure events only meaning a quantitative evaluation of the variety of failure exposure situations would not be necessary. Moreover, non-trivial comparisons between different types of risks are possible. The loss of accuracy identified from the case studies is assessed as a not satisfactory result. It is an open issue whether such inaccuracy is inherent, when applying a common and simple model for data analysis addressing various risk environments, or it can be reduced by a refined version of a common model or by improved scaling.

Keywords Data analysis · Applied probability · Risk analysis · Bayesian inference · Uncertainty · Variability · Performability · Feasibility study · Context and failure

29.1 Introduction

Numbers for probabilities or rates concerning failure events are needed to inform decisions in various areas such as risk assessment addressing complex production and transportation systems. The performability of analyses to obtain such numbers is often degraded because of the scarcity of input data and imperfect knowledge about the relevant impacts.

B. Reer (✉)
Villigen, Switzerland
e-mail: bernhardreer@yahoo.de

K. B. Misra (ed.), *Handbook of Advanced Performability Engineering*,
https://doi.org/10.1007/978-3-030-55732-4_29

Table 29.1 Incomplete data on occupational accident rates and worker's risk behaviour

Risk behaviour	Violation and No risk Compensation effort (VCN)	Violation and risk Compensation effort (VC)	No Violation (VN)	Total
Number of accidents	?	?	?	31
Number of task performances	?	?	?	336,000

Accident rate data from Rehhahn [1]; violation sub-division based on risk homeostasis theory [2].

As a hypothetical and simple example, let us assume a data analysis with the objective to estimate for given steel production task the probability of an occupational accident (e.g. getting burned) specific to the worker's risk behaviour concerning: violation from safety rules (e.g. not wearing hand protection device) and risk compensation effort (i.e. working with increased caution in case of a violation). Initially, however, the available data are incomplete as shown in Table 29.1.

Bayesian analyses—mostly in combination with expert judgment—provide options to overcome the problem of incomplete data as presented in a large number of publications (e.g. [3–5]). Respective applications show that rather sophisticated work is required for data collection or evaluation, e.g. when applying the Gibbs sampling procedure [6]. Component failure probabilities or rates and initiating event frequencies are the subjects of the handbook on data analysis for probabilistic risk assessment of nuclear power plants [7]. Human error probabilities (HEPs) have been addressed a feasibility study on the employment of Bayesian methods by Hallbert and Kolaczkowski [8]. In the context of safety in the chemical industry, Duijm and Goossens [9] propose the utilization of the Bayesian theorem for the assessment of failure probabilities of safety barriers in case of management deficiencies.

This chapter presents a framework serving to reduce the effort of data collection and evaluation. Variables associated with the Bayesian formula are used to distinguish classes of incomplete input data, and simplifying assumptions are made serving a rather fast generation of the output of interest. The basic concept of failure modelling and inference is presented in Sect. 29.2. The purpose of Sect. 29.2 is to show the mathematical background of the framework and to provide interpretations of the parameters of the mathematical model.

Due to the simplifying assumptions made, the scope of the framework is limited to applications with low accuracy standards in predicting failure probabilities (or rates). This is often the case for failure exposure in environments associated with

- a variety of contextual factors on the one hand, and
- limited knowledge about the relevant factors and their impacts, lack of impact data, variability of the impact or practical constraints to model a manageable number of factors.

The sources of uncertainty characterized above are likely to arise for failure events depending on human behaviour. As outlined by Dougherty [10], the factors affecting

Table 29.2 Laboratory experiments addressing risk-taking behaviour

Risk-taking behaviour (RTB)	Study	pr(RTB)	α
(I) Accepting a 0.001 probability loss of 5000$ instead of paying 5$ to avoid it	(a) Kahneman and Tversky [12], 72 subjects	0.17	1.7E−08
	(b) Slovic et al. [13], 36 subjects	0.72	
(II) Accepting a 0.5 probability loss of 1000$ instead of paying 500$ to avoid it	(a) Kahneman and Tversky [12], 68 subjects	0.69	0.028
	(b) Slovic et al. [13], 36 subjects	0.47	

α—Probability that the decision to reject hypothesis $H_0\{\Pr(\text{RTB}_{(a)}) = \Pr(\text{RTB}_{(b)})\}$ is false

human decision-making depend strongly on the context, and the impact of the context is not necessarily obvious. Consequently, the prediction of failures of type *human error* is a typical field of applications with low accuracy standards as reflected in the study by Kirwan et al. [11] on the validation of methods for HEP prediction: validation criteria with respect to precision are defined in terms of factors of three and ten in relation to an empirical HEP value.

Even under well-defined laboratory conditions—meaning the subjects are isolated to a wide extent from the various impacts expected in the real-life situations—human behaviour is difficult to predict. Laboratory experiments addressing risk-taking behaviour (RTB), for instance, indicate contradictory results as shown in Table 29.2. For a low probability risk (RTB I) as well as for a high probability risk (RTB II), the point estimate pr(RTB) (0.17 or 0.69, respectively) of the probability of risk-taking behaviour observed in the study of Kahneman and Tversky [12] is different from this point estimate (0.72 or 0.47, respectively) observed in the study of Slovic et al. [13]. Assuming a binominal distribution (approximated by a normal distribution) of the number of risk-taking subjects, (29.1) returns that each difference is significant at least on a significance level of 5%.

$$\alpha = 2 - 2\Phi\left(\frac{\left|\frac{n_{\text{RT}(a)}}{n_{(a)}} - \frac{n_{\text{RT}(b)}}{n_{(b)}}\right|}{\sqrt{\frac{n_{\text{RT}(a)}+n_{\text{RT}(b)}}{n_{(a)}+n_{(b)}}\left(1 - \frac{n_{\text{RT}(a)}+n_{\text{RT}(b)}}{n_{(a)}+n_{(b)}}\right)\left(\frac{1}{n_{(a)}} + \frac{1}{n_{(b)}}\right)}}\right) \tag{29.1}$$

In (29.1), $n_{(a)}$ and $n_{(b)}$ are a number of subjects in study (*a*) or (*b*), respectively, $n_{RT(a)}$ and $n_{RT(b)}$ the number of risk-taking subjects in study (*a*) or (*b*), respectively, $\Phi(\ldots)$ the function of the cumulative standard normal distribution, and α the probability that the decision to *reject* the *Null hypothesis* (H_0) of an equal RTB probability (in both studies) is false. The equation is taken from the basic literature (e.g. [14]) of statistical testing. Of course, the returned α value provides a coarse orientation only, since the underlying assumption (binominal distribution of the number of risk-taking subjects) is a strong simplification of the outcome of human decision-making and thus associated with particular uncertainty. Especially for the high probability risk, it might be reasonable to accept H_0 on a significance level of 1%.

Kahneman and Tversky [12] explain the results with the subjective weighting of outcome probabilities: small loss probabilities are over-weighted (risk aversion), while high loss probabilities are under-weighted. The risk-taking tendency towards the low probability risk is explained by Slovic et al. [13] with the threshold theory meaning a risk is neglected if the loss probability is below a certain value.

Failure events involving human behaviour are chosen here as the subjects of the feasibility case studies presented in Sect. 29.3. The purpose of this section and the main topic of this chapter is to illustrate how a simple but common model would overcome the problem of incomplete input data in various types of risk environments and to assess the loss of accuracy resulting from such kind of simplified data analysis.

29.2 Framework for Data Analysis

The framework is a high-level tool for the analysis of incomplete data. It guides to coarsely estimate distribution characteristics of probabilities or rates concerning an undesired event of interest. The undesired event is shortly denoted here as *failure* (*F*). It could be a specific accident, a human error, a damage or the like.

29.2.1 Failure and Condition in the Light of the Bayesian Theorem

Real-life situations comprise a variety of conditions under which a failure may occur, and each condition represents a constellation of values or levels from a set of variables. The selection of a limited set of variables to be addressed depends on the objective of the data analysis and branch-specific expertise. In principle, such conditional aspects of failure are represented by the *Bayesian theorem* and the related model of the *total probability*. For conditions resulting from a set of variables with discrete levels, the respective equations are

$$p_{C(j)|F} = \frac{p_{C(j)} p_{F|C(j)}}{p_F} \tag{29.2}$$

$$p_F = \sum_{C(j)} \left(p_{C(j)} p_{F|C(j)} \right) \tag{29.3}$$

where $p_{C(j)}$ is the probability of condition j, $p_{F|C(j)}$ the failure probability under this condition, $p_{C(j)|F}$ the probability of condition j under the condition of a failure and p_F the total failure probability. All $p_{C(j)}$ values sum up to one:

$$\sum_{C(j)} p_{C(j)} = 1 \tag{29.4}$$

According to the analysis objective in the example introduced in Table 29.1, *an occupational accident is the failure of interest, and the addressed conditions correspond to the three discrete levels (VCN, VC, VN) defined for the variable risk behaviour. Except for $p_F = 31/336{,}000$, none to the quantities in* (29.2) *can be derived directly.*

29.2.2 Scaling and Discretization

The inference lines (presented in Sect. 29.2.6) assume that conditions are represented on an *interval scale* with a limited number of discrete points. Each point represents a prevention index (i) as a quantity; a high numerical value of it corresponds to a low failure probability. Consequently, the input preparation comprises to transfer from (29.2) through (29.4) presented above to (29.5) through (29.7) presented below.

$$p_{i|F} = \frac{p_i \, p_{F|i}}{p_F} \tag{29.5}$$

$$p_F = \sum_i (p_i \, p_{F|i}) \tag{29.6}$$

In (29.5) and (29.6), p_i is the probability of a condition with a prevention index of i, $p_{F|i}$ the failure probability under this condition, $p_{i|F}$ the probability of a condition with a prevention index of i under the condition of a failure and p_F the total failure probability. All p_i values sum up to one:

$$\sum_i p_i = 1 \tag{29.7}$$

Various approaches exist for the preparation of an index scale. Many practical applications are based on the concepts of *Paired Comparison* (c.f.: [5, 15, 16]), *Multi-Attribute Utility Theory* (c.f.: [17, 18]) or *Psychological Test Theory* (c.f.: [19, 20]). Implementation details associated with scaling are subjects of the research on HEP assessment (e.g. [21]). Simplified examples of scaling are presented in Sect. 29.3.

Besides better transparency, the arguments for the use of a discrete scale are the limited ability to estimate an exact value on a continuous scale related to failure exposure in real-life situations, and incomplete knowledge about both the set and the impact of prevention and risk factors. In other words, a continuous scale would suggest a level of accuracy that is unlikely to achieve. A discrete scale is often entirely adequate for the purpose of risk or reliability analysis and, considering the scarcity of data, often necessary (c.f. [22], Chap. 6, p. 9).

In continuation of the hypothetical Table 29.1 *example, it is assumed that a team of occupational safety experts agrees—under consideration of the controversial discussion* (c.f. [23]) *of validity of the risk homeostasis theory—on a simple scale of tree points (0, 1, 2) with the prevention index assignments of* $i_{(\mathrm{VCN})} = 0$ *for violation and no risk compensation effort,* $i_{(\mathrm{VC})} = 1$ *for violation and risk compensation effort, and* $i_{(\mathrm{VN})} = 2$ *for no violation when performing the task.*

29.2.3 Parameterization

Parameterization concerns the specification—by means of a small number of constant quantities denoted as parameters—of the functional relationship between the probabilities ($p_{F|i}, p_{i|F}, p_i$) of interest and the prevention index (i). Respective assumptions made here about the types of functional relationships are listed and justified next.

1. Geometrical (exponential) relationship between $p_{F|i}$ and i:

$$p_{F|i} = p_{F|0}q^{i} = \mathrm{e}^{i \ln q + \ln p_{F|0}} = p_{F|m}q^{-(m-i)} = \mathrm{e}^{-(m-i)\ln q + \ln p_{F|m}} \tag{29.8}$$

2. Binominal distribution of i:

$$p_i = \binom{m}{i} v^i (1-v)^{m-i} \tag{29.9}$$

3. Binominal distribution of i under the condition of a failure:

$$p_{i|F} = \binom{m}{i} w^i (1-w)^{m-i} \tag{29.10}$$

(29.9) and (29.10) are probability mass functions. The five parameters in (29.8) through (29.10) are

(I) m as the size of a prevention index scale starting with $i = 0$;
(II) $p_{F|0}$ or $p_{F|m}$ as the position calibration parameter of this scale, i.e. the failure probability under the worst ($i = 0$) or best ($i = m$) condition, respectively, covered by the scale;
(III) q as the gradient parameter of the geometrical model, which is as well the gradient parameter for scale calibration;
(IV) v as the average degree of prevention, indicating (relatively) the average position ($\bar{\imath} = vm$) of the prevention index;
(V) w indicating (relatively) the average position ($\bar{\imath}|F = wm$) of the prevention index under the condition of a failure.

Assumption 1 is typical for applications dealing with failure probabilities associated with increased uncertainties (c.f.: [24, 17, 15, 5]). As outlined in Sect. 29.2.5, assumption 1 is furthermore in line with the frequently used model of a log-normal distribution of a failure probability.

Assumptions 2 and 3 are reasonable, since they imply assumption 1 because of the Bayesian formula (29.5). The proof is presented below.

$$p_{F|i} = \frac{p_F p_{i|F}}{p_i} = \frac{p_F \binom{m}{i} w^i (1-w)^{m-i}}{\binom{m}{i} v^i (1-v)^{m-i}} = \frac{p_F w^i (1-w)^{m-i}}{v^i (1-v)^{m-i}} \tag{29.11}$$

$$p_{F|(i+1)} = \frac{p_F p_{(i+1)|F}}{p_{i+1}} = \frac{p_F \binom{m}{i+1} w^{i+1} (1-w)^{m-(i+1)}}{\binom{m}{i+1} v^{i+1} (1-v)^{m-(i+1)}}$$
$$= \frac{p_F w^{i+1} (1-w)^{m-(i+1)}}{v^{i+1} (1-v)^{m-(i+1)}} \tag{29.12}$$

$$q = \frac{p_{F|(i+1)}}{p_{F|i}} = \frac{\frac{p_F w^{i+1} (1-w)^{m-(i+1)}}{v^{i+1} (1-v)^{m-(i+1)}}}{\frac{p_F w^i (1-w)^{m-i}}{v^i (1-v)^{m-i}}}$$
$$= \frac{w^{i+1} (1-w)^{m-i-1} v^i (1-v)^{m-i}}{w^i (1-w)^{m-i} v^{i+1} (1-v)^{m-i-1}} = \frac{(1-v)w}{(1-w)v} = \frac{w - vw}{v - vw} \tag{29.13}$$

It can be seen from (29.13) that the gradient parameter q, defined as the quotient $p_{F|(i+1)}/p_{F|i}$ for $i = 0$ to $i = (m-1)$, is constant, i.e. it does not depend on i. It can be seen as well that $v > w$ is the condition for a decreasing trend (over i) of the conditional failure probability.

To proceed with the hypothetical Table 29.1 *example, it is assumed that a team of occupational safety experts assesses that an observation (89% of the tasks carried without violations of safety rules)—available from another, smaller sample—is applicable to the sample (31 accidents per 336,000 tasks) in question. With* $p_{VN} = p_2 = 0.89$ *as input, resolving* (29.9) would yield then $v = 0.943$ *as the average degree of prevention. Moreover, it is assumed that the team concluded from the review of the 31 accident reports that is not explicitly documented (due to several reasons including liability) in each case whether a violation was involved but that is unlikely that each of the accidents would have happened without a violation. The team concludes therefore that the probability that no violation was involved under the condition of an accident shall return from a formula suitable for zero-event data* (c.f. [25]), *i.e.* $p_{VN|F} = p_{2|F} = 1 - 0.5^{1/31} = 0.0221$. *With this estimate as input, resolving* (29.10) *would yield* $w = 0.149$ *as the average degree of prevention under the condition of an accident.*

29.2.4 Relation to the Probability-Generating Function

The parametrizations for the conditional failure probability $p_{F|i}$ in (29.8) and the index probability p_i in (29.9) imply for the total failure probability p_F in (29.6):

$$p_F = \sum_{i=0}^{m} p_{F|0} q^i \binom{m}{i} v^i (1-v)^{m-i} \tag{29.14}$$

$$=> \frac{p_F}{p_{F|0}} = \sum_{i=0}^{m} q^i \binom{m}{i} v^i (1-v)^{m-i} \tag{29.15}$$

The expression on the right-hand side of (29.15) is equal to the probability-generating function of the binominal distribution and can therefore be substituted as shown in (29.16).

$$\frac{p_F}{p_{F|0}} = (vq + 1 - v)^m \tag{29.16}$$

Substituting q in (29.16) by (29.13) yields

$$\frac{p_F}{p_{F|0}} = \left(\frac{1-v}{1-w}\right)^m \tag{29.17}$$

In analogy, if $p_{F|m}$ (instead of $p_{F|0}$) is used as parameter (II), the related equations are

$$\frac{p_F}{p_{F|m}} = \left(\frac{1}{q} - \frac{v}{q} + v\right)^m \tag{29.18}$$

$$\frac{p_F}{p_{F|m}} = \left(\frac{v}{w}\right)^m \tag{29.19}$$

Insights on risk data analysis from the relationship in (29.17) are shown in Fig. 29.1. Given the evaluation of failure events returns a high average degree of prevention (w), a low value of the relative total failure probability ($p_F/p_{F|0}$) is expected only if the average degree of prevention in failure exposure situations (v) is very high. If one relative position parameter (v or w) is known, the range of the other relative position parameter can be bounded, e.g. if $w = 0.5$ is determined from the evaluation of failure events, v (indicating the average index position, $\bar{\imath} = vm$, of the failure exposure situations) is expected between 0.5 and 1. The size (m) of the index scale can be, by-and-large, interpreted as an indicator of the number of prevention opportunities, since a high number of m corresponds to a low value of the relative failure probability.

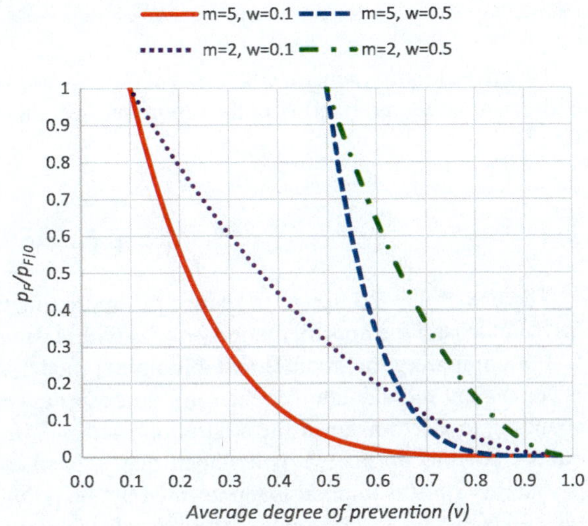

Fig. 29.1 Total to largest failure probability ratio ($p_F/p_{F|0}$) as a function of the average degree of prevention (v) for selected parameters of the size of the prevention index scale (m) and the average degree of prevention under the condition of a failure (w)

29.2.5 Relation to the Log-Normal Distribution

This section outlines how to derive the parameters of the geometrical–binominal model expressed by (29.8) and (29.9) from a known log-normal distribution of the failure probability of interest. This issue might be relevant if the data on the variation of the failure probability allow to estimate the parameters of a log-normal distribution.

The log-normal distribution is commonly used for the quantification of the uncertainty of a small failure probability treated as random variable Y. This means that $X = \ln Y$ has a normal distribution, which in turn would be suitable for the approximation of a binominal distribution assumed in (29.9) for i, i.e. the realization of random variable I (prevention index).

Equation (29.20) for the transformation yields from (29.8):

$$i = \frac{\ln p_{F|i} - \ln p_{F|0}}{\ln q} \tag{29.20}$$

If $\ln p_{F|i}$ is treated as a realization of random variable X, the mean (μ_I) and standard deviation (σ_I) of I can be expressed as a function of the mean (μ_X) and standard deviation (σ_X) of X:

$$\mu_I = \frac{\mu_X - \ln p_{F|0}}{\ln q} = vm \tag{29.21}$$

$$\sigma_I^2 = \frac{\sigma_X^2}{(\ln q)^2} = (1 - v)vm \tag{29.22}$$

From (29.21) and (29.22), q returns from (29.23):

$$q = e^{-\frac{\sigma_X}{\sqrt{(1-v)vm}}} = e^{-\frac{\sigma_X}{\sigma_I}} \tag{29.23}$$

Given the requirement that the mean of the log-normal distribution should be equal to the total failure probability of the binominal–geometric model, $p_{F|0}$ is derived from (29.24):

$$p_{F|0} = \frac{e^{\mu_X + 0.5\sigma_X^2}}{(vq + 1 - v)^m} \tag{29.24}$$

The use of $v = 0.5$ serves to reduce the approximation error, and for a scale size (m) of at least 5 the approximation error is within reasonable limits.

For comparing the geometrical–binominal model with the log-normal model, it is convenient to calculate for each $p_{F|i}$ an exceedance probability $p_{ex,B}(p_{F|i})$ corresponding to a retransfer to the continuous and infinite scale of a log-normal distribution. For this purpose, it is assumed that p_i is an interval probability, the failure probability for this interval is represented by the point estimate $p_{F|i}$, and within this interval 50% of p_i corresponds to the exceedance of $p_{F|i}$. With these assumptions, an estimate of the exceedance probability returns from (29.25).

$$p_{ex,B}(p_{F|i}) = 0.5p_i + \sum_{p_{F|j} > p_{F|i}} p_j = \begin{cases} 0.5p_i, \, i = 0 \\ 0.5p_i + \sum_{j=0}^{i-1} p_j, \, i > 0 \end{cases} \tag{29.25}$$

In this equation, p_i and p_j are determined from (29.9).

The estimate from (29.25) can be compared with the exceedance probability estimate (29.26) returning for $p_{F|i}$ from the log-normal distribution, where $p_{F|i}$ returns from (29.8), with the parameters estimated in (29.23) and (29.24).

$$p_{ex,L}(p_{F|i}) = 1 - \Phi\left(\frac{\ln p_{F|i} - \mu_X}{\sigma_X}\right) \tag{29.26}$$

For instance, a log-normal distribution of Y with a median of 0.0001 and an error factor (EF) of 5 has the parameters $\mu_X = -9.210$ (mean of $X = \ln Y$) and $\sigma_X = 0.978$ (standard deviation of X). For the ($v = 0.5$, $m = 5$) binominal distribution, the parameters—returning from (29.23) and (29.24)—of the geometrical model (29.8) are $q = 0.417$ and $p_{F|0} = 9.05E{-}04$. Figure 29.2 displays the comparison of the estimates returned from (29.25) and (29.26). It can be seen that the approximation has a fair accuracy.

Fig. 29.2 Exceedance probabilities calculated from the ($\mu_X = -9.210$, $\sigma_X = 0.978$) log-normal distribution and approximated by the ($v = 0.5$, $m = 5$) binominal distribution linked with the ($q = 0.417$, $p_{F|0} = 9.05\text{E}{-04}$) geometrical model

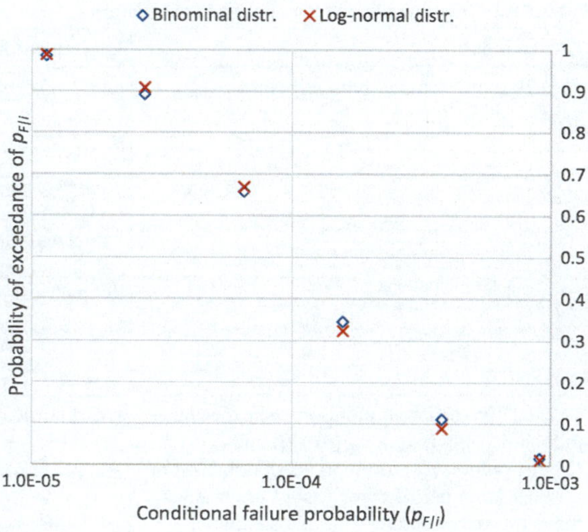

29.2.6 *Inference Lines*

The application of each inference line presented here requires that branch-specific expertise yielded a prevention index interval scale with a known size *m* of at least 1 and with substantial descriptions—by means of prevention or/and risk factors—of the scaling points.

In principle, three types of inference lines are of interest for risk analysis:

(A) Calibration of the index scale, i.e. estimation of the conditional failure probability ($p_{F|i}$) for each point (i) of it.
(B) Estimation of the total value (p_F) and the distribution (over i) of the failure probability.
(C) Prediction of the expected constellation of risk and prevention factors under the condition of a failure ($p_{i|F}$).

Table 29.3 illustrates the inference lines of type (A) for a set of situations with incomplete input data. The related inference lines of types (B) and (C) would work in a similar manner on the basis of the formula presented in Sects. 29.2.2 through 29.2.5.

With the additional estimates (p_{VN}, $p_{VN|F}$) *presented in* Sect. 29.2.3, *the input data situation for the hypothetical* Table 29.1 *example would correspond to* Table 29.3 *line (A5), i.e. known are (i)* $p_F = 31/336{,}000$, *(ii)* $p_{i(c)} = p_{VN} = p_2 = 0.89$ *and (iii)* $p_{i(a)|F} = p_{VN|F} = p_{2|F} = 0.0221$. *The calculation of parameters* $w = 0.149$ *and* $v = 0.943$ *in steps (A5.1) and (A5.2), respectively, is documented so far in* Sect. 29.2.3. *The scale calibration parameters returning from (A5.3) are* $q = 0.0105$ *and* $p_{F|0} = 0.0209$. *Finally returning from* (29.8), *the accident probabilities (per task) specific to the worker's risk behaviour are* $p_{F|0} = 2.09\text{E}{-02}$ *in case of violation and no risk*

Table 29.3 Inference lines for index scale calibration

Known input	Desired output generation
(A1) Two different calibration points, $p_{F\|i(a)}$, $p_{F\|i(b)}$	(A1.1) Calculate: $q = \left(\frac{p_{F\|i(a)}}{p_{F\|i(b)}}\right)^{1/(i_{(a)}-i_{(b)})}$ $p_{F\|0} = p_{F\|i(a)} q^{-i_{(a)}}$ (A1.2) If p_F or v or w is known, calculate—based on (29.13) and (29.16)—for validation the remaining parameters of the geometrical–binominal model; e.g. $v = \frac{(p_F/p_{F\|0})^{1/m}-1}{q-1} = w/(q-qw+w)$ $w = vq/(1-v+vq)$
(A2) (i) Two different points of the exceedance probability $\{p_{ex}(p_{F\|a}), p_{ex}(p_{F\|b})\}$, and (ii) one point ($p_{i(c)}$) of the index distribution or z values from paired comparisons or an index mean estimate $\bar{\imath}$ obtained from a sample of failure exposure situations	(A2.1) Calculate: $\sigma_X = \frac{\ln \frac{p_{F\|a}}{p_{F\|b}}}{\Phi^{-1}(1-p_{ex}(p_{F\|a}))-\Phi^{-1}(1-p_{ex}(p_{F\|b}))}$ $\mu_X = \ln, p_{F\|a} - \sigma_X \Phi^{-1}\left(1 - p_{ex}(p_{F\|a})\right)$ (A2.2) Determine v from $p_{i(c)} = \left(\frac{m}{i_{(c)}}\right) v^{i_{(c)}} (1-v)^{m-i_{(c)}}$ or $v = (1-\Phi(z_{max}))^{1/m}$ or $v = \bar{i}\ /m$, respectively (A2.3) Determine q from (29.23) (A2.4) If p_F is available, calculate $p_{F\|0}$ as in (A3.4); else from (29.24)
(A3) (i) One calibration point ($p_{F\|i(a)}$), (ii) the total failure probability p_F and (iii) input (A2) (ii)	(A3.1) Determine v as in (A2.2) (A3.2) Calculate: $p_{i(a)} = \left(\frac{m}{i_{(a)}}\right) v^{i_{(a)}} (1-v)^{m-i_{(a)}}$ $p_{i(a)\|F} = \frac{p_{i(a)} p_{F\|i(a)}}{p_F}$ (A3.3) Determine w from $p_{i(a)\|F} = \left(\frac{m}{i_{(a)}}\right) w^{i_{(a)}} (1-w)^{m-i_{(a)}}$ (A3.4) Calculate: $q = \frac{w-vw}{v-vw}$ $p_{F\|0} = (vq+1-v)^{-m} p_F$
(A4) (i) One calibration point ($p_{F\|i(b)}$), (ii) input (A2) (ii) and (iii) one point ($p_{i(a)\|F}$) of the conditional index distribution or a conditional index mean estimate $\bar{\imath}\|F$ obtained from a sample of failure cases	(A4.1) Determine v as in (A2.2) (A4.2) Determine w from (A3.3), or $w = (\bar{i}\|F)/m$, respectively (A4.3) Calculate: $p_{i(b)} = \left(\frac{m}{i_{(b)}}\right) v^{i_{(b)}} (1-v)^{m-i_{(b)}}$ $p_{i(b)\|F} = \left(\frac{m}{i_{(b)}}\right) w^{i_{(b)}} (1-w)^{m-i_{(b)}}$ $p_F = \frac{p_{i(b)} p_{F\|i(b)}}{p_{i(b)\|F}}$ (A4.3) Calculate q and $p_{F\|0}$ as in (A3.4)

(continued)

Table 29.3 (continued)

Known input	Desired output generation
(A5) (i) The total failure probability p_F, (ii) input (A2) (ii) or one calibration point ($p_{F|i(b)}$), and (iii) input (A4) (iii)	(A5.1) Determine w as in (A4.2) (A5.2) Determine v as in (A3.2), or from $w^{i(b)}(1-w)^{m-i(b)}\,p_F/p_{F|i(b)} = v^{i(b)}(1-v)^{m-i(b)}$, respectively (A5.3) Calculate q and $p_{F|0}$ as in (A3.4)

compensation effort, $p_{F|1} = 2.19E{-}04$ in case of violation and risk compensation effort, and $p_{F|2} = 2.32E{-}06$ in case of no violation. Note this result reflects as well the occupational safety expertise hypothetically assumed here for the purpose of illustration.

29.3 Feasibility Case Studies on Inference and Loss of Accuracy

Diverse examples of risk data analysis are presented, in order to illustrate how the framework would serve to deal with incomplete input data and to check the loss of accuracy induced by the simple geometrical–binominal model presented in Sect. 29.2. To meet the basic requirement for applying the inference lines, an approach for simplified scaling is applied in each case study. The approach is outlined in Sect. 29.3.1. A detailed illustration, covering each inference line from Table 29.3, is presented in Sect. 29.3.2. Brief illustrations covering selected lines are presented in Sects. 29.3.3, 29.3.4, 29.3.5 and 29.3.6.

29.3.1 Simplified Scaling

As presented in Sect. 29.3.6, the inference according to Table 29.3 requires the availability of a prevention index interval scale. To obtain such a scale for the illustration of inference and loss of accuracy, a simplified variant of the *Paired Comparisons* method is applied.

The result of a comparison between two conditions, say between $C(k)$ and $C(j)$, is expressed by a pair-wise value denoted as $c_{k:j}$, which can have three possible values (1, 0.5, 0) according to the notation below.

- $c_{k:j} = 1$ if the prevention index under $C(k)$ is greater than under $C(j)$,
- $c_{k:j} = 0.5$ if the prevention index is equal under both conditions, or
- $c_{k:j} = 0$ if the prevention index under $C(k)$ is smaller than $C(j)$.

The $c_{k:j}$ estimations carried out here (Sects. 29.3.2, 29.3.3, 29.3.4, 29.3.5 and 29.3.6) assume a level of branch-specific expertise that allows to reflect the ranking of the conditional failure probabilities. From the set of pair-wise values ($c_{k:j}, j = 1, \ldots, n_C$) of the comparisons, the overall value (c_k) for each condition $C(k)$ is calculated as follows:

$$c_k = \sum_{j=1}^{n_C} c_{k:j} \tag{29.27}$$

In this equation, n_C is the number of conditions. As the next step of the overall evaluation, a so-called z value is calculated for each condition:

$$z(k) = \Phi^{-1}\left(\frac{c_k}{n_C}\right) \tag{29.28}$$

The z value represents a point on an interval scale. It is up to the discretion of the analyst to choose the zero point of such a scale.

As expressed in (29.29), two options are considered for the estimation of the size (m) of the scale, in order to account for the uncertainty due to rounding. And finally, each prevention index returns from (29.30).

$$m \in \{z_{\max} - z_{\min}; \lceil z_{\max} - z_{\min} \rceil\} \tag{29.29}$$

$$i = \left[m \frac{z_k - z_{\min}}{z_{\max} - z_{\min}} \right] \tag{29.30}$$

In (29.29) and (29.30), $z_{\min}$ and $z_{\max}$ are the smallest or highest z value, respectively, calculated from (29.28), [...] the function that rounds up to the next highest natural number, and [...] the function that rounds to the nearest natural number.

29.3.2 *Non-survival per Severe Work-Related Traffic Crash*

29.3.2.1 Subject and Data

The first feasibility case study uses data from an analysis by Boufous and Williamson [26] of severe work-related traffic crashes in New South Wales, Australia, 1998–2002. The data comprise 13,124 traffic crashes that resulted in injury or death. The data concerning fatal outcome (non-survival), which is defined here as the failure of interest, are compiled in Table 29.4. They are specified by the factors (i.e. levels of the variables *Gender* and *Status of Driving*): GM (Male Gender), GF (Female Gender), SD (on-duty Status of driving) and SC (Commuting Status of driving). The data from [26] are the probability (p_C) of each condition (C) generated by

Table 29.4 Conditional fatality probability ($p_{F|C}$) per severe work-related traffic crash, condition probability (total: p_{CC}; conditional: $p_{CC|F}$) and exceedance probability ($p_{ex}(p_{F|C})$); specific to GM (Male Gender), GF (Female Gender), SD (on-duty Status of driving) and SC (Commuting Status of driving)

Condition (C)	GM SD	GM SC	GF SD	GF SC	Overall
$p_{F\|C}$*	2.84E−02	1.64E−02	5.47E−03	9.48E−03	1.80E−02
p_C*	2.94E−01	4.44E−01	3.77E−02	2.24E−01	1
$p_{C\|F}$	4.65E−01	4.06E−01	1.15E−02	1.18E−01	1
$p_{ex}(p_{F\|C})$	1.47E−01	5.16E−01	9.81E−01	8.50E−01	

*From [26]. Reading example: For females driving on-duty (GF SD), the probability of non-survival of a severe work-related traffic crash is 5.47E−03

these factors and the conditional fatal outcome probability ($p_{F|C}$) per severe crash. For the purpose of inference illustration, the table is completed by the conditional probabilities ($p_{C|F}$) calculated from (29.2) with input from (29.3), and the exceedance probabilities ($p_{ex}(p_{F|C})$) calculated according to the first line of (29.25).

Table 29.5 presents the results of the paired comparisons and scaling. According to the assumption in Sect. 29.3.1, the $c_{k:j}$ estimations reflect the $p_{F|CC}$ ranking shown in Table 29.4. The impact of factors SD and SC depends on the gender, e.g. on-duty status of driving (SD) is a risk factor for males, but a prevention factor for females. From the z values, the prevention indices are calculated for a scale (starting with $i = 0$) with a size of $m = 3$.

Table 29.5 Paired comparisons and scaling (addressing the conditions defined in Table 29.4) for non-survival (fatal outcome) per severe work-related traffic accident

	$c_{k:j}$			
	$C(k)$			
$C(j)$	GM SD	GM SC	GF SD	GF SC
GM SD	0.5	1	1	1
GM SC	0	0.5	1	1
GF SD	0	0	0.5	0
GF SC	0	0	1	0.5
c_k	0.5	1.5	3.5	2.5
$z(k)$	−1.1503	−0.3186	1.1503	0.3186
i	**0**	**1**	**3**	**2**

Reading example: During on-duty driving, fatal outcome prevention per severe crash is more likely for females (GF SD) than for males (GM SD), and the fatality prevention indices are 3 for females and 0 for males, respectively

29.3.2.2 Inference and Loss of Accuracy

Table 29.6 presents the results from applying inference lines (A1) through (A5) addressing the data on non-survival of a severe work-related traffic crash. The loss of accuracy is below a factor of 2 (compared with the observed values in Table 29.4) in the prediction of both the conditional failure (fatality) probability ($p_{F|C}$) and the condition probability (p_C). Figure 29.3 shows the loss of accuracy concerning $p_{F|C}$ when applying inference line A2. Except for $i = 1$, the predicted values are within the bounds of the 95% confidence intervals calculated by Boufous and Williamson [26].

As shown in Sect. 29.2.5, the geometrical–binominal model represents approximately a discretization of a log-normal distribution, which corresponds to a linear

Table 29.6 Inferences (according to Table 29.3), based on incomplete input data, for non-survival (F) per severe work-related traffic crash

	i	0	1	2	3
Inference line and Input	Condition (C) Output	GM SD	GM SC	GF SC	GF SD
(A1) $p_{F\|0}, p_{F\|3}, p_F$	$p_{F\|i}$ ($q = 0.54$)	2.84E−02	1.64E−02	9.48E−03	5.47E−03
	p_i ($v = 0.34$)	2.94E−01	4.44E−01	2.24E−01	3.77E−02
	$p_{i\|F}$ ($w = 0.29$)	4.65E−01	4.06E−01	1.18E−01	1.15E−02
	$p_{ex}(p_{F\|i})$	1.47E−01	5.16E−01	8.50E−01	9.81E−01
(A2) $p_{ex}(p_{F\|(GM\ SD)})$, $p_{ex}(p_{F\|(GF\ SD)})$, p_0	$p_{F\|i}$ ($q = 0.54$)	3.14E−02	1.71E−02	9.28E−03	5.04E−03
	p_i ($v = 0.43$)	1.82E−01	4.18E−01	3.19E−01	8.12E−02
	$p_{i\|F}$ ($w = 0.29$)	3.53E−01	4.39E−01	1.82E−01	2.52E−02
	$p_{ex}(p_{F\|i})$	9.11E−02	3.91E−01	7.59E−01	9.59E−01
(A3) $p_{F\|0}, p_0, p_F$	$p_{F\|i}$ ($q = 0.67$)	2.84E−02	1.91E−02	1.29E−02	8.66E−03
	p_i ($v = 0.43$)	1.82E−01	4.18E−01	3.19E−01	8.12E−02
	$p_{i\|F}$ ($w = 0.34$)	2.88E−01	4.44E−01	2.28E−01	3.91E−02
	$p_{ex}(p_{F\|i})$	9.11E−02	3.91E−01	7.59E−01	9.59E−01
(A4) $p_{F\|0}, p_0, p_{0\|F}$	$p_{F\|i}$ ($q = 0.67$)	2.84E−02	1.91E−02	1.29E−02	8.66E−03
	p_i ($v = 0.43$)	1.82E−01	4.18E−01	3.19E−01	8.12E−02
	$p_{i\|F}$ ($w = 0.34$)	2.88E−01	4.44E−01	2.28E−01	3.91E−02
	$p_{ex}(p_{F\|i})$	9.11E−02	3.91E−01	7.59E−01	9.59E−01
(A5) $p_3, p_{3\|F}, p_F$	$p_{F\|i}$ ($q = 0.55$)	3.34E−02	1.83E−02	1.00E−02	5.47E−03
	p_i ($v = 0.41$)	2.04E−01	4.28E−01	2.99E−01	6.96E−02
	$p_{i\|F}$ ($w = 0.28$)	3.78E−01	4.34E−01	1.66E−01	2.12E-02
	$p_{ex}(p_{F\|i})$	1.02E−01	4.18E−01	7.81E−01	9.65E−01

Reading example: With $p_{F|0} = p_{F|(GM\ SD)} = 2.84E{-}02$, $p_0 = p_{GM\ SD} = 2.94E{-}01$ and $p_F = p_F = 1.80E{-}02$ as inputs, inference line (A3) predicts for females driving on-duty (GF SD) a non-survival probability per work-related severe crash of $p_{F|3} = 8.66E{-}03$

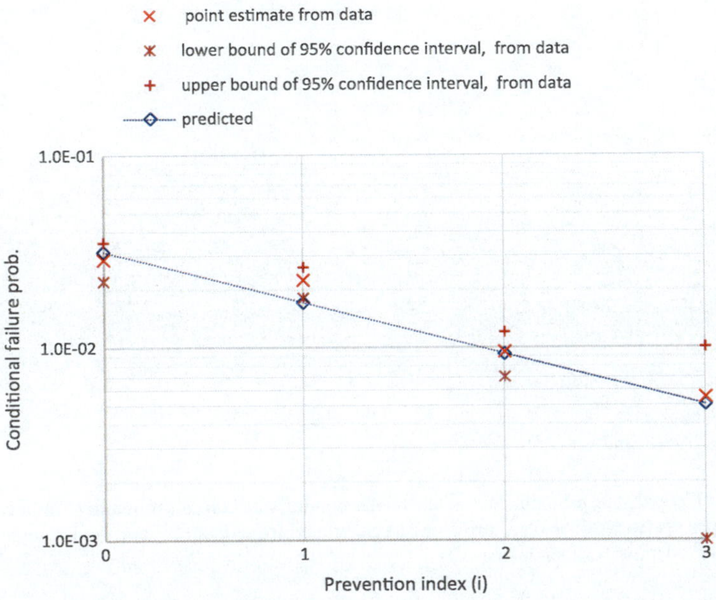

Fig. 29.3 Conditional failure probability (failure = fatal outcome per severe work-related traffic crash) calculated from data [26] and from geometrical – binominal modelling (inference line A2)

relationship between the conditional failure probability on a log scale and the z value, i.e. $z(p_{ex}) = \Phi^{-1}(p_{ex})$, of its exceedance probability, where p_{ex} is calculated from the p_i values (i.e. the predicted p_C values). Consequently, the loss of accuracy is reflected by the deviation from the linear relationship, as shown in Fig. 29.4. Except for $p_{F|1}$, the data-based points fit the linear relation quite well. However, Chi^2 testing indicates that the deviation between observation and prediction is significant concerning both the number of cases (exposures) and the number of failures (fatal outcomes).

29.3.3 Non-survival per Motorcycle Accident

Accident severity for motorcyclists in large French urban areas in the year 2003 is addressed in a study by De Lapparent [27]. 6348 accidents are evaluated. Severity per accident is quantified by means of four discrete classes: *No Injury*, *Slight Injury*, *Severe Injury* and *Fatal Injury*. The impact variables concerning groups of individuals involved in the accidents are *Age* (of the motorcyclist), *Gender* (of the motorcyclist) and *Engine Capacity* (of the motorcycle). The non-group relevant variables include the *Day Time*, *Helmet Wearing* (data assessed as unreliable in [27]) and the *Collision* (e.g. with a bicycle) if any. The results of the study include conditional probabilities

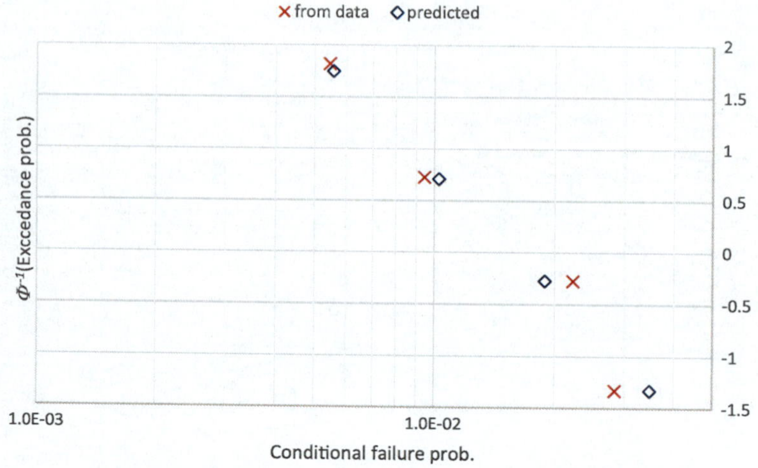

Fig. 29.4 Exceedance probability z value of the conditional failure probability (failure = fatal outcome per severe work-related traffic crash) calculated from data [26] and from geometrical – binominal modelling (inference line A2)

(denoted as "posterior probabilities") of severity specific to levels of the group variables (*Age*, *Gender*, *Engine Capacity*) as shown in Table 29.7 for *Fatal Injury*. The related condition probabilities are not presented in the paper [27].

The second illustration of inference and loss accuracy addresses the conditional fatality probabilities in Table 29.7. Paired comparisons are carried out as explained in Sect. 29.3.1. From the resulting z values, the prevention indices are calculated for a scale (starting with $i = 0$) with a size of $m = 3$. Table 29.8 presents both the condition assignments to the scale and the results from applying inference lines (A1) and (A2). The loss of accuracy is below a factor of 2.5 in the prediction of the conditional failure (fatality) probability (compared with the observed $p_{F|C}$ values in Table 29.7). The check of accuracy for the condition probabilities is degraded, since p_C values are not presented by De Lapparent [27]. However, the portion (about 1.6%) of accidents involving females (GF) with small motorcycles (ES) predicted by line (A1) deserves discussions, since it would yield an absolute number of 100 accidents only, and it is unlikely that a fatality probability ($p_{F|(\text{GF ES})}$) of 0.001 (per accident) has been empirically derived on the basis of 100 accidents only. However, it is possible that the probability of 0.001 is based on an extrapolation (from Bayesian Gibbs sampling in [27]). Nevertheless, line (A2) produces for $i = 3$ a result that would better comply with an empirical calculation of $p_{F|(\text{GF ES})}$.

29.3.4 *Teen Driver Night-Time Fatality per Young Resident*

The third case study uses data from a study [28] on teen driver fatality per young resident collected from 410 US state years. In Table 29.9, the data for night-time

Table 29.7 Conditional fatality probabilities ($p_{F|C}$) per motorcycle accident; specific to AL (driver's Age Low: younger than 30 years), AM (driver's Age Moderate: from 30 to <50 years), AH (driver's Age High: ≥50 years), GM (Male Gender), GF (Female Gender), ES (Engine capacity Small: <125 cm^3) and EL (Engine capacity Large: ≥25 cm^3). *Source* [27]

C	AL GM ES	AL GM EL	AL GF ES	AL GF EL	AM GM ES	AM GM EL	AM GF ES	AM GF EL	AH GM ES	AH GM EL	AH GF ES	AH GF EL	Total
$p_{F\|C}$	0.006	0.01	0.001	0.002	0.007	0.012	0.001	0.003	0.005	0.009	0.001	0.002	0.0078

Table 29.8 Inference according to Table 29.3 lines (A1) and (A2), based on incomplete input data for non-survival (F) per motorcycle accident

	i	0	1	2	3
	Condition (C)	AM GM EL	AL GM ES	AL GF EL	AL GF ES
			AL GM EL	AM GF EL	AM GF ES
			AM GM ES	AH GM ES	AH GF ES
Line and input	Output		AH GM EL	AH GF EL	
(A1) $p_{F\|0}, p_{F\|3}, p_F$	$p_{F\|i}$ ($q = 0.44$)	1.20E−02	5.24E−03	2.29E−03	1.00E−03
	p_i ($v = 0.25$)	4.21E−01	4.22E−01	1.41E−01	1.58E−02
	$p_{i\|F}$ ($w = 0.13$)	6.64E−01	2.91E−01	4.26E−02	2.07E−03
(A2) $p_{ex}(p_{F\|(AM\ GM\ EL)})$, $p_{ex}(p_{F\|(AL\ GF\ ES)})$, p_3	$p_{F\|i}$ ($q = 0.37$)	2.37E−02	8.75E−03	3.23E−03	1.19E−03
	p_i ($v = 0.5$)	1.25E−01	3.75E−01	3.75E−01	1.25E−01
	$p_{i\|F}$ ($w = 0.27$)	3.89E−01	4.32E−01	1.59E−01	1.96E−02

Table 29.9 Conditional night-time teen driver fatality probability ($p_{F|C}$) per young resident; specific to QG (Good Quality of graduated driver's license programs), QF (Fair Quality of graduated driver's license programs), LI (Law of graduated driver's license Implemented) and LM (Law of graduated driver's license (still) Missing)

Condition (C)	QG LI	QG LM	QF LI	QF LM	Total
Number of state years of the data*	15	72	96	227	410
$p_{F\|C}$*	6.13E−05	9.41E−05	9.26E−05	1.02E−04	9.70E−05
p_C	3.66E−02	1.76E−01	2.34E−01	5.54E−01	1
$p_{C\|F}$	2.31E−02	1.70E−01	2.23E−01	5.83E−01	1

*From Morrisey et al. [28]

are specified by the factors (variable levels) QG (Good Quality of graduated driver's license programs), QF (Fair Quality of graduated driver's license programs), LI (Law of graduated driver's license Implemented) and LM (Law of graduated driver's license (still) Missing). The condition probabilities in Table 29.9 are calculated from the numbers of state years of the data.

Due to the tiny $p_{F|C}$ difference between conditions QG LM and QF LI, the risk for each is assessed as equal in the paired comparisons. From the resulting z values, the prevention indices are calculated for a scale (starting with $i = 0$) with a size of $m = 2$. Table 29.10 presents the results from applying inference line (A5). The loss of accuracy is below a factor of 1.5 (compared with the observed values in Table 29.9) in the prediction of both the conditional teen driver fatality per young resident ($p_{F|C}$) and the condition probability (p_C). The deviation would be greater than a factor of 2, if $m = 3$ (instead of $m = 2$) is used as the scale size.

Table 29.10 Inference according to Table 29.3 line (A5) based on incomplete input data (p_2, $p_{2|F}$, p_F) for teen driver fatality per young resident

Condition (C)	QF LM	QG LM QF LI	QG LI
i	0	1	2
$p_{F\|i}$ ($q = 0.76$)	1.07E−04	8.09E−05	6.13E−05
p_i ($v = 0.19$)	6.54E−01	3.09E−01	3.66E−02

29.3.5 Crash per Driver Kilometre

From a Dutch travel survey (47,502 drivers, 4324 crashes of any severity), Langford et al. [29] investigated the crash rate by driver kilometre (km) as a function of five levels of the age of the driver and three levels of the driver's practice (concerning the annual distance driven). In the fourth case study presented here, these data are pooled for simplification as shown in Table 29.11. From the z values obtained from paired comparisons, the prevention indices are calculated for a scale (starting with $i = 0$) with a size of $m = 2$.

Table 29.12 presents the results from applying inference line (A3). The loss of accuracy is below a factor of 1.5 (compared with the observed values in Table 29.11) in the prediction of both the conditional crash rate per driver km ($p_{F|C}$) and the condition probability (p_C). However, the loss of accuracy would be greater than a factor of 2, if $m = 3$ (instead of $m = 2$) is used as the size of the scale.

Table 29.11 Conditional crash rate per driver kilometre ($p_{F|C}$); specific to AE (Extreme Age of the driver: younger than 20 or older than 74 years), AN (Non-extreme Age of driver: from 20 to 74 years), PL (Little driving Practice: 3000 km or less per year) and PE (Extended driving practice: >3000 km per year); pooled from the data presented in [29]

Condition (C)	AE PL	AE PE	AN PL	AN PE	Total
$p_{F\|C}$	5.30E−05	1.38E−05	2.79E−05	6.33E−06	1.05E−05
p_C	1.43E−02	2.53E−02	1.55E−01	8.05E−01	1

Table 29.12 Inference according to Table 29.3 line (A3) based on incomplete input data (p_0, $p_{F|0}$, p_F) for crash rate per driver km

Condition (C)	AE PL	AE PE AN PL	AN PE
i	0	1	2
$p_{F\|i}$ ($q = 0.37$)	5.30E−05	1.96E−05	7.28E−06
p_i ($v = 0.85$)	1.43E−02	2.10E−01	7.75E−01

29.3.6 Accident per Task in Steel Production

Rehhahn [1] investigated occupational accident data for various tasks in steel production. For block pull-off, the most dangerous task, 31 accidents (e.g. getting burned by contact to heated equipment) per 336,000 task performances were observed in the first year of the investigation. The observations of a subset of 7126 performances of various steel production tasks (not only block pull-off) identified that 11% of them were carried out with violations (e.g. not wearing hand protection device) of safety rules; such violation behaviour is shortly denoted here as risk-taking behaviour. Rehhahn attributes each accident to a violation and calculates therefore a rate of one accident per 1200 block pull-off tasks carried out with a violation; this rate returns from 31/(0.11 × 336,000). After improvements in the design of work conditions [e.g. replacing the hanging pliers by ones with self-inhibition (of uncontrolled pending movements)], it was observed that the number of accidents for this task reduced from 31 to 14 per year. The related data for the fifth case study on inference and accuracy are presented in Table 29.13. The value for $p_{F|(\mathrm{RA\ DP})}$ matches Rehhahn's [1] implicit assumption of no accident in case of risk-avoiding behaviour (i.e. compliance with safety rules). A point estimate suitable for a sample with no failure (c.f. [25]) has been used for the calculation of this value:

$$p_{F|(\mathrm{RA\ DP})} = 1 - 0.5^{1/((1-p_{\mathrm{RT}})n_{\mathrm{DP}})} = 1 - 0.5^{1/((1-0.11)336000)} = 2.32\mathrm{E} - 06 \quad (29.31)$$

No further specification of the data for the condition with improved design of work conditions (DI) is presented by Rehhahn [1]. Consequently, accuracy is checked against the data as they are. From the z values obtained from paired comparisons, the prevention indices are calculated for a scale (starting with $i = 0$) with a size of $m = 2$.

Table 29.14 presents the results from applying inference line (A3). The loss of accuracy is below a factor of 2 (compared with the observed values in Table 29.13) in the prediction of both the conditional accident probability per block pull-off task in steel production ($p_{F|C}$) and the condition probability (p_C).

Table 29.13 Conditional accident probability ($p_{F|C}$) per steel production task (block pull-off); specific to RT (Risk-Taking behaviour), RA (Risk-Avoiding behaviour), DP (Poor Design of work conditions) and DI (Improved Design of work conditions). *Source* [1], except $p_{F|(\mathrm{RA\ DP})}$, estimated from (29.31)

Condition (C)	RT DP	RA DP	DI	Total
$p_{F\|C}$	8.39E−04	2.32E−06	4.17E−05	6.80E−05
p_C	5.50E−02	4.45E−01	5.00E−01	1

Table 29.14 Inference according to Table 29.3 line (A3) based on incomplete input data (p_0, $p_{F|0}$, p_F) for accident per block pull-off task in steel production

Condition (C)	RT DP	DI	RA DP
i	0	1	2
$p_{F|i}$ ($q = 0.0656$)	8.39E−04	5.50E−05	3.61E−06
p_i ($v = 0.765$)	5.50E−02	3.59E−01	5.86E−01

29.4 Conclusions

The framework presented here provides options for the generation of probabilities concerning failures of interest in cases of incomplete input data. Its key features for achieving this goal are (a) utilizing the Bayesian theorem for both structuring cases of incomplete input data and providing options to overcome this problem; (b) scaling on the basis of branch-specific expertise regarding prevention and risk factors; and (c) discretization and parameterization.

Five diverse types of risk (as summarized in Table 29.15) are addressed to illustrate how the framework serves inference in case of incomplete input data. The risks concern failure events mainly influenced by human behaviour in complex transportation and production systems. The related prevention factors with potential from improvements are quality of training, practice, risk-avoiding behaviour and design of work conditions (if applicable). Of course, there are interrelations, e.g. the quality of training can influence the individual risk behaviour.

A key parameter returning from inference represents the average degree of prevention (v) in failure exposure situations. The parameter is bounded in the range from 0 to 1. This parameter serves non-trivial comparisons between risks and provides insights into the potential for the reduction of the average risk as shown in Fig. 29.5. For instance, the average night-time teen driver fatality per young resident (◆) has a promising margin for improvement by forcing prevention factors, since v is small, and the average risk is reducible by factor of 1.5.

Due to its bounded range, from 0 to 1, the average degree of prevention could be a suitable subject of expert judgement. As illustrated in Fig. 29.6, the advantage of the Bayesian structure of the framework is that such a judgement can be verified or falsified by the quantitative evaluation of failure events only meaning a quantitative evaluation of the variety of failure exposure situations would not be necessary. For instance, an expert assesses for the fatality per motorcycle accident an average degree of prevention of $v = 0.6$; this would correspond to a conditional average degree of prevention of $w = 0.4$ for an index scale (of any size) calibrated by $q = 0.437$. If a quantitative evaluation of a sample of fatality cases returns (via (A4.2) in Table 29.3) $w \ll 0.4$, this assessment ($v = 0.6$) would be debatable. However, given a scale calibrated with the dominant involvement of a factor with a very strong impact (e.g. risk-avoiding behaviour for the prevention of an accident in steel production), $w = 0.1$ would be in line with the estimate of $v = 0.6$. In turn, a small value of the scale calibration parameter q is an indicator of such dominant involvement. The

Table 29.15 Summary of geometrical–binominal inference and loss of accuracy

Average risk (p_F)	Prevention factors	Average degree of prevention (v)	m	q	Loss of accuracy due to simplification and incomplete input data
(■)1.8E−02 fatality per severe work-related traffic crash	Female gender	0.433	3	0.543	<Factor 2
(●)1.7E−03 fatality per motorcycle accident	Female gender. Small engine capacity. Daylight[a]. Collision with small mobile mass (pedestrian, bicycle or other motorcycle)[a]	0.251	3	0.437	<Factor 2.5
(◆)9.7E−05 night-time teen driver fatality per young resident	Law of graduated driver's license implemented. Good quality of graduated driver's license programs	0.191	2	0.758	<Factor 1.5
(X) 1.05E−05 crash rate per driver km	Extended driving practice (>3000 km per year). Non-extreme age of driver (from 20 to 74 years)	0.881	2	0.371	<Factor 2
(Δ) 6.80E−05 accident rate per block pull-off task in steel production	Risk-avoiding behaviour (compliance with safety rules). Improved design of work conditions	0.765	2	0.066	<Factor 2

[a]Not explicitly addressed in inference illustration due to missing data, but significant according to [27]

conclusion is that it would be desirable to develop a prevention index interval scale with a calibration parameter q not much smaller than 0.5.

The points marked in Figs. 29.5 and 29.6 correspond to the risks summarized in Table 29.15. The non-trivial interpretations outlined in the preceding two paragraphs are visible from these figures, which in turn represent the application of the simple framework outlined in Sect. 29.2.

However, loss of accuracy in the prediction of probabilities of conditions—concerning the constellations of risk and prevention factors—and conditional failure probabilities is the price to pay for such kind of attractive features when applying a

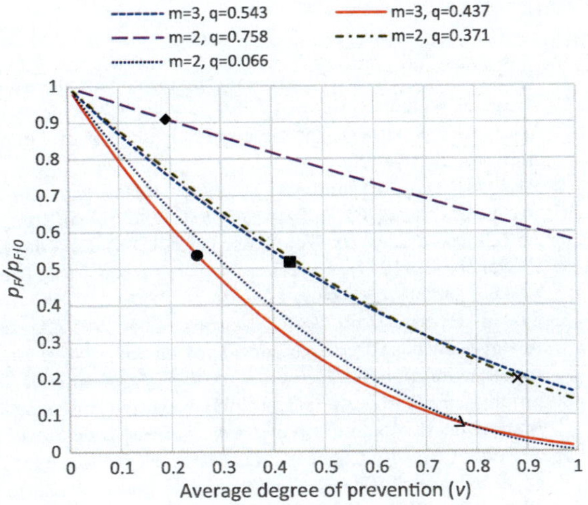

Fig. 29.5 Total to largest (for a given scale) failure probability ratio ($p_F/p_{F|0}$) as a function (29.16) of the average degree of prevention (v) for diverse values of the index scale size (m) and the gradient parameter (q) for prevention index scale calibration

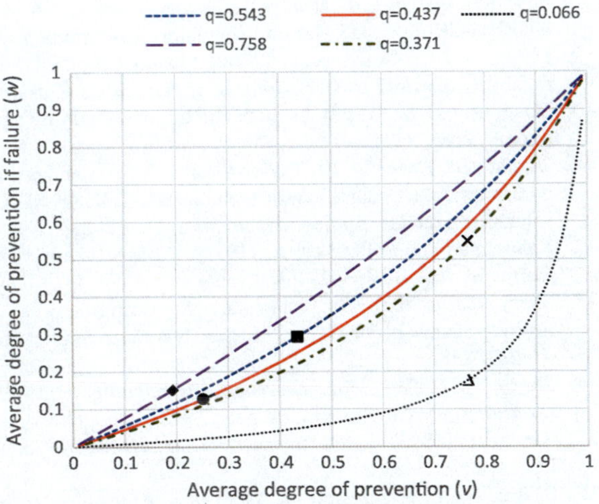

Fig. 29.6 Total (v) and conditional (w) average degree of prevention for diverse values of the gradient parameter (q) for prevention index scale calibration

common and simple model instead of applying branch-specific and more sophisticated models (like logistic regression or Gibbs sampling). The feasibility case studies carried out here for different types of risk indicate that overestimations or underestimations, respectively, by a factor of 1.5 through 2.5 are expected. This is not satisfactory, although large uncertainties are expected in predicting the elements of risks influenced by human behaviour (see Table 29.2 in the introduction). It is an open issue whether such inaccuracy is inherent, when applying a common and simple model for data analysis addressing various risk environments, or it can be reduced by a refined version of a common model or by improved scaling.

References

1. Rehhahn, H. (1973). Ergebnisse einer Fehlverhaltensuntersuchung zur Verbesserung der Arbeitssicherheit. *Arbeit und Leben, 3,* 44–47. ((in German)).
2. Wilde, G. J. S. (1982). The theory of risk homeostasis: Implications for safety and health. *Risk Analysis, 2,* 209–225.
3. Atwood, C. L., & Gentillon, C. D. (1996). *Bayesian treatment of uncertainty in classifying data.* In P. C. Cacciabue & I. A. Papazoglou (Eds.), *Probabilistic safety assessment and management: ESREL'96—PSAM-III* (Vol. 2, pp. 1283–1288). London: Springer.
4. Pukkinen, U. (1994). Bayesian analysis of consistent paired comparisons. *Reliability Engineering and System Safety, 43,* 1–16.
5. Szwed, P., van Dorp, J. R., Merrick, J. R. W., Mazzuchi, T. A., & Singh, A. (2006). A Bayesian paired comparison approach for relative accident probability assessment with covariate information. *European Journal of Operational Research, 169,* 157–177.
6. Smith, A. F., & Roberts, G. O. (1993). Bayesian computation via the Gibbs sampler and related Markov chain Monte Carlo methods. *Journal of the Royal Statistical Society, B, 55,* 3–23.
7. Atwood, C. L., La Chance, J. L., Martz, H. F., Anderson, D. J., Englehardt, M., Whitehead, D., & Wheeler, T. (2003) *Handbook of parameter estimation for probabilistic risk assessment.* Washington, DC: NUREG/CR-6823, US Nuclear Regulatory Commission.
8. Hallbert, B., & Kolaczkowski, A. (Eds.). (2007) *The employment of empirical data and Bayesian methods in human reliability analysis: A feasibility study.* Washington DC: NUREG/CR-6949, US Nuclear Regulatory Commission.
9. Duijm, N. J., & Goossens, L. (2006). Quantifying the influence of safety management on the reliability of safety barriers. *Journal of Hazardous Materials, 130,* 284–292.
10. Dougherty, E. M. (1993). Context and human reliability analysis. *Reliability Engineering and System Safety, 41,* 25–47.
11. Kirwan, B., Kennedy, R., Taylor-Adams, S., & Lambert, B. (1997). The validation of three human reliability quantification techniques—THERP, HEART and JHEDI: Part II—Results of validation exercise. *Applied Ergonomics, 28,* 17–25.
12. Kahneman, D., & Tversky, A. (1979). Prospect theory: An analysis of decision under risk. *Econometrica, 47,* 263–291.
13. Slovic, P., Fischhoff, B., Lichtenstein, S., Corrigan, B., & Combs, B. (1977). Preferences for insuring against probable small losses: Insurance implications. *Journal of Risk and Insurance, 44,* 237–258.
14. Storm, R. (1979). *Wahrscheinlichkeitsrechnung, mathematische Statistik und statistische Qualitätskontrolle* (7th ed.). Leipzig: VEB Fachbuchverlag (in German).
15. Kirwan, B. (1994). *A guide to practical human reliability assessment.* London: Taylor & Francis Ltd.
16. Seaver, D. A., & Sizewell, W. G. (1983). *Procedures for using expert judgment to estimate human error probabilities in nuclear power plant operations.* Washington, DC: NUREG/CR-2743, US Nuclear Regulatory Commission.
17. Embrey, D. E., Humphreys, P., Rosa, E. A., Kirwan, B., & Rea, K. (1984). *SLIM-MAUD: An approach to assessing human error probabilities using structured expert judgment. Vol. I: Overview of the SLIM-MAUD.* Washington, DC: NUREG/CR-3518, US Nuclear Regulatory Commission.
18. Von Winterfeldt, D., & Fischer, G. W. (1975). *Multi-attribute utility theory: Models and assessment procedures.* In D. Wendt, & C. Vlek (Eds.), *Utility, probability, and human decision making. Theory and Decision Library (An International Series in the Philosophy and Methodology of the Social and Behavioral Sciences)* (Vol. 11) Dordrecht: Springer.
19. Rasch, G. (1960). *Probabilistic models of some intelligence and attainment tests.* Copenhagen: Nielson & Lydiche.
20. Sträter, O. (2004). Considerations on the elements of quantifying human reliability. *Reliability Engineering and System Safety, 83,* 255–264.

21. Forester, J., Bley, D., Cooper, S., Lois, E., Siu, N., Kolaczkowski, A., & Wreathall, J. (2004). Expert elicitation approach for performing ATHEANA quantification. *Reliability Engineering and System Safety, 83,* 207–220.
22. Swain, A. D., & Guttmann, H. E. (1983). *Handbook of human reliability analysis with emphasis on nuclear power plant applications, Final report.* Washington, DC: NUREG/CR-1278, US Nuclear Regulatory Commission.
23. Trimpop, R. M. (1994). *The psychology of risk taking behaviour.* In *Advances in Psychology* (Vol. 107, pp. iii–xxv, 1–386). North Holland: Elsevier.
24. Duffey, R. (2004) *A new general accident theory.* In C. Spitzer et al. (Eds.), *International Conference on Probabilistic Safety Assessment and Management (7th: 2004: Berlin, Germany)* (Vol. 4, pp. 2371–2377). London: Springer.
25. Bailey, R. T. (1997). Estimation from zero-failure data. *Risk Analysis, 17,* 375–380.
26. Boufous, S., & Williamson, A. (2006). Work-related traffic crashes: A record linkage study. *Accident Analysis and Prevention, 38,* 14–21.
27. De Lapparent, M. (2006). Empirical Bayesian analysis of accident severity for motorcyclist in large French urban areas. *Accident Analysis and Prevention, 38,* 260–268.
28. Morrisey, M. A., Grabowski, D. C., Dee, T. S., & Campbell, C. (2006). The strength of graduated drivers license programs and fatalities among teen drivers and passengers. *Accident Analysis and Prevention, 38,* 135–141.
29. Langford, J., Methorst, R., & Hakarnies-Blomquist, L. (2006). *Older drivers do not have a high crash risk—A replication of low mileage bias. Accident Analysis and Prevention, 38,* 574–578.

Bernhard Reer, Ph.D. started 1986 his professional career in the field of Human Reliability Analysis (HRA) and Probabilistic Safety Assessment at the Jülich Research Center, Germany. He joined the Swiss Paul Scherrer Institute (PSI) in 1997 and since 2007, he is with the Swiss Federal Nuclear Safety Inspectorate (ENSI). At PSI, he led the development of the CESA method for the HRA of errors of commission. At ENSI, he works as a senior expert in various safety assessment areas including HRA, accident management (AM), aircraft crash hazards and operational event analysis. In the post-Fukushima EU Stress Test of nuclear power plants (2012), he was a leading author of country-specific AM review reports.

Chapter 30
Combining Domain-Independent Methods and Domain-Specific Knowledge to Achieve Effective Risk and Uncertainty Reduction

Michael Todinov

Abstract The common domain-specific approach to reliability improvement and risk reduction created the false perception that effective risk reduction can be successfully delivered solely by using methods offered by the specific domain. In standard textbooks on mechanical engineering and design of machine components, for example, there is no mention of general methods for improving reliability and reducing the risk of failure of engineering products. Accordingly, the chapter demonstrates the benefits from combining domain-independent methods and domain-specific knowledge for achieving effective risk and uncertainty reduction. In this respect, the chapter focuses on the domain-independent methods for reducing risk based on segmentation and algebraic inequalities and demonstrates that combining these methods with domain-specific knowledge helps to identify new simple and effective solutions in such mature fields like strength of components, kinematic analysis of mechanisms and electrical engineering. The meaningful interpretation of algebraic inequalities led to the discovery of new physical properties of electrical circuits and mechanical assemblies. These properties have never been suggested in standard textbooks and research literature covering the mature fields of electrical and mechanical engineering which demonstrates that the lack of knowledge of domain-independent methods for reducing risk and uncertainty made these properties invisible to domain experts.

Keywords Domain-independent methods · Reliability improvement · Risk reduction · Uncertainty reduction · Algebraic inequalities · Segmentation

---

M. Todinov (✉)
School of Engineering, Computing and Mathematics, Oxford Brookes University, Oxford, UK
e-mail: mtodinov@brookes.ac.uk

K. B. Misra (ed.), *Handbook of Advanced Performability Engineering*,
https://doi.org/10.1007/978-3-030-55732-4_30

30.1 Introduction

While reliability and risk assessment are truly domain-independent areas, this cannot be stated about the equally important areas of reliability improvement and risk reduction. For decades, the reliability and risk science failed to appreciate and emphasise that reliability improvement, risk reduction and uncertainty reduction are underpinned by general principles that work in many unrelated domains.

As a consequence, *methods for measuring and assessing reliability, risk and uncertainty were developed, not domain-independent methods for improving reliability, reducing risk and uncertainty which could provide direct input to the design process.* Indeed, in standard textbooks on mechanical engineering and design of machine components [1–10], for example, there is no mention of generic (domain-independent) methods for reliability improvement and risk and uncertainty reduction.

It needs to be pointed out that even the available methods for measuring and assessing reliability and risk cannot always be fully implemented in the design for the obvious reason that for new products and processes reliability data are simply unavailable.

In the rare cases where reliability data for the components and parts building the systems are available, they are relevant for a particular environment and duty cycle and their mechanical application to another environment and duty cycle, as experience has shown, is of highly questionable value. The lack of predictive capability of the existing reliability tools caused many engineers to lose faith in the tools and discard them as not adding real value to their work.

Why is engineering design so slow in exploiting the achievements of the reliability and risk science to improve reliability and reduce risk? This is certainly not due to the complexity of the reliability improvement and risk reduction methods. In this respect, the contrast with the complex generic mathematical methods for stress analysis, kinematic and dynamic analysis of solid bodies and fluids is striking. These mathematical modelling methods are penetrating all aspects of the engineering design.

The problem is that the current approach to reliability improvement and risk reduction almost solely relies on knowledge from a specific domain and is conducted exclusively by experts in that domain. This creates the incorrect perception that effective risk reduction can be delivered solely by using methods offered by the specific domain, without resorting to a general risk reduction methods and principles.

This incorrect perception resulted in ineffective reliability improvement and risk reduction across the entire industry, the loss of valuable opportunities for reducing risk and 'repeated reinvention of the wheel'. Current technology changes so fast that the domain-specific knowledge related to reliability improvement and risk reduction is outdated almost as soon as it is generated. In contrast, the domain-independent methods for reliability improvement, risk and uncertainty reduction are higher order methods that permit application in new, constantly changing situations and circumstances.

The development of the domain-specific, physics-of-failure approach for reliability improvement [11] has been prompted by the deficiencies of the data-driven approach. Although the physics-of-failure approach was very successful in addressing the underlying causes of failure and eliminating failure modes, it contributed to the widespread view among many reliability practitioners that only physics-of-failure models can deliver real reliability improvement.

It is necessary to point out that building accurate physics-of-failure models of the time to failure is not always possible because of the complexity of the physical mechanisms underlying the failure modes, the complex nature of the environment and the operational stresses. Physics-of-failure modelling certainly helps, for example, to increase the strength of a component by conducting research on the link between microstructure and mechanical properties of the material. However, this approach requires arduous and time-consuming research, special equipment and human resource. More importantly, physics-of-failure models *are not capable of capturing principles and invariants underlying reliability improvement and risk reduction in unrelated domains*. Despite their success and popularity, physics-of-failure models *cannot transcend the narrow domains they serve and cannot be used for improving reliability and reducing risk in unrelated domains*.

A central theme in the new domain-independent approach for reliability improvement and risk reduction introduced in [12] is the concept that risk reduction is underlined by common domain-independent principles which, combined with knowledge from the specific domain, are capable of generating effective risk-reducing solutions.

The domain-independent methods do not rely on the availability of past failure data or detailed knowledge of the underlying mechanisms of failure. As a result, they are particularly well suited for developing new designs, with unknown failure mechanisms and failure history. In many cases, these methods reduce risk at no extra cost or at a relatively small cost.

Establishing universally accepted theoretical principles for risk assessment requires a common definition of risk, valid in unrelated domains of human activity [13]. Similarly, establishing universally accepted theoretical principles for risk and uncertainty reduction goes through formulating domain-independent principles for reducing risk and uncertainty, valid in unrelated domains of human activity. Establishing the risk research as a mainstream science requires solid and universally accepted theoretical principles for the two fundamental components of risk management: *risk assessment* and *risk and uncertainty reduction*. The domain-independent principles and methods for risk and uncertainty reduction:

- Add value to decisions related to reliability improvement, risk and uncertainty reduction.
- Provide key input to the design process by improving the reliability of the designed product rather than measuring its performance only.
- Provide effective risk and uncertainty reduction across unrelated domains of human activity. Avoid loss of opportunities for reducing risk and 'reinvention of the wheel'.

- Deeply impact the current understanding of available methods and techniques for risk and uncertainty reduction.

It is important to point out that the domain-independent methods for reliability improvement and risk and uncertainty reduction are not a substitute for the domain-specific approach for risk reduction. Combined with knowledge from the specific domain, the domain-independent methods and principles help to obtain superior solutions. Accordingly, this chapter demonstrates that combining domain-specific knowledge from different areas of engineering with the domain-independent methods of the algebraic inequalities and segmentation leads to reliability improvement and uncertainty reduction.

30.2 Method of Segmentation to Improve Reliability and Develop Light-Weight Design

The underlying idea of the method of segmentation is to prevent failure modes and reduce the vulnerability to a single failure, by dividing an entity into a number of distinct parts. A large number of applications of the domain-independent method of segmentation have already been presented in [12].

There are numerous cases where design-engineers have control over the points of application of external loads. For the simply supported beam with length a in Fig. 30.1a, the concentrated load F is applied in the middle and results in a bending moment $M(x)$. The maximum bending moment $M_{1,\max}$ is attained at $x = a/2$ and is equal to $M_{1,\max} = Fa/4$ (Fig. 30.1b). Segmenting the concentrated load F into two loads with magnitude $F/2$ (Fig. 30.1c) reduces the maximum bending moment three times, from $M_{1,\max} = Fa/4$ to $M_{2,\max} = Fa/12$ (Fig. 30.1d). The reduction of the bending moment reduces the bending stress in the beam and increases its resistance to overstress failure.

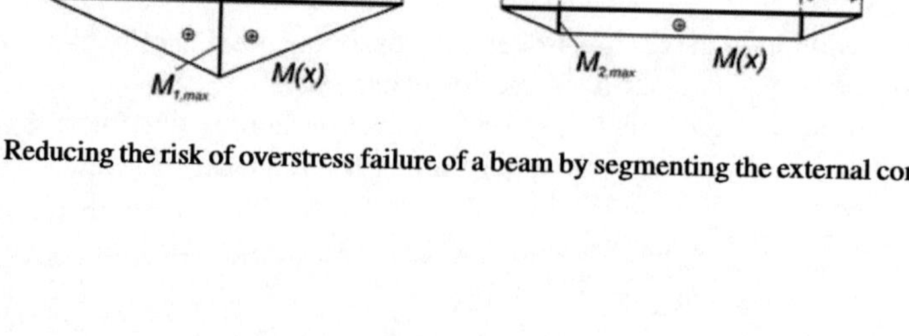

Fig. 30.1 Reducing the risk of overstress failure of a beam by segmenting the external concentrated load F

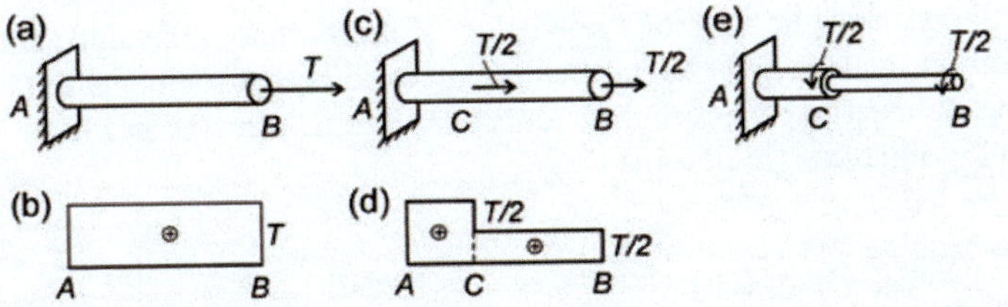

Fig. 30.2 Reducing the risk of overstress failure of a shaft by segmenting the external concentrated torque T

In some design applications (e.g. in motorsport design), the focus is often on obtaining a light-weight design, not on increasing the resistance to overstress failure. A light-weight design translates directly into enhanced performance, reduced fuel consumption and reduced emissions. As a result of the segmented external load and the reduced tensile stresses from bending in Fig. 30.1, the cross section of the loaded beam can be reduced which results in a light-weight design.

Indeed, the bending stress σ_b in a beam with a circular cross section with diameter d is given by the well-known formula [14]: $\sigma_b = 32M/(\pi d^3)$ where M is the bending moment acting in the particular section. Reducing the bending moment 3 times by preserving the bending stress σ_b, results in a significant reduction of the cross-sectional diameter of the beam. From $\sigma_b = 32M/(\pi d^3) = 32(M/3)/(\pi d_1^3)$, the diameter of the light-weight design is evaluated to be $d_1 = 0.693\,d$, which, for a uniform cross section, results in volume of the material per unit length of the beam equal to $\pi(0.693d)^2/4 = 0.48 \times \pi d^2/4$. As a result, the light-weight design carries the same bending stress σ_b with only 48% of the material of the original beam. The weight saving from segmenting the loading force is impressive.

The load segmentation also improves reliability and results in light-weight designs in the case of a concentrated external torque (Fig. 30.2a).

Segmenting the concentrated torque T into two torques of magnitude $T/2$ reduces the maximum shear stress from $\tau_{\max} = 16T/(\pi d^3)$ along the length AB in Fig. 30.2a, to $\tau_{\max,1} = 8T/(\pi d^3)$ along the section CB in Fig. 30.2c. Similarly, preserving the same shear stress $\tau_{\max}$ along the sections AC and CB yields the light-weight design in Fig. 30.2e with reduced cross section along the section CB.

These simple solutions for reducing the stresses in loaded structures, based on segmentation of external concentrated loads, have never been suggested in standard textbooks in the mature fields of stress analysis and strength of components [1, 2, 5, 14–16].

A primary objective of the topology optimisation of structural design is removing and redistributing a material in specified design spaces, for specified loads, constraints and boundary conditions so that a light-weight design is attained while preserving the required functionality. No solutions based on a segmentation of external loads have been suggested in the literature related to topological optimisation [17] despite that segmentation of external loads often leads to light-weight designs.

This shows that the lack of knowledge of the domain-independent method of segmentation made it invisible to the domain experts that segmentation of external loads can be used to reduce significantly the internal stresses in loaded structures and develop light-weight designs.

30.2.1 Improvement of Reliability of Computations

The next application of chain-rule segmentation to reduce the risk of computational errors is related to differentiating a very complex function $f(t)$ with respect to the parameter t.

The complex function $f(t)$ is first presented as a composition of nested continuous functions

$$f(t) = f(\varphi_1(\varphi_2(\ldots\varphi_n(t))))$$

where $f(\varphi_1), \varphi_1(\varphi_2), \varphi_2(\varphi_3), \ldots, \varphi_n(t)$ are simpler differentiable functions.

Consequently, the derivative $\mathrm{d}f(t)/\mathrm{d}t$ can be found by applying the chain rule for differentiation:

$$\frac{\mathrm{d}f(t)}{\mathrm{d}t} = \frac{\mathrm{d}f}{\mathrm{d}\varphi_1} \times \frac{\mathrm{d}\varphi_1}{\mathrm{d}\varphi_2} \times \ldots \times \frac{\mathrm{d}\varphi_n}{\mathrm{d}t}$$

The reduction of the risk of computational errors comes from the circumstance that each of the derivatives, $\mathrm{d}f/\mathrm{d}\varphi_1, \mathrm{d}\varphi_1/\mathrm{d}\varphi_2, \ldots, \mathrm{d}\varphi_n/\mathrm{d}t$, is much easier to evaluate than the derivative $\mathrm{d}f(t)/\mathrm{d}t$.

Consider an example from kinematics analysis of mechanisms. The mechanism whose kinematics is to be analysed incorporates three sliders *B*, *D* and *E* (Fig. 30.3). Sliders *B* and *D* move along the *x*-axis while slider *E* moves along the axis ET, which

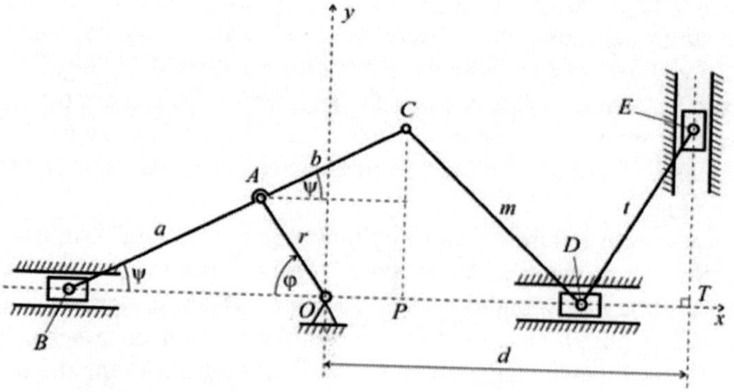

Fig. 30.3 A mechanism whose kinematics is analysed

is perpendicular to the x-axis and at a distance d from the origin O of the coordinate system Oxy.

The crank OA rotates in the clockwise direction, with a uniform angular velocity of $\omega = 1.5$ rad/s and subtends an angle φ with the horizontal x-axis which varies within the interval $[0, 2\pi]$. Note that the angle CDE is not fixed and varies as the links CD and ED rotate around the pin joint D. The values of the parameters fully specifying the mechanism are as follows: $\mathrm{OA} = r = 0.35\,\mathrm{m}$; $\mathrm{AB} = a = 0.65\,\mathrm{m}$; $AC = b = 0.50\,\mathrm{m}$; $CD = m = 0.80\,\mathrm{m}$; $DE = t = 0.75\,\mathrm{m}$ and $d = 1.3\,\mathrm{m}$.

The point of interest is the velocity of slider E.

Denoting $x_D = OD$ $y_E = TE$ and applying trigonometry yields

$$\sin\psi = r\sin\varphi/a \tag{30.1}$$

$$\cos\psi = \sqrt{1-\sin^2\psi} \tag{30.2}$$

$$x_D = b\cos\psi - r\cos\varphi + \sqrt{m^2-(a+b)^2\sin^2\psi} \tag{30.3}$$

$$y_E = \sqrt{t^2-(d-x_D)^2} \tag{30.4}$$

Substituting expressions (30.1) and (30.2) in (30.3), followed by substituting expression (30.3) in (30.4) expresses y_E as a function of the crank angle φ and by using the relationship $\varphi = \omega t$, y_E can also be expressed as a function of the time t. Once y_E has been presented as a function of time, it can be differentiated to obtain the velocity v_E of slider E: $v_E = \mathrm{d}y_E(t)/\mathrm{d}t$. However, this approach requires differentiating a very complex expression. During this process, the likelihood of making an error is very high. The risk of computational error can be reduced greatly if the method of segmentation is applied, by using the chain rule for differentiation. As a result, the initial problem of determining $v_E = \mathrm{d}y_E(t)/\mathrm{d}t$ is replaced by the simpler problem of determining the three derivatives:

$$v_E = [\mathrm{d}y_E/\mathrm{d}x_D]\times[\mathrm{d}x_D/\mathrm{d}\varphi]\times[\mathrm{d}\varphi/\mathrm{d}t] \tag{30.5}$$

Indeed,

$$\frac{\mathrm{d}y_E}{\mathrm{d}x_D} = \frac{d-x_D}{\sqrt{t^2-(d-x_D)^2}} \tag{30.6}$$

$$\begin{aligned}\frac{\mathrm{d}x_D}{\mathrm{d}\varphi} = &-\frac{br^2\sin\varphi\cos\varphi}{a^2\sqrt{1-(r/a)^2\sin^2\varphi}} + r\sin\varphi\\ &-\frac{(a+b)^2r^2\sin\varphi\cos\varphi}{a^2\sqrt{m^2-(a+b)^2(r/a)^2\sin^2\varphi}}\end{aligned} \tag{30.7}$$

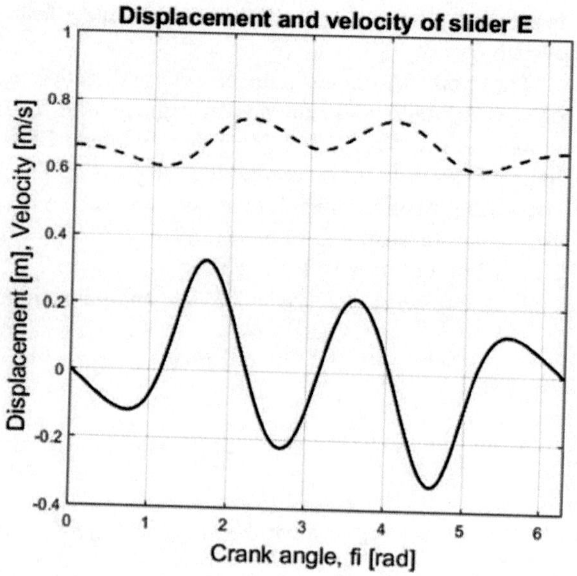

Fig. 30.4 Velocity v_E (Continuous line) and displacement y_E (Dashed line) of point E on slider E

$$d\varphi/dt = \omega \tag{30.8}$$

The velocity and displacement of slider E, as a function of the crank angle φ in radians, are shown in Fig. 30.4 with a continuous and dashed line, respectively. To test the chain-rule segmentation method, the velocity of slider E has also been calculated by using numerical differentiation.

$$v_{E,i} \approx \frac{y_{E,i} - y_{E,i-1}}{h} \times \omega \tag{30.9}$$

where $h = 0.001$ rad is a small step of the crank angle, $y_{E,i}$ and $y_{E,i-1}$ are the displacements of point E corresponding to crank angles φ_i and φ_{i-1}, $i = 1, \ldots, n$.

The velocity dependence obtained from the numerical differentiation and the velocity dependence obtained from the chain-rule segmentation coincide.

In the literature related to kinematic analysis of mechanisms [18–20], no solutions based on segmentation through the chain rule have been suggested, despite that segmentation based on the chain rule clearly leads to a significantly reduced likelihood of errors. The lack of knowledge of the domain-independent method of segmentation made it invisible to domain experts in the mature field of kinematic analysis of mechanisms that chain-rule segmentation yields a significantly reduced likelihood of computational errors.

30.3 Reducing Risk and Uncertainty by Using Algebraic Inequalities

In textbooks on reliability engineering [21–25] and in papers related to risk, reliability and uncertainty, there is a lack of discussion related to reducing risk and uncertainty by using algebraic inequalities. This is a surprising omission considering the power of algebraic inequalities in reducing risk and uncertainty and the existence of a significant number of publications covering the theory of algebraic inequalities [26–32]. It was only recently that some applications of the domain-independent method of algebraic inequalities for reducing risk and uncertainty have been presented in [12, 33].

A formidable advantage of the algebraic inequalities is their capacity to reduce aleatory and epistemic uncertainty and produce tight upper and lower bounds related to uncertain reliability-critical design parameters such as material properties, dimensions, loads and component reliabilities. Algebraic inequalities are capable of ranking systems, processes and decisions in terms of reliability in the absence of any knowledge related to the values of the reliability-critical parameters. In addition, algebraic inequalities can be interpreted in a meaningful way and this interpretation can be attached to real systems and processes. This yields not only to uncertainty reduction but also to the discovery of new fundamental properties of systems and processes.

By establishing tight bounds related to properties and parameters, algebraic inequalities can be applied to improve the robustness of designs, by complying them with the worst possible variation of the output parameters. As a result, a number of failure modes can be avoided.

30.3.1 Ranking Systems with Unknown Reliability of Components

Often, the reliabilities of the components building the system are unknown and the epistemic uncertainty associated with the reliabilities of the components building the system translates into epistemic uncertainty related to which system is superior.

An important way of using inequalities to improve reliability and reduce risk is to derive and prove an algebraic inequality which ranks systems performance. For two competing systems (a) and (b), built on components whose reliabilities are unknown, the steps for establishing which system is superior can be summarised as follows.

- For each of the competing systems, build the reliability network from its functional diagram.
- By using methods from system reliability analysis, determine the reliabilities R_a and R_b of the systems or the probabilities of system failure F_a,F_b.

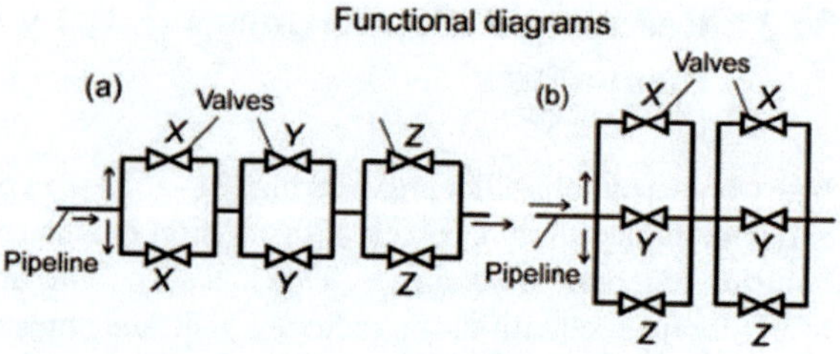

Fig. 30.5 Two competing systems with different topology, built with the same type of components

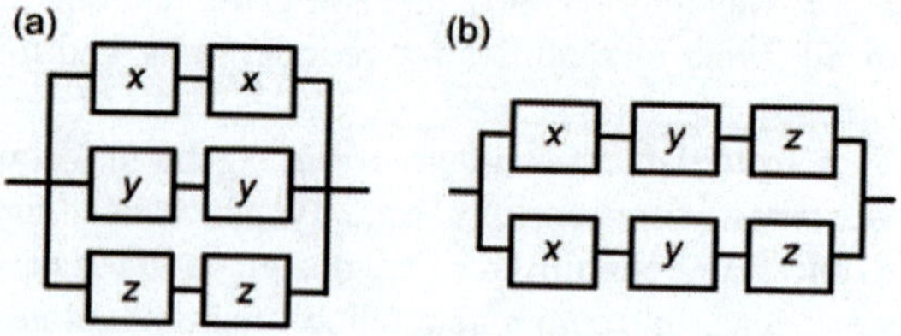

Fig. 30.6 The reliability networks of the systems from Fig. 30.5

- Subtract the reliabilities of the competing systems or the probabilities of system failure and prove any of the inequalities: $R_a - R_b > 0$, $R_a - R_b < 0$, $F_a - F_b > 0$, $F_a - F_b < 0$.
- Select the system with superior reliability or the system with the smaller probability of failure.

Consider two systems with different topologies, including the same type of valves (denoted by X, Y and Z) shown in Fig. 30.5. The valves are working independently from one another and all of them are initially open. The question of interest is which system is more reliable with respect to the function 'stopping the flow of fluid in the pipeline'. The signal for closing is issued to all valves simultaneously.

Figures 30.6a and b represent the reliability networks of the systems from Fig. 30.5a and b, correspondingly. The reliability values x, y and z characterising the separate valves are unknown. The only available information about the reliabilities of the valves are the obvious constraints: $0 < x < 1$; $0 < y < 1$; $0 < z < 1$.

Expressing the probabilities of failure characterising the competing systems as a function of the unknown reliabilities of the valves yields

$$F_a(x, y, z) = (1 - x^2)(1 - y^2)(1 - z^2) \text{ and } F_b(x, y, z) = (1 - xyz)^2$$

Ranking the systems' performance consists of proving $F_a(x, y, z) - F_b(x, y, z) < 0$ or $F_a(x, y, z) - F_b(x, y, z) > 0$. Proving $F_a(x, y, z) - F_b(x, y, z) < 0$, for example, is equivalent to proving the inequality

$$(1 - x^2)(1 - y^2)(1 - z^2) < (1 - xyz)^2 \tag{30.10}$$

To prove inequality (30.10), it suffices to prove the inequality $\sqrt{(1 - x^2)(1 - y^2)(1 - z^2)} < (1 - xyz)$ or the equivalent inequality

$$\sqrt{(1-x^2)(1-y^2)(1-z^2)} + xyz < 1 \tag{30.11}$$

Indeed, if inequality (30.11) is true, inequality (30.10) follows from it by squaring both sides of the inequality $\sqrt{(1-x^2)(1-y^2)(1-z^2)} < 1 - xyz$. The squaring operation will not change the direction of the inequality because $0 < x < 1$; $0 < y < 1$; $0 < z < 1$, and the following quantities are positive: $(1 - xyz) > 0$, $(1-x^2)(1-y^2)(1-z^2) > 0$.

To prove inequality (30.11), a combination of a substitution technique and a technique based on proving a simpler, intermediate inequality will be used.

Because the reliability r_i of a component is a number between zero and unity, the trigonometric substitutions $r_i = \sin \alpha_i$ where $\alpha_i \in (0, \pi/2)$ are appropriate. Making the substitutions: $x = \sin\alpha$; $y = \sin\beta$ and $z = \sin\gamma$ for the reliabilities of the components, transforms the left side of inequality (30.11) into

$$\sqrt{(1-x^2)(1-y^2)(1-z^2)} + xyz = \cos\alpha \times \cos\beta \times \cos\gamma + \sin\alpha \times \sin\beta \times \sin\gamma \tag{30.12}$$

Next, the positive quantity $\cos\alpha \times \cos\beta \times \cos\gamma + \sin\alpha \times \sin\beta \times \sin\gamma$ is replaced by the larger quantity $\cos\alpha \times \cos\beta + \sin\alpha \times \sin\beta$. Indeed, because $0 < \cos\gamma < 1$ and $0 < \sin\gamma < 1$, the inequality

$$\cos\alpha \times \cos\beta \times \cos\gamma + \sin\alpha \times \sin\beta \times \sin\gamma < \cos\alpha \times \cos\beta + \sin\alpha \times \sin\beta \tag{30.13}$$

holds. If the intermediate inequality $\cos\alpha \times \cos\beta + \sin\alpha \times \sin\beta \le 1$ can be proved, this will imply the inequality

$$\cos\alpha \times \cos\beta \times \cos\gamma + \sin\alpha \times \sin\beta \times \sin\gamma < 1 \tag{30.14}$$

Since $\cos\alpha \times \cos\beta + \sin\alpha \times \sin\beta = \cos(\alpha - \beta)$, and $\cos(\alpha - \beta) \le 1$, we finally get

$$\cos\alpha \times \cos\beta \times \cos\gamma + \sin\alpha \times \sin\beta \times \sin\gamma < \cos\alpha \times \cos\beta + \sin\alpha \times \sin\beta = \cos(\alpha - \beta) \le 1$$

from which inequality (30.11) follows.

Inequality (30.11) has been proved and from it, inequality (30.10) follows. The system in Fig. 30.5a is characterised by a smaller probability of failure compared to the system in Fig. 30.5b, therefore, the system in Fig. 30.5a is the more reliable system.

30.3.2 *Inequality of Negatively Correlated Random Events*

There is another, alternative way of using algebraic inequalities for risk and uncertainty reduction which consists of moving in the opposite direction: starting from existing abstract inequality and moving towards the real system or a process. An important step in this process is creating relevant meaning for the variables entering the algebraic inequality, followed by a meaningful interpretation of the different parts of the inequality which links it with a real physical system or process.

Consider m independent events $A_1, A_2, \ldots, A_m$ that are not mutually exclusive. This means that there are at least two events A_i and A_j for which $P(A_i \cap A_j) \neq \varnothing$. It is known with certainty, that if any particular event A_k of the set of events does not occur ($k = 1, \ldots, m$), then at least one of the other events occurs. In other words, the relationship

$$P(A_1 \cup \ldots \cup \bar{A}_k \cup \ldots \cup A_m) = 1$$

holds for the set of m events.

Under these assumptions, it can be shown that the following inequality holds

$$P(A_1) + P(A_2) + \ldots + P(A_m) > 1 \tag{30.15}$$

This inequality will be referred to as *the inequality of negatively correlated events*.

To prove this inequality, consider the number of outcomes $n_1, n_2, \ldots, n_m$ leading to the separate events $A_1, A_2, \ldots, A_m$, correspondingly. Let n denote the total number of possible outcomes. From the definition of inversely correlated events, it follows that any of the n possible outcomes corresponds to the occurrence of at least one event A_i. Since at least two events A_i and A_j can occur simultaneously, the sum of the outcomes leading to the separate events $A_1, A_2, \ldots, A_m$ is greater than the total number of outcomes n:

$$n_1 + n_2 + \ldots + n_m > n \tag{30.16}$$

This is because of the condition that at least two events A_i and A_j can occur simultaneously. Then, at least one outcome must be counted twice: once for event A_i and once for event A_j. Dividing both sides of (30.16) by the positive value n does not alter the direction of inequality (30.16) and the result is the inequality

$$n_1/n + n_2/n + \ldots + n_m/n > 1 \tag{30.17}$$

which is inequality (30.15).

Consider the reliability networks in Fig. 30.7, of two systems. Despite the deep uncertainty related to the components building the systems, the reliabilities of the systems can still be ranked, by a meaningful interpretation of the inequality of negatively correlated events.

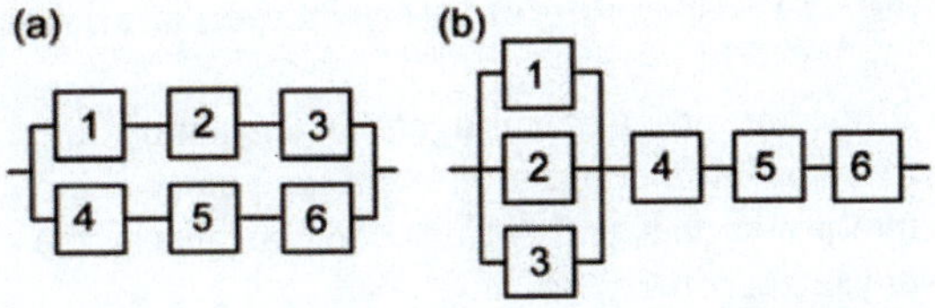

Fig. 30.7 Ranking the reliabilities of two systems with unknown reliability of components

The power of the simple inequality (30.15) can be demonstrated even if only two events $A_1 \equiv A$ and $A_2 \equiv \bar{B}$ are considered. Event $A_1 \equiv A$ stands for 'system (a) is working at the end of a specified time interval' while event $A_2 \equiv \bar{B}$ stands for 'system (b) is not working at the end of the specified time interval' $(P(\bar{B}) + P(B) = 1)$ (Fig. 30.7). The conditions of inequality (30.15) are fulfilled for events A and $\bar{B}$ related to the systems in Fig. 30.7.

Indeed, if event $\bar{B}$ does not occur, this means that system (b) is working. This can happen only if all components 4, 5 and 6 in Fig. 30.7b are working, which means that system (a) is working. As a result, if event $\bar{B}$ does not occur then event A occurs. Conversely, if event A does not occur then at least one of the components 4, 5, 6 in Fig. 30.7a does not work, which means that system (b) does not work (the event $\bar{B}$ occurs). At the same time, both events can occur simultaneously $(P(A \cap \bar{B}) \neq 0)$. This is, for example, the case if components 1, 2, 3 are in working state at the end of the time interval (0, t) and component 5 is in a failed state.

The conditions of inequality (30.15) are fulfilled, therefore

$$P(A) + P(\bar{B}) > 1 \tag{30.18}$$

holds, which is equivalent to

$$P(A) > 1 - P(\bar{B}) = P(B)$$

As a result, it follows that $P(A) > P(B)$ irrespective of the reliabilities $r_1, r_2, r_3, r_4, r_5, r_6$ of components (1–6) building the systems. The meaningful interpretation of the inequality of negatively correlated events helped to reveal the intrinsic reliability of competing design solutions and rank these in terms of reliability, in the absence of knowledge related to the reliabilities of their building parts.

In other cases, knowledge about the age of the components is available which can be used in proving the inequalities related to the system reliabilities. For example, it is known that the functional diagrams of the competing systems are built with three valves (A, B and C) with different ages. Valve A is a new valve, followed by valve B with an intermediate age and valve C which is an old valve. If the reliabilities of the valves are denoted by a, b and c, the reliabilities of the valves can be ranked: $a > b > c$ and this ranking can be used in proving the inequalities related to the reliabilities of the competing systems [12].

30.3.2.1 Meaningful Interpretation of an Abstract Algebraic Inequality

While the proof of an algebraic inequality does not normally pose problems, the meaningful interpretation of an inequality is not a straightforward process. Such an interpretation usually brings deep insights, some of which stand at the level of a new physical property/law.

Consider the abstract algebraic inequality

$$(x_1 + x_2 + \ldots + x_n) \geq n^2 \left(\frac{1}{1/x_1 + 1/x_2 + \ldots + 1/x_n} \right), \tag{30.19}$$

which is valid for any set of n non-negative quantities x_i.

A proof of Inequality (30.19) can be obtained by transforming the inequality to the classical Cauchy–Schwarz inequality

$$(a_1 b_1 + a_2 b_2 + \ldots + a_n b_n)^2 \leq (a_1^2 + a_2^2 + \ldots + a_n^2)(b_1^2 + b_2^2 + \ldots + b_n^2) \tag{30.20}$$

which is valid for any two sequences of real numbers $a_1, a_2, \ldots, a_n$ and $b_1, b_2, \ldots, b_n$.

Note that the transformation $a_i = \sqrt{x_i}$ $(i = 1, \ldots, n)$ and $b_i = 1/\sqrt{x_i}$ $(i = 1, \ldots, n)$, substituted in the Cauchy–Schwarz inequality (30.20) leads to inequality (30.19).

Appropriate meaning can now be attached to the variables entering inequality (30.19) and the two sides of the inequality can be interpreted in various meaningful ways.

A relevant meaning for the variables in the inequality can be created, for example, if each x_i stands for 'electrical resistance of element i'. The equivalent resistances $R_{e,s}$ and $R_{e,p}$ of n elements arranged in series and parallel are given by [34]

$$R_{e,s} = x_1 + x_2 + \ldots + x_n \tag{30.21}$$

$$R_{e,p} = \frac{1}{1/x_1 + 1/x_2 + \ldots + 1/x_n} \tag{30.22}$$

where x_i is the resistance of the ith element $(i = 1, \ldots, n)$. In this case, expression (30.21) on the left side of the inequality (30.19) can be meaningfully interpreted as the equivalent resistance of n elements arranged in series. The expression (30.22), on the right side of inequality (30.19), can be meaningfully interpreted as the equivalent resistance of n elements arranged in parallel. Inequality (30.19) now expresses a new physical property: the equivalent resistance of n elements arranged in parallel is at least n^2 times smaller than the equivalent resistance of the same elements arranged in series, irrespective of the individual resistance values of the elements. Equality is attained for $x_1 = x_2 = \ldots = x_n$.

It needs to be pointed out that for resistors of equal values, the fact that the equivalent resistance in parallel is exactly n^2 times smaller than the equivalent resistance of the resistors in series is a trivial result, easily derived and known for a long period of time [35].

Indeed, For the resistance of n resistors arranged in series $x_1 = x_2 = \ldots = x_n = r$, the value nr is obtained from Eq. (30.21), while for the same n resistors arranged in parallel, the value r/n is obtained from Eq. (30.22). As can be seen, the value r/n is exactly n^2 times smaller than the value nr. However, the bound provided by inequality (30.19) is a much deeper result. It is valid for *any possible values of the resistances. The bound given by inequality* (30.19) *does not require equal resistances.*

The meaning created for the variables x_i in inequality (30.19) is not unique and can be altered. Suppose that x_i in inequality (30.19) stands for electrical capacity. The equivalent capacitances $C_{e,p}$, $C_{e,s}$ of n capacitors arranged in parallel and series are given by [34]:

$$C_{e,p} = x_1 + x_2 + \ldots + x_n \tag{30.23}$$

and

$$C_{e,s} = \frac{1}{1/x_1 + 1/x_2 + \ldots + 1/x_n} \tag{30.24}$$

correspondingly, where x_i is the capacitance of the ith capacitor ($i = 1, \ldots, n$). The expression (30.23) on the left side of inequality (30.19) can now be meaningfully interpreted as the equivalent capacitance of n capacitors arranged in parallel. The expression (30.24) on the right side of inequality (30.19) can be meaningfully interpreted as the equivalent capacitance $C_{e,s}$ of n capacitors arranged in series. Inequality (30.19) now expresses another physical property: the equivalent capacitance of n capacitors arranged in parallel is at least n^2 times larger than the equivalent capacitance of the same capacitors arranged in series, irrespective of the values of the individual capacitors.

Suppose that another meaning for the variables x_i in Inequality (30.19) is created, for example, each x_i now stands for the stiffness of the elastic element i ($i = 1, \ldots, n$) Consider the equivalent stiffness $k_{e,s}$ of n elastic elements in series and the equivalent stiffness $k_{e,p}$ of n elastic elements in parallel. The stiffness values of the separate elastic elements, denoted by $x_1, x_2, \ldots, x_n$, are unknown. The equivalent stiffness of n elastic elements in series is given by the well-known relationship:

$$k_{e,s} = \frac{1}{1/x_1 + 1/x_2 + \ldots + 1/x_n} \tag{30.25}$$

and for the same elastic elements in parallel, the equivalent stiffness is

$$k_{e,p} = x_1 + x_2 + \ldots + x_n \tag{30.26}$$

Now, the two sides of inequality (30.19) can be meaningfully interpreted in the following way. The expression (30.25) on the right-hand side of the inequality (30.19) can be interpreted as the equivalent stiffness of n elastic elements arranged in series. The left side of inequality (30.19) can be interpreted as the equivalent stiffness of n elastic elements arranged in parallel. The inequality now expresses a different physical property: the equivalent stiffness of n elastic elements arranged in parallel is at least n^2 times larger than the equivalent stiffness of the same elements arranged in series, irrespective of the individual stiffness values characterising the separate elements. These are examples of different physical properties derived from a meaningful interpretation of a single abstract algebraic inequality.

The considered examples illustrate new physical properties predicted from interpreting a correct algebraic inequality which give the basis for *the principle of non-contradiction: If a correct algebraic inequality permits meaningful interpretation that can be related to a real process, the process realization yields results that do not contradict the algebraic inequality.*

Further details regarding the principle of non-contradiction will be presented elsewhere.

Inequality (30.19) is domain-independent. It provides tight bounds for electrical and mechanical properties. At the same time, the uncertainty associated with the relationship between the equivalent parameters characterising elements arranged in series and parallel (due to the epistemic uncertainty related to the values of the building elements) is reduced.

These properties have never been suggested in standard textbooks and research literature covering the mature fields of mechanical and electrical engineering, which demonstrates that the lack of knowledge of the domain-independent method of algebraic inequalities made these properties invisible to the domain experts.

30.4 Conclusions

1. The benefit from combining the domain-independent method of segmentation with domain-specific knowledge in strength of components was demonstrated in reducing the risk of overstress failure by segmenting concentrated external loads. It was demonstrated that the domain-independent method of segmentation also achieves light-weight design.
2. The capability of the chain-rule segmentation to reduce the risk of computational errors has been demonstrated in the area of kinematic analysis of complex mechanisms.
3. The domain-independent method of algebraic inequalities has been used to reduce uncertainty, reveal the intrinsic reliability of competing designs and rank these in terms of reliability, in the absence of knowledge related to the reliabilities of their building parts.
4. The meaningful interpretation of an algebraic inequality led to the discovery of new physical properties.

Thus, the equivalent resistance of n elements arranged in parallel is at least n^2 smaller than the equivalent resistance of the same elements arranged in series, irrespective of the resistances of the elements.
Another physical property discovered by a meaningful interpretation of an algebraic inequality is that the equivalent capacity of n capacitors arranged in series is at least n^2 times smaller than the equivalent capacity of the same capacitors arranged in parallel, irrespective of the actual capacities of the separate capacitors.
5. The inequality of negatively correlated random events was introduced and its meaningful interpretation was used to reveal the intrinsic reliability of competing design solutions and to rank them in the absence of knowledge related to the reliabilities of the building parts.
6. The domain-independent method of segmentation and the domain-independent method based on algebraic inequalities combined with knowledge from specific domains achieved effective risk reduction solutions.

References

1. Collins, J. A. (2003). *Mechanical design of machine elements and machines*. New York: Wiley.
2. Norton, R. L. (2006). *Machine design, an integrated approach* (3rd ed). Pearson International edition.
3. Pahl, G., Beitz, W., Feldhusen, J., & Grote, K. H. (2007). *Engineering design*. Berlin: Springer.
4. Childs, P. R. N. (2014). *Mechanical design engineering handbook*. Amsterdam: Elsevier.
5. Budynas, R. G., & Nisbett, J. K. (2015). *Shigley's mechanical engineering design* (10th ed.). New York: McGraw-Hill.
6. Mott, R. L, Vavrek, E. M., & Wang, J. (2018). *Machine elements in mechanical design* (6th ed). Pearson Education Inc.
7. Gullo, L. G., & Dixon, J. (2018). *Design for safety*. Chichester: Wiley.
8. French, M. (1999). *Conceptual design for engineers* (3rd ed.). London: Springer.
9. Samuel, A., & Weir, J. (1999). *Introduction to engineering design: Modelling, synthesis and problem solving strategies*. London: Elsevier.
10. Thompson, G. (1999). *Improving maintainability and reliability through design*. London: Professional Engineering Publishing Ltd.
11. Pecht, M., Dasgupta, A., Barker, D., & Leonard, C. T. (1990). The reliability physics approach to failure prediction modelling. *Quality and Reliability Engineering International, 4*, 267–273.
12. Todinov, M. T. (2019). *Methods for reliability improvement and risk reduction*. Wiley.
13. Aven, T. (2016). Risk assessment and risk management: review of recent advances on their foundation. *European Journal of Operational Research, 253*, 1–13.
14. Gere, J., & Timoshenko, S. P. (1999). *Mechanics of materials* (4th edn). Stanley Thornes Ltd.
15. Hearn, E. J. (1985). *Mechanics of materials, vol. 1 and 2*, (2nd edition). Butterworth.
16. Budynas, R. G. (1999). *Advanced strength and applied stress analysis* (2nd ed.). New York: McGraw-Hill.
17. Bendsøe, M. P., & Sigmund, O. (2003). *Topology optimization—theory, methods and applications*. Berlin Heidelberg: Springer.
18. Sandor, G. N., & Erdman, A. G. (1984). *Advanced mechanism design: analysis and synthesis* (Vol. 2). Englewood Cliffs NJ: Prentice-Hall Inc.
19. Dicker, J. J., Pennock, G. R., & Shigley, J. E. (2003). *Theory of machines and mechanisms*. Oxford: Oxford University Press.

20. Uicker, J. J. Jr, & Pennock, G. R., & Shigley, J. E. (2017). *Theory of machines and mechanisms* (5th ed). New York: Oxford University Press.
21. Modarres, M., Kaminskiy, M. P., & Krivtsov, V. (2017). *Reliability engineering and risk analysis, a practical guide* (3rd ed). CRC Press.
22. Dhillon, B. S. (2017). *Engineering systems reliability, safety, and maintenance*. New York: CRC Press.
23. Ebeling, C. E. (1997). *Reliability and maintainability engineering*. Boston: McGraw-Hill.
24. Lewis, E. E. (1996). *Introduction to reliability engineering*. New York: Wiley.
25. O'Connor, P. D. T. (2002). *Practical Reliability Engineering* (4th ed.). New York: Wiley.
26. Bechenbach, E., & Bellman, R. (1961). *An introduction to inequalities*. New York: The L.W.Singer company.
27. Cloud, M., Byron, C., & Lebedev, L. P. (1998). *Inequalities: with applications to engineering*. New York: Springer.
28. Engel, A. (1998). *Problem-solving strategies*. New York: Springer.
29. Hardy, G., Littlewood, J. E., & Pólya, G. (1999). *Inequalities*. New York: Cambridge University Press.
30. Pachpatte, B. G. (2005). *Mathematical inequalities. North Holland Mathematical Library* (vol. 67). Amsterdam: Elsevier.
31. Steele, J. M. (2004). *The cauchy-schwarz master class: An introduction to the art of mathematical inequalities*. New York: Cambridge University Press.
32. Kazarinoff, N. D. (1961). *Analytic inequalities*. New York: Dover Publications Inc.
33. Todinov, M. T. (2018). Improving reliability and reducing risk by using inequalities. *Safety and reliability, 38*(4), 222–245.
34. Floyd, T. (2004). *Electronics fundamentals: circuits, devices and applications* (6th ed.). New Jersey: Pearson Education Inc.
35. Rozhdestvenskaya, T. B., & Zhutovskii, V. L. (1968). High-resistance standards. *Measurement techniques, 11*, 308–313.

Michael Todinov has a background in mechanical engineering, applied mathematics and computer science. He received his Ph.D. and his higher doctorate (DEng) from the University of Birmingham and is currently a professor in mechanical engineering in Oxford Brookes University, UK. Prof. Todinov pioneered research on reliability analysis based on the cost of failure, repairable flow networks and networks with disturbed flows, domain-independent methods for reliability improvement and reducing risk by using algebraic inequalities. In the area of reliability and risk, Prof. Todinov authored five books with reputable academic publishers and numerous research papers. In 2017, he received the prestige award of the Institution of Mechanical Engineers (UK) in the area of risk reduction in mechanical engineering.

Chapter 31
Stochastic Effort Optimization Analysis for OSS Projects

Yoshinobu Tamura, Adarsh Anand, and Shigeru Yamada

Abstract It is very important to produce and maintain a reliable system structured from several open-source software, because many open-source software (OSS) have been introduced in various software systems. As for the OSS development paradigm, the bug tracking systems have been used for software quality management in many OSS projects. It will be helpful for OSS project managers to assess the reliability and effort management of OSS, if many fault data recorded on the bug tracking system are analyzed for software quality improvement. In this chapter, we focus on a method of stochastic effort optimization analysis for OSS projects by using the OSS fault big data. Then, we discuss the method of effort estimation based on stochastic differential equation and jump-diffusion process. In particular, we use the OSS development effort data obtained from fault big data. Then, deep learning is used for the parameter estimation of jump-diffusion process model. Also, we discuss the optimal maintenance problem based on our methods. Moreover, several numerical examples of the proposed methods are shown by using the effort data in the actual OSS projects. Furthermore, we discuss the results of numerical examples based on our methods of effort optimization.

Keywords Open-source software · Software effort · Optimization · Stochastic model · Deep learning

---

Y. Tamura (✉)
Tokyo City University, Tokyo 1588557, Japan
e-mail: tamuray@tcu.ac.jp

A. Anand
University of Delhi, Delhi 110007, India

S. Yamada
Tottori University, Tottori 6808550, Japan

K. B. Misra (ed.), *Handbook of Advanced Performability Engineering*,
https://doi.org/10.1007/978-3-030-55732-4_31

31.1 Introduction

Many open-source software (OSS) are embedded in many software systems and services. Also, most software systems have used any OSS components because of quick delivery, standardization, cost reduction, etc. Then, many software reliability growth models for OSS systems have been actively proposed in the past (Yamada [1]—Zhou [2]).

This chapter discusses the flexible effort prediction models based on deep learning considering the Wiener process and AI for the complexed several external factors of OSS development project. In particular, we discuss the flexible jump-diffusion process model. Then, it is difficult to make a decision about the unknown parameters of flexible jump terms in the proposed models because of the complexity in likelihood function included in the multiple distributions based on the Wiener process and jump-diffusion one. In this chapter, the methods of deep learning and maximum likelihood are used as the parameter estimation methods based on AI for our models.

Moreover, several numerical examples based on the actual fault big data on the projects of OSS are shown by using the flexible effort prediction models proposed in this chapter. Then, numerical illustrations of parameter estimation based on deep learning are shown. Finally, we show that the proposed flexible effort prediction models will be useful to predict the total developmental effort and optimal maintenance time of OSS developed under the open-source project.

31.2 Flexible Effort Estimation Model

We discuss stochastic maintenance effort modeling of jump-diffusion processes to control the OSS project operation. Let $X(t)$ be the cumulative OSS maintenance effort expenditures up to time $t|(t \geq 0)$ in the operation. $X(t)$ is the real values continuously. Then, $X(t)$ gradually grows with the OSS project operation. By using stochastic modeling technique of classical software reliability models (Yamada [3]—Kapur [4]), we consider the following equation considering the OSS maintenance effort:

$$\frac{dX(t)}{dt} = b(t)\{a - X(t)\}, \tag{31.1}$$

where $b(t)$ is the effort expenditure rate of the OSS project at time t. The parameter a in Eq. (31.1) means the expected cumulative maintenance effort of OSS expended for the specified version.

We expand Eq. (31.1) to the following stochastic differential equation process (Arnold [5]—Yamada [6]) with Brownian motion:

$$\frac{dX(t)}{dt} = \{b(t) + cn(t)\}\{a - X(t)\}, \tag{31.2}$$

where c is added as a positive value meaning a level of the irregular continuous fluctuation, and $n(t)$ a standardized Gaussian white noise. Then, Eq. (31.2) is extended to the following stochastic differential equation process of an It ô type:

$$\mathrm{d}X(t) = \left\{b(t) - \frac{1}{2}\sigma^2\right\}\{a - X(t)\}\mathrm{d}t + c\{a - X(t)\}\mathrm{d}\omega(t), \tag{31.3}$$

where $\omega(t)$ means a one-dimensional Wiener process. $\omega(t)$ can be represented as the white noise $n(t)$.

The jump term is inserted to the stochastic differential equation process model of Eq. (31.3) by considering the unexpected irregular situation at time t with many external complicated project factors. The process of jump diffusion is obtained as follows:

$$\begin{aligned}\mathrm{d}X_j(t) &= \left\{b(t) - \frac{1}{2}c^2\right\}\{a - X_j(t)\}\mathrm{d}t,\\ &+ c\{a - X_j(t)\}\mathrm{d}\omega(t) + d\left\{\sum_{i=1}^{N_t(\lambda)}(J_i - 1)\right\},\end{aligned} \tag{31.4}$$

where a Poisson point process with frequency λ at time t is represented as $N_t(\lambda)$, i.e., the number of jumps, and λ the rate of jump. J_i means the range of i-th jump. We assume that $\omega(t)$, $N_t(\lambda)$, and J_i are mutually independent. Moreover, the increasing rates of OSS maintenance effort for $b(t)$ are shown as

$$b(t) \doteq \frac{\frac{\mathrm{d}C_*(t)}{\mathrm{d}t}}{\alpha - C_*(t)}, \tag{31.5}$$

$$C_e(t) = \alpha\left(1 - \mathrm{e}^{-\beta t}\right), \tag{31.6}$$

$$C_s(t) = \alpha\left\{1 - (1 + bt)\mathrm{e}^{-\beta t}\right\}. \tag{31.7}$$

In this chapter, $b(t)$ is assumed to be the intensity functions in Eqs. (31.6) and (31.7) from nonhomogeneous Poisson process (NHPP) models as the OSS effort expenditure function of our model, where $a \doteq \alpha$ is the expected cumulative number of latent faults (the total amount of maintenance effort), and $b \doteq \beta$ the detection rate per fault (the effort consumption rate) in terms of software reliability growth models.

Based on It ô's formula (Tamura [7]), $X_{j*}(t)$ in Eq. (31.4) can be derived as

$$X_{je}(t) = a\left[1 - \exp\left\{-bt - c\omega(t) - \sum_{i=1}^{N_t(\lambda)} \log J_i\right\}\right], \tag{31.8}$$

$$X_{js}(t) = a\left[1 - (1 + bt)\exp\left\{-bt - c\omega(t) - \sum_{i=1}^{N_t(\lambda)} \log J_i\right\}\right]. \tag{31.9}$$

Moreover, we extend the existing jump-diffusion process models obtained from Eq. (31.4) to the following time-delay jump-diffusion processes:

$$\mathrm{d}X_{fj}(t) = \left\{b(t) - \frac{1}{2}c^2\right\}\{a - X_{fj}(t)\}\mathrm{d}t + c\{a - X_{fj}(t)\}\mathrm{d}\omega(t)$$
$$+ d\left\{\sum_{i=0}^{N_t(\lambda_1)} (J_i^1 - 1)\right\}. \quad (t \geq 0) \tag{31.10}$$

$$\mathrm{d}X_{fj}(t) = \left\{b(t) - \frac{1}{2}c^2\right\}\{a - X_{fj}(t)\}\mathrm{d}t + c\{a - X_{fj}(t)\}\mathrm{d}\omega(t)$$
$$+ d\left\{\sum_{i=0}^{N_t(\lambda_1)} (J_i^1 - 1)\right\} + d\left\{\sum_{i=0}^{N_{t'}(\lambda_2)} (J_i^2 - 1)\right\}, (t \geq 0, t' \geq t_1) \tag{31.11}$$

where $N_t(\lambda_1)$ and $N_{t'}(\lambda_2)$ are Poisson point processes with parameter λ_1 and λ_2 at each operation time $(t \geq 0)$ and $\left(t^{'} \geq t_1\right)$, respectively. Moreover, J_i^1 and J_i^2 are i-th jump ranges in each operation time $(t \geq 0)$ and $\left(t' \geq t_1\right)$, respectively. We assume that $N_t(\lambda_1)$, $N_t(\lambda_2)$, J_i^1, and J_i^2 are mutually independent in this chapter.

From It ô's formula (Tamura [7]), the solution of Eqs. (31.10) and (31.11) can be obtained as follows.

In the cases $(t \geq 0)$:

$$X_{fje}(t) = a\left[1 - \exp\left\{-bt - c\omega(t) - \sum_{i=1}^{N_t(\lambda_1)} \log J_i^1\right\}\right], \tag{31.12}$$

$$X_{fjs}(t) = a\left[1 - (1 + bt)\cdot\exp\left\{-bt - c\omega(t) - \sum_{i=1}^{N_t(\lambda_1)} \log J_i^1\right\}\right]. \tag{3.13}$$

In the cases $\left(t \geq 0, |t' \geq t_1\right)$:

$$X_{fje}(t) = a\left[1 - \exp\left\{-bt - c\omega(t) - \sum_{i=1}^{N_t(\lambda_1)} \log J_i^1 - \sum_{i=1}^{N_{t'}(\lambda_2)} \log J_i^2\right\}\right] \tag{31.14}$$

$$X_{fjs}(t) = a\left[1 - (1 + bt)\cdot\exp\left\{-bt - c\omega(t) - \sum_{i=1}^{N_t(\lambda_1)} \log J_i^1 - \sum_{i=1}^{N_{t'}(\lambda_2)} \log J_i^2\right\}\right]. \tag{31.15}$$

Considering the time delay over $t_2|(t_2 \geq t_1)$, we can formulate the flexible jump-diffusion process models as follows:

$$X_{fje}(t) = a\left[1 - \exp\left\{-bt - c\omega(t) - \sum_{k=1}^{K}\sum_{i=1}^{N_{t^k}(\lambda_k)} \log J_i^k\right\}\right], \tag{31.16}$$

$$X_{fje}(t) = a\left[1 - (1+bt)\exp\left\{-bt - c\omega(t) - \sum_{k=1}^{K}\sum_{i=1}^{N_{t^k}(\lambda_k)} \log J_i^k\right\}\right], \tag{31.17}$$

where $t^k|(k = 1, 2, \cdots, K)$ means k-th specific time for major version upgrade, and K is the number of the major version upgrades of OSS (Tamura [8]—Tamura [9]).

31.3 Optimal Maintenance Problem Based on Flexible Effort Estimation Models

It is well known that the development cycle of OSS project is proceeded as follows:

1. Upload of OSS,
2. The usage of OSS by users,
3. The record of bug contents on the bug tracking system, and
4. The modification and development of OSS.

Then, we assume that the maintenance of OSS is the major version upgrade from a standpoint of developers. On the other hand, we assume that the maintenance of OSS means the reboot and the software shift in the new version from a standpoint of OSS users. In fact, the maintenance of OSS means the project operation associated with the OSS system halt from a standpoint of OSS users. Our method will be useful for both sides: the developer and users to assess the maintenance time.

Considering the characteristics of several version upgrade of OSS, it is interesting for the software developers to predict and estimate the time when we should stop project operating in order to maintain the OSS system efficiently. We formulate the total effort based on flexible jump-diffusion process model by using the classical software release problems. We define the following parameters in terms of effort:

r_1 : the importance rate of effort per unit time during OSS operation,
r_2 : the fixing cost per fault during OSS operation,
r_3 : the maintenance cost per fault during OSS operation.

Then, the software efforts in the operation can be formulated as

$$\phi_1^e(t) = r_1 t + r_2 X_{fje}(t), \tag{31.18}$$

$$\phi_1^s(t) = r_1 t + r_2 X_{fjs}(t). \tag{31.19}$$

Also, the software effort expenditures after the maintenance of cloud software system are represented as follows:

$$\phi_2^e(t) = r_3 X_{fjre}(t), \tag{31.20}$$

$$\phi_2^s(t) = r_3 X_{fjrs}(t). \tag{31.21}$$

From the above equations, the total software maintenance effort expenditures are given by

$$\emptyset^e(t) = \phi_1^e(t) + \phi_2^e(t), \tag{31.22}$$

$$\emptyset^s(t) = \phi_1^s(t) + \phi_2^s(t). \tag{31.23}$$

The optimum maintenance time t^* is obtained by minimizing $\emptyset^e(t)$ and $\emptyset^s(t)$ in the above equations.

31.4 Steps of Parameter Estimation

We apply a method of maximum likelihood to the estimation of unknown parameters a, b, and c. In particular, it is very difficult to make a decision about the unknown parameters of jump terms in our models because of the complexity in likelihood function included in the multiple distributions based on the Wiener process and jump-diffusion one. Several estimation methods for jump parameters in jump-diffusion process model have been proposed by several specified researchers (Tamura [8]). However, there are no effective methods of such parameter estimation. We discuss the estimation method of parameters in terms of jump terms. Then, the algorithm of deep learning is used in order to estimate the jump parameters of the discussed model.

As an example, we assume that our jump-diffusion process models include the parameters λ_1 and λ_2 for Y_t and $Y_{t'}$, similarly, μ_1, μ_2, τ_1, and τ_2 for J_i^1 and J_i^2 in Eqs. (31.16) and (31.17). Then, the set parameters $\boldsymbol{J}$ in terms of λ_1, μ_1, and τ_1 are estimated by a deep learning algorithm in case of $(t \geq 0)$. Similarly, the set parameters $\boldsymbol{J}'$ in terms of λ_2, μ_2, and τ_2 make a decision by using the deep learning algorithm in case of $(t' \geq t_1)$. We apply a deep feedforward neural network as the algorithms of deep learning in order to learn the OSS fault big data on bug tracking systems. We apply the following input data sets in terms of each unit on the input layer. Then, the unknown parameters as the objective variable are given as the parameter set $\boldsymbol{J}$ in terms of λ_1, μ_1, and τ_1. The following nine items as explanatory variables are set to the units of input layer:

- Opened and Changed date, time
- Product name
- Component name
- Name of version
- Nickname for Reporter
- Nickname for Assignee
- Status of faults
- Operating system
- Level of severity

We show the steps of parameter estimation in our method as follows:

Step 1 The method of maximum likelihood is used for the parameters a, b, and c.

Step 2 Then, the data used in the method of maximum likelihood is the cumulative maintenance effort.

Step 3 The unknown parameters included in $N_t(\lambda_*)$ and J_i^* of the jump terms are estimated by using deep learning.

Step 4 Then, the data used in the method of deep learning is the fault big data structured by nine items as explanatory variables. The fault big data sets have 10,000 fault lines. The weight parameters are learned by using the past actual learning data. Moreover, the output values are estimated by using the learned weight parameters and testing fault data. Furthermore, the output values are used as the estimated parameters included in $N_t(\lambda_*)$ and J_i^* of jump terms.

31.5 Numerical Examples

The Apache HTTP server (The Apache Software Foundation [10]) is well known as OSS. We show several numerical examples based on our method by using the fault big data of Apache HTTP server.

Figure 31.1 is the estimated cumulative OSS operation effort based on the exponential effort prediction model by using deep learning. The long-dash line shows the start line of version 4.1.31 major-version-upgraded from version 3.x line. Moreover, the dot-dash line shows the start line of beta version 7.0.0 major-version-upgraded from version 6.x line. After 1826 days, we found that the jump becomes as shown in Fig. 32.1. Similarly, Fig. 31.2 shows the estimated cumulative OSS project operation effort based on S-shaped type effort prediction model by using deep learning. From Fig. 31.2, we find that the S-shaped effort prediction model given by Eq. (31.17) fits better than the exponential type effort prediction model given by Eq. (31.16) for the actual data sets. In our method, we estimate the unknown parameters of jump terms based on the deep learning by using the data sets until version 4.1.31. Moreover, the unknown parameters of jump terms are estimated by using the data sets from version 4.1.31 to beta version 7.0.0.

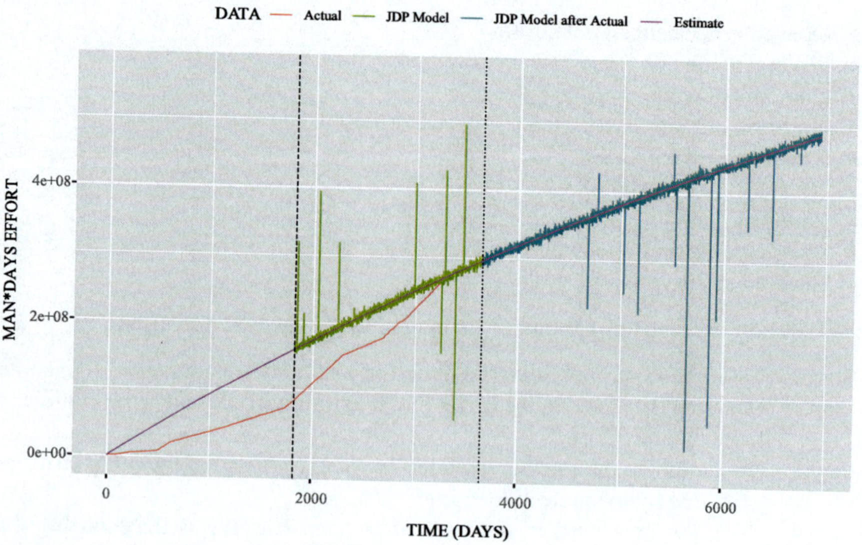

Fig. 31.1 The estimated cumulative OSS project operation effort based on exponential type effort prediction model by using the deep learning

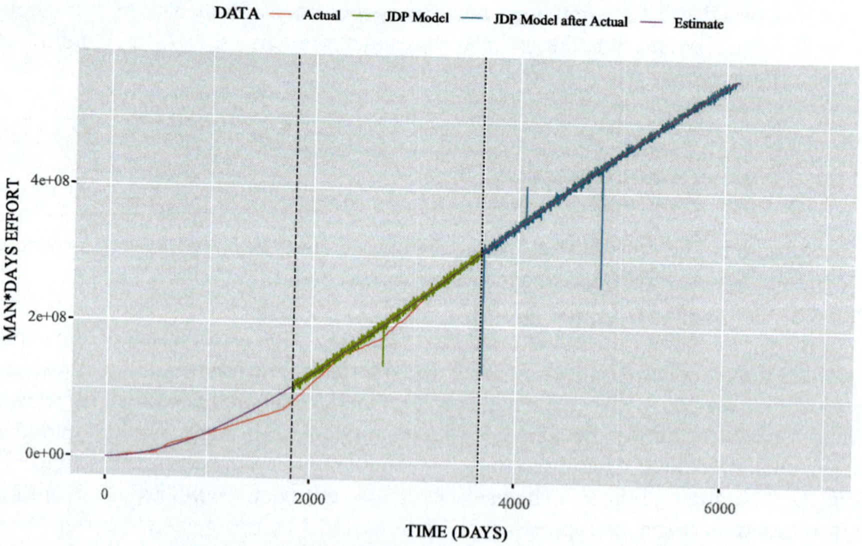

Fig. 31.2 The estimated cumulative OSS project operation effort based on S-shaped type effort prediction model by using deep learning

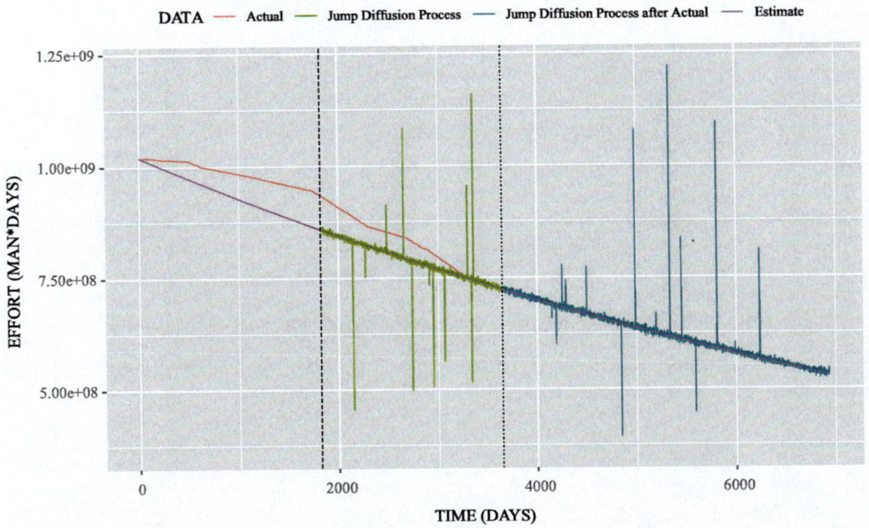

Fig. 31.3 The estimated cumulative OSS project operation effort expenditures based on exponential type effort prediction model by using deep learning

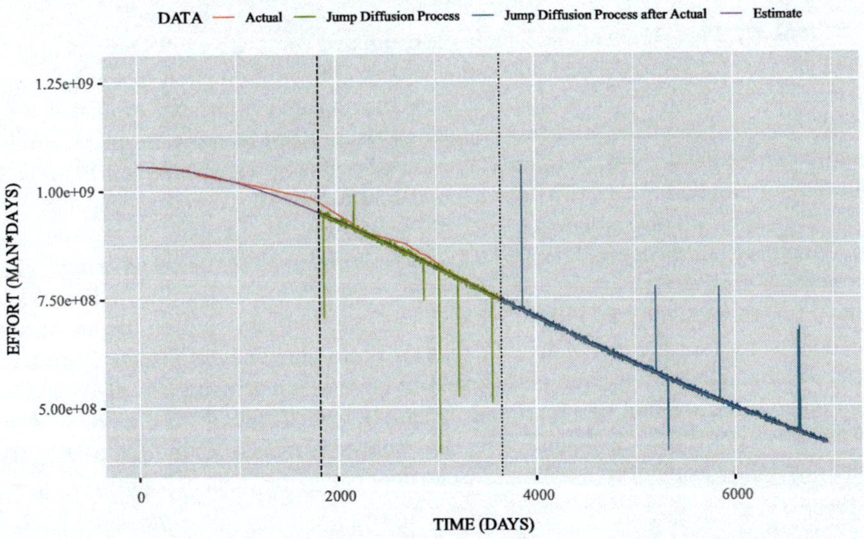

Fig. 31.4 The estimated cumulative OSS project operation effort expenditures based on S-shaped type effort prediction model by using deep learning

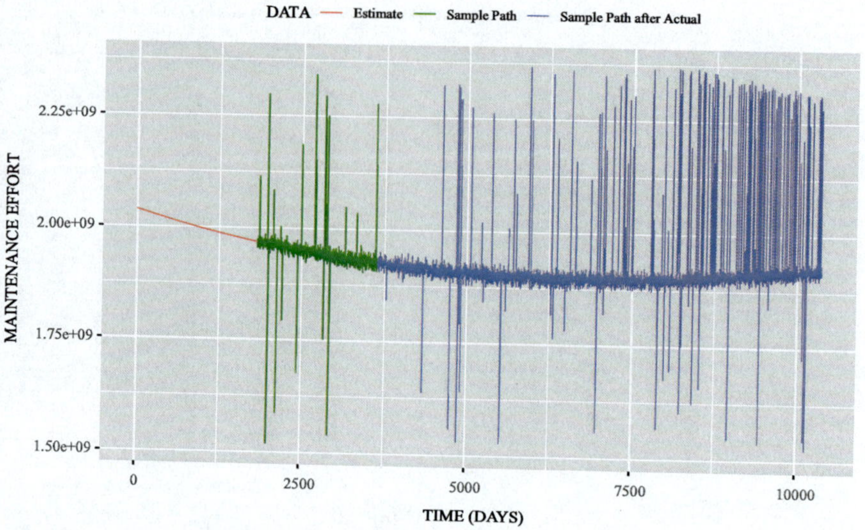

Fig. 31.5 The estimated cumulative software effort based on exponential type effort prediction model by using deep learning

Similarly, Figs. 31.3 and 31.4 show the estimated cumulative OSS project operation effort expenditures based on exponential type effort prediction model by using deep learning, and the estimated cumulative maintenance effort expenditures based on S-shaped type effort prediction model by using deep learning, respectively. In particular, the data sets of specified phases are estimated by deep learning in Figs. 31.3 and 31.4. From Figs. 31.3 and 31.4, we found that the estimates based on deep learning can show for each phase in detail.

Moreover, we show the estimated cumulative software effort based on exponential type and S-shaped type effort prediction models by using deep learning in Figs. 31.5 and 31.6, respectively. In Figs. 31.5 and 31.6, the blue line means the estimated sample paths after the end of actual data sets. From blue lines of Fig. 31.5, we find that the characteristic of jump noise becomes large as the progress of operation. On the other hand, we find that the trend of jump noise changes with the progress of operation from Fig. 31.6. Finally, Fig. 31.7 shows estimated J_{i}^{k} in case of $X_{fje}(\mathrm{t})$ and $X_{fjs}(t)$ by using deep learning. We can understand the trend of jump noise from Fig. 31.7.

31.6 Concluding Remarks

This chapter focuses on the stochastic maintenance effort optimization based on the jump-diffusion process and deep learning for OSS projects. The optimal maintenance time depends on the cumulative software effort expenditures under the operation. In

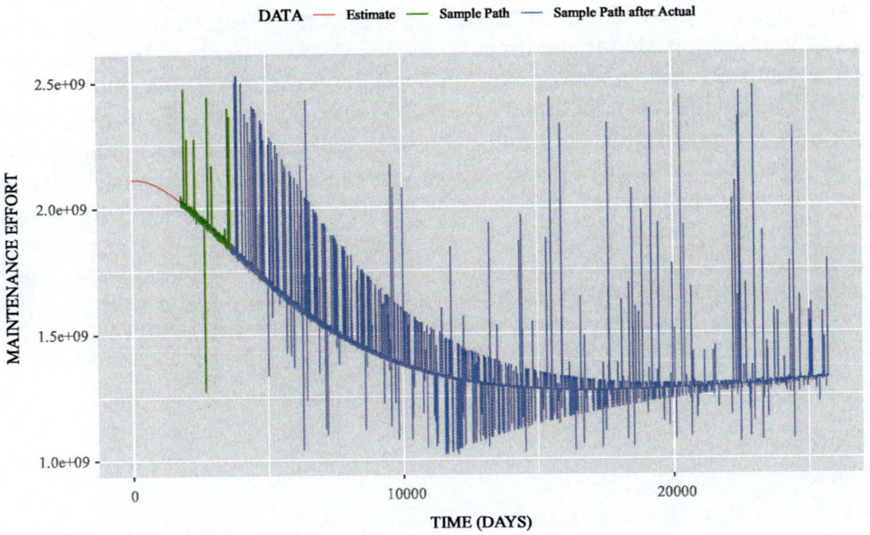

Fig. 31.6 The estimated cumulative software effort based on S-shaped type effort prediction model by using deep learning

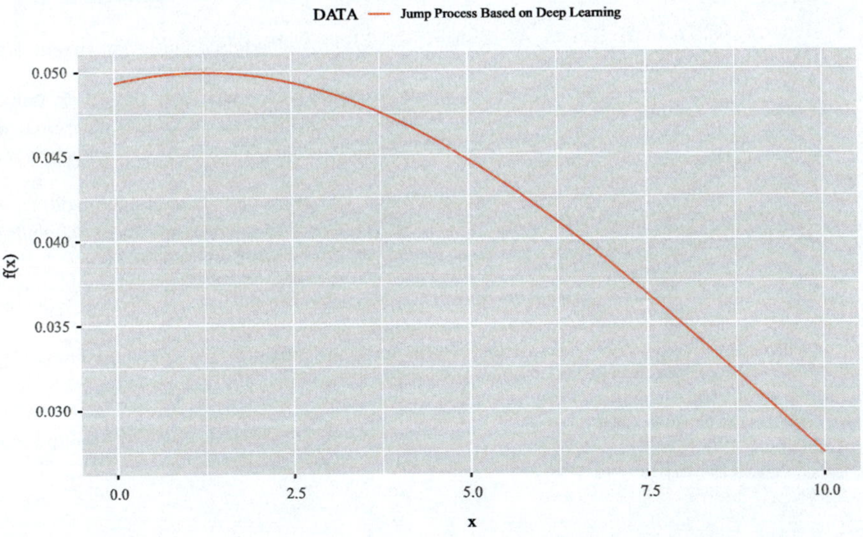

Fig. 31.7 The estimated $J_{\mathrm{i}}^{\mathrm{k}}$ in case of $X_{fje}(\mathrm{t})$ and $X_{fjs}(t)$ by using deep learning

this chapter, we have discussed the method of OSS project effort analysis based on deep learning considering the irregular situations with jump term from the trends of several OSS version upgrade. It is difficult for us to estimate the many parameters of jump terms because of the complicated stochastic processes. Then, we have discussed several methods of parameter estimation based on deep learning in order to comprehend the trends of OSS version upgrade in our effort prediction models. The proposed parameter estimation methods will be useful as the estimation method considering the progress trends with OSS version upgrade.

Acknowledgements This work was supported in part by the JSPS KAKENHI Grant No. 20K11799 in Japan.

References

1. Yamada, S., & Tamura, Y. (2016). *OSS reliability measurement and assessment*. Switzerland: Springer International Publishing.
2. Zhou, Y., & Davis, J. (2005). OSS reliability model: an empirical approach. In *Proceedings of the fifth workshop on OSS engineering* (pp. 67–72).
3. Yamada, S. (2014). *Software reliability modeling: Fundamentals and applications*. Tokyo: Springer.
4. Kapur, P. K., Pham, H., Gupta, A., & Jha, P. C. (2011). *Software reliability assessment with OR applications*. London: Springer.
5. Arnold, L. (1974). *Stochastic differential equations-theory and applications*. New York: Wiley.
6. Yamada, S., Kimura, M., Tanaka, H., & Osaki, S. (1994). Software reliability measurement and assessment with stochastic differential equations. *IEICE Transactions on Fundamentals, E77–A*(1), 109–116.
7. Tamura, Y., Sone, H., & Yamada, S. (2019). Productivity assessment based on jump diffusion model considering the effort management for OSS project. *International Journal of Reliability, Quality and Safety Engineering, 26*(5), 1950022-1-1950022-22. World Scientific.
8. Tamura, Y. & Yamada, S. (2019). Maintenance effort management based on double jump diffusion model for OSS project, *Annals of Operations Research*. 1–16, 10.1007/s10479-019-03170-w, Springer US, Online First.
9. Tamura, Y., & Yamada, S. (2017). Fault identification and reliability assessment tool based on deep learning for fault big data. *Journal of Software Networking, 2017*(1). pp 161–176. https://doi.org/10.13052/jsn2445-9739.2017.008.
10. The Apache Software Foundation, The Apache HTTP Server Project, https://www.httpd.apache.org/.

Chapter 32
Should Software Testing Continue After Release of a Software: A New Perspective

P. K. Kapur, Saurabh Panwar, and Vivek Kumar

Abstract Software reliability is a highly active and thriving field of research. In the past, various software reliability growth models have been suggested to analyze the reliability and security of the software systems. The present chapter seeks to focus on analyzing the software release policy under different modeling frameworks. This study discusses both the conventional policy where testing stop time and software release times coincide and the modern release time policy wherein software time-to-market and testing termination time are treated as two distinct time-points. The modern release policy represents the situation in which the software developers release the software early to capture maximum market share and continue the testing process for an added period to maximize the reliability of the software product. In addition, the concept of change-point with two different schemes is addressed in the present study. Change-point indicates the time-point at which the model parameters experience a discontinuity in time. In one scenario, the change-point is considered occurring before the release of the software and in the second scenario, the release time is treated as a change-point for the tester's fault detection process. The study further provides numerical illustrations to test the different release time policies and analyze the practical applicability of the optimization problem to minimize the cost function and maximize the reliability attribute.

Keywords Software reliability assessment · Release time decisions · Reliability growth models · Change-point · Post-release testing

P. K. Kapur (✉)
Amity Center for Interdisciplinary Research, Amity University, Noida, Uttar Pradesh, India
e-mail: pkkapur1@gmail.com

S. Panwar · V. Kumar
Department of Operational Research, University of Delhi, Delhi, India

K. B. Misra (ed.), *Handbook of Advanced Performability Engineering*,
https://doi.org/10.1007/978-3-030-55732-4_32

32.1 Introduction

Software systems are logical instructions that control the function of hardware in the digitalized domain [1]. The tremendous involvement of the software system in the day-to-day human activity has spawned the importance of delivering a reliable and assured software system in the market. The reliability of the software products has turned into an essential attribute of the system quality. In software reliability engineering, reliability is defined as the probability of failure-free operation of a system for a given period in an intended environment [2]. According to ISO/IEC 9001, quality is stated as the effectiveness of a software system to conform to requirements [3]. As reliability is an important aspect of software quality, the evaluation of software reliability is necessary for the current software production process. Software reliability is enhanced by carefully debugging the faults present in the source code of the software system [4].

The principal aim of software producers is to achieve a desirable level of reliability and satisfy the users to attain long-term earnings and strengthen the brand value in the market for a protracted period. To debug the underlying defects, the software goes through a comprehensive testing process prior to its commercial release. Based on the data of defects encountered during the testing phase, the software reliability can be predicted using suitable Software Reliability Growth Models (SRGMs) [5]. SRGMs are analytical models that explain the failure observation and defect identification phenomena during the software testing and debugging phases. The accurate prediction of software failure using SRGMs can help software engineers to devise proper quality support and steady resource planning [6].

In software engineering literature, the development of reliability growth models has been a long-standing study for researchers and practitioners. The majority of the SRGMs are parametric models based on the Non-homogenous Poisson Process (NHPP). In NHPP-based reliability growth models, the mean value function is applied to calculate the expected number of defects detected, isolated, and corrected from the software system at any point in time [5]. The first contribution to SRGM was made by Goel and Okumoto [7] who have developed the model under the fundamental assumption that the discovery of defects follows an exponential path. Later, Yamada et al. [8] considered the failure observation phenomenon to have an S-shaped growth pattern. Thereafter, various analytical studies have been carried out to develop plausible software reliability prediction tools [9–13]. Many authors have extended the conventional models either by adding new dimensions (e.g., test coverage or testing efforts) [14–16] or by relaxing some of the assumptions (e.g., incorporating the concept of imperfect debugging or change-point) [17–22].

In the past, several models have also been proposed focusing on assessing the optimal release time of the software system. The release time of software products depends on many attributes, viz., size, reliability level, the competence of developers, testing environment, user requirements, software development cost, market opportunity cost, warranty cost, risk, etc. Users require fast deliveries, an inexpensive product with high quality. On the contrary, software producers aim at delivering a reliable product to its clients with low development cost and high-profit margins. Thus, software analyst needs to find a trade-off between these conflicting objectives of user's requirements with that of software producers. The problem of evaluating optimal software release time and testing termination time complying both users and developer's requirements are generally known as Software release time decisions (SRTD). As every firm has limited testing resources for software development, project developers need to examine when to stop the testing process and when to deliver the software to its users. The early delivery of the software will reduce the development cost and provide competitive advantages but insufficient testing may result in the user's disappointment and may impede the goodwill of the company. Nevertheless, prolong testing without releasing the software may escalate the development cost and leads to market opportunity loss. Consequently, software release time and testing stop time forms a complex problem.

As release time decisions are a comprehensive part of software development, several studies have been conducted by various researchers in the past to suggest plausible release time policy. These optimization problems are formulated by using different techniques depending on the model parameters and application taken into consideration. Okumoto and Goel [23] formulated two basic optimal release time problems, one with an objective of minimizing the total software development cost and another with an aim of maximizing the reliability function. Later, Yamada and Osaki [24] developed the constrained cost minimization and reliability maximization release time problems. Kapur and Garg [25] examined the release time planning using testing effort-based reliability growth models. In addition, several studies have been carried out for determining the optimal release planning for successive releases of a software product.

In addition, few academicians have planned the optimal release time for software systems based on the criterion that testing stop time and software release time should be treated as two separate decision variables [26–31]. These studies are based on the strategy of delivering the software early to clients and continuing the testing process in the user environment to upgrade the overall reliability and quality of the software. The advantage of such practises is that it facilitates software producers to avoid market manipulation by the competitors. Moreover, continued testing in the operational phase will support developers to achieve the desired level of product reliability. The focus of this chapter is to study and analyze different classes of Software Release Time Decisions (SRTD) using different optimization techniques. The practical applicability of various release time policies is further established in the present chapter.

32.2 Assumptions

The various SRGMs described in the present study are based upon the following assumptions:

1. The fault removal process is modeled using a Non-Homogenous Poison Process (NHPP)
2. Faults causing failure are removed as soon as they are detected.
3. Software consists of a finite number of faults.
4. All the faults are removed perfectly without generating any additional faults.
5. During post-release testing phase, some proportion of the remaining faults are detected by the testing team and some are detected by users and reported to the testing team.
6. After detecting the faults, customers immediately report it to the testing team for rectification.
7. The cost of providing the patch or updates to users for fault rectification is negligible.

32.3 Unified Approach

In general, the software reliability model states that the fault detection at any instant of time is directly proportional to the remaining number of defects present in the software, i.e.,

$$\frac{dm(t)}{dt} = h(t)(a - m(t)) \tag{32.1}$$

where $h(t)$ denotes the hazard rate function or conditional probability of a software failure because of the fault present in the system given that no failure has occurred due to that fault before time t. Mathematically, the conditional probability or hazard rate function $h(t)$ can be described as

$$h(t) = \frac{f(t)}{1 - F(t)} \tag{32.2}$$

where $F(t)$ represents the non-decreasing cumulative probability function of fault removal by time t and $f(t)$ denotes the non-cumulative distribution function of software failure.

Thus, the instantaneous failure observation at any time t can be expressed using the following differential equation:

$$\frac{dm(t)}{dt} = \frac{f(t)}{1 - F(t)}(a - m(t)) \tag{32.3}$$

The above differential equation can be solved to get a closed-form solution by using the initial condition, at $t = 0$, $m(0) = 0$ and $F(0) = 0$:

$$m(t) = aF(t); \qquad t \geq 0 \tag{32.4}$$

Equation (32.4) represents the expected number of defects identified by time t where distribution function $F(t)$ can take different functional forms.

32.4 Software Release Time Decisions

Software release time determination in the testing phase is a typical application of software reliability models. The prior evaluation of software release time is important for developers to reduce the dual losses related to both early and late releases. The optimization problem of determining the optimal time of software release is mainly formulated based on goals set by the management in terms of cost, reliability, failure intensity, etc. subject to constraints.

32.4.1 *Conventional Release Time Decisions*

The software release time problems have been studied in many different ways and using different methodologies. The traditional release time decision is based on an implicit assumption that the formal testing stops completely at the time of release of the software. After release, the task of identifying software bugs is shifted to the users, and bugs are identified and fixed only if they cause problems to users. Figure 32.1 depicts the software lifecycle under two phases. Following are some most commonly used release time policies.

32.4.1.1 Unconstrained Release Time Decisions

Since the earlier work, the release time determination primarily focused on cost minimization or reliability maximization problems. In software engineering literature, Okumoto and Goel [23] were the first to develop optimal release time policy. They have proposed two unconstrained problems (a) cost minimization and (b) reliability maximization. According to their study, the total expected cost incurred during testing and operational phases is expressed as

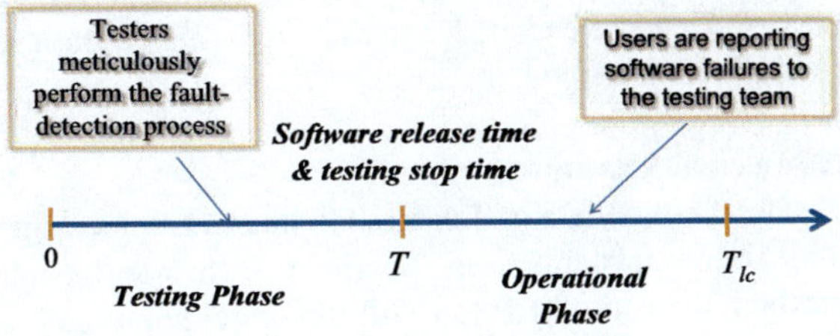

Fig. 32.1 Different phases of software lifecycle under conventional release time policy

$$C(T) = C_1T + C_2m(T) + C_3(m(T_{lc}) - m(T)) \tag{32.5}$$

In the cost model given in Eq. (32.5), T is the release time of the software system, C_1 denotes the testing cost per unit time, C_2 represents the cost of debugging a fault in the testing phase, and C_3 signifies the cost of debugging a fault in the operational phase. Moreover, $m(T)$ denotes the expected number of faults detected by time T following exponential distribution function, i.e.,

$$m(T) = a(1 - \mathrm{e}^{-bt}) \tag{32.6}$$

Equation (32.6) describes the mean value function of exponential SRGM [7].

The focus of the optimization problem was on minimizing the cost function. Thus, the first release time policy (Policy 1) is simply unconstrained minimization problem of expected cost function:

$$\textbf{Policy 1}: \ \text{Minimize}\, C(T) = C_1T + C_2m(T) + C_3(m(T_{lc}) - m(T)) \tag{32.7}$$

The release time is obtained by differentiating the cost function with respect to time, T and computing the time-point where the first derivative is zero based on the method of calculus. However, the aim of simply minimizing the cost attribute for evaluating the release time is solely a developer-oriented policy. The consideration of a client's requirement of high quality and a secure software system cannot be ignored while determining the software release time. Accordingly, Okumoto and Goel [23] developed another release time policy (Policy 2) wherein the aim is to maximize the reliability function of the software products, i.e.:

$$\textbf{Policy 2}: \ \text{Maximize}\, R(x|T) = \mathrm{e}^{-[m(T+x)-m(T)]} \tag{32.8}$$

where x is a small time duration; $R(x|0) = \mathrm{e}^{-m(x)}$ and $R(x|\infty) = 1$.

The reliability is a necessary attribute of the software system that affects the optimal decisions involved with software release time and testing stop time. Reliability function is defined as a probability of failure-free operation performed by the software with desirable output in a specific period under certain environmental conditions [32]. It may be noted that software producers do not have infinite resources to carry on the immeasurable testing process for achieving the maximum reliability function with value 1. Therefore, companies specify a desirable level of reliability (say, R_0) and release the software when aspiration level of reliability is achieved regardless of the cost incurred.

Numerical Example

The fault count data of Tandem Computers software projects [33] is obtained to estimate the model parameters. By applying the nonlinear least square (NLLS) regression method, the estimated result of the model parameters is $a = 130.30$ and $b = 0.083$.

Furthermore, the cost parameters are considered as $C_1 = \$100$, $C_2 = \$10$, and $C_3 = \$50$. The software lifecycle is further assumed to be $T_{lc} = 3$ years. On solving the Policy 1 using these parameters, the optimal release time is obtained as $T* = 17.65$ weeks with minimum cost $C(T*) = \$4272.46$ and the reliability level achieved is 0.0909. Such an optimal result cannot be accepted by the software engineers because at the release time extremely low reliability is achieved. Now, consider the Policy 2 that aims at maximizing the reliability function of the software system at the release time with minimum reliability aspiration level of $R_0 = 0.80$. Then, the optimal release time is obtained as $T* = 46.27$ with \$6035.64 budget consumption. When testing cost increases to $C_1 = \$500$, then total budget consumed is \$24547.53 to achieve 0.80 level of reliability at software release time. Thus, when only reliability is considered as an objective function, then no check is given on the budget allocated for testing. Consequently, results from Policy 1 and 2 suggest that unconstrained optimization of either cost minimization or reliability maximization is insufficient to ascertain the optimal release time decisions.

32.4.1.2 Constrained Release Time Decisions

The release time policy suggested by Okumoto and Goel [23] was either unconstrained cost minimization or unconstrained reliability maximization problem. In view of this, Yamada and Osaki [24] discussed the constrained optimization problem to calculate the software release time. They developed the optimization model with the aim of cost minimization under reliability aspiration constraint (Policy 3) and optimization model with the aim of reliability maximization under cost constraint (Policy 4), i.e.

$$\textbf{Policy 3:}\quad \text{Minimize } C(T) \tag{32.9}$$

$$\text{Subject to}: \ R(x|T) \geq R_0 \tag{32.10}$$

$$\textbf{Policy 4:}\quad \text{Maximize } R(x|T) \tag{32.11}$$

$$\text{Subject to}: \ C(T) \leq C_b \tag{32.12}$$

where $C(T)$ is the expected software development cost, R_0 is the predefined reliability level, and C_b is the total budget available for testing the software. The release time policy is based on three SRGMs, namely, exponential, modified exponential, and S-shaped SRGM.

Numerical Example

The exponential SRGM-based release time decision is evaluated by considering the example of Tandem Computers. The estimated values of parameters are $a = 130.30$

and $b = 0.083$. The cost parameters are taken as $C_1 = \$100$, $C_2 = \$10$, and $C_3 = \$50$, and reliability aspiration level as $R_0 = 0.8$. According to Policy 3, the optimal software time-to-market is $T* = 46.26$ weeks with minimum cost $C(T*) = \$6041.08$. Now, if the per unit testing cost changes to $C_1 = \$500$, then the optimal result is $T* = 46.26$ weeks with minimum cost $C(T*) = \$24515.64$. Again if $R_0 = 0.85$, then $T^* = 50.08$ weeks and $C(T^*) = \$26424.63$. Now, for deducing the release time using Policy 4, the total budget is considered as $C_b = \$6000$ with cost parameters as $C_1 = \$100$, $C_2 = \$10$, and $C_3 = \$50$, then optimal result is $T^* = 49.811$ weeks and achieved reliability level is $R^* = 0.7939$. If the allocated budget is $C_b = \$80,000$ with cost parameters as $C_1 = \$500$, $C_2 = \$10$, and $C_3 = \$50$ then optimal result is $T^* = 39.245$ weeks and achieved reliability level is $R^* = 0.6707$.

32.4.2 *Recent Release Time Decisions*

The reliability aspiration level of the device dictates that testing should continue until all faults are identified and removed. However, prolong testing without releasing the software may cause excessive development cost and loss in market share due to the competitive environment. Therefore, the recent release time decisions are based on the principle that software release time and testing stop time should be treated as two separate decision variables. In the succeeding sections, the generalized framework to determine the optimal release and testing stop time of software under reliability and budgetary constraints are discussed.

32.4.2.1 Release Time Decision When Software Time-to-Market and Testing Stop Time Are Different Time-Points

Based on the concept of post-release testing practiced in the software industry, Kapur et al. [30] proposed a software scheduling policy wherein the benefits of early software release with continued testing for a specific period in the operational phase has been examined. The prime advantage of early delivery of software is to have a competitive edge, while prolong testing in the user environment ensures high software reliability. Moreover, in the operational phase, the efficiency of the testing is improved as both testers and users meticulously identify faults in the system. According to the present release time policy, the software lifecycle is divided into three phases: *pre-release testing* phase $[0, \tau]$, *post-release testing* phase $[\tau, T]$ and *operational* phase $[T, T_{lc}]$ as depicted in Fig. 32.2. Furthermore, mean value function for fault detection phenomenon in these phases is subsequently described.

Phase 1: Pre-release Testing Phase $[0, \tau]$

In this phase, testing team rigorously identify faults lying latent in the software system. Thus, using the unified approach, the mean value function for faults identified by testers is given as

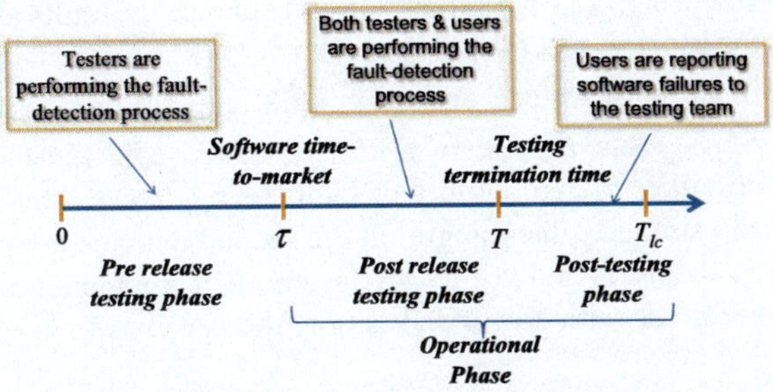

Fig. 32.2 Different phases of software lifecycle with Post-release testing

$$m_{\text{pre}}(t) = aF_{\text{pre}}(t); \quad 0 \le t \le \tau \tag{32.13}$$

where $F_{\text{pre}}(t)$ denotes the cumulative distribution of fault detection during pre-release testing phase. When fault identification function follows an exponential distribution function, then

$$m_{\text{pre}}(t) = a(1 - \mathrm{e}^{-bt}) \tag{32.14}$$

where a is the initial faults present in the system; b is the rate parameter of exponential distribution.

Phase 2: Post-Release Testing Phase $[\tau, T]$

It is considered that the software is released for the commercial purpose at time τ and testing continues for an extra period $[\tau, T]$. During this phase, the faults are identified simultaneously by both the testing team and the end users. However, the efficiency of the testers in detecting the faults is higher compared to the users. This is because of the varied proficiency and testing skill differences between the two groups. Additionally, the intensity of testing depends on the time spent on testing per day. Usually, the time spent by users on the software is lower than the testing team whose only intention of using the software is to debug faults. Let $F^{*}_{\text{post}}(T - \tau)$ and $F^{**}_{\text{post}}(T - \tau)$ be the failure observation function of testers and users during post-release testing phase. Then, the expected number of faults identified during this phase can be expressed as

$$m_{\text{post}}(t - \tau) = (a - m_{\text{pre}}(\tau)) \int_{0}^{t-\tau} f^{*}_{\text{post}}(x)\bar{F}^{**}_{\text{post}}(x)\,\mathrm{d}x$$
$$+ (a - m_{\text{pre}}(\tau)) \int_{0}^{t-\tau} \bar{F}^{*}_{\text{post}}(x) f^{**}_{\text{post}}(x)\mathrm{d}x; \quad \tau < t \le T \tag{32.15}$$

$$m_{\text{post}}(t - \tau) = m^{*}_{\text{post}}(t - \tau) + m^{**}_{\text{post}}(t - \tau); \quad \tau < t \le T \tag{32.16}$$

where $m^{*}_{\text{post}}(t-\tau)$ and $m^{**}_{\text{post}}(t-\tau)$ denotes the faults identified by the testing team and the users, respectively; $f^{*}_{\text{post}}(x)$ and $f^{**}_{\text{post}}(x)$ represents the non-cumulative distribution function of tester's and users fault detection process, respectively: $\bar{F}^{*}_{\text{post}}(x) = 1 - F^{*}_{\text{post}}(x)$, and $\bar{F}^{**}_{\text{post}}(x) = 1 - F^{**}_{\text{post}}(x)$. In Eq. (32.15), the integral in the first component describes the probability of faults being detected by the testers and not by the users and similarly, the integral in the second denotes the probability of failures observed by the users and not by the testers. When exponential SRGM is used to model the failure observation phenomenon, then

$$m^{*}_{\text{post}}(t-\tau) = ae^{-b\tau}\left(1 - \mathrm{e}^{-(b+b')(t-\tau)}\right)(1/1+r) \tag{32.17}$$

$$m^{**}_{\text{post}}(t-\tau) = ae^{-b\tau}\left(1 - \mathrm{e}^{-(b+b')(t-\tau)}\right)(r/1+r) \tag{32.18}$$

where $1/1+r$ is the conditional probability of fault detection in the second phase by the testing team; b' is the fault detection rate of users.

Phase 3: Operational Phase $[T, T_{lc}]$

After the testing process is stopped at T, the faults are detected only by the users who report it to the testing team. The software testers then rectify the faults and send a patch to its users to update the system. The expected number of faults detected during this phase is given as

$$m_{\text{op}}(t-T) = (a - m_{\text{pre}}(\tau))\left(1 - \left(\int_{0}^{T-\tau} f^{*}_{\text{post}}(x)\bar{F}^{**}_{\text{post}}(x)\mathrm{d}x + \int_{0}^{T-\tau} \bar{F}^{*}_{\text{post}}(x) f^{**}_{\text{post}}(x)\mathrm{d}x\right)\right) F_{\text{op}}(t-T);\ T < t \le T_{lc} \tag{32.19}$$

where $F_{\text{op}}(t-T)$ denotes the fault detection function of users during the operational phase. Using exponential distribution function, the expected number of defects identified by the users is given as

$$m_{\text{op}}(t-T) = ae^{-b\tau}\mathrm{e}^{-(b+b')(T-\tau)}\left(1 - \mathrm{e}^{-b''(t-T)}\right) \tag{32.20}$$

where b'' is the rate parameter of user detection function in the operational phase.

Furthermore, to evaluate the two decision variables, namely, software release time and testing stop time, the cost functions that influence the optimal decision are considered in the optimization problem. Costs functions: testing cost, market opportunity cost, and software debugging cost during three phases have been identified as major cost components.

Testing cost: It includes the cost of resources such as CPU hours consumed and workforce involved in the testing and execution of the software system. As per the software engineering literature [16], the cost of testing is assumed as a linear function of testing duration.

Market Opportunity Cost: It involves the loss incurred by the firm due to the delay in the market entry of their software product. Market opportunity cost is an imperative factor that reflects the manipulation of the market by competitor firms. As per prior studies [27], the cost function is assumed a quadratic function of the software release time.

Faults Debugging Cost during Pre-release Testing Phase: It comprises the resources consumed by the testing team in removing the faults identified during this phase. This cost function is directly dependent on the number of faults detected [23].

Faults Debugging Cost during Post-release Testing Phase: It includes the cost associated with faults debugging after software release but before testing termination time. It is worth noting that when the software is in the user environment, various unanticipated cost components such as liability costs, user disapproval cost, revenue losses, and indirect costs due to damaged reputation increase the overall debugging cost. This cost component is linearly dependent on the faults identified by both the testers and users during this phase.

Faults Debugging Cost during Operational Phase: It comprises the resources consumed in debugging the faults reported by the users during the operational phase.

Thus, the overall cost function for the present release policy is given as

$$\begin{aligned} C(\tau, T) = C_1 T + C_2 \tau^2 + C_3 m_{\text{pre}}(\tau) + C_4 m^*_{\text{post}}(T - \tau) \\ + C_5 m^{**}_{\text{post}}(T - \tau) + C_6 m_{\text{op}}(T_{lc} - T) \end{aligned} \tag{32.21}$$

where C_1 is the testing cost per unit time; C_2 denotes the release time-dependent market opportunity cost; C_3 and C_4 are the cost of debugging a fault detected by the testing team in the pre-release and post-release testing period, respectively; C_5 and C_6 signifies the debugging cost of a fault reported by the customer in the post-release and post-testing period, respectively.

Going ahead, the reliability function of the software system can be defined. Reliability is the necessary attribute that influences the optimal decisions involved with software release time and testing termination time. The conditional reliability function is defined as [2]:

$$R(x|t) = \mathrm{e}^{-[m(t+x)-m(t)]} \tag{32.22}$$

where $R(x|0) = \mathrm{e}^{-m(x)}$ and $R(x|\infty) = 1$.

So, the reliability at release time τ is given as

$$R(x_1|\tau) = \exp(-(m(\tau + x_1) - m(\tau))) \tag{32.23}$$

The expected number of faults identified by time $(\tau + x_1)$ is expressed as

$$m(\tau + x_1) = aF_1(\tau) + a(1 - F_{pre}(\tau))$$
$$\left(\int_0^{\tau + x_1 - \tau} f^*_{post}(x) \overline{F}^{**}_{post}(x)\, dx + \int_0^{\tau + x_1 - \tau} \overline{F}^*_{post}(x) f^{**}_{post}(x) dx \right) \tag{32.24}$$

Hence, the reliability function at τ is given as

$$R(x_1|\tau) = \exp\left(-a(1 - F_{pre}(\tau)) \left(\int_0^{\tau + x_1 - \tau} f^*_{post}(x) \bar{F}^{**}_{post}(x)\, dx \right.\right.$$
$$\left.\left. + \int_0^{\tau + x_1 - \tau} \bar{F}^*_{post}(x) f^{**}_{post}(x) dx \right)\right) \tag{32.25}$$

Likewise, the reliability attribute at testing termination time, T is given as

$$R(x_2|T) = \exp(-(m(T + x_2) - m(T))) \tag{32.26}$$

The expected number of faults detected by time $(T + x_2)$ is given as

$$m(T + x_2) = m(T) + \left[a(1 - F_{pre}(\tau)) \left(1 - \left(\int_0^{T-\tau} f^*_{post}(x) \bar{F}^{**}_{post}(x)\, dx \right.\right.\right.$$
$$\left.\left.\left. + \int_0^{T-\tau} \bar{F}^*_{post}(x) f^{**}_{post}(x) dx \right)\right)\right] F_{op}(T + x_2 - T) \tag{32.27}$$

where $m(T) = aF_{pre}(\tau) + a(1 - F_{pre}(\tau)) \left(\int_0^{T-\tau} f^*_{post}(x).\overline{F}^{**}_{post}(x)\, dx \right.$
$\left. + \int_0^{T-\tau} \overline{F}^*_{post}(x) f^{**}_{post}(x) dx \right)$

Therefore, reliability function can be re-written as

$$R(x_2|T) = \exp\left(- \left[a(1 - F_{pre}(\tau)) \left(1 - \left(\int_0^{T-\tau} f^*_{post}(x) \overline{F}^{**}_{post}(x)\, dx \right.\right.\right.\right.$$

$$+\int_{0}^{T-\tau} \overline{F}^{*}_{\text{post}}(x) f^{**}_{\text{post}}(x)\mathrm{d}x \Bigg)\Bigg)\Bigg] F_{\text{op}}(T + x_2 - T)\Bigg) \quad (32.28)$$

Consequently, the optimal release time policy (Policy 5) of minimizing the total cost function subject to reliability requirement and budgetary constraint can be stated as

Policy 5:

$$\text{Minimize } C(\tau, T) = C_1T + C_2\tau^2 + C_3m_{\text{pre}}(\tau) + C_4m^{*}_{\text{post}}(T-\tau) + C_5m^{**}_{\text{post}}(T-\tau) + C_6m_{\text{op}}(T_{lc} - T) \quad (32.29)$$

$$\text{Subject to } R(x_1|\tau) \geq R_0,\ R(x_2|T) \geq R_1 \text{ and } C(\tau, T) \leq C_b \quad (32.30)$$

Numerical Example

To gain a better insight into the above cost model, we presented a numerical using the first release data set of Tandem computers [33]. The parameter values of exponential SRGM is $a = 130.30$ and $b = 0.083$. It is further assumed that the fault detection rate of testers is the same in both the testing phases. The efficiency of the users in identifying faults is low as compared to the testing team. Consider $b' = rb$ as the combined fault detection rate of the customers, where r is the ratio of fault detection rate under user's usage with respect to testers testing in the second phase $[\tau, T]$. Besides fault identification rate of users in the third phase, $[T,\ T_{lc}]$ is $b'' = sb$ where s is the ratio of fault detection rate under user's usage with respect to developers testing in this phase. It may be noted that $s \geq r$ because in the third phase user's base is increased; therefore, the chance of failure observation by the customer in the third phase is more.

To obtain the optimal result, following values of model parameters is considered: $r = 0.5$, $s = 0.6$, $T_{lc} = 100$, $C_1 = \$50$, $C_2 = \$5$, $C_3 = \$25$, $C_4 = \$350$, $C_5 = \$25$, $C_6 = \$350$, $x_1 = 0.1$, $x_2 = 0.1$, $R_0 = 0.80$, $R_1 = 0.85$, and $C_b = \$15000$. The optimal result of Policy 5 using the abovementioned parameter values is $\tau = 20.23$ weeks, $T = 40.48$ weeks and $C(\tau, T) = \$10,342.63$. However, when no post-testing is considered, that is, testing is stopped as soon as software is released, then the cost function becomes $C(\tau) = C_1\tau + C_2\tau^2 + C_3m_{\text{pre}}(\tau) + C_4m_{\text{op}}(T_{lc} - \tau)$. The optimal result in such a case is: $\tau = T = 28.11$ weeks and $C(\tau) = \$12490.11$. From the aforementioned results, it is evident that minimum cost is attained when post-testing is considered, i.e., it is more beneficial for a company to release software early and continue the testing process for an added period in the operational phase. The convexity plot of the objective function with and without post-release testing is illustrated using Fig. 32.3a and 32.3b, respectively. In addition, Table 32.1 summarizes the phase-wise description of failure observation.

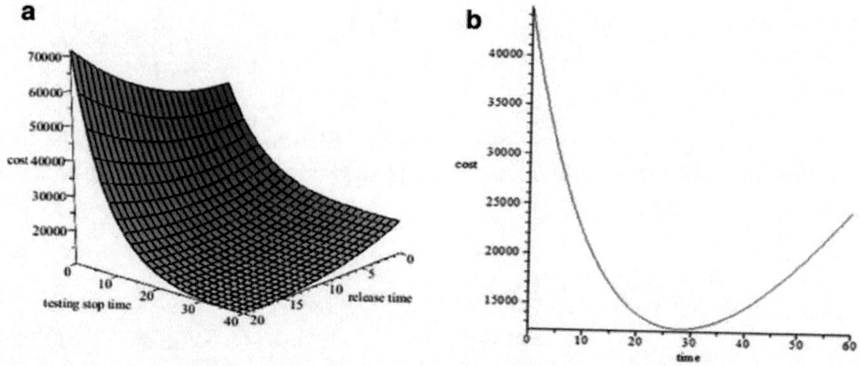

Fig. 32.3 **a** Convexity plot of cost function with post-release testing. **b** Convexity plot of cost function without post-release testing

Table 32.1 Phase-wise fault detection description under policy 5

Lifecycle phase		Mean value function	Number of faults identified
Pre-release testing phase $[0, \tau]$		$m_{pre}(\tau)$	105.89 (106 approx.)
Post-release testing phase $[\tau, T]$	Total	$m_{post}(T-\tau)$	22.35 (22 approx.)
	Tester	$m^{*}_{post}(T-\tau)$	14.90 (15 approx.)
	User	$m^{**}_{post}(T-\tau)$	7.45 (7 approx.)
Post-testing phase $[T, T_{lc}]$		$m_{op}(T_{lc}-T)$	1.85 (2 approx.)

32.4.2.2 Release Time Decision by Incorporating Change-Point and Post-Release Testing

Shrivastava et al. [34] proposed a release policy by incorporating the change-point concept in the software failure observation phenomenon. It has been established in the prior literature that developers intensify their testing strategy after certain time-point to attain the desired level of reliability early. The time instance at which fault detection rate alters due to change in the testing strategy is known as change-point. For modeling, the mean value function of fault detection, the software lifecycle is divided into four phases as illustrated in Fig. 32.4.

Phase 1: Pre-release Testing Phase before Change-point $[0, \delta)$

As described in the earlier section, during this period testers meticulously identify any faults lying dormant in the software system. The differential equation for failure observation phenomenon under this phase can be stated as

$$\frac{dm_{pre\_1}(t)}{dt} = \frac{f_{pre\_1}(t)}{1 - F_{pre\_1}(t)}\left(a - m_{pre\_1}(t)\right); \quad 0 \le t \le \delta \tag{32.31}$$

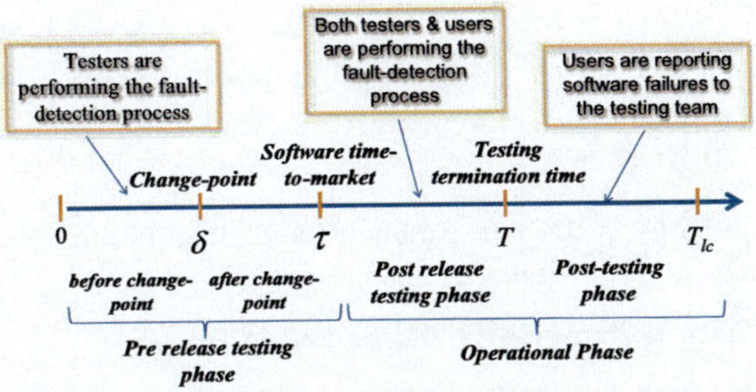

Fig. 32.4 Different phases of software life cycle with Change-point and Post-release testing

where $F_{\text{pre\_1}}(t)$ is the cumulative distribution function of the testers and δ represents the change-point. On solving above equation under the initial condition $t = 0$, $m_{\text{pre\_1}}(t) = 0$, following mean value function is obtained:

$$m_{\text{pre\_1}}(t) = aF_{\text{pre\_1}}(t); \quad 0 \le t < \delta \tag{32.32}$$

Now, if the fault detection phenomenon follows a logistic distribution function (Kapur et al. [1]), then mean value function becomes

$$m_{\text{pre\_1}}(t) = a\left(\frac{1 - e^{-b_1 t}}{1 + \beta_1 e^{-b_1 t}}\right); \quad 0 \le t < \delta \tag{32.33}$$

where b_1 and β_1 are parameters of logistic distribution function.

Phase 2: Post-release Testing Phase after Change-point $[\delta,\ \tau]$

After change-point, δ is the rate of fault detection by the testing team modifies. Thus, the differential equation for fault identification during this phase becomes

$$\frac{dm_{\text{pre\_2}}(t-\delta)}{dt} = \frac{f_{\text{pre\_2}}(t)}{1 - F_{\text{pre\_2}}(t)}\left(a(1 - F_{\text{pre\_1}}(\delta)) - m_{\text{pre\_2}}(t-\delta)\right); \quad \delta < t \le \tau \tag{32.34}$$

where $F_{\text{pre\_2}}(t)$ is the distribution function for fault detection after change-point. Above equation can be further solved under the boundary condition, i.e., at $t = \delta,\ \ m_{\text{pre\_2}}(t-\delta) = 0$, to get the following closed-form solution:

$$m_{\text{pre\_2}}(t-\delta) = a\left(1 - F_{\text{pre\_1}}(\delta)\right)\left[1 - \frac{\left(1 - F_{\text{pre\_2}}(t)\right)}{\left(1 - F_{\text{pre\_2}}(\delta)\right)}\right]; \quad \delta < t \le \tau \tag{32.35}$$

Using logistic distribution function to model the fault detection phenomenon, mean value function of failure observation after change-point becomes

$$m_{\mathrm{pre\_2}}(t-\delta)=a\left(1-\left(\frac{1-\mathrm{e}^{-b_1\delta}}{1+\beta_1\mathrm{e}^{-b_1\delta}}\right)\right)\left(1-\frac{1-\left(\frac{1-\mathrm{e}^{-b_2t}}{1+\beta_2\mathrm{e}^{-b_2t}}\right)}{1-\left(\frac{1-\mathrm{e}^{-b_2\delta}}{1+\beta_2\mathrm{e}^{-b_2\delta}}\right)}\right);\quad \delta<t\le\tau \tag{32.36}$$

where b_2 is the rate parameter and β_2 is the learning parameter of tester's fault detection function after change-point δ.

Phase 3: Post-release Testing Period $[\tau,\ T]$

During this phase, both users and testers are observing software failures. Therefore, the expected number of faults identified in post-release testing phase will be the sum of faults detected by the testers and reported by the users. However, the debugging of faults is done by the testers alone. After successfully rectifying the failure, the testing team sends a patch to its users to update their software system. It is considered that a fixed portion say λ, of the remaining faults from the previous phase will be detected by the testers and remaining (1-λ) will be identified by the users who then immediately report it to the developers for correcting it. Thus, the expected number of faults detected during this phase by testers and users, respectively, is given as

$$m^{*}_{\mathrm{post}}(t-\tau)=\lambda(a-m(\tau))F^{*}_{\mathrm{post}}(t-\tau);\quad \tau<t\le T \tag{32.37}$$

$$m^{**}_{\mathrm{post}}(t-\tau)=(1-\lambda)(a-m(\tau))F^{**}_{\mathrm{post}}(t-\tau);\quad \tau<t\le T \tag{32.38}$$

where $m(\tau)=m_{\mathrm{pre\_1}}(\delta)+m_{\mathrm{pre\_2}}(\tau-\delta)$ is the expected number of faults debugged during the pre-release testing phase; $F^{*}_{\mathrm{post}}(t-\tau)$ and $F^{**}_{\mathrm{post}}(t-\tau)$ is the cumulative distribution function of testers and users, respectively, during the post-release testing phase.

Phase 4: Post-testing Stop Time Phase $[T,\ T_{lc}]$

When developers stop the testing process at time T, the task of fault detection entirely shifts to the users. Therefore, in the post-testing phase, clients may encounter failure due to undetected faults from the previous phases and will report it to the testing team for removal. This process of fault detection will continue until the end of the software lifecycle. Thus, the instantaneous fault detection by users in this phase becomes

$$\frac{\mathrm{d}m_{\mathrm{op}}(t-T)}{\mathrm{d}(t-T)}=\frac{f_{\mathrm{op}}(t-T)}{1-F_{\mathrm{op}}(t-T)}\left(a-m_3(T)-m_{\mathrm{op}}(t-T)\right);\quad T<t\le T_{lc} \tag{32.39}$$

where $m_3(T)=m_{\mathrm{pre\_1}}(\delta)+m_{\mathrm{pre\_2}}(\tau-\delta)+m^{*}_{\mathrm{post}}(T-\tau)+m^{**}_{\mathrm{post}}(T-\tau)$ denotes the expected number of faults detected by time T.

On further solving Eq. (32.39) using the initial condition, $t=T,\ m_{\mathrm{op}}(t-T)=0$, the following solution is obtained:

$$m_{op}(t - T) = (a - m_3(T))F_{op}(t - T); \quad T < t \le T_{lc} \tag{32.40}$$

Besides, the cost components involved in the present optimal policy is given as

$$C(\tau, T) = C_1 T + C_2 \tau^2 + C_3 m_{pre\_1}(\delta) + C_4 m_{pre\_2}(\tau - \delta) + C_5 m^*_{post}(T - \tau) + C_6 m^{**}_{post}(T - \tau) + C_7 m_{op}(T_{lc} - T) \tag{32.41}$$

where C_1 is the testing cost per unit time; C_2 denotes the release time-dependent market opportunity cost; C_3 and C_4 are the cost of debugging a fault detected by the testing team in the pre-release testing period before and after the change-point, respectively; C_5 and C_6 signifies the debugging cost of a fault detected by the tester and user respectively in the post-release testing period; C_7 represents the cost of removing the fault reported by the user during post-testing period.

Mathematically, the optimization problem for the cost minimization problem under the budgetary constraint (Policy 6) is stated as

Policy 6:

$$\text{Minimize } C(\tau, T) = C_1 T + C_2 \tau^2 + C_3 m_{pre\_1}(\delta) + C_4 m_{pre\_2}(\tau - \delta) + C_5 m^*_{post}(T - \tau) + C_6 m^{**}_{post}(T - \tau) + C_7 m_{op}(T_{lc} - T) \tag{32.42}$$

$$\text{Subject to } C(\tau, T) \le C_b \tag{32.43}$$

where C_b is the total budget available to the software firm for software development.

Numerical Example

The actual failure data of Tandem computers from the testing period is used to estimate the parameter of the described reliability growth model. The change-point for this data occurs at the 8th week of the testing period. The parameter estimation is carried out using nonlinear regression performed using SPSS software. The estimated value of parameters are $a = 104$, $b_1 = 0.02$, $b_2 = 0.2$, $\beta_1 = 1.2$, and $\beta_2 = 2.1$. Furthermore, it is assumed that the fault detection rate of tester's increases by 50% after software release, i.e., the hazard rate after τ will be $\frac{b_2}{1+\beta_2 e^{-b_2 t}} + \frac{1}{2}\frac{b_2}{1+\beta_2 e^{-b_2 t}} = \frac{3}{2}\frac{b_2}{1+\beta_2 e^{-b_2 t}}$ and the corresponding distribution function will be $F^*_{post}(t - \tau) = \left(1 - \left(\frac{1+\beta_2}{\beta_2 + e^{b_2(t-\tau)}}\right)^{3/2}\right)$. Besides, it is assumed that the failure observation rate of clients is 40% of that of the tester's fault detection rate, i.e., hazard rate will be $\frac{2}{5}\left(\frac{b_2}{1+\beta_2 e^{-b_2 t}}\right)$ and the corresponding distribution function will be $F^{**}_{post}(t - \tau) = \left(1 - \left(\frac{1+\beta_2}{\beta_2 + e^{b_2(t-\tau)}}\right)^{2/5}\right)$. Moreover, the user's distribution function of fault detection remains same in the post-testing phase, i.e., $F^{**}_{post}(t - \tau) = F_{op}(t - T)$.

In addition, cost parameters considered for the given optimization problem is $C_1 = \$100$, $C_2 = \$10$, $C_3 = \$40$, $C_4 = \$70$, $C_5 = \$70$, $C_6 = \$150$, $C_7 = \$200$ and $C_B = \$15000$. Also, the efficiency of testers in identifying the faults is more

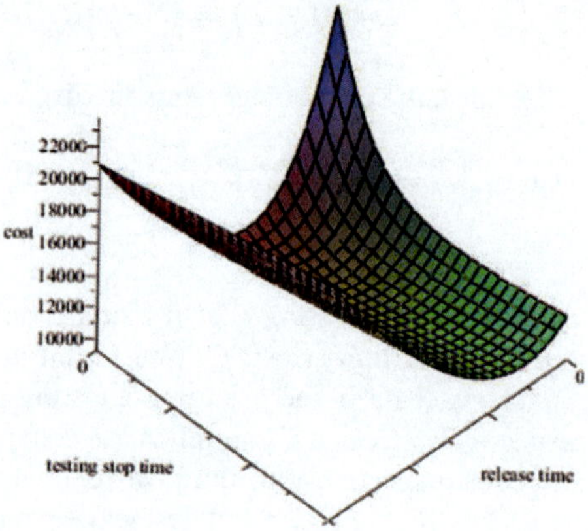

Fig. 32.5 Convexity plot of the cost function

Table 32.2 Phase-wise faults detection description under policy 6

Lifecycle phase	Mean value function	Number of faults detected (Approx.)
Pre-release phase	$m_{pre\_1}(\delta)$	8 (by testers before change-point)
	$m_{pre\_1}(\tau - \delta)$	65 (by testers after change-point)
Post-release phase	$m^{*}_{post}(T - \tau)$ $m^{**}_{post}(T - \tau)$	17 (by testers after release and before testing stops)
	$m^{**}_{post}(T - \tau)$	6 (reported by users for removal)
Post-testing phase	$m_{op}(T_{lc} - T)$	8 (reported by users for removal)

as compared to the users, i.e., $\lambda = 0.6$. The optimal result of Release Policy 6 obtained using the above parameter values is $\tau = 12$ weeks, $T = 25$ weeks with minimum cost as $C(\tau, T) = \$13023$. The pictorial representation of the objective function of Policy 6 is presented in Fig. 32.5. Additionally, Table 32.2 summarizes the phase-wise description of the number of faults detected in the software by testers and users.

32.4.2.3 Release Time Decision When Change-Point and Release Time Coincides

Kapur et al. [28] developed the failure observation phenomenon by considering software release time and testing stop time as two distinct time-points. According to their approach, software lifecycle is categorized into three phases: pre-release testing phase, post-release testing phase, and post-testing phase. Figure 32.6 illustrates the different phases of software lifecycle.

The modeling framework describing the fault detection process under different phases of software lifecycle is provided below.

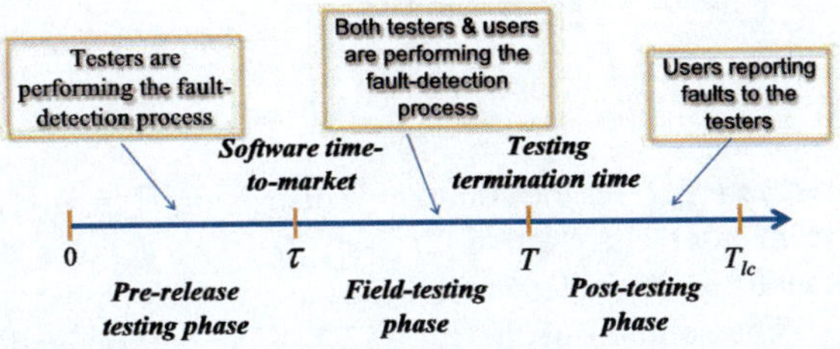

Fig. 32.6 Different phases of failure occurrence phenomenon

Phase 1: Pre-release Testing Period $[0, \tau)$

The differential equation describing the instantaneous fault detection during pre-release testing period is given as

$$\frac{\mathrm{d}m_{\mathrm{pre}}(t)}{\mathrm{d}t} = \frac{f_{\mathrm{pre}}(t)}{1 - F_{\mathrm{pre}}(t)}\left(a - m_{\mathrm{pre}}(t)\right) \tag{32.44}$$

where $F_{\mathrm{pre}}(t)$ is the cumulative distribution function of fault detection by the testing team before change-point τ.

On solving Eq. (32.44), using the boundary condition at $t = 0$, $m_{\mathrm{pre}}(t) = 0$, following closed-form solution representing the expected number of faults identified by time t is obtained:

$$m_{\mathrm{pre}}(t) = aF_{\mathrm{pre}}(t); \quad 0 \le t < \tau \tag{32.45}$$

Equation (32.45) describes the mean value function of failure observation in pre-release testing phase.

Furthermore, the fault detection function is considered to follow an S-shaped curve. Therefore, the NHPP-based SRGM model with delayed S-shaped curve is used to represent the fault identification process with distribution function [8], i.e.,

$$m_{\mathrm{pre}}(t) = a\left(1 - (1 + b_1 t)\mathrm{e}^{-b_1 t}\right) \tag{32.46}$$

where b_1 is the rate parameter of developers' fault detection rate before change-point τ.

Phase 2: Field Testing or Post-release Testing Phase $[\tau, T]$

During this phase, both testing team and clients are detecting remaining defects in the system that left undetected in the previous phase of the software lifecycle. Let λ be the proportion of the remaining faults identified by the developers during post-release phase and $(1 - \lambda)$ be the failures reported by the users in this phase. Therefore, the expected number of faults detected by the developers during this phase is expressed as

$$\frac{\mathrm{d}m^{*}_{post}(t-\tau)}{\mathrm{d}t}=\frac{f^{*}_{post}(t)}{1-F^{*}_{post}(t)}\left(\lambda a(1-F_{pre}(\tau))-m^{*}_{post}(t-\tau)\right);\quad \tau<t\leq T \tag{32.47}$$

where $F^{*}_{post}(t)$ is the cumulative distribution function of fault detection by the testing team after change-point τ; $m^{*}_{post}(t-\tau)$ denotes the expected number of faults identified by the testers in post-release testing phase.

The solution of the Eq. (32.47) can be obtained by solving it using the initial condition $t=\tau,\quad m^{*}_{post}(t-\tau)=0$:

$$m^{*}_{post}(t-\tau)=\lambda a\big(1-F_{pre}(\tau)\big)\left[1-\frac{\left(1-F^{*}_{post}(t)\right)}{\left(1-F^{*}_{post}(\tau)\right)}\right];\quad \tau<t\leq T \tag{32.48}$$

Equation (32.48) describes the mean value function of defects identified by the testing team during post-release testing phase. Now, the differential equation expressing the instantaneous failure observation by the clients in this phase is given as

$$\frac{\mathrm{d}m^{**}_{post}(t-\tau)}{\mathrm{d}(t-\tau)}=\frac{f^{**}_{post}(t-\tau)}{1-F^{**}_{post}(t-\tau)}\left((1-\lambda)a(1-F_{pre}(\tau))-m^{**}_{post}(t-\tau)\right) \tag{32.49}$$

where $F^{**}_{post}(t)$ is the cumulative distribution function of fault detection by users; $m^{**}_{post}(t-\tau)$ denotes the expected number of faults identified by the users in post-release testing phase.

Using the boundary condition, $t=\tau,\quad m^{**}_{post}(t-\tau)=0, F^{**}_{post}(t-\tau)=0$, Eq. (32.51) can be solved to get the following mean value solution:

$$m^{**}_{post}(t-\tau)=(1-\lambda)a\big(1-F_{pre}(\tau)\big)F^{**}_{post}(t-\tau);\quad \tau<t\leq T \tag{32.50}$$

Equation (32.50) provides the mean value function of faults detected by the users in the post-release testing phase. If failure observation phenomenon follows a delayed S-shaped curve then,

$$m^{*}_{post}(t-\tau)=\lambda a(1+b_1\tau)\mathrm{e}^{-b_1\tau}\left[1-\left(\frac{1+b_2t}{1+b_2\tau}\right)\mathrm{e}^{-b_2(t-\tau)}\right];\quad \tau<t\leq T \tag{32.51}$$

where b_2 is the defect detection parameter for tester's after change-point τ.

$$m^{**}_{post}(t-\tau)=(1-\lambda)a(1+b_1\tau)\mathrm{e}^{-b_1\tau}\left(1-(1+b_3(t-\tau))\mathrm{e}^{-b_3(t-\tau)}\right);\quad \tau<t\leq T \tag{32.52}$$

where b_3 is the users' rate parameter of fault detection function.

Phase 3: Post-testing Period $[T,\ T_{lc}]$

After the testing has been stopped at time T, the users may still encounter failure due to undetected faults. Thus, during post-testing period, users report the failure to the testing team for correction. The fault detection rate of users remains the same as that in the previous phase. Accordingly, the differential equation for the failure observation phenomenon during this phase becomes

$$\frac{dm_{op}(t-T)}{d(t-\tau)} = \frac{f^{**}_{post}(t-\tau)}{1-F^{**}_{post}(t-\tau)}\left(\mathrm{A}-m_{op}(t-T)\right);\quad T < t \le T_{lc} \tag{32.53}$$

where $\mathrm{A} = a(1-F_{pre}(\tau))\left(1-\lambda\left\{1-\frac{(1-F^{*}_{post}(T))}{(1-F^{*}_{post}(\tau))}\right\}-(1-\lambda)F^{**}_{post}(T-\tau)\right)$ denotes the amount of unidentified faults from the previous phases of software lifecycle.

Equation (32.53) can be further solved using the initial condition, $t = T$, $m_{op}(t-T) = 0$ to obtain the following solution:

$$m_{op}(t-T) = a(1-F_{pre}(\tau))\begin{pmatrix} 1-\lambda\left\{1-\dfrac{\left(1-F^{*}_{post}(T)\right)}{\left(1-F^{*}_{post}(\tau)\right)}\right\} \\ -(1-\lambda)F^{**}_{post}(T-\tau)\end{pmatrix}$$

$$\left[1-\left(\frac{1-F^{**}_{post}(t-\tau)}{1-F^{**}_{post}(T-\tau)}\right)\right];\quad T < t \le T_{lc} \tag{32.54}$$

Equation (32.54) describes the expected number of faults identified by the users in post-testing period. Now, if the failure observation phenomenon follows delayed S-shaped distribution function, then

$$m_{op}(t-T) = a(1+b_1\tau)e^{-b_1\tau}\begin{pmatrix} 1-\lambda\left[1-\left(\dfrac{1+b_2T}{1+b_2\tau}\right)e^{-b_2(T-\tau)}\right] \\ -(1-\lambda)\left(1-(1+b_3(T-\tau))e^{-b_3(T-\tau)}\right)\end{pmatrix}$$

$$\left(1-\left(\frac{1+b_3(t-\tau)}{1+b_3(T-\tau)}\right)e^{-b_3(t-T)}\right) \tag{32.55}$$

Going ahead, Table 32.3 summarizes the functional form of major cost components, which are essential to evaluate the software release time and the testing duration.

Thus, the overall cost function is expressed as

$$\begin{aligned} C(\tau,T) = {} & C_1T + C_2\tau^2 + C_3m_{pre}(\tau) + C_4m^{*}_{post}(T-\tau) \\ & + C_5m^{**}_{post}(T-\tau) + C_6m_{op}(T_{lc}-T) \end{aligned} \tag{32.56}$$

Table 32.3 Cost components for the optimization problem

Cost components	Cost function
Testing cost	$C_{testing}(t) = C_1 T$
Market opportunity cost	$C_{market\_opp}(t) = C_2 \tau^2$
Faults debugging cost during pre-release testing phase	$C_{phase\_I}(t) = C_3 m_{pre}(\tau)$
Faults debugging cost during post-release testing phase	$C_{Phase\_II}(t) = C_4 m^{*}_{post}(T - \tau) + C_5 m^{**}_{post}(T - \tau)$
Faults debugging cost during post- testing phase	$C_{Phase\_III}(t) = C_6 m_{op}(T_{lc} - T)$

Optimal Release Time Decisions Using MAUT

Multi-attribute utility theory is a well-established methodology to solve the optimization problem concerning multiple factors with a conflicting objective function to provide the best solution [35]. In software engineering, the MAUT has been progressively used to assess the trade-off between the conflicting factors for developing the optimal release time policies [36]. Kapur et al. [29] considered two vital attributes, namely, cost and reliability functions to evaluate the optimal software release time and testing termination time. This technique comprises the following four steps.

Step 1: Selection of Suitable Attributes

The release time strategy should be computed based on crucial attributes. These attributes must be measurable and have practical relevance. The prime concern of software engineers is to deliver a reliable and secure software system to its clients. Thus, *reliability* is the important attribute that influence the optimal decision of software release. Thus, the first attribute included in the proposed optimization problem is

$$\text{Maximize}\, R(x|\tau, T) = e^{-[m(\tau+x_1)-m(\tau)]-[m(T+x_2)-m(T)]} \tag{32.57}$$

where x_1 and x_2 are small time durations; $m(\tau + x_1)$ and $m(T + x_2)$ denotes the expected number of faults detected by time $(\tau + x_1)$ and $(T + x_2)$, respectively.

The mean value function of fault detection in small interval $[\tau, \ \tau + x_1]$ is given by

$$m(\tau + x_1) - m(\tau) = \lambda a\big(1 - F_{pre}(\tau)\big)\left[1 - \left(\frac{1 - F^{*}_{post}(\tau + x_1)}{1 - F^{*}_{post}(\tau)}\right)\right] + (1-\lambda)a\big(1 - F_{pre}(\tau)\big)F^{**}_{post}(\tau + x_1 - \tau) \tag{32.58}$$

Similarly, the mean value function of fault detection in small interval $[T, \ T + x_2]$ is given as

$$m(T + x_2) - m(T) = a\big(1 - F_{\text{pre}}(\tau)\big)\left(1 - \lambda\left\{1 - \frac{\big(1 - F^{*}_{\text{post}}(T)\big)}{\big(1 - F^{*}_{\text{post}}(\tau)\big)}\right\}\right.$$
$$\left. - (1-\lambda)F^{**}_{\text{post}}(T-\tau)\right)\left[1 - \left(\frac{1 - F^{**}_{\text{post}}(T + x_2 - \tau)}{1 - F^{**}_{\text{post}}(T-\tau)}\right)\right] \quad (32.59)$$

The second critical attribute for strategic release time decision is the cost function. The determination of cost budget is essential for the software producers to develop good quality software at minimal cost. Therefore, cost function for the given problem is given as

$$\text{Minimize } C = \frac{C(\tau,\ T)}{C_b} \quad (32.60)$$

where $C(\tau, T) = C_1T + C_2\tau^2 + C_3m_{\text{pre}}(\tau) + C_4m^{*}_{\text{post}}(T-\tau) + C_5m^{**}_{\text{post}}(T-\tau) + C_6m_{\text{op}}(T_{lc} - T)$ and C_b is the total budget allocated for fault debugging process.

Step 2: Elicitation of SAUF for each Attribute

Utility functions are employed to describe the goal of each attribute. Single Attribute Utility Function (SAUF) gives the aspiration level of management for each attribute. Two most commonly used SAUF functional forms are linear and exponential. For the present problem, linear function is applied to represent the utility function for both attributes. The elicitation of SAUF for each attribute is carried out based on the management decision. This process requires a subjective assessment and may not be specifically precise. The utility function of two attributes, namely, reliability and cost function are expressed as

$$u(C) = l_c + u_cC \text{ and } u(R) = l_r + m_rR \quad (32.61)$$

Additionally, the utility function is bounded with the best, $u(y^{\text{best}}) = 1$ and the worst, $u(y^{\text{worst}}) = 0$ value. Furthermore, the bounds are calculated based on the management and decision-maker's aspirations:

(a) For the reliability function, minimum 60% of the faults should be detected and maximum of 100% must be identified.
(b) For the cost function, the minimum budget prerequisite is 90% and the maximum requirement is 100%.

Therefore, the bounds for these attributes is $C^{\text{worst}} = 0.9$, $C^{\text{best}} = 1$, $R^{\text{worst}} = 0.6$ and $R^{\text{best}} = 1$. Under these boundary conditions, the SAUF for the two attributes takes the following functional form:

$$U(C) = 10C - 9 \text{ and } U(R) = 2.5R - 1.5 \quad (32.62)$$

Step 3: Estimation of Weight Parameters

The weight or scale parameter describes the relative importance of an attribute over another. For deciding which attribute should be given priority, the lottery method or management's discretion is taken into consideration when the number of attributes is less. For the present study, the weight has been allotted on the management's judgment. The value of the weight parameter lies between zero and one, where the value closer to 1 denotes the higher significance. Moreover, the sum of weight parameters should be equal to 1, i.e., $W_r + W_c = 1$. For the present problem, the weight given by the software development management to the reliability attribute is $W_r = 0.6$ and consequently weight assigned to cost attribute is $W_c = 0.4$.

Step 4: Formulation of MAUF

Finally, the Multi-attribute Utility Function (MAUF) is developed by arithmetically summing all the Single Attribute Utility Functions (SAUF) using the weight parameters. Therefore, the MAUF (U) (Policy 7) for the proposed framework is expressed as

Policy 7:

$$\text{Maximize } U(R, C) = W_r U(R) - W_C U(C) \tag{32.63}$$

where $W_r + W_c = 1$ where $U(R)$ and $U(C)$ represent the single utility functions for reliability and cost attribute, respectively. In the present study, the focus of the software producers is to maximize the overall utility function. Therefore, the utility of the cost attribute is multiplied by a negative sign to synchronize it with the reliability attribute and to obtain the maximum value of the MAUF. After substituting the values from previous steps, the MAUF function can be re-written as

$$\text{Maximize } U(R, C) = 0.6 \times (2.5R - 1.5) - 0.4 \times (10C - 9) \tag{32.64}$$

where $w_R + w_c = 1$ and $C(\tau,\ T)/C_b \leq 1$

Numerical Example

In this section, the practical applicability of the proposed problem is illustrated through an example by using the actual fault discovery data. The parameters of the pre-release testing phase are estimated by fitting the model to the actual data of the second release of Tandem computers [33]. The estimated parameter values are $a = 127.3989$ and $b_1 = 0.241689$. Besides, it has been considered that the efficiency of testers in detecting the defects increases by 50% after the change-point. Therefore, the fault detection rate of testers during post-release testing phase is $b_2 = 0.3625335$. In addition, the ability of customers in identifying the fault is less as compared to the testers. Therefore, the fault detection rate of users is taken as 60% of that of developers, i.e., $b_3 = 0.1450134$. The rest of the model parameter values are set based on the previous studies: $C_1 = \$100,\ C_2 = \$16,\ C_3 = \$40,\ C_4 = \$70,\ C_5 = \$120,$ $C_6 = \$150,\ C_b = \$23,500,\ x_1 = 2,\ x_2 = 2,\ \lambda = 0.6,$ and $T_{lc} = 100$ *weeks*

Table 32.4 Optimal results

Release policies	$U(\tau^*,T^*)$	τ^* *(in weeks)*	T^* *(in weeks)*
Recent release time policy	0.844	10.231	16.864
Traditional release time policy	0.727	12.500	–

The formulated MAUF (Policy 7) is solved using the parameter values to obtain the optimal results. Besides, the developed release time policy is compared with the traditional release time policy with no post-release testing. Under conventional release time decision, reliability attribute is

$$\text{Maximize}\, R(x|\tau) = e^{-[m(\tau+x_1)-m(\tau)]} \tag{32.65}$$

where τ is the software release time as well as testing stop time under conventional release time policy.

In addition, the cost attribute under traditional release time policy takes the following functional form:

$$\text{Minimize}\, C = C(\tau)\big/ C_b \tag{32.66}$$

where $C(\tau) = C_1\tau + C_2\tau^2 + C_3 m_{\text{pre}}(\tau) + C_6 m_{\text{op}}(T_{lc} - \tau)$ is the total cost components associated with software development.

The optimal results for both release time policy are summarized in Table 32.4. The concavity plot of the utility functions under two release time policies is provided in Figs. 32.7a, b. It is evident from the findings that maximum utility is attained when post-release testing is performed. To maximize the reliability of the system with the aim of minimum cost consumption, it is suggested that firms should expedite the software release for commercial use but continue to conduct the testing process in the field environment for a specific period. By following the modern release time strategy, the company benefits in two ways. Firstly, by speeding up the software release, they will be able to avoid the manipulation of the market by their competitors. Secondly, the continuous testing process in the user environment with better efficiency and the client's contribution to failure observation improves the reliability of the software system and thus satisfies the customer requirements.

32.5 Concluding Remarks

In today's highly competitive market, every firm seeks to attain maximum market share by satisfying the client's requirement of a high-quality software system. Therefore, the reliability of software products holds vital importance for software engineers. Moreover, delivering the software on time with minimum development cost is

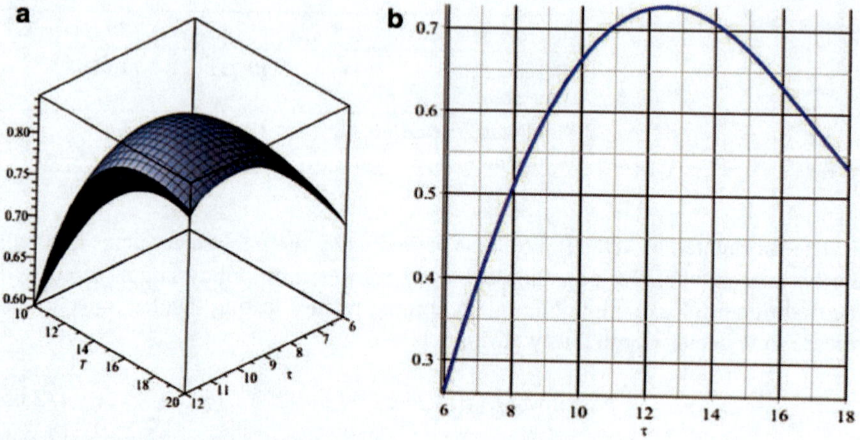

Fig. 32.7 **a** Utility function plot under recent release time policy. **b** Utility function plot under traditional release time policy

imperative for developers to avoid manipulation of the market by its competitors. The release time decisions are examined using either continuous- or discrete-time models. The former kind makes use of the execution time (i.e., CPU time) or calendar time to express the software failure observation phenomena. The second category utilizes the number of test cases (i.e., computer test run) executed as a unit for measuring the testing process. The focus of this chapter is to discuss various release time polices carried out using continuous-time models and are suggested by the researchers and practitioners from the past many decades. In this study, the Software Release Time Decisions (SRTD) are classified under two categories, specifically, conventional or traditional release time policy and recent or modern release time policy. Earlier, software producers followed a policy of terminating the testing process as soon as the software is released for commercial purposes. However, according to a new perspective on software release time, software should be released early to capture the market and continue the testing process for an additional period to improve the reliability of the systems. In this chapter, seven release time polices are described using the numerical illustration. The practical application of all the release polices is established by fitting the model to the real-life failure data. Findings of optimal release policies 5, 6, and 7 indicate that it is beneficial for a firm to follow recent release time decisions to maximize the reliability of the system and minimize the overall cost function.

References

1. Lee, S. H., Lee, S. J., Shin, S. M., Lee, E. C., & Kang, H. G. (2020). Exhaustive testing of safety-critical software for reactor protection system. *Reliability Engineering & System Safety, 1*(193), 106667.
2. Kapur, P. K., Pham, H., Gupta, A., & Jha, P. C. (2011). *Software reliability assessment with OR applications*. London: Springer.
3. Fonseca, L. M. (2015). ISO 9001 quality management systems through the lens of organizational culture. *Calitatea, 16*(148), 54.
4. Ivanov, V., Reznik, A., & Succi, G. (2018). Comparing the reliability of software systems: A case study on mobile operating systems. *Information Sciences, 1*(423), 398–411.
5. Kapur, P. K., Kumar, S., & Garg, R. B. (1999) Contributions to hardware and software reliability. *World Scientific*.
6. Jabeen, G., Luo, P., & Afzal, W. (2019). An improved software reliability prediction model by using high precision error iterative analysis method. *Software Testing, Verification and Reliability, 29*(6–7), e1710.
7. Goel, A. L., & Okumoto, K. (1979). Time-dependent error-detection rate model for software reliability and other performance measures. *IEEE Transactions on Reliability, 28*(3), 206–211.
8. Yamada, S., Ohba, M., & Osaki, S. (1983). S-shaped reliability growth modeling for software error detection. *IEEE Transactions on Reliability, 32*(5), 475–484.
9. Kapur, P. K., & Garg, R. B. (1992). A software reliability growth model for an error-removal phenomenon. *Software Engineering Journal, 7*(4), 291–294.
10. Kapur, P. K., Anand, S., Yamada, S., & Yadavalli, V. S. (2009). Stochastic differential equation-based flexible software reliability growth model. *Mathematical Problems in Engineering*.
11. Li, Q., & Pham, H. (2017). NHPP software reliability model considering the uncertainty of operating environments with imperfect debugging and testing coverage. *Applied Mathematical Modelling, 1*(51), 68–85.
12. Nagaraju, V., Fiondella, L., & Wandji, T. (2019). A heterogeneous single changepoint software reliability growth model framework. *Software Testing, Verification and Reliability., 29*(8), e1717.
13. Yamada, S., Ohba, M., & Osaki, S. (1984). S-shaped software reliability growth models and their applications. *IEEE Transactions on Reliability, 33*(4), 289–292.
14. Huang, C. Y., & Kuo, S. Y. (2002). Analysis of incorporating logistic testing-effort function into software reliability modeling. *IEEE Transactions on Reliability, 51*(3), 261–270.
15. Kapur, P. K., Goswami, D. N., Bardhan, A., & Singh, O. (2008). Flexible software reliability growth model with testing effort dependent learning process. *Applied Mathematical Modelling, 32*(7), 1298–1307.
16. Pham, H., & Zhang, X. (2003). NHPP software reliability and cost models with testing coverage. *European Journal of Operational Research, 145*(2), 443–454.
17. Huang, C. Y., & Lyu, M. R. (2011). Estimation and analysis of some generalized multiple change-point software reliability models. *IEEE Transactions on Reliability, 60*(2), 498–514.
18. Kapur, P. K., Gupta, A., Shatnawi, O., & Yadavalli, V. S. (2006). Testing effort control using flexible software reliability growth model with change point. *International Journal of Performability Engineering, 2*(3), 245–263.
19. Kapur, P. K., Pham, H., Anand, S., & Yadav, K. (2011). A unified approach for developing software reliability growth models in the presence of imperfect debugging and error generation. *IEEE Transactions on Reliability, 60*(1), 331–340.
20. Wang, J., Wu, Z., Shu, Y., & Zhang, Z. (2015). An imperfect software debugging model considering log-logistic distribution fault content function. *Journal of Systems and Software, 1*(100), 167–181.
21. Xie, M., & Yang, B. (2003). A study of the effect of imperfect debugging on software development cost. *IEEE Transactions on Software Engineering, 29*(5), 471–473.

22. Zhao, J., Liu, H. W., Cui, G., & Yang, X. Z. (2006). Software reliability growth model with change-point and environmental function. *Journal of Systems and Software, 79*(11), 1578–1587.
23. Okumoto, K., & Goel, A. L. (1979). Optimum release time for software systems based on reliability and cost criteria. *Journal of Systems and Software, 1*(1), 315–318.
24. Yamada, S., & Osaki, S. (1987). Optimal software release policies with simultaneous cost and reliability requirements. *European Journal of Operational Research, 31*(1), 46–51.
25. Kapur, P. K., & Garg, R. B. (1991). Optimal release policies for software systems with testing effort. *International Journal of Systems Science, 22*(9), 1563–1571.
26. Arora, A., Caulkins, J. P., & Telang, R. (2006). Research note—Sell first, fix later: Impact of patching on software quality. *Management Science, 52*(3), 465–471.
27. Jiang, Z., Sarkar, S., & Jacob, V. S. (2012). Postrelease testing and software release policy for enterprise-level systems. *Information Systems Research, 23*(3-part-1), 635–657.
28. Kapur, P. K., Panwar, S., Singh, O., & Kumar, V. (2019). Joint optimization of software time-to-market and testing duration using multi-attribute utility theory. *Annals of Operations Research, 2*, 1–28.
29. Kapur, P. K., Panwar, S., Singh, O., & Kumar, V. (2019). Joint release and testing stop time policy with testing-effort and change point. In *Risk Based Technologies* (pp. 209-222). Singapore: Springer.
30. Kapur, P. K., Shrivastava, A. K., & Singh, O. (2017). When to release and stop testing of a software. *Journal of the Indian Society for Probability and Statistics., 18*(1), 19–37.
31. Singh, O., Panwar, S., & Kapur, P. K. (2020). Determining software time-to-market and testing stop time when release time is a change-point. *International Journal of Mathematical, Engineering and Management Sciences., 5*(2), 208–224.
32. Yamada, S. (2014). *Software reliability modeling: Fundamentals and applications*. Tokyo: Springer.
33. Wood, A. (1996). Predicting software reliability. *Computer, 29*(11), 69–77.
34. Shrivastava, A. K., Kumar, V., Kapur, P. K., & Singh, O. (2020). Software release and testing stop time decision with change point. *International Journal of System Assurance Engineering and Management*.
35. Keeney, R. L. (1971). Utility independence and preferences for multiattributed consequences. *Operations Research, 19*(4), 875–893.
36. Li, X., Xie, M., & Ng, S. H. (2012). Multi-objective optimization approaches to software release time determination. *Asia-Pacific Journal of Operational Research, 29*(03), 1240019.

P. K. Kapur is Director, Center for Interdisciplinary Research, Amity University, Noida and Former Dean of the Faculty of Mathematical Sciences and Former Head of the Department of Operational Research, University of Delhi. He has supervised 40 PhDs and 25 M.Phil dissertations in the areas of Innovation Diffusion in Marketing, Software Reliability, Reliability-based optimization, and Multi-Criteria Decision-Making (MCDM) as a tool for interdisciplinary research in Human Resource Development (HRD), Marketing of Brands, Big data projects adoption and other areas of management. He is the author of two world-renowned books "Software Reliability Assessment with O.R. Applications," Springer UK (2011) and "Contributions to Hardware and Software Reliability," (1999), World Scientific, Singapore. He has executed various research projects from UGC, DRDO in the field of Mathematical Modeling in Marketing and Software Reliability. He has been the President of Society for Reliability Engineering, Quality and Operations Management (Regd.) since 2000 and former President of Operational Research Society of India. He is the Editor-in-Chief of International Journal of Systems Assurance Engineering and Management (IJSAEM) published by Springer. He has edited 5 volumes of Conference Proceedings published by leading publishers of India and has been the Guest Editor for special issues of IJRQSE (USA), IJSAEM (Springer India), CDQM (Serbia), International Journal of Modeling and Optimization (Singapore), OPSEARCH (India), International Journal of Performability Engineering (India). He obtained his Ph.D. degree in Reliability Theory (Operational Research) from University of Delhi in 1977. He has published extensively in Indian journals and abroad in the areas of Marketing, MCDM, Hardware Reliability, Optimization, Queuing Theory and Maintenance and Software Reliability (more than 300 papers).

Mr. Saurabh Panwar is a research scholar in the Department of Operational Research, University of Delhi, Delhi, India. He is currently pursuing his Ph.D. degree in Marketing Research. He has received his M.Phil degree in Operational Research from the University of Delhi in 2015. He has his Master's degree in Operational Research and a Bachelor's degree in Computer Science from the University of Delhi, India. He is a lifetime member of the Society for Reliability Engineering, Quality, and Operations Management (SREQOM) since 2015. He is also an Associate Editor of International Journal of Systems Assurance Engineering and Management (IJSAEM), Springer. He has published quality research papers in international journals and proceedings of high repute. His research interests include mathematical modeling, new product development, software reliability, and optimization.

Mr. Vivek Kumar is pursuing Ph.D. in the area of Software Reliability Modeling from Department of Operational Research, University of Delhi, Delhi, India. He did his M.Phil. and M.Sc. in Operational Research from Department of Operational Research, University of Delhi. He has published papers in International Journals and Conference Proceedings of repute. He is a lifetime member of the Society for Reliability Engineering, Quality, and Operations Management (SREQOM). He is an associate editor of International Journal of Systems Assurance Engineering and Management (IJSAEM), Springer.

Chapter 33
Data Resilience Under Co-residence Attacks in Cloud Environment

Gregory Levitin and Liudong Xing

Abstract The virtualization technology, particularly virtual machines (VMs) used in cloud computing systems have raised unique security and reliability risks for cloud users. This chapter focuses on the resilience to one of such risks, co-residence attacks where a user's information in one VM can be accessed/stolen or corrupted through side channels by a malicious attacker's VM co-residing on the same physical server. Both users' and attackers' VMs are distributed among cloud servers at random. We consider different users' data protection policies with the aim to make the data resilient to the co-residence attacks, including data partition with and without replication of the parts, and attack detection through the early warning mechanism. Probabilistic models are suggested to derive the overall probabilities of an attacker's success in data theft and data corruption. Based on the suggested probabilistic evaluation models, optimization problems of obtaining the data partition/replication policy to balance data security, data reliability, and a user's overheads are formulated and solved, leading to the optimal data protection policy to achieve data resilience. The possible user's uncertainty about the number of attacker's VMs is taken into account. Numerical examples demonstrating the influence of different constraints on the optimal policy are presented.

Keywords Cloud system · Co-residence attack · Data replication/partition · Data security · Data reliability · Virtual machine

G. Levitin (✉)
The Israel Electric Corporation, P. O. Box 10, Haifa 31000, Israel
e-mail: levitin@iec.co.il

L. Xing
University of Massachusetts, Dartmouth, MA 02747, USA
e-mail: lxing@umassd.edu

K. B. Misra (ed.), *Handbook of Advanced Performability Engineering*,
https://doi.org/10.1007/978-3-030-55732-4_33

33.1 Introduction

Cloud computing systems are vulnerable to co-resident attacks (CRAs) [1, 2]. Empowered by the virtualization technology, a cloud system provides on-demand services to its users through constructing and running virtual machines (VMs). VMs from different cloud users may be hosted on the same physical server, raising security and reliability concerns. Specifically, a malicious cyber attacker may launch a CRA through co-locating its VM with a target user's VM; a side channel can then be built between these VMs to enable unauthorized access or corruption of user's data. Intuitively, the VM co-residence probability, thus the data theft and corruption probabilities are decreasing functions of the number of cloud servers. Therefore, these probabilities can be negligibly small for large-scale cloud systems or cloud services involving a large number of physical servers. However, the CRA success probability can become significantly large for cloud systems dedicated to an institute, like private and community clouds that may contain only a few to a couple of dozen physical servers [3, 4]. In this case, certain mitigation mechanisms are desired to make data resilient to CRAs. Various mechanisms have been suggested, including for example schemes based on side-channel handling [5, 6], malicious VM detection [7, 8], VM migrations [9], VM allocations [10], virtual private cloud [11], the game theory [1, 12]. Refer to Section II of [13] for a review of the existing CRA mitigation techniques. Based on our work in [14–18], this chapter presents data resilience techniques utilizing the data partition with or without replication to combat the negative effects of CRAs. Further, attack detection through the early warning mechanism is investigated to enhance the data resilience to CRAs.

Note that cyber and physical resilience has received significant research attentions in the past decade [19–23]. Resilient techniques against different hazards such as malicious cyberattacks (e.g., distributed denial of service attack, insider attack) [24, 25], natural disasters (e.g., hurricanes, earthquakes) [26, 27], and human errors [28, 29] have been proposed. This chapter focuses on the resilience to one particular type of cyberattacks that is the CRA. Different from the existing CRA mitigation techniques that either requires a modification to the cloud architecture or actions from the cloud service providers [13], the resilience techniques presented in this chapter are based on data partition and replication and can be easily implemented from the cloud user's perspective. The problem of optimal data partition and replication balancing the information survivability/reliability and security was addressed in [30] with the restrictive assumption that all the data parts are located in the same server. This chapter extends the single-server model to the case of cloud systems with distributed servers where data parts may be allocated to VMs hosted on different servers, and each data part (replica) can be attacked only if the user's and attacker's VMs co-reside on the same server.

The remainder of this chapter is arranged as follows: Sect. 33.2 focuses on modeling and optimizing the data partition policy to balance data theft and data corruption threats caused by CRAs. Section 33.3 discusses data resilience based on data replication in clouds subject to dynamic CRAs. Section 33.4 considers the combined data partition and replication policy. Section 33.5 models and optimizes the data protection policy based on data partition coupled with attack detection mechanism. Section 33.6 summarizes the chapter and discusses threats to validity of the model and future research directions.

33.2 Balancing Theft and Corruption Threats by Data Partition

Consider a cloud system with n servers. To store or process data a user sends a service request to the cloud resource management system (RMS), which responds to the request by creating a VM and allocating it to an available cloud server. To protect the data against being stolen by malicious attackers, the data partition policy can be adopted [14]. The data is assumed to be useful only in its integrity. With the partition, data is divided into multiple separate blocks (corresponding to different VMs) and stored in different places. To access the data an attacker must get access to all the blocks. Thus, the partition can enhance the data security (decrease the data theft probability). However, the partition may lower the data reliability because if any block is corrupted the entire data becomes useless to its user. Therefore, it is crucial to design the number of blocks used in the data partition policy. More blocks can make data theft more difficult, but data corruption easier. In this section, the optimal data partition policy is investigated, which aims to strike a balance between data theft and data corruption probabilities.

33.2.1 System and CRA Model

A user divides the sensitive data into k non-overlapped blocks and sends k requests to the RMS to construct k VMs (one for each data block). These VMs are referred to as user's VMs (UVMs) hereinafter. The UVMs are then allocated to available servers randomly and freely. It is assumed that the RMS chooses any server with equal probability. Note that due to technical reasons or budget constraints, the maximum value of k can be specified.

To get access to the UVMs, an attacker sends m requests containing malware to the same RMS. The RMS cannot distinguish malicious requests from normal requests; it simply responds by constructing m VMs (one for each malicious request). These VMs are referred to as attacker's VMs (AVMs) hereinafter. The AVMs are also randomly allocated among the n available servers. In the case of any AVM co-locating with a UVM in the same cloud server, a side channel may be established by this AVM to the co-resident UVM, enabling unauthorized accesses to the user's data.

There are two possible types of AVMs allocations among servers: free allocation (FA) and different servers allocation (DSA). Under the FA, one server may be

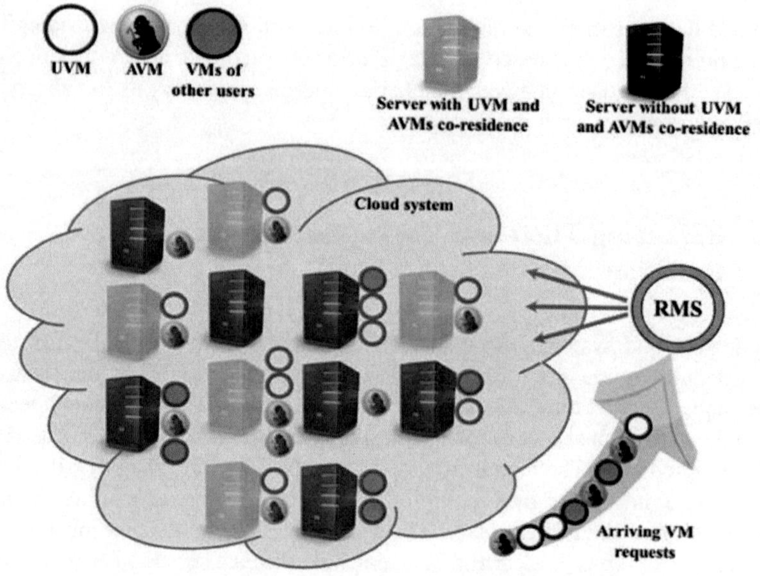

Fig. 33.1 Example of free VMs allocation leading to a failed data theft CRA

assigned any number of AVMs ranging from 0 to m; the AVMs may be allocated to from 1 to min(n, m) servers. Under the DSA, the RMS allocates the m AVMs to m different servers. Since the DSA can guarantee that the AVMs reside on the maximal number of cloud servers, it is more beneficial to the attacker from the perspective of the co-residence probability; it may, however, increase the chance to expose the attacker's malicious intention. Unlike legitimate users who are ready to be checked, the attacker may try to avoid attracting attention. In this case, the attacker may prefer the FA policy for the AVMs allocation.

Figure 33.1 illustrates an example of free VMs allocation in a cloud system where the data theft attack fails because not all the UVMs co-reside with at least one AVM.

In the case of any UVM co-residing with at least one AVM on the same physical server, the probabilities that in this server the AVMs succeed to corrupt and steal the UVM data are u and v, respectively. When all the servers use the same data protection measures, the event of attacker's success in building the side channel and accessing data is common for all servers where UVMs and AVMs co-reside; if an AVM succeeds to build the side channel, it happens in all servers. When the servers use individual independent protection measures, the data corruption/theft events for different servers are independent. We assume that probabilities u and v are the same in all the servers and independent of the number of UVMs and AVMs residing on the same physical server. However, values of v and u are not necessarily equal. For example, $u > v$ takes place in cases where an attacker penetrates to certain encrypted data and corrupts it, but cannot decrypt the data and thus steal it; $u < v$ takes place in cases where the data is write protected, making data corruption far more difficult than data theft.

33.2.2 Probabilities of Data Theft and Corruption

Consider the model with n cloud servers, k UVMs and m AVMs. Among n cloud servers there exist $\binom{n}{h}$ different groups of h servers, where $1 \le h \le \min(n, k)$. For each h-server group, applying the inclusion–exclusion principle [16], the number of possible allocations where any server belonging to the group hosts at least one UVM and no servers not belonging to the group host UVMs is

$$\sum_{i=0}^{h-1} (-1)^i \binom{h}{i} (h-i)^k. \tag{33.1}$$

Thus, the probability that exactly h out of n servers host UVMs is given by (33.2), where n^k is the number of possible ways to allocate k UVMs among the n available servers.

$$q(n,k,h) = n^{-k} \binom{n}{h} \sum_{i=0}^{h-1} (-1)^i \binom{h}{i} (h-i)^k. \tag{33.2}$$

Consider a h-server group hosting the UVMs ($1 \le h \le \min(n, k)$). In the case of the free AVM allocation, according to the inclusion–exclusion principle, the number of AVM allocations where AVMs co-reside with UVMs in a fixed subset of x servers selected from the h-server group is

$$\sum_{i=0}^{x} (-1)^{x-i} \binom{x}{i} (n-h+i)^m. \tag{33.3}$$

Thus, the conditional probability that AVMs and UVMs co-reside in exactly x cloud servers given that the UVMs are allocated among a fixed set of h cloud servers is given by (33.3), where n^m is the number of possible ways to allocate m AVMs among the n servers under the FA policy.

$$g(n,m,h,x) = \begin{cases} n^{-m} \binom{h}{x} \sum_{i=0}^{x} (-1)^{x-i} \binom{x}{i} (n-h+i)^m & \text{if } x \le \min(h,m), \\ 0, & \text{otherwise.} \end{cases} \tag{33.4}$$

In the case of the DSA policy, the total number of the possible AVM allocations is $\binom{n}{m}$ and the conditional probability that UVMs and AVMs co-reside in exactly x servers given that the UVMs are distributed among a fixed set of h servers is

$$g(n,m,h,x) = \begin{cases} \binom{h}{x}\binom{n-h}{m-x}\binom{n}{m}^{-1} & \text{if } x \le \min(m,h) \\ 0, & \text{otherwise.} \end{cases} \tag{33.5}$$

As the case of the DSA is relatively rare, below we consider only the more computationally complicated FA case.

With $q(n, k, h)$ and $g(n, m, h, x)$, the total probability that AVMs and UVMs co-reside in exactly x cloud servers can be computed as (33.6), where h runs from x (when UVMs and AVMs co-reside in all $h = x$ servers) to its maximum value $\min(n, k)$ and x cannot exceed the total number of servers n, or the total number of UVMs k, or the number of AVMs m.

$$p(n,k,m,x) = \begin{cases} \sum_{h=x}^{\min(n,k)} q(n,k,h)g(n,m,h,x), & x \le \min(n,k,m) \\ 0, & \text{otherwise} \end{cases} \tag{33.6}$$

Having (33.6) the probability that at least one UVM co-resides with AVMs can be obtained as

$$\begin{aligned} w(n,k,m) &= 1 - p(n,k,m,0) \\ &= \sum_{h=1}^{\min(n,k)} \binom{n}{h}\left(1-\left(1-\frac{h}{n}\right)^m\right)\left(\sum_{i=0}^{h-1}(-1)^i\binom{h}{i}\left(\frac{h-i}{n}\right)^k\right). \end{aligned} \tag{33.7}$$

and the probability that all the UVMs co-reside with AVMs is

$$\begin{aligned} z(n,k,m) &= \sum_{h=1}^{\min(n,k)} q(n,k,h)g(n,m,h,h) \\ &= \sum_{h=1}^{\min(n,k)} \binom{n}{h}\left(\sum_{i=0}^{h-1}(-1)^i\binom{h}{i}\left(\frac{h-i}{n}\right)^k\right)\left(\sum_{i=0}^{h}(-1)^i\binom{h}{i}\left(1-\frac{i}{n}\right)^m\right). \end{aligned} \tag{33.8}$$

When the data protection for all servers is common, all AVMs co-residing with UVMs either succeed to get access or not simultaneously. As the probability of data corruption in all co-resident UVMs is u, the overall probability that AVMs corrupt user's data in at least one server (i.e., the data corruption probability) is

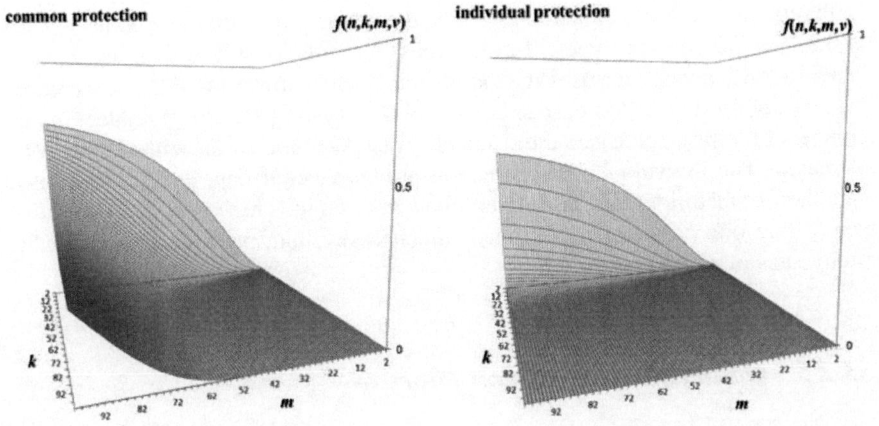

Fig. 33.2 Data theft probability $f(n, k, m, v)$ for $v = 0.8$, $n = 30$ under common and individual data protections

$$s(n, k, m, u) = u \cdot w(n, k, m). \tag{33.9}$$

As the probability of data theft in all co-resident UVMs is v, the overall probability that AVMs succeed in stealing user's data (all UVMs co-reside with AVMs and the data theft succeeds) is

$$f(n, k, m, v) = v \cdot z(n, k, m). \tag{33.10}$$

In the case of individual independent protections in different servers, when AVMs and UVMs co-reside in exactly x cloud servers, the conditional probability that AVMs succeed in corrupting UVM data in at least one of the servers is 1-(1-u)^x. Thus, the overall data corruption probability can be obtained as

$$\begin{aligned} s(n, k, m, u) &= \sum_{x=1}^{\min(n,k,m)} p(n, k, m, x)(1 - (1 - u)^x) \\ &= \sum_{h=1}^{\min(n,k)} q(n, k, h) \sum_{x=1}^{\min(h,m)} g(n, m, h, x)(1 - (1 - u)^x). \end{aligned} \tag{33.11}$$

The probability that all UVMs co-reside with AVMs and the attack succeeded in all the co-resident servers (i.e., the data theft probability) is

$$f(n, k, m, v) = \sum_{h=1}^{\min(n,k,m)} q(n, k, h)g(n, m, h, h)v^h. \tag{33.12}$$

Figure 33.2 illustrates examples of the data theft probability $f(n, k, m, v)$ for $v = 0.8$, $n = 30$ and two types of protections. It can be observed that $f(n, k, m, v)$ increases with m but decreases with k. Indeed, as the number of UVMs k increases, the probability that AVMs co-reside with all the UVMs (data theft) reduces; as the number of AVMs m increases, the probability that AVMs co-reside with all the UVMs increases. The individual data protection provides better data defense and lower data theft probability than the common data protection. These results are intuitive: penetrating one common protection is much easier than penetrating x individual protections in x different servers.

33.2.3 Optimal Data Partition Policy

In the case of the number of AVMs m being known, the optimization problem

$$k = \arg\min_{k} f(n, k, m, v) \text{ subject to } s(n, k, m, u) < s^*. \quad (33.13)$$

aims to determine the number of data blocks k minimizing the data theft probability (i.e., maximizing the data security) subject to meeting a certain level of data corruption probability s^*. Note that the value of the maximum allowable data corruption probability s^* can be specified, for example, based on the formal requirement of military or government applications. However, in some practical scenarios, the value of s^* may be unspecified before investigating data security and reliability issues. In this case, the optimal solutions for different values of s^* may be obtained for the tradeoff analysis of data corruption and theft probabilities, as demonstrated in this section.

In the case of the number of AVMs m being uncertain but with known range $[m_{\min}, m_{\max}]$ and distribution $\mu(l) = \Pr(m = l)$ for $m_{\min} \le l \le m_{\max}$, the optimization problem is formulated as

$$k = \arg\min_{k} \sum_{l=m_{\min}}^{m_{\max}} \mu(l) f(n, k, l, v) \text{subject to} \sum_{l=m_{\min}}^{m_{\max}} \mu(l) s(n, k, l, u) < s^*. \quad (33.14)$$

The optimization problems of (33.13) and (33.14) can be solved using the brute-force approach, which enumerates all possible integer values of k in a given range.

Figures 33.3, 33.4 and 33.5 demonstrate the optimal value of k and corresponding values of s (data corruption probability) and f (data theft probability) as functions of s^* for $u = 0.1$ and $v = 0.8$ under the case of individual and independent data protection in different servers.

Two values (10 and 50) for the number of servers n are considered. Three cases for m are considered: $m = 10$, $m = 30$, and the case when m is uncertain with the uniform distribution in the range (10, 30). It can be observed that the optimal value of k increases as n increases, and decreases as m increases. The minimum value of

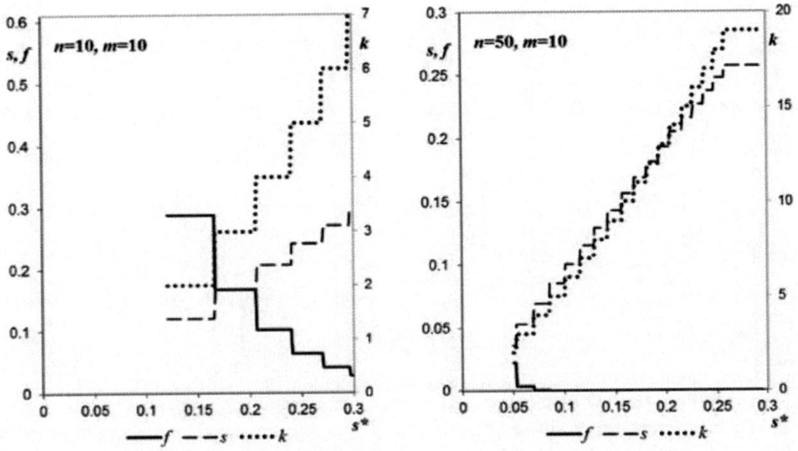

Fig. 33.3 Optimal value of k and corresponding values of f and s as functions of s^* for $u = 0.1$, $v = 0.8$, and $m = 10$ with individual independent data protection in different servers

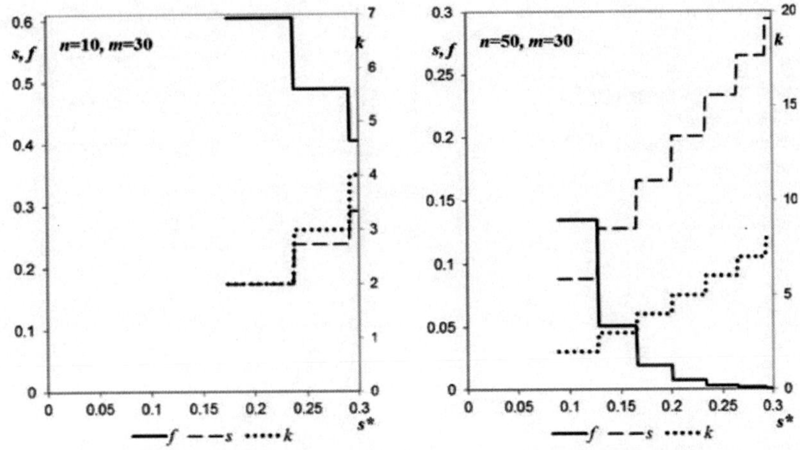

Fig. 33.4 Optimal value of k and corresponding values of f and s as functions of s^* for $u = 0.1$, $v = 0.8$, and $m = 30$ with individual independent data protection in different servers

s^* so that constraint $s < s^*$ can be met decreases with n and increases with m. The minimum values of f decrease with n and increase with m.

Figure 33.6 presents the optimal value of k as a function of u for different n with $s^* = 0.1$ and m being uniformly distributed in the range (10, 30) for the case of individual independent data protection in different servers. With an increase in the corruption success probability u the number of data blocks (UVMs) should be reduced to keep the value of corruption probability within the specified limit $s < 0.1$.

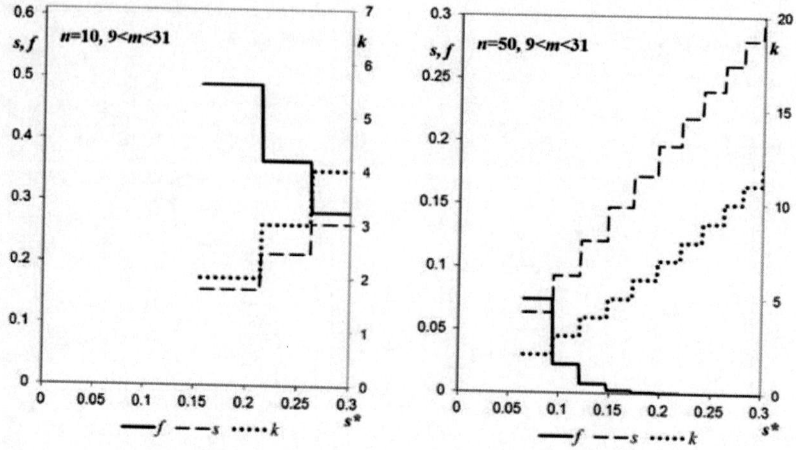

Fig. 33.5 Optimal value of k and corresponding values of f and s as functions of s^* for $u = 0.1$, $v = 0.8$, and uncertain m with individual data protection in different servers

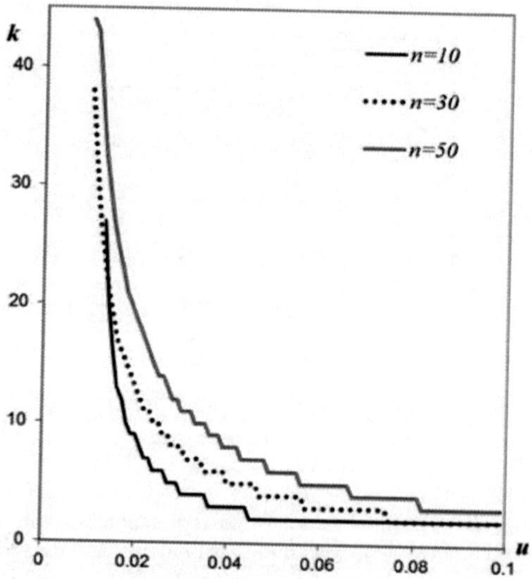

Fig. 33.6 Optimal value of k as a function of u and n for $s^* = 0.1$ and m with uniform distribution in the range (10, 30) and individual independent data protection in different servers

This reduction should be more considerable when the number of servers is low and the co-residence probability is high.

Figure 33.7 demonstrates the values of f corresponding to the optimal k for five different values of v (1, 0.8, 0.6, 0.4, 0.2) and two different values of n (30, 50). Since the data corruption probability is not dependent on v (the constraint in problem (33.14) can be met for a certain value of k for any value of v), the optimal value of k is not dependent on v. However, as demonstrated in Fig. 33.7, the data theft probability f is strongly dependent on v. As probability u increases, the number of UVMs k for

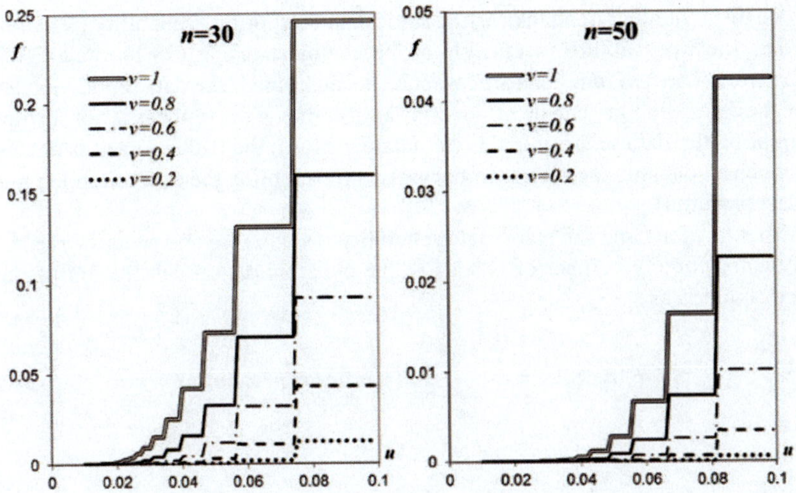

Fig. 33.7 Minimal value of f as a function of u and v for $s^* = 0.1$, $n = 30$, 50 and m uniformly distributed in the range (10, 30) for the case of individual independent data protection in different servers

which the overall data corruption probability still satisfies the constraint decreases, causing an increase in f.

33.3 Data Resilience to Dynamic CRAs by Data Replication

Different from the study in the previous section that assumes static CRAs, this section considers attacks that are random and constitute a stochastic process. Instead of using the data partition, this section models the data replication policy where a user's information is replicated k times and stored on different VMs to reduce the chance of data corruption by attackers.

We assume that CRAs (requests to create AVMs) constitute a nonhomogeneous Poisson process with a time-varying attack rate $a(t)$. Thus, the probability that m CRAs occur in time interval [0, t) is

$$\pi(t, a, m) = \frac{1}{m!} \exp\left\{ - \int_0^t a(x)\mathrm{d}x \right\} \left(\int_0^t a(x)\mathrm{d}x \right)^m . \qquad (33.15)$$

In the case of $a(t) \equiv A$ (a constant), we have a homogeneous Poisson attack process and the probability in (33.15) becomes

$$\pi(t, A, m) = \frac{\exp\{-At\}}{m!}(At)^m . \qquad (33.16)$$

For a given number of attacks *m*, the static data corruption probability (reliability) and data theft probability (security) can be evaluated using (33.11) and (33.12), respectively. The two functions are swapped because under the data replication technique the data reliability can be achieved as long as one of the replicas is not corrupted (resembling the data security under the data partition); the data security is achieved only when all the replicas are not compromised (resembling the data reliability under the data partition).

With $\pi(t, a, m)$ and the static data reliability $s(n, k, m, v)$ and security $f(n, k, m, v)$ evaluated for any number of attacks *m*, the dynamic data reliability and security can be obtained as

$$R(n, k, t) = 1 - \sum_{m=1}^{\infty} \pi(t, a, m) f(n, k, m, v), \tag{33.17}$$

and

$$\Theta(n, k, t) = 1 - \sum_{m=1}^{\infty} \pi(t, a, m) s(n, k, m, u), \tag{33.18}$$

respectively.

As *m* increases, $\pi(t, a, m)$ defined in (33.15) converges to zero. Hence, the summation in (33.17) and (33.18) should be conducted until $\pi(t, a, m) \leq \varepsilon$ (a predefined small value). $\varepsilon = 10^{-8}$ is used in the example below to provide a desired precision for $R(t)$ and $\Theta(t)$ obtained.

Figure 33.8 demonstrates values of $R(t)$ and $\Theta(t)$ for $n = 30$ servers, $k = 10$ UVMs/replicas, $u = 0.03$, $v = 0.8$, and three different constant attack rates $a(t) = A = 1, 3$, or 5. It can be observed that as mission time *t* increases, both $R(t)$ and $\Theta(t)$ converge to constant values, and as the attack rate *A* increases, the convergence rate increases. However, the values to which $R(t)$ and $\Theta(t)$ converge are not dependent on the attack rate. Indeed, as time proceeds, the number of AVMs increases causing the probability of any UVM co-locating with at least one AVM to approach 1. In this case, the data theft and corruption probabilities are not dependent on the number of AVMs anymore; they only depend on the number of servers hosting UVMs, and probabilities *v* and *u*.

In the case of the attack rate being time-varying, for example, $a(t) = c + gt$, according to (33.15) we have $\pi(t, a, m) = \frac{1}{m!} \exp\{-(c + 0.5gt)t\}((c + 0.5gt)t)^m$. Figure 33.8 presents examples of $R(t)$ and $\Theta(t)$ for the cloud system with the same parameters but three different cases of increasing attack rates. As parameter *g* increases, the same convergence behavior can be observed.

Similar to the optimization study in Sect. 33.2.3, with the data reliability and security functions, optimal decisions can be made about time during which the user's data can be stored in the cloud and about the dynamic change of the number of data replicas while achieving a balance between data reliability and security.

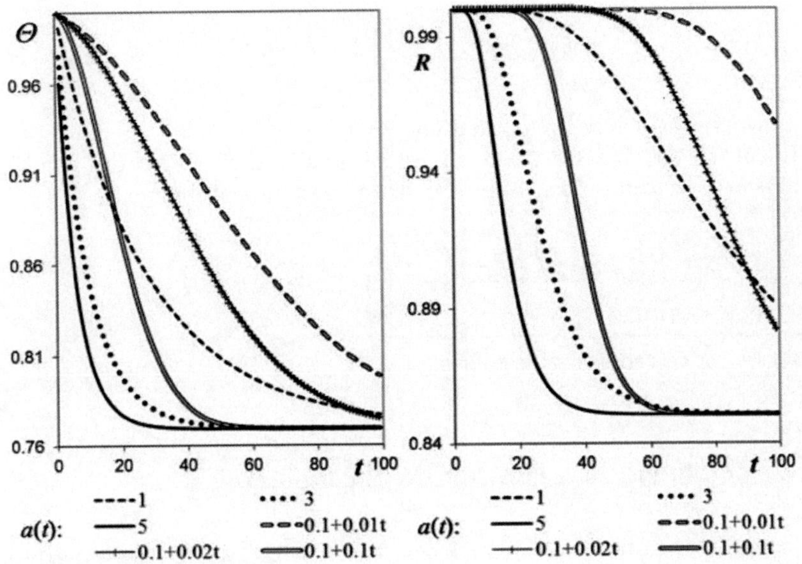

Fig. 33.8 $\Theta(t)$ and $R(t)$ for $n = 30$, $k = 10$, $u = 0.03$, $v = 0.8$ and different attack rates $a(t)$

33.4 Data Resilience by Combined Data Partition and Replication

As shown in Sect. 33.2, the data partition makes the theft difficult as an attacker must get access to all the k blocks to steal the information. On the other hand, the information is corrupted and becomes useless to its user if the attacker succeeds in corrupting any of the blocks. To decrease the data corruption probability (or enhance the data reliability), y_i replicas can be created for each data block i ($1 \leq i \leq k$). With the combined data partition and replication, the attacker must destroy all y_i replicas of any block i to corrupt the information; the attacker should get access to at least one replica of each data block to steal the information.

More data blocks make information theft harder, but data corruption easier; more replicas for each block make data corruption harder, but data theft easier. The optimal solution should strike the balance between data corruption and theft probabilities.

33.4.1 Data Theft and Corruption Probabilities

Consider the case when the data protection for all servers is common. To access the full data, an attacker has to get access to at least one replica for each data block, in particular, one out of y_i UVMs storing replicas of the i-th block for $i = 1,\ldots, k$. For any fixed m and the data partition-replication policy $\mathfrak{R} = (k, y_1, \ldots, y_k)$, the data theft probability can be obtained as

$$f(n, \mathfrak{R}, m, v) = v\prod_{i=1}^{k} w(n, y_i, m). \tag{33.19}$$

where $w(n, y_i, m)$ can be evaluated using (33.7).

To corrupt the data, the attacker must get access to all UVMs that contain all replicas of at least one block. Thus, the data corruption probability is

$$s(n, \mathfrak{R}, m, u) = u\left(1 - \prod_{i=1}^{k} (1 - z(n, y_i, m))\right). \tag{33.20}$$

where $z(n, y_i, m)$ can be evaluated using (33.8).

33.4.2 Optimal Data Partition-Replication Policy

Under the data partition-replication policy $\mathfrak{R} = (k, y_1, \ldots, y_k)$, the user's overhead is

$$O(\mathfrak{R}) = c\sum_{i=1}^{k} y_i, \tag{33.21}$$

where c is the overhead of creating one VM, $\sum_{i=1}^{k} y_i$ gives the total number of UVMs created under the policy $\mathfrak{R}$.

The optimal data partition-replication policy $\mathfrak{R}$ is determined by solving the multi-objective optimization problem formulated in (33.22).

$$\begin{aligned} &\mathfrak{R} = \arg\min_{\mathfrak{R}}\{f(n, \mathfrak{R}, m, v), s(n, \mathfrak{R}, m, v), O(\mathfrak{R})\} \\ &\text{subject to } 1 \le k \le k_{\max}, 1 \le y_i \le y_{\max} \text{ for } i = 1, \ldots, k, \end{aligned} \tag{33.22}$$

where k and y_i ($i = 1, \ldots, k$) are integer variables, k_{max} denotes an upper limit of the number of blocks the original data can be separated into, and y_{max} denotes an upper limit of the number of replicas that can be made for any data block. k_{max} and y_{max} can be decided based on specific data structure and are typically known in advance.

For uncertain m with a known distribution μ the problem (33.22) takes the form

$$\begin{aligned} &\mathfrak{R} = \arg\min_{\mathfrak{R}}\{f(n, \mathfrak{R}, \mu, v), s(n, \mathfrak{R}, \mu, v), O(\mathfrak{R})\} \\ &\text{subject to } 1 \le k \le k_{\max}, 1 \le y_i \le y_{\max} \text{for } i = 1, \ldots, k, \end{aligned}$$

where

$$f(n, \mathfrak{R}, \mu, v) = v \sum_{l=m_{\min}}^{m_{\max}} \mu(l) \prod_{i=1}^{k} w(n, y_i, l),$$

$$s(n, \mathfrak{R}, \mu, u) = u \sum_{l=m_{\min}}^{m_{\max}} \mu(l) \left(1 - \prod_{i=1}^{k} (1 - z(n, y_i, l))\right). \tag{33.23}$$

The multi-objective optimization problems (33.22), (33.23) are usually solved in practice as one of the following constrained single-objective optimization problems:

$$\min f(n, \mathfrak{R}, \mu, v) \text{ s. t.} s(n, \mathfrak{R}, \mu, u) \leq s^*, O(\mathfrak{R}) \leq O^*, 1 \leq k \leq k_{\max}, 0 < y_i \leq y_{\max} \text{ for } i = 1, \ldots, k; \tag{33.24}$$

$$\min s(n, \mathfrak{R}, \mu, u) \text{ s. t.} f(n, \mathfrak{R}, \mu, v) \leq f^*, O(\mathfrak{R}) \leq O^*, 1 \leq k \leq k_{\max}, 0 < y_i \leq y_{\max} \text{ for } i = 1, \ldots, k; \tag{33.25}$$

$$\min O(\mathfrak{R}) \text{ s. t.} s(n, \mathfrak{R}, \mu, u) \leq s^*, f(n, \mathfrak{R}, \mu, v) \leq f^*, 1 \leq k \leq k_{\max}, 0 < y_i \leq y_{\max} \text{ for } i = 1, \ldots, k. \tag{33.26}$$

To obtain the solutions of these problems one can use any optimization algorithm, which minimizes a single criterion

$$\begin{aligned} F = {} & \eta_f \max(f(n, \mathfrak{R}, \mu, v) - f^*, 0) \\ & + \eta_s \max(s(n, \mathfrak{R}, \mu, u) - s^*, 0) \\ & + \eta_O \max(O(\mathfrak{R}) - O^*, 0). \end{aligned} \tag{33.27}$$

When $f^* = 0, \eta_s \gg \eta_f, \eta_O \gg \eta_f$, the problem min F reduces to (33.24); when $s^* = 0, \eta_f \gg \eta_s, \eta_O \gg \eta_s$, it reduces to (33.25); and when $O^* = 0, \eta_s \gg \eta_O, \eta_f \gg \eta_O$, it reduces to (33.26).

Table 33.1 demonstrates examples of solutions to the optimization problem (33.24) for $s^* = 0.05$, $n = 30$, $v = 0.2$, $u = 0.6$, $c = 1$, $k_{\max} = 10$, and $y_{\max} = 10$.

Different overhead constraints are considered for different values of m. For example, in the case of m being uncertain and uniformly distributed in the range (10, 30), the optimal partition-replication policy for overhead constraint $O^* = 20$ is (2, 4, 4) meaning two data blocks and four replicas of each data block; for $O^* = 40$ is (6, 7, 5, 6, 6, 6, 6) meaning six blocks, seven replicas of the first block, five replicas of the second block, and six replicas for each of the remaining four data blocks.

It can be observed from Table 33.1 that with an increase in the number of AVMs m the optimal number of data blocks decreases, which is necessary to meet the constraint $s(n, \mathfrak{R}, \mu, u) < 0.05$ when the co-residence probability increases.

Table 33.1 Examples of solutions to optimization problem (33.24)

Overhead constraint	O	s	f	$\Re$
$m = 10$				
$5 \le O^* \le 7$	5	0.0125	0.0383	2, 3, 2
$12 \le O^* \le 14$	12	0.0059	0.0485	4, 4, 3, 3, 2
$15 \le O^* \le 18$	15	0.0042	0.0422	5, 3, 3, 3, 3, 3
$23 \le O^* \le 26$	23	0.0023	0.0469	7, 4, 4, 3, 3, 3, 3, 3
$32 \le O^* \le 35$	32	0.0014	0.0498	9, 5, 4, 4, 4, 3, 3, 3, 3, 3
$36 \le O^* \le 50$	36	0.0011	0.0483	10, 4, 4, 4, 4, 4, 4, 3, 3, 3, 3
$m = 30$				
$5 \le O^* \le 13$	5	0.0397	0.0412	1, 5
$14 \le O^* \le 50$	14	0.0396	0.0486	2, 6, 5
$10 \le m \le 30$				
$5 \le O^* \le 7$	3	0.0336	0.0473	1, 3
$8 \le O^* \le 20$	8	0.0330	0.0496	2, 4, 4
$36 \le O^* \le 43$	36	0.0325	0.0498	6, 7, 5, 6, 6, 6, 6
$44 \le O^* \le 50$	44	0.0324	0.0499	7, 8, 6, 6, 6, 6, 6, 6

33.5 Systems with Attack Detection by Early Warning

To enhance the resilience of data from CRAs, attack detection using the early warning agents [18, 31] can be implemented. Specifically, the cloud RMS attaches an early warning agent (EWA) to each VM. The EWA aims to detect the attempts of AVMs to create side channels, and then inform other UVMs in the case of an attack being detected. Thus, the attack fails as long as at least one EWA can detect the attack before AVMs access all the k UVMs. It is assumed that the attack detection and data theft events in different servers are independent.

Both the time T_D required by any EWA to successfully detect the attack and the time T_A required by the AVM to have access to data of co-resident UVMs are random, respectively, following known distributions with *pdf* s $f_D(t)$ and $f_A(t)$. It is assumed that these times are identical for all the servers and independent of the number of UVMs and AVMs co-locating on the same server. We also assume that the CRA attacks (request to create AVMs) constitute a Poisson process with constant rate A.

33.5.1 Evaluating Data Theft Probability

We start considering competing data theft and attack detection processes in cloud system with data partition from the first co-residence event (FCE) when the first of AVMs is allocated by the RMS in the server containing UVM. It is assumed that the co-resident AVM starts creating the side channel to steal the data (occurs) immediately after the FCE (at time 0).

The probability $q(n, k, h)$ that after random allocation of k UVMs, they are located in h out of n servers is determined in (33.2). Given all the UVMs are located in h specific servers, the AVMs created in the cloud system after the FCE must be allocated in each of the remaining h-1 servers to get access to all UVMs. If in time t the number of new AVMs created after the FCE in the cloud is m, the probability that all remaining UVMs co-reside with newly created AVMs in all h-1 remaining servers is $g(n, m, h\text{-}1, h\text{-}1)$ (see eq. (33.4)). The total probability that all UVMs co-reside with AVMs given that m AVMs are created after the FCE is (similar to eq. (33.8))

$$\tilde{z}(n, k, m) = \sum_{h=1}^{\min(n,k)} q(n, k, h)g(n, m, h-1, h-1). \tag{33.28}$$

Given the attack rate A, the probability $\pi(t, A, m)$ that m attacks happen in time interval $[0, t)$ is determined in eq. (33.16). Having the functions $\pi(t, A, m)$, $q(n, k, h)$, and $g(n, m, h\text{-}1, h\text{-}1)$ one can obtain the probability that during time t since the FCE the attacker succeeds to get its AVMs co-residing with all the UVMs as

$$\psi(n, k, A, t) = \sum_{h=1}^{\min(n,k)} q(n, k, h) \sum_{m=h-1}^{\infty} \pi(t, A, m)g(n, m, h-1, h-1). \tag{33.29}$$

The attack is successful if AVMs co-reside with all the UVMs in time t after the FCE, and time $t + T_A$ required for completing the data theft from all UVMs is less than the time T_D required for detecting the attack after the FCE. Hence, the data theft success probability is

$$\Omega(n, k, A) = \int_{d_{\min}}^{d_{\max}} \int_{\alpha_{\min}}^{\min\{t_D,\alpha_{\max}\}} \psi(n, k, A, t_D - t_A) f_A(t_A) f_D(t_D) dt_A dt_D. \tag{33.30}$$

where $d_{\min}$, $d_{\max}$ are minimum, maximum possible realizations of T_D, and $\alpha_{\min}$, $\alpha_{\max}$ are minimum, maximum possible realizations of T_A.

33.5.2 *Optimal Data Partition-Protection Policy*

Cloud providers suggest different data protection options and different types of EWAs at different prices to users. Let $c_A(i)$ denote the cost of using data protection option i on each UVM, and $f_A(i, t)$ be the AVM data access time *pdf* under protection option i. Let $c_D(j)$ denote the cost of using EWA type j on each UVM, and $f_D(i, t)$ be the attack detection time *pdf* under EWA type j. Let c_U denote the cost of constructing

Table 33.2 Parameters of available data protection and EWA options

Option	d_{min}	d_{max}	d_{mean}	d_σ	c_D	α_{min}	α_{max}	α_{mean}	α_σ	c_A
1	10	50	15	10	15	5	30	10	4	5
2	10	50	12	10	22	7	30	12	4	11
3	10	40	10	8	34	7	33	15	5	19
4	7	40	10	8	61	7	38	18	6	31
5	7	30	9	6	75	9	40	22	6	54

a single UVM. In the case of k data blocks corresponding to k UVMs with data protection option i and EWA type j, the total cost of constructing and defending the UVMs is

$$C(i, j, k) = k(c_A(i) + c_D(j) + c_U). \tag{33.31}$$

The optimization problem solved in this section is formulated in (33.32), which determines values of i, j, and k minimizing the total cost subject to meeting a desired level of data theft probability.

$$\text{minimize}\, C(i, j, k) \text{s.t.}\, \Omega(i, j, k) \le \Omega^* \tag{33.32}$$

where Ω^* is the maximum allowed level of data theft probability. The brute-force enumeration can be used to solve the problem.

Table 33.2 presents parameter values for a case where the available data protection and EWA options provide truncated normal distributions of T_A with parameters (α_{min}, α_{max}, α_{mean}, α_σ) and T_D with parameters (d_{min}, d_{max}, d_{mean}, d_σ). The costs under each option are also presented in Table 33.2. The single UVM creation cost is c_U=12. The attack rate is $A = 1$.

Table 33.3 presents some example solutions to (33.32) for different values of n (the number of servers) and different data security requirements Ω^*. It can be observed that when n is small ($n = 10$), the user should prefer the most costly and effective protection and EWA options, whereas when $n = 20$ inexpensive options are preferable and the desired data security requirement can be met through increasing the number of UVMs. When n increases further, fewer UVMs and less costly data protection and EWA options can be utilized because as the number of servers increases, the AVM and UVM co-residence probability decreases. Hence the overall user cost C reduces as n increases, as shown in Fig. 33.9. The sensitivity of the cost to the value of n increases as Ω^* decreases (the required data security level increases).

Table 33.3 Examples of optimal data protection and detection policies

n	Ω^*	Protection option	EWA option	k	Ω	C
10	0.02	5	5	3	0.01257	423
	0.005	5	5	5	0.00425	705
	0.001	5	5	10	0.00098	1410
	0.0005	5	5	15	0.00047	2115
20	0.02	5	5	2	0.01587	282
	0.005	5	5	3	0.00433	423
	0.001	5	5	5	0.00062	705
	0.0005	4	5	7	0.00049	826
50	0.02	1	1	4	0.01551	128
	0.005	2	1	5	0.00408	190
	0.001	1	1	7	0.00095	224
	0.0005	1	1	8	0.00043	256
100	0.02	1	1	3	0.01439	96
	0.005	1	1	4	0.00262	128
	0.001	1	1	5	0.00055	160
	0.0005	2	1	5	0.00039	190

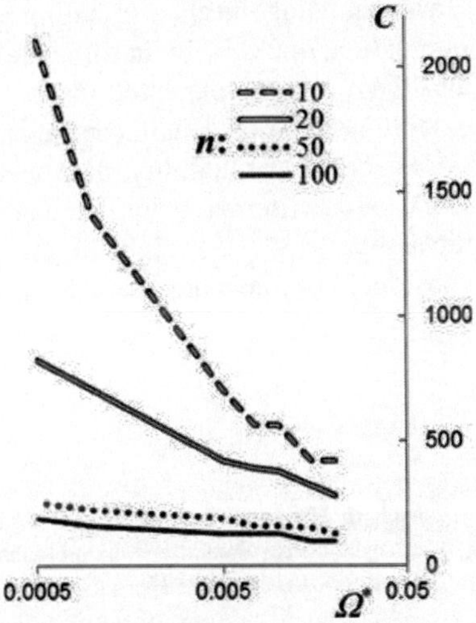

Fig. 33.9 User cost C as functions of the number of cloud servers and threshold of the data theft probability Ω^*

33.6 Summary and Conclusions

This chapter presents different data resilient techniques based on data partition, data replication, combined data partition and replication, and data partition coupled with attack detection/EWAs to combat CRAs in the cloud systems. The data theft probability (security) and data corruption probability (reliability) are evaluated using probabilistic models. Under the different data resilience techniques, the optimal data protection policy is investigated to balance data reliability, data security, and user's overhead. The cases where the number of attacker's VMs is fixed or uncertain with given distribution as well as the dynamic attacks with the number of AVMs increasing in time are considered.

It should be noted that some assumptions made in the suggested models can affect their validity. First, it is assumed that the VMs are distributed among the servers totally at random. In fact, the RMS can distribute the VMs according to a predetermined schedule or based on cost, energy consumption, or load balance considerations. However, in some cases, when the cloud management decides to use specific subsets of servers, it may be still assumed that within this subset the VMs are randomly distributed. It is also assumed that the probabilities of data theft and corruption as well as the probability of attack detection do not depend on servers hosting the VMs. In practice, different servers may have different protections and attackers can use different types of AVMs, which results in different attack detection probabilities. However, using the corresponding probabilities averaged over the servers' and VMs population, one can obtain some realistic estimate of influence of the number of VMs and EWAs on the expected losses. Relaxing the assumptions discussed above and extending the model should be a subject of further research.

Besides data reliability, data security and user's overhead are considered in this work, other criteria (such as data accessibility, energy efficiency, response time, and throughput) [32, 33] might be critical to the cloud service performed and can be considered to make the analysis and optimization more realistic and thorough.

References

1. Mehedi Hasan, M. G. M., & Ashiqur Rahman, M. (2020). A signaling game approach to mitigate co-resident attacks in an IaaS cloud environment. *Journal of Information Security and Applications, 50*, 102397.
2. Levitin, G., Xing, L., &, Xiang, Y. (2020). Optimization of time constrained N-version programming service components with competing task execution and version corruption processes. *Reliability Engineering & System Safety, 193*, 106666.
3. Cáliz Ospino, R., Pérez Arteaga, P., & Pérez Castillo, J. (2015). Lessons learned in the design and implementation of a private cloud for high-performance computing using OpenStack in existing university infrastructure. In: *Proceedings of the tenth computing Colombian conference (10CCC)*. https://doi.org/10.1109/columbiancc.2015.7333473.
4. VirtualDCS. (2017). *Example of a private cloud implementation.* VMware vCloud TM document. https://www.virtualdcs.co.uk/files/example-of-a-private-cloud.pdf.

5. Zhang, Y., & Reiter, M. K. (2013). Düppel: Retrofitting commodity operating systems to mitigate cache side channels in the cloud. In *Proceedings of ACM SIGSAC conference* computing communications security (pp. 827–838).
6. Varadarajan, V., Ristenpart, T., & Swift, M. Scheduler-based defenses against cross-VM side-channels. In *Procedings of 23rd USENIX* security symposium (pp. 687–702).
7. Zhang, Y., Juels, A., Oprea, A., & Reiter, M. K. (2011). HomeAlone: Co-residency detection in the cloud via side-channel analysis. In *Proceedings of IEEE symposium on security and privacy* (pp. 313–328). DC, USA: IEEE Computer Society Washington.
8. Bates, A., Mood, B., Pletcher, J., Pruse, H., Valafar, M., & Butler, K. (2014). On detecting co-resident cloud instances using network flow watermarking techniques. *International Journal of Information Security, 13*(2), 171–189.
9. Atya, A. O. F., Qian, Z., Krishnamurthy, S. V. Porta, T. L. McDaniel P., & Marvel, L. (2017). Malicious co-residency on the cloud: Attacks and defense. In *Proceedings of IEEE INFOCOM 2017—IEEE conference on computer communications* (pp. 1–9). Atlanta, GA.
10. Han, Y., Chan, J., Alpcan, T., & Leckie, C. (2017). Using virtual machine allocation policies to defend against co-resident attacks in cloud computing. *IEEE Transactions on Dependable and Secure Computing, 14*(1), 95–108.
11. Varadarajan, V., Zhang, Y., Ristenpart, T. & Swift, M. (2015). A placement vulnerability study in multi-tenant public clouds. In *Proceedings of the 24th USENIX Conference on Security Symposium* (pp. 913–928). CA, USA: USENIX Association Berkeley.
12. Han, Y., Chan, J., Alpcan, T., & Leckie, C. (2015). A game theoretical approach to defend against co-resident attacks in cloud computing: Preventing co-residence using semi-supervised learning. *IEEE Transactions on Information Forensics and Security, 11*(3), 556–570.
13. Xing, L., Levitin, G., & Xiang, Y. Defending N-version programming service components against Co-resident Attacks in IoT Cloud Systems, *IEEE Transactions on Services Computing, in press.* https://doi.org/10.1109/tsc.2019.2904958.
14. Levitin, G., Xing, L., & Dai, Y. (2017). Optimal data partitioning in cloud computing system with random server assignment. *Future Generation Computer Systems, 70,* 17–25.
15. Levitin, G., Xing, L., & Dai, Y. (2018). Co-residence based data vulnerability vs. security in cloud computing system with random server assignment. *European Journal of Operational Research, 267*(2), 676–686.
16. Xing, L., & Levitin, G. (2017). Balancing theft and corruption threats by data partition in cloud system with independent server protection. *Reliability Engineering and System Safety, 167,* 248–254.
17. Luo, L., Xing, L., & Levitin, G. (2019). Optimizing dynamic survivability and security of replicated data in cloud systems under co-residence attacks. *Reliability Engineering & System Safety, 192,* 106265.
18. Levitin, G., Xing, L., & Huang, H.-Z. (2019). Security of separated data in cloud systems with competing attack detection and data theft processes. *Risk Analysis, 39*(4), 846–858.
19. Harrop, W., & Matteson, A. (2013). Cyber resilience: A review of critical national infrastructure and cyber security protection measures applied in the UK and USA. *Journal of Business Continuity & Emergency Planning, 7*(2), 149–162.
20. Linkov, I., Roslicky, L., & Trump, B. D. (2020). *Resilience and hybrid threats: Security and integrity for the digital world.* Amsterdam: IOS Press, Incorporated.
21. Herrington, L., & Aldrich, R. (2013). The future of cyber-resilience in an age of global complexity. *Politics, 33*(4), 299–310.
22. Kott, A., &, Linkov, I. (2019). Cyber Resilience of Systems and Networks (1st ed.). Springer International Publishing: Imprint: Springer: Cham.
23. Flammini, F. (2019). Resilience of cyber-physical systems: From risk modelling to threat counteraction (1st ed. 2019. ed.). Cham: Springer International Publishing: Imprint: Springer.
24. Dsouza, G., Hariri, S., Al-Nashif, Y., & Rodriguez, G. (2013). Resilient dynamic data driven application systems (rDDDAS). *Procedia Computer Science, 18,* 1929–1938.
25. Osanaiye, O., Raymond Choo, K., & Dlodlo, M. (2016). Distributed denial of service (DDoS) resilience in cloud: Review and conceptual cloud DDoS mitigation framework. *Journal of Network and Computer Applications, 67,* 147–165.

26. Fang, Y. P., & Zio, E. (2019). An adaptive robust framework for the optimization of the resilience of interdependent infrastructures under natural hazards. *European Journal of Operational Research, 276*(3), 1119–1136.
27. Bostick, T. P., Connelly, E. B., Lambert, J. H., & Linkov, I. (2018). Resilience science, policy and investment for civil infrastructure. *Reliability Engineering and System Safety, 175,* 19.
28. Anderson, T., Busby, J., Gouglidis, A., Hough, K., Hutchison, D. & Rouncefield, M. (2020). Human and organizational issues for resilient communications. In *Guide to disaster-resilient communication networks. Computer communications and networks.* Cham: Springer.
29. Borsci, S., David, L. Z. (2020). Chapter 117—uman factors and system thinking for medical device. In I. Ernesto (Ed.), *Clinical engineering handbook (Second Edition)* pp. 829–831. Academic Press.
30. Levitin, G., Hausken, K., Taboada, H. A., & Coit, D. W. (2012). *Data Survivability versus* Security in Information Systems. *Reliability Engineering & System Safety, 100,* 19–27.
31. Chen, D., Xu, M., & Shi, W. (2018). Defending a cyber system with early warning mechanism. *Reliability Engineering & System Safety, 169,* 224–234.
32. Naidu, P., & Bhagat, B. (2017). Emphasis on cloud optimization and security gaps: A literature review. *Cybernetics and Information Technologies, 17*(3), 165–185.
33. Qiu, X., Dai, Y., Xiang, Y., & Xing, L. (2019). Correlation modeling and resource optimization for cloud service with fault recovery. *IEEE Transactions on Cloud Computing, 7*(3), 693–704.

Gregory Levitin received the B.S. and M.S. degrees in electrical engineering from Kharkov Polytechnic Institute, Kharkov, Ukraine. B.S. degree in mathematics from Kharkov State University, Kharkov, Ukraine, and the Ph.D. degree in industrial automation from the Moscow Research Institute of Metalworking Machines Moscow, Russia, in 1982, 1986, and 1989, respectively. He is currently a Distinguished Visiting Professor with the University of Electronic Science and Technology of China, Chengdu, China, and a Senior Expert at the Reliability Department of the Israel Electric Corporation, Haifa, Israel. His current research interests include operations research and artificial intelligence applications in reliability and security. He has published over 300 papers and five books. Prof. Levitin is the Chair of the European Safety and Reliability Association Technical Committee on System Reliability. He was an Associate Editor of the IEEE Transactions on Reliability and an Area Coordinator of the International Journal of Performability Engineering. He is an Associate Editor of Reliability Engineering and System Safety and IISE Transactions, and an Editorial Board Member of the Journal of Risk and Reliability, and Reliability and Quality Performance.

Liudong Xing received her M.S. and Ph.D. degrees in electrical engineering from the University of Virginia in 2000 and 2002, respectively. She is currently a Professor and Graduate Program director with the Department of Electrical and Computer Engineering, University of Massachusetts (UMass) Dartmouth, USA. Her current research interests include reliability and resilience modeling, analysis and optimization of complex systems and networks. Prof. Xing was the recipient of the 2014 Leo M. Sullivan Teacher of the Year Award, the 2010 Scholar of the Year Award, and the 2011 Outstanding Women Award of UMass Dartmouth. She was the recipient of the 2018 IEEE Region 1 Outstanding Teaching in an IEEE Area of Interest (University or College) Award, the 2015 Chang Jiang Scholar award by the Ministry of Education of China, and the 2007 IEEE Region 1 Technological Innovation (Academic) Award. She was also co-recipient of the Best (Student) Paper Award at several conferences and journals. She has published two books titled "Binary Decision Diagrams and Extensions for System Reliability Analysis" and "Dynamic System Reliability: Modeling and Analysis of Dynamic and Dependent Behaviors." She is an Associate Editor or Editorial Board member of multiple journals including Reliability Engineering & System Safety, IEEE Internet of Things Journal, and International Journal of Systems Science. She is a senior member of IEEE and fellow of The International Society of Engineering Asset Management.

Chapter 34
Climate Change Causes and Amplification Effects with a Focus on Urban Heat Islands

Alec Feinberg

Abstract Global Warming has man-made root causes and amplification effects. As reliability engineers, we know that without understanding real root causes, you may not be addressing the main part of a problem. This will be a key issue in this chapter as the global warming emphasis has been on CO_2 reduction. Therefore, our focus in this chapter will be to look at a key root cause that is not currently being addressed enough often termed Urban Heat Islands (UHI) effect. This is the heat created from cities and their area coverage. We will focus on this primarily because at the present time, the International Panel on Climate Change (IPCC), the world's governing body on the subject, is not providing any guidance on "albedo" goals similar to the way they have made suggestions for CO_2 reduction. This is important as most countries look to the IPCC for guidance and it is the author's opinion that UHI do provide a reasonable contribution to global warming, as well as it is known that they cause health-related problems from their excess heat.

Keywords Global warming · Urban heat islands · Greenhouse gas · Impermeable surfaces · Hydro-hotspots · Hotspots

34.1 Introduction

Before we delve into this subject, in this section, we will address the primary mechanisms associated with climate change root causes and amplification effects summarized in Table 34.1.

In climate change, it makes sense to talk about two dates. In this chapter, we will look at 1950 to present day (2019–2020).

As Table 34.1 shows, population is a key driver in expanding urban heat islands, roads, and greenhouse gases. This means that population increase is the main root cause and its growth is directly correlated to global warming. Most people are familiar with how well the greenhouse gas CO2 is correlated to population growth and

A. Feinberg (✉)
dba DfRSoft.com, 9510 Centerwood Drive, Raleigh, NC 27617, USA
e-mail: dfrsoft@gmail.com

K. B. Misra (ed.), *Handbook of Advanced Performability Engineering*,
https://doi.org/10.1007/978-3-030-55732-4_34

Table 34.1 Global warming cause and effects

Global warming causes →	Population → Expanding urban heat islands (uhi), roads & increases in greenhouse gas
Global warming feedback amplification effects →	Water Vapor Feedback, Land Albedo Change Due to Cities & Roads, Ice and Snow—Albedo Feedback, Lapse Rate Feedback, Cloud Feedback, etc.
Urban heat Island amplification effects →	UHI Solar Heating Area (Building Areas), UHI Building Heat Capacities, Humidity Effects, and Hydro-Hot Spots, Reduced Wind Cooling, Solar Canyons, Loss of Wetlands, Increase in Impermeable Surfaces, Loss of Evapotranspiration Natural Cooling.

global warming temperature. In this chapter, we will demonstrate the same type of correlation for UHI growth. The population since 1950 has increased by a factor of 3.

Population is the main driver

- 1950 2,580 Million
- 2019 7,800 Million
- Three times greater than 1950

The population growth rate is currently about 1.2% per year and has varied as high as 2.1% in the 1960s and early 1970s. This is shown in Appendix.

34.1.1 Global Warming Root Causes

One interesting point is that more than half of the world's population now lives in urban areas. Therefore, right away, we see that cities are one key suspect causes of man-made global warming. This is included in our table under Urban Heat Islands and Roads.

Urban Heat Islands (UHI): Typically, cities are roughly 1–4°C warmer than surrounding rural areas (Fig. 34.1). Thus, the temperature profile mimics that of an island. We are all familiar with how hot pavements can get on a sunny day due to solar heating. This makes cities really a hot spot place to live. So, this is the first concept we need to understand. Climatologists including the Intergovernmental Panel on Climate Change (IPCC) have recognized that cities are much hotter than the rural areas often called Urban Heat Islands (UHI) effect. This is due to their low albedo absorbing effect of cities. Furthermore, it is found that cities in humid climates are even warmer than the cities in dry climates [1]. Therefore, humidity provides an amplification effect. We discuss how some authors have found that 1/3–1/2 of global warming is due to UHI [2–6]. This is supported by the author as well

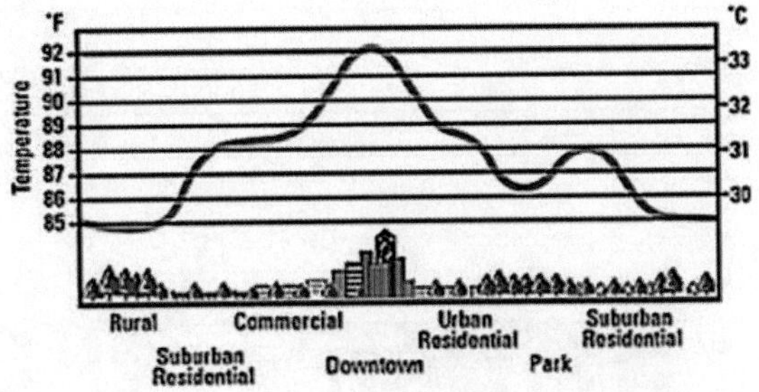

Fig. 34.1 Urban heat island profile

in an modeling [7–9] which is presented here. One might ask why climatologists (specifically the IPCC) have not called for regulations on albedo (higher reflectivity) city design requirements compared to their CO_2 emissions effort in global meetings. As we mentioned, this will be discussed in the chapter.

Population also drives Greenhouse (GH) gases.

Greenhouse gases can be thought of as blankets on your bed, you need enough to keep you warm, but too many can get too hot. The main GH gases are

- Water vapor (H_2O, 25,000 PPM), CO_2 (414 PPM), Methane (CH_4), Ozone (O_3)

Water vapor is the most abundant GH gas with a high value of around 50,000 ppm and an average value of very roughly 25,000PPM [10, 11], while CO_2 was around 300 PPM in 1950 and at the end of 2019 has increased to 414 PPM. The other GH gases are less influential. The increase in CO_2 is what climatologists point to as the main reason for global warming.

- We note that GH gases do not absorb sunlight which is often referred to as short-wave (SW) electromagnetic radiation; so sunlight is essentially transparent to all of the GH gases.
- However, when the Earth absorbs sunlight, it re-radiates Infrared (IR) which is referred to as the longwave (LW) electromagnetic radiation (heat) or IR radiation. IR radiation is in a different part of the spectrum. Therefore, this radiation given off as heat has a long wavelength compared to the short wavelengths found in sunlight. In this area of the spectrum, the greenhouse gases can absorb LW radiation. GH gases absorb a certain long wavelength, and then re-emit this radiation of the same long wavelength in arbitrary directions, some back to Earth and some to outer space.

34.1.2 The Root Causes Result in Warming Since 1950

These root causes and amplification effects create warming. Before we explain the amplification effects, let's look at the temperature data. How do we know that climate

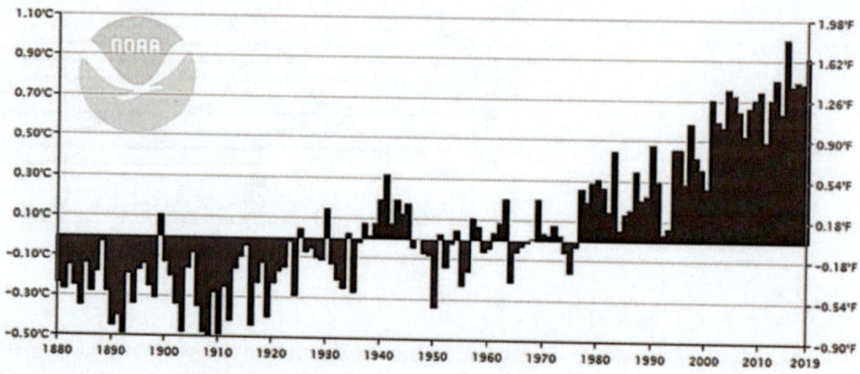

Fig. 34.2 NOAA [9] global land and ocean november temperature anomalies 1880–2019

change is occurring? It is logical not to trust just one data set. The data signs of global warming are

- Surface thermometers—climatologists look at surface thermometers from across the globe and take the averages. The reported temperature rise is about 1.6°F from 1950 to 2019.
- Satellite temperatures support temperature increases
- Ice is melting
- Ocean heat is increasing about the same rate as atmospheric temperatures
- Sea levels are rising

Figure 34.2 obtained from NOAA, 2019 [12] shows the warming trend from 1880 to 2019. The crossover warming period is usually considered to occur around 1950. The average temperature change may be summarized

- A 0.95°C rise corresponds to a 1.7°F rise since 1950 to the beginning of 2020
- 1950 average temperature of 57°F
- 2019–2020 average temperature of 58.73°F

34.1.3 Proof the Global Warming is Due to Man

We know there are many skeptics that global warming is due to mankind. Everyone agrees on the fact that our Earth is warming. But a few people will claim it is a natural trend. The best illustration that it is unnatural is to look at the warming over the last 20,000 years and observe the trend.

We can immediately see from Fig. 34.3 that the warming occurring at present day has never been observed before, even since human civilization started (Holocene area on the graph) and is totally different, unlike any other warming trend. This unfortunately is pretty much irrefutable evidence that warming is due to mankind.

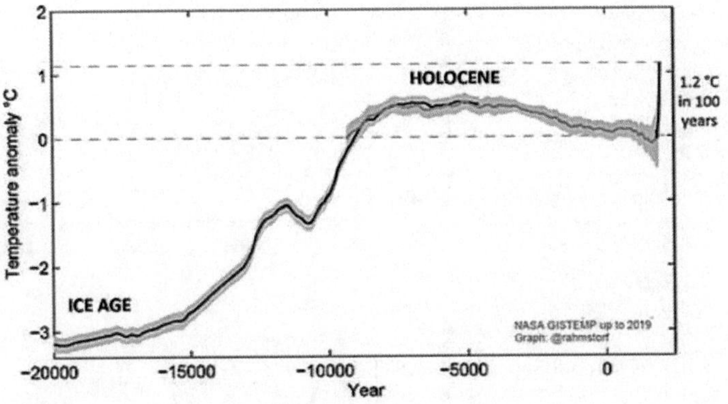

Fig. 34.3 Irrefutable evidence that global warming is due to mankind

34.2 Global Warming Feedback Amplification Effects

Climate feedback is similar to electrical engineering operational amplifier feedback gain. As the root cause warms the planet, we get a number of feedbacks such as sea ice and snow melting. In this section, we will review some of the major feedback problems.

34.2.1 Water Vapor Feedback

In climate change, one of the key issues is that warmer air causes air to expand so it can hold more water vapor. Thus, the root causes, UHI and greenhouse gases, create warmer air which in turn unfortunately cause one of the most problematic climate feedback mechanisms. The main problem is that warmer air

- ***Increases the specific humidity and decreases % Relative Humidity (RH).*** Both of these facts are illustrated in Fig. 34.4a, b. Moisture content in the atmosphere is a major concern since water vapor is known to be the most potent of all the greenhouse gases. Therefore, when evaporation does not all go into clouds, some of it is held in the atmosphere, it becomes a major amplification feedback effects simply because as mentioned above, water vapor is itself a potent GH gas.

The amount of global warming created by water vapor feedback is thought to be a factor of 2.

The percent of global warming created by this feedback is close to that of CO1.

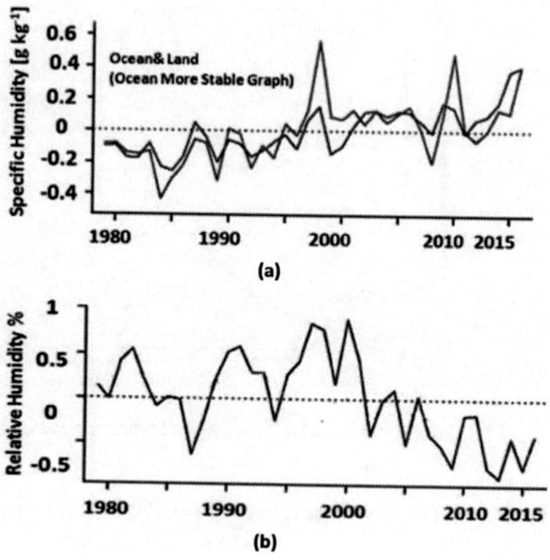

Fig. 34.4 **a** Increasing specific humidity **b** decrease trend in % relative humidity (Warmer air holds more moisture) graph by NOAA climate.gov [12], Fdez-Sevilla [13]

34.2.2 Global Warming Due to Arctic Sea Ice Loss

While the Antarctic is relatively stable in the amount of sea ice, the arctic sea ice is currently melting at an alarming rate. The primary root cause is:

- Loss of albedo (reflectivity) from a decreasing amount of snow and ice.

Albedos and rough area estimates are provided in Table 34.2. Albedo [14] is the solar reflectivity of objects. This is illustrated in Table 34.2. For example, ice is 0.6 which is 60% reflective and 40% absorbing. If an area of ice melts, then it likely becomes part of the open ocean having an albedo of 0.06 or 94% absorbing a factor of 10 more shortwave radiation (see Table 34.2). This global warming feedback is troubling in the last two decades; the sea ice loss is about 12.85% per decade [15]. Fortunately, the Arctic areas receive only about 40% as much solar radiation [16] reducing the feedback effect.

- Feinberg [7] using an albedo model showed that this change may result in about 0.15°C of global warming. As global warming is close to 1 °C this represents about 15% of the problem.

34.2.3 Quantifying Global Warming

To get a crude estimate of global warming and its feedback, some suggested values can be provided. Feinberg [7] illustrated that if greenhouse gases were responsible for 40% of global warming, then Table 34.3 illustrates some examples.

Table 34.2 Albedo of different surfaces and estimated areas

Surface	% of Earth Area	Albedo (0–1)
Water type	71	
Sea Ice	15	0.66
Open ocean	56	0.06
Land Type	29.006	
Roads (0.04)	0.09	0.04
Urban cov (0.12)	0.316	0.12
Forest (0.17)	3.3	0.17
Forest (snow)	5	0.81
Grass lands (0.26)	3.7	0.26
Grass lands snow	7	0.81
Desert (0.4)	9.6	0.4
Sum % of earth area	100.006	
Weighted earth		
Clouds (0.47)	60	0.472

Table 34.3 Some crude estimate of global warming

Warming component	Percent of GW	% of GW root cause
Urbanization	1.9–22%	4.6%–43%
Greenhouse gases (40%)	40.00%	95%–57%
Sea ice melting feedback	15%	
Water vapor feedback	42%–51%	
X-Other)	1.5% to -24%	

We note that X-Other is other feedback mechanisms such as cloud coverage, snow melting, lapse rate, etc. Cloud coverage can cause negative feedback which reverses some of the global warming trends as cloud albedo reduces shortwave radiation onto the Earth's surface. Note that Urbanization creates an albedo effect that increases shortwave radiation absorption and longwave radiation also warming the Earth. Note in the last column shows the percent of the root cause for global warming. We see that Urbanization can be quite problematic, which leads us to the next topic. The large variation associated with this value is related to the difficulty in estimating how much land has been urbanized and the estimate for what is described below as UHI amplification values.

34.3 UHI Additional Amplification Effects

Urban heat islands have amplification effects that extend the effective area. From Table 34.1, we see that there are three main UHI amplification effects due to humidity, heat capacity, and hydro-hot spots.

- ***The humidity amplification effect***: This effect has been observed. For example, Zhao et al. [1] noted that UHI temperature increases in daytime ΔT by 3.0°C in humid climates but decreases by 1.5°C in dry climates. They noted that such relationships imply UHI will exacerbate heat wave stress on human health in wet UHI climates. One explanation is how heat dissipates through convection which is more difficult in humid climates. Another explanation is that warmer air holds more water vapor. This can increase local specific humidity so that there could be local greenhouse effects.
- ***The heat capacity and solar heating area amplification effect***: This effect contributes to the day–night UHI cycle. In most cities, it is observed that daytime atmospheric temperatures are actually cooler compared to night. For example, in a study by Basara et al. [17] in Oklahoma city UHI, it was found that at just 9 m height, the UHI was consistently 0.5–1.75°C greater in the urban core than the surrounding rural locations at night. Further, in general, the UHI impact was strongest during the overnight hours and weakest during the day. This inversion effect can be the result of massive UHI buildings acting like heat sinks, having giant heat capacities, and storing heat in their reservoir via convection as solar radiation is absorbed during the day. This occurrence often reduces the UHI day effect, but at night, buildings cool down, giving off their stored heat that increases local temperatures to the surrounding atmosphere. This effect increases with city growth as buildings have gotten substantially taller since 1950 [18].
- ***The hydro-hots pot amplification effect***: This effect is not well addressed. Atmospheric moisture source is a complex issue due to Hydro-Hot Spots (HHS). HHS occur when buildings are hot due to sun exposure. Then, during precipitation periods, the hot evaporation surfaces increase localized water vapor as warm air holds more moisture. This increase in local greenhouse gas could blanket city heat and increase infrared radiation during these periods, providing another UHI humidity amplification source.
- ***Reduced wind cooling and solar canyons***: In UHI, reduced wind is a known effect due to building wind friction that inhibits cooling by convection. Tall buildings also create solar canyons and trap sunlight, reducing the average albedo, although some benefits occur from shading. In general, both have the effect of amplifying the temperature profile of UHI.

The main problem is the difficulty in figuring out a way to quantify the above amplification effect. For the interested reader, the Feinberg [7] paper provides two estimates.

34.4 The Problem with the IPCC Guidelines and Our Focus in This Chapter

The International Panel on Climate Change (IPCC) report SYR_AR5 [19] guidelines are to meet a goal of less than 2°C rise. This to be achieved by focusing on CO_2 reduction:

> Multi-model results show that limiting total human-induced warming to less than 2°C relative to the period 1861–1880 with a probability of > 66% would require total CO_2 emissions from all anthropogenic sources since 1870 to be limited to about 2900 Gt CO_2 when accounting for non-CO_2 forcing as in the RCP2.6 scenario, with a range of 2550–3150 Gt CO_2 arising from variations in non-CO_2 climate drivers across the scenarios considered by WGIII. About 1900 [1650–2150] $GtCO_2$ were emitted by 2011, leaving about 1000 $GtCO_2$ to be consistent with this temperature goal.

The main contention of this author is that there are ***no albedo guidelines for cities and roads similar to what is said regarding*** $\boldsymbol{CO_2}$. As we have stated, UHI must be a source of global warming. The question is, by how much? Therefore, our focus on this chapter is primarily on UHI. We will see there is a need to

- Raise awareness that UHI are the root cause of global warming and many other local effects
- Have the IPCC and world leaders to set Albedo guidelines for Cities & Roads
- Have the IPCC and world leaders to have guidelines for Zero Population Growth
- Have the IPCC and world leaders to set more guidelines for Eco-Friendly Urban Design
- Have governments to measure city hot spots and find ways to mitigate them in each city such as taxing buildings with low albedos
- Recommend an agency like NASA be tasked with finding applicable solutions to cool down UHI.

34.4.1 Lack of Albedo Guidelines for UHI Similar to CO_2

If UHI are likely significantly contributing to global warming, as such, it would be important to have both CO_2 and albedo goals for cities and roads. Even if UHI are not significant contributors on a global scale, they definitely are problematic to human health due to the increased warming in cities from the UHI effect. That is, it makes sense that we need to increase the reflectivity of our cities and roads to slow down the warming trend and possibly reverse some of the global warming as we show below in our albedo solar model. The knowledge that UHI may be causing significant issues to global warming was pointed out as early as 2007 by McKitrick et al. [3]. We will discuss this in the next section. We see that this poses a major risk. For example, even if the IPCC is 99% confident that global warming is only formed CO_2, the risk can be quantified as follows:

- Risk = Probability of Failure x Severity
- Risk Quantification:
  - If you are a 99% GW is only caused by CO_2 (Prob. of Fail = 1%)
  - Severity = World Population 7.7×10^9
- GW Risk = 1% × 7.7 Billion People = 77 Million People
- This has been the risk where the IPCC has failed to address UHI global warming issues
- Conclusion: **Better Safe than Sorry!**

Unfortunately, the IPCC is really the only group capable of making such guidelines that would help on a global scale with the UHI albedo climate problems because they are the global climate leaders tasked with this responsibility.

34.5 Some Key Publications on How Much Do UHI Contributes to Global Warming?

There have been a good handful of key papers (besides the author's recent publication) on UHI and their findings have shown anticipated significance to global warming. Here is a list of some key papers:

- In 2007, McKindrick and Michaels published a highly controversial paper [3]. Their research showed that UHI may be causing as much ***as half of the global warming from the period they studied from 1999 to 2005***. They used gridded Earth Temperature stations in their assessment. This paper became a major issue through the years with many IPCC authors and Mckindrick has strongly defended his work on his website. I suggest that you go to his website to research his responses and his additional publications that he made in defending his work.
- Huang & Lu [6], Yang et al. [5], UHI China Studies: "Our results on the relative contribution of the UHI to climate warming are consistent with previous studies. Ren et al. [4] found that urbanization-induced warming for Beijing (Wuhan) was significant and accounted for 80.4% (64.5%) of the warming over 1961–2000 and 61.3% (39.5%) of the warming over 1981–2000.... The warming rate due to the UHI and its contributions to the climate warming in the ***fifth report of the IPCC can still be regarded as conservative in the urban agglomeration region.*** Some studies [6] have ***suggested that "significant" contribution of urbanization to temperature changes might be comparable to that of GHG emission for metropolises and large cities."***
- Yang, Hou & Chen China Study 2011 [5]: "For metropolises and large cities in east China, the significant contribution of urbanization to temperature change may be comparable to that of GHG concentration.... The increasing divergence between urban and rural surface temperature trends highlights the limitations of the response policy to climate change; these policies focus only on GHG reduction

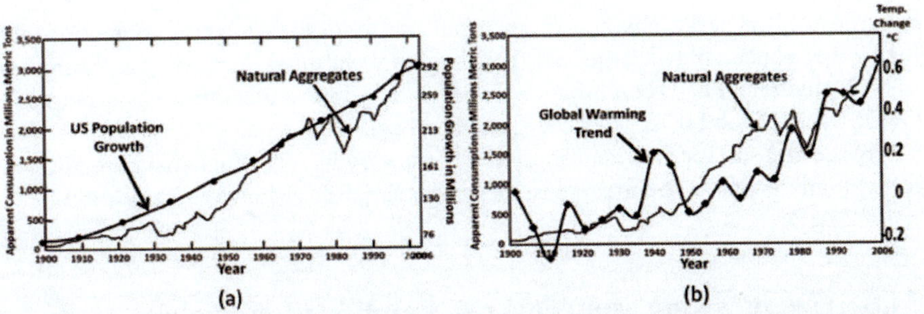

Fig. 34.5 **a** Natural aggregates (USGS 1900–2006) [20] correlated to U.S. population growth (World Bank 2018) **b** Natural aggregates correlated to global warming (NASA 2020) [21]

….***Policymakers need to address the impact of land use such as urbanization*** and deforestation on climate change in addition to that of GHG emissions. ***Serious measures for broadening the range of management strategies beyond GHG reductions and a land-based mitigation framework should be included in the scheme for mitigating climate change***."

34.6 Correlation Assessment of Urban Heat Islands to Global Warming

Often, people refer to correlation assessments as proof of events occurring. This is especially true for CO_2. It is easy to Google CO_2 correlation to population growth and to global warming. We see correlations are helpful. But keep in mind while they are necessary, they are typically not a sufficient condition to prove what is occurring. For example, we can just as easily show correlations to UHI. Figure 34.5 shows the correlation of Cities and Road building material (called natural aggregates) to U.S. population growth and to Global warming. After all, it is the building materials that get hot from solar radiation.

Now as we said it is necessary but it's not sufficient condition that cities and roads contribute significantly to global warming. We now need some more scientific proof as in Sect. 34.4. In the next sections, we will approach this issue from a novel perspective.

34.7 How Much Area Do UHI and Their Urban Areas Cover?

Many climatologist have the contention that UHI effects are basically only of local significance. This is most likely related to urban area estimates. For example, IPCC [22] AR5 report references Schneider et al. [23] study that resulted in urban coverage of 0.148% of the Earth. This seemingly small area tends to dismiss the contention that the UHI effect can play a large scale role in global warming. Furthermore, estimates of how much land has been urbanized vary widely in the literature. For example, a GRUMP [24] found 0.783% of the Earth had been urbanized.

We see that surface area land approximations vary widely which makes prediction difficult and is one of the main reasons for the variability shown in Table 34.3.

34.8 Urbanization Surface Area Amplification Factors

Authors have found that the UHI effect has what is called a footprint area effect that is larger than the UHI and its coverage itself. This is due to all the amplification factors shown in Table 34.1. Therefore, the temperature of the UHI effectively extends beyond the area of land that is covered by the core and its urbanization. For example, Zhang et al. [25] found the ecological footprint of the urban land cover extends beyond the perimeter of urban areas, and the footprint of urban climates on vegetation phenology they found was 2.4 times the size of the actual urban land cover. In a more recent study by Zhou et al. [26], found that the "footprint" of UHI effect, including urban areas, was 2.3 and 3.9 times of urban size for the day and night, respectively.

From the items in Table 34.1, we see that the Amplification Factor (AF) is some function of all the components listed

$$\mathrm{AF}_{\mathrm{UHI\ for\ 2019}} = f\left(\overline{\mathrm{Build}}_{\mathrm{Area}} \times \overline{\mathrm{Build}}_{C_P} \times \overline{R}_{\mathrm{wind}} \times \overline{\mathrm{LossE}}_{\mathrm{vtr}} \times \overline{\mathrm{Hy}} \times \overline{S}_{\mathrm{canyon}}\right) \tag{34.1}$$

In a recent study by Feinberg [7], he was able to illustrate how this factor likely varies from 3.1 to 8.4. If we established a reference year such as 1950, this means that any new area would effectively be amplified by this factor which would affect the heat over the amplified area.

34.9 UHI Global Warming Estimates

There have been numerous studies on UHI effects. We have discussed the key publication in Sect. 34.4. This included the McKitrick and Michaels [3] paper that half of the global warming trend from 1979 to 2002 is caused by UHI. We also noted that research in China [6] indicates that UHI effects contribute to climate warming by about 30%. There is an apparent pushback as the handful of papers have been unsuccessful in getting the world leaders (including IPCC authors) to date on making city albedo guidelines similar to what they have established for CO_2.

A simplistic feasibility model has its strength in

- Supporting estimate from these authors

- Corrective action assessment using "what if" scenarios for changes to the albedo

In a recent paper by the author [7, 8], a nominal coverage area found by Schneider [23] and the worst case by GRUMP [24] was used. Both estimates were on data taken mainly from satellites around 2000. Then we extrapolated up to 2019 and down to 1950. The area amplification climate factor as discussed in the footprint section was then applied. The compiled results were inserted into a Weighted Amplification Albedo Solar Urbanization (WAASU) Model (from Feinberg 2020 [7]) with the results in Table 34.3 that

- *Urbanization likely has contributed to global warming between 1.9% and 22% with the large variable due to uncertainty in area estimates. Furthermore, urbanization contributes to feedback so as a root cause it contributes almost two times this, 4.6%–43% of the global warming temperature rise.*

We see these results vary widely as estimates on what percent of the Earth is urbanized also vary by a large amount along with the difficulty in estimating amplification factors, as this result is also based on UHI amplification effects. There are also global amplification effects due to feedback issues such as albedo decrease due to ice and snow melting, and an increase in specific humidity due to the fact that warm air holds more moisture. The increase in water vapor is considered to double the warming. If we use just this effect, then the Feinberg [7] results yield an increase of about 3–27% of global warming due to Urbanization. Note that this somewhat supports the finding of these other authors from a totally different perspective using an albedo model and estimating feedback temperature rises (see Table 34.3).

34.10 Basics Physics of Global Warming

Understanding global warming physics is not that complicated. We can provide a brief overview of what climatologists call the Earth's Energy Budget (EEB). This is a common term that one can easily lookup. In general, when sunlight shines on the Earth, energy absorbed by the sun that warms the Earth, this energy is the main concept for the EEB. The important aspect of this is the fact that we can make assessments without taking into account the Greenhouse Gas (GHG) effect. This allows what portion is just due to the albedo effect. Here, we can understand how much heating is created by the sun's power absorption. This really helps to make things simple. The amount of solar power that falls on the Earth is well established and is given by 1361 W/m^2. This is shown in Fig. 34.6.

However, we need to take into account that only half of the Earth is illuminated at any time and that the Earth is spherical so we lose another half so this factor is a quarter shown in the formula. Then the EEB formula for sun absorption is

$$P_{\text{EDergy Budget}} = 1361W/\text{m}^2\{0.25 \times 1 - \text{Albedo}) = \sigma T^4 \tag{34.2}$$

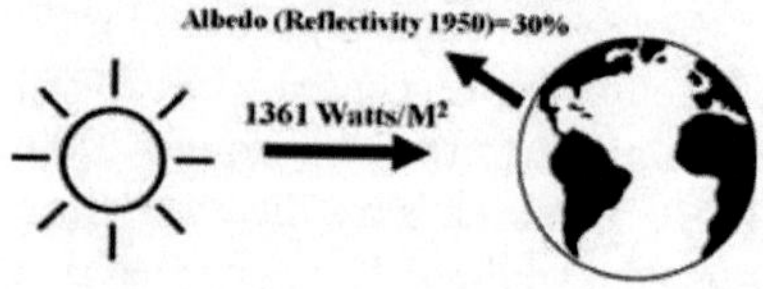

Fig. 34.6 Showing the power of the sunlight falling on the earth with its 1950 albedo of 30%

Table 34.4 IPCC Earth's energy budget values (Hartmann et al., 2013)

IPCC Item	Incident and reflected radiation (W/m^2)	Albedo %	Absorbed (W/m^2)
Earth	100/340	**29.4118**	$240 = 340 \times (1-.294)$
Atmosphere & clouds	76/340	**22.3529**	**79**
Earth surface albedo	24/340	**7.0588**	**161**

This is the shortwave radiation incoming that is absorbed by the Earth, must be equal to the longwave radiation that leaves the Earth when we are in equilibrium. Equilibrium is a difficult term in global warming as there is transient warming occurring all the time. That is, the world stopped for a moment so no increase in GHGs or building construction, for example, warming would still increase due to the transient effects such as snow and ice melting. However, at some point, we would establish a new equilibrium. The table 34.4 is taken from an IPCC report [27] on the EEB.

In Table 34.4, we see that the albedo of the Earth is 29.4118%, this is close to the 30% (Albedo Science Direct) shown in Fig. 34.6. We are concerned with the energy the Earth absorbs. You might notice the σT^4 term on the RHS of Eq. 34.2. This is the Stefam-Boltzmann formula. That is, once we solve for the absorbed power on the LHS of the equation, we can convert that to the Earth's average temperature by solving for T. Here, σ is a constant and is given by $\sigma = 5.670367 \times 10^{-8}$ W. m^{-2}.K^{-4}. Notice that if you multiply 1361 by 0.25 in Eq. 34.2, we obtain about 340 Watts/m^2 shown in the table.

34.11 Physics of Global Temperature for 1950

We are now in a position to use Eq. 34.2 to estimate the Earth's energy budget power absorbed and temperatures. For the 1950 power absorbed using Eq. 34.2, we have for an albedo of 30%

$$P_{\text{Energy Budget 1950}} = 1361\ \text{W/m}^2\{0.25 \times 1 - 0.30) = 238.15\ W/M^2 \qquad (34.3)$$

We see this value is close to the 240 W/m^2 shown in the IPCC Table 34.4. Now we need to convert this to temperature. What we use is the Stefam-Boltzmann formula also shown in Eq. 34.2 as

$$P_{\text{Energy Budget}} = \sigma T^4 \text{ Stefam-Boltzmann Formula,}$$
$$\text{where } \sigma = 5.6, 70, 367 \times 10^{-8} W.m^{-2} \cdot K^{-4} \quad (34.4)$$

Inserting values and solving for temperature T, we find the

$$T_{1950} = (P_{\text{EFB}}/\sigma)^{1/4} = \left(238.15/5.6, 70, 367 \times 10^{-8}\right) = 284.58^{\circ} K$$
$$\sim -1.43^{\circ} F\left(-18.572^{\circ}\text{C}\right) \quad (34.5)$$

This is fairly cold because we have not introduced any greenhouse gases. What greenhouse gases do is allow the sun's rays to pass through to the Earth. We know about 70% of the sun's power is absorbed and 30% reflected. That is according to the IPCC Table about 100 W/m^2 is reflected by clouds, the atmosphere, and the Earth.

That is from Table 0.294118 × 340 W/m^2 incoming is reflected which is 100 W/m^2 shown in the IPCC table. Furthermore, the remaining 240 W/m^2 is absorbed with 79 by the atmosphere and clouds, and 161 W/m^2 by the Earth.

The absorb sun rays turn into infrared radiation which is a different area of the electromagnetic spectrum, then the greenhouse gases in this area of the spectrum can absorb this energy and re-emit it. Some back to Earth and some to outer space.

If for example, 147.74 W/M^2 of the 238.15 W/M^2 power given off is re-emitted back to Earth (about 61.6%) then the total power absorbed by the Earth is 384.9 W/m^2.

$$T_{1950} = (P_{\text{EEB}}/\sigma)^{1/4} = \left(384.91/5.670367 \times 10^{-8}\right) = 287^{\circ} K \sim 57^{\circ} F\left(-13.89^{\circ} C\right) \quad (34.6)$$

This shows the temperature in 1950.

34.11.1 Global Warming Due to One-Fifth Percent Albedo Change

What if in 2019 urbanization caused the reflectivity of the Earth to drop by 0.2%? This would reduce the IPCC albedo from 29.4118 to 29.3519. We can use the equation

$$\%GW = \left\{(P/\sigma)^{0.25}_{2019} - (P/\sigma)^{0.25}_{1950}\right\}/0.95^{\circ} C \quad (34.7)$$

where $P =$ **340 W/m^2** x (1-Albedo) and we note that there is about a 0.95°C global temperature increase since 1950.

Inserting this value, we find

$$\%GW = 5.7\% \quad (34.8)$$

$$T_{2020} - T_{1950} = 0.057 \times 0.95^{\circ} C = 0.068^{\circ} C \quad (34.9)$$

Therefore, this calculation indicates that

- UHI are responsible for 5.7% of global warming.

This is basically how one can estimate the effect related to albedo change. Unfortunately, the albedo of the Earth is hard to measure due to cloud coverage as well even more difficult to assess the effect of urbanization. In the Feinberg [7] model though, it is possible to make such assessments. In a similar manner, the WAASU model only was able to assess a maximum contribution of about 15% of global warming from urbanization (Table 34.3) compared with the McKitrick and Michaels [3] contention that half of the global warming trend from 1979 to 2002 is caused by UHI.

34.12 Implication of Ignoring the Urbanization Effect on Global Warming

We find that both CO_2 and UHI global warming estimates are very difficult to make. It puts climatologists in a difficult spot. Currently, it is clear that the IPCC authors do not account in their reports the radiation forcing due to urbanization change that this author could find. Therefore, their models, which are likely highly complex and computer-driven are built to match current day warming trends, having numerous adjustable parameters. One can see the difficulty of incorporating something like a WAASU model which incorporates amplification factors along with their CO_2 predictions. However, such an addition should be considered.

The underlying truth is that, if they do not add in the urbanization factor, this makes the CO2 estimates even more inaccurate. What we can say is that they are likely both significant and both need to be addressed. Unfortunately, global warming models do not account for UHI. This adds risk. We have pointed out that numerous authors including this one have found that the UHI effect is at least partly responsible for global warming trends. There are partially easily implemented solutions like cool roofs for example that are not being used to a large degree. However, cities for the most part are continuing to use absorbing colors in roofs, roads, and designs with non-reflective architecture with other mitigating methods as well as not being addressed on a global scale. This is highly risky if it turns out the McKitrick and Michael's [3] estimates along with other such authors including this one are correct. It means we are going to do everything that we can in terms of corrective action. Therefore, it would be very helpful if the IPCC and its authors would start to work on this part of global warming in their models and add it to their guidelines. We provide further suggestion and conclusion at the end of this chapter.

34.13 Highly Evaporation Surfaces and Rainwater Management HHS Feedback Mechanisms

In this section, we briefly review UHI-related global warming issues by summarizing issues with the aid of Figs. 34.7a, b. Figure 34.7a which shows HHS from Highly Evaporating Surfaces (HES) feedback and Fig. 34.7b illustrates Rain Water Management (RWM) feedback contributions to global warming.

Figure 34.7a shows HHS–HEHS feedback that may be summarized:

- Low albedo cities and roads emitting infrared radiation (IR), increased warming (approx. 1/3)
- Precipitation occurs, followed by evaporation of HHS–HES moisture, lower %RH increase specific humidity Greenhouse gas in the warmed city area
- Local heat amplification, less local cooling with increased specific humidity amplifies heat index
- Local warming radiates heat increasing Global warming (with the 1/3 estimate)
- Evaporation increases in cities and ocean primarily from UHI and roads creates lower %RH and higher specific humidity globally along with CO_2 increase creating more humidity issues

Figure 34.7b Shows HHS–RWM feedback that may be summarized:

- Higher temperature stormwater is collected off of HHS buildings, streets, and hot cars
- A large percentage is drained to nearby rivers, lakes, or ocean
- Warmer air allows for an increase in specific humidity
- The impermeable city building and roads have replaced vegetative land creating a lost area that would have stored cooler water in soil keeping the land moist with less generated heat compared to HHS runoff.
- This increases land dryness and can mean less land evaporation and more ocean rain since precipitation often follows evaporation areas as discussed below.

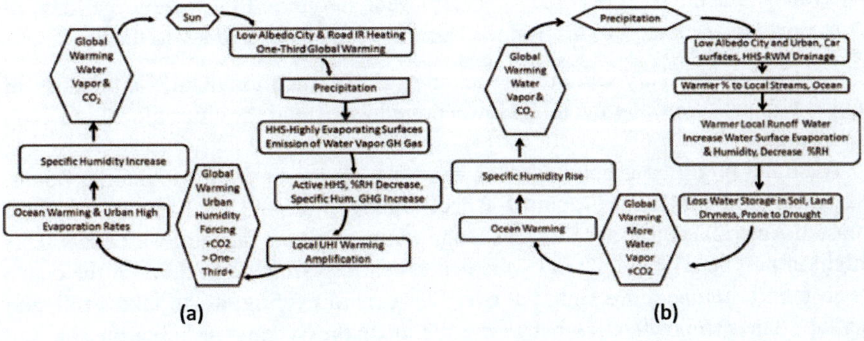

Fig. 34.7 **a** HHS–HES feedback view of contribution to global warming, **b** HHS rainwater management (rwm) high-temperature water cycling in climate change [28]

- The RWM is often warmer from HHS activity raising stormwater temperatures from hot city buildings and street cycling each year billions of gallons of rainwater to local streams, lakes, and ocean contributing to local surface water temperature increases depending on location. These runoffs affect atmospheric warming trends and GH gases.

34.14 Some Data Information on Rainwater Management (RWM) Trends

Another important aspect not addressed by the IPCC is high-temperature stormwater runoff. Rainwater management is an important factor in UHI as it too can influence global warming trends and should be included in their reports. It can also impact where it rains. Rain sometimes follows local evapotranspiration. Apart from precipitation, evapotranspiration is the major component in the hydrologic budget.

When it rains in a city, much of the land in urban areas is covered by pavement or asphalt. These impermeable surfaces in urban cities commonly estimated around 55% runoff, with 30% for evapotranspiration, 10% shallow soil infiltration, and 5% deep soil infiltration. Water temperatures from runoffs are often hotter due to HHS. For example,

- The New York Environment Report, in 2014 [29], "Every year, old sewers flooded by stormwater release more than 27 billion gallons of untreated sewage into New York Harbor."
- Fry [30] reported that in February California estimated that 18 trillion gallons of rain in February alone had most of the water going to the Pacific Ocean. The article goes on to point out the LA dept. of water captured 22 billion gallons of water during the recent storm.
- In August 2001, rains over Cedar Rapids, Iowa, led to a 10.5 C rise in the nearby stream within one hour, which led to the killing of fishes. Similar events have been documented across the American Midwest, as well as Oregon and California [10]
- Sydney Paper reported [31] "Every year around 132 billion gallons of stormwater—enough to fill Sydney Harbor—runs from Sydney to the sea."

It is of course very difficult to tell the global thermodynamic influences of higher temperature water cycling. However, it does extend the cities global warming footprint.

Australia might be a good extreme example, on the Sydney–Melbourne South-East side, the Tasman Sea is about 1–2 deciles range warmer [32] than the Southwest coast of Australia and about 5 deciles range warmer that the far southwest coast. This might in part be an example of cyclic ocean heating. We tend to think of the ocean as an infinite temperature sink, but over 70 years of cycling, it can take a toll, and perhaps this is somewhat of what we are seeing on the Sydney—Melbourne side and coastal issues.

34.14.1 Some Data Information on RWM Causing Dry Day Increases

As an example of the importance in losing wetland (water storage), Cao et al. [33] did a study on wetland reduction in China and correlation to drought with the following conclusion:

- "The wetland distributions and areas of the five provinces of southwestern China in the 1970s, 1990, 2000 and 2008 show that the total reduction of wetland area was 3553.21 km^2 in the five provinces of southwestern China from 1970 to 2008, accounting for about 17% of the ground area, and thus the average annual reduction area is about 88.83 km^2. The reduction rate was comparatively fast from 2000 to 2008 with an average annual reduction of 329.31 km^2. The changes to the wetland area show a negative correlation with temperature (i.e. wetland decrease, increase in temperature), and a positive correlation with precipitation (i.e. wetland decrease, precipitation decrease)."

  Hirshi et al. [34] did the following study:

- "We analyzed observational indices based on measurements at 275 meteorological stations in central and southeastern Europe, and on publicly available gridded observations. We find a relationship between soil-moisture deficit, as expressed by the standardized precipitation index, and summer hot extremes in southeastern Europe. This relationship is stronger for the high end of the distribution of temperature extremes. We compare our results with simulations of current climate models and find that the models correctly represent the soil-moisture impacts on temperature extremes in southeastern Europe, but overestimate them in central Europe."

In Hirshi et al. [34] study, they observed a negative linear relationship between wetland decrease and dry days increase.

Wetland issues are recognized by the IPCC in Chap. 2 [27] 2019, "warming trends over dry lands are twice the global average [35]. However, there is little connection to UHI rainwater runoff being dumped into oceans and this in part causing some of the dry lands."

34.15 Conclusions and Suggestions

From our review of data and its analysis presented, it is our opinion that the IPCC guidelines focused solely on CO_2 reduction appear not to be enough to stop global warming trends from occurring. Our conclusion is that the albedo reduction of UHI is needed to help stop global warming anomalies. This will also reduce HHS contribution to atmospheric moisture issues. Of course, we also feel more studies are needed to assess these impacts such as better estimates of global UHI solar surface areas.

Below we provide suggestions and corrective actions related to albedo and HHS reduction that includes:

- Creating new IPCC goals to include and recognize albedo forcing the issue of UHI and roads
- Recommending changes for the albedo of roads and cities to reducing HHS and the area effect dramatically, i.e. paint roads and building with reflective colors (have minimally albedo requirements, 0.25–0.5)
- Sugawara et al. [36] estimate UHI have an albedo of about 0.12, while Feinberg [7] found that this must increase to 0.2 in order to evenly offset the warming increase
- Mandating future albedo design requirements of city and roads
- Roads to be more HHS eco-friendly
- Recommendation for cars to be more reflective. Here, although worldwide cars likely do not embody much of the Earth's area, recommending that all newly manufactured cars are higher in reflectivity (e.g., silver or white) would help raise awareness of this issue similar to electric cars that help improve CO_2 emissions
- Thoroughly assess and making goals for rainwater management issues including evapotranspiration and rainwater runoff allowed temperatures released into streams, rivers, lakes, and oceans
- Requiring negative population growth to reduce increase HHS–HES surfaces and fossil fuel use
- Improve HHS–HES irrigation to soil
- Improving vegetation in runoff areas
- Adopting Low-Impact Development in city planning and improvements for design approach aiming to mimic naturalized water balances with semi-permeable surfaces
- Requiring severe HHS–RWM changes to reduce runoff into the ocean worldwide that can cause loss of wetlands and local increase in dry days and an increase in evaporation rates
- Providing new studies on albedo and humidity forcing from UHI to better understand their effects, address conflicts with CO_2 theory. Providing updated UHI radiative forcing contribution to GW. Provide a modern microclimate doubling experiment if possible to verify doubling claims.

Appendix

Figure 34.8 shows a plot of the world population growth rate that varies from about 2.1–1.2%. This is used to make a growth rate estimate of urban coverage.

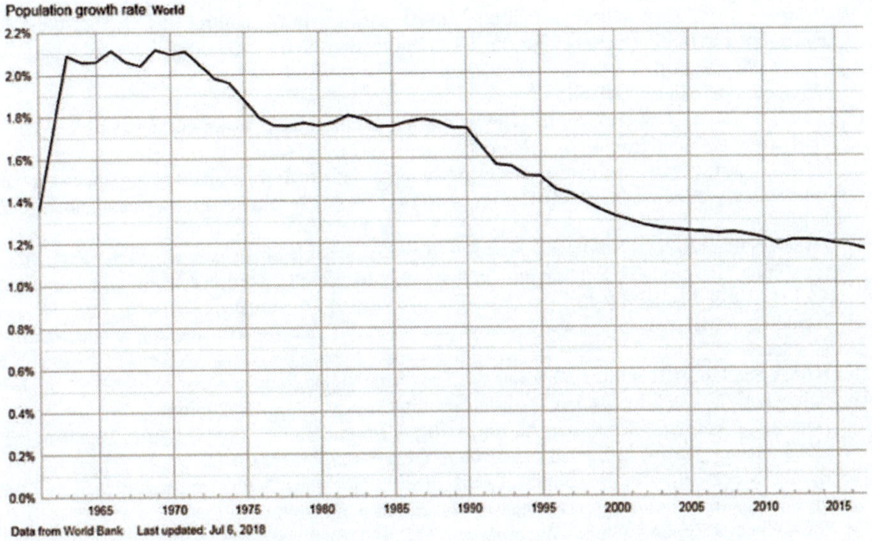

Fig. 34.8 Population growth rate by year from 1960 to 2018 (World Bank 2018) [37]

References

1. Zhao, L., Lee, X., Smith, R. B., & Oleson, K. (2014). Strong contributions of local background climate to urban heat islands. *Nature, 10;511*(7508), 216–219. https://doi.org/10.1038/nature13462.
2. McKitrick, R., & Michaels, J. (2004). A test of corrections for extraneous signals in gridded surface temperature data. *Climate Research.*
3. McKitrick, R., & Michaels, P. (2007). Quantifying the influence of anthropogenic surface processes and inhomogeneities on gridded global climate data. *Journal of Geophysical Research-Atmospheres.* See also McKitrick R., webpage, https://www.rossmckitrick.com/temperature-data-quality.html
4. Ren, G., Chu, Z., Chen, Z., & Ren, Y. (2007). Implications of temporal change in urban heat island intensity observed at Beijing and Wuhan stations. *Geophysical Reseach Letters, 34,* L05711. https://doi.org/10.1029/2006GL027927.
5. Yang, X., Hou, Y., & Chen, B. (2011). Observed surface warming induced by urbanization in east China. *Journal of Geophysical Research Atmospheres, 116.* https://doi.org/10.1029/2010jd015452.
6. Huang, Q., & Lu, Y. (2015). Effect of urban heat island on climate warming in the Yangtze river delta urban agglomeration in China. *International Journal of Environmental Research and Public Health, 12*(8), 8773.
7. Feinberg, A. (2020). Urban heat island amplification estimates on global warming using an albedo model, Preprint in *viXra:2003.0088,* https://vixra.org/abs/2003.0088?ref=11151864 *Researchgate* https://doi.org/10.13140/RG.2.2.32758.14402, (Currently under peer review in the *Journal of SN Applied Science*).
8. Feinberg, A. On geoengineering and implementing an albedo solution with UHI GW and cooling estimates. vixra 2006.0198, https://doi.org/10.13140/RG.2.2.26006.37444/6 (Currently in Peer Review in the *Journal of Mitigation and Adaptation Strategies for Global Change*).

9. Feinberg, A. (2020). Albedo solution to global warming in the control of CO_2, hotspots, & hydro-hotspot forcing and their albedo-GHG interaction, Vixra 2008.0098, https://doi.org/10.13140 (Submitted).
10. Wikipedia, Urban heat island.
11. Kiehl, J. T., & Trenberth, K. E. (1997). Earth's annual global mean energy budget (PDF). *Bulletin of the American Meteorological Society, 78*(2), 197–208.
12. NOAA. (2019). National centers for environmental information, climate at a glance: Global time series, published Dec 2019. Retrieved on Dec 25, 2019 from https://www.ncdc.noaa.gov/cag/.
13. Fdez-Sevilla. (2014). New theory proposal to assess possible changes in atmospheric circulation. Blog "Filling in or Finding out". https://diegofdezsevilla.wordpress.com https://doi.org/10.13140/RG.2.1.4859.3440.
14. Albedo 30%, https://www.sciencedirect.com/topics/earth-and-planetary-sciences/earth-albedo.
15. NASA Sea Ice. (2019). https://climate.nasa.gov/vital-signs/arctic-sea-ice/.
16. Sciencing. (2018). https://sciencing.com/sun-intensity-vs-angle-23529.html.
17. Basara, J., Hall Jr, P., Schroeder, A., Illston, B., & Nemunaitis, K. (2008). Diurnal cycle of the Oklahoma City urban heat island. *Journal of Geophysical Research.*
18. Barr, J. M. (2019). The economics of skyscraper height (Part IV): Construction costs around the world, https://buildingtheskyline.org/skyscraper-height-iv/.
19. IPCC Special Reports, Global Warming of 1.5°C. (2018). 2019 Refinement of the 2006 IPVV guidelines for National Greenhouse Gas Inventories, https://www.ipcc.ch/2019/, 2007 IPCC Fourth Assessment Report, AR4 (2007), AR5 (2014), Gensuo Jia et. Al. Chapter 2: *Land-Climate Interactions* (2019), AR5 Chapter 8 (2014) *Urban Areas*, Aromar Revi et. Al.
20. USGS. (1900–2006). Materials in Use in U.S. Interstate Highways. https://pubs.usgs.gov/fs/2006/3127/2006-3127.pdf.
21. NASA. (1900–2006). Updated, 2020. https://climate.nasa.gov/vital-signs/global-temperature/.
22. Satterthwaite, D. E., Aragón-Durand, F., Corfee-Morlot, J., Kiunsi, R. B. R., Pelling, M., Roberts, D.C., & Solecki, W. (2014). Urban areas. In: Climate change 2014: Impacts, adaptation, and vulnerability. part a: global and sectoral aspects. Contribution of working group ii to the fifth assessment report of the intergovernmental panel on climate change (IPCC).
23. Schneider, A., Friedl, M., & Potere, D. (2009). A new map of global urban extent from MODIS satellite data. *Environmental Research Letters, 4*(4), 044003. https://doi.org/10.1088/1748-9326/4/4/044003.
24. GRUMP. (2005). Global rural urban mapping project, Columbia university socioeconomic data and applications center, Gridded population of the world and the global rural-urban mapping project.
25. Zhang, X., Friedl, M. A., Schaaf, C. B., Strahler, A. H., & Schneider, A. (2004). The footprint of urban climates on vegetation phenology. *Geophysical Reseach Letters, 31,* L12209.
26. Zhou, D., Zhao, S., Zhang, L., Sun, G., & Liu, Y. (2015). The footprint of urban heat island effect in China. *Scientific Reports, 5,* 11160.
27. Hartmann, D. L., Klein Tank, A. M. G., Rusticucci, M., Alexander, L. V., Bronnimann, S., Charabi, Y., et al. (2013). Observations: Atmosphere and surface. In: Climate Change 2013: The Physical Science Basis. In: T.F. Stocker, D. Qin, G.-K. Plattner, M. Tignor, S.K. Allen, J. Boschung, A. Nauels, et al. (Eds.). Contribution of Working Group I to the Fifth Assessment Report of the Intergovernmental Panel on Climate Change. Cambridge: Cambridge University Press, United Kingdom and New York, NY, USA.
28. Feinberg, A. (2020). Review of global warming urban heat island forcing issues unaddressed by IPCC suggestions including CO_2 doubling estimates, viXra:2001.0415.
29. New York Environment Report. (2014), It's been raining in NYC: Where does all that water go?, https://www.nyenvironmentreport.com/its-been-raining-in-nyc-where-does-all-that-water-go/.
30. Fry, H. (2019). *California wastes most of its rainwater, which simply goes down the drain.* Times: LA.

31. Cormack, L., (2015). Where does all the stormwater go after the Sydney weather clears? *The Sydney Morning Hearald*, May. https://www.smh.com.au/environment/where-does-all-the-stormwater-go-after-the-sydney-weather-clears-20150430–1mx4ep.html.
32. BOM. (2018). Bureau of meteorology, annual climate statement, sea surface temperatures very much warmer than average for the Australian region as a whole, issues Jan 2019. http://www.bom.gov.au/climate/current/annual/aus/.
33. Cao, C. X., Zhao, J., Gong, P., Ma, G. R., Bao, D. M., & Tian, K. (2011). Wetland changes and droughts in southwestern China, *Geomatics, Natural Hazards and Risk*, Oct 2011. https://www.tandfonline.com/doi/full/10.1080/19475705.2011.588253.
34. Hirshi, M., Seneviratne, S. I., Alexandrov, V., Boberg, F., Boroneant, C., Christensen, O. B., et al. (2011). Observational evidence for soil-moisture impact on hot extremes in southeastern Europe. *Nature Geoscience, 4,* 17–21.
35. Lickley, M., & Solomon, S. (2018). Drivers, timing and some impacts of global aridity change. *Environmental Research Letters, 13,* 104010. https://doi.org/10.1088/1748-9326/.
36. Sugawara, H., & Takamura, T. (2014). Surface albedo in cities (0.12): Case study in sapporo and Tokyo, Japan. *Boundary-Layer Meteorol, 153,* 539–553. https://doi.org/10.1007/s10546-014-9952-0.
37. World Bank. (2018). Population growth rate, worldbank.org.

Alec Feinberg is the founder of DfRSoft. He has a Ph.D. in Physics and is the principal author of the books, Design for Reliability and Thermodynamic Degradation Science: Physics of Failure, Accelerated Testing, Fatigue, and Reliability Applications. He has presented numerous technical papers and won the 2003 RAMS best tutorial award for the topic, "Thermodynamic Reliability Engineering." He has studied degradation systems for his entire professional career.

Chapter 35
On the Interplay Between Ecology and Reliability

Ali Muhammad Ali Rushdi and Ahmad Kamal Hassan

Abstract This chapter attempts to enhance the interplay between the ecology and reliability fields by employing Boolean-based reliability language and techniques to quantify ecological metrics related to connectivity and redundancy. We emphasize the question of connectivity in models of probabilistic networks as a common area of interest for both fields. The chapter borrows techniques from mainstream reliability theory to treat a prominent problem of ecology, namely that of survivability (of a species), defined here as the probability of successful migration of a certain organism escaping from critical source habitat patches and seeking refuge in specific destination habitat patches via heterogeneous deletable ecological corridors, possibly with uninhabitable stepping stones en route. This problem might be reformulated in contexts other than that of migration, including those of (a) dynamics of metapopulations, colonization, or invasion, (b) gene flow, (c) spread of infectious diseases, epidemics, or pandemics, and (d) energy transfer within food webs. Indicators of network connectivity in classical reliability theory are probabilities that might be designated according to the set of source nodes and the set of destination nodes as one to one, one to many, many to many, or all to all. Our present notion of survivability (of a species) is also a probability of connectivity, now measured from any node (among many nodes) to any node (among many nodes). We explore methods for computing the survivability (of a species) by adapting switching-algebraic techniques that are usually employed in the reliability field. In addition to this survivability metric, we comment on some other connectivity indicators that are currently used in ecology. We stress two recent contributions to the ecology literature, one employing analogy with electric circuit theory, and another concerning the most reliable (or minimum-lag) dispersal paths.

A. M. A. Rushdi (✉)
Department of Electrical and Computer Engineering, King Abdulaziz University, P.O. Box 80204, Jeddah 21589, Saudi Arabia
e-mail: arushdi@kau.edu.sa

A. K. Hassan
Faculty of Electrical Engineering, Ghulam Ishaq Khan Institute of Engineering Sciences and Technology, Topi 23640, Pakistan

K. B. Misra (ed.), *Handbook of Advanced Performability Engineering*,
https://doi.org/10.1007/978-3-030-55732-4_35

Keywords Habitat patch · Ecological corridor · Stepping stone · Survivability of species · Migration-like contexts · Probability-ready expressions

35.1 Introduction

This chapter employs a perspective and a methodology of system reliability in handling the ecological problem of survivability (of a specific species), i.e. the problem of computing the probability that the concerned species migrates successfully between various habitat patches through imperfect heterogeneous ecological corridors and perfect stepping stones. We compute this survivability measure through a minimal adaptation of the techniques used in system reliability evaluation. The chapter represents a modest contribution to ongoing attempts for achieving a profitable interplay between ecology and reliability. Constructive interactions and unifications are sought between these two branches of science, with an ambitious goal of creating a new interdisciplinary paradigm that fruitfully exploits similarities and common interests shared by the reliability and ecology disciplines, as well as inherits strengths and capabilities possessed by any of them. To this end, ecologists are encouraged to become more knowledgeable about essential reliability measures and techniques that are significant to ecology. Similarly, reliability practitioners are advised to acquaint themselves with ecology problems that are amenable to solution by reliability techniques and to seek familiarity with ecological indicators that measure connectivity and redundancy. We confine ourselves herein to the issue of connectivity, which might be singled out as the most prominent common thread between reliability and ecology. In fact, we deal with a single problem of connectivity, but this problem has a multitude of beneficial interpretations.

From an ecology perspective, the issue of landscape connectivity is of paramount importance, since it is a decisive factor for the survival of a species in its habitat, wherein the species normally secures its basic needs (food, water, shelter from weather and predators, a breeding place, etc.) The literature abounds with ecological definitions for landscape connectivity, as well as with metrics for measuring it [1–17]. Taylor et al. [2] define landscape connectivity as "*the degree to which the landscape facilitates or impedes movement among resource patches*," while With and King [3] define it as "*the functional relationship among habitat patches owing to the spatial contagion of habitat and the movement responses of organisms to landscape structure*." According to Tischendorf and Fahrig [4], these definitions indirectly or implicitly support the notion that connectivity is dependent on both species and landscape. Common approaches of predicting landscape connectivity metrics include deducing landscape pattern indexes, individual-based movement simulations, analytic measures such as those based on graph theory or on least cost path models, as well as models utilizing a purported analogy of certain ecological parameters with the electric circuit quantities of current, voltage, and resistance [18]. Notable among the metrics suggested for landscape connectivity is the survivability metric, which is presented herein and is directly borrowed from reliability theory.

This metric, introduced earlier in [19–21] under the name of 'survival reliability' employs minimal realistic assumptions to produce effective predictions and explanations. It was first introduced by Jordán in his seminal work [22], which is based on the use of a metapopulation landscape graph; whose nodes stand for perfect patches and whose edges represent identical corridors that can be deleted independently. Though the work in [22] attempted to apply reliability theory in the study of landscape connectivity, its tools were limited to those of elementary probability. Later work in [19–21] generalized the work in [22] by relaxing the assumption of equal corridor deletion probabilities, and by employing advanced tools and extended measures of Boolean-based reliability theory.

We are using the term 'survivability' herein to denote the probability that the species evades extinction through successful migration, thereby surviving harsh conditions in its local environment. This usage is in line with the general utilization of the term 'system survivability' [23–30], defined in [24] as "*the ability of a network to withstand and recover from failures,*" and more precisely in [26] as "*the system's ability to continuously deliver services in compliance with the given requirements in the presence of failures and other undesired events.*" The main issue here is that critical services in a system (such as a telecommunication network) should be continuously provided even during undesirable events like criminal attacks, earthquakes, floods, hurricanes, other natural disasters, or network own failures. If we view a species as a system, then its survivability would mean that it continues its normal existence as a living organism despite adverse and detrimental events in the surrounding environment.

To set the stage for our work in this chapter, we note that many populations of species in nature are fragmented: they consist of local or insular populations occupying separate isolated habitat patches [31], which are surrounded by less suitable or uninhabitable areas where the species is almost absent or virtually non-existent. An important ecological phenomenon is the migration of the species from its original habitat patch, which might influence population stability, population cycles, and density and other demographic, genetic, and behavioral variables [32]. In fact, such migration might be the only way for a species to avoid the threat of local extinction in a critical habitat patch, whose conditions are moving in the direction of becoming totally impossible for living and breeding. Throughout this chapter, we will make frequent reference to the following ecology terms [33–35]:

- **Habitat patch**: a discrete area where the local population of a specific species may survive and continue to breed and obtain necessary resources for a long term. In our current migration problem, there are some source habitat patches (with critical conditions), in which the species is threatened by imminent local extinction, and there are some destination habitat patches (with favorable conditions), in which the species might seek refuge so as to secure its survivability (as a species). Habitat patches are called simply "habitats" by the seminal paper on migration reliability by Jordán [22]. However, other prominent ecologists use this term in a general sense to refer to the whole ecological network and employ the narrow-sense term "habitat patch" for fragments or parts of the whole network.

- **Stepping stone**: a relatively small place that helps the local population of the pertinent species achieve its goal of migration to a safer or more prosperous habitat patch. A stepping stone is not considered inhabitable for the species; it allows temporary or transitory residence but it is not suitable for long-term survival. Besides source and destination habitat patches, stepping stones serve as nodes in the graph representation of the migration problem.
- **Ecological corridor**: a physical area which connects patches (habitat patches and stepping stones) and makes migration possible for the given species between the source and destination habitat patches. However, a corridor supports short-term (rather than long-term) survival for the concerned species. Ecological corridors have many beneficial functions such as minimizing extinction and supporting biodiversity, but they also occasionally have detrimental effects such as spreading diseases. Corridors serve as edges or links in the graph representation of the migration problem.

The problem considered in this chapter might be reformulated in many ecological contexts other than that of migration of species, including those of (see, e.g. Hock & Mumby [16]):

(a) **Metapopulations**, which depict a wide variety of models describing a group of spatially separated local populations of the same species that are connected by colonization or invasion. Growth, development, and change within the whole metapopulation are stimulated and influenced by colonization and extinction of the sub-populations. A metapopulation might be stable, though its constituent populations possess finite lifespans. Immigrants from a sub-population experiencing a population boom might re-colonize an open habitat (whose original population has gone extinct), or they may join a too small population, thereby rescuing it from imminent local extinction,
(b) **Gene flow (allele flow)**, which is the transfer (through interbreeding) of genetic variation (subtle differences in DNA sequences among individuals) from one population of a species to a different one. Such transfer introduces new alleles (gene forms) to the genetic information or gene pool of the receiving population,
(c) **Infectious diseases, epidemics, or pandemics**, which are spread by pathogenic microorganisms, such as bacteria, viruses, parasites, or fungi. Models of transmission of such microorganisms from one host to another might mimic migration models, and
(d) **A food web (a consumer-producer system)**, which constitutes an interconnection of all the food chains (what-eats-what relations) in a single ecosystem. Again, the migration model can be used to study the transfer of nutrients and energy from their source in producers (typically green plants) through herbivores to multiple levels of carnivores.

For space limitation, we will not give much background of the system reliability techniques that we use herein for the computation of survivability of species, though we will provide a detailed and expository treatment for the examples considered. The reader might consult any of the available excellent treatises on system reliability

[36–38] or refer to some of our earlier papers on the present topic [19–21]. Here, it might suffice to say that we will employ switching algebra (usually known as two-valued Boolean algebra) by utilizing the indicator variables for probabilistic events instead of the events themselves [41]. Our work starts in the switching or Boolean domain, wherein an expression for the indicator variable of a successful migration is obtained as a switching function of the indicator variables of corridor successes. Subsequently, our work moves from the Boolean domain to the probability domain through the use of the real transform (also known as the probability transform) [39, 40], which is a probability expression of survivability as a multi-affine function of corridor reliabilities. We observe that probability formulas are considerably simplified when probabilistic events in union are mutually exclusive (their indicators are disjoint) and/or when intersected events or their indicators are statistically independent. Attempting to make the most of this observation, many efficient algorithms have been written for converting the switching expression for the indicator variable of system success into a special form that we call a probability-ready expression (PRE). This expression is a switching formula enjoying the advantageous property that it is directly convertible, on a one-to-one basis, to the corresponding probability transform, which stands for a reliability or survivability expression. For a PRE to enjoy such a desirable property, it must possess two characteristics [41–43], which are in accordance with our aforementioned observation:

(a) 'Disjointness' (that all Oared terms (products) are mutually exclusive or disjoint), a characteristic of paramount importance in system reliability, since it can be induced via procedures generalizing the Reflection Law in switching algebra [41], and
(b) 'Independence' (that all ANDed terms (sums) are statistically independent), a characteristic that cannot be deliberately created, but might be enhanced through choosing between success and failure expressions, and that should be preserved once observed to exist. This characteristic is not overly stressed in the study of general networks, but it should be noted well and taken care of in ecology networks, which are frequently of parallel or almost-parallel logical structures, and usually involve products of independent expressions that should preferably not be multiplied out. We emphasize that it is not necessary to look always at success expressions since it might be of advantage to look at the problem from a failure perspective. In fact, it is better to look at the failure of a structure of better redundancy (parallel or almost parallel) and to consider the success of a structure of poorer redundancy (series or almost series).

The conversion from a PRE to a probability expression is trivially achieved by replacing switching variables by their probabilistic expectations, AND operations by arithmetic multiplications, and OR operations by arithmetic additions.

For space limitations also, we do not address herein the simple survivability problem that arises in an ecological network with a single source habitat patch and a single destination habitat patch. Techniques for handling this problem are the old techniques used to handle two-terminal (source-to-terminal or st) reliability in classical reliability theory [36–45]. These techniques can, in a sense, be viewed as advanced (usually Boolean-based) applications of probability theory. In fact, it is clear that the migration problem had not been attacked by ecologists via methods beyond that of elementary probability (see, e.g. Jordán [22] and Hock & Mumby [16]).

The organization of the rest of this chapter is as follows. Section 35.2 lists our assumptions, and notation pertaining to the ecology and reliability domains. Sections 35.3 and 35.4 treat ecological networks with several source habitat patches and/or several destination habitat patches. Section 35.3 demonstrates that the survivability analysis in an ecological network that has several such habitat patches with paths sharing no edges in common reduces to the reliability analysis of several ecological networks each having a single source habitat patch and a single destination habitat patch (i.e. to standard source-to-terminal reliability problems). Section 35.4 deals with the problem of networks with several source habitat patches and/or several destination habitat patches with paths that share some edges in common. This problem is similar to (albeit with a subtle difference from) the problem of broadcast reliability in classical reliability theory, which considers connectivity from one node to all the nodes in a set of other nodes. The problem in Sect. 35.4 considers connectivity from at least one node among the nodes in a given set of nodes to at least one node among the nodes in a set of other nodes. The problem is solved via enumeration of cutsets, pathsets, partitioned groups of paths. This enumeration reveals certain particular features that are specific to ecology networks. The problem is solved once more by a divide-and-conquer method relying on the Boole–Shannon expansion. The symbolic survivability–unsurvivability expressions obtained herein are all checked via the exhaustive tests set by [42]. Section 35.5 investigates survivability polynomials when the assumption of corridor heterogeneity is relaxed. Section 35.6 concludes the chapter and points out new directions for further research with a stress on the warranted utilization of more reliability techniques in ecology.

35.2 Assumptions and Notation

35.2.1 Assumptions

- We consider one particular organism or species of very general unspecified characteristics.
- Initially, the pertinent species is in a critical habitat patch, in which it is imminently threatened by local extinction. To avoid this fate, it attempts migration to one out of a few potential perfect destination habitat patches with favorable conditions. Along the way, it traverses imperfect heterogeneous corridors and perfect uninhabitable stepping stones.
- Corridors are considered 'soft edges' or boundaries that are permeable to emigrating or dispersing individuals of the organism. Permeability of an edge means that a good proportion of potential emigrants can reach the patch boundary

and then cross over it. By contrast, other patch boundaries are impenetrable so that dispersing individuals virtually never cross due to physical or psychological barriers (see, e.g. Stamps et al. [32]).

- Each of the corridors is in one of two states, either good (permeable) or failed (deleted or destroyed), and the migration process is also in one of two states, either successful or unsuccessful.
- Corridor states are statistically independent.

35.2.2 Notation

N	Number of ecological corridors in the investigated network, $n \geq 0$.
X_i	Success of corridor $i =$ indicator variable, which reveals that the pertinent species successfully migrates through that corridor = a switching (two-valued Boolean) random variable that takes only one of the two discrete values 0 and 1; ($X_i = 1$ if and only if corridor i is permeable (good), while $X_i = 0$ if and only if corridor i is deleted or destroyed (failed)).
$\bar{X}_i$	Failure or deletion of corridor $i =$ indicator variable, which denotes that the concerned species fails to migrate through corridor i, where $\bar{X}_i = 0$ if and only if corridor i is good, while $\bar{X}_i = 1$ if and only if corridor i is deleted/destroyed. The success X_i and the failure $\bar{X}_i$ are complementary variables ($\bar{X}_i = 1 - X_i$).
$\boldsymbol{X}$	A vector of n indicator variables denoting corridor successes, $\boldsymbol{X} = [X_1\ X_2 \ldots X_n]^T$.
$S(\boldsymbol{X})$	Indicator variable for the successful migration of the pertinent species, called migration success or network success. Its complement $\bar{S}(\boldsymbol{X})$ is called migration or network failure.
Pr[...]	Probability of the event [...].
E[...]	Probabilistic expectation of the random variable [...].
q_i, p_i	Reliability and unreliability of corridor i. Both q_i and p_i are real values in the closed real interval [0.0,1.0]. These two variables usually have the opposite meanings in reliability circles, where p_i refers to reliability and q_i denotes unreliability.
P_i	A random switching variable denoting the success of minimal path i (indicating an irredundant connection between the source habitat patch and a destination one. This is a prime implicant of the migration success S. It is a conjunction of the successes of corridors *en route*. A minimal path does not subsume any other path.
C_j	A random switching variable denoting the failure of minimal cutset j (indicating that the source habitat patch is disconnected irredundantly from all destination habitat patches). This is a prime implicant of the migration failure $\bar{S}$. It is a conjunction of failures of corridors belonging to the cutset. A minimal cutset does not subsume any other cutset.
q_i	$\Pr[X_i = 1] = E[X_i] = 1.0 - p_i$.

$\boldsymbol{q}$ A vector of n elements representing the corridor reliabilities, $\boldsymbol{q}= [q_1\ q_2 \ldots q_n]^T$.

$\boldsymbol{p}$ A vector of n elements representing the corridor unreliabilities = **1.0**−$\boldsymbol{q}$, where **1.0** is a vector of n elements each equal to 1.0.

$R(\boldsymbol{q}), U(\boldsymbol{q})$ Survivability and unsurvivability of the species. Both $R(\boldsymbol{q})$ and $U(\boldsymbol{q})$ are real values in the closed real interval [0.0,1.0]. For convenience, we retain for these two quantities the conventional notation for system reliability and unreliability.

$R(\boldsymbol{q})$ $\Pr[S(X) = 1] = E[S(X)] = 1 - U(\boldsymbol{q})$

35.3 Networks with Several Source/Destination Habitat Patches with Paths Having no Edges in Common

When the ecological network under consideration has several source/destination habitat patches with paths sharing no edges in common, then it is obvious that this network comprises several independent sub-networks. The success of any of these sub-networks depicts the connectivity of an individual source habitat patch to an individual destination habitat patch and hence is statistically independent of the successes of other sub-networks. Each of these successes is evaluated herein (via any of the standard techniques for *st* reliability) and used as the success of an equivalent single composite edge connecting the original source habitat patch to the original destination habitat patch. The original overall network is a logically parallel structure (albeit not a physical parallel one) of the constituent sub-networks (each replaced by its equivalent single composite edge). In fact, this network is characterized as having a success equal to the ORing of the successes of these equivalent edges. More economically, it is characterized as having a single minimal cutset so that its failure is easily expressed as the failure of this cutset, i.e. as a conjunction (ANDing) of the equivalent edge failures. Solving the present problem should be done in terms of failure rather than success, with care to avoid the tempting habit of multiplying out products of independent sub-expressions.

For example, let us consider the small ecological network in Fig. 35.1a, which is essentially similar to networks in [19–22]. Sub-networks of this network (involving paths to different destination habitat patches) do not share edges in common. This network is equivalent to the one in Fig. 35.1b, where the edge successes $S_A - S_D$ are directly evaluated in terms of pertinent link or corridor successes. Now, Fig. 35.1b shows that the network has a single minimal cutset (i.e. it is a logically parallel structure). The network failure is exactly the failure of this cutset, namely

$$\bar{S} = \overline{S_A} \wedge \overline{S_B} \wedge \overline{S_C} \wedge \overline{S_D}. \tag{35.1}$$

Equation (35.1) expresses the network failure $\bar{S}$ as the ANDing of statistically independent sub-expressions. We preserve statistical independence by keeping these sub-expressions in factored form and avoiding to multiply them out. We now write

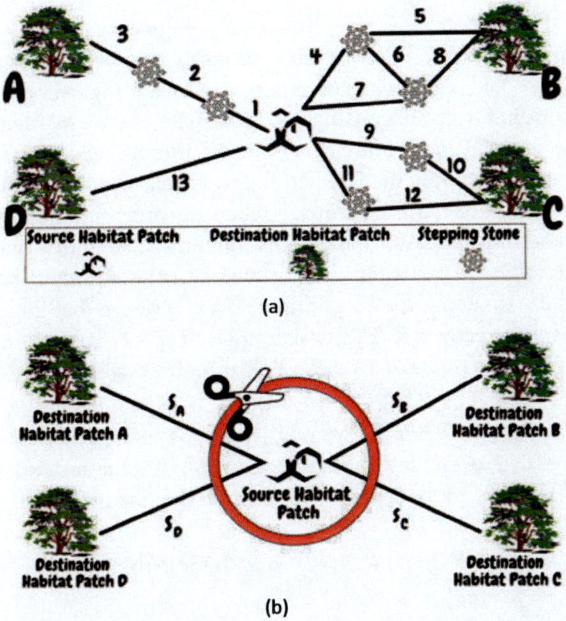

Fig. 35.1 **a** A toy landscape network that has only one source habitat patch but multiple destination habitat patches with source–destination paths that share **no** edges in common, **b** Replacing each individual sub-network to produce an equivalent network, which is a logically parallel structure that has a unique minimal cutset marked in red

these sub-expressions, initially in minimal form, simply by inspecting the pertinent sub-networks in Fig. 35.1a. Since the expression for $\bar{S}$ will be a Probability-ready expression (PRE) if these sub-expressions are PRE, we convert each of them from its minimal form to a PRE form (when the two forms differ) as follows:

$$\overline{S_A} = \overline{X_1} \vee \overline{X_2} \vee \overline{X_3} = \overline{X_1} \vee X_1\left(\overline{X_2} \vee X_2\overline{X_3}\right). \tag{35.2}$$

$$\begin{aligned}\overline{S_B} &= \overline{X_4}\,\overline{X_7} \vee \overline{X_5}\,\overline{X_8} \vee \overline{X_4}\,\overline{X_6}\,\overline{X_8} \vee \overline{X_5}\,\overline{X_6}\,\overline{X_7} \\ &= \overline{X_4}\,\overline{X_7} \vee \left(X_4 \vee \overline{X_4}X_7\right)\left(\overline{X_5}\,\overline{X_8} \vee \left(X_5 \vee \overline{X_5}X_8\right)\right. \\ &\quad \left(\overline{X_4}\,\overline{X_6}\,\overline{X_8} \vee \left(X_4 \vee \overline{X_4}X_6 \vee \overline{X_4}\,\overline{X_6}X_8\right)\right)\overline{X_5}\,\overline{X_6}\,\overline{X_7})) \\ &= \overline{X_4}\,\overline{X_7} \vee \overline{X_5}\,\overline{X_8}\left(X_4 \vee \overline{X_4}X_7\right) \vee \overline{X_4}X_5\overline{X_6}X_7\overline{X_8} \vee X_4\overline{X_5}\,\overline{X_6}\,\overline{X_7}X_8\end{aligned} \tag{35.3}$$

$$\overline{S_C} = \overline{X_9}\,\overline{X_{10}} \vee \overline{X_{11}}\,\overline{X_{12}} = \overline{X_9}\,\overline{X_{10}} \vee \overline{X_{11}}\,\overline{X_{12}}\left(X_9 \vee \overline{X_9}X_{10}\right). \tag{35.4}$$

$$\overline{S_D} = \overline{X_{13}}. \tag{35.5}$$

We have tried to give the reader a glimpse of how to conceptually convert a sum-of-products expression into a PRE by disjointing its products. In a sum (ORing) of products, we apply the Reflection Law repeatedly by adding every product complemented as a multiple for its succeeding products. The complemented product could be written as the sum (ORing) of its complemented literals according to De Morgan's law. However, we again apply the Reflection law repeatedly within this sum so as to render it disjoint. Of course, this disjointing scheme is rather inefficient as it generates many unwarranted multiplications and vanishing terms. Actually, one may resort to any of the more efficient variations of this scheme [41]. Another issue here is that we might avoid the disjointing effort entirely if we observe that the complement of a sub-expression is already a PRE. For example, in (35.2) we obtained the disjoint subexpression $\overline{S_A} = \overline{X_1} \vee X_1(\overline{X_2} \vee X_2\overline{X_3})$, which corresponds to the unsurvivability of sub-network A as $E\{\overline{S_A}\} = p_1 + q_1(p_2 + q_2 p_3)$. However, if we note that $S_A = X_1 \wedge X_2 \wedge X_3$ is a PRE as is, we might skip the task of disjointing and simply get $E\{\overline{S_A}\} = 1 - q_1 q_2 q_3$, which is not only easier to obtain but also more compact. Equations (35.1)–(35.5) transform now directly to the unsurvivability expression

$$U = p_{13}(1 - q_1 q_2 q_3)(p_4 p_7 + p_5 p_8(q_4 + p_4 q_7) + p_4 q_5 p_6 q_7 p_8 + q_4 p_5 p_6 p_7 q_8) \times (p_9 p_{10} + p_{11} p_{12}(q_9 + p_9 q_{10})). \tag{35.6}$$

In short, we stress that the survivability analysis of an ecological network that has paths (between source and destination habitat patches) that share *no* edges in common reduces to the survivability analysis of several ecological networks each having *a single* source habitat patch and *a single* destination habitat patch (i.e. to standard *st* reliability problems).

35.4 Networks with Several Source/Destination Habitat Patches with Some Shared Edges Among Source–Destination Paths

This section deals with an ecological network that has several source/destination habitat patches such that there are some shared edges among the source–destination paths. Such a network can be analyzed via a variety of ways. We explore herein the methods of (a) enumeration of minimal cutsets (which minimally destroy connectivity from source habitat patches to destination habitat patches), (b) enumeration of minimal pathsets (which minimally ensure connectivity from source habitat patches to destination habitat patches), and (c) Boole–Shannon Expansion of the network success or failure with respect to the successes or failures of some edges (preferably the shared edges). To expose and compare the various methods, we introduce a small ecological network in Fig. 35.2a. This is one of the simplest possible examples for our current purpose since it has merely a single source habitat patch and only two destination habitat patches. Hopefully, this toy example might trigger work on larger

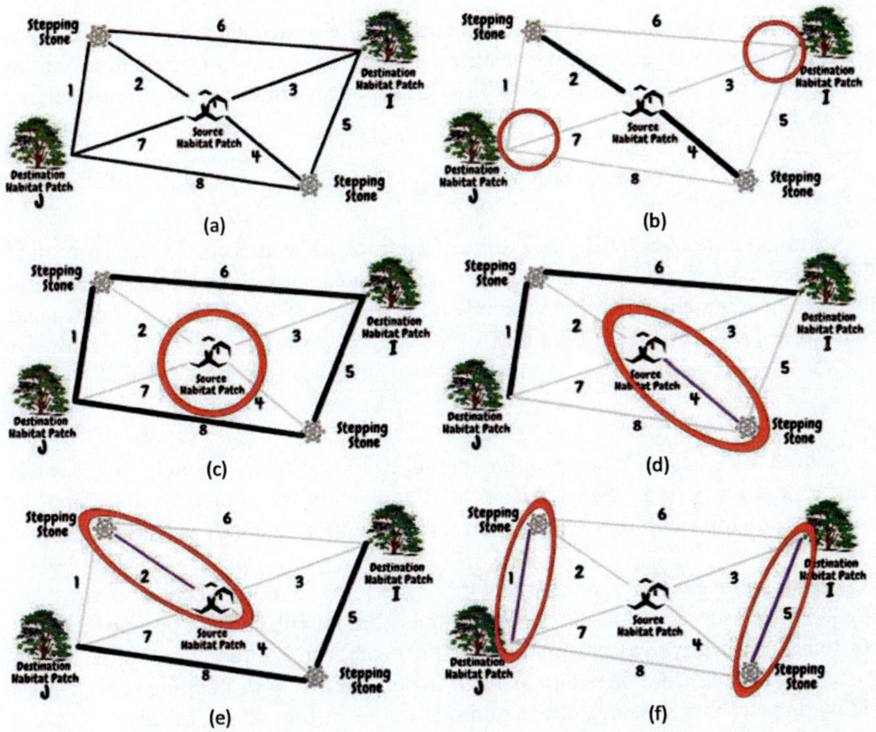

Fig. 35.2 **a** A toy landscape network with one-to-any-out-of-many survivability, **b–e** Enumeration of the minimal cutsets for a successful migration in the network in (a), **f** A cutset that is not minimal, being a *superset* of the minimal cutset in Fig. 35.2c

more sophisticated networks and ultimately lead to the formulation of algorithmic solutions of the problem in the general case.

Starting with the method of cutset enumeration, we use Figs. 35.2b–e to enumerate all the minimal cutsets for the considered network, which turn out to be four. Of course, none of these minimal cutsets is a subset or a superset of another, which means that their failures are only partially overlapping so that none of these failures totally covers another. Figure 35.2b illustrates the first of these cutsets, showing that its failure is a conjunction of the failures of its two partitions

$$C_1 = \left(\overline{X_3}\,\overline{X_5}\,\overline{X_6}\right) \wedge \left(\overline{X_1}\,\overline{X_7}\,\overline{X_8}\right) = \overline{X_1}\,\overline{X_3}\,\overline{X_5}\,\overline{X_6}\,\overline{X_7}\,\overline{X_8}. \quad (35.7)$$

Here, C_1 is the ANDing of failures for the two vertex cutsets of the two destination habitat patches. Likewise, Fig. 35.2c shows the second cutset and asserts that its failure is

$$C_2 = \overline{X_2}\,\overline{X_3}\,\overline{X_4}\,\overline{X_7}. \quad (35.8)$$

Clearly, C_2 is the failure of the vertex cutset of the source habitat patch. We might alternatively view C_2 as a cutset that surrounds simultaneously *all* destination habitat patches and *all* stepping stones. Now, Fig. 35.2d portrays the third cutset and declares that its failure is

$$C_3 = \overline{X_2}\,\overline{X_3}\,\overline{X_5}\,\overline{X_7}\,\overline{X_8}. \tag{35.9}$$

We might interpret C_3 to be the failure of a purported "vertex cutset" that surrounds the source habitat patch in conjunction with stepping stone 1 under the assumption that these two patches constitute a single node. Figure 35.2e displays the fourth and final cutset, and identify its failure as

$$C_4 = \overline{X_1}\,\overline{X_3}\,\overline{X_4}\,\overline{X_6}\,\overline{X_7} \tag{35.10}$$

Note that C_4 has an interpretation analogous to that of C_3, namely that it is the failure of a purported "vertex cutset" that surrounds the source habitat patch in conjunction with stepping stone 2 under the assumption that they constitute a single node.

Thorough investigation reveals that there are no more minimal cutsets. However, Fig. 35.2f is used to propose a fifth potential minimal cutset. This purported cutset surrounds *separately* (a) destination habitat patch I in conjunction with stepping stone 1 and (b) destination habitat patch II in conjunction with stepping stone 2. This claimed cutset is a *superset* of the minimal cutset in Fig. 35.2c. In fact, it is not a minimal cutset since its failure

$$C_5 = \left(\overline{X_6}\,\overline{X_3}\,\overline{X_4}\,\overline{X_8}\right) \wedge \left(\overline{X_6}\,\overline{X_2}\,\overline{X_7}\,\overline{X_8}\right) = \overline{X_2}\,\overline{X_3}\,\overline{X_4}\,\overline{X_6}\,\overline{X_7}\,\overline{X_8}, \tag{35.11}$$

subsumes that of the second cutset $C_2 = \overline{X_2}\,\overline{X_3}\,\overline{X_4}\,\overline{X_7}$. Here, subsumption means literal-set containment [41], i.e. the set of literals in C_2 is contained as a subset in the set of literals of C_5. A well-known consequence of the subsumption relation is that the subsumed term C_2 absorbs the subsuming term C_5 when ORed together [41], namely

$$C_2 \vee C_5 = C_2, \tag{35.12}$$

This absorption suggests that during (algorithmic) cutset enumeration, the analyst does need to take pains avoiding the inclusion of non-minimal cutset failures such as C_5 in an expression for $\bar{S}$. Such an inclusion does not make the resulting expression a wrong one. It only causes this expression to be longer, but it yields the same PRE result since PRE generation includes a mechanism for detecting and absorbing any subsuming terms. The final expression for the overall network failure is

$$\bar{S} = \overline{X_3}\,\overline{X_7}\left(\overline{X_2}\,\overline{X_4} \vee \overline{X_2}\,\overline{X_5}\,\overline{X_8} \vee \overline{X_1}\,\overline{X_4}\,\overline{X_6} \vee \overline{X_1}\,\overline{X_5}\,\overline{X_6}\,\overline{X_8}\right), \tag{35.13}$$

where we factored out the product $\overline{X_3}\,\overline{X_7}$ and used an arrangement with shorter terms ahead of longer ones, which is more likely to produce a more compact PRE. In fact, the resulting PRE might be written via any disjointness procedure [41] as

$$\bar{S}_{\mathrm{PRE}} = \overline{X_3}\,\overline{X_7}(\overline{X_2}\,\overline{X_4} \vee \overline{X_2}X_4\overline{X_5}\,\overline{X_8} \vee \overline{X_1}X_2\overline{X_4}\,\overline{X_6} \vee \overline{X_1}X_2X_4\overline{X_5}\,\overline{X_6}\,\overline{X_8}) \tag{35.14}$$

This PRE is transformed, on a one-to-one basis, to the unsurvivability expression

$$U = p_3p_7(p_2p_4 + p_2q_4p_5p_8 + p_1q_2p_4p_6 + p_1q_2q_4p_5p_6p_8). \tag{35.15}$$

In our second solution method, we express system success S as the ORing of the minimal paths for the network in Fig. 35.3. First, we observe that a minimal path connects the source habitat to one of the destinations without traversing the other destination. Therefore, we partition system success into two sums of products, namely

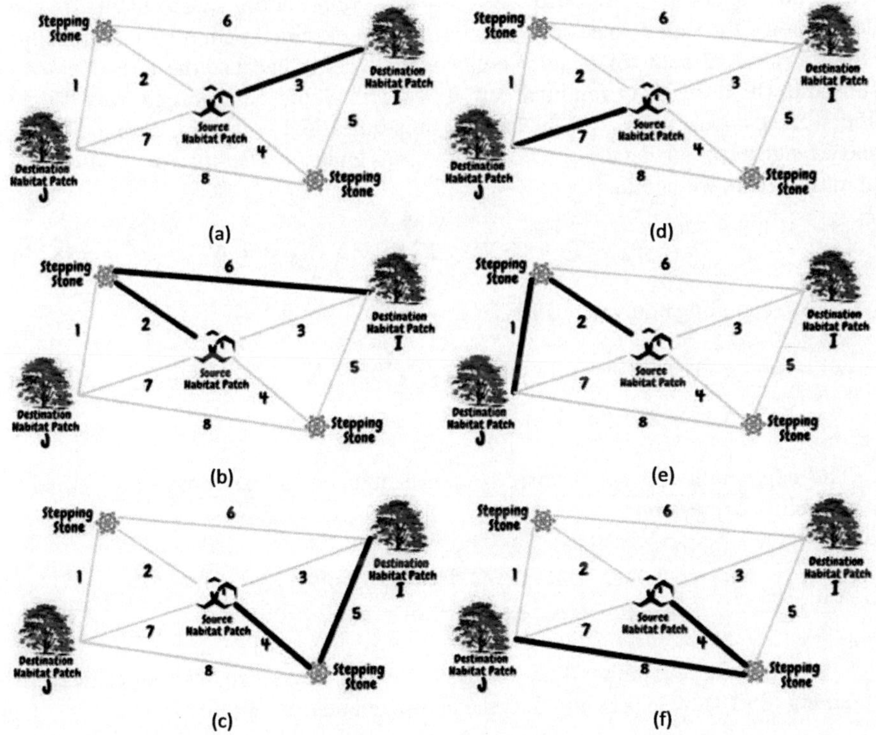

Fig. 35.3 Enumeration of all the minimal paths for a successful migration in the network in Fig. 35.2a. Paths (a), (b), and (c) are from the source habitat to destination I, while Paths (d), (e), and (f) are from the source habitat to destination J only

$$S = S_I \vee S_J, \tag{35.16}$$

where S_I is the disjunction of all minimal paths from the source habitat to destination habitat I that do not pass through destination J. These paths are portrayed in Fig. 35.3a–c, and their terms are ORed to express S_I as

$$S_I = X_3 \vee X_2X_6 \vee X_4X_5. \tag{35.17}$$

Likewise, S_J is the disjunction of all minimal paths from the source habitat to destination habitat J that do not pass through destination I. These paths are displayed in Fig. 35.3d–f, and their ORing produces the following expression for S_J

$$S_J = X_7 \vee X_1X_2 \vee X_4X_8. \tag{35.18}$$

We stress that the two expressions S_I and S_J are neither independent nor disjoint. They would have been statistically independent (with the corresponding sets of paths being disjoint), had it not been for the shared edges 2 and 4. It is remarkable that we do not include in S_I the path $X_5X_7X_8$ that connects the source habitat to the destination I through J since such a path (though a minimal path for I individually) is not a minimal path for I and J collectively. This is because the term $X_5X_7X_8$ subsumes the term X_7 (a minimal path in S_J) and would be absorbed by it if they both were to appear in $(S_I \vee S_J)$. Now, we substitute (35.17) and (35.18) into (35.16) and arrange terms so that those with fewer literals appear earlier than ones with more literals. Hence, we obtain

$$S = X_3 \vee X_7 \vee X_2X_6 \vee X_1X_2 \vee X_4X_5 \vee X_4X_8. \tag{35.19}$$

A corresponding equivalent PRE is

$$\begin{aligned} S_{\mathrm{PRE}} &= X_3 \vee X_7\overline{X_3} \vee X_2X_6\overline{X_3}\,\overline{X_7} \vee X_1X_2\overline{X_3}\,\overline{X_7}\,\overline{X_6} \vee X_4X_5\overline{X_3}\,\overline{X_7}(\overline{X_2} \vee \overline{X_1}\,\overline{X_2}\,\overline{X_6}) \\ &= X_4X_8\overline{X_3}\,\overline{X_7}(\overline{X_2} \vee \overline{X_1}X_2\overline{X_6})\overline{X_5}, \end{aligned} \tag{35.20}$$

The expression (35.20) is directly convertible, on a one-to-one basis to the survivability expression

$$\begin{aligned} R = q_3 + q_7q_3 + q_2q_6p_3p_7 + q_1q_2p_3p_7p_6 + q_4q_5p_3p_7(p_2 \\ + p_1q_2p_6) + q_4q_8p_3p_7(p_2 + p_1q_2p_6)p_5. \end{aligned} \tag{35.21}$$

Looking at failure rather than success, we can devise an alternative way of obtaining (35.13) by expressing $\bar{S}$ as the complementation of (35.16), i.e.

$$\bar{S} = \overline{S_I} \wedge \overline{S_J}, \tag{35.22}$$

where $\overline{S_I}$ and $\overline{S_J}$ are the failures to connect the source habitat to destination habitat I alone and destination habitat J alone, respectively. These are given by direct graph enumeration, or preferably in factored form by complementation of (35.17) and (35.18) as

$$\overline{S_I} = \overline{X_3}\,\overline{X_5}\,\overline{X_6} \vee \overline{X_2}\,\overline{X_3}\,\overline{X_5} \vee \overline{X_3}\,\overline{X_4}\,\overline{X_6} \vee \overline{X_2}\,\overline{X_3}\,\overline{X_4} = \overline{X_3}\big(\overline{X_2} \vee \overline{X_6}\big)\big(\overline{X_4} \vee \overline{X_5}\big), \tag{35.23}$$

$$\overline{S_j} = \overline{X_1}\,\overline{X_7}\,\overline{X_8} \vee \overline{X_2}\,\overline{X_7}\,\overline{X_8} \vee \overline{X_1}\,\overline{X_4}\,\overline{X_7} \vee \overline{X_2}\,\overline{X_4}\,\overline{X_7} = \overline{X_7}\big(\overline{X_2} \vee \overline{X_1}\big)\big(\overline{X_4} \vee \overline{X_8}\big). \tag{35.24}$$

We now perform the ANDing of (35.23) and (35.24) to obtain (35.13) after factoring out the product $\overline{X_3}\,\overline{X_7}$ and absorbing subsuming terms, or preferably to obtain its equivalent factored form through 'intelligent multiplication.'

$$\bar{S} = \overline{X_3}\,\overline{X_7}\big(\overline{X_2} \vee \overline{X_1}\,\overline{X_6}\big)\big(\overline{X_4} \vee \overline{X_5}\,\overline{X_8}\big), \tag{35.25}$$

Equation (35.25) leads immediately to the following PRE and corresponding unsurvivability

$$\bar{S}_{\mathrm{PRE}} = \overline{X_3}\,\overline{X_7}\big(\overline{X_2} \vee \overline{X_1}X_2\overline{X_6}\big)\big(\overline{X_4} \vee X_4\overline{X_5}\,\overline{X_8}\big), \tag{35.26}$$

$$U = p_3 p_7 (p_2 + p_1 q_2 p_6)(p_4 + p_5 q_4 p_8). \tag{35.27}$$

We now explore the possibility of improving the path set solution (35.21) to obtain one of computational complexity comparable to that of the cutset solution (35.27). We obtain the following partially factored expression of system success through complementation of both sides of (35.25)

$$S = X_3 \vee X_7 \vee X_4(X_5 \vee X_8) \vee X_2(X_1 \vee X_6). \tag{35.28}$$

Equation (35.28) might be obtained directly by inspecting the network's graph, provided we avoid the exhaustive enumeration of the individual paths that was done in Fig. 35.3, and instead enumerate groups of minimal paths as shown in Fig. 35.4. The success expression can now be converted into PRE form, with care taken not to spoil the form of partial factoring therein by using the techniques in [41, 46], namely

$$\begin{aligned} S_{\mathrm{PRE}} = {} & X_3 \vee \overline{X_3}\big(X_7 \vee \overline{X_7}(X_4\big(X_5 \vee \overline{X_5}X_8\big) \vee X_2 \\ & \big(X_1 \vee \overline{X_1}X_6\big)\big(\overline{X_5}\,\overline{X_8} \vee \big(X_5 \vee \overline{X_5}X_8\big)\overline{X_4}\big)\big). \end{aligned} \tag{35.29}$$

This PRE can be transformed immediately to the following survivability expression (which is slightly better than (35.21) but not competitive to (35.27))

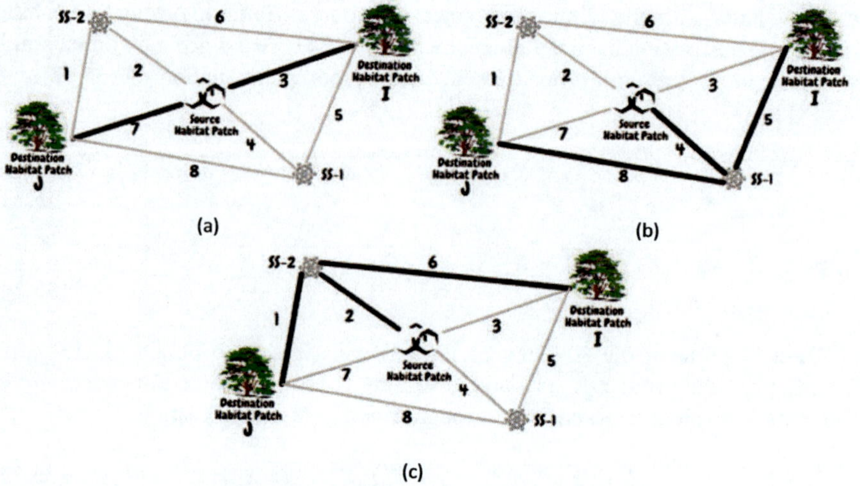

Fig. 35.4 Enumeration of groups of minimal paths for a successful migration in the network in Fig. 35.2a: **a** Direct paths from the source to one of the destinations, **b** Paths via the first stepping stone (SS-1), **c** Paths via the second stepping stone (SS-2)

$$R = q_3 + p_3(q_7 + p_7(q_4(q_5 + p_5q_8) + q_2(q_1 + p_1q_6)(p_5p_8 + (q_5 + p_5q_8)p_4))). \quad (35.30)$$

Finally, we explore a method based on the Boole–Shannon Expansion (BSE) of the ecological network success with respect to the successes of a few edges [41]. This Boolean-domain method is a divide-and-conquer technique, which is effective and fast in generating a PRE from the outset. It amounts to an application of the Total Probability Law [47, 48] in the probability domain, and is effectively the same as the Factoring theorem [49, 50] in the graph domain. If the expansion is performed about successes of all edges that are shared in the source-destination paths, the original network is replaced with several sub-networks, each of which has no edges common among paths and hence is solved directly as in Sect. 35.2.

A single BSE application means that the switching function $S(\boldsymbol{X})$ (representing migration success) is expanded about one of its arguments X_i (representing corridor success) as [41]

$$S(\boldsymbol{X}) = (\bar{X}_i \wedge S(\boldsymbol{X}|0_i)) \vee (X_i \wedge S(\mathbf{X}|1_i)). \quad (35.31)$$

Here, function $S(\boldsymbol{X})$ is expressed in terms of its two subfunctions $S(\boldsymbol{X}|0_i)$ and $S(\boldsymbol{X}|1_i)$, which represent the Boolean quotients, subfunctions or ratios $S(\boldsymbol{X})/\bar{X}_i$ and $S(\boldsymbol{X})/X_i$, and are obtained by assigning the expansion variable X_i the two values 0 and 1, respectively in the function $S(\boldsymbol{X})$. In our current case, we expand the migration success S (twice) with respect to the successes X_2 and X_4 of the two common edges (for paths from the critical habitat to the two destinations), namely

$$S = (\bar{X}_2\bar{X}_4 \wedge (S/\bar{X}_2\bar{X}_4)) \vee (\bar{X}_2 X_4 \wedge (S/\bar{X}_2 X_4)) \vee$$
$$(X_2\bar{X}_4 \wedge (S/X_2\bar{X}_4)) \vee (X_2 X_4 \wedge (S/X_2 X_4)) \tag{35.32}$$

The four Boolean quotients in (35.32) denote network success subject to the four combinations that each of corridors 2 and 4 is shorted (perfect) or open (failed). They are obtained from the sub-networks resulting subject to these four combinations as

$$S/\bar{X}_2\bar{X}_4 = X_3 \vee X_7, \tag{35.33}$$

$$S/\bar{X}_2 X_4 = X_3 \vee X_7 \vee X_5 \vee X_8, \tag{35.34}$$

$$S/X_2\bar{X}_4 = X_3 \vee X_7 \vee X_1 \vee X_6, \tag{35.35}$$

$$S/X_2 X_4 = X_3 \vee X_7 \vee X_5 \vee X_8 \vee X_1 \vee X_6. \tag{35.36}$$

Substituting (35.33)–(35.36) into (35.329), we obtain an expression that can be immediately reduced to (35.28). Each of the expressions in (35.33)–(35.36) has a PRE complement. Therefore, we can replace (35.32) by its expectation (using the abbreviation $p_{ij} = p_i p_j$)

$$R = p_2 p_4(1 - p_3 p_7) + p_2 q_4(1 - p_3 p_7 p_5 p_8) + q_2 p_4(1 - p_3 p_7 p_1 p_6)$$
$$+ q_2 q_4(1 - p_3 p_7 p_5 p_8 p_1 p_6) = p_2 p_4(1 - p_{37}) + p_2 q_4(1 - p_{37} p_{58})$$
$$+ q_2 p_4(1 - p_{37} p_{16}) + q_2 q_4(1 - p_{37} p_{58} p_{16}). \tag{35.37}$$

The various unsurvivability and survivability expressions obtained in this section vary in the competitiveness of their computational complexity. The numbers of multiplications in the expressions (35.15), (35.21), (35.27), (35.30), and (35.37) are 14, 21, 7, 10 and 14, respectively. Table 35.1 demonstrates that the five expressions are equivalent since each of them passes the test for correctness in Rushdi [39]. We observe that each of these expressions is a multi-affine function in its arguments (an algebraic function depicting a straight line relation in each of the arguments) and that each expression has a correct "truth table." Note that all R expressions share the same 'truth table' which is identical to the genuine truth table of the corresponding S and S_{PRE} expressions, and is line-wise complementary to the 'truth tables' of the corresponding U, $\bar{S}$ and $\bar{S}_{PRE}$ expressions. A full truth table of eight variables would have $2^8 = 256$ lines, while the much-reduced version given in Table 35.1 has 10 'lines' only, which correspond to the six minimal paths together with the four minimal cutsets, with the (–) entries in the table depicting "don't-cares." To understand what the table does, consider the 'line' $C_2 = \overline{X_2}\ \overline{X_3}\ \overline{X_4}\ \overline{X_7}$. This line assigns 0 to each of q_2, q_3, q_4, and q_7 and leaves q_1, q_5, q_6, and q_8 unspecified, and hence it covers $2^4 = 16$ truth table lines. For this 'line,' Eq. (35.15) reduces to an identity (equal to 1) in the four variables q_1, q_5,q_6, and q_8, namely

Table 35.1 A compact "truth table" for the complete verification of the multi-affine formulas U in (35.15) or (35.27), and R in either of (35.21), (35.30), or (35.34)

Line Depicts	q_1	q_2	q_3	q_4	q_5	q_6	q_7	q_8	U	R
$P_1 = X_3$	–	–	1	–	–	–	–	–	0	1
$P_2 = X_7$	–	–	–	–	–	–	1	–	0	1
$P_3 = X_2X_6$	–	1	–	–	–	1	–	–	0	1
$P_4 = X_1X_2$	1	1	–	–	–	–	–	–	0	1
$P_5 = X_4X_5$	–	–	–	1	1	–	–	–	0	1
$P_6 = X_4X_8$	–	–	–	1	–	–	–	1	0	1
$C_1 = \overline{X}_1\,\overline{X}_3\,\overline{X}_5\,\overline{X}_6\,\overline{X}_7\,\overline{X}_8$	0	–	0	–	0	0	0	0	1	0
$C_2 = \overline{X}_2\,\overline{X}_3\,\overline{X}_4\,\overline{X}_7$	–	0	0	0	–	–	0	–	1	0
$C_3 = \overline{X}_2\,\overline{X}_3\,\overline{X}_5\,\overline{X}_7\,\overline{X}_8$	–	0	0	–	0	–	0	0	1	0
$C_4 = \overline{X}_1\,\overline{X}_3\,\overline{X}_4\,\overline{X}_6\,\overline{X}_7$	0	–	0	0	–	0	0	–	1	0

$$U = (1)(1)((1)(1) + (1)(0)p_5p_8 + p_1(0)(1)p_6 + p_1(0)(0)p_5p_6p_8), \qquad (35.38)$$

while Eq. (35.27) reduces to an identity (equal to 0) in these four variables, namely

$$R = 0 + (1)(0 + (1)(0(q_5 + p_5q_8) + 0(q_1 + p_1q_6)(p_5p_8 + (q_5 + p_5q_8)(1)))). \qquad (35.39)$$

Each of the expressions (35.15), (35.21), (35.27), (35.30), and (35.37) is a symbolic expression and hence can be used in computing several measures of classical reliability such as (a) the region of useful redundancy (computed via equation solving) [51], (b) the life expectancy or mean-time-to-failure (MTTF) (computed via integration), and (c) various importance measures (computed via differentiation or differencing) [52]. An interesting finding is that the MTTF of a classical landscape network of an appropriate configuration and a sufficient number of redundant corridors can be made more than double that of a single corridor [19].

In passing, we remark that a prominent metric of landscape connectivity (reported by Hock & Mumby [16]) concerns the most reliable (or minimum-lag) dispersal paths. The analysis to obtain this metric is quite related to (and much simpler than) our current analysis, as it pertains to the enumeration of all minimal paths and the computation of the success of each path as a product of the successes of its links. The network in Fig. 35.3 has six paths of path successes and path reliabilities, given respectively by $\{X_3, X_7, X_2X_6, X_1X_2, X_4X_5, X_4X_8\}$ and $\{q_3, q_7, q_2q_6, q_1q_2, q_4q_5, q_4q_8\}$. The paths have time lags that are inversely proportional to their reliabilities. The metric M used in [16] is the maximum value among the six path reliabilities, i.e.

$$M = \text{Max}\{q_3, q_7, q_2q_6, q_1q_2, q_4q_5, q_4q_8\}. \qquad (35.40)$$

This metric sets a lower bound on our survivability metric R above, since

$$R \geq M. \tag{35.41}$$

This result asserts that while measures that are currently used in ecology still avoid the somewhat more involved computations of reliability theory, they gain such an advantage by settling down for approximate less encompassing values. To supplement the result in (35.41), we suggest the utilization of the concept of duality to deduce an upper bound of R in a similar fashion as well. We note that the aforementioned network has four cutsets whose failures are given by

$$\{\overline{X_1}\,\overline{X_3}\,\overline{X_5}\,\overline{X_6}\,\overline{X_7}\,\overline{X_8},\ \overline{X_2}\,\overline{X_3}\,\overline{X_4}\,\overline{X_7},\ \overline{X_2}\,\overline{X_3}\,\overline{X_5}\,\overline{X_7}\,\overline{X_8},\ \overline{X_1}\,\overline{X_3}\,\overline{X_4}\,\overline{X_6}\,\overline{X_7}\}. \tag{35.42}$$

Hence, we can obtain a lower bound on the unsurvivability metric U, which corresponds to an upper bound on the survivability metric R, namely

$$1 - R = U \leq \text{Max}\{p_1p_3p_5p_6p_7p_8,\ p_2p_3p_4p_7,\ p_2p_3p_5p_7p_8,\ p_1p_3p_4p_6p_7\}. \tag{35.43a}$$

$$R \geq \text{Min}\{1 - p_1p_3p_5p_6p_7p_8,\ 1 - p_2p_3p_4p_7,\ 1 - p_2p_3p_5p_7p_8,\ 1 - p_1p_3p_4p_6p_7\}. \tag{35.43b}$$

We advocate the combined use of (35.41) and (35.43b) to establish a range for the possible value of the survivability R. If this range is tight enough, then we have established a simple and fast method of approximating R.

35.5 The Case of All Corridors Having the Same Probability of Deletion

This section deals with the *i.i.d.* case when all corridor unreliabilities are equal to the same value p. In this case, the unsurvivability U for our first network in Fig. 35.1a is obtained from Eq. (35.6) as

$$U = -2p^{13} + 11p^{12} - 19p^{11} - 3p^{10} + 46p^9 - 44p^8 + 12p^6, \tag{35.44}$$

While its value for our second network in Fig. 35.2a is obtained from any of Eqs. (35.15), (35.21), (35.27), (35.30) and (35.37) as

$$U = p^8 - 2p^7 - p^6 + 2p^5 + p^4. \tag{35.45}$$

Figure 35.5 displays plots of the survivability polynomials (35.44) and (35.45) versus corridor unreliability p. Each of the two plots is a type III curve that lies entirely below the simplex line p versus p for $p \in (0.0, 1.0)$. Each plot depicts a case of total redundancy, for which the region of useful redundancy extends over

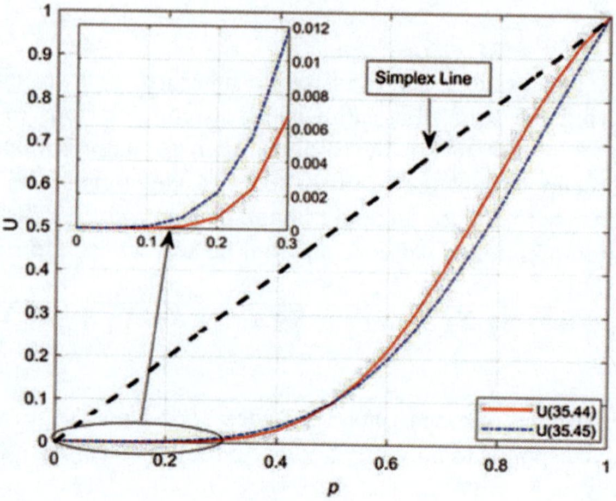

Fig. 35.5 Plots of two survivability polynomials versus corridor unreliability

the entire doubly open-ended unit interval (0.0, 1.0). The two plots intersect once within this interval. Both networks secure high species survivability (with the first surpassing the second) for highly reliable corridors. For poor corridor reliability, the species survivability in each network deteriorates, with a performance of the first network becoming inferior to that of the second.

35.6 Conclusions and Potential Future Work

This chapter dealt with the migration problem in ecology by computing the survivability of the migrating species in the somewhat involved situation when there are definitely several source/destination habitat patches connected with paths that share some edges (corridors). The species survivability under these circumstances was identified with the connectivity from at least one node within a specified set of nodes to at least one node within another specified set of nodes. Survivability (so defined) differs slightly from broadcast reliability in classical reliability theory, which is equivalent to connectivity from one node to all nodes among a specified set of nodes. This concept of survivability is useful in a variety of ecological contests other than that of migration. It is more suitable than broadcast reliability in many other contexts as well. In fact, it should be the measure to use in communication systems when the sent signal forwarded to a set of potential receivers is required to reach at least one of these receivers. An obvious example of this is the situation of a sinking ship, which sends a distress SOS signal (widely believed to stand for Save-Our-Ship) to all ships around it. Sailors aboard the sinking ship are likely to be rescued if the SOS signal is received by any of the surrounding ships. Another related example in telemedicine applications arises when surgeons operating on a patient need immediate advice from one out of several experts.

Work in this paper is based on a few simplifying assumptions, whose partial relaxation might lead to useful extensions in a variety of ways. Modeling accuracy would improve if we assume that the stepping stones (which act as intermediate nodes in the graph) are not perfect. The present analysis assumes that corridor reliabilities are specified exactly and deterministically. A more realistic analysis should take uncertainty into consideration [21, 52], for example, by employing a doubly stochastic model that assigns probabilistic rather than deterministic values to corridor probabilities of success. Alternatively, these probabilities might be generated by fuzzy logic, instead of being assigned crisp values [53]. The current notion of dichotomy can be generalized by using multiple rather than binary states for the system and its components [54].

The concept of mere connectivity obviously does not correspond exactly to (or exactly capture the essence of) the ecological notion of corridor permeability, and might be enhanced by considering capacity as another corridor attribute, thereby leading to a more encompassing model of a capacitated or flow network [55–57]. Such a network might mimic any type of many engineering networks such as computer, transportation, or telecommunication networks, but it should adhere to ecological requirements and constraints. Algorithms written for these engineering networks might be adapted or tailored to ecological networks by making the most utilization of specific features of these latter networks such as their being logically almost-parallel structures. We expect the capacitated network model of the ecological notion of landscape connectivity to be a serious competitor to the now-dominating electric-circuit model. Both models are probabilistic (rather than deterministic), and both model organism movement via a 'through' quantity (flow or electric current). However, the electric-circuit model makes use of an 'across' quantity (electric voltage) without identifying an ecological counterpart of exactly the same physical nature. In fact, an 'across' quantity should not be specified in absolute terms (only the 'difference' of its values at two points matter). Moreover, when the 'across' quantity is divided by the associated 'through' quantity, it should give some measure of impedance or resistance, and when these two quantities are multiplied together, energy should emerge.

The availability of the present reliability-borrowed techniques in ecology is expected to facilitate the handling of problems of design of ecological corridors at various scales. Since these corridors are mainly intended as a means of ensuring long-term wildlife conservation and biodiversity persistence, the life expectancy of the overall landscape network should be a major concern, and the corridors themselves should not be species specific but are to be designed for multiple species, with some emphasis on fragmentation-sensitive species for which corridors are likely to be proposed.

References

1. Fahrig, L., & Merriam, G. (1985). Habitat patch connectivity and population survival: Ecological archives E066-008. *Ecology, 66*(6), 1762–1768.
2. Taylor, P. D., Fahrig, L., Henein, K., & Merriam, G. (1993). Connectivity is a vital element of landscape structure. *Oikos, 1*(68), 571–573.
3. With, K. A., & King, A. W. (1997). The use and misuse of neutral landscape models in ecology. *Oikos, 79*(2), 219–229.
4. Tischendorf, L., & Fahrig, L. (2000). On the usage and measurement of landscape connectivity. *Oikos, 90*(1), 7–19.
5. Tischendorf, L., & Fahrig, L. (2000). How should we measure landscape connectivity? *Landscape Ecology, 15*(7), 633–641.
6. Moilanen, A., & Hanski, I. (2001). On the use of connectivity measures in spatial ecology. *Oikos, 95*(1), 147–151.
7. Moilanen, A., & Nieminen, M. (2002). Simple connectivity measures in spatial ecology. *Ecology, 83*(4), 1131–1145.
8. Brooks, C. P. (2003). A scalar analysis of landscape connectivity. *Oikos, 102*(2), 433–439.
9. Jordán, F. (2003). Quantifying landscape connectivity: Key patches and key corridors. *WIT Transactions on Ecology and the Environment, 28*(64), 883–891.
10. Pe'er, G., Heinz, S. K., & Frank, K. (2006). Connectivity in heterogeneous landscapes: Analyzing the effect of topography. *Landscape Ecology, 21*(1), 47–61.
11. Pascual-Hortal, L., & Saura, S. (2006). Comparison and development of new graph-based landscape connectivity indices: Towards the priorization of habitat patches and corridors for conservation. *Landscape Ecology, 21*(7), 959–967.
12. Saura, S., & Pascual-Hortal, L. (2007). A new habitat availability index to integrate connectivity in landscape conservation planning: Comparison with existing indices and application to a case study. *Landscape and Urban Planning, 83*(2–3), 91–103.
13. Kindlmann, P., & Burel, F. (2008). Connectivity measures: A review. *Landscape Ecology, 23*(8), 879–890.
14. Minor, E. S., & Urban, D. L. (2008). A graph-theory framework for evaluating landscape connectivity and conservation planning. *Conservation Biology, 22*(2), 297–307.
15. Luque, S., Saura, S., & Fortin, M. J. (2012). Landscape connectivity analysis for conservation: Insights from combining new methods with ecological and genetic data. *Landscape Ecology, 27*(2), 153–157.
16. Hock, K., & Mumby, P. J. (2015). Quantifying the reliability of dispersal paths in connectivity networks. *Journal of the Royal Society, Interface, 12*(105), 20150013.
17. Zhang, Z., Meerow, S., Newell, J. P., & Lindquist, M. (2019). Enhancing landscape connectivity through multifunctional green infrastructure corridor modeling and design. *Urban Forestry and Urban Greening, 1*(38), 305–317.
18. McRae, B. H., Dickson, B. G., Keitt, T. H., & Shah, V. B. (2008). Using circuit theory to model connectivity in ecology, evolution, and conservation. *Ecology, 89*(10), 2712–2724.
19. Rushdi, A. M., & Hassan, A. K. (2015). Reliability of migration between habitat patches with heterogeneous ecological corridors. *Ecological Modelling, 24*(304), 1–10.
20. Rushdi, A. M., & Hassan, A. K. (2016). An exposition of system reliability analysis with an ecological perspective. *Ecological Indicators, 1*(63), 282–295.
21. Rushdi, A. M. (2016). Quantification of uncertainty in the reliability of migration between habitat patches. *Computational Ecology and Software, 6*(3), 66–82.
22. Jordán, F. (2000). A reliability-theory approach to corridor design. *Ecological Modelling, 128*(2–3), 211–220.
23. Ellison, R. J., Fisher, D. A., Linger, R. C., Lipson, H. F., Longstaff, T. A., & Mead, N. R. (1999). Survivability: Protecting your critical systems. *IEEE Internet Computing, 3*(6), 55–63.

24. Zhou, D., & Subramaniam, S. (2000). Survivability in optical networks. *IEEE Network, 14*(6), 16–23.
25. Jindal, V., Dharmaraja, S., & Trivedi, K. S. (2006). Analytical survivability model for fault tolerant cellular networks supporting multiple services. *Simulation Series, 38*(3), 505–512.
26. Heegaard, P. E., & Trivedi, K. S. (2009). Network survivability modeling. *Computer Networks, 53*(8), 1215–1234.
27. Jindal, V., Dharmaraja, S., & Trivedi, K. S. (2011). Markov modeling approach for survivability analysis of cellular networks. *International Journal of Performability Engineering, 7*(5), 429–440.
28. Menasché, D. S., Avritzer, A., Suresh, S., Leão, R. M., de Souza e Silva, E., Diniz, M., et al. (2014). Assessing survivability of smart grid distribution network designs accounting for multiple failures. *Concurrency and Computation: Practice and Experience, 26*(12), 1949–1974.
29. Trivedi, K. S., & Xia, R. (2015). Quantification of system survivability. *Telecommunication Systems, 60*(4), 451–470.
30. Koziolek, A., Avritzer, A., Suresh, S., Menasché, D. S., Diniz, M., e Silva, E. D., et al. (2016). Assessing survivability to support power grid investment decisions. *Reliability Engineering and System Safety, 155,* 30–43.
31. Khasin, M., Meerson, B., Khain, E., & Sander, L. M. (2012). Minimizing the population extinction risk by migration. *Physical Review Letters, 109*(13), 138104.
32. Stamps, J. A., Buechner, M., & Krishnan, V. V. (1987). The effects of edge permeability and habitat geometry on emigration from patches of habitat. *The American Naturalist, 129*(4), 533–552.
33. Collinge, S. K. (1998). Spatial arrangement of habitat patches and corridors: Clues from ecological field experiments. *Landscape and Urban Planning., 42*(2–4), 157–168.
34. Tulloch, A. I., Barnes, M. D., Ringma, J., Fuller, R. A., & Watson, J. E. (2016). Understanding the importance of small patches of habitat for conservation. *Journal of Applied Ecology, 53*(2), 418–429.
35. Wintle, B. A., Kujala, H., Whitehead, A., Cameron, A., Veloz, S., Kukkala, A., et al. (2019). Global synthesis of conservation studies reveals the importance of small habitat patches for biodiversity. *Proceedings of the National Academy of Sciences, 116*(3), 909–914.
36. Misra, K. B. (2012). *Reliability analysis and prediction: A methodology oriented treatment.* Elsevier.
37. Misra, K. B. (Ed). (2012). *New trends in system reliability evaluation.* Elsevier.
38. Misra, K. B. (Ed.). (2008). *Handbook of performability engineering.* Springer Science and Business Media.
39. Rushdi, A. M., & Goda, A. S. (1985). Symbolic reliability analysis via Shannon's expansion and statistical independence. *Microelectronics and Reliability, 25*(6), 1041–1053.
40. Rushdi, A. M., & Abdulghani, A. A. (1993). A comparison between reliability analyses based primarily on disjointness or statistical independence: The case of the generalized INDRA network. *Microelectronics and Reliability, 33*(7), 965–978.
41. Rushdi, A. M., & Rushdi, M. A. (2017). Switching-algebraic analysis of system reliability. Chapter 6 in Ram, M., & Davim, P. (Eds.). *Advances in Reliability and System Engineering* (pp. 139–161). Switzerland: Springer International Publishing, Cham.
42. Rushdi, A. M. (1983). How to hand-check a symbolic reliability expression. *IEEE Transactions on Reliability, 32*(5), 402–408.
43. Rushdi, A. M., & Ghaleb, F. A. (2014). The Walsh spectrum and the real transform of a switching function: A review with a Karnaugh-map perspective. *Journal of Qassim University: Engineering and Computer Sciences, 7*(2), 73–112.
44. Misra, K. B. (1970). An algorithm for the reliability evaluation of redundant networks. *IEEE Transactions on Reliability, 19*(4), 146–151.
45. Aggarwal, K. K., Misra, K. B., & Gupta, J. S. (1975). Reliability evaluation a comparative study of different techniques. *Microelectronics Reliability, 14*(1), 49–56.

46. Rushdi, A. M. (1984). On reliability evaluation by network decomposition. *IEEE Transactions on Reliability, 33*(5), 379–384.
47. Rushdi, R. A., Rushdi, A. M., & Talmees, F. A. (2018). Novel pedagogical methods for conditional-probability computations in medical disciplines. *Journal of Advances in Medicine and Medical Research., 25*(10), 1–15.
48. Jiang, Y., & Yang, W. (2013). An approach based on theorem of total probability for reliability analysis of RC columns with random eccentricity. *Structural Safety, 1*(41), 37–46.
49. Satyanarayana, A., & Chang, M. K. (1983). Network reliability and the factoring theorem. *Networks, 13*(1), 107–120.
50. Page, L. B., & Perry, J. E. (1988). A practical implementation of the factoring theorem for network reliability. *IEEE Transactions on Reliability, 37*(3), 259–267.
51. Rushdi, A. M., & Al-Hindi, K. A. (1993). A table for the lower boundary of the region of useful redundancy for k-out-of-n systems. *Microelectronics and Reliability, 33*(7), 979–992.
52. Rushdi, A. M., Hassan, A. K., & Moinuddin, M. (2020). System reliability analysis of small-cell deployment in heterogeneous cellular networks. *Telecommunication Systems, 73*(3), 371–381.
53. Liu, K. R., Kuo, J. Y., Yeh, K., Chen, C. W., Liang, H. H., & Sun, Y. H. (2015). Using fuzzy logic to generate conditional probabilities in Bayesian belief networks: A case study of ecological assessment. *International Journal of Environmental Science and Technology, 12*(3), 871–884.
54. Nichols, J. D., & Kendall, W. L. (1995). The use of multi-state capture-recapture models to address questions in evolutionary ecology. *Journal of Applied Statistics, 22*(5–6), 835–846.
55. Rushdi, A. M. (1988). Indexes of a telecommunication network. *IEEE Transactions on Reliability, 37*(1), 57–64.
56. Jane, C. C., Lin, J. S., & Yuan, J. (1993). Reliability evaluation of a limited-flow network in terms of minimal cutsets. *IEEE Transactions on Reliability, 42*(3), 354–361.
57. Lin, J. S., Jane, C. C., & Yuan, J. (1995). On reliability evaluation of a capacitated-flow network in terms of minimal pathsets. *Networks, 25*(3), 131–138.

Ali Muhammad Ali Rushdi has a B.Sc. in Electrical Engineering from Cairo University, Giza, Arab Republic of Egypt, in 1974, and an M.S. and a Ph.D. in Electrical Engineering from the University of Illinois at Urbana-Champaign (UIUC), USA, in 1977 and 1980, respectively. He maintained a perfect GPA of 5.0/5.0 throughout his entire study. Since 1980, he has been with King Abdulaziz University (KAU) in Jeddah, Saudi Arabia, where he is now Professor of Electrical and Computer Engineering and Head of the Computer Engineering Programs. He has served (and is serving) as an Associate Editor for several prestigious scientific journals. His research interests and contributions over the past 45 years spanned the areas of Electromagnetic Communications, Computer Engineering, Reliability, Digital Design, Engineering Pedagogy and Education, Switching Networks, Boolean Algebras and Equations, Engineering Design, Dimensional Analysis, Diagnostic Testing, Ecological Modelling, Futurology, Translation, Inferential Thinking, and Innovative Problem-Solving. He is an initiated member of the Honorary Societies: *Eta Kappa Nu* and *Phi kappa Phi*. He is a Life Senior Member of the Institute of Electrical and Electronics Engineers (IEEE).

Ahmad Kamal Hassan has a B.Eng. in electrical and electronics engineering with honors from the University of Bradford, UK, in 2007, an M.S. in electrical engineering from Karlstad University, Sweden, in 2012, and a Ph.D. in electrical engineering at King Abdulaziz University, Saudi Arabia, in 2017. He was a Consultant for Radio Access Networks with Nokia Siemens Networks, Ericsson AB., and Advance Communication and Electronics Systems in the western region of Saudi Arabia from 2007 to 2010. He was a Research Associate with the Ghulam Ishaq Khan Institute of Engineering Sciences and Technology from 2013 to 2014. He was associated with the Center of Excellence in Intelligent Engineering Systems, Pakistan. Since 2017, he is an Assistant Professor at Ghulam Ishaq Khan Institute of Engineering Sciences and Technology, Pakistan.

His research interests include reliability modeling and analysis of networks in general, and system modeling, performance analysis, and beamforming design of MIMO communication networks in particular. He is a Chartered Engineer (C.Eng.) and a member of the Institution of Engineering and Technology.